Lecture Notes in Computer Science 8236

Commenced Publication in 1973
Founding and Former Series Editors:
Gerhard Goos, Juris Hartmanis, and Jan van Leeuwen

Ivan Dimov   István Faragó
Lubin Vulkov (Eds.)

Numerical Analysis and Its Applications

5th International Conference, NAA 2012
Lozenetz, Bulgaria, June 15-20, 2012
Revised Selected Papers

 Springer

Volume Editors

Ivan Dimov
Bulgarian Academy of Sciences
Institute of Information and Communication Technologies
Acad. G. Bonchev, bl. 25A, 1113 Sofia, Bulgaria
E-mail: ivdimov@bas.bg

István Faragó
Eötvös Loránd University, Department of Applied Analysis
Pázmány Péter sétány 1/C, 1117 Budapest, Hungary
E-mail: faragois@cs.elte.hu

Lubin Vulkov
University of Rousse
Department of Applied Mathematics and Statistics
8 Studentska str., 7017 Rousse, Bulgaria
E-mail: lvalkov@uni-ruse.bg

ISSN 0302-9743   e-ISSN 1611-3349
ISBN 978-3-642-41514-2   e-ISBN 978-3-642-41515-9
DOI 10.1007/978-3-642-41515-9
Springer Heidelberg New York Dordrecht London

Library of Congress Control Number: 2013950214

CR Subject Classification (1998): G.1, F.2.1, G.4, I.6, G.2, J.2

LNCS Sublibrary: SL 1 – Theoretical Computer Science and General Issues

Typesetting: Camera-ready by author, data conversion by Scientific Publishing Services, Chennai, India

Printed on acid-free paper

Springer is part of Springer Science+Business Media (www.springer.com)

Preface

The 5th International Conference Numerical Analysis and Applications was held in the hotel 'Ambelitz', Lozenetz, Bulgaria, June 15–20, 2012. The conference was organized by the Department of Applied Mathematics and Statistics at the University of Ruse in cooperation with Union of Mathematicians of Bulgaria, section Ruse. The conference continued the tradition of 4 previous ones (1996, 2000, 2004 and 2008) as a forum, where scientists from leading research groups from the "East" and "West" were provided with the opportunity to meet and exchange ideas and establish research cooperation. More than 100 scientists from all over the world participated the conference. A wide range of problems concerning recent achievement on numerical analysis and its applications in physics, chemistry, engineering, and economics were discussed. The key lectures reviewed some of the advanced achievements in the field of numerical methods and their applications. Four special sessions were organized:

- G. Shishkin, Computational Methods for Boundary and Interior Layers;
- A. Smirnov, Numerical Modeling of Shell-like Structures and their Applications;
- N. Nefedov, Asymptotic Approximation Methods for Non-local Reaction-Diffusion-Advection Problems;
- N. Yamanaka, Verified Numerical Computations.

The conference was opened by the lecture of Prof. M.J. Gander: From Euler, Ritz and Galerkin to Modern Computing.

The success of the conference and the present volume in particular are the outcome of the joint effort of many colleagues from various institutions and organizations. First, we are indebted to all members of the Scientific Committee for their valuable contribution forming the scientific level of the conference, as well as their help in reviewing contributed papers. We thank the Local Organizing Committee and especially M. Koleva for the help in putting together the book.

The 6th International Conference on Numerical Analysis and Its Applications will be held in June 2016.

June 2013

Ivan Dimov
István Faragó
Lubin Vulkov

Table of Contents

I Invited Papers

II Contributed Talks

Note on the Convergence of the Implicit Euler Method

István Faragó

Eötvös Loránd University, Institute of Mathematics
and
MTA-ELTE "Numerical Analysis and Large Networks" Research Group
Pázmány P. s. 1/c, 1117 Budapest, Hungary

Abstract. For the solution of the Cauchy problem for the first order ODE, the most popular, simplest and widely used method are the Euler methods. The two basic variants of the Euler methods are the explicit Euler methods (EEM) and the implicit Euler method (IEM). These methods are well-known and they are introduced almost in any arbitrary textbook of the numerical analysis, and their consistency is given. However, in the investigation of these methods there is a difference in concerning the convergence: for the EEM it is done almost everywhere but for the IEM usually it is missed. (E.g., [1,2,6–9].)The stability (and hence, the convergence) property of the IEM is usually shown as a consequence of some more general theory. Typically, from the theory for the implicit Runge-Kutta methods, which requires knowledge of several basic notions in numerical analysis of ODE theory, and the proofs are rather complicated. In this communication we will present an easy and elementary prove for the convergence of the IEM for the scalar ODE problem. This proof is direct and it is available for the non-specialists, too.

Keywords: Numerical solution of ODE, implicit and explicit Euler method, Runge-Kutta methods, finite difference method.

1 Introduction

Many different problems (physical, chemical, etc.) can be described by the initial-value problem for first order ordinary differential equation (ODE) of the form

$$\frac{du}{dt} = f(t, u), \quad t \in (0, T), \tag{1}$$

$$u(0) = u_0, \tag{2}$$

We note that, using the semidiscretization, the time-dependent partial differential equations also lead to the problem (1)-(2). Hence, the solution of such problem plays a crucial role in mathematical modelling. (For simplicity, in sequel we consider only the scalar problem, i.e., when $f : \mathbb{R}^2 \to \mathbb{R}$.) We know that under the the Lipshitz conditon, i.e., in case

$$|f(t, s_1) - f(t, s_2)| \leq L|s_1 - s_2|, \text{ for all } (t, s_1), (t, s_2) \in dom(f) \tag{3}$$

I. Dimov, I. Faragó, and L. Vulkov (Eds.): NAA 2012, LNCS 8236, pp. 1–11, 2013.
© Springer-Verlag Berlin Heidelberg 2013

with the Lipschitz constant $L > 0$, the problem (1)-(2) has unique solution. However, for general form of function f the solution cannot be defined analytically, it can be done only for a very few f. Therefore, we apply some numerical method. Hence, the numerical integration of the problem (1)-(2) -under the condition (3)- is one of the most typical tasks in the numerical modelling of real-life problems.

For the solution of the above Cauchy problem, the most popular, simplest and widely used method are the Euler methods. The two basic variants of the Euler methods are the explicit Euler methods (EEM) and the implicit Euler method (IEM). These methods are well-known and in the next section we give their description. These methods are introduced almost in any arbitrary textbook of the numerical analysis, and their consistency is given. Since the direct proof of convergence is typically given for the EEM, only, therefore our aim is to give an easy and elementary prove for the convergence of the IEM, too. This proof is direct and it is available for the non-specialists, too.

The paper is organized as follows.

In Section 2 we formulate the basic Euler methods, namely, the EEM and IEM, and we analyze their consistency. In Section 3 we give the proof of the convergence of the EEM. In Section 4 we give an overview of the traditional proof of the convergence for the IEM. The Section 5 contains the simple and compact proof of the convergence of the IEM, and we define the order of its convergence, too. Finally, we finish the paper with giving some comments.

2 The Euler Methods

Our aim is to define some numerical solution at some fixed point $t^\star \in (0, T)$ to the Cauchy problem (1)-(2). Therefore, we construct the sequence of the uniform meshes with the mesh-size $h = t^\star/N$ of the form

$$\omega_h = \{t_n = n \cdot h, n = 0, 1, \ldots, N\},$$

and our aim is to define at the mesh-point $t^\star = t_N$ a suitable approximation y_N on each fixed mesh.

This requires to give the rule how to define the mesh-function $y_h : \omega_h \to \mathbb{R}$. Using the notation $y_h(t_n) = y_n$, we introduce the following two methods, called *Euler methods*:

1. *Explicit Euler method (EEM):*

$$
\begin{aligned}
y_{n+1} &= y_n + hf(t_n, y_n), \quad n = 0, 1, \ldots, N \\
y_0 &= u_0.
\end{aligned}
\tag{4}
$$

2. *Implicit Euler method (IEM):*

$$
\begin{aligned}
y_{n+1} &= y_n + hf(t_{n+1}, y_{n+1}), \quad n = 0, 1, \ldots, \\
y_0 &= u_0.
\end{aligned}
\tag{5}
$$

In mathematics and computational science, these methods are most basic explicit and implicit method for numerical integration of ordinary differential equations and they are the simplest Runge-Kutta methods.

Remark 1. The Euler methods are named after Leonhard Euler (1707-1783), who treated it in his book Institutionum calculi integralis (published 1768-70) [4]. We note that Euler developed the explict method to prove that the initial value problem (1)-(2) had a solution.

Let us define the local truncation error for these methods, under the assumption that f continuously differentiable, and hence the solution $u(t)$ has second continuous derivative.

For EEM, the local truncation error $l(h)$ can be defined as

$$l(h) = u(t_{n+1}) - u(t_n) - hf(t_n, u(t_n)), \tag{6}$$

where $u(t)$ stands for the solution of the problem (1)-(2). Therefore, the relation

$$u(t_{n+1}) = u(t_n + h) = u(t_n) + hu'(t_n) + \frac{h^2}{2}u''(\vartheta_n^{EEM}) \tag{7}$$

holds with some $\vartheta_n^{EEM} \in (t_n, t_{n+1})$. Using the equation (1) at the point $t = t_n$, and then the truncated Taylor expansion (7) for the local truncation error $l(h)$ in (6), we obtain

$$l(h) = u(t_{n+1}) - u(t_n) - hu'(t_n) = \frac{h^2}{2}u''(\vartheta_n^{EEM}). \tag{8}$$

For the local truncation error $l(h)$ of IEM we have the relation

$$l(h) = u(t_{n+1}) - u(t_n) - hf(t_{n+1}, u(t_{n+1})), \tag{9}$$

and hence the relation

$$u(t_n) = u(t_{n+1} - h) = u(t_{n+1}) - hu'(t_{n+1}) + \frac{h^2}{2}u''(\vartheta_n^{IEM}) \tag{10}$$

holds with some $\vartheta_n^{IEM} \in (t_n, t_{n+1})$. Using the equation (1) at the point $t = t_{n+1}$, and then the truncated Taylor expansion (10) for the local truncation error $l(h)$ in (9), we obtain

$$l(h) = u(t_{n+1}) - u(t_n) - hu'(t_{n+1}) = -\frac{h^2}{2}u''(\vartheta_n^{IEM}). \tag{11}$$

The order of a numerical method is defined by the local truncation error: when $l(h) = \mathcal{O}(h^{p+1})$ then the method is called consistent of order p. This means that for both Euler methods the order of consistency is equal to one.

We also introduce the notation $e_n = u(t_n) - y_n$ for the global error at the mesh-point t_n. The convergence at the point $t = t^*$ means the relation $\lim_{h \to 0} |e_N| = 0$, where $N \cdot h = t^*$. (Obviously, $n = N(h)$ and $\lim_{h \to 0} N(h) = \infty$.) When $e_N = \mathcal{O}(h^p)$ with some $p > 0$ then the convergence is of p-th order. Our aim is to show the convergence and its order.

3   Convergence of the Explicit Euler Method

For sake of completeness, in the sequel we consider the convergence for the EEM. We note that in the textbooks the proof is given only for this case, and it is not considered for the IEM directly. (See e.g. [1, 2, 6–9] and many others.)

First we formulate a useful statement.

Lemma 1. *Let $a > 0$ and $b > 0$ given numbers, and s_i $(i = 0, 1, \ldots, k-1)$ such numbers that the inequalities*

$$|s_i| \leq a|s_{i-1}| + b, \quad i = 1, 2, \ldots, k - 1 \tag{12}$$

hold. Then the estimations

$$|s_i| \leq a^i|s_0| + \frac{a^i - 1}{a - 1}b, \quad i = 1, 2, \ldots, k \tag{13}$$

are valid.

(The proof is done by induction, and, as an easy exercise, it is left to the Reader.)

Corollary 1. *When $a > 1$ then obviously we have*

$$\frac{a^i - 1}{a - 1} = a^{i-1} + a^{i-2} + \cdots + 1 \leq ia^{i-1}.$$

Hence, for this case the Lemma 1, instead of (13) yields the estimation

$$|s_i| \leq a^i|s_0| + ia^{i-1}b, \quad i = 1, 2, \ldots, k. \tag{14}$$

Let us consider the EEM, which means that the values y_n at the mesh-points ω_h defined by the one-step recursion (4).

Since for the local truncation error $l(h)$ of EEM we have the relation (6). Re-arranging this formula, we have

$$u(t_{n+1}) = u(t_n) + hf(t_n, u(t_n)) + l(h). \tag{15}$$

Based on the expressions (4) and (11), we get the error equation in the form

$$e_{n+1} = e_n + h\left(f(t_n, u(t_n)) - f(t_n, y_n)\right) + l(h). \tag{16}$$

Hence, using the Lipschitz property (3), we obtain

$$|e_{n+1}| \leq |e_n| + hL|e_n| + |l(h)|, \tag{17}$$

which implies the relation

$$|e_{n+1}| \leq (1 + Lh)|e_n| + |l(h)|. \tag{18}$$

Therefore for any index $n = 0, 1, \ldots N$ we have the relation

$$|e_n| \le (1 + Lh)|e_{n-1}| + |l(h)| < \exp(hL)|e_{n-1}| + |l(h)|. \tag{19}$$

Let us apply Lemma 1 to the global error e_n. Choosing $a = \exp(hL) > 1$, $b = |l(h)| > 0$, due to (19) we can use Corrolary 1. Hence, based on (14) we get

$$|e_n| \le [\exp(hL)]^n |e_0| + n [\exp(hL)]^{n-1} |l(h)|. \tag{20}$$

Due to the obvious relations

$$[\exp(hL)]^n = \exp(Lhn) = \exp(Lt_n),$$

and

$$n [\exp(hL)]^{n-1} |l(h)| < nh \exp(Lhn)\frac{|l(h)|}{h} = t_n \exp(Lt_n)\frac{|l(h)|}{h},$$

the relation (20) results in the estimation

$$|e_n| \le \exp(Lt_n) \left[|e_0| + t_n \frac{|l(h)|}{h} \right], \tag{21}$$

which is hold for every $n = 1, 2, \ldots, N$.

Using the expression for the local truncation error in the form (7), the estimation (21) can be rewritten as

$$|e_n| \le \exp(Lt_n) \left[|e_0| + t_n \frac{h}{2} u''(\vartheta_n^{EEM}) \right], \tag{22}$$

where, as before $\vartheta_n^{EEM} \in (t_n, t_{n+1})$ is some fixed number. Hence, introducing the notation $M_2^{(n)} = \max_{[t_{n-1}, t_n]} |u''(t)|$, for the global error at any mesh-point $t = t_n$ we get the estimation

$$|e_n| \le \exp(Lt_n) \left[|e_0| + \frac{M_2^{(n)}}{2} h t_n \right], \tag{23}$$

for any $n = 1, 2, \ldots, N$.

Putting $n = N$ into (23), and taking into the account the equality $t^* = Nh$, with the notation $M_2 = \max_{[0, t^*]} |u''(t)|$ we get

$$|e_N| \le \exp(Lt^*) \left[|e_0| + \frac{M_2}{2} t^* h \right]. \tag{24}$$

Since $e_0 = 0$, therefore, finally we get

$$|e_N| \le C_{EEM} \cdot h, \tag{25}$$

where $C_{EEM} = \frac{M_2}{2} t^* \exp(Lt^*) = constant$. This proves the first order convergence of the EEM.

4   Convergence of the Implicit Euler Method–Traditional Approach

The convergence of the IEM cannot be proven directly as it was done in the Section 3. The main reason is that the basic inequality (12) of the Lemma 1 cannot be obtained for any h. (For the EEM we have got it in the form (19) for arbitrary h.)

First we summarize the usual way of proving the convergence of the IEM. Let us consider the so called *linear multi-step methods*, which are defined by the k-step recursion:

$$\sum_{j=0}^{k} \alpha_j y_{n+j} = h \sum_{j=0}^{k} \beta_j f(t_{n+j}, y_{n+j}), \tag{26}$$

where the coefficients $\alpha_0, \ldots, \alpha_k$ and and $\beta_0, \ldots, \beta_k$ are real fixed constants.

Remark 2. We always assume that $\alpha_k \neq 0$. When $\beta_k = 0$ then y_{n+k} is obtained explicitly from previous values of y_j and $f(t_j, y_j)$, and the k-step method is then said to be explicit. On the other hand, if $\beta_k \neq 0$ then y_{n+k} appears not only on the left-hand side but also on the right, within $f(t_{n+k}, y_{n+k})$, and due to this implicit dependence on y_{n+k} the method is then called implicit. We note that the numerical method (26) is called linear because it involves only linear combinations of the y_n and the $f(t_n, y_n)$.

The starting values for the method (26) are defined as follows: y_0 is given from the initial condition, $y_1, \ldots, y_{k-1}$ are computed by some suitable RungeKutta methods. This means that the starting values will contain numerical errors and it is important to know how these will affect further approximations y_n, $n > k$, which are calculated by means of (26). Thus, we wish to consider the "stability" of the numerical method with respect to "small perturbations" in the starting conditions.

Definition 1. *A linear k-step linear multi-step method for the ordinary differential equation (1) is said to be zero-stable if there exists a constant K such that, for any two sequences (y_n) and $(\tilde{y}_n)$ which have been generated by the same formulae (26) but different initial data $y_0, y_1, \ldots, y_{k-1}$ and $\tilde{y}_0, \tilde{y}_1, \ldots, \tilde{y}_{k-1}$, respectively, we have*

$$|y_n - \tilde{y}_n| \leq K \max\{|y_0 - \tilde{y}_0|, |y_1 - \tilde{y}_1|, , \ldots, |y_{k-1} - \tilde{y}_{k-1}|\}, \tag{27}$$

for $t_n \leq t^*$ and as h tends to 0.

Given the linear k-step method (26) we consider its first characteristic polynomial, defined as $\varrho(z) = \sum_{j=0}^{k} \alpha_j z^{k-j}$.

Theorem 1. *A linear multi-step method is zero-stable for any ordinary differential equation of the form (1) where f satisfies the Lipschitz condition (3), if and only if its first characteristic polynomial has zeros roots inside the closed unit disc, with any which lie on the unit circle being simple.*

The proof of the Theorem 1 is complicated, and it can be found in [5, 10]. We mention that the algebraic stability condition contained in this theorem, namely that the roots of the first characteristic polynomial lie in the closed unit disc and those on the unit circle are simple, is often called the root condition.

For the convergence of the linear k-step linear multi-step method we have the following statement.

Theorem 2. *The linear k-step linear multi-step method* (26) *applied to the initial value problem* (1)-(2) *the zero stability and the consistency are necessary and sufficient condition of the convergence.*

The proof of this theorem is can be found in [3].

Let us notice that IEM, defined by (5) can be considered as a one-step ($k = 1$ linear multi-step method of the form (26), where $\alpha_0 = -1$, $\alpha_1 = 1$, $\beta_0 = 0$ and $\beta_1 = 1$. Therefore, its characteristic polynomial has the unique root $z = 1$. Hence, according to the Theorem 1, this method is zero-stable. However, we already know that it is consistent, too. Hence, based on Theorem 2, we get

Theorem 3. *The implicit Euler method is convergent.*

However, we emphasize that the "proof" of Theorem 3 is incomplete: it uses two statements (Theorem 1 and Theorem 2) which were not proven, and their proofs are rather difficult. Moreover, this classical statement doesn't give information about the rate of the convergence. Therefore, we are going to prove the convergence of the IEM in another way, which is compact and short.

5   Convergence of the Implicit Euler Method- Another Approach

We consider the IEM, which means that the values y_n at the mesh-points ω_h defined by the one-step recursion (5). Rearranging the local truncation error for IEM of the form (9) we have

$$u(t_{n+1}) = u(t_n) + hf(t_{n+1}, u(t_{n+1})) + l(h). \tag{28}$$

Based on the expressions (5) and (11), we get the error equation in the form

$$e_{n+1} = e_n + h\left(f(t_{n+1}, u(t_{n+1})) - f(t_{n+1}, y_{n+1})\right) + l(h). \tag{29}$$

Hence, using the Lipschitz property (3), we obtain

$$|e_{n+1}| \leq |e_n| + hL|e_{n+1}| + |l(h)|, \tag{30}$$

which implies the relation

$$(1 - Lh)|e_{n+1}| \leq |e_n| + |l(h)|. \tag{31}$$

Assume that $h < h_0 := \dfrac{1}{2L}$. Then (31) implies the relation

$$|e_{n+1}| \le \frac{1}{1-hL}|e_n| + \frac{1}{1-hL}|l(h)|. \tag{32}$$

We give an estimation for $\dfrac{1}{1-hL}$. Based on the assumption, $hL \le 0.5$, therefore we can write this expression as

$$1 < \frac{1}{1-hL} = 1 + hL + (hL)^2 + \cdots + (hL)^n + \cdots =$$
$$1 + hL + (hL)^2\left(1 + hL + (hL)^2 + \ldots\right) = 1 + hL + (hL)^2\frac{1}{1-hL}. \tag{33}$$

Obviously, for the values $Lh < 0.5$ the estimation $\dfrac{(hL)^2}{1-hL} < hL$ holds. Therefore, we have the upper bound

$$\frac{1}{1-hL} < 1 + 2hL < \exp(2hL). \tag{34}$$

Since for the values $Lh \in [0, 0.5]$ obviously $\dfrac{1}{1-hL} \le 2$, therefore for the global error the substitution (34) into (32) results in the recursive relation

$$|e_n| \le \exp(2hL)|e_{n-1}| + 2|l(h)|. \tag{35}$$

Let us apply Lemma 1 to the global error e_n. Choosing $a = \exp(2hL) > 1$, $b = 2|l(h)| > 0$, and owning the relation (35), based on (14) we get

$$|e_n| \le [\exp(2hL)]^n |e_0| + n[\exp(2hL)]^{n-1} 2|l(h)|. \tag{36}$$

Due to the obvious relations

$$[\exp(2hL)]^n = \exp(2Lhn) = \exp(2Lt_n),$$

and

$$n[\exp(2hL)]^{n-1} 2|l(h)| < 2nh\exp(2Lhn)\frac{|l(h)|}{h} = 2t_n \exp(2Lt_n)\frac{|l(h)|}{h},$$

the relation (36) results in the estimation

$$|e_n| \le \exp(2Lt_n)\left[|e_0| + 2t_n\frac{|l(h)|}{h}\right], \tag{37}$$

which is hold for every $n = 1, 2, \ldots, N$.

Using the expression for the local truncation error in the form (10), the estimation (37) can be rewritten as

$$|e_n| \le \exp(2Lt_n)\left[|e_0| + 2t_n\frac{h}{2}u''(\vartheta_n^{IEM})\right], \tag{38}$$

where, as before $\vartheta_n^{IEM} \in (t_n, t_{n+1})$ is some fixed number. Hence, introducing the notation $M_2^{(n)} = \max\limits_{[t_{n-1}, t_n]} |u''(t)|$, for the global error at any mesh-point $t = t_n$ we get the estimation

$$|e_n| \leq \exp(2Lt_n) \left[|e_0| + M_2^{(n)} h t_n \right], \tag{39}$$

for any $n = 1, 2, \ldots, N$.

Putting $n = N$ into (39), and taking into the account the relation $t^\star = Nh$, with the notation $M_2 = \max\limits_{[0, t^\star]} |u''(t)|$ we get

$$|e_N| \leq \exp(2Lt^\star) \left[|e_0| + M_2 t^\star h \right]. \tag{40}$$

Since $e_0 = 0$, therefore, finally we get

$$|e_N| \leq C_{IEM} \cdot h, \tag{41}$$

where $C_{IEM} = M_2 t^\star \exp(2Lt^\star) = constant$. This proves the following statement.

Theorem 4. *The implicit Euler method is convergent and then rate of convergence is one.*

6 Concluding Remarks

Finally, we give some comments.

◊ The convergence on the interval $[0, t^\star]$ yields the relation $\lim\limits_{h \to 0} max_{n=1,2,\ldots,N} |e_n| = 0$. As one can easily see, based on the relations (21) (for the EEM) and (37) (for the IEM) the local truncation error $|e_n|$ can be bounded by the expression $C_{EEM} \cdot h$ (for the EEM) and $C_{IEM} \cdot h$ (for the IEM). This means that both methods are convergent on the interval $[0, t^\star]$ in the first order.

◊ The general (not necessarily linear) multi-step methods can be written in the form

$$\sum_{j=0}^{k} \alpha_j y_{n+j} = h \Phi_f(y_{n+k}, y_{n+k-1}, \ldots, y_n, t_n; h), \tag{42}$$

where the subscript f on the right-hand side indicates that the dependence of Φ on $y_{n+k}, y_{n+k-1}, \ldots, y_n, t_n$ is through the function $f(t, u)$. We impose the following two conditions on (42):

$$\left. \begin{array}{c} \Phi_{f \equiv 0}(y_{n+k}, y_{n+k-1}, \ldots, y_n, t_n; h) \equiv 0, \\ \\ |\Phi_f(y_{n+k}, y_{n+k-1}, \ldots, y_n, t_n; h) - \Phi_f(\tilde{y}_{n+k}, \tilde{y}_{n+k-1}, \ldots, \tilde{y}_n, t_n; h)| \\ \\ \leq M \sum_{j=0}^{k} |y_{n+j} - \tilde{y}_{n+j}|, \end{array} \right\} \tag{43}$$

where M is a constant. (These conditions are not very restrictive, e.g., the second one is automatically satisfied if the initial value problem to be solved satisfies a Lipschitz condition.)

Theorem 5. *The necessary and sufficient conditions for the method (42) to be convergent are that it be both consistent and zero-stable.*

The necessary and sufficient conditions for consistency can be expressed by the first characteristic polynomial: the method (42) is consistent if and only if

$$\varrho(1) = 0 \qquad (44)$$

and

$$\Phi_f(y(t_n), y(t_n), \ldots, y(t_n), t_n; 0)/\varrho'(1) = f(t_n, y(t_n)), \qquad (45)$$

see [7], p. 30.

To verify the zero-stability, we refer to root condition in Theorem 1. Hence, we can check the convergence directly for both the EEM and IEM. Clearly, for these methods $k = 1$, and

$$\Phi_f^{EEM}(y_{n+1}, y_n, t_n; h) = f(t_n, y_n), \quad \text{for the EEM,}$$

and

$$\Phi_f^{IEM}(y_{n+1}, y_n, t_n; h) = f(t_n + h, y_{n+1}), \quad \text{for the IEM.}$$

Moreover, $\varrho(z) = z - 1$. Hence the properties (43), (44), (45) and the root condition can be checked directly. These together result in the convergence of the Euler methods.

◇ As (44) shows, for the consistent methods $z = 1$ is always the root of the first characteristic polynomial. Since for the one-step methods this polynomial is of first order, therefore for this case the polynomial has only one root, thus, such methods are always zero-stabile. Hence, we can conclude the following statement.

Theorem 6. *The consistent one-step methods are convergent.*

◇ For the general linear multi-step method we can get more sharp result than the statement of Theorem 3. Namely,

Theorem 7. *For a linear multi-step method that is consistent with the ordinary differential equation (1) where f is assumed to satisfy a Lipschitz condition, and starting with consistent initial data, zero-stability is necessary and sufficient for convergence. Moreover if the solution has continuous derivative of order $p + 1$, and the local truncation error $\mathcal{O}(h^{p+1})$, then the global error e_n is of order $\mathcal{O}(h^p)$.*

This theorems is due to Dahlquist (the proof can be found e.g. in [10]), and it says that the orders of consistency and convergence are coincided.

◇ In our paper we did not consider roundoff error,which is always present in computer calculations. At the present time there is no universally accepted method to analyze roundoff error after a large number of time steps. The three main methods for analyzing roundoff accumulation are the analytical method, the probabilistic method and the interval arithmetic method, each of which has both advantages and disadvantages.

◇ In the IEM is implicit, in each step we must solve a -usually non-linear-equations, namely, the root of the equation of the form $g(s) := s - hf(t_n, s) - y_n = 0$. This can be done by using some iterative method such as Newton method.

◇ In this paper we have been concerned with the stability and accuracy properties of the Euler methods in the asymptotic limit of $h \to 0$ and $n \to \infty$ here $n \cdot h$ fixed. However, it is of practical significance to investigate the performance of methods in the case of $h > 0$ and fixed and $n \to \infty$. Specifically, we would like to ensure that when applied to an initial value problem whose solution decays to zero as $t \to \infty$, the Euler methods exhibit a similar behavior, for $h > 0$ fixed and $t_n \to \infty$. This problem is investigated on the famous Dalquaist scalar test equation, and it requires the so called A-stability property. As it is known, from this point of view the IEM is much more better: it is A-stable without any restriction w.r.t. h ("absolute stable"), however the EEM is bounded only under some strict condition for h. The latter makes the EEM unusable for several class of the problem, like stiff problems.

Acknowledgement. The work supported by the Hungarian Research Grant OTKA K 67819.

References

1. Ascher, U.M., Petzold, L.R.: Computer methods for ordinary differential equations and differential-algebraic equations. SIAM, Philadelphia (1998)
2. Bachvalov, N. S.: Numerical methods, Nauka, Moscow (1975) (in Russian)
3. Dahlquist, G.: Convergence and stability in the numerical integration of ordinary differential equations. Math. Scand. 4, 33–53 (1956)
4. Collected Works of Euler, L. vol. 11 (1913), 12 (1914) pp. 549–553(1984)
5. Isaacson, E., Keller, H.B.: Analysis of numerical methods. Wiley, New York (1966)
6. LeVeque, R.: Finite difference mehods for ordinary and partial differential equations. SIAM, Philadelphia (2007)
7. Lambert, J.D.: Numerical methods for ordinary differential systems: The initial value problem. John Wiley and Sons, Chicester (1991)
8. Molchanov, I.N.: Computer methods for solving applied problems. Differential equations. Naukova Dumka, Kiew(1988) (in Russian)
9. Samarskij, A.A., Gulij, A.V.: Numerical methods, Nauka, Moscow (1989) (in Russian)
10. Suli, E.: Numerical solution of ordinary differential equations, Oxford (2010)

Uniform Shift Estimates for Transmission Problems and Optimal Rates of Convergence for the Parametric Finite Element Method

Hengguang Li[1], Victor Nistor[2], and Yu Qiao[3,*]

[1] Department of Mathematics, Wayne State University, Detroit, MI 48202, USA
hli@math.wayne.edu
[2] Department of Mathematics, Pennsylvania State University
University Park, PA, 16802, USA
Inst. Math. Romanian Acad.
PO BOX 1-764, 014700, Bucharest, Romania
nistor@math.psu.edu
[3] College of Mathematics and Information Science
Shaanxi Normal University, Xi'an, Shaanxi, 710062, P.R. China
yqiao@snnu.edu.cn

Abstract. Let $\Omega \subset \mathbb{R}^d$, $d \geqslant 1$, be a bounded domain with piecewise smooth boundary $\partial\Omega$ and let U be an open subset of a Banach space Y. Motivated by questions in "Uncertainty Quantification," we consider a parametric family $P = (P_y)_{y \in U}$ of uniformly strongly elliptic, second order partial differential operators P_y on Ω. We allow jump discontinuities in the coefficients. We establish a regularity result for the solution $u : \Omega \times U \to \mathbb{R}$ of the parametric, elliptic boundary value/transmission problem $P_y u_y = f_y$, $y \in U$, with mixed Dirichlet-Neumann boundary conditions in the case when the boundary and the interface are smooth and in the general case for $d = 2$. Our regularity and well-posedness results are formulated in a scale of broken weighted Sobolev spaces $\hat{\mathcal{K}}^{m+1}_{a+1}(\Omega)$ of Babuška-Kondrat'ev type in Ω, possibly augmented by some locally constant functions. This implies that the parametric, elliptic PDEs $(P_y)_{y \in U}$ admit a shift theorem that is uniform in the parameter $y \in U$. In turn, this then leads to h^m-quasi-optimal rates of convergence (i. e., algebraic orders of convergence) for the Galerkin approximations of the solution u, where the approximation spaces are defined using the "polynomial chaos expansion" of u with respect to a suitable family of tensorized Lagrange polynomials, following the method developed by Cohen, Devore, and Schwab (2010).

1   Introduction

Recently, questions related to differential equations with random coefficients have received a lot of attention due to the practical applications of these problems

* H. Li was partially supported by the NSF Grant DMS-1158839. V. Nistor was partially supported by the NSF Grant DMS-1016556.

I. Dimov, I. Faragó, and L. Vulkov (Eds.): NAA 2012, LNCS 8236, pp. 12–23, 2013.

[2,7,9,16,17]. Our paper is motivated by the approach in [6,13], where families of differential operators on polyhedral domains indexed by $y \in U$, were studied. As in those papers, U is an open subset of a Banach space Y, which allows us to study the analyticity of the solution in terms of $y \in U$.

We here study the properties of solutions to a family of strongly elliptic, mixed boundary value/transmission problems

$$P_y u_y(x) = Pu(x, y) = f_y(x) = f(x, y), \quad x \in \Omega, \ y \in U \tag{1}$$

on a domain $\Omega \subset \mathbb{R}^d$, $d \geq 1$. The domain Ω is assumed to be piecewise smooth and bounded. Thus, for each $y \in U$, we are given a second order, uniformly strongly positive, parametric partial differential operator P_y on Ω whose coefficients are functions of $(x, y) \in \Omega \times U$ and are allowed to have jump discontinuities across a fixed interface Γ. More precisely, we assume that $\overline{\Omega} = \cup_{k=1}^K \overline{\Omega}_k$, where Ω_k are disjoint domains with piecewise smooth boundaries and $\Gamma := \left(\cup_{k=1}^K \partial\Omega_k \right) \smallsetminus \partial\Omega$.

Under suitable regularity assumptions on the coefficients of P and on the source term $f : \Omega \times U \to \mathbb{R}$, we establish in Section 5 a regularity and well-posedness result for the solution $u : \Omega \times U \to \mathbb{R}$ of the parametric, elliptic boundary value/transmission problem (1) with mixed Dirichlet-Neumann boundary conditions. Our regularity result is formulated in a scale of broken weighted Sobolev spaces $\hat{\mathcal{K}}_{a+1}^{m+1}(\Omega) = \oplus_{k=1}^K \mathcal{K}_{a+1}^{m+1}(\Omega_k)$ of Babuška-Kondrat'ev type in Ω, for which we prove that our elliptic PDEs $(P_y)_{y \in U}$ admit a shift theorem that is uniform in the parameter $y \in U$. We deal completely in this paper with the cases when the boundary $\partial\Omega$ and the interface Γ are smooth and disjoint. We also indicate how to proceed in the general case for $d = 2$. Our results generalize the results of [13] by allowing jump discontinuities in the coefficients and by allowing adjacent edges to be endowed with Neumann-Neumann boundary conditions. We will be therefore brief in our presentation, referring to [13], as well as [6,7] for more details.

The main contribution of this paper is to study the regularity of the solution of a (non-parametric) transmission/boundary value problem with rather weak smoothness assumptions on the coefficients. As far as we know, this paper is the only place where a complete proof for the regularity of transmission problems is given, even in the case of smooth coefficients. The results are general enough so that one can use the approach in [6,13] to obtain regularity results for families and then to obtain optimal rates of convergence for the Galerkin method. An abstract version of this method is explained in [3]. These issues will be discussed in more detail in a forthcoming paper.

The paper is organized as follows. In Section 2 we formulate our parametric partial differential boundary value/transmission problem and introduce some of our main assumptions. We also discuss the needed notions of positivity for families of operators and derive some simple consequences. In Section 3, we review the "broken" version of usual Sobolev spaces, and then formulate and prove the main results, Theorem 1, which is a regularity and well-posedness result for non-parametric solutions in smooth case. In Section 4, we recapitulate regularity and well-posedness results for the non-parametric, elliptic problem from [4,5,10], the main result being Theorem 2. This theorem is then generalized to families in Section 5, thus yielding our main regularity and well-posendess result for parametric families of uniformly strongly boundary value/transmission problems, namely Theorem 3. As mentioned above, this result is formulated in broken weighted Sobolev spaces (the so called "Babuška-Kondrat'ev" spaces). As in [6,7,13] these results lead to h^m-quasi-optimal rates of convergence for a suitable Galerkin method for the approximation of our parametric solution u.

Acknowledgements. VN acknowledges the support of the Hausdorff Institute for Mathematics (HIM) in Bonn during the HIM Trimester "High dimensional approximation," where this work was initiated. We also thank Christoph Schwab for several useful discussions.

2 Ellipticity, Positivity, Solvability for Parametric Families

We now formulate our parametric partial differential boundary value/ transmission problem and introduce some of our main assumptions.

2.1 Notation and Assumptions

By $\Omega \subset \mathbb{R}^d$, $d \geq 1$, we shall denote a connected, bounded piecewise smooth domain, which we assume is decomposed into finitely many subdomains Ω_k with piecewise smooth boundary, $\overline{\Omega} = \cup_{k=1}^K \overline{\Omega}_k$. We obtain results on the spatial regularity of PDEs whose data depend on a parameter vector $y \in U \subset Y$, where U is an open subset of a Banach space Y. By $a_{pq}^{ij}, b_{pq}^i, c_{pq} : \Omega \times U \to \mathbb{R}$, $1 \leq i,j \leq d$, we shall denote bounded, measurable functions satisfying smoothness and other assumptions to be made precise later. We denote by $A = (a_{pq}^{ij}, b_{pq}^i, c_{pq})$. Let us denote by $\partial_i = \frac{\partial}{\partial x_i}$, $i = 1, \ldots, d$. We shall then denote by $P^A = [P_{pq}^A]$ a $\mu \times \mu$ matrix of parametric differential operators in divergence form

$$P_{pq}^A u(x,y) := \left(-\sum_{i,j=1}^d \partial_i \big(a_{pq}^{ij}(x,y)\partial_j\big) + \sum_{i=1}^d b_{pq}^i(x,y)\partial_i + c_{pq}(x,y) \right) u(x,y), \quad (2)$$

where $x \in \Omega$ and $y \in U$. Note that the derivatives act only in the x-direction and y is just a parameter. The matrix case is needed in order to handle the case of systems, such as that of (anisotropic) linear elasticity.

A matrix $P = [P_{pq}]_{p,q=1}^\mu$ of differential operators acts on vector-valued functions $u = (u_q)_{q=1}^\mu$ in the usual way $(Pu)_p = \sum_{q=1}^\mu P_{pq}u_q$, for $u = (u_q) \in \mathcal{C}^\infty(\Omega \times U)^\mu$. We recall that $H^{-1}(\Omega)$ is defined as the dual of $H_0^1(\Omega) := \{u \in H^1(\Omega), u|_{\partial\Omega} = 0\}$ with pivot $L^2(\Omega)$. Occasionally, we shall need to specialize a family P for a particular value of y, in which case we shall write $P_y : \mathcal{C}^\infty(\Omega)^\mu \to H^{-1}(\Omega)^\mu$ for the induced operator. We emphasize that we allow P to have non-smooth coefficients, so that Pu may be non-smooth in general.

2.2   Boundary and Interface Conditions

We impose mixed Dirichlet and Neumann boundary conditions. To this end, we assume there is given a closed set $\partial_D \Omega \subset \partial \Omega$, which is a union of polygonal subsets of the boundary and we let $\partial_N \Omega := \partial \Omega \smallsetminus \partial_D \Omega$. The set $\partial_D \Omega$ will be referred to as "Dirichlet boundary" and $\partial_N \Omega$ as "Neumann boundary," according to the type of boundary conditions that we associate to these parts of the boundary. The case of cracks is also allowed, provided that one treats different sides of the crack as *different* parts of the boundary, as in [10], for instance, but we choose not to treat this case explicitly in this paper. We then define the *conormal derivatives*

$$(\nabla_\nu^A u)_p(x,y) = \sum_{q=1}^{\mu} \sum_{i,j=1}^{d} \nu_i a_{pq}^{ij}(x,y) \partial_j u_q(x,y), \quad x \in \partial_N \Omega, \ y \in U, \qquad (3)$$

where $\nu = (\nu_i)$ is the outward unit normal vector at $x \in \partial_N \Omega$. The conormal derivatives $\nabla_\nu^A u^\pm$ at the interface Γ are defined similarly, using an arbitrary but fixed labeling of the two sides of the interface into a positive and a negative part.

We shall also need the spaces $H_D^1(\Omega)$ and $H_D^{-1}(\Omega)$ for vector-valued functions: $H_D^1(\Omega) := \{u \in H^1(\Omega)^\mu, \ u = 0 \text{ on } \partial_D \Omega\}$ and $H_D^{-1}(\Omega)$, defined to be the dual of $H_D^1(\Omega)$ with pivot space $L^2(\Omega)$. Note that we assume here as in [13] that we have *the same type of boundary conditions for all components u_q of the solution vector u.*

Recall that our domain Ω is decomposed into K subdomains of the same type (with piecewise smooth boundary), $\overline{\Omega} = \cup_{k=1}^K \overline{\Omega}_k$. We then denote by $\Gamma :=$ $\left(\cup_{k=1}^K \partial \Omega_j\right) \smallsetminus \partial \Omega$ the interface of our problem. We also fix arbitrarily the sides of the interface, and we thus denote by u^+, respectively u^- the non-tangential limits of u at the two sides of the interface. We define similarly the conormal derivatives $\nabla_\nu^A u^+$ and $\nabla_\nu^A u^-$ at the interface, but using the two sided limits of the coefficients a_{pq}^{ij} at Γ. We consider the parametric family of boundary value/interface problems

$$\begin{cases} P^A u(x,y) = f(x,y) & x \in \Omega, \\ u(x,y) = 0 & x \in \partial_D \Omega, \\ \nabla_\nu^A u(x,y) = g(x,y) & x \in \partial_N \Omega \\ u^+(x,y) - u^-(x,y) = 0 & x \in \Gamma \\ \nabla^A u^+(x,y) - \nabla^A u^-(x,y) = h(x,y) & x \in \Gamma \end{cases} \qquad (4)$$

where P^A is as in Equation (2), ∇_ν^A is as in Equation (3), and $y \in U$. We stress that for us the dependence of P^A on its coefficients, that is on A, is important, which justifies our notation.

2.3   Ellipticity and Positivity for Differential Operators

In this subsection we recall the definition of the positivity property for parametric families of differential operators. Let us therefore consider, for any $y \in U$, the parametric bilinear form $B(y; \cdot, \cdot)$ defined by

$$B(y; v, w) := \int_{x \in \Omega} \sum_{p,q=1}^{\mu} \left(\sum_{i,j=1}^{d} a_{pq}^{ij}(x,y) \partial_i v_p(x,y) \partial_j w_q(x,y) \right.$$

$$\left. \sum_{i=1}^{d} b_{pq}^{i}(x,y) \partial_i v_p(x,y) w_q(x,y) + c_{pq}(x,y) v_p(x,y) w_q(x,y) \right) dx, \quad y \in U. \quad (5)$$

Definition 1. *The family $(P_y)_{y \in U}$ is called* uniformly strictly positive definite *on $H_0^1(\Omega)^\mu \subset V \subset H^1(\Omega)^\mu$ if the coefficients a_{pq}^{ij} are symmetric in i, j and in p, q (that is, $a_{pq}^{ij} = a_{pq}^{ji} = a_{qp}^{ij}$, for all i, j, p, and q), and if there exist $0 < r < R < \infty$ such that for all $y \in U$, and $v, w \in V$, we have*

$$|B(y; v, w)| \le R \|v\|_{H^1(\Omega)} \|w\|_{H^1(\Omega)} \quad and \quad r \|v\|_{H^1(\Omega)}^2 \le B(y; v, v).$$

If U is reduced to a single point, that is, if we deal with the case of a single operator instead of a family, then we say that P is strictly positive definite. Throughout this paper, we shall assume that $(P_y)_{y \in U}$ *is uniformly strictly positive definite.* Positivity is closely related to ellipticity.

Definition 2. *The family $(P_y)_{y \in U}$ is called* uniformly strongly elliptic *if the coefficients a_{pq}^{ij} are symmetric in i, j and in p, q and if there exist $0 < r_e < R_e < \infty$ such that for all $x \in D$, $y \in U$, $\xi \in \mathbb{R}^d$, and $\eta \in \mathbb{R}^\mu$*

$$r_e |\xi|^2 |\eta|^2 \le \sum_{p,q=1}^{\mu} \sum_{i,j=1}^{d} a_{pq}^{ij}(x,y) \xi_i \xi_j \eta_p \eta_q \le R_e |\xi|^2 |\eta|^2. \quad (6)$$

In case one is interested only in scalar equations (not in systems), then for $V = H_D^1(\Omega)$, the assumption that our family P_y is uniformly positive definite can be replaced with the (slightly weaker) assumption that the family P_y is uniformly strongly elliptic, that $\sum_{i=1}^{d} \partial_i b^i = 0$ in Ω, $\sum_{i=1}^{d} \nu_i b^i = 0$ on $\partial_N \Omega$, $c \ge 0$, and $\partial_D \Omega \ne \emptyset$ (in which case it also follows that $\partial_D \Omega$ has a non-empty measure). In general, a uniformly strictly positive family P will also be uniformly strongly elliptic.

2.4 Consequences of Positivity

The usual Lax-Milgram lemma gives the following result as in [6]. Recall the constant r from Definition 1.

Proposition 1. *Assume that $f_y := f(\cdot, y) \in H_D^{-1}(\Omega)$, for any $y \in U$. Also, assume that the family P is uniformly strictly positive definite. Then our family of boundary value problems $P_y u_y = f_y$, $u_y \in H_D^1(\Omega)$, i.e., Equation (4), admits a unique solution $u_y = P_y^{-1} f_y$. Moreover, $\|P_y^{-1}\|_{\mathcal{L}(H_D^{-1}; H_D^1)} \le r^{-1}$, for all $y \in U$.*

The parametric solution $u_y \in H_D^1(\Omega)$ of Proposition 1 is then obtained from the usual weak formulation: given $y \in U$, find $u_y \in V := H_D^1(\Omega)$ such that

$$B(y; u_y, w) = (f_y, w) + \int_{\partial_N \Omega} g_y w dS + \int_{\Gamma} h_y w dS, \quad \forall w \in V, \quad (7)$$

where (f_y, w) denotes the $L^2(\Omega)$ inner product and dS is the surface measure on $\partial \Omega$ or on Γ. Also, $f_y(x) = f(x, y)$, and similarly for u_y, g_y, and h_y.

3   Broken Sobolev Spaces and Higher Regularity of Non-parametric Solutions in the Smooth Case

One of our main goals is to obtain regularity of the solution u both in the space variable x and in the parameter y. It is convenient to split this problem into two parts: regularity in x and regularity in y. We first address regularity in x in the case when the boundary $\partial\Omega$ and the interface Γ are smooth and disjoint. We also assume that each connected component of the boundary is given a single type of boundary conditions: either Dirichlet or Neumann. This leads to Theorem 1, which states the regularity and well-posedness of Problem (4) in this smooth case ($\partial\Omega$ and Γ smooth and disjoint). This is the main result of this paper, and, as far as we know, no complete proof was given before. We also consider coefficients with lower regularity than it is usually assumed, which is needed to treat the truly parametric case. (We are planning to address this question in a future paper.)

We assume throughout this and the following section that we are dealing with a single, non-parametric equation (not with a family), that is, that U is reduced to a single point in this subsection. We also assume that $\mu = 1$, to simplify the notation.

We shall need the "broken" version of the usual Sobolev spaces to deal with our interface problem. Recall the subdomains $\Omega_k \subset \Omega$, $1 \le k \le K$, we define:

$$\hat{H}^m(\Omega) := \{v : \Omega \to \mathbb{R}, v \in H^m(\Omega_k), \forall 1 \le k \le K\} \tag{8}$$
$$\hat{W}^{m,\infty}(\Omega) := \{v : \Omega \to \mathbb{R}, \partial^\alpha v \in L^\infty(\Omega_k), \forall 1 \le k \le K, |\alpha| \le m\}.$$

For further reference we note that the definitions of these spaces imply that the multiplication and differentiation maps $\hat{W}^{m,\infty}(\Omega) \times \hat{H}^m(\Omega) \to \hat{H}^m(\Omega)$ and $\partial_i : \hat{H}^m(\Omega) \to \hat{H}^{m-1}(\Omega)$ are continuous.

One of the difficulties of dealing with interface problems is the more complicated structure of the domains and ranges of our operators. When $m = 0$, we define $\mathcal{D}_m = \mathcal{D}_0 = H_D^1(\Omega) = V$ and $\mathcal{R}_m = \mathcal{R}_0 = H_D^{-1}(\Omega) = V^*$. Then we define $\tilde{P}^A$ in a weak sense using the bilinear form B introduced in Equation (5) (see the discussion around Equation (2.12) in [10] for more details or the discussion around Equation (20) in [11]). Assume now that $m \ge 1$. We then define

$$\mathcal{D}_m := \hat{H}^{m+1}(\Omega) \cap \{u = 0 \text{ on } \partial_D\Omega\} \cap \{u^+ - u^- = 0 \text{ on } \Gamma\} \text{ and}$$
$$\mathcal{R}_m := \hat{H}^{m-1}(\Omega) \oplus H^{m-1/2}(\partial_N\Omega) \oplus H^{m-1/2}(\Gamma).$$

In particular, $\mathcal{D}_m = \hat{H}^{m+1}(\Omega) \cap H_D^1(\Omega)$. Let $A = (a^{ij}, b^i, c) \in \hat{W}^{m,\infty}(\Omega)^{d^2+d+1}$ and $P^A u = \sum_{i,j=1}^2 \partial_i(a^{ij}\partial_j u) + \sum_{i=1}^2 b^i\partial_i u + cu$, as before. Then the family of partial differential operators $\tilde{P}_m^A : \mathcal{D}_m \to \mathcal{R}_m$,

$$\tilde{P}_m^A u = \left(P^A u, \nabla_\nu^A u|_{\partial_N\Omega}, (\nabla_\nu^A u^+ - \nabla_\nu^A u^+)|_\Gamma\right) \tag{9}$$

is well defined. Note that the domain $\mathcal{D}_m$ and codomain $\mathcal{R}_m$ are independent of $y \in U$, which justifies why we do not consider homogeneous Neumann boundary

conditions. We are now ready to state and prove our main theorem. Let us denote
$$\|u\|_{\hat{H}^m(\Omega)} := \left(\sum_{k=1}^{K} \|u\|_{H^m(\Omega_k)^2}\right)^{1/2} \text{ and } \|v\|_{\hat{W}^{m,\infty}(\Omega)} := \sum_{k=1}^{K} \|v\|_{W^{m,\infty}(\Omega_k)}$$
the resulting natural norms on the spaces introduced in Equation (8).

Theorem 1. *Let us assume that $\Omega \subset \mathbb{R}^d$ is smooth and bounded, that the interface Γ is smooth and does not intersect the boundary, and that to each component of the boundary it is associated a single type of boundary conditions (either Dirichlet or Neumann). Assume that $A = (a^{ij}, b^i, c) \in \hat{W}^{m,\infty}(\Omega)^{d^2+d+1}$ and that P^A is strictly positive definite on $H_D^1(\Omega)$, then $\tilde{P}_m^A : \mathcal{D}_m \to \mathcal{R}_m$ is invertible.*

Moreover, let $\|P^{-1}\|$ denote norm of the inverse of the map $P : H_D^1(\Omega) \to H_D^1(\Omega)^* =: H_D^{-1}(\Omega)$. Then there exists a constant $\tilde{C}_1 > 0$ such that the solution u of (4) satisfies

$$\|u\|_{\hat{H}^{m+1}(\Omega)} + \|u\|_{H^1(\Omega)} \leq \tilde{C}_1 \big(\|f\|_{\hat{H}^{m-1}(\Omega)} \|g\|_{H^{m-\frac{1}{2}}(\partial_N \Omega)} + \|h\|_{H^{m-\frac{1}{2}}(\Gamma)}\big), \quad (10)$$

with the constant $\tilde{C}_1 = \tilde{C}_1(m, \|P^{-1}\|, \|A\|_{\hat{W}^{m,\infty}}).$

Proof. In the case of the pure Dirichlet boundary conditions for an equation and without the explicit bounds in Equation (10), this lemma is a classical result, which is proved using divided differences and the so called "Nirenberg's trick" (see [8,12]). Since we consider transmission problems and want the more explicit bounds in the above Equation (10), let us now indicate the main steps to treat the interface regularity following the classical proof and [13]. The boundary conditions (*i. e.*, regularity at the boundary) were dealt with in [13]. In all the calculations below, all the constants C in this proof will be generic constants that will depend only on the variables on which $\tilde{C}_1$ depends (*i. e.*, on the order m, the norms $\|P^{-1}\|$ and $\|A\|_{\hat{W}^{m,\infty}(\Omega)}$). We split the proof into several steps.

Step 1. We first use Proposition 1 to conclude that $P : H_D^1(\Omega) \to H_D^1(\Omega)^*$ is indeed invertible. This provides the needed estimate for $m = 0$ (in which case, we recall, our problem (4) has to be interpreted in a weak sense).

Step 2. For $m > 0$ we can assume $g = 0$ and $h = 0$ by using the extension theorem as in [13].

Step 3. We also notice that, in view of the invertibility of P for $m = 0$ and since $u \in H_D^1(\Omega)$, it suffices to prove

$$\|u\|_{\hat{H}^{m+1}(\Omega)} \leq C\big(\sum_k \|f\|_{H^{m-1}(\Omega_k)} + \|u\|_{\hat{H}^m(\Omega)}\big). \quad (11)$$

Indeed, the desired inequality (10) will follow from Equation (11) by induction on m. Since Equation (11) holds for P if, and only if, it holds for $\lambda + P$, in order to prove Equation (11), it is also enough to assume that $\lambda + P$ is strictly positive for some $\lambda \in \mathbb{R}$. In particular, Equation (11) will continue to hold–with possibly different constants–if we change the lower order terms of P.

Step 4. Let us assume that $\Omega = \mathbb{R}^d$ with the interface given by $\Gamma = \{x_d = 0\}$. Let $\Omega_1 = \mathbb{R}_+^d$ and $\Omega_2 = \mathbb{R}_-^d$ be the two halves into which $\mathbb{R}^d$ is divided (so $K = 2$). Then we prove Equation (11) for these particular domains and for

$g = 0$ and $h = 0$ by induction on m. As we have noticed, the Equation (11) is true for $m = 0$, since the stronger relation (10) is true in this case. Thus, we shall assume that Equation (11) has been proved for m and for smaller values and we will prove it for $m + 1$. That is, we want to prove

$$\|u\|_{\hat{H}^{m+2}(\mathbb{R}^d)} \leq C\big(\|f\|_{\hat{H}^m(\mathbb{R}^d)} + \|u\|_{\hat{H}^{m+1}(\mathbb{R}^d)}\big). \tag{12}$$

To this end, let us first write

$$\|u\|_{\hat{H}^{m+2}(\mathbb{R}^d)} \leq \sum_{j=1}^{d} \|\partial_j u\|_{\hat{H}^{m+1}(\mathbb{R}^d)} + \|u\|_{L^2(\mathbb{R}^d)}. \tag{13}$$

We then use our estimate (11) for m (using the induction hypothesis) applied to the function $\partial_j u$ for $j < d$. This gives

$$\|\partial_j u\|_{\hat{H}^{m+1}(\mathbb{R}^d)} \leq \|P\partial_j u\|_{\hat{H}^{m-1}(\mathbb{R}^d)} \leq \|\partial_j f\|_{\hat{H}^{m-1}(\mathbb{R}^d)} + \|[P, \partial_j]u\|_{\hat{H}^{m-1}(\mathbb{R}^d)}$$
$$\leq \|f\|_{\hat{H}^m(\mathbb{R}^d)} + C\|u\|_{\hat{H}^{m+1}(\mathbb{R}^d)} \tag{14}$$

since the commutator $[P, \partial_j] = P\partial_j - \partial_j P$ is an operator of order ≤ 2 whose coefficients can be bounded in terms of $\|A\|_{\hat{W}^{m,\infty}(\Omega)}$. We now only need to estimate $\|\partial_d u\|_{H^{m+1}}$, we do that on each half subspace.

$$\|\partial_d u\|_{\hat{H}^{m+1}(\mathbb{R}^d)} \leq \sum_{j=1}^{d} \|\partial_j \partial_d u\|_{\hat{H}^m(\mathbb{R}^d)} + \|\partial_d u\|_{L^2(\mathbb{R}^d)}$$
$$\leq \sum_{j=1}^{d-1} \|\partial_j u\|_{\hat{H}^{m+1}(\mathbb{R}^d)} + \|\partial_d^2 u\|_{\hat{H}^m(\mathbb{R}^d)} + \|u\|_{\hat{H}^1(\mathbb{R}^d)}. \tag{15}$$

The right hand side of the above equation contains only terms that have already been estimated in the desired way, except for $\|\partial_d^2 u\|_{H^m}$. Since $m \geq 0$, we can use the relation $Pu = f$ to estimate this term as follows. Let us write $Pu = \sum \partial_i(a^{ij}\partial_j u) + \sum b^i \partial_i u + cu$. This gives $a^{dd}\partial_d^2 u = f - \sum_{(i,j)\neq(d,d)} a^{ij}\partial_i\partial_j u + Qu$, where Q is a first order differential operator. Next we notice that a^{dd} is uniformly bounded from below by the uniform strong positivity property (which implies uniform strong ellipticity): $(a^{dd})^{-1} \leq r^{-1}$. Note that by Proposition 1, we have $\|P^{-1}\| \leq r$, and hence r^{-1} is an admissible constant. This gives $\partial_d^2 u = (a^{dd})^{-1}f - \sum_{j=1}^{d-1} B^j\partial_j u + Q_1 u$ where B^j and Q_1 are first order differential operators with coefficients bounded by admissible constants, which then gives

$$\|\partial_d^2 u\|_{\hat{H}^m(\mathbb{R}^d)} \leq C\big(\|f\|_{\hat{H}^m(\mathbb{R}^d)} + \sum_{j=1}^{d-1} \|\partial_j u\|_{\hat{H}^{m+1}(\mathbb{R}^d)} + \|u\|_{\hat{H}^{m+1}(\mathbb{R}^d)}\big) \tag{16}$$
$$\leq C\big(\|f\|_{\hat{H}^m(\mathbb{R}^d)} + \|u\|_{\hat{H}^{m+1}(\mathbb{R}^d)}\big)$$

by Equation (14). Equation (15) and (16) then give

$$\|\partial_d u\|_{\hat{H}^{m+1}(\mathbb{R}^d)} \leq C\big(\|f\|_{\hat{H}^m(\mathbb{R}^d)} + \|u\|_{\hat{H}^{m+1}(\mathbb{R}^d)}\big). \tag{17}$$

Combining Equations (17) and (14) with Equation (13) gives then the desired Equation (12) for m replaced with $m + 1$.

Step 5. We finally reduce to the case of a half-space or a full space using a partition of unity as in the classical case, as follows. We choose a smooth partition of unity (ϕ_j) on Ω consisting of functions with small supports. The supports should be small enough so that if the support of ϕ_j intersects the boundary of Ω or the interface Γ, then the boundary or the interface can be straightened in a small neighborhood of the support of ϕ_j. We arrange that the resulting operators are positive and we complete the proof as in [13].

See also [14,15].

4   Weighted Sobolev Spaces and Higher Regularity of Non-parametric Solutions

We now assume $d = 2$, so Ω is a plane domain. We allow however Ω to be *piecewise* smooth. We also consider coefficients with lower regularity than the ones considered in [10]. This leads to Theorem 2, which will be then generalized to families in a forthcoming paper, which will contain also full details for the remaining results. *We continue to assume that we are dealing with a single, non-parametric equation and that $\mu = 1$.*

To formulate further assumptions on our problem and to state our results, we shall need weighted Sobolev spaces, both of L^2 and of L^∞ type. Let $\mathcal{V}$ be the set of singular points, where $Q \in \mathcal{V}$ if one of the following is satisfied: (1) it is a vertex, (2) it is a point where the boundary condition changes type (from Dirichlet to Neumann), (3) it is a point where the interface meets the boundary, or (4) it is a non-smooth point on the interface of the subdomains Ω_k. Let us denote from now on by $\rho : \mathbb{R}^2 \to [0, 1]$ a continuous function that is smooth outside the set $\mathcal{V}$ and is such that $\rho(x)$ is equal to the distance from $x \in \mathbb{R}^2$ to $\mathcal{V}$ when x is close to the singular set $\mathcal{V}$. The function ρ will be called the *smoothed distance to the set of singular points*. We can also assume $\|\nabla \rho\| \leq 1$, which will be convenient in later estimates, since it will reduce the number of constants (or parameters) in our estimates. We first define the *Babuška-Kondrat'ev spaces*

$$\mathcal{K}_a^m(\Omega) := \{v : \Omega \to \mathbb{R}, \, \rho^{|\alpha|-a}\partial^\alpha v \in L^2(\Omega), \, \forall \, |\alpha| \leq m\} \tag{18}$$

$$\mathcal{W}^{m,\infty}(\Omega) := \{v : \Omega \to \mathbb{R}, \, \rho^{|\alpha|}\partial^\alpha v \in L^\infty(\Omega), \, \forall \, |\alpha| \leq m\}. \tag{19}$$

We shall denote by $\| \cdot \|_{\mathcal{K}_a^m(\Omega)}$ and $\| \cdot \|_{\mathcal{W}^{m,\infty}(\Omega)}$ the resulting natural norms on these spaces. We shall need also the "broken" version of these Babuška-Kondrat'ev spaces for our interface problem. Recall the subdomains $\Omega_k \subset \Omega$, $1 \leq k \leq K$. In analogy with the smooth case, we then define: $\hat{\mathcal{K}}_a^m(\Omega) := \{v : \Omega \to \mathbb{R}, \, v \in \mathcal{K}_a^m(\Omega_k), \, \forall 1 \leq k \leq K\}$, and $\hat{\mathcal{W}}^{m,\infty}(\Omega) := \{v : \Omega \to \mathbb{R}, \, v \in \mathcal{W}^{m,\infty}(\Omega_k), \, \forall 1 \leq k \leq K\}$. If $\mathcal{V}$ is empty (that is, if the domain Ω is smooth and the interface is also smooth and does not touch the boundary), then we set $\rho \equiv 1$ and our spaces reduce to the broken Sobolev spaces $\hat{H}^m(\Omega)$ and $\hat{W}^{m,\infty}$ introduced in the previous section, Equation (8). As in the smooth case, the multiplication and differentiation maps $\hat{\mathcal{W}}^{m,\infty}(\Omega) \times \hat{\mathcal{K}}_a^m(\Omega) \to \hat{\mathcal{K}}_a^m(\Omega)$ and $\partial_i : \hat{\mathcal{K}}_a^m(\Omega) \to \tilde{\mathcal{K}}_{a-1}^{m-1}(\Omega)$ are continuous. Let $S \subset \partial\Omega_k$. Also as in the smooth case, we define the spaces $\mathcal{K}^{m+1/2}_{a+1/2}(S)$ as the restrictions to S of the functions $u \in \mathcal{K}^{m+1}_{a+1}(\Omega_k)$. These spaces have intrinsic descriptions [1,11] similar to the usual Babuška-Kondrat'ev spaces. Note that no "hat" is needed for the boundary version of the spaces $\hat{\mathcal{K}}$. Also $\mathcal{K}^{m+1/2}_{a+1/2}(S_1 \cup S_2) = \mathcal{K}^{m+1/2}_{a+1/2}(S_1) \oplus \mathcal{K}^{m+1/2}_{a+1/2}(S_2)$, if S_1 and S_2 are disjoint.

We need to consider the subset $\mathcal{V}_s$ of $\mathcal{V}$ consisting of Neumann-Neumann corners (i. e., corners where two edges endowed with Neumann boundary conditions meet) and non-smooth points of the interface, which can be described as $\mathcal{V}_s := \mathcal{V} \smallsetminus \{Q \in \mathcal{V}, Q \in \overline{\partial_D \Omega}\}$. Note that, if a point Q at the intersection of the interface Γ and the boundary falls on an edge with Neumann boundary conditions, then Q is also included in $\mathcal{V}_s$. In order to deal with the singularities arising at the points in $\mathcal{V}_s$ (which behave differently than the singularities at the other points of $\mathcal{V}$), we also need to augment our weighted Sobolev spaces with a suitable finite-dimensional space. Namely, for each point $Q \in \mathcal{V}_s$, we choose a function $\chi_Q \in \mathcal{C}^\infty(\bar{\Omega})$ that is constant equal to 1 in a neighborhood of Q. We can choose these functions to have disjoint supports. Let W_s be the linear span of the functions χ_Q for any $Q \in \mathcal{V}_s$. We now define the domains and ranges of our operators. Assume first that $m \geq 1$.

$$\mathcal{D}_{a,m} := (\hat{\mathcal{K}}^{m+1}_{a+1}(\Omega) + W_s) \cap \{u = 0 \text{ on } \partial_D \Omega\} \cap \{u^+ - u^- = 0 \text{ on } \Gamma\}$$
$$\mathcal{R}_{a,m} := \hat{\mathcal{K}}^{m-1}_{a-1}(\Omega) \oplus \mathcal{K}^{m-1/2}_{a-1/2}(\partial_N \Omega) \oplus \mathcal{K}^{m-1/2}_{a-1/2}(\Gamma).$$

Let us observe that, by definition, the functions in W_s satisfy the interface and boundary conditions (so $W_s \subset V := H^1_D(\Omega)$). Moreover, for $a \geq 0$, we have $\mathcal{D}_{a,m} = (\hat{\mathcal{K}}^{m+1}_{a+1}(\Omega) + W_s) \cap H^1_D(\Omega)$. Denote $A = (a^{ij}, b^i, c) \in \hat{\mathcal{W}}^{m,\infty}(\Omega)^{d^2+d+1}$, $d = 2$, and $P^A u = \sum^2_{i,j=1} \partial_i(a^{ij}\partial_j u) + \sum^2_{i=1} b^i \partial_i u + cu$, as before. Then the family of partial differential operators $\tilde{P}^A_{a,m} : \mathcal{D}_{a,m} \to \mathcal{R}_{a,m}$

$$\tilde{P}^A_{a,m} u = \left(Pu, \nabla^A_\nu u|_{\partial_N \Omega}, (\nabla^A_\nu u^+ - \nabla^A_\nu u^+)|_\Gamma\right) \tag{20}$$

is well defined and the induced map $\hat{\mathcal{W}}^{m,\infty}(\Omega)^{(d^2+d+1)} \ni A = (a^{ij}, b^i, c) \to P^A_{a,m} \in \mathcal{L}(\mathcal{D}_{a,m}, \mathcal{R}_{a,m})$ is continuous (recall that $d = 2$). The continuity of this map motivates the use of the spaces $\hat{\mathcal{W}}^{m,\infty}(\Omega)$.

When $m = 0$, we define

$$\mathcal{D}_{a,m} = \mathcal{D}_{a,0} = \mathcal{K}^1_{a+1}(\Omega) \cap \{u = 0 \text{ on } \partial_D \Omega\} + W_s$$
$$\mathcal{R}_{a,m} = \mathcal{R}_{a,0} = (\mathcal{K}^1_{-a+1}(\Omega) \cap \{u = 0 \text{ on } \partial_D \Omega\})^*,$$

where in the last equation the dual is defined as the dual with pivot $L^2(\Omega)$. Then we define $\tilde{P}_{a,0}$ in a weak sense using the bilinear form B introduced in Equation (5), as in the smooth case.

Recall the constant $0 < r$ in the definition of the uniform strict positivity (Definition 1) and Proposition 1. We now state the main result of this section. Recall that U is reduced to a point in this section.

Theorem 2. *Assume that* $A = (a^{ij}, b^i, c) \in \hat{\mathcal{W}}^{m,\infty}(\Omega)^{d^2+d+1}$, $d = 2$, *and that* P^A *is strictly positive definite on* $V = H_D^1(\Omega)$. *Then there exists* $0 < \eta$ *such that for any* $m \in \mathbb{N}_0$ *and for any* $0 < a < \eta$, *the map* $\tilde{P}_{a,m}^A : \mathcal{D}_{a,m} \to \mathcal{R}_{a,m}$ *is boundedly invertible and* $\|(\tilde{P}_{a,m}^A)^{-1}\| \leq \tilde{C}$, *where* $\tilde{C} = \tilde{C}(m, r, a, \|A\|_{\hat{\mathcal{W}}^{m,\infty}(\Omega)})$ *depends only on the indicated variables.*

A more typical formulation is given in the following corollary.

Corollary 1. *We use the notation and the assumptions of Theorem 2. If* $f \in \hat{\mathcal{K}}_{a-1}^{m-1}(\Omega)$, $g \in \mathcal{K}_{a-1/2}^{m-1/2}(\partial_N \Omega)$, *and* $h \in \mathcal{K}_{a-1/2}^{m-1/2}(\Gamma)$, *then the solution* $u \in H_D^1(\Omega)$ *of Problem (4) can be written* $u = u_r + u_s$, *with* $u_r \in \hat{\mathcal{K}}_{a+1}^{m+1}(\Omega)$ *and* $u \in W_s$, *such that*

$$\|u_r\|_{\hat{\mathcal{K}}_{a+1}^{m+1}} + \|u_s\|_{L^2} \leq \tilde{C}\big(\|f\|_{\hat{\mathcal{K}}_{a-1}^{m-1}} + \|g\|_{\mathcal{K}_{a-1/2}^{m-1/2}(\partial_N \Omega)} + \|h\|_{\mathcal{K}_{a-1/2}^{m-1/2}(\Gamma)}\big),$$

with $\tilde{C}$ *as in Theorem 2.*

5   Applications

We keep the settings and notations of the previous section. In particular, $d = 2$ and we are dealing with equations (not systems). One can proceed as in [6,7,13] to obtain h^m-quasi-optimal rates of convergence for the Galerkin u_n approximations of u. Namely, under suitable additional regularity in the $y \in U$ variable one can construct a sequence of finite dimensional subspaces $S_n \subset L^2(U; V)$ such that

$$\|u - u_n\|_{L^2(U;V)} \leq C \dim(S_n)^{-m/2} \|f\|_{H^{m-1}(\Omega)}. \tag{21}$$

This is based on a holomorphic regularity in U and on the approximation properties in [10]. We now state a uniform shift theorem for our families of boundary value/transmission problems.

Let us denote by $\mathcal{C}_b^k(U; Z)$ the space of k-times boundedly differentiable functions defined on U with values in the Banach space Z. By $\mathcal{C}_b^\omega(U; Z)$ we shall denote the space of *analytic* functions with bounded derivatives defined on U with values in the Banach space Z. Recall that r is the constant appearing in the definition of uniform positivity of the family $(P_y)_{y \in U}$. Theorem 2 extends to families of boundary value problems as in [13] as follows. Let us denote by $\eta(y)$ the best constant appearing in Theorem 2 for $P = P_y$ and $\eta = \inf_{y \in U} \eta(y)$.

Theorem 3. *Let* $m \in \mathbb{N}_0$ *and* $k_0 \in \mathbb{N}_0 \cup \{\infty, \omega\}$ *be fixed. Assume that* $A = (a^{ij}, b^i, c) \in \mathcal{C}_b^{k_0}(U; \hat{\mathcal{W}}^{m,\infty}(\Omega))^{d^2+d+1}$, $d = 2$, *and that the family* P_y^A *is uniformly positive definite. Then* $\eta = \inf_{y \in U} \eta(y) > 0$. *Let* $f \in \mathcal{C}_b^{k_0}(U; \hat{\mathcal{K}}_{a-1}^{m-1}(\Omega))$, $g \in \mathcal{C}_b^{k_0}(U; \mathcal{K}_{a-1/2}^{m-1/2}(\partial_N \Omega))$, $h \in \mathcal{C}_b^{k_0}(U; \mathcal{K}_{a-1/2}^{m-1/2}(\Gamma))$, *and* $0 < a < \eta$. *Then the solution* u *of our family of boundary value problems (4) satisfies* $u \in \mathcal{C}_b^{k_0}(U; \mathcal{D}_{a,m})$. *Moreover, for each finite* $k \leq k_0$, *there exists a constant* $C_{a,m} > 0$ *such that*

$$\|u\|_{\mathcal{C}_b^k(U;\mathcal{D}_{a,m})} \leq C_{a,m}\big(\|f\|_{\mathcal{C}_b^k(U;\hat{\mathcal{K}}_{a-1}^{m-1}(\Omega))}$$
$$+ \|g\|_{\mathcal{C}_b^k(U;\mathcal{K}_{a-1/2}^{m-1/2}(\partial_N \Omega))} + \|h\|_{\mathcal{C}_b^k(U;\mathcal{K}_{a-1/2}^{m-1/2}(\Gamma))}\big).$$

The constant $C_{a,m}$ depends only on r, m, a, k, and the norms of the coefficients a^{ij}, b^i, c in $C_b^k(U; \hat{W}^{m,\infty}(\Omega))$, but not on f or g.

References

1. Ammann, B., Ionescu, A., Nistor, V.: Sobolev spaces on Lie manifolds and regularity for polyhedral domains. Documenta Math (Electronic) 11, 161–206 (2006)
2. Babuska, I., Nobile, F., Tempone, R.: A stochastic collocation method for elliptic partial differential equations with ramdom input data. SIAM Rev. 52(2), 317–355 (2010)
3. Bacuta, C., Nistor, V.: A priori estimates and high order Galerkin approximation for parametric partial differential equations (work in progress)
4. Bacuta, C., Nistor, V., Zikatanov, L.: Improving the rate of convergence of 'high order finite elements' on polygons and domains with cusps. Numerische Mathematik 100, 165–184 (2005)
5. Bacuta, C., Nistor, V., Zikatanov, L.: Improving the rate of convergence of high order finite elements on polyhedral II: Mesh refinements and interpolation. Numer. Funct. Anal. Optim. 28(7-8), 775–842 (2007)
6. Cohen, A., DeVore, R., Schwab, C.: Convergence rates of best N-term Galerkin approximations for a class of elliptic PDEs. Found. Comput. Math. 10(6), 615–646 (2010)
7. Cohen, A., DeVore, R., Schwab, C.: Analytic regularity and polynomial approximation of parametric and stochastic elliptic PDEs. Anal. Appl. (Singap.) 9(1), 11–47 (2011)
8. Fichera, F.: Linear elliptic differential systems and eigenvalue problems. Lecture Notes in Mathematics, vol. 8. Springer, Berlin (1965)
9. Gerritsma, M., van der Steen, J., Vos, P., Karniadakis, G.: Time-dependent generalized polynomial chaos. J. Comput. Phys. 229(22), 8333–8363 (2010)
10. Li, H., Mazzucato, A., Nistor, V.: Analysis of the finite element method for transmission/mixed boundary value problems on general polygonal domains. Electron. Trans. Numer. Anal. 37, 41–69 (2010)
11. Mazzucato, A., Nistor, V.: Well-posedness and regularity for the elasticity equation with mixed boundary conditions on polyhedral domains and domains with cracks. Arch. Ration. Mech. Anal. 195(1), 25–73 (2010)
12. Morrey, C.: Jr.: Second-order elliptic systems of differential equations. In: Contributions to the Theory of Partial Differential Equations, Annals of Mathematics Studies, vol. (33), pp. 101–159. Princeton University Press, Princeton (1954)
13. Nistor, V.: Schwab, Ch.: High order Galerkin approximations for parametric second order elliptic partial differential equations. ETH Seminar for Applied Mathematics Report 2012-21, to appear in M3AS (2012)
14. Roĭtberg, J., Šeftel', Z.: On equations of elliptic type with discontinuous coefficients. Dokl. Akad. Nauk SSSR 146, 1275–1278 (1962).
15. Roĭtberg, J., Šeftel, Z.: General boundary-value problems for elliptic equations with discontinuous coefficients. Dokl. Akad. Nauk SSSR 148, 1034–1037 (1963).
16. Schwab, C., Todor, R.: Sparse finite elements for elliptic problems with stochastic loading. Numer. Math. 95(4), 707–734 (2003)
17. Wan, X., Karniadakis, G.: Error control in multi-element generalized polynomial chaos method for elliptic problems with random coefficients. Commun. Comput. Phys. 5(2-4), 793–820 (2009)

Galerkin FEM for Fractional Order Parabolic Equations with Initial Data in H^{-s}, $0 \le s \le 1$

Bangti Jin, Raytcho Lazarov, Joseph Pasciak, and Zhi Zhou

Department of Mathematics, Texas A&M University, College Station, TX 77843, USA

Abstract. We investigate semi-discrete numerical schemes based on the standard Galerkin and lumped mass Galerkin finite element methods for an initial-boundary value problem for homogeneous fractional diffusion problems with non-smooth initial data. We assume that $\Omega \subset \mathbb{R}^d$, $d = 1, 2, 3$ is a convex polygonal (polyhedral) domain. We theoretically justify optimal order error estimates in L_2- and H^1-norms for initial data in $H^{-s}(\Omega)$, $0 \le s \le 1$. We confirm our theoretical findings with a number of numerical tests that include initial data v being a Dirac δ-function supported on a $(d-1)$-dimensional manifold.

1 Introduction

We consider the initial–boundary value problem for the fractional order parabolic differential equation for $u(x,t)$:

$$
\begin{aligned}
\partial_t^\alpha u(x,t) + \mathcal{L}u(x,t) &= f(x,t) && \text{in } \Omega, \quad T \ge t > 0, \\
u(x,t) &= 0 && \text{in } \partial\Omega, \quad T \ge t > 0, \\
u(x,0) &= v(x) && \text{in } \Omega,
\end{aligned}
\tag{1}
$$

where $\Omega \subset \mathbb{R}^d$ $(d = 1, 2, 3)$ is a bounded convex polygonal domain with a boundary $\partial\Omega$, and $\mathcal{L}$ is a symmetric, uniformly elliptic second-order differential operator. Integrating the second order derivatives by parts (once) gives rise to a bilinear form $a(\cdot, \cdot)$ satisfying

$$
a(v,w) = (\mathcal{L}v, w) \forall v \in H^2(\Omega), w \in H_0^1(\Omega),
$$

where $(\cdot, \cdot)$ denotes the inner product in $L_2(\Omega)$. The form $a(\cdot, \cdot)$ extends continuously to $H_0^1(\Omega) \times H_0^1(\Omega)$ where it is symmetric and coercive and we take $\|u\|_{H^1} = a(u,u)^{1/2}$, for all $u \in H_0^1(\Omega)$. Similarly, $\mathcal{L}$ extends continuously to an operator from $H_0^1(\Omega)$ to $H^{-1}(\Omega)$ (the set of bounded linear functionals on $H_0^1(\Omega)$) by

$$
\langle \mathcal{L}u, v \rangle = a(u,v) \forall u, v \in H_0^1(\Omega).
\tag{2}
$$

Here $\langle \cdot, \cdot \rangle$ denotes duality pairing between $H^{-1}(\Omega)$ and $H_0^1(\Omega)$. We assume that the coefficients of $\mathcal{L}$ are smooth enough so that solutions $v \in H_0^1(\Omega)$ satisfying

$$
a(v,\phi) = (f,\phi) \forall \phi \in H_0^1(\Omega)
$$

with $f \in L_2(\Omega)$ are in $H^2(\Omega)$.

I. Dimov, I. Faragó, and L. Vulkov (Eds.): NAA 2012, LNCS 8236, pp. 24–37, 2013.
© Springer-Verlag Berlin Heidelberg 2013

Here $\partial_t^\alpha u$ $(0 < \alpha < 1)$ denotes the left-sided Caputo fractional derivative of order α with respect to t and it is defined by (cf. [1, p. 91] or [2, p. 78])

$$\partial_t^\alpha v(t) = \frac{1}{\Gamma(1-\alpha)} \int_0^t (t-\tau)^{-\alpha} \frac{d}{d\tau} v(\tau)\, d\tau,$$

where $\Gamma(\cdot)$ is the Gamma function. Note that as the fractional order α tends to unity, the fractional derivative $\partial_t^\alpha u$ converges to the canonical first order derivative $\frac{du}{dt}$ [1], and thus (1) reproduces the standard parabolic equation. The model (1) captures well the dynamics of subdiffusion processes in which the mean square variance grows slower than that in a Gaussian process [3] and has found a number of practical applications. A comprehensive survey on fractional order differential equations arising in viscoelasticity, dynamical systems in control theory, electrical circuits with fractance, generalized voltage divider, fractional-order multipoles in electromagnetism, electrochemistry, and model of neurons is provided in [4]; see also [2].

The goal of this study is to develop, justify, and test a numerical technique for solving (1) with non-smooth initial data $v \in H^{-s}(\Omega)$, $0 \le s \le 1$, an important case in various applications and typical in related inverse problems; see e.g., [5], [6, Problem (4.12)] and [7,8]. This includes the case of v being a delta-function supported on a $(d-1)$–dimensional manifold in $\mathbb{R}^d$, which is particularly interesting from both theoretical and practical points of view.

The weak form for problem (1) reads: find $u(t) \in H_0^1(\Omega)$ such that

$$(\partial_t^\alpha u, \chi) + a(u, \chi) = (f, \chi) \quad \forall \chi \in H_0^1(\Omega),\ T \ge t > 0, \quad u(0) = v. \tag{3}$$

The folowing two results are known, cf. [6]: (1) for $v \in L_2(\Omega)$ the problem (1) has a unique solution in $C([0,T]; L_2(\Omega)) \cap C((0,T]; H^2(\Omega) \cap H_0^1(\Omega))$ [6, Theorem 2.1]; (2) for $f \in L_\infty(0,T; L_2(\Omega))$, problem (1) has a unique solution in $L_2(0,T; H^2(\Omega) \cap H_0^1(\Omega))$ [6, Theorem 2.2].

To introduce the semidiscrete FEM for problem (1) we follow standard notation in [9]. Let $\{\mathcal{T}_h\}_{0<h<1}$ be a family of regular partitions of the domain Ω into d-simplexes, called finite elements, with h denoting the maximum diameter. Throughout, we assume that the triangulation $\mathcal{T}_h$ is quasi-uniform, i.e., the diameter of the inscribed disk in the finite element $\tau \in \mathcal{T}_h$ is bounded from below by h, uniformly on $\mathcal{T}_h$. The approximation u_h will be sought in the finite element space $X_h \equiv X_h(\Omega)$ of continuous piecewise linear functions over $\mathcal{T}_h$:

$$X_h = \left\{ \chi \in H_0^1(\Omega) : \ \chi \ \text{is a linear function over} \ \tau \ \forall \tau \in \mathcal{T}_h \right\}.$$

The semidiscrete Galerkin FEM for problem (1) is: find $u_h(t) \in X_h$ such that

$$(\partial_t^\alpha u_h, \chi) + a(u_h, \chi) = (f, \chi) \quad \forall \chi \in X_h,\ T \ge t > 0, \quad u_h(0) = v_h, \tag{4}$$

where $v_h \in X_h$ is an approximation of v. The choice of v_h will depend on the smoothness of v. For smooth data, $v \in H^2(\Omega) \cap H_0^1(\Omega)$, we can choose v_h to be either the finite element interpolant or the Ritz projection $R_h v$ onto X_h. In the case of non-smooth data, $v \in L_2(\Omega)$, following Thomée [9], we shall take $v_h = P_h v$, where P_h is the L_2-orthogonal projection operator $P_h : L_2(\Omega) \to X_h$, defined by $(P_h \phi, \chi) = (\phi, \chi)$, for all $\chi \in X_h$. In the intermediate case, $v \in H_0^1(\Omega)$, we can choose either $v_h = P_h v$ or $v_h = R_h v$. The goal of this paper is to study the convergence rates of the semidiscrete Galerkin method (4) for initial data $v \in H^{-s}(\Omega)$, $0 \leq s \leq 1$ when $f = 0$.

The rest of the paper is organized as follows. In Section 2 we briefly review the regularity theory for problem (1). In Section 3 we motivate our study by considering an 1-D example with initial data being a δ–function. Then in Theorem 2 we prove the main result: for $0 \leq s \leq 1$, the following error bound holds

$$\|u(t) - u_h(t)\| + h\|\nabla(u(t) - u_h(t))\| \leq Ch^{2-s}t^{-\alpha}\ell_h\|v\|_{-s}, \quad \ell_h = |\ln h|.$$

Further, in Section 4 we show a similar result for the lumped mass Galerkin method. Finally, in Section 5 we present numerical results for test problems with smooth, intermediate, non-smooth initial data and initial data that is a δ–function, all confirming our theoretical findings.

2 Preliminaries

For the existence and regularity of the solution to (1), we need some notation and preliminary results. For $s \geq -1$, we denote by $\dot{H}^s(\Omega) \subset H^{-1}(\Omega)$ the Hilbert space induced by the norm

$$|v|_s^2 = \sum_{j=1}^{\infty} \lambda_j^s \langle v, \varphi_j \rangle^2 \tag{5}$$

with $\{\lambda_j\}_{j=1}^{\infty}$ and $\{\varphi_j\}_{j=1}^{\infty}$ being respectively the Dirichlet eigenvalues and the L_2-orthonormal eigenfunctions of $\mathcal{L}$. As usual, we identify functions f in $L_2(\Omega)$ with the functional F in $H^{-1}(\Omega)$ defined by $\langle F, \phi \rangle = (f, \phi)$, for all $\phi \in H_0^1(\Omega)$. The set $\{\varphi_j\}_{j=1}^{\infty}$, respectively, $\{\lambda_j^{\frac{1}{2}} \varphi_j\}_{j=1}^{\infty}$, forms an orthonormal basis in $L_2(\Omega)$, respectively, $H^{-1}(\Omega)$. Thus $|v|_0 = \|v\| = (v, v)^{\frac{1}{2}}$ is the norm in $L_2(\Omega)$ and $|v|_{-1} = \|v\|_{H^{-1}(\Omega)}$ is the norm in $H^{-1}(\Omega)$. It is easy to check that $|v|_1 = a(v, v)^{\frac{1}{2}}$ is also the norm in $H_0^1(\Omega)$. Note that $\{\dot{H}^s(\Omega)\}$, $s \geq -1$ form a Hilbert scale of interpolation spaces. Motivated by this, we denote $\|\cdot\|_{H^s}$ to be the norm on the interpolation scale between $H_0^1(\Omega)$ and $L_2(\Omega)$ when s is in $[0, 1]$ and $\|\cdot\|_{H^s}$ to be the norm on the interpolation scale between $L_2(\Omega)$ and $H^{-1}(\Omega)$ when s is in $[-1, 0]$. Thus, $\|\cdot\|_{H^s}$ and $|\cdot|_s$ provide equivalent norms for $s \in [-1, 1]$.

We further assume that the coefficients of the elliptic operator $\mathcal{L}$ are sufficiently smooth and the polygonal domain Ω is convex, so that $|v|_2 = \|\mathcal{L}v\|$ is equivalent to the norm in $H^2(\Omega) \cap H_0^1(\Omega)$ (cf. the proof of Lemma 3.1 of [9]).

Now we introduce the operator $E(t)$ by

$$E(t)v = \sum_{j=1}^{\infty} E_{\alpha,1}(-\lambda_j t^\alpha)\,(v, \varphi_j)\,\varphi_j, \quad \text{where } E_{\alpha,\beta}(z) = \sum_{k=0}^{\infty} \frac{z^k}{\Gamma(k\alpha + \beta)}. \tag{6}$$

Here $E_{\alpha,\beta}(z)$ is the Mittag-Leffler function defined for $z \in \mathbb{C}$ [1]. The operator $E(t)$ gives a representation of the solution u of (1) with a homogeneous right hand side, so that for $f(x,t) \equiv 0$ we have $u(t) = E(t)v$. This representation follows from eigenfunction expansion [6]. Further, we introduce the operator $\bar{E}(t)$ defined for $\chi \in L_2(\Omega)$ as

$$\bar{E}(t)\chi = \sum_{j=1}^{\infty} t^{\alpha-1} E_{\alpha,\alpha}(-\lambda_j t^\alpha)\,(\chi,\varphi_j)\,\varphi_j. \tag{7}$$

The operators $E(t)$ and $\bar{E}(t)$ together give the following representation of the solution of (1):

$$u(t) = E(t)v + \int_0^t \bar{E}(t-s)f(s)ds. \tag{8}$$

Motivated by [5,6], we will study the convergence of semidiscrete Galerkin methods for problem (1) with very weak initial data, i.e., $v \in H^{-s}(\Omega), 0 \le s \le 1$. Then the following question arises naturally: in what sense should we understand the solution for such weak data? Obviously, for any $t > 0$ the function $u(t) = E(t)v$ satisfies equation (1). Moreover, by dominated convergence we have

$$\lim_{t\to 0+} |E(t)v - v|_{-s} = \left(\lim_{t\to 0+} \sum_{j=1}^{\infty} (E_{\alpha,1}(-\lambda_j t^\alpha) - 1)^2 \lambda_j^{-s}(v,\varphi_j)^2 \right)^{\frac{1}{2}} = 0,$$

provided that $v \in H^{-s}(\Omega)$. Here $(v,\varphi_j) = \langle v,\varphi_j \rangle_{H^{-s},H^s}$ is well defined since $\varphi_j \in H_0^1(\Omega)$. Therefore, the function $u(t) = E(t)v$ satisfies (1) and for $t \to 0$ it converges to v in H^{-s}–norm. That is, it is a weak solution to (1); see also [5, Proposition 2.1].

For the solution of the homogeneous equation (1), which is the object of our study, we have the following stability and smoothing estimates.

Theorem 1. *Let $u(t) = E(t)v$ be the solution to problem (1) with $f \equiv 0$. Then for $t > 0$ we have the the following estimates:*

(a) for $\ell = 0$, $0 \le q \le p \le 2$ and for $\ell = 1$, $0 \le p \le q \le 2$ and $q \le p+2$

$$|(\partial_t^\alpha)^\ell u(t)|_p \le Ct^{-\alpha(\ell + \frac{p-q}{2})}|v|_q, \tag{9}$$

(b) for $0 \le s \le 1$ and $0 \le p+s \le 2$

$$|\partial_t^\alpha u(t)|_{-s} \le Ct^{-\alpha}|v|_{-s} \quad \text{and} \quad |u(t)|_p \le Ct^{-\frac{p+s}{2}\alpha}|v|_{-s}. \tag{10}$$

Proof. Part (a) can be found in [6, Theorem 2.1] and [10, Theorem 2.1]. Hence we only show part (b). Note that for $t > 0$,

$$|u(t)|_p^2 \leq \sum_{j=1}^{\infty} \lambda_j^p |E_{\alpha,1}(-\lambda_j t^\alpha)|^2 |(v, \varphi_j)|^2 \leq C \sum_{j=1}^{\infty} \frac{\lambda_j^p}{(1 + \lambda_j t^\alpha)^2} |(v, \varphi_j)|^2$$

$$\leq C t^{-(p+s)\alpha} \sum_{j=1}^{\infty} \frac{(\lambda_j t^\alpha)^{p+s}}{(1 + \lambda_j t^\alpha)^2} \lambda_j^{-s} |(v, \varphi_j)|^2$$

$$\leq C t^{-(p+s)\alpha} \sum_{j=1}^{\infty} \lambda_j^{-s} |(v, \varphi_j)|^2 = C t^{-(p+s)\alpha} |v|_{-s}^2,$$

which proves the second inequality of case (b). The first estimate follows similarly by noticing the identity $\partial_t^\alpha E_{\alpha,1}(-\lambda t^\alpha) = -\lambda E_{\alpha,1}(-\lambda t^\alpha)$ [1].

We shall need some properties of the L_2-projection P_h onto X_h.

Lemma 1. *Assume that the mesh is quasi–uniform. Then for $s \in [0,1]$,*

$$\|(I - P_h)w\|_{H^s} \leq C h^{2-s} \|w\|_{H^2} \forall w \in H^2(\Omega) \cap H_0^1(\Omega),$$

and

$$\|(I - P_h)w\|_{H^s} \leq C h^{1-s} \|w\|_{H^1} \forall w \in H_0^1(\Omega).$$

In addition, P_h is stable on $H^s(\Omega)$ for $s \in [-1, 0]$.

Proof. Since the mesh is quasi-uniform, the L_2–projection operator P_h is stable in $H_0^1(\Omega)$ [11]. This immediately implies its stability in $H^{-1}(\Omega)$. Thus, stability on $H^{-s}(\Omega)$ follows from this, the trivial stability of P_h on $L_2(\Omega)$ and interpolation.

Let I_h be the finite element interpolation operator and C_h be the Clement or Scott-Zhang interpolation operator. It follows from the stability of P_h in $L_2(\Omega)$ and $H_0^1(\Omega)$ that

$$\|(I - P_h)w\|_{L_2} \leq \|(I - I_h)w\|_{L_2} \leq C h^2 \|w\|_{H^2} \forall w \in H^2(\Omega) \cap H_0^1(\Omega),$$

$$\|(I - P_h)w\|_{H^1} \leq C \|(I - I_h)w\|_{H^1} \leq C h \|w\|_{H^2} \forall w \in H^2(\Omega) \cap H_0^1(\Omega),$$

$$\|(I - P_h)w\|_{L_2} \leq \|(I - C_h)w\|_{L_2} \leq C h \|w\|_{H^1} \forall w \in H_0^1(\Omega),$$

$$\|(I - P_h)w\|_{H^1} \leq C \|w\|_{H^1} \forall w \in H_0^1(\Omega).$$

The inequalities of the lemma follow by interpolation.

Remark 1. All the norms appearing in Lemma 1 can be replaced by their corresponding equivalent dotted norms.

3   Galerkin Finite Element Method

To motivate our study we shall first consider the 1-D case, i.e., $\mathcal{L}u = -u''$, and take initial data the Dirac δ-function at $x = \frac{1}{2}$, $\langle \delta, v \rangle = v(\frac{1}{2})$. It is well known that $H_0^{\frac{1}{2}+\epsilon}(0,1)$ embeds continuously into $C_0(0,1)$, hence the δ-function is a bounded linear functional on the space $H_0^{\frac{1}{2}+\epsilon}(\Omega)$, i.e., $\delta \in H^{-\frac{1}{2}-\epsilon}(\Omega)$.

In Tables 1 and 2 we show the error and the convergence rates for the semidiscrete Galerkin FEM and semidiscrete lumped mass FEM (cf. Section 4) for initial data v being a Dirac δ-function at $x = \frac{1}{2}$. The results suggest an $O(h^{\frac{1}{2}})$ and $O(h^{\frac{3}{2}})$ convergence rate for the H^1- and L_2-norm of the error, respectively. Below we prove that up to a factor $|\ln h|$ for fixed $t > 0$, the convergence rate is of the order reported in Tables 1 and 2. In Table 3 we show the results for the case that the δ-function is supported at a grid point. In this case the standard Galerkin method converges at the expected rate in H^1-norm, while the convergence rate in the L_2-norm is $O(h^2)$. This is attributed to the fact that in 1-D the solution with the δ-function as the initial data is smooth from both sides of the support point and the finite element spaces have good approximation property.

Table 1. Standard FEM with initial data $\delta(\frac{1}{2})$ for $h = 1/(2^k + 1)$, $\alpha = 0.5$

time	k	3	4	5	6	7	ratio	rate
$t = 0.005$	L_2-norm	3.95e-2	1.59e-2	6.00e-3	2.19e-3	7.89e-4	≈ 2.75	$O(h^{\frac{3}{2}})$
	H^1-norm	1.21e0	8.99e-1	6.52e-1	4.66e-1	3.33e-1	≈ 1.40	$O(h^{\frac{1}{2}})$
$t = 0.01$	L_2-norm	2.85e-2	1.13e-2	4.26e-3	1.55e-3	5.58e-4	≈ 2.77	$O(h^{\frac{3}{2}})$
	H^1-norm	8.66e-1	6.39e-1	4.62e-1	3.31e-1	2.35e-1	≈ 1.40	$O(h^{\frac{1}{2}})$

Table 2. Lumped mass FEM with initial data $\delta(\frac{1}{2})$, $h = 1/2^k$ $\alpha = 0.5$

time	k	3	4	5	6	7	ratio	rate
$t = 0.005$	L_2-norm	7.24e-2	2.66e-2	9.54e-3	3.40e-3	1.21e-3	≈ 2.79	$O(h^{\frac{3}{2}})$
	H^1-norm	1.51e0	1.07e0	7.60e-1	5.40e-1	3.81e-1	≈ 1.41	$O(h^{\frac{1}{2}})$
$t = 0.01$	L_2-norm	5.20e-2	1.89e-2	6.77e-3	2.40e-3	8.54e-4	≈ 2.79	$O(h^{\frac{3}{2}})$
	H^1-norm	1.07e0	7.59e-1	5.37e-1	3.80e-1	2.70e-1	≈ 1.41	$O(h^{\frac{1}{2}})$

The numerical results in Tables 1–3 motivate our study on the convergence rates of the semidiscrete Galerkin and lumped mass schemes for initial data $v \in H^{-s}(\Omega)$, $0 \le s \le 1$.

Table 3. Standard semidiscrete FEM with initial data $\delta(\frac{1}{2})$, $h = 1/2^k$, $\alpha = 0.5$

Time	k	3	4	5	6	7	ratio	rate
$t = 0.005$	L_2-norm	5.13e-3	1.28e-3	3.21e-4	8.03e-5	2.01e-5	≈ 3.99	$O(h^2)$
	H^1-norm	4.29e-1	3.09e-1	2.21e-1	1.56e-1	1.11e-1	≈ 1.41	$O(h^{\frac{1}{2}})$
$t = 0.01$	L_2-norm	3.07e-3	7.70e-4	1.93e-4	4.82e-5	1.21e-5	≈ 3.98	$O(h^2)$
	H^1-norm	3.04e-1	2.19e-1	1.56e-1	1.11e-1	7.87e-2	≈ 1.41	$O(h^{\frac{1}{2}})$

Theorem 2. *Let u and u_h be the solutions of* (1) *and the semidiscrete Galerkin finite element method* (4) *with $v_h = P_h v$, respectively. Then there is a constant $C > 0$ such that for $0 \le s \le 1$*

$$\|u_h(t) - u(t)\| + h\|\nabla(u_h(t) - u(t))\| \le Ch^{2-s} \ell_h \, t^{-\alpha} |v|_{-s}. \tag{11}$$

Remark 2. Note that for any fixed ϵ there is a $C_\epsilon > 0$ such that $|\delta|_{-\frac{1}{2}-\epsilon} \le C_\epsilon$. Thus, modulo the factor $\ell_h = |\ln h|$, the theorem confirms the computational results of Table 1, namely convergence in the L_2–norm with a rate $O(h^{\frac{3}{2}})$ and in H^1–norm with a rate $O(h^{\frac{1}{2}})$.

Proof. We shall need the following auxiliary problem: find $u^h(t) \in H_0^1(\Omega)$, s.t.

$$(\partial_t^\alpha u^h(t), \chi) + a(u^h(t), \chi) = (f(t), \chi) \quad \forall \chi \in H_0^1(\Omega), \; t > 0, \quad u^h(0) = P_h v. \tag{12}$$

We note that the initial data $u^h(0) = P_h v \in H_0^1(\Omega)$ is smooth.

Now we consider the semidiscrete Galerkin method for problem (12), i.e., equation (4) with $v_h = P_h v$. By Theorem 3.2 of [10] we have

$$\|u_h(t) - u^h(t)\| + h\|\nabla(u_h(t) - u^h(t))\| \le Ch^2 \ell_h \, t^{-\alpha} \|P_h v\|. \tag{13}$$

Now, using the inverse inequality $\|P_h v\| \le Ch^{-s}\|P_h v\|_{-s}$, for $0 \le s \le 1$, and the stability of P_h in $H^{-s}(\Omega)$ (cf. Lemma 1), we get

$$\|u_h(t) - u^h(t)\| + h\|\nabla(u_h(t) - u^h(t))\| \le Ch^{2-s} \ell_h \, t^{-\alpha} \|v\|_{-s}. \tag{14}$$

Now we estimate $u(t) - u^h(t) = E(t)(v - P_h v)$. To this end, let $\{v_n\} \subset L_2(\Omega)$ be a sequence converging to v in $H^{-s}(\Omega)$. Noting that the operators P_h and $E(t)$ are self-adjoint in $(\cdot, \cdot)$ and using the smoothing property (9) of $E(t)$ with $\ell = 0$, $q = 0$ and $p = 2$, we obtain for any $\phi \in L_2(\Omega)$

$$|(E(t)(I - P_h)v_n, \phi)| = |(v_n, (I - P_h)E(t)\phi)| \le |v_n|_{-s}|(I - P_h)E(t)\phi|_s$$
$$\le Ch^{2-s}|v_n|_{-s}|E(t)\phi|_2 \le Ch^{2-s}t^{-\alpha}|v_n|_{-s}\|\phi\|.$$

Taking the limit as n tends to infinity gives

$$\|u(t) - u^h(t)\| = \sup_{\phi \in L_2(\Omega)} \frac{|(E(t)(I - P_h)v, \phi)|}{\|\phi\|} \le Ch^{2-s}t^{-\alpha}|v|_{-s}. \tag{15}$$

Then by the triangle inequality we arrive at the L_2-estimate in (11).

Next, for the gradient term $\|\nabla(u(t) - u^h(t))\|$, we observe that for any $\phi \in \dot{H}^1(\Omega)$, by the coercivity of $a(\cdot, \cdot)$, we have

$$C_0 \|\nabla(E(t)(I - P_h)v_n)\|^2 \leq a(E(t)(I - P_h)v_n, E(t)(I - P_h)v_n)$$

$$\leq \sup_{\phi \in H_0^1(\Omega)} \frac{a(E(t)(I - P_h)v_n, \phi)^2}{a(\phi, \phi)}. \tag{16}$$

Meanwhile we have

$$|a(E(t)(I - P_h)v_n, \phi)| = |((I - P_h)v_n, E(t)\mathcal{L}\phi)| = |(v_n, (I - P_h)E(t)\mathcal{L}\phi)|$$

$$\leq C|v_n|_{-s}|(I - P_h)E(t)\mathcal{L}\phi|_s \leq Ch^{1-s}|v_n|_{-s}|E(t)\mathcal{L}\phi|_1$$

$$\leq Ch^{1-s}t^{-\alpha}|v_n|_{-s}|\mathcal{L}\phi|_{-1} \leq Ch^{1-s}t^{-\alpha}|v_n|_{-s}|\phi|_1.$$

Passing to the limit as n tends to infinity and combining with (16) gives

$$\|\nabla(u(t) - u^h(t))\| \leq Ch^{1-s}t^{-\alpha}|v|_{-s}. \tag{17}$$

Thus, (15) and (17) lead to the following estimate for $0 \leq s \leq 1$:

$$\|u(t) - u^h(t)\| + h\|\nabla(u(t) - u^h(t))\| \leq Ch^{2-s}t^{-\alpha}|v|_{-s}. \tag{18}$$

Finally, (14), (18), and the triangle inequality give the desired estimate (11) and this completes the proof.

4 Lumped Mass Method

In this section, we consider the lumped mass FEM in planar domains (see, e.g. [9, Chapter 15, pp. 239–244]). An important feature of the lumped mass method is that when representing the solution $\bar{u}_h$ in the nodal basis functions, the mass matrix is diagonal. This leads to a simplified computational procedure. For completeness we shall briefly describe this approximation. Let z_j^τ, $j = 1, \ldots, d+1$ be the vertices of the d-simplex $\tau \in \mathcal{T}_h$. Consider the following quadrature formula and the induced inner product in X_h:

$$Q_{\tau,h}(f) = \frac{|\tau|}{d+1} \sum_{j=1}^{d+1} f(z_j^\tau) \approx \int_\tau f \, dx, \quad (w, \chi)_h = \sum_{\tau \in \mathcal{T}_h} Q_{\tau,h}(w\chi)$$

Then lumped mass finite element method is: find $\bar{u}_h(t) \in X_h$ such that

$$(\partial_t^\alpha \bar{u}_h, \chi)_h + a(\bar{u}_h, \chi) = (f, \chi) \quad \forall \chi \in X_h, \ t > 0, \quad \bar{u}_h(0) = P_h v. \tag{19}$$

To analyze this scheme we shall need the concept of *symmetric* meshes. Given a vertex $z \in \mathcal{T}_h$, the patch Π_z consists of all finite elements having z as a vertex. A mesh $\mathcal{T}_h$ is said to be symmetric at the vertex z, if $x \in \Pi_z$ implies $2z - x \in \Pi_z$, and $\mathcal{T}_h$ is symmetric if it is symmetric at every interior vertex.

In [10, Theorem 4.2] it was shown that if the mesh is symmetric, then the lumped mass scheme (19) for $f = 0$ has an almost optimal convergence rate in L_2-norm for nonsmooth data $v \in L_2(\Omega)$.

Now we prove the main result concerning the lumped mass method:

Theorem 3. *Let $u(t)$ and $\bar{u}_h(t)$ be the solutions of the problems (1) and (19), respectively. Then for $t > 0$ the following error estimate is valid:*

$$\|\bar{u}_h(t) - u(t)\| + \|\nabla(\bar{u}_h(t) - u(t))\| \le Ch^{1-s}\ell_h t^{-\alpha}\|v\|_{-s}, \quad 0 \le s \le 1. \quad (20)$$

Moreover, if the mesh is symmetric then

$$\|\bar{u}_h(t) - u(t)\| \le Ch^{2-s}\ell_h t^{-\alpha}\|v\|_{-s}, \quad 0 \le s \le 1. \quad (21)$$

Proof. We split the error into $\bar{u}_h(t) - u(t) = \bar{u}_h(t) - u^h(t) + u^h(t) - u(t)$, where $u^h(t) - u(t)$ was estimated in (18). The term $\bar{u}_h(t) - u^h(t)$ is the error of the lumped mass method for the auxiliary problem (12). Since the initial data $P_h v \in L_2(\Omega)$, we can apply known estimates on $\bar{u}_h(t) - u^h(t)$ [10, Theorem 4.2]. Namely,
(a) If the mesh is globally quasiuniform, then

$$\|\bar{u}_h(t) - u^h(t)\| + h\|\nabla(\bar{u}_h(t) - u^h(t))\| \le Cht^{-\alpha}\ell_h\|P_h v\|;$$

(b) If the mesh is symmetric, then

$$\|\bar{u}_h(t) - u^h(t)\| \le Ch^2 t^{-\alpha}\ell_h\|P_h v\|.$$

These two estimates, the inequality $\|P_h v\| \le Ch^{-s}\|v\|_{-s}$, $0 \le s \le 1$, and estimate (14) give the desired result. This completes the proof of the theorem.

Remark 3. The H^1-estimate is almost optimal for any quasi-uniform meshes, while the L_2-estimate is almost optimal for symmetric meshes. For the standard parabolic equation with initial data $v \in L_2(\Omega)$, it was shown in [12] that the lumped mass scheme can achieve at most an $O(h^{\frac{3}{2}})$ convergence order in L_2-norm for some nonsymmetric meshes. This rate is expected to hold for fractional order differential equations as well.

5   Numerical Results

Here we present numerical results in 2-D to verify the error estimates derived herein and [10]. The 2-D problem (1) is on the unit square $\Omega = (0,1)^2$ with $\mathcal{L} = -\Delta$. We perform numerical tests on four different examples:

(a) Smooth initial data: $v(x,y) = x(1-x)y(1-y)$; in this case the initial data v is in $H^2(\Omega) \cap H_0^1(\Omega)$, and the exact solution $u(x,t)$ can be represented by a rapidly converging Fourier series:

$$u(x,t) = \sum_{n=1}^{\infty}\sum_{m=1}^{\infty} \frac{4c_n c_m}{m^3 n^3 \pi^6} E_{\alpha,1}(-\lambda_{n,m}t^\alpha)\sin(n\pi x)\sin(m\pi y),$$

where $\lambda_{n,m} = (n^2 + m^2)\pi^2$, and $c_l = 4\sin^2(l\pi/2) - l\pi\sin(l\pi)$, $l = m, n$.

(b) Initial data in $H_0^1(\Omega)$ (case of intermediate smoothness):

$$v(x) = (x - \tfrac{1}{2})(x - 1)(y - \tfrac{1}{2})(y - 1)\chi_{[\frac{1}{2},1]\times[\frac{1}{2},1]},$$

where $\chi_{[\frac{1}{2},1]\times[\frac{1}{2},1]}$ is the characteristic function of $[\frac{1}{2},1] \times [\frac{1}{2},1]$.

(c) Nonsmooth initial data: $v(x) = \chi_{[\frac{1}{4},\frac{3}{4}]\times[\frac{1}{4},\frac{3}{4}]}$.

(d) Very weak data: $v = \delta_\Gamma$ with Γ being the boundary of the square $[\frac{1}{4},\frac{3}{4}]\times[\frac{1}{4},\frac{3}{4}]$ with $\langle \delta_\Gamma, \phi \rangle = \int_\Gamma \phi(s)ds$. One may view (v,χ) for $\chi \in X_h \subset \dot{H}^{\frac{1}{2}+\epsilon}(\Omega)$ as duality pairing between the spaces $H^{-\frac{1}{2}-\epsilon}(\Omega)$ and $\dot{H}^{\frac{1}{2}+\epsilon}(\Omega)$ for any $\epsilon > 0$ so that $\delta_\Gamma \in H^{-\frac{1}{2}-\epsilon}(\Omega)$. Indeed, it follows from Hölder's inequality

$$\|\delta_\Gamma\|_{H^{-\frac{1}{2}-\epsilon}(\Omega)} = \sup_{\phi \in \dot{H}^{\frac{1}{2}+\epsilon}(\Omega)} \frac{|\int_\Gamma \phi(s)ds|}{\|\phi\|_{\frac{1}{2}+\epsilon,\Omega}} \le |\Gamma|^{\frac{1}{2}} \sup_{\phi \in \dot{H}^{\frac{1}{2}+\epsilon}(\Omega)} \frac{\|\phi\|_{L_2(\Gamma)}}{\|\phi\|_{\frac{1}{2}+\epsilon,\Omega}},$$

and the continuity of the trace operator from $\dot{H}^{\frac{1}{2}+\epsilon}(\Omega)$ to $L_2(\Gamma)$.

The exact solution for each example can be expressed by an infinite series involving the Mittag-Leffler function $E_{\alpha,1}(z)$. To accurately evaluate the Mittag-Leffler functions, we employ the algorithm developed in [13]. To discretize the problem, we divide the unit interval $(0,1)$ into $N = 2^k$ equally spaced subintervals, with a mesh size $h = 1/N$ so that $[0,1]^2$ is divided into N^2 small squares. We get a symmetric mesh for the domain $[0,1]^2$ by connecting the diagonal of each small square. All the meshes we have used are symmetric and therefore both semidiscrete Galerkin FEM and lumped mass FEM have the same theoretical accuracy. Unless otherwise specified, we have used the lumped mass method.

To compute a reference (replacement of the exact) solution we have used two different numerical techniques on very fine meshes. The first is based on the exact representation of the semidiscrete lumped mass solution $\bar{u}_h$ by

$$\bar{u}_h(t) = \sum_{n,m=1}^{N-1} E_{\alpha,1}(-\lambda_{n,m}^h t^\alpha)(v,\varphi_{n,m}^h)\varphi_{n,m}^h,$$

where $\varphi_{n,m}^h(x,y) = 2\sin(n\pi x)\sin(m\pi y)$, $n,m = 1,\dots,N-1$, with x,y being grid points, are the discrete eigenfunctions and

$$\lambda_{n,m}^h = \frac{4}{h^2}\left(\sin^2\frac{n\pi h}{2} + \sin^2\frac{m\pi h}{2}\right)$$

are the corresponding eigenvalues.

The second numerical technique is based on fully discrete scheme, i.e., discretizing the time interval $[0,T]$ into $t_n = n\tau$, $n = 0,1,\dots$, with τ being the time step size, and then approximating the fractional derivative $\partial_t^\alpha u(x,t_n)$ by finite difference [14]:

$$\partial_t^\alpha u(x,t_n) \approx \frac{1}{\Gamma(2-\alpha)}\sum_{j=0}^{n-1} b_j \frac{u(x,t_{n-j}) - u(x,t_{n-j-1})}{\tau^\alpha}, \tag{22}$$

where the weights $b_j = (j+1)^{1-\alpha} - j^{1-\alpha}$, $j = 0, 1, \ldots, n-1$. We have used this approximation on very fine meshes in both space and time to compute a reference solution. This fully discrete solution is denoted by U_h. Throughout, we have set $\tau = 10^{-6}$ so that the error incurred by temporal discretization is negligible (see Table 6).

We measure the accuracy of the approximation $u_h(t)$ by the normalized error $\|u(t) - u_h(t)\|/\|v\|$ and $\|\nabla(u(t) - u_h(t))\|/\|v\|$. The normalization enables us to observe the behavior of the error with respect to time in case of nonsmooth initial data.

Smooth initial data: example (a). In Table 4 we show the numerical results for $t = 0.1$ and $\alpha = 0.1$, 0.5, 0.9. Here `ratio` refers to the ratio between the errors as the mesh size h is halved. In Figure 1, we plot the results from Table 4 in a log-log scale. The slopes of the error curves are 2 and 1, respectively, for L_2- and H^1-norm of the error. This confirms the theoretical result from [10].

Table 4. Numerical results for smooth initial data, example (a), $t = 0.1$

α	h	1/8	1/16	1/32	1/64	1/128	ratio	rate
0.5	L_2-norm	1.45e-3	3.84e-4	9.78e-5	2.41e-5	5.93e-6	≈ 4.02	$O(h^2)$
	H^1-norm	5.17e-2	2.64e-2	1.33e-2	6.67e-3	3.33e-3	≈ 1.99	$O(h)$
0.9	L_2-norm	1.88e-3	4.53e-4	1.13e-4	2.82e-5	7.06e-6	≈ 4.00	$O(h^2)$
	H^1-norm	6.79e-2	3.43e-2	1.73e-2	8.63e-3	4.31e-3	≈ 2.00	$O(h)$

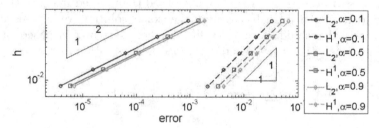

Fig. 1. Error plots for smooth initial data, Example (a): $\alpha = 0.1, 0.5, 0.9$ at $t = 0.1$

Table 5. Intermediate case (b) with $\alpha = 0.5$ at $t = 0.1$

h	1/8	1/16	1/32	1/64	1/128	ratio	rate
L_2-error	3.04e-3	8.20e-4	2.12e-4	5.35e-5	1.32e-5	≈ 3.97	$O(h^2)$
H^1-error	5.91e-2	3.09e-2	1.56e-2	7.88e-3	3.93e-3	≈ 1.98	$O(h)$

Intermediate smooth data: example (b). In this example the initial data $v(x)$ is in $H_0^1(\Omega)$ and the numerical results are shown in Table 5. The slopes of the error curves in a log-log scale are 2 and 1 respectively for L_2- and H^1-norm of the errors, which agrees well with the theory for the intermediate case [10].

Nonsmooth initial data: example (c). First in Table 6 we compare fully discrete solution U_h via the finite difference approximation (22) with the semidiscrete lumped mass solution $\bar{u}_h$ via eigenexpansion to study the error incurred by time discretization. We observe that for each fixed spatial mesh size h, the difference between $\bar{u}_h$, the lumped mass FEM solution, and U_h decreases with the decrease of τ. In particular, for time step $\tau = 10^{-6}$ the error incurred by the time discretization is negligible, so the fully discrete solutions U_h could well be used as reference solutions. In Table 7 and Figure 2 we present the numerical results for problem (c). These numerical results fully confirm the theoretically predicted rates for nonsmooth data.

Table 6. The difference $\bar{u}_h - U_h$, nonsmooth initial data, example (c): $\alpha = 0.5$, $t = 0.1$

Time step	h	1/8	1/16	1/32	1/64	1/128
$\tau = 10^{-2}$	L_2-norm	2.03e-3	2.01e-3	2.00e-3	2.00e-3	2.00e-3
	H^1-norm	9.45e-3	9.17e-3	9.10e-3	9.08e-3	9.07e-3
$\tau = 10^{-4}$	L_2-norm	1.81e-5	1.79e-5	1.79e-5	1.79e-5	1.79e-5
	H^1-norm	8.47e-5	8.22e-5	8.15e-5	8.13e-5	8.13e-5
$\tau = 10^{-6}$	L_2-norm	1.80e-7	1.78e-7	1.78e-7	1.78e-7	1.78e-7
	H^1-norm	8.42e-7	8.17e-7	8.10e-7	8.10e-7	8.09e-7

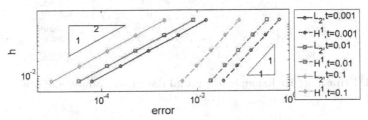

Fig. 2. Error plots for lumped FEM for nonsmooth initial data, Example (c): $\alpha = 0.5$

Table 7. Error for the lumped FEM for nonsmooth initial data, example (c): $\alpha = 0.5$

Time	h	1/8	1/16	1/32	1/64	1/128	ratio	rate
$t = 0.001$	L_2-norm	1.55e-2	3.99e-3	1.00e-3	2.52e-4	6.26e-5	≈ 4.01	$O(h^2)$
	H^1-norm	6.05e-1	3.05e-1	1.48e-1	7.29e-2	3.61e-2	≈ 2.00	$O(h)$
$t = 0.01$	L_2-norm	8.27e-3	2.10e-3	5.28e-4	1.32e-4	3.29e-5	≈ 4.01	$O(h^2)$
	H^1-norm	3.32e-1	1.61e-1	7.90e-2	3.90e-2	1.93e-2	≈ 2.02	$O(h)$
$t = 0.1$	L_2-norm	2.12e-3	5.36e-4	1.34e-4	3.36e-5	8.43e-6	≈ 3.99	$O(h^2)$
	H^1-norm	8.23e-2	4.01e-2	1.96e-2	9.72e-3	4.84e-3	≈ 2.01	$O(h)$

Table 8. Error for standard FEM: initial data Dirac δ-function, $\alpha = 0.5$

Time	h	1/8	1/16	1/32	1/64	1/128	ratio	rate
$t = 0.001$	L_2-norm	5.37e-2	1.56e-2	4.40e-3	1.23e-3	3.41e-4	≈ 3.57	$O(h^{1.84})$
	H^1-norm	2.68e0	1.76e0	1.20e0	8.21e-1	5.68e-1	≈ 1.45	$O(h^{\frac{1}{2}})$
$t = 0.01$	L_2-norm	2.26e-2	6.20e-3	1.67e-3	4.46e-4	1.19e-4	≈ 3.74	$O(h^{1.90})$
	H^1-norm	9.36e-1	5.90e-1	3.92e-1	2.65e-1	1.84e-1	≈ 1.46	$O(h^{\frac{1}{2}})$
$t = 0.1$	L_2-norm	8.33e-3	2.23e-3	5.90e-3	1.55e-3	4.10e-4	≈ 3.77	$O(h^{1.91})$
	H^1-norm	3.08e-1	1.91e-1	1.26e-1	8.44e-2	5.83e-2	≈ 1.46	$O(h^{\frac{1}{2}})$

Table 9. Error for lumped mass FEM: initial data Dirac δ-function, $\alpha = 0.5$

Time	h	1/8	1/16	1/32	1/64	1/128	ratio	rate
$t = 0.001$	L_2-norm	1.98e-1	7.95e-2	3.00e-2	1.09e-2	3.95e-3	≈ 2.75	$O(h^{\frac{3}{2}})$
	H^1-norm	5.56e0	4.06e0	2.83e0	2.02e0	1.41e0	≈ 1.42	$O(h^{\frac{1}{2}})$
$t = 0.01$	L_2-norm	6.61e-2	2.56e-2	9.51e-3	3.47e-3	1.25e-3	≈ 2.78	$O(h^{\frac{3}{2}})$
	H^1-norm	1.84e0	1.30e0	9.10e-1	6.40e-1	4.47e-1	≈ 1.42	$O(h^{\frac{1}{2}})$
$t = 0.1$	L_2-norm	2.15e-2	8.13e-3	3.01e-3	1.09e-3	3.95e-4	≈ 2.75	$O(h^{\frac{3}{2}})$
	H^1-norm	5.87e-1	4.14e-1	2.88e-1	2.03e-1	1.41e-1	≈ 1.43	$O(h^{\frac{1}{2}})$

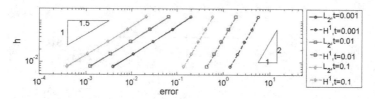

Fig. 3. Error plots for Example (d): initial data Dirac δ-function, $\alpha = 0.5$

Very weak data: example (d). The empirical convergence rate for the weak data δ_Γ agrees well with the theoretically predicted convergence rate in Theorem 2, which gives a ratio of 2.82 and 1.41, respectively, for the L_2- and H^1-norm of the error; see Table 9. Interestingly, for the standard Galerkin scheme, the L_2-norm of the error exhibits super-convergence; see Table 8.

Acknowledgements. The research of R. Lazarov and Z. Zhou was supported in parts by US NSF Grant DMS-1016525 and J. Pasciak has been supported by NSF Grant DMS-1216551. The work of all authors has been supported also by Award No. KUS-C1-016-04, made by King Abdullah University of Science and Technology (KAUST)

References

1. Kilbas, A., Srivastava, H., Trujillo, J.: Theory and Applications of Fractional Differential Equations. Elsevier, Amsterdam (2006)

2. Podlubny, I.: Fractional Differential Equations. Academic Press, San Diego (1999)
3. Bouchaud, J.P., Georges, A.: Anomalous diffusion in disordered media: statistical mechanisms, models and physical applications. Phys. Rep. 195(4-5), 127–293 (1990)
4. Debnath, L.: Recent applications of fractional calculus to science and engineering. Int. J. Math. Math. Sci. 54, 3413–3442 (2003)
5. Cheng, J., Nakagawa, J., Yamamoto, M., Yamazaki, T.: Uniqueness in an inverse problem for a 1-d fractional diffusion equation. Inverse Problems 25(11), 115002, 1–16 (2009)
6. Sakamoto, K., Yamamoto, M.: Initial value/boundary value problems for fractional diffusion-wave equations and applications to some inverse problems. J. Math. Anal. Appl. 382(1), 426–447 (2011)
7. Jin, B., Lu, X.: Numerical identification of a Robin coefficient in parabolic problems. Math. Comput. 81(279), 1369–1398 (2012)
8. Keung, Y.L., Zou, J.: Numerical identifications of parameters in parabolic systems. Inverse Problems 14(1), 83–100 (1998)
9. Thomée, V.: Galerkin Finite Element Methods for Parabolic Problems. Springer Series in Computational Mathematics, vol. 25. Springer, Berlin (1997)
10. Jin, B., Lazarov, R., Zhou, Z.: Error estimates for a semidiscrete finite element method for fractional order parabolic equations. SIAM J. Numer. Anal. 51(1), 445–466 (2013)
11. Bramble, J.H., Xu, J.: Some estimates for a weighted L^2 projection. Math. Comp. 56(194), 463–476 (1991)
12. Chatzipantelidis, P., Lazarov, R., Thomee, V.: Some error estimates for the finite volume element method for a parabolic problem. Computational Methods in Applied Mathematics 13(3), 251–275 (2013); also arXiv:1208-3219
13. Seybold, H., Hilfer, R.: Numerical algorithm for calculating the generalized Mittag-Leffler function. SIAM J. Numer. Anal. 47(1), 69–88 (2008)
14. Lin, Y., Xu, C.: Finite difference/spectral approximations for the time-fractional diffusion equation. J. Comput. Phys. 225(2), 1533–1552 (2007)

Boundary Value Problems for Fractional PDE and Their Numerical Approximation

Boško S. Jovanović[1], Lubin G. Vulkov[2], and Aleksandra Delić[1]

[1] University of Belgrade, Faculty of Mathematics
Studentski trg 16, 11000 Belgrade, Serbia
{bosko,adelic}@matf.bg.ac.rs
[2] University of Rousse, Center of Applied Mathematics,
Studentska Str. 8, Rousse 7017, Bulgaria
vulkov@ami.ru.acad.bg

Abstract. Fractional order partial differential equations are considered. The main attention is devoted to fractional in time diffusion equation. An interface problem for this equation is studied and its well posedness in the corresponding Sobolev like spaces is proved. Analogous results are obtained for a transmission problem in disjoint intervals.

Keywords: Fractional derivative, fractional PDE, initial-boundary value problem, interface problem, transmission problem, finite differences.

1 Introduction

The use of fractional partial differential equations in mathematical models has become increasingly popular in recent years. Such equations are used for the description of large classes of physical and chemical processes (anomalous diffusion, turbulent flow, chaotic dynamics etc.) that occur in media with fractal geometry, disordered materials, amorphous semiconductors, viscoelastic media, as well as in the mathematical modelling of economic, biological and social phenomena (see e.g. [2, 3, 5, 11, 16–20, 23]).

Because of the integral in the definition of the fractional order derivatives, it is apparent that these derivatives are nonlocal operators. This explains one of their most significant uses in applications: non-integer derivatives possess a memory effect which it shares with several materials such as viscoelastic materials or polymers. On the other side, this feature of the fractional derivatives makes the design of accurate and fast numerical methods difficult.

In this paper we present some examples of fractional PDE and highlight the main theoretical and numerical problems appearing. In particular, we emphasize some interface and transmission problems related with fractional PDE.

2 Fractional Integrals and Derivatives

The most often used definition for integration of arbitrary real positive order comes from the extension of the formula for n-fold integration

I. Dimov, I. Faragó, and L. Vulkov (Eds.): NAA 2012, LNCS 8236, pp. 38–49, 2013.
© Springer-Verlag Berlin Heidelberg 2013

$$\int_a^x dx_1 \int_a^{x_1} dx_2 \cdots \int_a^{x_{n-1}} f(x_n)\, dx_n = \frac{1}{(n-1)!} \int_a^x (x-\xi)^{n-1} f(\xi)\, d\xi\,, \quad n \in \mathbb{N}.$$

Replacing n by real positive parameter α and factorial by Gamma function one obtains so-called left Riemann-Liouville fractional integral:

$$(I_{a+}^\alpha f)(x) = \frac{1}{\Gamma(\alpha)} \int_a^x (x-\xi)^{\alpha-1} f(\xi)\, d\xi\,, \quad x > a.$$

The right Riemann-Liouville fractional integral is defined analogously:

$$(I_{b-}^\alpha f)(x) = \frac{1}{\Gamma(\alpha)} \int_x^b (\xi-x)^{\alpha-1} f(\xi)\, d\xi\,, \quad x < b.$$

For $\alpha = 0$ one sets: $I_{a+}^0 f = I_{b-}^0 f = f$.

Fractional integral possess the following basic properties [12]:

$$Q I_{a+}^\alpha = I_{b-}^\alpha Q\,, \quad Q I_{b-}^\alpha = I_{a+}^\alpha Q\,, \qquad (Qf)(x) = f(a+b-x),$$

$$I_{a+}^\alpha I_{a+}^\beta = I_{a+}^{\alpha+\beta}\,, \quad I_{b-}^\alpha I_{b-}^\beta = I_{b-}^{\alpha+\beta}.$$

Analogously, inspired by the formula

$$\frac{d^n f(x)}{dx^n} = \frac{d^{n+1}}{dx^{n+1}}\left(\int_a^x f(\xi)\, d\xi \right),$$

the left and right Riemann-Liouville fractional derivatives are defined by

$$D_{a+}^\alpha = \frac{d^{[\alpha]+1}}{dx^{[\alpha]+1}} I_{a+}^{[\alpha]+1-\alpha}\,, \quad D_{b-}^\alpha = (-1)^{[\alpha]+1} \frac{d^{[\alpha]+1}}{dx^{[\alpha]+1}} I_{b-}^{[\alpha]+1-\alpha}\,,$$

i.e.

$$(D_{a+}^\alpha f)(x) = \frac{1}{\Gamma(n-\alpha)} \frac{d^n}{dx^n}\left(\int_a^x (x-\xi)^{n-\alpha-1} f(\xi)\, d\xi \right), \quad x > a,$$

$$(D_{b-}^\alpha f)(x) = \frac{(-1)^n}{\Gamma(n-\alpha)} \frac{d^n}{dx^n}\left(\int_x^b (\xi-x)^{n-\alpha-1} f(\xi)\, d\xi \right), \quad x < b,$$

where $n-1 \le \alpha < n$ and $n \in \mathbb{N}$.

Notice that if function $f(x)$ has n-order continuous derivative in $[a,b]$, then as α tends to n or $n-1$, the left (right) Riemann-Liouville derivative becomes a conventional n- or $(n-1)$-order derivative of $f(x)$.

An alternative definition of fractional derivatives (named after Caputo) one obtains by commuting $d^{[\alpha]+1}/dx^{[\alpha]+1}$ and $I_{a+}^{[\alpha]+1-\alpha}$ ($I_{b-}^{[\alpha]+1-\alpha}$):

$$^cD_{a+}^\alpha = I_{a+}^{[\alpha]+1-\alpha} \frac{d^{[\alpha]+1}}{dx^{[\alpha]+1}}, \qquad ^cD_{b-}^\alpha = (-1)^{[\alpha]+1} I_{b-}^{[\alpha]+1-\alpha} \frac{d^{[\alpha]+1}}{dx^{[\alpha]+1}}.$$

For $n - 1 < \alpha < n$ and $n \in \mathbb{N}$ we have

$$(D_{a+}^\alpha f)(x) = (^cD_{a+}^\alpha f)(x) + \sum_{j=0}^{n-1} f^{(j)}(a+0) \frac{(x-a)^{j-\alpha}}{\Gamma(j-\alpha+1)}.$$

In the framework of generalized functions from $\mathcal{D}_+'$, fractional integrals and fractional derivatives are interpreted as convolutions [25]:

$$I_{0+}^\alpha f = f * \psi_\alpha, \quad D_{0+}^\alpha f = f * \psi_{-\alpha}, \quad \alpha \geq 0,$$

where $\psi_\alpha \in \mathcal{D}_+'$ is defined in the following way:

$$\psi_\alpha(x) = \begin{cases} \dfrac{\theta(x)}{\Gamma(\alpha)} x^{\alpha-1}, & \alpha > 0, \\ \psi_{\alpha+1}'(x), & \alpha \leq 0. \end{cases}$$

In particular, $\psi_1(x) = \theta(x)$ is the Heaviside function, while $\psi_0(x) = \delta(x)$ is Dirac distribution.

Fractional derivatives satisfies semigroup property in $\mathcal{D}_+'$:

$$D_{0+}^\alpha D_{0+}^\beta f = D_{0+}^\beta D_{0+}^\alpha f = D_{0+}^{\alpha+\beta} f, \quad f \in \mathcal{D}_+'.$$

For continuous functions the same is valid under some additional assumptions, for example [21]:

$$(D_{a+}^\alpha D_{a+}^\beta f)(x) = (D_{a+}^{\alpha+\beta} f)(x), \quad 0 < \alpha, \beta < 1, \quad f(a) = 0, \quad x > a.$$

For the functions of many variables partial derivatives of fractional order are defined in analogous manner, for example:

$$(D_{x,a+}^\alpha f)(x,t) = \frac{1}{\Gamma(n-\alpha)} \frac{d^n}{dx^n} \left(\int_a^x (x-\xi)^{n-\alpha-1} f(\xi,t)\, d\xi \right), \quad x > a,$$

$$(D_{t,c+}^\beta f)(x,t) = \frac{1}{\Gamma(m-\beta)} \frac{d^m}{dt^m} \left(\int_c^t (t-\tau)^{m-\beta-1} f(x,\tau)\, d\tau \right), \quad t > c,$$

where $n - 1 < \alpha < n$, $m - 1 < \beta < m$ and $n, m \in \mathbb{N}$.

3   Some Function Spaces

First, we introduce some notations and define some function spaces, norms and inner products that are used thereafter. Let Ω be an open domain in $\mathbb{R}^n$. As usual, by $C^k(\Omega)$ and $C^k(\overline{\Omega})$ we denote the spaces of k-fold differentiable functions. By $\dot{C}^\infty(\Omega) = C_0^\infty(\Omega)$ we denote the space of infinitely differentiable functions with compact support in Ω. The space of measurable functions whose square is Lebesgue integrable in Ω is denoted by $L^2(\Omega)$. The inner product and norm of $L^2(\Omega)$ are defined by

$$(u,v)_\Omega = (u,v)_{L^2(\Omega)} = \int_\Omega uv\, d\Omega, \quad \|u\|_\Omega = \|u\|_{L^2(\Omega)} = (u,u)_\Omega^{1/2}.$$

We also use $H^\alpha(\Omega)$ and $\dot{H}^\alpha(\Omega) = H_0^\alpha(\Omega)$ to denote the usual Sobolev spaces [14], whose norms are denoted by $\|u\|_{H^\alpha(\Omega)}$.

For $\alpha > 0$ let us set

$$|u|_{H_+^\alpha(a,b)} = \|D_{a+}^\alpha u\|_{L^2(a,b)}, \quad |u|_{H_-^\alpha(a,b)} = \|D_{b-}^\alpha u\|_{L^2(a,b)}$$

and

$$\|u\|_{H_\pm^\alpha(a,b)} = \left(\|u\|_{L^2(a,b)}^2 + |u|_{H_\pm^\alpha(a,b)}^2 \right)^{1/2}.$$

Then we define the spaces $H_\pm^\alpha(a,b)$ and $\dot{H}_\pm^\alpha(a,b)$ as the closure of $C^\infty[a,b]$ and $\dot{C}^\infty(a,b)$, respectively, with respect to the norm $\|\cdot\|_{H_\pm^\alpha(a,b)}$. Because for $\alpha = n \in \mathbb{N}$ fractional derivative reduces to standard integer order derivative, we have $H_\pm^n(a,b) = H^n(a,b)$.

From Theorem 2 in [15] follows:

Lemma 1. Let $0 < \alpha < 1$, $u \in H_+^\alpha(a,b)$ and $v \in H_-^\alpha(a,b)$. Then

$$(D_{a+}^\alpha u, v)_{L^2(a,b)} = (u, D_{b-}^\alpha v)_{L^2(a,b)}.$$

From Lemma 2.4 in [4] immediately follows:

Lemma 2. Let $\alpha > 0$, $u \in \dot{C}^\infty(\mathbb{R})$ and $\operatorname{supp} u \subset (a,b)$. Then

$$(D_{a+}^\alpha u, D_{b-}^\alpha u)_{L^2(a,b)} = \cos\pi\alpha \, \|D_{a+}^\alpha u\|_{L^2(a,+\infty)}^2.$$

For $\alpha > 0$, $\alpha \neq n + 1/2$, $n \in \mathbb{N}$, we set

$$|u|_{H_c^\alpha(a,b)} = |(D_{a+}^\alpha u, D_{b-}^\alpha u)_{L^2(a,b)}|^{1/2}, \quad \|u\|_{H_c^\alpha(a,b)} = \left(\|u\|_{L^2(a,b)}^2 + |u|_{H_c^\alpha(a,b)}^2 \right)^{1/2}$$

and define the space $\dot{H}_c^\alpha(a,b)$ as the closure of $\dot{C}^\infty(a,b)$ with respect to the norm $\|\cdot\|_{H_c^\alpha(a,b)}$.

Lemma 3. (see [13]) For $\alpha > 0$, $\alpha \neq n + 1/2$, $n \in \mathbb{N}$, the spaces $\dot{H}_+^\alpha(a,b)$, $\dot{H}_-^\alpha(a,b)$, $\dot{H}_c^\alpha(a,b)$ and $\dot{H}^\alpha(a,b)$ are equal and their seminorms as well as norms are equivalent.

For the functions of two variables, x and t, defined in the rectangle $Q = (0,1) \times (0,T)$, we introduce anisotropic Sobolev spaces $H^{\alpha,\beta}(Q)$, $\alpha, \beta \geq 0$, in the usual manner [14]:

$$H^{\alpha,\beta}(Q) = L^2((0,T), H^\alpha(0,1)) \cap H^\beta((0,T), L^2(0,1)).$$

Analogously we define

$$H_\pm^{\alpha,\beta}(Q) = L^2((0,T), H^\alpha(0,1)) \cap H_\pm^\beta((0,T), L^2(0,1)).$$

Notice that for $0 \leq \beta < 1/2$

$$H_+^{\alpha,\beta}(Q) = H_-^{\alpha,\beta}(Q) = H^{\alpha,\beta}(Q).$$

4   Time Fractional Diffusion Equation

Let $\alpha \in (0,1)$. In $Q = (0,1) \times (0,T)$ we consider the following fractional in time diffusion equation

$$D_{t,0+}^\alpha u - \frac{\partial^2 u}{\partial x^2} = f(x,t), \quad (x,t) \in Q, \tag{1}$$

subject to homogeneous initial and boundary conditions

$$u(x,0) = 0, \quad x \in (0,1), \tag{2}$$

$$u(0,t) = u(1,t) = 0, \quad t \in (0,T). \tag{3}$$

Problem (1)-(3) is often called subdiffusion problem.

Taking inner product of equation (1) with function v and using Lemma 1 and properties of fractional derivatives one obtains the following weak formulation of the problem (1)-(3) (see [13]): find $u \in \dot{H}^{1,\alpha/2}(Q)$ such that

$$a(u,v) = l(v), \quad \forall v \in \dot{H}^{1,\alpha/2}(Q), \tag{4}$$

where

$$\dot{H}^{1,\alpha/2}(Q) = L^2((0,T), \dot{H}^1(0,1)) \cap \dot{H}^{\alpha/2}((0,T), L^2(0,1)),$$

the bilinear form $a(\cdot,\cdot)$ is defined by

$$a(u,v) = \left(D_{t,0+}^{\alpha/2} u, D_{t,T-}^{\alpha/2} v\right)_Q + \left(\frac{\partial u}{\partial x}, \frac{\partial v}{\partial x}\right)_Q,$$

and the linear functional $l(\cdot)$ is given by

$$l(v) = (f,v)_Q.$$

Now, from lemmae 1-3 and Lax-Milgram lemma, we immediately obtain the following result:

Theorem 1. (see [13]) *For all $\alpha \in (0,1)$ and $f \in L^2(Q)$ the problem (4) is well posed and its solution satisfies a priory estimate*

$$\|u\|_{H^{1,\alpha/2}(Q)} \leq C \|f\|_{L^2(Q)} .$$

Here and in the sequel C denotes positive generic constant which may take different values in different formulas.

Analogous result holds in the case when the right-side part of (1) has the form

$$f(x,t) = f_0(x,t) + \frac{\partial f_1}{\partial x} + D_{t,0+}^{\alpha/2} f_2 ,$$

where $f_i \in L^2(0,1)$. In this case the linear functional $l(\cdot)$ in weak formulation (4) is given by

$$l(v) = (f_0, v)_Q - \left(f_1, \frac{\partial v}{\partial x} \right)_Q + \left(f_2, D_{t,T-}^{\alpha/2} v \right)_Q . \tag{5}$$

Using Lax-Milgram lemma, Cauchy-Schvarz inequality and lemmae 1-3 one obtains the following result:

Theorem 2. *Let $\alpha \in (0,1)$ and $f_i \in L^2(Q)$, $i = 0, 1, 2$. Then the problem (4)-(5) is well posed and its solution satisfies a priory estimate*

$$\|u\|_{H^{1,\alpha/2}(Q)} \leq C \left(\|f_0\|_{L^2(Q)} + \|f_1\|_{L^2(Q)} + \|f_2\|_{L^2(Q)} \right) .$$

In analogous way one obtains the following result:

Theorem 3. *Let $\alpha \in (0,1)$ and $f \in L^2(Q)$. Then the problem (1)-(3) is well posed in the space $H_+^{2,\alpha}(Q) \cap \dot{H}^{1,\alpha/2}(Q)$ and its solution satisfies a priory estimate*

$$\|u\|_{H_+^{2,\alpha}(Q)} \leq C \|f\|_{L^2(Q)} .$$

5   Finite Difference Approximation

In the rectangle $\bar{Q} = [0,1] \times [0,T]$ we introduce the uniform mesh $\bar{Q}_{h\tau} = \bar{\omega}_h \times \bar{\omega}_\tau$, where $\bar{\omega}_h = \{x_i = ih \mid i = 0, 1, \ldots, n; \ h = 1/n\}$ and $\bar{\omega}_\tau = \{t_j = j\tau \mid j = 0, 1, \ldots, m; \ \tau = T/m\}$. We will use standard notation from the theory of finite difference schemes [22]:

$$v = v(x,t), \quad \hat{v} = v(x,t+\tau); \quad v^j = v(x,t_j), \quad x \in \bar{\omega}_h ,$$

$$v_x = \frac{v(h+h,t) - v(x,t)}{h} = v_{\bar{x}}(x+h,t), \quad v_t = \frac{v(h,t+\tau) - v(x,t)}{\tau} = v_{\bar{t}}(x,t+\tau).$$

We approximate the initial-boundary value problem (1)-(3) with the following weighted finite difference scheme:

$$\Delta^\alpha_{t,0+}v^j - \sigma\, v^{j+1}_{\bar{x}x} - (1-\sigma)\, v^j_{\bar{x}x} = \varphi^j, \quad x \in \omega_h, \quad j = 1, 2, \dots, m-1 \qquad (6)$$

$$v(x,0) = 0, \quad x \in \omega_h, \qquad (7)$$

$$v(0,t) = v(1,t) = 0, \quad t \in \bar{\omega}_\tau, \qquad (8)$$

where $\sigma \in [0,1]$ is free weight parameter, $\varphi^j = f(x, t_j + \tau/2)$, $\omega_h = \bar{\omega}_h \cap (0,1)$ and

$$\Delta^\alpha_{t,0+}v^j = \frac{1}{\Gamma(2-\alpha)} \left(\sum_{k=0}^{j} \left(t^{1-\alpha}_{j-k+1} - t^{1-\alpha}_{j-k} \right) v^k \right)_t$$

is difference analogue of the left Riemann-Liouville fractional time derivative.

Notice that the finite difference scheme (6)-(8) on the each time level t_j reduces to a three-diagonal system of linear equations. On the other hand, the solution v^j on time level t_j explicitly depends on the solutions at all previous time levels t_k, $k < j$. Thus, numerical effort is $O(n\, m^2)$ (instead of $O(n\, m)$ for $\alpha = 1$). All values v^k must be permanently stored, which can be expensive, especially in the multidimensional case.

Let us define discrete inner products and norms

$$(v,w)_h = (v,w)_{L^2(\omega_h)} = h \sum_{i=1}^{n-1} v(x_i)w(x_i), \quad \|v\|_h = \|v\|_{L^2(\omega_h)} = (v,v)^{1/2}_h,$$

$$(v,w]_h = (v,w]_{L^2(\omega_h)} = h \sum_{i=1}^{n} v(x_i)w(x_i), \quad \|v]_h = \|v]_{L^2(\omega_h)} = (v,v]^{1/2}_h,$$

$$\|v\|_{L^2(Q_{h\tau})} = \left(\tau \sum_{j=0}^{m-1} \|v^j\|^2_h \right)^{1/2}, \quad \|v]_{L^2(Q_{h\tau})} = \left(\tau \sum_{j=0}^{m-1} \|v^j]^2_h \right)^{1/2},$$

$$\|v\|_{B^{1,\alpha/2}(Q_{h\tau})} = \left(\|\sigma\hat{v}_{\bar{x}} + (1-\sigma)v_{\bar{x}}]\|^2_{L^2(Q_{h\tau})} + \sum_{j=0}^{m-1}(t^{1-\alpha}_{m-j} - t^{1-\alpha}_{m-j-1})\|\hat{v}^j\|^2_h \right)^{1/2}.$$

The following result holds true (see [1]):

Theorem 4. *Let $0 < \alpha < 1$ and $\sigma \geq 1/(3 - 2^{1-\alpha})$. Then finite difference scheme (6)-(8) is absolutely stable and its solution satisfies the following a priori estimate:*

$$\|v\|_{B^{1,\alpha/2}(Q_{h\tau})} \leq C\, \|\varphi\|_{L^2(Q_{h\tau})}.$$

Let us assume that the solution of the initial-boundary value problem (1)-(3) belongs to the space $C^{4,3}(\bar{Q})$. Then from the theorem 4 one immediately obtains the following convergence rate estimate for finite difference scheme (6)-(8)

$$\|u - v\|_{B^{1,\alpha/2}(Q_{h\tau})} \leq C\,(h^2 + \tau). \qquad (9)$$

In [24] it is proved that the finite difference scheme (6)-(8) is stable in discrete C-norm

$$\|v\|_{C(\bar{Q}_{h\tau})} = \max_{(x,t)\in\bar{Q}_{h\tau}} |v(x,t)|$$

if

$$\tau^\alpha \leq \frac{C(\alpha)\,h^2}{1-\sigma},$$

where $C(\alpha)$ is computable constant depending on α. In this case the estimate holds

$$\|u-v\|_{C(\bar{Q}_{h\tau})} \leq C\,\frac{\tau+h^2}{\tau^{1-\alpha}},$$

whereby, assuming $\tau^\alpha \asymp h^2$, follows convergence for $1/2 < \alpha < 1$.

Norm $\|v\|_{B^{1,\alpha/2}(Q_{h\tau})}$ is the discrete analogue of the norm

$$\|u\|_{B^{1,\alpha/2}(Q)} = \left(\left\|\frac{\partial u}{\partial x}\right\|^2_{L^2(Q)} + \int_0^T (T-t)^{-\alpha}\|u(\cdot,t)\|^2_{L^2(0,1)} \right)^{1/2}$$

which is weaker than $\|u\|_{H^{1,\alpha/2}(Q)}$. It would be interesting to prove discrete analogues of theorems 1-3 for the finite difference scheme (6)-(8) and to derive convergence rate estimates consistent with the smoothness of the solution of boundary value problem (1)-(3).

6   Interface Problem for Fractional in Time Diffusion Equation

Let $\alpha,\ \xi \in (0,1)$ and $K = \mathrm{const} > 0$. In $Q = (0,1) \times (0,T)$ we consider the following fractional in time diffusion equation

$$[1 + K\delta(x-\xi)]\,D^\alpha_{t,0+}u - \frac{\partial^2 u}{\partial x^2} = f(x,t), \quad (x,t)\in Q, \tag{10}$$

where $\delta(x-\xi)$ is Dirac distribution concentrated at $x=\xi$. We assume that the solution of (10) satisfies homogeneous initial and boundary conditions (2)-(3).

Notice that in the case when $f(x,t)$ does not contain singular terms equation (10) reduces to

$$D^\alpha_{t,0+}u - \frac{\partial^2 u}{\partial x^2} = f(x,t), \quad (x,t)\in Q_1\cap Q_2, \tag{11}$$

where $Q_1 = (0,\xi)\times(0,T)$, $Q_2 = (\xi,1)\times(0,T)$, subject to conjugation condition

$$K\left(D^\alpha_{t,0+}u\right)(\xi,t) = \left[\frac{\partial u}{\partial x}\right]_{x=\xi} \equiv \frac{\partial u}{\partial x}(\xi+0,t) - \frac{\partial u}{\partial x}(\xi-0,t). \tag{12}$$

Problems of similar type are usually called interface problems. Analogous problem for integer order diffusion equation is considered in [7, 8]. Another interface problem for fractional in space diffusion equation is studied in [6].

Let $\tilde{L}^2(0,1)$ be the space of functions defined on the interval $(0,1)$, with the inner product and norm

$$(v,w)_{\tilde{L}^2(0,1)} = \int_0^1 v(x)w(x)\,dx + v(\xi)w(\xi), \quad \|v\|_{\tilde{L}^2(0,1)} = (v,v)_{\tilde{L}^2(0,1)}^{1/2}.$$

For the functions defined in the rectangle $Q = (0,1) \times (0,T)$, we define the space $\tilde{L}^2(Q) = L^2((0,T), \tilde{L}^2(0,1))$, with inner product and associated norm:

$$(v,w)_{\tilde{L}^2(Q)} = \iint_Q v(x,t)w(x,t)\,dxdt + \int_0^T v(\xi,t)w(\xi,t)\,dt, \quad \|v\|_{\tilde{L}^2(Q)} = (v,v)_{\tilde{L}^2(Q)}^{1/2}.$$

Finally, for $\alpha, \beta \geq 0$, we introduce anisotropic Sobolev type spaces:

$$\tilde{H}^{\alpha,\beta}(Q) = L^2((0,T), H^\alpha(0,1)) \cap H^\beta((0,T), \tilde{L}^2(0,1))$$

and

$$\tilde{H}_\pm^{\alpha,\beta}(Q) = L^2((0,T), H^\alpha(0,1)) \cap H_\pm^\beta((0,T), \tilde{L}^2(0,1)).$$

Notice that for $0 \leq \beta < 1/2$: $\tilde{H}_+^{\alpha,\beta}(Q) = \tilde{H}_-^{\alpha,\beta}(Q) = \tilde{H}^{\alpha,\beta}(Q)$.

The weak formulation of the problem (10), (2), (3) is: find $u \in \dot{\tilde{H}}^{1,\alpha/2}(Q)$ such that

$$a(u,v) = l(v), \quad \forall v \in \dot{\tilde{H}}^{1,\alpha/2}(Q), \tag{13}$$

where

$$\dot{\tilde{H}}^{1,\alpha/2}(Q) = L^2((0,T), \dot{H}^1(0,1)) \cap \dot{H}^{\alpha/2}((0,T), \tilde{L}^2(0,1)),$$

the bilinear form $a(\cdot,\cdot)$ is defined by

$$a(u,v) = \left(D_{t,0+}^{\alpha/2}u, D_{t,T-}^{\alpha/2}v \right)_{L^2(Q)} + K \left(D_{t,0+}^{\alpha/2}u(\xi,\cdot), D_{t,T-}^{\alpha/2}v(\xi,\cdot) \right)_{L^2(0,T)}$$

$$+ \left(\frac{\partial u}{\partial x}, \frac{\partial v}{\partial x} \right)_{L^2(Q)},$$

and the linear functional $l(\cdot)$ is given by

$$l(v) = (f,v)_{L^2(Q)}.$$

Similarly as in the previous section, one obtains the following results:

Theorem 5. *For all $\alpha \in (0,1)$ and $f \in L^2(Q)$ the problem (13) is well posed and its solution satisfies a priory estimate*

$$\|u\|_{\tilde{H}^{1,\alpha/2}(Q)} \leq C \|f\|_{L^2(Q)}.$$

Theorem 6. *Let $\alpha \in (0,1)$ and $f \in L^2(Q)$. Then the problem (10), (2), (3) is well posed in the space $H^{2,0}(Q_1) \cap H^{2,0}(Q_2) \cap \tilde{H}_+^{0,\alpha}(Q) \cap \dot{\tilde{H}}^{1,\alpha/2}(Q)$ and its solution satisfies a priory estimate*

$$\left\| \frac{\partial^2 u}{\partial x^2} \right\|_{L^2(Q_1)} + \left\| \frac{\partial^2 u}{\partial x^2} \right\|_{L^2(Q_2)} + \left\| D_{t,0+}^\alpha u \right\|_{\tilde{L}^2(Q)} + \|u\|_{L^2(Q)} \leq C \|f\|_{L^2(Q)}.$$

7   Transmission Problem in Disjoint Intervals

Let $Q_i = (a_i, b_i) \times (0, T)$, $i = 1, 2$, where $-\infty < a_1 < b_1 < a_2 < b_2 < +\infty$ and $\alpha \in (0, 1)$. We consider the following system of fractional in time diffusion equations

$$D^{\alpha}_{t,0^+} u_i - \frac{\partial^2 u_i}{\partial x^2} = f_i(x, t), \quad (x, t) \in Q_i, \quad i = 1, 2, \tag{14}$$

subject to nonlocal internal boundary conditions of Robin-Dirichlet type

$$\frac{\partial u_1}{\partial x}(b_1, t) + p_1 u_1(b_1, t) = q_1 u_2(a_2, t), \quad t \in (0, T), \tag{15}$$

$$-\frac{\partial u_2(a_2, t)}{\partial x} + p_2 u_2(a_2, t) = q_2 u_1(b_1, t), \quad t \in (0, T), \tag{16}$$

and homogeneous initial and external boundary conditions

$$u_i(x, 0) = 0, \quad x \in (a_i, b_i), \quad i = 1, 2, \tag{17}$$

$$u_1(a_1, t) = 0, \quad u_2(b_2, t) = 0. \quad t \in (0, T). \tag{18}$$

We also assume that

$$p_i > 0, \quad q_i > 0, \quad i = 1, 2 \quad \text{and} \quad q_1 q_2 \leq p_1 p_2. \tag{19}$$

Analogous problem for integer order diffusion equation is considered in [9, 10]. Problems of such type are known as transmission problems.

Let us introduce the product space

$$L^2 = L^2(Q_1) \times L^2(Q_2) = \left\{ v = (v_1, v_2) \mid v_i \in L^2(Q_i), \ i = 1, 2 \right\},$$

endowed with the inner product and norm

$$(v, w)_{L^2} = q_2 (v_1, w_1)_{L^2(Q_1)} + q_1 (v_2, w_2)_{L^2(Q_2)}, \quad \|v\|_{L^2} = (v, v)^{1/2}_{L^2}.$$

Analogously, we introduce the spaces

$$H^{\alpha, \beta} = H^{\alpha, \beta}(Q_1) \times H^{\alpha, \beta}(Q_2) \quad \text{and} \quad H^{\alpha, \beta}_{\pm} = H^{\alpha, \beta}_{\pm}(Q_1) \times H^{\alpha, \beta}_{\pm}(Q_2).$$

In particular, we set

$$H^{1, \alpha/2}_0 = \left\{ v = (v_1, v_2) \mid v_1 \in L^2((0, T), H^1(a_1, b_1)) \cap \dot{H}^{\alpha/2}((0, T), L^2(a_1, b_1)), \right.$$

$$v_2 \in L^2((0, T), H^1(a_2, b_2)) \cap \dot{H}^{\alpha/2}((0, T), L^2(a_2, b_2)), \ v_1(a_1, t) = v_2(b_2, t) = 0 \right\}.$$

The weak formulation of the problem (14)-(18) has the form: find $u \in H^{1, \alpha/2}_0$ such that

$$a(u, v) = l(v), \quad \forall v \in H^{1, \alpha/2}_0, \tag{20}$$

where

$$a(u, v) = a_1(u_1, v_1) + a_2(u_2, v_2) + a_3(u, v),$$

$$a_i(u_i, v_i) = q_{3-i} \left(D_{t,0+}^{\alpha/2} u_i,\, D_{t,T_-}^{\alpha/2} v_i \right)_{Q_i} + q_{3-i} \left(\frac{\partial u_i}{\partial x},\, \frac{\partial v_i}{\partial x} \right)_{Q_i},\quad i = 1, 2,$$

$$a_3(u, v) = \int_0^T \left[q_2 p_1 u_1(b_1, t) v_1(b_1, t) + q_1 p_2 u_2(a_2, t) v_2(a_2, t) \right.$$

$$\left. - q_1 q_2 u_2(a_2, t) v_1(b_1, t) - q_1 q_2 u_1(b_1, t) v_2(a_2, t) \right] dt,$$

and

$$l(v) = (f, v)_{L^2} = q_2 (f_1, v_1)_{Q_1} + q_1 (f_2, v_2)_{Q_2}.$$

Now, using lemmae 1-3, Cauchy-Schwarz and Poincaré inequalities and condition (19) we convince that bilinear form $a(\cdot, \cdot)$ and linear functional $l(\cdot)$ satisfy requirements of Lax-Milgram lemma. Hence, one obtains the following assertion:

Theorem 7. *Let $\alpha \in (0, 1)$, $f \in L^2$ and let the conditions (19) are satisfied. Then the problem (20) is well posed and its solution satisfies a priory estimate*

$$\|u\|_{H^{1,\alpha/2}} \le C \|f\|_{L^2}.$$

In analogous way one obtains the following results:

Theorem 8. *Let the assumptions of Theorem 7 are satisfied. Then the initial-boundary value problem (14)-(18) is well posed in $H_+^{2,\alpha} \cap H_0^{1,\alpha/2}$ and its solution satisfies a priory estimate*

$$|u|_{H_+^{2,\alpha}} \equiv \left[\sum_{i=1}^2 \left(\left\| \frac{\partial^2 u_i}{\partial x^2} \right\|_{L^2(Q_i)}^2 + \left\| D_{t,0+}^{\alpha} u_i \right\|_{L^2(Q_i)}^2 \right) \right]^{1/2} \le C \|f\|_{L^2}.$$

Acknowledgement. The research of the first and third author was supported by Ministry of Education and Science of Republic of Serbia under project 174015, while the research of the second author was supported by Bulgarian National Fund of Sciences under project VU-MI-106/2005.

References

1. Alikhanov, A.A.: Boundary value problems for the diffusion equation of the variable order in differential and difference settings (May 10, 2011); arXiv:1105.2033v1 [math.NA]
2. Benson, D.A., Wheatcraft, S.W., Meerschaeert, M.M.: The fractional order governing equations of Levy motion. Water Resour. Res. 36, 1413–1423 (2000)
3. Carreras, B.A., Lynch, V.E., Zaslavsky, G.M.: Anomalous diffusion and exit time distribution of particle tracers in plasma turbulence models. Phys. Plasmas 8(12), 5096–5103 (2001)
4. Ervin, V.J., Roop, J.P.: Variational formulation for the stationary fractional advection dispersion equation. Numer. Methods Partial Differential Equations 23, 558–576 (2006)

5. Henry, B., Wearne, S.: Existence of turing instabilities in a two-species fractional reactiondiffusion system. SIAM J. Appl. Math. 62, 870–887 (2002)
6. Ilic, M., Turner, I.W., Liu, F., Anh, V.: Analytical and numerical solutions to one-dimensional fractional-in-space diffusion in a composite medim. Appl. Math. Comp. 216, 2248–2262 (2010)
7. Jovanović, B.S., Vulkov, L.G.: Operator's approach to the problems with concentrated factors. In: Vulkov, L.G., Waśniewski, J., Yalamov, P. (eds.) NAA 2000. LNCS, vol. 1988, pp. 439–450. Springer, Heidelberg (2001)
8. Jovanović, B.S., Vulkov, L.G.: On the convergence of finite difference schemes for the heat equation with concentrated capacity. Numer. Math. 89(4), 715–734 (2001)
9. Jovanović, B.S., Vulkov, L.G.: Numerical solution of a two-dimensional parabolic transmission problem. Int. J. Numer. Anal. Model. 7(1), 156–172 (2010)
10. Jovanović, B.S., Vulkov, L.G.: Numerical solution of a parabolic transmission problem. IMA J. Numer. Anal. 31, 233–253 (2011)
11. Kirchner, J.W., Feng, X., Neal, C.: Fractal stream chemistry and its implications for contaminant transport in catchments. Nature 403, 524–526 (2000)
12. Kiryakova, V.: A survey on fractional calculus, fractional order differential and integral equations and related special functions. Rousse (2012)
13. Li, X., Xu, C.: A space-time spectral method for the time fractional diffusion equation. SIAM J. Numer. Anal. 47(3), 2108–2131 (2009)
14. Lions, J.L., Magenes, E.: Non homogeneous boundary value problems and applications. Springer, Berlin (1972)
15. Love, E.R., Young, L.C.: On fractional integration by parts. Proc. London Math. Soc., Ser. 2 44, 1–35 (1938)
16. Mainardi, F.: Fractional diffusive waves in viscoelastic solids. In: Wegner, J.L., Norwood, F.R. (eds.) Nonlinear Waves in Solids, pp. 93–97. Fairfield (1995)
17. Mainardi, F.: Fractional calculus. In: Carpinteri, A., Mainardi, F. (eds.) Fractals and Fractional Calculus in Continuum Mechanics. Springer, New York (1997)
18. Metzler, R., Klafter, J.: The random walks guide to anomalous diffusion: A fractional dynamics approach. Phys. Rep. 339, 1–77 (2000)
19. Müller, H.P., Kimmich, R., Weis, J.: NMR flow velocity mapping in random percolation model objects: Evidence for a power-law dependence of the volume-averaged velocity on the probe-volume radius. Phys. Rev. E 54, 5278–5285 (1996)
20. Nigmatullin, R.R.: The realization of the generalized transfer equation in a medium with fractal geometry. Physica Status Solidi. B 133(1), 425–430 (1986)
21. Podlubny, I.: Fractional Differential Equations. Academic Press (1999)
22. Samarskii, A.A.: Theory of difference schemes. Nauka, Moscow (1989) Russian; English edition: Pure and Appl. Math., vol. 240. Marcel Dekker, Inc., (2001)
23. Sokolov, I.M., Klafter, J., Blumen, A.: Fractional kinetics. Physics Today 55(11), 48–54 (2002)
24. Taukenova, F.I., Shkhanukov-Lafishev, M.K.: Difference methods for solving boundary value problems for fractional differential equations. Comput. Math. Math. Phys. 46(10), 1785–1795 (2006)
25. Vladimirov, V.S.: Equations of mathematical physics. Nauka, Moscow (1988) (Russian)

Numerical Study of Maximum Norm a Posteriori Error Estimates for Singularly Perturbed Parabolic Problems

Natalia Kopteva[1] and Torsten Linß[2]

[1] Department of Mathematics and Statistics,
University of Limerick, Limerick, Ireland
natalia.kopteva@ul.ie

[2] Fakultät für Mathematik und Informatik, FernUniversität in Hagen,
Universitätsstr. 11, 58095 Hagen, Germany
torsten.linss@fernuni-hagen.de

Abstract. A second-order singularly perturbed parabolic equation in one space dimension is considered. For this equation, we give computable a posteriori error estimates in the maximum norm for two semidiscretisations in time and a full discretisation using P_1 FEM in space. Both the Backward-Euler method and the Crank-Nicolson method are considered. Certain critical details of the implementation are addressed. Based on numerical results we discuss various aspects of the error estimators in particular their effectiveness.

Keywords: a posteriori error estimate, maximum norm, singular perturbation, elliptic reconstruction, backward Euler, Crank-Nicolson, parabolic equation, reaction-diffusion.

1 Introduction

The authors' recent paper [4] gives certain maximum norm a posteriori error estimates for time-dependent semilinear reaction-diffusion equations in 1-3 space dimensions, applicable in both regular and singularly perturbed regimes. The purpose of the present paper is to numerically investigate the sharpness and robustness of the theoretical results [4] when applied to a relatively simple equation. Our test problem will be a singularly perturbed equation in the form

$$\mathcal{M}u := u_t + \mathcal{L}u = f \quad \text{in} \ \ \Omega \times (0,T], \quad \Omega := (0,1), \quad \mathcal{L}u := -\varepsilon^2 u_{xx} + ru, \quad \text{(1a)}$$

with a small positive perturbation parameter ε and functions $r : \bar{\Omega} \to \mathbb{R}, r \geq \varrho^2$, $\varrho > 0$, $f : \bar{\Omega} \times [0,T] \to \mathbb{R}$, subject to the initial and Dirichlet boundary conditions

$$u(x,0) = \varphi(x) \quad \text{for} \ \ x \in \bar{\Omega}, \quad u(0,t) = u(1,t) = 0 \quad \text{for} \ \ t \in [0,T]. \quad \text{(1b)}$$

Solutions to (1) typically exhibit sharp layers of width $\mathcal{O}\left(\varepsilon \ln \varepsilon^{-1}\right)$ at the two end points of the spatial domain. Interior layers may also be present depending

I. Dimov, I. Faragó, and L. Vulkov (Eds.): NAA 2012, LNCS 8236, pp. 50–61, 2013.
© Springer-Verlag Berlin Heidelberg 2013

on the right-hand side and on the initial condition. These layers form challenges for any numerical method; see [9] for an overview.

Recently much attention has been paid to the design of adaptive methods for partial differential equations that automatically adapt the discretisation to the features of the solution, see e.g. [10]. The main ingredients of such methods are reliable a posteriori error estimators.

In [4,5] error estimators in the maximum norm for singularly perturbed parabolic problems like (1) have been derived. The crucial issue when analysing methods for such problems is to carefully monitor any dependence of constants on the perturbation parameter. In the present paper we shall numerically investigate the sharpness of the a posteriori error bounds derived in [4].

The outline of the paper is as follows. In §2 we review properties of the Green's function of (1) which are the basis for the analysis in [4]. Semidiscretisations in time are studied in §3. Both the implicit Euler method and the Crank-Nicolson method will be considered. §4 is concerned with full discretisations which are obtained by applying a FEM to the semidiscretisations. Finally, in §5 the effects of changing the spatial mesh are studied.

2   The Green's Function

The main tool for deriving a posteriori error estimators in [4] is the Green's function $\mathcal{G}$ associated with the differential operator $\mathcal{M}$ of (1). It can be used to express the error of a numerical approximation in terms of its residual.

For definitions and properties of fundamental solutions and Green's functions of parabolic operators, we refer the reader to [2, Chap. 1 and §7 of Chap. 3]. Any given function v of sufficient regularity can be represented as

$$v(x,t) = \int_\Omega \mathcal{G}(x,t;\xi,0)\, v(\xi,0)\, \mathrm{d}\xi + \int_0^t \int_\Omega \mathcal{G}(x,t;\xi,s)\, (\mathcal{M}v)(\xi,s)\, \mathrm{d}\xi\, \mathrm{d}s. \quad (2)$$

Theorem 1 ([5, Th. 2.1]). *Let* $r \in C^1(\bar{\Omega})$. *Assume* $\varrho^2 \le r$ *on* $\bar{\Omega}$ *with some constant* $\varrho > 0$. *Then, for the Green's function* $\mathcal{G}$ *one has*

$$\int_\Omega |\mathcal{G}(x,t;\xi,s)|\, \mathrm{d}\xi \le \mathrm{e}^{-\varrho^2(t-s)},$$

$$\int_\Omega |\partial_\xi^k \mathcal{G}(x,t;\xi,s)|\, \mathrm{d}\xi \le \frac{\gamma_k\, \mathrm{e}^{-\varrho^2(t-s)}}{\varepsilon^k (t-s)^{k/2}} + \mathcal{O}\left(\varepsilon^{k-1}\right), \quad \text{for } k=1,2,$$

and

$$\int_\Omega |\partial_s \mathcal{G}(x,t;\xi,s)|\, \mathrm{d}\xi \le \left(\frac{\gamma_2}{t-s} + \|r\|_\infty\right) \mathrm{e}^{-\varrho^2(t-s)} + \mathcal{O}\left(\varepsilon\right)$$

with constants $\gamma_1 = 1/\sqrt{\pi}$ *and* $\gamma_2 = \sqrt{2/(\pi\mathrm{e})}$.

Let $\tilde{u}$ be an approximation of the exact solution u of (1). Replacing v by $u - \tilde{u}$ in (2), we get the error representation

$$
\begin{aligned}
(u - \tilde{u})\,(x,t) = &\int_\Omega \mathcal{G}(x,t;\xi,0)\,(\varphi - \tilde{u})\,(\xi,0)\,\mathrm{d}\xi \\
&+ \int_0^t \int_\Omega \mathcal{G}(x,t;\xi,s)\,(f - \tilde{u}_s + \mathcal{L}\tilde{u})(\xi,s)\,\mathrm{d}\xi\,\mathrm{d}s.
\end{aligned}
\tag{3}
$$

The main idea when deriving a posteriori error estimates is the use of the Hölder inequality, the L_1-norm bounds for the Green's function in Theorem 1 and maximum-norm bounds for the residuum.

In the case of the backward-Euler discretisation the approximation $\tilde{u}$ is considered to be piecewise constant in time. Therefore, $\tilde{u}$ will be discontinuous in time and $\tilde{u}_s$ has to be read in the context of distributions. Further discontinuities will occur when the spatial discretisation mesh changes between time levels.

To deal with these discontinuities, integration by parts is applied to the second integral in (3):

$$
\begin{aligned}
\int_0^t \mathcal{G}(x,t;\xi,s)\,\tilde{u}_s(\xi,s)\,\mathrm{d}s = &\; \mathcal{G}(x,t;\xi,s)\,\tilde{u}(\xi,s)\Big|_{s=0}^{t-\tau} \\
&- \int_0^{t-\tau} \mathcal{G}_s(x,t;\xi,s)\,\tilde{u}(\xi,s)\,\mathrm{d}s + \int_{t-\tau}^t \mathcal{G}(x,t;\xi,s)\,\tilde{u}_s(\xi,s)\,\mathrm{d}s.
\end{aligned}
$$

For the L_1 norm of $\mathcal{G}_s$, Theorem 1 yields the bound

$$
\int_0^t \int_\Omega |\mathcal{G}_s(x,t;\xi,s)|\,\mathrm{d}\xi\,\mathrm{d}s \le \gamma_2 \ell(\tau,t) + \bar{\varrho} + \mathcal{O}\,(\varepsilon), \quad 0 < \tau < t \le T,
$$

where $\ell(\tau,t) := \int_\tau^t s^{-1} e^{-\varrho^2 s/2}\,\mathrm{d}s \le \ln(t/\tau)$ and $\bar{\varrho} := \varrho^{-2}\|r\|_\infty$.

3   Semidiscretisation in Time

Let $\omega_t : 0 = t_0 < t_1 < t_2 < \cdots < t_M = T$, be an arbitrary nonuniform mesh in time direction with step sizes $\tau_j := t_j - t_{j-1}$ and mesh intervals $J_j := (t_{j-1}, t_j)$, $j = 1, \dots, M$. Set $f^j := f(\cdot, t_j)$.

Given an arbitrary function $v : \omega_t \to H_0^1(\Omega) : t_j \mapsto v^j$, we introduce its standard piecewise linear interpolant

$$
(I_{1,t}v)\,(\cdot,t) := \frac{t_j - t}{\tau_j}\,v^{j-1} + \frac{t - t_{j-1}}{\tau_j}\,v^j \quad \text{for } t \in \bar{J}_j, \quad j = 1, \dots, M,
$$

and the piecewise-constant interpolant

$$
(I_{0,t}v)\,(\cdot,t) := v^j \quad \text{for } t \in (t_{j-1}, t_j], \quad j = 1 \dots, M; \quad (I_{0,t}v)\,(\cdot,0) := v^1;
$$

so $I_{0,t}v$ is continuous on $[t_0, t_1]$. Furthermore, introduce the difference quotient

$$
\delta_t v^j := \frac{v^j - v^{j-1}}{\tau_j}
$$

as an approximation of the first-order time derivative.

3.1   Backward Euler

We associate an approximate solution $U^j \in H_0^1(\Omega)$ with the time level t_j and require it to satisfy

$$\delta_t U^j + \mathcal{L}U^j = f^j \quad \text{in } \Omega, \quad j = 1, \ldots, M; \quad U^0 = \varphi. \tag{4}$$

Using (3) with $\tilde{u} = I_{0,t}U$, the following a posteriori error estimate was obtained in [4].

Theorem 2. *For $m = 1, \ldots, M$, the maximum-norm error satisfies*

$$\left\| U^m - u(\cdot, t_m) \right\|_{\infty, \Omega} \leq \eta^{\mathrm{bE}} := \eta_{\mathrm{osc}}^{\mathrm{bE}} + \eta_t^{\mathrm{bE}} + \eta_{t,*}^{\mathrm{bE}} \tag{5}$$

with

$$\eta_{\mathrm{osc}}^{\mathrm{bE}} := \sum_{j=1}^m \tau_j e^{-\varrho^2 (t_m - t_j)} \left\| f - I_{0,t}f \right\|_{\infty, \Omega \times J_j}, \quad \eta_{t,*}^{\mathrm{bE}} := 2\tau_m \left\| \delta_t U^m \right\|_{\infty, \Omega},$$

$$\eta_t^{\mathrm{bE}} := \left(\gamma_2 \ln \frac{t_m}{\tau_m} + \bar{\varrho} + \mathcal{O}(\varepsilon) \right) \max_{j=1,\ldots,m-1} \tau_j \left\| \delta_t U^j \right\|_{\infty, \Omega}.$$

Remark 1. In practice, for a singularly perturbed problem the $\mathcal{O}(\varepsilon)$ term is small (compared to $\bar{\varrho}$). Therefore, it will be neglected.

The term $\eta_{\mathrm{osc}}^{\mathrm{bE}}$ captures the data oscillations. Therefore, it cannot be evaluated exactly and needs to be approximated. In our experiments this is done as follows:

$$\left\| f - I_{\nu,t}f \right\|_{\infty, \Omega \times J_j} \approx \max_{k=0,\ldots,3} \left\| (f - I_{\nu,t}f)(\cdot, t_{j-1} + k\tau_j/4) \right\|_{\infty, \Omega}, \quad \nu = 0, 1, \tag{6}$$

i.e., the difference between the right-hand side f and it piecewise constant (and later linear) interpolant is sampled at 4 equally spaced points per time interval.

We present numerical results for the following *test problem*:

$$u_t - \varepsilon^2 u_{xx} + (1+x)u = 1 - \cos 10xt^2, \quad \text{in } \Omega \times (0, T],$$
$$u(x, 0) = \sin \pi x, \quad x \in [0, 1], \quad u(0, t) = u(1, t) = 0, \quad t \in (0, T], \tag{7}$$

with $\varepsilon = 10^{-6}$. This is a sufficiently small value to bring out the singular-perturbation nature of the problem. The exact solution is not available. Instead the true errors are approximated by means of a numerical solution on a very fine layer-adapted mesh. Errors arising from the spatial discretisation can be neglected.

In Table 1 we present results for the semi-discretisation error at final time $T = 1$ and compare it with the a posteriori error estimator of Theorem 2. In time we use a mesh with M mesh intervals and varying step sizes:

$$\tau_j = \begin{cases} \frac{2}{3M} & \text{if } j \text{ is odd,} \\ \frac{4}{3M} & \text{if } j \text{ is even.} \end{cases}$$

Table 1. Semidiscretisation by the backward Euler method, $\varepsilon = 10^{-6}$

M	$\|u - U\|_\infty$	rate	η^{bE}	$\frac{C_{\mathrm{eff}}}{\ln(1/\tau_M)}$	$\eta_{\mathrm{osc}}^{\mathrm{bE}}$	η_t^{bE}	$\eta_{t,*}^{\mathrm{bE}}$
2^{11}	1.401e-4	1.00	3.793e-3	2.35	1.091e-4	*2.767e-3*	9.171e-4
2^{12}	7.006e-5	1.00	1.981e-3	2.46	5.455e-5	*1.468e-3*	4.586e-4
2^{13}	3.503e-5	1.00	1.032e-3	2.56	2.727e-5	*7.758e-4*	2.293e-4
2^{14}	1.751e-5	1.00	5.371e-4	2.67	1.363e-5	*4.088e-4*	1.147e-4
2^{15}	8.757e-6	1.00	2.790e-4	2.77	6.817e-6	*2.149e-4*	5.733e-5
2^{16}	4.379e-6	—	1.447e-4	2.88	3.408e-6	*1.127e-4*	2.867e-5

The table contains the maximum errors at time $T = 1$, the error bounds obtained by the error estimator η^{bE}, the efficiency index

$$C_{\mathrm{eff}} := \eta^{\mathrm{bE}} \Big/ \left\| U^M - u(\cdot, T) \right\|_{\infty, \Omega}$$

and the various parts of the error estimator. The dominant term η_t^{bE} in the estimator is highlighted in the table. It does not converge with first order because of the presence of the $\ln(1/\tau_m)$ term (which also appears in [5]). Also, note that the efficiency slightly deteriorates with increasing M. C_{eff} is approximately proportional to $\ln(1/\tau_M)$. We conjecture that the factor $\ln(1/\tau_m)$ appearing in $\eta_{t,*}^{\mathrm{bE}}$ is merely an artifact of the analysis. Apart from this the estimator is quite efficient with $\frac{C_{\mathrm{eff}}}{\ln(1/\tau_M)} \approx 2.5 \ldots 3.0$.

3.2   Crank-Nicolson

An approximate solution $U^j \in H_0^1(\Omega)$ is associated with the time level t_j. It satisfies

$$\delta_t U^j + \frac{\mathcal{L}U^j + \mathcal{L}U^{j-1}}{2} = \frac{f^{j-1} + f^j}{2} \quad \text{in } \Omega \quad j = 1, \ldots, M; \quad U^0 = \varphi.$$

For the Crank-Nicolson method the following error bound is given in [4]:

$$\left\| U^m - u(\cdot, t_m) \right\|_{\infty, \Omega} \leq \eta^{\mathrm{CN}} := \eta_{\mathrm{osc}}^{\mathrm{CN}} + \eta_t^{\mathrm{CN}} + \eta_{t,*}^{\mathrm{CN}} \tag{8}$$

where

$$\eta_{\mathrm{osc}}^{\mathrm{CN}} := \sum_{j=1}^{m} \tau_j e^{-\varrho^2(t_m - t_j)} \left\| f - I_{1,t} f \right\|_{\infty, \Omega \times J_j}, \quad \eta_{t,*}^{\mathrm{CN}} := \frac{5\tau_m}{8} \left\| \delta_t \psi^m \right\|_{\infty, \Omega},$$

$$\eta_t^{\mathrm{CN}} := \frac{1}{8} \left(\gamma_2 \ln \frac{t_m}{\tau_m} + \bar{\varrho} + \mathcal{O}(\varepsilon) \right) \max_{j=1,\ldots,m-1} \tau_j \left\| \delta_t \psi^j \right\|_{\infty, \Omega}$$

and $\psi^j := \mathcal{L}U^j - f(\cdot, t_j)$.

The term $\eta_{\mathrm{osc}}^{\mathrm{CN}}$ captures the data oscillations and needs to be approximated. This is done by means of (6). The results of our test computations can be found in Table 2. They are in agreement with the theoretical results. Again, we have highlighted the dominant term of the estimator in the table. Its second order convergence is affected by the presence of the $\ln(1/\tau_m)$ term.

Table 2. Semidiscretisation by the Crank-Nicolson method, $\varepsilon = 10^{-6}$

M	$\|u - U\|_\infty$	rate	η^{CN}	$\frac{C_{eff}}{\ln(1/\tau_M)}$	η_{osc}^{CN}	η_t^{CN}	$\eta_{t,*}^{CN}$
2^7	4.108e-6	2.00	2.231e-4	6.23	2.486e-5	9.239e-5	1.058e-4
2^8	1.028e-6	2.00	5.755e-5	6.42	6.189e-6	2.494e-5	2.642e-5
2^9	2.570e-7	2.00	1.484e-5	6.62	1.544e-6	6.695e-6	6.600e-6
2^{10}	6.427e-8	2.00	3.824e-6	6.82	3.856e-7	1.789e-6	1.649e-6
2^{11}	1.607e-8	2.00	9.846e-7	7.02	9.635e-8	4.760e-7	4.123e-7
2^{12}	4.018e-9	—	2.533e-7	7.23	2.408e-8	1.262e-7	1.031e-7

4 Full Discretisations

In this section we describe our results for full discretisations of the parabolic
problem (1). To this end we apply piecewise linear P_1 finite elements to the
semidiscrete backward Euler and Crank-Nicolson methods.

4.1 The Spatial Discretisation

Consider a steady-state version of the abstract parabolic problem (1):

$$\mathcal{L}v = -\varepsilon^2 v_{xx} + rv = g \quad \text{in } by, \quad v(0) = v(1) = 0, \qquad (9)$$

with $0 < \varepsilon \ll 1$ and $r \geq \varrho^2$ on Ω, $\varrho > 0$. The corresponding variational formulation is: Find $u \in H_0^1(0,1)$ such that

$$a(u,v) := \varepsilon^2 \langle u_x, v_x \rangle + \langle ru, v \rangle = \langle f, v \rangle \quad \forall w \in H_0^1(\Omega),$$

where $\langle \cdot, \cdot \rangle$ is the standard inner product in $L_2(\Omega)$.

An approximate solution of (9) is obtained by means of the P_1-Galerkin FEM.
Let V_h be the space of continuous piecewise-linear finite element functions on
an arbitrary nonuniform mesh $\bar{\omega}_x = \{x_i\}_{i=0}^N$ with $0 = x_0 < x_1 < \cdots < x_N = 1$,
$h_i := x_i - x_{i-1}$ and $I_i := (x_{i-1}, x_i)$. Note that here we make absolutely no mesh
regularity assumptions. As solutions of our problem typically exhibit sharp layers
so a suitable mesh is expected to be highly-nonuniform; see, e.g., [7].

Our discretisation of (9) is: Find $v_h \in \mathring{V}_h := V_h \cap H_0^1(\Omega)$ such that

$$a_{V_h}(v_h, w_h) := \varepsilon^2 \langle v_h', w_h' \rangle + \langle rv_h, w_h \rangle_{V_h} = \langle g, w_h \rangle_{V_h} \quad \forall w_h \in \mathring{V}_h, \qquad (10)$$

where $\langle \psi, w \rangle_{V_h} := \langle I_{1,x}\psi, w \rangle$ with the standard piecewise-linear nodal interpolation $I_{1,x} : C(\bar{\Omega}) \to V_h$. For the resulting FEM we cite the following a posteriori
error bound from [7].

Theorem 3. *Let v be the solution of (9) and v_h its finite element approximation
defined by (10). Then the maximum-norm error satisfies.*

$$\|v - v_h\|_{\infty,\Omega} \leq \eta_\varepsilon(V_h, rv_h - g)$$

with the a posteriori error estimator

$$\eta_\varepsilon(V_h, q) := \max_{i=1,\ldots,N} \left\{ \frac{h_i^2}{4\varepsilon^2} \|I_{1,x}q\|_{\infty,I_i} \right\} + \varrho^{-2} \|q - I_{1,x}q\|_{\infty,\Omega}. \qquad (11)$$

4.2   Fully Discrete Backward Euler Method

A spatial mesh $\bar{\omega}_x^j : 0 = x_0^j < x_1^j < \cdots < x_{N_j}^j = 1$, a finite-element space V_h^j of piecewise-linear functions and a computed solution $u_h^j \in \mathring{V}_h^j := V_h^j \cap H_0^1(\Omega)$ are associated with the time level t_j. By $I_{1,x}^j : C(\bar{\Omega}) \to V_h^j$ we denote the nodal interpolation in V_h^j and by $P_h^j : L^2(\Omega) \to V_h^j$ the L_2 projection onto V_h^j. Given the computed solution u_h^j, we set $\hat{u}_h^{j-1} := P^j u_h^{j-1}$ and

$$\delta_t^* u_h^j := \frac{u_h^j - \hat{u}_h^{j-1}}{\tau_j}.$$

Note that $\hat{u}_h^{j-1} = u_h^{j-1}$ if $V_h^{j-1} \subset V_h^j$, i.e. when the mesh is purely refined. Otherwise, when parts of the mesh are coarsend, one typically has $\hat{u}_h^{j-1} \neq u_h^{j-1}$.

We apply the FEM (10) to (4) and obtain the full discretisation: Find $u_h^j \in \mathring{V}_h^j$, $j = 0, \ldots, M$, such that

$$\langle \delta_t^* u_h^j, w_h \rangle + a_{V_h^j}(u_h^j, w_h) = \langle f^j, w_h \rangle_{V_h} \quad \forall w_h \in \mathring{V}_h^j, \ j = 1, \ldots, M, \qquad (12)$$

with some initial value u_h^0, for example $u_h^0 = I_{1,x}^0 \varphi$.

Elliptic Reconstruction. For each time level t_j, $j = 1, \ldots, M$, we follow an idea from [8] and introduce the *elliptic reconstruction* $R^j \in H_0^1(\Omega)$ of u_h^j, which is uniquely defined by

$$a(R^j, w) = \langle f^j - \delta_t^* u_h^j, w \rangle \quad \forall w \in H_0^1(\Omega). \qquad (13)$$

In view of (12), u_h^j can be interpreted as the finite-element approximation of R^j obtained by (10). Therefore, Theorem 3 applies and yields

$$\left\| u_h^j - R^j \right\|_{\infty,\Omega} \leq \eta^j := \eta_\varepsilon(V_h^j, q^j), \qquad j = 1, \ldots, M, \qquad (14a)$$

with

$$q^j := r u_h^j - f^j + \delta_t^* u_h^j, \qquad (14b)$$

Remark 2. (i) The second term in the error estimator η_ε, see (11), simplifies to

$$\varrho^{-2} \left\| q^j - I_{1,x} q^j \right\|_{\infty,\Omega} = \varrho^{-2} \left\| r u_h^j - f^j - I_{1,x}(r u_h^j - f^j) \right\|_{\infty,\Omega},$$

because $I_{1,x}^j \delta_t^* u_h^j = \delta_t^* u_h^j$.

(ii) The first term of η_ε requires to evaluate q^j in the mesh nodes x_i^j. For small ε, its evaluation using (14b) is numerically unstable because rounding errors are amplified. A stable alternative is to determine $q^j \in V_h^j$ such that $q^j(0) = f^j(0)$, $q^j(1) = f^j(1)$ and

$$\langle q^j, v_h \rangle_{V_h^j} = -\varepsilon^2 \langle u_{h,x}^j, v_{h,x} \rangle \quad \forall v_h \in \mathring{V}_h^j.$$

This requires to invert the standard mass matrix.

A Posteriori Estimator for the Parabolic Problem. Consider the error at time t_m. By the triangle inequality, we have

$$\|u_h^m - u(\cdot, t_m)\|_{\infty,\Omega} \leq \|u_h^m - R^m\|_{\infty,\Omega} + \|R^m - u(\cdot, t_m)\|_{\infty,\Omega}$$
$$\leq \eta^m + \|R^m - u(\cdot, t_m)\|_{\infty,\Omega}.$$

The difference $R^m - u(\cdot, t_m)$ is represented using (3) with $\tilde{u} = I_{0,t}R$. The reconstruction R can be completely eliminated using (13) and (14a). We arrive at the following a posteriori error bound [4].

Theorem 4. *For* $m = 1, \ldots, M$, *the maximum-norm error satisfies*

$$\|u_h^m - u(\cdot, t_m)\|_{\infty,\Omega} \leq \eta^{\mathrm{bE}} := \eta_{\mathrm{init}} + \eta_{\mathrm{osc}}^{\mathrm{bE}} + \eta_{\mathrm{proj}} + \eta_t^{\mathrm{bE}} + \eta_{t,*}^{\mathrm{bE}} + \eta_{\mathrm{ell}}^{\mathrm{bE}} + \eta_{\mathrm{ell},*}^{\mathrm{bE}}$$

with

$$\eta_{\mathrm{init}} := e^{-\gamma^2 t_m} \|u_h^0 - \varphi\|_{\infty,\Omega}, \quad \eta_{\mathrm{osc}}^{\mathrm{bE}} := \sum_{j=1}^{m} \tau_j e^{-\varrho^2(t_m - t_j)} \|f - I_{0,t}f\|_{\infty, \Omega \times J_j},$$

$$\eta_{\mathrm{proj}} := \sum_{j=1}^{m} e^{-\gamma^2(t_m - t_j)} \|\hat{u}_h^{j-1} - u_h^{j-1}\|_{\infty,\Omega}, \quad \eta_{t,*}^{\mathrm{bE}} := 2\tau_m \|\delta_t^* u_h^m\|_{\infty,\Omega},$$

$$\eta_t^{\mathrm{bE}} := \left(\gamma_2 \ln \frac{t_m}{\tau_m} + \bar{\varrho} + \mathcal{O}(\varepsilon) \right) \max_{j=1,\ldots,m-1} \tau_j \|\delta_t^* u_h^j\|_{\infty,\Omega},$$

$$\eta_{\mathrm{ell}}^{\mathrm{bE}} := \left(\gamma_2 \ln \frac{t_m}{\tau_m} + \bar{\varrho} + \mathcal{O}(\varepsilon) \right) \max_{j=1,\ldots,m-1} \eta^j \quad \text{and} \quad \eta_{\mathrm{ell},*}^{\mathrm{bE}} := 2\eta^m.$$

Remark 3. Comparing with Theorem 2, we notice four new terms.

- η_{init}: the error in approximating the initial condition,
- η_{proj}: the accumulated erros due to projections when the mesh is coarsend,
- $\eta_{\mathrm{ell}}^{\mathrm{bE}}$ and $\eta_{\mathrm{ell},*}^{\mathrm{bE}}$: elliptic error estimates for the spatial discretisation.

Numerical Results. In order of balancing the accuracy in space and time, we use a Bakhvalov mesh with $N = [\sqrt{8M}]$ mesh points in space. We do so, because the method is formally 1st order in time and 2nd order in space. The Bakhvalov mesh [1] is given by

$$x_i^j = x_i = \mu(\xi_i), \quad \xi_i = i/N$$

with the mesh generating function

$$\mu(\zeta) = \begin{cases} \vartheta(\zeta) := \dfrac{\sigma\varepsilon}{\varrho} \ln \dfrac{\alpha}{\alpha - \zeta} & \zeta \in [0, \zeta^*], \\ \vartheta(\zeta^*) + \vartheta'(\zeta^*)(\zeta - \zeta^*) & \zeta \in [\zeta^*, 1/2], \\ 1 - \mu(1 - \zeta) & \zeta \in [1/2, 1]. \end{cases}$$

Table 3. Backward Euler and linear FEM, $\varepsilon = 10^{-6}$

M	N	error η^{bE}	rate $\frac{C_{\mathrm{eff}}}{\ln(1/\tau_M)}$	η_{init} $\eta_{\mathrm{osc}}^{\mathrm{bE}}$	rate rate	η_t^{bE} $\eta_{\mathrm{ell}}^{\mathrm{bE}}$	rate rate	$\eta_{t,*}^{\mathrm{bE}}$ $\eta_{\mathrm{ell},*}^{\mathrm{bE}}$	rate rate
2^{11}	362	1.401e-4	1.00	2.283e-5	1.00	*2.767e-3*	*0.91*	9.171e-4	1.00
		1.062e-2	6.59	1.091e-4	1.00	*4.998e-3*	*0.92*	1.803e-3	1.00
2^{12}	512	7.006e-5	1.00	1.142e-5	1.00	*1.468e-3*	*0.92*	4.586e-4	1.00
		5.534e-3	6.87	5.455e-5	1.00	*2.643e-3*	*0.92*	8.985e-4	1.00
2^{13}	724	3.503e-5	1.00	5.709e-6	1.00	*7.758e-4*	*0.92*	2.293e-4	1.00
		2.881e-3	7.15	2.727e-5	1.00	*1.394e-3*	*0.93*	4.483e-4	1.00
2^{14}	1024	1.751e-5	1.00	2.854e-6	1.00	*4.088e-4*	*0.93*	1.147e-4	1.00
		1.497e-3	7.44	1.363e-5	1.00	*7.335e-4*	*0.93*	2.238e-4	1.00
2^{15}	1448	8.757e-6	1.00	1.427e-6	1.00	*2.149e-4*	*0.93*	5.733e-5	1.00
		7.775e-4	7.72	6.817e-6	1.00	*3.852e-4*	*0.93*	1.118e-4	1.00
2^{16}	2048	4.379e-6	—	7.135e-7	—	*1.127e-4*	—	2.867e-5	—
		4.031e-4	8.01	3.408e-6	—	*2.018e-4*	—	5.584e-5	—

The transition point ζ^* satisfies $(1 - 2\zeta^*)\vartheta'(\zeta^*) = 1 - 2\vartheta(\zeta^*)$ which implies $\mu \in C^1[0,1]$. For the mesh parameters are chosen we take $\sigma = 4$ and $\alpha = 1/4$.

Table 3 displays the results of our test computations for (7). It contains the error at final time $T = 1$, the a posteriori error estimator, the efficiency index and the various components of the error estimator together with their respective rate of convergence.

While the results are in aggreement with Theorem 4, we observe that the terms η_t^{bE} and $\eta_{\mathrm{ell}}^{\mathrm{bE}}$ (highlighted in the table) dominate and converge slower than all other terms. This is because of the factor $\ln \frac{t_m}{\tau_m}$ in their definition. As for the semidiscretisation we conjecture that this factor is an artifact of the error analysis in [4].

Note that $\eta_{\mathrm{proj}} \equiv 0$ because the mesh does not change with time. The effect of mesh adaptivity will be discussed in more detail in §5.

In Table 4 we present computational results for a uniform mesh in space. The method does not converge. This has to be expected because the mesh is not adapted to the layer structure. Examining the various terms of the error estimator, we see that the terms $\eta_{\mathrm{ell}}^{\mathrm{bE}}$ and $\eta_{\mathrm{ell},*}^{\mathrm{bE}}$ dominate. Thus, the source of the bad behaviour is correctly attributed to a wrong spatial resolution.

4.3 Fully Discrete Crank-Nicolson Method

With each time level t_j, $j = 0, \ldots, M$, we associate an approximation $u_h^j \in \mathring{V}_h^j$ of $u(\cdot, t_j)$ that satisfies

$$\langle \delta_t^* u_h^j, w_h \rangle + \tfrac{1}{2} a_{V_h^j}\left(u_h^j + \hat{u}_h^{j-1}, w_h\right) = \tfrac{1}{2}\langle f^j + f^{j-1}, w_h \rangle_{V_h}, \quad \forall w_h \in \mathring{V}_h^j,$$

$$j = 1, \ldots, M,$$

with some initial value u_h^0, e.g., $u_h^0 = I_{1,x}^0 \varphi$.

Using elliptic reconstructions and piecewise *linear* interpolation in time, the following a posteriori error bound was derived in [4].

Table 4. Backward Euler and linear FEM, $\varepsilon = 10^{-6}$, uniform mesh

M	N	error η^{bE} / $\frac{C_{\mathrm{eff}}}{\ln(1/\tau_M)}$	rate	η_{init} / $\eta_{\mathrm{osc}}^{\mathrm{bE}}$	rate	$\eta_{\mathrm{t}}^{\mathrm{bE}}$ / $\eta_{\mathrm{ell}}^{\mathrm{bE}}$	rate	$\eta_{\mathrm{t},*}^{\mathrm{bE}}$ / $\eta_{\mathrm{ell},*}^{\mathrm{bE}}$	rate
2^{11}	362	5.371e-2	0.00	5.710e-6	1.00	3.208e-3	0.91	1.157e-3	1.00
		33.98	55.03	1.091e-4	1.00	24.97	-0.09	9.006	0.00
2^{12}	512	5.363e-2	0.00	2.854e-6	1.00	1.705e-3	0.92	5.796e-4	1.00
		35.50	57.58	5.455e-5	1.00	26.49	-0.08	9.006	0.00
2^{13}	724	5.359e-2	0.00	1.428e-6	1.00	9.021e-4	0.92	2.901e-4	1.00
		37.01	60.08	2.727e-5	1.00	28.01	-0.08	9.006	0.00
2^{14}	1024	5.357e-2	0.00	7.136e-7	1.00	4.758e-4	0.93	1.451e-4	1.00
		38.53	62.57	1.363e-5	1.00	29.52	-0.07	9.006	0.00
2^{15}	1448	5.356e-2	0.00	3.569e-7	1.00	2.502e-4	0.93	7.261e-5	1.00
		40.04	65.03	6.817e-6	1.00	31.03	-0.07	9.006	0.00
2^{16}	2048	5.355e-2	—	1.784e-7	—	1.312e-4	—	3.632e-5	—
		41.55	67.50	3.408e-6	—	32.54	—	9.006	—

Theorem 5. *For $m = 1, \ldots, M$, the maximum-norm error satisfies*

$$\left\| u_h^m - u(\cdot, t_m) \right\|_{\infty, \Omega} \le \eta^{\mathrm{CN}} := \eta_{\mathrm{init}} + \eta_{\mathrm{osc}}^{\mathrm{CN}} + \eta_{\mathrm{proj}} + \eta_{\mathrm{t}}^{\mathrm{CN}} + \eta_{\mathrm{t},*}^{\mathrm{CN}} + \eta_{\mathrm{ell}}^{\mathrm{CN}} + \eta_{\mathrm{ell},*}^{\mathrm{CN}}$$

with η_{init} and η_{proj} as in Theorem 4, and

$$\eta_{\mathrm{osc}}^{\mathrm{CN}} := \sum_{j=1}^{m} \tau_j e^{-\varrho^2(t_m - t_j)} \left\| f - I_{1,t} f \right\|_{\infty, \Omega \times J_j}, \qquad \eta_{\mathrm{t},*}^{\mathrm{CN}} := \frac{5\tau_m^2}{8} \left\| \delta_t^* \psi^m \right\|_{\infty, \Omega},$$

$$\eta_{\mathrm{t}}^{\mathrm{CN}} := \frac{1}{8} \left(\gamma_2 \ln \frac{t_m}{\tau_m} + \bar{\varrho} + \mathcal{O}(\varepsilon) \right) \max_{j=1, \ldots, m-1} \tau_j^2 \left\| \delta_t^* \psi^j \right\|_{\infty, \Omega},$$

$$\eta_{\mathrm{ell}}^{\mathrm{CN}} := 2 \left(\gamma_2 \ln \frac{t_m}{\tau_m} + \bar{\varrho} + \mathcal{O}(\varepsilon) \right) \max_{j=1, \ldots, m-1} \eta^j \quad and \quad \eta_{\mathrm{ell},*}^{\mathrm{CN}} := 5\eta^m,$$

where $\psi^j, \hat{\psi}^{j-1} \in \mathring{V}_h^j$ solve

$$\langle \psi^j, w_h \rangle_{V_h^j} = a_{V_h^j}(u_h^j, w_h) - \langle f^j, w_h \rangle_{V_h^j} \quad \forall\, w_h \in V_h^j, \tag{15a}$$

$$\langle \hat{\psi}^{j-1}, w_h \rangle_{V_h^j} = a_{V_h^j}(\hat{u}_h^{j-1}, w_h) - \langle f^{j-1}, w_h \rangle_{V_h^j} \quad \forall\, w_h \in V_h^j. \tag{15b}$$

Furthermore,

$$\eta^j := \max \left\{ \eta_\varepsilon \left(V_h^j, r u_h^j - f^j + \psi^j \right), \eta_\varepsilon \left(V_h^j, r \hat{u}_h^{j-1} - f^{j-1} + \hat{\psi}^{j-1} \right) \right\}.$$

Remark 4. The evaluation of the error estimator requires the solutions of the two auxiliary problems (15a) and (15b). With regard to the numerical stability of computing η, Remark 2(ii) applies.

Numerical Results for our test problem (7) are contained in Table 5. The Crank-Nicolson method with linear FEM in space is formally 2nd order both in

Table 5. Crank-Nicolson and linear FEM, $\varepsilon = 10^{-6}$

M	N	error η^{CN}	rate $\frac{C_{eff}}{\ln(1/\tau_M)}$	η_{init} η_{osc}	rate rate	η_t^{CN} η_{ell}^{CN}	rate rate	$\eta_{t,*}^{CN}$ $\eta_{ell,*}^{CN}$	rate rate
2^7	2^{11}	1.362e-5	1.99	7.135e-7	2.00	9.239e-5	1.89	1.058e-4	2.00
		5.938e-4	5.00	2.486e-5	2.01	2.304e-4	1.88	1.396e-4	2.00
2^8	2^{12}	3.421e-6	2.00	1.784e-7	2.00	2.494e-5	1.90	2.642e-5	2.00
		1.554e-4	5.21	6.189e-6	2.00	6.277e-5	1.89	3.486e-5	2.00
2^9	2^{13}	8.570e-7	2.00	4.459e-8	2.00	6.695e-6	1.90	6.600e-6	2.00
		4.052e-5	5.42	1.544e-6	2.00	1.692e-5	1.90	8.712e-6	2.00
2^{10}	2^{14}	2.145e-7	2.00	1.115e-8	2.00	1.789e-6	1.91	1.649e-6	2.00
		1.054e-5	5.64	3.856e-7	2.00	4.532e-6	1.91	2.177e-6	2.00
2^{11}	2^{15}	5.365e-8	2.00	2.787e-9	2.00	4.760e-7	1.92	4.123e-7	2.00
		2.739e-6	5.85	9.635e-8	2.00	1.207e-6	1.91	5.443e-7	2.00
2^{12}	2^{16}	1.342e-8	—	6.967e-10	—	1.262e-7	—	1.031e-7	—
		7.103e-7	6.07	2.408e-8	—	3.202e-7	—	1.361e-7	—

Table 6. Euler method on two nested Bakhvalov meshes

M	error η^{bE}	rate $\frac{C_{eff}}{\ln(1/\tau_M)}$	η_{init} η_{osc}	rate rate	η_t^{bE} η_{ell}^{bE}	rate rate	$\eta_{t,*}^{bE}$ $\eta_{ell,*}^{bE}$	rate rate	η_{proj}^{bE}	rate
2^{11}	2.577e-4	0.91	5.709e-6	1.00	2.767e-3	0.91	9.171e-4	1.00	2.765e-4	0.98
	9.609e-3	3.24	1.091e-4	1.00	4.998e-3	0.51	5.357e-4	0.97		
2^{12}	1.369e-4	0.93	2.854e-6	1.00	1.468e-3	0.92	4.586e-4	1.00	1.401e-4	0.99
	5.914e-3	3.76	5.455e-5	1.00	3.516e-3	1.33	2.744e-4	0.97		
2^{13}	7.183e-5	0.95	1.427e-6	1.00	7.758e-4	0.92	2.293e-4	1.00	7.069e-5	0.99
	2.638e-3	3.20	2.727e-5	1.00	1.394e-3	0.93	1.396e-4	0.98		
2^{14}	3.711e-5	0.97	7.135e-7	1.00	4.088e-4	0.93	1.147e-4	1.00	3.556e-5	0.99
	1.378e-3	3.23	1.363e-5	1.00	7.335e-4	0.93	7.065e-5	0.99		
2^{15}	1.900e-5	0.97	3.568e-7	1.00	2.149e-4	0.93	5.733e-5	1.00	1.786e-5	1.00
	7.181e-4	3.29	6.817e-6	1.00	3.852e-4	0.93	3.564e-5	0.99		
2^{16}	9.708e-6	—	1.784e-7	—	1.127e-4	—	2.867e-5	—	8.958e-6	—
	3.736e-4	3.35	3.408e-6	—	2.018e-4	—	1.795e-5	—		

space and in time. Therefore, N should be chosen proportional to M. We have chosen $N = 16M$ to balance the accuracy. The results are in agreement with our theoretical findings. Again they suggest that the logarithmic term is a mere artifact of the error analysis.

5 Mesh Adaptivity, Projection Errors

So far we have considered discretizations where the spatial mesh remains unchanged while we integrate in time. In this final section of the paper we will investigate some effects of changing the spatial discretisation.

A typical approach in mesh adaptivity is enrichment of the spatial discretisation by adding mesh points whenever required. At stages when the discretisation becomes too big it is reduced by removing mesh points where they are not needed anymore. Thus, typically the refinement steps outnumbers the coarsening steps. There are two adavantage of this approach

- no projection errors are introduced during refinement and
- the L_2 projection has to be computed only when the mesh is coarsened.

We model this strategy in our next experiment.

Starting from a Bakhvalov mesh with $2N$ mesh points we coarsen the mesh at time $T/8$ by removing every other mesh point. At time $T/4$ we switch back the mesh with $2N$ points then coarsen again at time $3T/8$ etc. Table 6 gives the results for the backward Euler method, $N = [\sqrt{8M}]$. Again, the estimator predicts the actual errors very well.

Acknowledgement. This publication has emanated from research conducted with the financial support of Science Foundation Ireland under the Research Frontiers Programme 2008; Grant 08/RFP/MTH1536.

References

1. Bakhvalov, N.S.: Towards optimization of methods for solving boundary value problems in the presence of boundary layers. Zh. Vychisl. Mat. i Mat. Fiz. 9, 841–859 (1969)
2. Friedman, A.: Partial differential equations of parabolic type. Prentice-Hall, Englewood Cliffs (1964)
3. Kopteva, N.: Maximum norm a posteriori error estimates for a 1D singularly perturbed semilinear reaction-diffusion problem. IMA J. Numer. Anal. 27, 576–592 (2007)
4. Kopteva, N., Linß, T.: A posteriori error estimation for parabolic problems using elliptic reconstructions. I: Backward-Euler and Crank-Nicolson methods. Univerisity of Limerick (2011) (preprint)
5. Kopteva, N., Linß, T.: Maximum norm a posteriori error estimation for a time-dependent reaction-diffusion problem. Comp. Meth. Appl. Math. 12(2), 189–205 (2012)
6. Linß, T.: Maximum-norm error analysis of a non-monotone FEM for a singularly perturbed reaction-diffusion problem. BIT Numer. Math. 47, 379–391 (2007)
7. Linß, T.: Layer-adapted meshes for reaction-convection-diffusion problems. Lecture Notes in Mathematics, vol. 1985. Springer, Berlin (2010)
8. Makridakis, C., Nochetto, R.H.: Elliptic reconstruction and a posteriori error estimates for parabolic problems. SIAM J. Numer. Anal. 41, 1585–1594 (2003)
9. Roos, H.-G., Stynes, M., Tobiska, L.: Robust numerical methods for singularly perturbed differential equations. Springer Series in Computational Mathematics, vol. 24. Springer, Berlin (2008)
10. Schmidt, A., Siebert, K.G.: Design of adaptive finite element software. Lecture Notes in Computational Science and Engineering, vol. 42. Springer, Berlin (1976)

Comparison Principle
for Reaction-Diffusion-Advection Problems
with Boundary and Internal Layers

Nikolay Nefedov

Department of Mathematics, Faculty of Physics,
Lomonosov Moscow State University, 119991 Moscow, Russia
nefedov@phys.msu.ru

Abstract. In the present paper we discuss father development of the general scheme of the asymptotic method of differential inequalities and illustrate it applying for some new important cases of initial boundary value problem for the nonlinear singularly perturbed parabolic equations, which are called in applications as reaction-diffusion-advection equations. The theorems which state front motion description and stationary contrast structures formation are proved for parabolic, parabolic-periodic and integro-parabolic problems.

Keywords: singularly perturbed problems, comparison principle, reaction-diffusion-advection equations.

1   Introduction

Nonlinear singularly perturbed PDE's which have solutions with boundary and internal layers are of increasing interest because of many applications of practical importance. This work is devoted to nonlinear singularly perturbed parabolic and integro-parabolic equations. In applications, these problems may be interpreted as models for local and non-local reaction-diffusion and reaction-diffusion-advection processes in chemical kinetics, synergetic, astrophysics, biology, et. al. The solutions of these problems often feature a narrow boundary layer region of rapid change as well as internal layers of different types (stationary internal layers - contrast structures, moving internal layers - fronts and moving spikes). It is well-known that such problems are extremely complicated for numerical treatment as well for asymptotic investigations and it needs to develop new asymptotic methods to investigate them formally as well as rigorously.

We present our recent extension of the well-known boundary layer functions method to construct the formal asymptotics of solutions of different classes of problems with internal layers. These results is father development of our investigations of contrast structures which were published in the review paper [1].

Our rigorous investigation is based on modern development of comparison principle for elliptic and parabolic problems. The basic ideas of this approach where suggested in the papers of H.Aman and D.Sattinger (see [2,3]) and recently got a father development in the works of P. Hess [4] and H.Aman [5]. These works

I. Dimov, I. Faragó, and L. Vulkov (Eds.): NAA 2012, LNCS 8236, pp. 62–72, 2013.
© Springer-Verlag Berlin Heidelberg 2013

are essentially using so called Krein-Ruthman theorem (see, for example, [4]) and basic results of M.A. Krasnoselskij on positive operators theory (see, for example, [6] and references threin).

In the present paper we discuss father development of the general scheme of asymptotic method of differential inequalities, the basic ideas of which were proposed in [7] and illustrate it applying for some new important cases of initial boundary value problem for the equation

$$\varepsilon^2 \Delta u - \frac{\partial u}{\partial t} = f(u, \nabla u, x, \varepsilon), \quad x \in \mathcal{D} \subset R^N, t > 0, \tag{1}$$

which plays important role in many applications and is called reaction-diffusion-advection equation. For these problems we state the conditions which imply the existence of contrast structures - solutions with internal layers. Particularly the cases when equation (1) is semilinear or quasylinear are considered. The results for equation (1) are extended for periodic parabolic problems and for some classes for nonlocal reaction-diffusion-advection equations. Among others we discuss the following problems:

1.Existence and Lyapunov stability of stationary solutions.

2. The analysis of local and global domain of stability of the stationary contrast structures.

3. The problem of stabilization of the solution of initial boundary value problem.

Our investigation uses so-called positivity property of the operators producing formal asymptotics and is based on some recent extensions of Krein-Ruthman theorem. The basic idea of this approach is to construct lower and upper solutions to the problem by using formal asymptotics. By using these we state the existence of the solutions, estimate the accuracy of the asymptotics. The new significant result of our work is that we propose a new approach to investigate asymptotic stability of the stationary and periodic solutions in the sense of Ljapunov.

Another aspects of this work is to emphasize the possibility of use this analytical treatment for numerical approaches. Some examples with moving fronts are presented.

2   General Scheme of Asymptotic Method of Differential Inequalities for Reaction-Advection-Diffusion Equations

We consider some cases of initial boundary value problem

$$\begin{aligned} \varepsilon^2 \Delta u - \frac{\partial u}{\partial t} &= f(u, \nabla u, x, \varepsilon), \quad x \in \mathcal{D} \subset R^N, t > 0, \\ Bu &= h(x), \quad x \in \partial \mathcal{D}, t > 0, \end{aligned} \tag{2}$$

where ε is a small parameter, f, h, and $\partial \mathcal{D}$ are sufficiently smooth, B is a boundary operator for Dirichlet, Neumann or third order boundary conditions.

Denote by N the nonlinear operator in (2)

$$Nu \equiv \varepsilon^2 \Delta u - f(u, \nabla u, x, \varepsilon)$$

We introduce the following definition for an upper and a lower solution, which is more strong than classical definition.

Definition. *We call an upper solution β and a lower solution α asymptotic of order $q > 0$ of problem (2) if they satisfy the inequalities*

$$N\beta \leq -c\varepsilon^q, \ N\alpha \geq c\varepsilon^q, \quad x \in D \tag{3}$$

$$B\alpha \leq h(x) \leq B\beta \tag{4}$$

where c is a positive constant.

Denote by L the linear operator which we get from N by linearizing f on the stationary solution, and by H the following characteristic of the nonlinearity

$$H \equiv f(\beta, \nabla\beta, x, \varepsilon) - f(\alpha, \nabla\alpha, x, \varepsilon) - L_f(\beta - \alpha),$$

where L_f is the linearization of f on the stationary solution. Suppose we know how construct the asymptotic lower and upper solutions. Note that one of the most important our achievements is the method to construct them by using the formal asymptotic expansion. In what follows we describe this approach. Our assumption is

(A_1). *There exist asymptotic of order q an upper solution β and a lower solution α such that $\beta > alpha$ and $|\beta - \alpha| \leq c\varepsilon^r$.*

From assumption (A_1) it follows the existence of the stationary solution $u(x, \varepsilon)$ of problem (2)satisfying the inequalities inequalities $\alpha \leq u(x, \varepsilon) \leq \beta$ and therefore we also have the asymptotic estimate for the solution. It differ from the upper or lower solution on the value of order $O(\varepsilon^r)$.

We also assume

(A_2). $|H| \leq c\varepsilon^p$

(A_3). $p \geq q$

It is clear that the estimates of the assumptions (A_2) (A_2) depend on the properties of the nonlinearity f and the lower and Under the assumptions above the following theorem take place.

Theorem 1. *Suppose the assumptions $(A_1) - (A_3)$ to be valid. Then, for sufficiently small ε there exists a solution $u(x, \varepsilon)$ of (2) which differ from the upper or lower solution on the value of order $O(\varepsilon^r)$ and is asymptotically stable in Lyapunov sense with the local domain of stability $[\alpha, \beta]$*

Proof

The proof of Theorem 1 is based on the revised maximum principal, which used Krein-Ruthman theorem.

¿From $(A_2), (A_3)$ it is follows that $L(\beta - \alpha) < 0$, $B(\beta - \alpha) > 0$ and therefore the principal eigenvalue which exists and real satisfy the estimate $\lambda_p < 0$, which imply the asymtotic stability of the stationary solution in the sense of Ljapunov.

The analogues of Theorem 1 are valid for periodic-parabolic problems and nonlocal reaction-advection-diffusion equations.

In order to get the upper and lower solutions satisfying the assumptions of Theorem 1 we use the formal asymptotics, which can be constructed in a lot of cases by our method, proposed in [1,8]. Under quite natural assumptions the formal asymptotics of internal layer solution is produced by the boundary layer operators L_B, regular expansion operators L_R and by the operators describing the location of transition layer A^Γ. To construct the formal asymptotic we assume that the operators are invertible.

For the construction of asymptotic lower and upper solutions we require that *these operators have positive inverse when they act in the same classes of functions in which we construct the asymptotic expansions by means of these operators.*

Finally we get $\alpha \equiv \alpha_n$, $\beta \equiv \beta_n$ – modified n-th order formal asymptotic.

We illustrate our approach by two examples.

3   Periodic Solutions with Boundary Layers

We consider the boundary value problem

$$N_\varepsilon(u) := \varepsilon \left(\frac{\partial^2 u}{\partial x^2} - \frac{\partial u}{\partial t} \right) -$$

$$A(u,x,t)\frac{\partial u}{\partial x} - B(u,x,t) = 0 \quad \text{for } x \in (0,1),\ t \in R \tag{5}$$

$$u(0,t,\varepsilon) = u^{(-)}(t), \quad u(1,t,\varepsilon) = u^{(+)}(t) \text{ for } t \in R,$$

$$u(x,t,\varepsilon) = u(x,t+T,\varepsilon) \quad \text{for} \quad t \in R,$$

where ε is a small parameter, A, B, $u^{(-)}$ and $u^{(+)}$ are sufficiently smooth and T-periodic in t.

If we put $\varepsilon = 0$ in equation (9) we get the so-called degenerate equation

$$A(u,x,t)\frac{\partial u}{\partial x} + B(u,x,t) = 0, \tag{6}$$

where t has to be considered as a parameter. Equation (6) is a first order ordinary differential equation and can be considered with one of the following initial conditions from problem (9)

$$u(0,t) = u^{(-)}(t), \tag{7}$$

$$u(1,t) = u^{(+)}(t). \tag{8}$$

(A_1). The problems (6),(7) and (6),(8) have the solutions $u = \varphi^{(-)}(x,t)$ and $u = \varphi^{(+)}(x,t)$, respectively, which are defined for $0 \le x \le 1, t \in R$ and which are T-periodic in t. Additionally we assume

$$\varphi^{(-)}(x,t) < \varphi^{(+)}(x,t) \text{ for } x \in [0,1],\ t \in R,$$

$$A(\varphi^{(+)}(x,t),x,t) < 0, \; A(\varphi^{(-)}(x,t),x,t) > 0,$$

$$x \in [0,1], \; t \in R.$$

To formulate the next assumptions we introduce the function $I(x,t)$ by

$$I(x,t) := \int_{\varphi^{(-)}(x,t)}^{\varphi^{(+)}(x,t)} A(u,x,t)du.$$

(A_2). The equation

$$I(x,t) = 0 \tag{9}$$

has a smooth solution $x = x_0(t)$ which is T-periodic and obeys the conditions

$$0 < x_0(t) < 1 \quad \text{for} \quad t \in R,$$

$$\int_{\varphi^{(-)}(x_0(t),t)}^{s} A(u,x_0(t),t)\,du > 0$$

$$\text{for any } s \in \left(\varphi^{(-)}(x_0(t),t), \varphi^{(+)}(x_0(t),t)\right)$$

$$\text{and for} \quad t \in R.$$

(A_3). The root $x_0(t)$ of equation (9) satisfies the condition

$$\frac{\partial I}{\partial x}(x_0(t),t) < 0 \quad \text{for} \quad t \in R,$$

that is, $x_0(t)$ is a simple root.

Construction of the Formal Asymptotics. To characterize the location of the interior layer we introduce the curve $x = x_*(t,\varepsilon)$ as locus of the intersection of the solution $u(x,t,\varepsilon)$ of (2) with the surface

$$u = \frac{1}{2}\left(\varphi^{(-)}(x,t) + \varphi^{(+)}(x,t)\right) =: \varphi(x,t).$$

In what follows we construct the asymptotic expansion of $x_*(t,\varepsilon)$ in the form

$$x_*(t,\varepsilon) = x_0(t) + \varepsilon \, x_1(t) + ..., \tag{10}$$

where $x_0(t)$ is the solution of equation (9) and $x_k(t), k = 1,2,...$, are T-periodic functions to be determined. For the following we use the notation

$$\xi := \frac{x - x_*(t,\varepsilon)}{\varepsilon},$$

$$\overline{\mathcal{D}}^{(-)} := \{(x,t) \in R^2 : 0 \le x \le x_*(t,\varepsilon), \, t \in R\},$$

$$\overline{\mathcal{D}}^{(+)} := \{(x,t) \in R^2 : x_*(t,\varepsilon) \le x \le 1, \, t \in R\}.$$

First we consider in $\overline{\mathcal{D}}^{(-)}$ the boundary value problem

$$\varepsilon\left(\frac{\partial^2 u}{\partial x^2} - \frac{\partial u}{\partial t}\right) - A(u,x,t)\frac{\partial u}{\partial x} - B(u,x,t) = 0$$

$$\text{for} \quad (x,t) \in \overline{\mathcal{D}}^{(-)},$$

$$u(0,t,\varepsilon) = u^0(t), \quad u(x(t,\varepsilon),t,\varepsilon) = \varphi(t,\varepsilon) \tag{11}$$

$$\text{for} \quad t \in R,$$

$$u(x,t,\varepsilon) = u(x,t+T,\varepsilon) \quad \text{for} \quad t \in R.$$

We look for the formal asymptotic expansion of the solution $U^{(-)}(x,t,\varepsilon)$ of this problem in the form

$$U^{(-)}(x,t,\varepsilon) = \bar{U}^{(-)}(x,t,\varepsilon) + Q^{(-)}(\xi,t,\varepsilon) =$$

$$\sum_{i=0}^{\infty} \varepsilon^i \left(\bar{U}_i^{(-)}(x,t) + Q_i^{(-)}(\xi,t)\right), \tag{12}$$

where $\bar{U}^{(-)}$ and $\bar{Q}^{(-)}$ denote the regular and the interior layer parts. Next we study in $\overline{\mathcal{D}}^{(+)}$ similar problem to construct $U^{(+)}(x,t,\varepsilon)$.

By using the standard procudure of boundary layer function method we can constuct these expansions and to show that operators $L_R^{\pm}$, produsing regular part of the asymptotics have the form

$$L_R^{\pm} \equiv -A(\varphi^{(\pm)}(x,t),x,t)\frac{\partial}{\partial x} - \left(A_u(\varphi^{(\pm)}(x,t),x,t)\frac{\partial \overline{U}_0^{(\pm)}}{\partial x}\right.$$

$$\left. + B_u(\varphi^{(\pm)}(x,t),x,t)\right), \tag{13}$$

and therefore positivelly invertible for a negative wrigte hand part - inequality $L_R^{\pm}\bar{u} < 0$ has a positive solution.

The periodic functions $x_i(t)$ are determined from C^1-matching conditions (see [8]).

$$\frac{\partial I}{\partial x}(x_0(t),t)x_k(t) = h_k(t), \, k = 1,2,..., \quad t \in R.$$

We see that in our case $A^\Gamma = \frac{\partial I}{\partial x}(x_0(t),t)$ and inequality $A^\Gamma(\delta x(t)) < 0$ has a positive solution.

Existence Results. We denote by $\mathcal{D}_n^{(-)}$ and $\mathcal{D}_n^{(-)}$ the domains

$$\mathcal{D}_n^{(-)} := \{(x,t) \in R^2 : 0 \leq x \leq \sum_{i=0}^{n+1} x_i(t)\varepsilon^i, \, t \in R\},$$

$$\mathcal{D}_n^{(+)} := \{(x,t) \in R^2 : \sum_{i=0}^{n+1} x_i(t)\varepsilon^i \leq x \leq 1, \, t \in R\}$$

and denote by $U_n^{(\pm)}$ the partial sums of order n of the expansions , where ξ is replaced by $\left(x - \sum_{i=0}^{n+1} x_i(t)\varepsilon^i\right)/\varepsilon$.

We introduce the notation

$$U_n(x,t,\varepsilon) = \begin{cases} U_n^{(-)}(x,t,\varepsilon) \text{ for } (x,t) \in \mathcal{D}_n^{(-)}, \\ U_n^{(+)}(x,t,\varepsilon) \text{ for } (x,t) \in \mathcal{D}_n^{(+)}. \end{cases}$$

Then we have the following existence theorem

Theorem 2. *Suppose the assumptions $(A_0) - (A_3)$ to be valid. Then, for sufficiently small ε there exists a solution $u(x,t,\varepsilon)$ of (9) which T-periodic in t, has an interior layer and satisfies*

$$|u(x,t,\varepsilon) - U_n(x,t,\varepsilon)| \le c\varepsilon^{n+1} \ (x,t) \in \overline{\mathcal{D}}$$

where the positive constant c does not depend on ε.

Construction of the Upper and Lower Solutions. The proof of the theorem presented in the previous section is based on the technique of lower and upper solutions.

The upper and lower solutions satisfying the definition above are constructed by means of the modification of the formal asymptotics. In order to describe them we introduce the periodic curves $x = x_\beta(t,\varepsilon)$ and $x = x_\alpha(t,\varepsilon)$ as the $n+1$-th partial sums of the asymptotics of $x^*(t,\varepsilon)$ with a small shifts at the last term

$$x_\beta(t,\varepsilon) = x_0(t) + \varepsilon x_1(t) + ... + \varepsilon^{n+1}(x_{n+1}(t) - \delta)$$

and

$$x_\alpha(t,\varepsilon) = x_0(t) + \varepsilon x_1(t) + ... + \varepsilon^{n+1}(x_{n+1}(t) + \delta)$$

where $\delta > 0$ is an independent of ε number. These curves divide our domain $\overline{\mathcal{D}}$ into two subdomains $\overline{\mathcal{D}}_\beta^{(-)}, \overline{\mathcal{D}}_\beta^{(+)}$ and $\overline{\mathcal{D}}_\alpha^{(-)}, \overline{\mathcal{D}}_\alpha^{(+)}$ where

$$\overline{\mathcal{D}}_\beta^{(-)} := \{(x,t) \in R^2 : 0 \le x \le x_\beta(t,\varepsilon), \ t \in R\},$$

$$\overline{\mathcal{D}}_\beta^{(+)} := \{(x,t) \in R^2 : x_\beta(t,\varepsilon) \le x \le 1, \ t \in R\}.$$

The domains $\overline{\mathcal{D}}_\alpha^{(\pm)}$ are defined similarly.

Now we can define the upper solution $\beta(x,t,\varepsilon) = \beta_n(x,t,\varepsilon)$ and the lower solution $\alpha(x,t,\varepsilon) = \alpha_n(x,t,\varepsilon)$ by the expressions

$$\beta_n(x,t,\varepsilon) = \beta_n^{(\pm)}(x,t,\varepsilon) = \bar{U}_0^{(\pm)}(x,t) + \varepsilon\bar{U}_1^{(\pm)}(x,t)$$

$$+ ... + \varepsilon^n\bar{U}_n^{(\pm)}(x,t) + \varepsilon^{n+1}(\bar{U}_{n+1}^{(\pm)}(x,t) + v(x))$$

$$+ Q_0^{(\pm)}(\xi_\beta,t) + \varepsilon Q_1^{(\pm)}(\xi_\beta,t) + ...$$

$$+ \varepsilon^{n+1}Q_{(n+1)\beta}^{(\pm)}(\xi_\beta,t) + \varepsilon^{n+2}Q_{(n+2)\beta}^{(\pm)}(\xi_\beta,t,\varepsilon)$$

and

$$a_n(x,t,\varepsilon) = a_n^{(\pm)}(x,t,\varepsilon) = \bar{U}_0^{(\pm)}(x,t) + \varepsilon \bar{U}_1^{(\pm)}(x,t)$$
$$+ ... + \varepsilon^n \bar{U}_n^{(\pm)}(x,t) + \varepsilon^{n+1}(\bar{U}_{n+1}^{(\pm)}(x,t) - v(x))$$
$$+ Q_0^{(\pm)}(\xi_\alpha, t) + \varepsilon Q_1^{(\pm)}(\xi_\alpha, t) + ...$$
$$+ \varepsilon^{n+1} Q_{(n+1)\alpha}^{(\pm)}(\xi_\alpha, t) + \varepsilon^{n+2} Q_{(n+2)\alpha}^{(\pm)}(\xi_\alpha, t, \varepsilon),$$

where $v(x) = exp(mx)$ and $m > 0$ is an independent of ε sufficiently large number, $\xi_\beta = (x - x_\beta)/\varepsilon$, $\xi_\alpha = (x - x_\alpha)/\varepsilon$,, the function $\beta_n(x,t,\varepsilon) = \beta_n^{(\pm)}(x,t,\varepsilon)$ in $\overline{\mathcal{D}}_\beta^{(\pm)}$ and similarly we define $a_n(x,t,\varepsilon)$.

The existence theorem and its estimate for the solution follows from the differential inequalities theorem and from the structure of the upper and lower solutions.

Lemma 1. *The functions $\beta_n(x,t,\varepsilon)$ and $a_n(x,t,\varepsilon)$ satisfies the following uniform in $\overline{\mathcal{D}}$ estimates:*

$$\beta_n(x,t,\varepsilon) - a_n(x,t,\varepsilon) = O(\varepsilon^n),$$
$$|a_n(x,t,\varepsilon) - u_p(x,t,\varepsilon)| = O(\varepsilon^n),$$
$$|\beta_n(x,t,\varepsilon) - u_p(x,t,\varepsilon)| = O(\varepsilon^n) \tag{14}$$
$$\frac{\partial a_n}{\partial x} = \frac{\partial u_p}{\partial x} + O(\varepsilon^{n-1}), \quad \frac{\partial \beta_n}{\partial x} = \frac{\partial u_p}{\partial x} + O(\varepsilon^{n-1}).$$

where $u_p(x,t,\varepsilon)$ is the periodic internal layer solution of problem (9), stated in the Theorem 2.

The estimates of this proposition follows from the structure of the functions $\beta_n(x,t,\varepsilon)$ and $a_n(x,t,\varepsilon)$.

Stability Results. In this section we investigate the stability (in the sense of Lyapunov) of the periodic solution $u_p(x,t,\varepsilon)$ established by Theorem 2. It is known that the nonlinear stability problem under consideration can be solved by means of the linearized problem. For this purpose we study the following linear eigenvalue problem

$$L_\varepsilon v := \varepsilon \left(\frac{\partial^2 v}{\partial x^2} - \frac{\partial v}{\partial t} \right) - A(u_p(x,t,\varepsilon), x, t) \frac{\partial v}{\partial x}$$
$$- \left[A_u(u_p(x,t,\varepsilon), x, t) \frac{\partial u_p}{\partial x} + B_u(u_p(x,t), x, t) \right] v = \mu v \tag{15}$$
$$\text{for} \quad x \in (0,1), \ t \in R,$$
$$v(0,t,\varepsilon) = v(1,t,\varepsilon) = 0 \quad \text{for} \quad t \in R,$$
$$v(x,t,\varepsilon) = v(x,t+T,\varepsilon) \quad \text{for} \quad t \in R,$$

If μ satisfies $\mu < 0$, then the periodic solution $u_p(x,t,\varepsilon)$ is asymptotically stable, if μ satisfies $\mu > 0$, then $u_p(x,t,\varepsilon)$ is unstable.

It is shown by P.Hess that under our smoothness assumptions the operator L_ε is such that the Krein-Rutman theory can be applied. With slight extension of the results of Hess we can show that if we consider the auxiliary problems for

$$L_\varepsilon w = -h \quad w(0,t) = w(1,t) = 0 \; t \in R, \tag{16}$$

then the following theorem holds.

Lemma 2.
  (i).If $\mu < 0$, then problem (16) has a unique strictly positive solution.
  (iI).If $\mu > 0$, then problem (16) has no positive solution.
  (iii).If $\mu = 0$, then problem (16) has no solution.

Corollary. *If problem (16) has for some positive h a positive solution, then the principal eigenvalue μ of (15) and is negative.*

We have from the construction of the lower and upper solutions

$$N_\varepsilon(\beta_n) \equiv \varepsilon \left(\frac{\partial^2 \beta_n}{\partial x^2} - \frac{\partial \beta_n}{\partial t} \right) - A(\beta_n, x, t)\frac{\partial \beta_n}{\partial x} - B(\beta_n, x, t)$$

$$= -g^{\beta_n}(x, t, \varepsilon)\varepsilon^{n+1}$$

$$N_\varepsilon(\alpha_n) = g^{\alpha_n}(x, t, \varepsilon)\varepsilon^{n+1},$$

where

$$g^{\beta_n}(x, t, 0) > 0 \text{ and } g^{\alpha_n}(x, t, 0) > 0 \; x \in [0,1], t \in R,$$

and therefore, α_n and β_n are the asymptotic lower and upper solutions order of $q = n + 1$.

Using the notation $w_n = \beta_n - \alpha_n$ we get

$$N_\varepsilon(\beta_n) - N_\varepsilon(\alpha_n) = \varepsilon \left(\frac{\partial^2 w_n}{\partial x^2} - \frac{\partial w_n}{\partial t} \right)$$

$$- \left[A(\beta_n, x, t)\frac{\partial \beta_n}{\partial x} - A(\alpha_n, x, t)\frac{\partial \alpha_n}{\partial x} \right] - \tag{17}$$

$$\left[B(\beta_n, x, t) - B(\alpha_n, x, t) \right] = -g(x, t, \varepsilon)\varepsilon^{n+1}$$

Taking into account the estimates of Lemma 1 we obtain

$$A(\beta_n, x, t)\frac{\partial \beta_n}{\partial x} - A(\alpha_n, x, t)\frac{\partial \alpha_n}{\partial x} =$$

$$A(u_p(x, t, \varepsilon), x, t)\frac{\partial w_n}{\partial t} + A_u(u_p(x, t, \varepsilon), x, t)\frac{\partial u_p}{\partial x}w_n$$

$$+ O(\varepsilon^{2n-1}) \; x \in [0,1], t \in R.$$

Furthermore, we have

$$B(\beta_n(x, t, \varepsilon), x, t) - B(\alpha_n(x, t, \varepsilon), x, t) =$$

$$B_u(u_p(x, t, \varepsilon), x, t)w_n + O(\varepsilon^{2n}).$$

From these estimates it follows that the estimate of the Assumption A_2 of the Theorem 1 is satisfied with $p = 2n - 1$

Therefore, from we get the auxiliary problem

$$L_\varepsilon w_n = h,$$

where $h = -g(x, t, \varepsilon)\varepsilon^{n+1} + O(\varepsilon^{2n-1}) + O(\varepsilon^{2n})$. If we choose n such that $n > 2$, i.e. the Assumption A_3 of the Theorem 1 $p > q$ is satisfied and we have $h < 0$.

Theorem 3. *Suppose the assumptions* $(A_0) - (A_2)$ *to be satisfied. Then for sufficiently small* ε *the periodic solution of problem (9) with interior layer is asymptotically stable with a local region of attraction* $[\alpha_3(x, t, \varepsilon), \beta_3(x, t, \varepsilon)]$.

4 Moving Fronts in Nonlocal Reaction-Diffusion-Advection Equations

Another classes of problems where our approach is successfully applicable are integro-parabolic equations. Recently the problem of asymptotic description of front motion for the problem

$$L[u] \equiv -\varepsilon \frac{\partial u}{\partial t}(x, t, \varepsilon) + \varepsilon^2 \frac{\partial^2 u}{\partial x^2}(x, t, \varepsilon) - \varepsilon A(x, \varepsilon) \frac{\partial u}{\partial x}(x, t, \varepsilon) -$$

$$- \int_a^b g(u(x, t, \varepsilon), u(s, t, \varepsilon), x, s, \varepsilon)\, ds = 0, \quad a < x < b, \tag{18}$$

$$\frac{\partial u}{\partial x}(a, t, \varepsilon) = 0, \quad \frac{\partial u}{\partial x}(b, t, \varepsilon) = 0, \quad u(x, 0, \varepsilon) = u^0(x, \varepsilon) \tag{19}$$

was investigated in [9]. Under some natural assumptions where the crucial is

Condition I. *There exist two functions*

$$\varphi^{(-)} \in C(\Omega^{(-)}), \quad where \quad \Omega^{(-)} \equiv \{(x, y) : a \le x \le y \le b\},$$

$$\varphi^{(+)} \in C(\Omega^{(+)}), \quad where \quad \Omega^{(+)} \equiv \{(x, y) : a \le y \le x \le b\},$$

which for every $y \in (a, b)$ *satisfy* $\varphi^{(-)}(y, y) < \varphi^{(+)}(y, y)$ *and the system of the two coupled integral equations*

$$\int_a^y g(\varphi^{(-)}(x, y), \varphi^{(-)}(s, y), x, s, 0)\, ds+$$

$$+ \int_y^b g(\varphi^{(-)}(x, y), \varphi^{(+)}(s, y), x, s, 0)\, ds = 0, \quad a < x < y,$$

$$\int_a^y g(\varphi^{(+)}(x, y), \varphi^{(-)}(s, y), x, s, 0)\, ds+$$

$$+ \int_y^b g(\varphi^{(+)}(x, y), \varphi^{(+)}(s, y), x, s, 0)\, ds = 0, \quad y < x < b.$$

We state the existence of moving fronts at this problem. This results have an important applicability to describe the formation of stationary contrast structures.

Acknowledgement. This work is supported by RFBR, pr. N 13-01-00200.

References

1. Vasilieva, A.B., Butuzov, V.F., Nefedov, N.N.: Contrast structures in singularly perturbed problems. Fundamentalnaja i Prikladnala Matemat. 3(4), 799–851 (1998) (in Russian)
2. Amann, H.: Periodic Solutions of Semilinear Parabolic Equations. In: Nonlinear Analysis: a Collection of Papers in Honor of Erich Rothe, pp. 1–29. Academic, New York (1978)
3. Sattinger, D.H.: Monotone Methods in Elliptic and Parabolic Boundary Value Problems. Indiana Univ. Math. J. 21(11), 979–1001 (1972)
4. Hess, P.: Periodic-Parabolic Boundary Value Problems and Positivity. Pitman Research Notes in Math. Series, vol. 247. Longman Scientific&Technical, Harlow (1991)
5. Amann, H.: Maximum priciples and principal eigenvalues. In: Ten Mahtmatical Essays in Analysis and Topology. Elsevier (2005)
6. Zabrejko, P.P., Koshelev, A.I., Krasnoseiskij, M.A et al.: Integral equations. M.: Nauka (1968) (in Russian)
7. Nefedov, N.N.: The Method of Differential Inequalities for Some Classes of Nonlinear Singularly Perturbed Problems with Internal Layers. Differ. Uravn. 31(7), 1142–1149 (1995)
8. Vasileva, A.B., Butuzov, V.F., Nefedov, N.N.: Singularly Perturbed problems with Boundary and Internal Layers. Proceedings of the Steklov Institute of Mathmatics 268, 258–273 (2010)
9. Nefedov, N.N., Nikitin, A.G., Petrova, M.A., Recke, L.: Moving fronts in integro-parabolic reaction-diffusion-advection equations. Differ. Uravn. 47(9), 1–15 (2011)

Multiscale Convection in One Dimensional Singularly Perturbed Convection–Diffusion Problems

E. O'Riordan and J. Quinn

School of Mathematical Sciences, Dublin City University, Dublin 9, Ireland

Abstract. Linear singularly perturbed ordinary differential equations of convection diffusion type are considered. The convective coefficient varies in scale across the domain which results in interior layers appearing in areas where the convective coefficient decreases from a scale of order one to the scale of the diffusion coefficient. Appropriate parameter-uniform numerical methods are constructed. Numerical results are given to illustrate the theoretical error bounds established.

1 Continuous Problem Class

In this paper, we examine singularly perturbed ordinary differential equations of the form

$$-\varepsilon u'' + a_\varepsilon(x)u' + b(x)u = f(x), \quad a_\varepsilon(x) > 0; \quad x \in (0,1),$$

where the magnitude of the convective coefficient $a_\varepsilon(x)$ varies in scale across the domain. Problems of this type may arise when linearizing certain nonlinear singularly perturbed problems or when generating approximations to the solution of a coupled system of singularly perturbed equations.

Define the subdomains $\Omega_i, i = 1, 2, 3$ of $[0,1]$ to be

$$\Omega_i := (d_{i-1}, d_i), \quad 0 = d_0 \le d_1 < d_2 \le d_3 = 1; \quad \Omega := \cup_{i=1}^3 \Omega_i.$$

The points d_1, d_2 are points where the convective coefficient is of the same order as the diffusion coefficient ε in the following class of singularly perturbed problems: find $u \in C^3(\Omega) \cap C^1(\bar{\Omega})$ such that

$$L_\varepsilon u := -\varepsilon u'' + a_\varepsilon(x)u' + b(x)u = f(x), \ x \in \Omega, \tag{1.1a}$$

$$u(0) = u(1) = 0, \tag{1.1b}$$

$$b(x) \ge \beta > 0, \ x \in \Omega, \tag{1.1c}$$

$$1 = \|a\| \ge a_\varepsilon(x) > \varepsilon, \ x \in \Omega_2, \tag{1.1d}$$

$$\varepsilon \ge a_\varepsilon(x) \ge \alpha\varepsilon > 0, \ |a'_\varepsilon(x)| \le C\varepsilon, \quad x \in \Omega_1 \cup \Omega_3, \tag{1.1e}$$

where $0 < \varepsilon \le 1$ is a singular perturbation parameter and a_ε, f, b are smooth. The points d_1, d_2 are assumed to be independent of the singular perturbation parameter ε. For all $x \in [d_1, d_2]$, define the limiting function $a_0(x)$ by

I. Dimov, I. Faragó, and L. Vulkov (Eds.): NAA 2012, LNCS 8236, pp. 73–85, 2013.

$$a_0(x) := \lim_{\varepsilon \to 0} a_\varepsilon(x), \ x \neq d_1, d_2;$$

$$a_0(d_1) := \lim_{x \to d_1^+} a_0(x), \quad \text{and} \quad a_0(d_2) := \lim_{x \to d_2^-} a_0(x).$$

Observe that $a_0(x) \equiv 0, x \in [0, d_1) \cup (d_2, 1], a_0(x) > 0, x \in (d_1, d_2)$. The different cases of $a_0(x)$ being either continuous or discontinuous at the point d_1 will be examined.

The nature of the growth (decay) of the convective coefficient in the vicinity of the point $d_1(d_2)$ will be restricted by the following constraints on the problem class: assume also that for a given constant $0 < \theta \leq 0.5$ independent of ε, there exists two points $\gamma_1, \gamma_2 \in \Omega_2, \ \gamma_1 < \gamma_2$ such that

$$\int_{\gamma_2}^{d_2} a_\varepsilon(t)dt \geq \theta(d_2 - d_1) > 0; \quad (1.1f)$$

$$a_\varepsilon(x) \geq \theta > 0, \quad |a_\varepsilon'(x)| \leq C, \quad \gamma_1 \leq x \leq \gamma_2; \quad (1.1g)$$

$$\varepsilon a_\varepsilon'(x) \leq \theta_1 a_\varepsilon^2(x), \ \theta_1 < 1, x \in (d_1, \gamma_1); \quad a_\varepsilon'(x) \leq 0, x \in (\gamma_2, d_2); \quad (1.1h)$$

$$a_\varepsilon(x) - a_0(x) = \sum_{i=1}^{3} \xi_i(x; \varepsilon), \quad x \in \Omega_2; \quad (1.1i)$$

where for $k = 0, 1$ and all $x \in \Omega_2$

$$|\xi_1(x)|_k \leq C\varepsilon^{-k/2}e^{-\theta_1(x-d_1)/\sqrt{\varepsilon}}, \quad |\xi_2(x)|_k \leq C\varepsilon^{-k}e^{-\theta(d_2-x)/\varepsilon}, \quad |\xi_3|_k \leq C\varepsilon^2.$$

Note that $\gamma_i, \theta_i, i = 1, 2$ are independent of ε. Specific choices for the convective coefficient are taken in the test examples examined in §5.

In this paper, our interest is focused on the interior layers appearing in the interior region Ω_2. To exclude the appearance of reaction-diffusion type layers in the region $\Omega_1 \cup \Omega_3$, we impose the following restriction on the forcing term f

$$|f(x)| \leq C_1 a_0(x), \quad x \in \bar{\Omega}. \quad (1.1j)$$

As $f \in C^2(0, 1)$ and $|f(x)| \leq Ca_0(x)$, then (1.1j) implies that $f(x) \equiv 0, x \in \bar{\Omega}_1 \cup \bar{\Omega}_3$. To exclude any layer emerging in the vicinity of the point d_1 and consequently being convected throughout the domain, we impose the additional constraint on the forcing term, that

$$f(x) \equiv 0, \quad x \in (d_1, 2\gamma_1]. \quad (1.1k)$$

A significant effect of interest in this problem is that the problem is singularly perturbed only in the subdomain $\Omega_2 \subset \Omega$. Within the literature on singularly perturbed convection-diffusion equations, it is normally assumed that $a(x) \geq \alpha > 0$ everywhere. In this paper, we examine the effect of $0 < a_\varepsilon(x) \leq C\varepsilon$ in an $O(1)$-neighbourhood of the outflow boundary point $x = 1$.

In [1,3], the case of a discontinuous coefficient $a(x)$ (independent of ε) was examined under the assumption that $a(x) > \alpha > 0$ away from points of discontinuity. Weak internal layers [3] will form in the vicinity of points of discontinuity in a if $u \in C^1(0, 1)$.

In §5 we present a numerical algorithm, which is applicable to problems of the form (1.1). However, in §4, we present an asymptotic error bound for this algorithm under the additional restriction that

$$a_\varepsilon(x) \geq \alpha_\varepsilon(x) + C\varepsilon, \gamma_2 \leq x \leq d_2 \quad \alpha_\varepsilon(x) := \frac{\theta}{2}(1 - e^{-\theta(d_2-x)/\varepsilon}). \quad (1.2)$$

In [4], a boundary turning point problem was studied, where an equivalent condition on a_ε was assumed such that a boundary layer of width $O(\varepsilon)$ was supported near the boundary.

Notation. Throughout the paper C denotes a generic constant that is independent of both the singular perturbation parameter ε and the discretization parameter N. The semi-noms $|\cdot|_k$ are defined by

$$|g|_k := \|g^{(k)}\| = \left\|\frac{d^k g}{dx^k}\right\|, \quad \text{where} \quad \|g\| := \max_{x \in \Omega} |g(x)|.$$

2   Solution Decompositions

We begin by stating a standard comparison principle associated with the differential operator in problem (1.1).

Lemma 1. *[1] Suppose that a function $\omega \in C^0(\bar{\Omega}) \cup C^2(\Omega \setminus \{p_1, p_2\})$, where $p_1, p_2 \in \Omega$, satisfies $\omega(0) \geq 0$, $\omega(1) \geq 0$ $L_\varepsilon \omega(x) \geq 0$, $x \in \Omega \setminus \{p_1, p_2\}$ and $[-\varepsilon \omega'](p_i) \geq 0, i = 1, 2$ then $\omega(x) \geq 0$, $x \in [0,1]$.*

Hence, the solution is uniformly bounded from the assumption $b(x) \geq \beta > 0$.

Lemma 2. *We have the following stability estimate*

$$\|u\|_\Omega \leq \frac{\|f\|}{\beta}.$$

From the argument in [2, Lemma 3.2], it follows that for all $x \in \Omega$

$$\|u^{(k)}\|_\Omega \leq C\varepsilon^{-k} \max\{\|f\|, \|u\|\}, \ k = 1, 2 \quad (2.1a)$$
$$\|u^{(3)}\|_\Omega \leq C\varepsilon^{-3} \max\{\|f\|, \|f'\|, \|u\|\}. \quad (2.1b)$$

The reduced solution v_0 is the potentially discontinuous solution of

$$a_0 v_0' + b v_0 = f; \ d_1 < x < d_2 \ v_0 \equiv 0, x \in \bar{\Omega} \setminus \Omega_2.$$

On the interior interval Ω_2 and noting (1.1j), we have that

$$v_0(x) = \int_{t=d_1}^x \frac{f(t)}{a_0(t)} e^{-\int_{s=t}^x \frac{b(s)}{a_0(s)} ds} dt, \quad |v_0| \leq C.$$

If $a_0(x) > 0$, $x \in \bar{\Omega}_2$, then it immediately follows that $|v_0|_k \leq C, k \leq 4$.

Lemma 3. *Based on the assumptions (1.1k), (1.1h), it follows that (for ε sufficiently small)*

$$|u(x)| \leq C\varepsilon^2, \quad x \leq d_1.$$

Proof. On the interval $[0, 2\gamma_1]$, consider the following barrier function

$$\psi(x) = e^{-\int_x^{2\gamma_1} \frac{(1-\theta_1)a_\varepsilon(t)}{\varepsilon} dt}, \quad 0 < \theta_1 < 1.$$

Then, using (1.1h),

$$-\varepsilon\psi'' + a_\varepsilon\psi' + b\psi = \frac{(1-\theta_1)}{\varepsilon}\left(a_\varepsilon^2\theta_1 - \varepsilon a_\varepsilon'\right)\psi + b\psi \geq 0.$$

Thus, by applying the maximum pronciple over the subinterval $[0, 2\gamma_1]$,

$$|u(x)| \leq |u(2\gamma_1)|e^{-\int_x^{2\gamma_1} \frac{(1-\theta_1)a_\varepsilon(t)}{\varepsilon} dt}, \quad x \in (0, 2\gamma_1);$$

and, so by (1.1g),

$$|u(d_1)| \leq Ce^{-\int_{\gamma_1}^{2\gamma_1} \frac{(1-\theta_1)a_\varepsilon(t)}{\varepsilon} dt} \leq Ce^{-\frac{(1-\theta_1)\theta\gamma_1}{\varepsilon}} \leq C\varepsilon^2.$$

The regular component v associated with (1.1) is defined to be the solution of the problem: find the function v such that

$$L_\varepsilon v = f(x), \quad x \in \Omega_1 \cup \Omega_3,$$
$$-\varepsilon v'' + a_0 v' + bv = f(x), \quad x \in \Omega_2,$$
$$v(0) = 0, \; v(d_1) = u(d_1), v(d_2^-) = v(d_2^-), v(d_2^+) = v(1) = 0;$$

where $v(d_2^-)$ is specified below. In general, this function will be discontinuous at the point d_2. Recall assumption (1.1e) and using the argument from [2, Lemma 3.2], we have that $v \equiv u, \; x \in \Omega_1$ and

$$|u| \leq C\varepsilon^2, |u'| \leq C\varepsilon, |u''| \leq C\varepsilon, |u'''| \leq C, \; x \in \Omega_1, \quad v \equiv 0, \; x \in \Omega_3.$$

In the interior region Ω_2, consider the further sub-decomposition of $v = v_0 + \varepsilon v_1 + \varepsilon^2 v_2 + u(d_1)$. The first correction v_1 to the reduced solution v_0 satisfies

$$a_0 v_1' + bv_1 = v_0'', v_1(d_1) = 0, \quad x \in \Omega_2;$$

and, the second correction v_2 satisfies

$$-\varepsilon v_2'' + a_0 v_2' + bv_2 = v_1'', v_2(d_1) = 0; \; v_2(d_2) = 0.$$

Taking $v(d_2^-) = v_0(d_2^-) + \varepsilon v_1(d_2^-) + u(d_1)$, one can then deduce (assuming (1.1k)) that for $k = 0, 1, 2, 3$,

$$|v(x)|_k \leq C(1 + \varepsilon^{2-k}), x \in \bar{\Omega}_2, \quad \text{if} \quad a_0(d_2) \geq \alpha > 0.$$

Assumption (1.1k) implies that $v_i^{(k)}(d_1^+) = 0, 1 \leq k \leq 5, i = 0, 1$. If $a_0(d_2) = 0$, then additional restrictions will be implicitly placed on the data by assuming that $|v_0|_k \leq C, |v_1|_k \leq C, \ k = 0, 1, 2, 3$. Hence, under assumption (1.2) (which implies that $a_0(d_2) \geq 0.5\theta_2 > 0$), we can deduce the following bounds on the derivatives of the regular component

$$|v(x)|_k \leq C(1 + \varepsilon^{2-k}); \qquad x \in \Omega_1 \cup \Omega_2 \cup \Omega_3. \tag{2.2}$$

In the case of problem (1.1), we identify two interior layer functions w, y defined respectively as the solutions of: Find a discontinuous w such that

$$L_\varepsilon w = 0, x \in \Omega, \tag{2.3a}$$
$$w(0) = w(d_1^-) = 0, \quad w(d_2^+) = u(d_2), w(1) = 0, \tag{2.3b}$$
$$w(d_1^+) = 0, \ w(d_2^-) = (u - v)(d_2^-). \tag{2.3c}$$

Find $y \in C^0(0, 1)$ such that

$$L_\varepsilon y = (a_0 - a_\varepsilon)v', x \in \Omega_2, \ y \equiv 0, x \in \bar{\Omega}_1 \cup \bar{\Omega}_3. \tag{2.4}$$

We can establish the following bounds on the first layer function w

$$|w(x)| \leq Ce^{-\int_x^{d_2} \frac{g_\varepsilon(t)}{\varepsilon} dt} =: C\phi_1(x), x \leq d_2 \tag{2.5a}$$
$$|w(x)| \leq Ce^{-\sqrt{\frac{\beta}{2\varepsilon}}(x - d_2)}, x > d_2, \tag{2.5b}$$

where $g_\varepsilon : [d_1, d_2] \to [0, 1]$ is defined so that

$$g_\varepsilon(x) = a_\varepsilon(x), \gamma_2 \leq x \leq d_2, \quad g_\varepsilon(x) = 0, d_1 \leq x \leq 0.5(\gamma_2 + \gamma_1);$$

and in the interval $[0.5(\gamma_2 + \gamma_1), \gamma_2)]$ the function $g_\varepsilon(x) = p(x)$ is a polynomial so that $\min\{\beta, \|a_\varepsilon'\|_{[\gamma_1, \gamma_2]}\} \geq p' \geq 0, \ p(0.5(\gamma_2 + \gamma_1)) = 0, \ p(\gamma_2) = a_\varepsilon(\gamma_2)$. Observe that by this choice, using assumption (1.1h), we have that

$$-\varepsilon g_\varepsilon' + g_\varepsilon(a_\varepsilon - g_\varepsilon) + \varepsilon b \geq 0, x \neq \gamma_2; \ [-\phi_1'](\gamma_2) = 0. \tag{2.5c}$$

By the Mean Value Theorem, there exists a point $z \in (d_1, d_1 + \varepsilon)$ such that

$$|w'(z)| \leq \frac{|w(d_1 + \varepsilon)|}{\varepsilon} \leq C\varepsilon^{-1}e^{-\int_{d_1+\varepsilon}^{d_2} \frac{g_\varepsilon(t)}{\varepsilon} dt}$$

and

$$\varepsilon w'(z) = \varepsilon w'(d_1) + a_\varepsilon(z)w(z) + \int_{d_1}^z (b - a_\varepsilon')wdt.$$

Hence,

$$|w'(d_1)| \leq C\varepsilon^{-1}e^{-\int_{d_1}^{d_2} \frac{g_\varepsilon(t)}{\varepsilon} dt}$$

as $z \in (d_1, d_1 + \varepsilon) \subset (d_1, \gamma_1)$ and

$$\int_{d_1}^z |a_\varepsilon'|dt = a_\varepsilon(z) - a_\varepsilon(d_1) \leq C.$$

We can then deduce that: $w \equiv 0$ on Ω_1 and

$$|w(x)|_k \leq C\varepsilon^{-k} e^{-\int_x^{d_2} \frac{g_\varepsilon(t)}{\varepsilon} dt}, \quad k = 1, 2, 3, \qquad x \in \Omega_2, \qquad (2.5d)$$

$$|w(x)|_k \leq C\varepsilon^{-k/2} e^{-\sqrt{\frac{\beta}{2\varepsilon}}(x - d_2)}, \quad k = 1, 2, 3, \qquad x \in \Omega_3. \qquad (2.5e)$$

In bounding the second layer function y we note (1.1i) and the following three cases:

(a) If $|a_0(x) - a_\varepsilon(x)|_k \leq C\varepsilon^2, x \in \Omega_2,\ k = 0, 1$ then

$$|y|_k \leq C\varepsilon^{2-k}, \quad 0 \leq k \leq 3. \qquad (2.6a)$$

(b) If

$$|a_0(x) - a_\varepsilon(x)|_k \leq C e^{-\theta_1(x - d_1)/\sqrt{\varepsilon}}, \qquad x \in \Omega_2,\ k = 0, 1;$$

then by assumption (1.1k), the decomposition $v'(x) = (v_0 + \varepsilon v_1)'(x) + \varepsilon^2 v_2'(x)$ and $(v_0 + \varepsilon v_1)^{(k)}(d_1^+) = 0, 0 \leq k \leq 5$, we can obtain the bound

$$|(a_0 - a_\varepsilon)v'(x)| \leq |(a_0 - a_\varepsilon)((v_0 + \varepsilon v_1)'(x) - (v_0 + \varepsilon v_1)'(d_1^+))| + C\varepsilon^2$$
$$\leq C(x - d_1)^4 |(a_0 - a_\varepsilon)| + C\varepsilon^2 \leq C\varepsilon^2, \quad x \in \Omega_2.$$

Using this bound, we again deduce that the derivatives of y satisfy the bounds

$$|y|_k \leq C\varepsilon^{2-k}, \quad 0 \leq k \leq 3 \qquad (2.6b)$$

(c) If

$$|a_0(x) - a_\varepsilon(x)|_k \leq C e^{-\theta(d_2 - x)/\varepsilon}, \ x \in \Omega_2,\ k = 0, 1$$

then for all $x \in \Omega_2$ and $0 \leq k \leq 3$,

$$|y(x)|_k \leq C\varepsilon^{-k} e^{-\theta(d_2 - x)/\varepsilon}, \qquad |y'(d_1)| \leq C\varepsilon. \qquad (2.6c)$$

Note that $v + w + y \in C^0(\bar{\Omega})$; $L(v + w + y) = f$ in Ω; by construction $v + w + y = u, x \in \bar{\Omega} \setminus \Omega$; and so $u = v + w + y, x \in \bar{\Omega}$. The bounds on the derivatives of these components are given in (2.2), (2.5) and (2.6)

3   Discrete Problem

For any mesh function $Z(x_i)$, we introduce the finite difference operators

$$D^+ Z(x_i) := \frac{Z(x_{i+1}) - Z(x_i)}{h_{i+1}}, \quad D^- Z(x_i) := \frac{Z(x_i) - Z(x_{i-1})}{h_i},$$

$$\delta^2 Z(x_i) := \frac{D^+ Z(x_i) - D^- Z(x_i)}{\bar{h}_i}.$$

Here the mesh step is $h_i := x_i - x_{i-1}$ and $\bar{h}_i := (h_i + h_{i+1})/2$ for each i.

Define the transition parameter τ implicitly using

$$\int_{d_2 - \tau}^{d_2} a_\varepsilon(t) dt = 2\varepsilon \ln N. \qquad (3.1)$$

On Ω a piecewise-uniform mesh of N mesh intervals is constructed as follows. The subdomain $\overline{\Omega}_2$, is further subdivided into

$$[d_1, d_2 - \sigma] \cup [d_2 - \sigma, d_2], \quad \sigma = \min\{\theta(d_2 - d_1), \tau\} \qquad (3.2)$$

Note that, by (1.1f), $\gamma_2 \leq d_2 - \tau$. In the subdomain Ω_3 a piecewise-uniform is also employed. The subdomain is subdivided into

$$[d_2, d_2 + \sigma^*] \cup [d_2 + \sigma^*, 1] \qquad (3.3a)$$

where

$$\sigma^* = \min\{0.5(1 - d_2), \tau^*\}, \quad \tau^* := \sqrt{\frac{2\varepsilon}{\beta}} \ln N. \qquad (3.3b)$$

On each of the five subintervals

$$[0, d_1] \cup [d_1, d_2 - \sigma] \cup [d_2 - \sigma, d_2] \cup [d_2, d_2 + \sigma^*] \cup [d_2 + \sigma^*, 1]$$

a uniform mesh with $\frac{N}{8}, \frac{N}{8}, \frac{N}{4}, \frac{N}{4}, \frac{N}{4}$ mesh-intervals is placed. The resulting fitted piecewise uniform mesh is denoted by Ω_ε^N. Note that the points d_1, d_2 are mesh points. Consider the following upwind finite difference method

$$L^N U := -\varepsilon\delta^2 U(x_i) + aD^- U(x_i) + bU(x_i) = f(x_i), \quad x_i \in \Omega_\varepsilon^N, \qquad (3.4a)$$

$$U(0) = U(1) = 0. \qquad (3.4b)$$

4 Error Analysis

In this section, we restrict the discussion to the case of (1.2), where the convective coefficient decreases exponentially in the vicinity of the point d_2.

Theorem 1. *Assume (1.2). Then,*

$$|(U - u)(x_i)| \leq CN^{-1}(\ln N) + C\varepsilon^2, \quad x_i \in \Omega_\varepsilon^N$$

where u is the solution of problem (1.1) and U is the numerical solution generated by the numerical method constructed in §3.

Proof. We outline the proof. To begin, note that $\|U\| \leq C$. We confine the argument to the case where the mesh is piecewise-uniform in each of the subdomains Ω_2, Ω_3. The case of a uniform mesh (when ε is sufficiently large relative to N^{-1}) is handled by a classical argument. In an analogous fashion to the decomposition of the continuous solution, the discrete solution can be decomposed into subcomponents. That is, $U = V + W + Y$ where the discrete regular component V is multi-valued at d_2 and satisfies

$$L^N V = f, x_i \in \Omega_1 \cup \Omega_3, \qquad (4.1a)$$

$$(-\varepsilon\delta^2 + a_0 D^- + b)V = f, x_i \in \Omega_2, \qquad (4.1b)$$

$$V(0) = V(d_1) = 0, \ V^-(d_2) = v(d_2^-), \ V^+(d_2) = V(1) = 0. \qquad (4.1c)$$

The discrete layer components are defined as follows:

$$L^N W^- = 0, x_i \in (0, d_2), \quad W^-(0) = 0, W^-(d_2) = (U - V^-)(d_2); \quad (4.1d)$$
$$L^N W^+ = 0, x_i \in \Omega_3, \quad W^+(d_2) = U(d_2), W^+(1) = 0; \quad (4.1e)$$
$$L^N Y = (a_0 - a_\varepsilon) D^- V(x_i), x_i \in \Omega_2, \quad (4.1f)$$
$$Y(d_1) = Y(d_2) = 0, \quad Y(x_i) = 0, x_i \in \Omega_1 \cup \Omega_3. \quad (4.1g)$$

Observe that the discrete layer component W is multivalued at d_2, where $W :=$ W^-, $0 \le x_i \le d_2$, $W := W^+$, $d_2 \le x_i \le 1$. Consider the following discrete barrier function for W^-

$$\Psi(x_i) := \frac{\Pi_{j=1}^{i}(1 + \frac{g(x_j)h_j}{2\varepsilon})}{\Pi_{j=1}^{3N/4}(1 + \frac{g(x_j)h_j}{2\varepsilon})}$$

which has the properties

$$D^- \Psi(x_i) = \frac{g(x_i)}{2\varepsilon(1 + \frac{g(x_j)h_j}{2\varepsilon})} \Psi(x_i), \quad D^+ \Psi(x_i) = \frac{g(x_{i+1})}{2\varepsilon} \Psi(x_i).$$

Then

$$L^N \Psi(x_i) = \left(\frac{g(x_i)}{4\varepsilon} \frac{(\bar{h}_i 2a_\varepsilon(x_i) - g(x_j)h_j)}{\bar{h}_i(1 + \frac{g(x_j)h_j}{2\varepsilon})} + \frac{g(x_i) - g(x_{i+1})}{2\bar{h}_i} + b(x_i) \right) \Psi(x_i).$$

Using this barrier function, the bound (2.5) and the definition (3.1), we deduce that

$$|W(x_i) - w(x_i)| \le CN^{-1}, \quad 0 \le x_i \le d_2 - \tau.$$

Using the bounds (2.2), a decomposition of the form $V = V_0 + \varepsilon V_1 + \varepsilon^2 V_2$ and a standard stability and consistency argument one can derive the error bound

$$|V - v| \le CN^{-1} + C\varepsilon^2.$$

As in [4, Lemma 2.2] one can also establish that $|D^- V| \le C$. Analagous arguments to bounding $|W - w|$ may be used to bound the errors $|Y - y|$ for $x_i \le d_2 - \tau$. Hence we have established that

$$|(U - u)(x_i)| \le CN^{-1} + C\varepsilon^2, \quad x_i \le d_2 - \tau.$$

From the bounds (2.5) and assumption (1.2), we deduce that

$$|w(x)|_k \le C\varepsilon^{-k} e^{-\int_x^{d_2} \frac{\alpha_\varepsilon(t)}{\varepsilon} dt}; \quad \gamma_2 \le x \le d_2.$$

Using standard truncation error analysis and the bounds on the derivatives of the components v, w and y given in §2, we have that

$$|L^N (U - u)(x_i)| \le CN^{-1} + C\frac{N^{-1}\log N}{\varepsilon} e^{-\int_{x_{i+1}}^{d_2} \left(\theta + \frac{\alpha_\varepsilon(t)}{\varepsilon}\right) dt}, \quad x_i \in (d_2 - \tau, d_2);$$

$$|L^N (U - u)(d_2)| \le C\frac{N^{-1}\log N}{\sqrt{\varepsilon}} + CN^{-1};$$

$$|L^N (U - u)(x_i)| \le CN^{-1}\log N, \quad x_i \in \Omega_3.$$

Consider the discrete barrier function,

$$\Phi(x_i) := \begin{cases} \dfrac{\Pi_{j=1}^{i}(1+\frac{\alpha_\varepsilon(x_{j-1})h_j}{2\varepsilon})}{\Pi_{j=1}^{3N/4}(1+\frac{\alpha_\varepsilon(x_{j-1})h_j}{2\varepsilon})}, & d_2 - \tau \le x_i \le d_2 \\[2ex] \dfrac{1-x_i}{1-d_2}, & d_2 \le x_i \le 1. \end{cases}$$

Then, using the arguments in [4, Lemma 2.3], we deduce that

$$L^N\Phi(x_i) \ge \frac{\theta^2}{4\varepsilon}\Phi(x_i), d_2 - \tau \le x_i \le d_2.$$

Combine these bounds with the truncation error bounds to complete the proof, using the barrier function

$$CN^{-1}\ln N + C\varepsilon^2 + CN^{-1}\ln N\Phi(x_i),$$

to bound the error, since $b \ge \beta > 0$.

5   Numerical Experiments

We present an algorithm for which to calculate the transition parameter τ in (3.1) in the following: Using a numerical integration routine (Simpson's rule used here), approximate the integral in (3.1) over increasing intervals whose end point are all d_2, increasing with an initial step size of $\delta := \varepsilon \ln N$, until the value $2\varepsilon \ln N$ is exceeded at some point a. From the point a, integrate over decreasing intervals whose end points are all d_2, decreasing with a step size $\frac{1}{2}\delta$, until a value less than $2\varepsilon \ln N$ is reached at some point b. From that point b, integrate over increasing intervals whose end points are all d_2, increasing with a step size $\frac{1}{4}\delta$, until the value of $2\varepsilon \ln N$ is exceeded at some point c. Continue this process, halving the interval step size each time, until a desired tolerance is reached ($10^{-6}\varepsilon \ln N$ used as the tolerance here).

For the purposes of constructing test problems, we split the definition of a_ε into three parts

$$a_\varepsilon(x) = \varepsilon x, \ x \le d_1, \quad a_\varepsilon(x) = \varepsilon(1-x), \ x \ge d_2; \tag{5.1}$$

and in the interval Ω_2 we examine two different choices. In each of the following examples $d_1 = 0.25$ and $d_2 = 0.75$.

Test Example 1: $a_0(d_1) = 0$, $a_0(d_2) \ne 0$

$$a_\varepsilon(x) = x - d_2 + (\varepsilon(d_1 + d_2 - 1) + d_2 - d_1)\frac{1 - e^{-(d_2-x)/\varepsilon}}{1 - e^{-(d_2-d_1)/\varepsilon}} + \varepsilon(1 - d_2),$$

$$a_0(x) = x - d_1, \ b(x) = 1, \ f(x) = 5a_0(x)(d_2 - x).$$

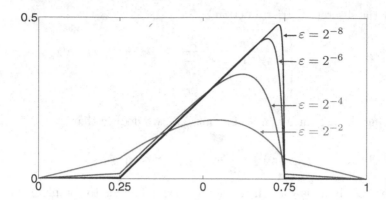

Fig. 1. Plots of the convective coefficient $a_\varepsilon(x)$ for several values of ε in the case of Example 1

Table 1. Computed double mesh rates of convergence p_ε^N and computed uniform rates of convergence p^N in the case of Example 1, where $\tau \approx 4.1\varepsilon \ln N$

ε	p_ε^N						
	N=32	N=64	N=128	N=256	N=512	N=1024	N=2048
2^{-0}	1.78	1.64	1.22	1.12	1.07	1.03	1.02
2^{-2}	1.67	1.21	1.10	1.05	1.03	1.01	1.01
2^{-4}	0.95	0.97	0.98	0.99	0.99	1.00	1.00
2^{-6}	1.05	0.99	0.98	0.98	0.99	0.99	1.00
2^{-8}	1.12	1.06	1.02	1.02	1.00	1.00	1.00
2^{-10}	1.04	1.02	1.00	1.00	1.00	1.00	1.00
2^{-12}	1.00	1.01	1.00	1.00	0.99	0.99	0.99
2^{-14}	1.00	1.00	1.00	1.00	1.00	0.99	0.99
2^{-16}	1.00	1.00	1.00	1.00	1.00	1.00	1.00
2^{-18}	1.00	1.00	1.00	1.00	1.00	1.00	1.00
2^{-20}	1.00	1.00	1.00	1.00	1.00	1.00	1.00
p^N	1.00	1.00	1.00	1.00	1.00	1.00	0.99

Test Example 2: $a_0(d_1) \neq 0$, $a_0(d_2) \neq 0$

$$a_\varepsilon(x) = C_1(x - d_1) + C_2(d_2 - x) - e^{-\frac{x - d_1}{\sqrt{\varepsilon}}} - e^{-\frac{d_2 - x}{\varepsilon}}, a_\varepsilon(d_1) = \varepsilon d_1,$$
$$a_\varepsilon(d_2) = \varepsilon(1 - d_2), b(x) = 1, \ f(x) = 10a_0(x)(x - d_1)(d_2 - x).$$

Plots of the convective coefficient $a_\varepsilon(x)$ in both test examples are displayed in Figures 1 and 3, respectively, for a range of values of ε. In both cases, the convective coefficient has a layer to the left of the point d_2. For test example 2, the convective coefficient has an additional layer to the right of d_1. The computed solutions for both test examples are displayed in Figures 2 and 4, with zooms of

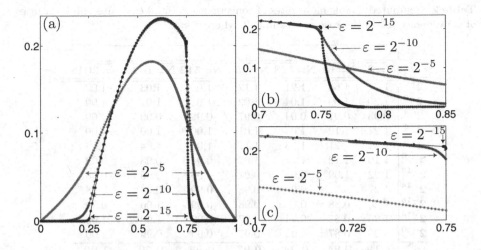

Fig. 2. Plot of the numerical solution, with $d_2 = 0.75$, for sample values of ε over the axis (a) $[0, 1]$, (b) $[0.7, 0.85]$ and (c) $[0.7, 0.75]$ using $N = 1024$ points for Example 1

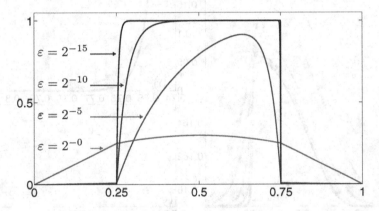

Fig. 3. Convective coefficient a_ε for sample values of ε for Example 2

the layer structure in the vicinity of d_2 given. Observe the absence of layers in the vicinity of the point d_1 in both figures. Table 1 and 2 display the computed rates of convergence p_ε^N and the uniform rates of convergence p^N, using the double mesh principle (see [2]). The numerical results for both test examples suggest that the method is first order uniformly convergent and that the error bound given in Theorem 4.1 is not sharp, as the term $C\varepsilon^2$ is not evident in these tables when ε is large and, in addition, the factor $\ln N$ is not visible.

Table 2. Computed double mesh rates of convergence p_ε^N and computed uniform rates of convergence p^N in the case of Example 2, where $\tau \approx 2.1\varepsilon \ln N$

ε	p_ε^N						
	N=32	N=64	N=128	N=256	N=512	N=1024	N=2048
2^{-0}	1.77	1.60	1.21	1.12	1.06	1.03	1.02
2^{-2}	1.20	1.09	1.04	1.02	1.01	1.01	1.00
2^{-4}	0.93	0.91	0.94	0.97	0.98	0.99	1.00
2^{-6}	1.27	1.12	1.04	1.01	1.00	1.00	1.00
2^{-8}	1.22	1.21	1.18	1.11	1.06	1.04	1.03
2^{-10}	1.19	1.12	1.10	1.07	0.94	0.98	0.99
2^{-12}	1.22	1.09	0.95	0.97	0.96	0.96	0.98
2^{-14}	1.27	0.91	0.96	0.98	0.99	0.98	0.98
2^{-16}	1.26	0.88	0.95	0.98	0.99	1.00	0.99
2^{-18}	1.25	0.87	0.94	0.97	0.99	1.00	1.00
2^{-20}	1.25	0.87	0.94	0.97	0.99	0.99	1.00
$\mathbf{p^N}$	**1.19**	**0.94**	**0.94**	**0.97**	**0.98**	**0.99**	**1.00**

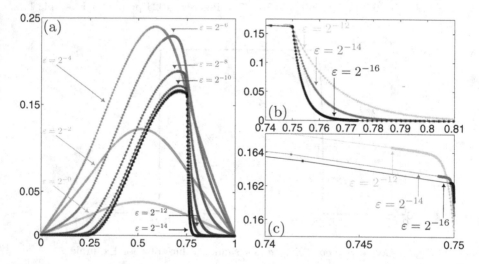

Fig. 4. Plots of numerical solution for sample values of ε over the axis (a) $[0,1]$, (b) $[0.74, 0.81]$ and (c) $[0.74, 0.75]$ using $N = 1024$ points in the case of Example 2

References

1. de Falco, C., O' Riordan, E.: A parameter robust Petrov-Galerkin scheme for advection-diffusion-reaction equations. Numerical Algorithms 5(1), 107–127 (2011)
2. Farrell, P.A., Hegarty, A.F., Miller, J.J.H., O'Riordan, E., Shishkin, G.I.: Robust computational techniques for boundary layers. Chapman and Hall/CRC Press, Boca Raton, U.S.A (2000)

3. Farrell, P.A., Hegarty, A.F., Miller, J.J.H., O' Riordan, E., Shishkin, G.I.: Singularly perturbed convection diffusion problems with boundary and weak interior layers. Journal Computational and Applied Mathematics 166(1), 133–151 (2004)
4. O' Riordan, E., Quinn, J.: Parameter-uniform numerical methods for some Linear and Nonlinear Singularly Perturbed Convection Diffusion Boundary Turning Point Problems. BIT Numerical Mathematics 51, 317–337 (2011)

Hybrid Functions for Nonlinear Differential Equations with Applications to Physical Problems

M. Razzaghi

Department of Mathematics and Statistics,
Mississippi State University, MS 39762, USA,
Currently, Fulbright Scholar, Department of Mathematics and Computer Sciences
Technical University of Civil Engineering, Bucharest, Romania

Abstract. A numerical method for solving nonlinear initial-value problems is proposed. The Lane-Emden type equations which have many applications in mathematical physics are then considered. The method is based upon hybrid function approximations. The properties of hybrid functions of block-pulse functions and Bernoulli polynomials are presented and are utilized to reduce the computation of nonlinear initial-value problems to a system of equations. The method is easy to implement and yields very accurate results.

Keywords: Block-pulse functions, Bernoulli polynomials, hybrid, nonlinear initial-value problems, Lane-Emden equation.

1 Introduction

The available sets of orthogonal functions can be divided into three classes. The first class includes sets of piecewise constant basis functions (e.g., block-pulse, Haar, Walsh, etc.). The second class consists of sets of orthogonal polynomials (e.g., Chebyshev, Laguerre, Legendre, etc.). The third class is the set of sine-cosine functions in the Fourier series. While orthogonal polynomials and sine-cosine functions together form a class of continuous basis functions, piecewise constant basis functions have inherent discontinuities or jumps.

Orthogonal functions have been used when dealing with various problems of the dynamical systems. The main advantage of using orthogonal functions is that they reduce the dynamical system problems to those of solving a system of algebraic equations. The approach is based on converting the underlying differential equation into an integral equation through integration, approximating various signals involved in the equation by truncated orthogonal functions, and using the operational matrix of integration to eliminate the integral operations. This matrix can be uniquely determined based on the particular orthogonal functions. Among piecewise constant basis functions, block-pulse functions are found to be very attractive, in view of their properties of simplicity and disjointedness and among orthogonal polynomials, the shifted Legendre polynomials is computationally more effective [1]. The Bernoulli polynomials and Taylor series are not

I. Dimov, I. Faragó, and L. Vulkov (Eds.): NAA 2012, LNCS 8236, pp. 86–94, 2013.

based on orthogonal functions, nevertheless, they possess the operational matrix of integration. However, since the integration of the cross product of two Taylor series vectors is given in terms of a Hilbert matrix [2], which are known to be ill-posed, the applications of Taylor series are limited.

In recent years the hybrid functions consisting of the combination of block-pulse functions with Chebyshev polynomials [3],[4] Legendre polynomials [5],[6] or Taylor series [7],[8] have been shown to be a mathematical power tool for discretization of selected problems.

Many problems in the literature of mathematical physics can be formulated as equations of the Lane-Emden type defined in the form

$$y'' + \frac{2}{x}y' + g(y(x)) = 0, \qquad 0 < x < \infty, \qquad (1)$$

subject to

$$y(0) = A, \qquad y'(0) = 0, \qquad (2)$$

where prime denotes differentiation with respect to x. The solution of the Lane-Emden equation, as well as those of a variety of nonlinear problems in quantum mechanics and astrophysics such as the scattering length calculations in the variable phase approach, is numerically challenging because of the singular point at the origin. Eqs. (1) and (2) with specializing $g(y(x))$ and A occurs in several models of mathematical physics and astrophysics [9]-[16]. For $g(y) = y^\alpha$, $h(x) = 0$, and $A = 1$ in Eqs. (1) and (2), we obtain the Lane-Emden equation of index α which has been the object of much study [13]-[15]. It was physically shown that interesting values of α lie in the interval $[0, 5]$, and this equation has analytical solutions for $\alpha = 0$, 1 and 5.

In the present work, we first consider the nonlinear ordinary differential equations of the form

$$f\left(t, x(t), \dot{x}(t), \ddot{x}(t)\right) = 0, \qquad 0 \le t < \infty, \qquad (3)$$

with initial conditions

$$x(0) = x_0, \qquad \dot{x}(0) = 0, \qquad (4)$$

where dots denote differentiation with respect to t. We assume that Eqs. (3) and (4) have a unique solution $x(t)$ to be determined. We then solve a variety of Lane-Emden equations which fall into this category. Here, we introduce a new direct computational method for solving Eqs. (3) and (4). This method consists of reducing the solution of Eqs. (3) and (4) to a set of algebraic equations by first expanding $\ddot{x}(t)$ in terms of hybrid functions with unknown coefficients. These hybrid functions, which consist of block-pulse functions and Bernoulli polynomials, are presented. The operational matrix of integration is given. This matrix, together with the properties of hybrid functions, are then utilized to evaluate the unknown coefficients for the solution of Eqs. (3) and (4).

This paper is organized as follows: in section 2, we describe the basic properties of the hybrid of block-pulse functions and Bernoulli polynomials required for our subsequent development. Section 3 is devoted to the solution of nonlinear initial-value problems given with Eqs. (3) and (4) by using hybrid functions. In section 4, the present method is applied to different Lane-Emden equations as well as Lane-Emden equation of index α. The numerical solutions are compared with some exact or approximate solutions in order to assess the accuracy of the proposed method.

2   Hybrid Functions

2.1   Properties of Bernoulli Polynomials

The Bernoulli polynomials of order m, are defined in [17] by

$$\beta_m(t) = \sum_{k=0}^{m} \binom{m}{k} \alpha_k t^{m-k},$$

where α_k, $k = 0, 1, ..., m$ are Bernoulli numbers. These numbers are a sequence of signed rational numbers, which arise in the series expansion of trigonometric functions [18] and can be defined by the identity

$$\frac{t}{e^t - 1} = \sum_{n=0}^{\infty} \alpha_n \frac{t^n}{n!}.$$

The first few Bernoulli numbers are

$$\alpha_0 = 1, \quad \alpha_1 = -\frac{1}{2}, \quad \alpha_2 = \frac{1}{6}, \quad \alpha_4 = -\frac{1}{30},$$

with $\alpha_{2k+1} = 0$, $k = 1, 2, 3, \ldots$.
The first few Bernoulli polynomials are

$$\beta_0(t) = 1, \quad \beta_1(t) = t - \frac{1}{2}, \quad \beta_2(t) = t^2 - t + \frac{1}{6}, \quad \beta_3(t) = t^3 - \frac{3}{2}t^2 + \frac{1}{2}t.$$

According to [19], Bernoulli polynomials form a complete basis over the interval $[0,1]$.

For approximating an arbitrary time function, the advantages of Bernoulli polynomials $\beta_m(t)$, $m = 0, 1, 2, ..., M$ where $0 \le t \le 1$, over shifted Legendre polynomials $p_m(t)$, $m = 0, 1, 2, ..., M$, where $0 \le t \le 1$, are given in [20].

2.2   Hybrid of Block-Pulse and Bernoulli Polynomials

Hybrid functions $b_{nm}(t)$, $n = 1, 2, \ldots, N$, $m = 0, 1, \ldots, M$ are defined on the interval $[0, t_f]$ as

$$b_{nm}(t) = \begin{cases} \beta_m(\frac{N}{t_f}t - n + 1), & t \in [\frac{n-1}{N}t_f, \frac{n}{N}t_f], \\ 0, & \text{otherwise,} \end{cases}$$

where n and m are the order of block-pulse functions and Bernoulli polynomials, respectively.

2.3 Function Approximation

Let $H = L^2[0,1]$, and assume that $\{b_{10}(t), b_{20}(t), ..., b_{NM}(t)\} \subset H$ be the set of hybrid of block-pulse and Bernoulli polynomials, and

$$Y = span\{b_{10}(t), b_{20}(t), ..., b_{N0}(t), b_{11}(t), b_{21}(t), ..., b_{N1}(t), ..., b_{1M}(t), b_{2M}(t), ..., b_{NM}(t)\},$$

and f be an arbitrary element in H. Since Y is a finite dimensional vector space, f has the unique best approximation out of Y such as $f_0 \in Y$, that is

$$\forall y \in Y, \| f - f_0 \| \leq \| f - y \|.$$

Since $f_0 \in Y$, there exist unique coefficients $c_{10}, c_{20}, ..., c_{NM}$ such that

$$f \simeq f_0 = \sum_{m=0}^{M} \sum_{n=1}^{N} c_{nm} b_{nm}(t) = C^T B(t),$$

where

$$B^T(t) = [b_{10}(t), b_{20}(t), ..., b_{N0}(t), b_{11}(t), b_{21}(t), ..., b_{N1}(t), ..., b_{1M}(t), b_{2M}(t), ..., b_{NM}(t)], \tag{5}$$

and

$$C^T = [c_{10}, c_{20}, ..., c_{N0}, c_{11}, c_{21}, ..., c_{N1}, ..., c_{1M}, c_{2M}, ..., c_{NM}].$$

2.4 Operational Matrix of Integration

The integration of the $B(t)$ defined in Eq. (5) is given by

$$\int_0^t B(t')dt' \simeq PB(t), \tag{6}$$

where P is the $N(M+1) \times N(M+1)$ operational matrix of integration. The matrix P for $t_f = 1$ is given in [20] by

$$P = \frac{1}{N} \begin{bmatrix} P_0 & I & O & ... & O \\ \frac{-1}{2}\alpha_2 I & O & \frac{1}{2}I & ... & O \\ \vdots & \vdots & \vdots & \ddots & \vdots \\ \frac{-1}{M}\alpha_M I & O & O & ... & \frac{1}{M}I \\ \frac{-1}{M+1}\alpha_{M+1}I & O & O & ... & O \end{bmatrix},$$

where I and O are $N \times N$ identity and zero matrices respectively, and

$$P_0 = \begin{bmatrix} -\alpha_1 & 1 & ... & 1 & 1 \\ 0 & -\alpha_1 & ... & 1 & 1 \\ \vdots & \vdots & \ddots & \vdots & \vdots \\ 0 & 0 & ... & -\alpha_1 & 1 \\ 0 & 0 & ... & 0 & -\alpha_1 \end{bmatrix}.$$

It is seen that P is more sparse than operational matrices of integration for the hybrid of block-pulse with Chebyshev polynomials [3], with Legendre polynomials [5] and Taylor series [7].

3   Solution of Nonlinear Initial-Value Problems

In order to solve Eqs. (3) and (4) with a hybrid of block-pulse functions and Bernoulli polynomials, we first choose an interval $[0, t_f)$. By expanding x_0 in terms of hybrid functions we get

$$x_0 = [x_0, 0, \ldots, 0, \ldots, x_0, 0, \ldots, 0] B(t) = e^T B(t). \tag{7}$$

Let

$$\ddot{x}(t) = C^T B(t), \tag{8}$$

Integrating Eq. (8) from 0 to t and using Eqs. (4) and (7), we obtain

$$\dot{x}(t) = C^T P B(t), \tag{9}$$

$$x(t) = \left(C^T P^2 + e^T\right) B(t), \tag{10}$$

where P is the operational matrix of integration given in Eq. (6). By substituting Eqs. (8)-(10) in Eq. (3), we get

$$f\left(t, \left(C^T P^2 + e^T\right) B(t), C^T P B(t), C^T B(t)\right) = 0. \tag{11}$$

We now collocate Eq. (11) at MN points t_{nm}, $n = 1, \ldots, N$, $m = 0, 1, \ldots, M-1$, as

$$f\left(t_{nm}, \left(C^T P^2 + e^T\right) B(t_{nm}), C^T P B(t_{nm}), C^T B(t_{nm})\right) = 0. \tag{12}$$

For a suitable collocation points t_{nm}, we use Gaussian nodes which are given in [21]. Using Eq. (12), we obtain a system of MN nonlinear equations which can be solved by the Newton's iterative method. It is well known that the initial guess for Newton's iterative method is very important especially for complicated problems. To choose the initial guess for our problem, in the first stage we set $N = 1$. When selecting M we choose an arbitrary low number depending on the problem and apply the Newton's iterative method for solving M nonlinear equations by choosing x_0 in Eq. (4) as our initial guess. We then set $N = 2$ and use the approximate solution in stage one as our initial guess in this stage. We continue this approach until the results are similar up to a required number of decimal places for the same N and two consecutive M values.

4   Illustrative Examples

In order to assess the applicability, efficiency and accuracy of our method, we applied the proposed method to a variety of singular Lane-Emden equations as indicated in the following examples.

4.1   Example 1 (Linear, Non-homogeneous Lane-Emden Equation)

This example corresponds to the following Lane-Emden equation [13]

$$y'' + \frac{8}{x}y' + xy = x^5 - x^4 + 44x^2 - 30x, \qquad 0 < x \le 1, \qquad (13)$$

with

$$y(0) = 0, \qquad y'(0) = 0. \qquad (14)$$

To solve Eq. (13) with initial conditions in Eq. (14) by the present method, we first rewrite it in the following equivalent form

$$f\left(t, x(t), \dot{x}(t), \ddot{x}(t)\right) = t\ddot{x}(t) + 8\dot{x}(t) + t^2 x(t) - t\left(t^5 - t^4 + 44t^2 - 30t\right) = 0, \quad (15)$$

$$x(0) = 0, \qquad \dot{x}(0) = 0.$$

We then apply the method presented in this paper and solve Eq. (15) with $t_f = 1$ for $N = 1$ and $M = 4$. For this equation we obtain

$$x(t) = t^4 - t^3,$$

which is the exact solution.

4.2   Example 2 (Linear, Non-homogeneous Lane-Emden Equation).

This example corresponds to the following Lane-Emden equation [13]

$$y'' + \frac{2}{x}y' = 2(2x^2 + 3)y, \qquad 0 < x \le 1,$$

with

$$y(0) = 1, \qquad y'(0) = 0.$$

The equivalent form is

$$f\left(t, x(t), \dot{x}(t), \ddot{x}(t)\right) = t\ddot{x}(t) + 2\dot{x}(t) - 2t(2t^2 + 3)x(t) = 0,$$

with

$$x(0) = 1, \qquad \dot{x}(0) = 0,$$

which has the following exact solution:

$$x_e(t) = e^{t^2}.$$

Table 1 shows that high accuracy that has been obtained by the proposed method for $N = 2$ and $M = 8$ with $t_f = 1$ together with the exact solution $x_e(t)$.

Table 1. Estimated and exact values of $x(t)$ for Example 1

t	$x(t)$	$x_e(t)$
0	0	0
0.1	1.01005016712	1.01005016712
0.2	1.04081077418	1.04081077419
0.3	1.09417428370	1.09417428371
0.4	1.17351087101	1.17351087099
0.5	1.28402541675	1.28402541669
0.6	1.43332941453	1.43332941456
0.7	1.63231621996	1.63231621996
0.8	1.89648087932	1.89648087930
0.9	2.24790798667	2.24790798668
1	2.71828182846	2.71828182846

4.3   Example 3 (Lane-Emden Equation of Index α)

Consider the Lane-Emden equation of index α given by

$$y'' + \frac{2}{x}y' + y^\alpha = 0, \qquad 0 < x < \infty,$$

with

$$y(0) = 1, \qquad y'(0) = 0.$$

The equivalent form is

$$f\left(t, x(t), \dot{x}(t), \ddot{x}(t)\right) = t\ddot{x}(t) + 2\dot{x}(t) + tx^\alpha(t) = 0, \tag{16}$$

with

$$x(0) = 1, \qquad \dot{x}(0) = 0. \tag{17}$$

This equation is linear for $\alpha = 0$ and 1, nonlinear otherwise, and exact solutions exist only for $\alpha = 0$, 1 and 5 and are given in [13], respectively, by

$$x(t) = 1 - \frac{1}{6}t^2, \quad x(t) = \frac{sin(t)}{t}, \quad x(t) = \left(1 + \frac{t^2}{3}\right)^{-\frac{1}{2}}.$$

It was physically shown that interesting value of α lie in the interval $[0, 5]$. Moreover, Bender et al. [12] determined the zeros of $x(t)$ asymptotically, here denoted by ξ, and found that

$$\xi = \pi + 0.885273956\,\delta + 0.24222\,\delta^2,$$

for $\delta = -0.5$, 0, 0.5, 1.0 and 1.5 which correspond to $\alpha = 0$, 1, 2, 3, and 4, respectively. We applied the method presented in this paper and solved Eqs. (16) and (17) and then evaluated the zeros of $x(t)$, which also are evaluated in [10], by using perturbation methods and a (1-1) Pade approximation, and in [15] by linearization technique. In Table 2, the resulting values of the zeros of $x(t)$ for $\alpha = 0, 0.5, \cdots, 4.5$ using the present method with $N = 2$ and $M = 4$, with the results obtained in [10] and [15], together with exact values given in [21] are presented. Table 2 shows that the present method provides more accurate predictions of the zeros of $x(t)$

Table 2. Comparison of numerical results for ξ ($x(\xi) = 0$) for Example 3

α	Method in [10]	Method in [15]	present method	Exact [21]
0.0	2.4465	2.44899	2.44948972	2.44948972
0.5	-	-	2.75269805	2.75269805
1.0	-	3.14048	3.14159265	3.14159265
1.5	-	-	3.65375373	3.65375373
2.0	4.3603	4.35086	4.35287459	4.35287459
2.5	-	-	5.35527545	5.35527545
3.0	7.0521	6.89312	6.89684861	6.89684861
3.5	-	-	9.53580534	9.53580534
4.0	17.976	14.96518	14.97154634	14.97154634
4.5	-	-	31.83646324	31.83646324

5   Conclusion

In the present work the hybrid of block-pulse functions and Bernoulli polynomials together with the operational matrix of integration P are used for solving nonlinear second-order initial value problems and the Lane-Emden equation. The matrix P has many zeros which make this method computationally very attractive without sacrificing the accuracy of the solution. Illustrative examples are given to demonstrate the validity and applicability of the proposed method.

References

1. Razzaghi, M., Elnagar, G.: Linear quadratic optimal control problems via shifted Legendre state parameterization. Int. J. Syst. Sci. 25, 393–399 (1994)
2. Razzaghi, M., Razzaghi, M.: Instabilities in the solution of a heat conduction problem using Taylor series and alternative approaches. J. Frank. Instit. 326, 683–690 (1989)
3. Razzaghi, M., Marzban, H.R.: Direct method for variational problems via hybrid of block-pulse and Chebyshev functions. Math. Prob. Eng. 6, 85–97 (2000)
4. Wang, X.T., Li, Y.M.: Numerical solutions of integro differential systems by hybrid of general block-pulse functions and the second Chebyshev polynomials. Appl. Math. Comput. 209, 266–272 (2009)
5. Marzban, H.R., Razzaghi, M.: Optimal control of linear delay systems via hybrid of block-pulse and Legendre polynomials. J. Frank. Instit. 341, 279–293 (2004)
6. Singh, V.K., Pandey, R.K., Singh, S.: A stable algorithm for Hankel transforms using hybrid of block-pulse and Legendre polynomials. Comput. Phys. Communications. 181, 1–10 (2010)

7. Marzban, H.R., Razzaghi, M.: Analysis of time-delay systems via hybrid of block-pulse functions and Taylor series. J. Vibra. Contr. 11, 1455–1468 (2005)
8. Marzban, H.R., Razzaghi, M.: Solution of multi-delay systems using hybrid of block-pulse functions and Taylor series. J. Sound. Vibra. 292, 954–963 (2006)
9. Davis, H.T.: Introduction to Nonlinear Differential and Integral Equations. Dover Publications, New York (1962)
10. Chandrasekhar, S.: Introduction to the Study of Stellar Structure. Dover Publications, New York (1967)
11. Davis, P.J., Rabinowitz, P.: Methods of Numerical Integration. Academic Press, New York (1975)
12. Bender, C.M., Pinsky, K.S., Simmons, L.M.: A new perturbative approach to nonlinear problems. J. Math. Phys. 30, 1447–1455 (1989)
13. Wazwaz, A.M.: A new algorithm for solving differential equations of Lane-Emden type. Appl. Math. Comput. 118, 287–310 (2001)
14. He, J.H.: Variational approach to the Lane-Emden equation. Appl. Math. Comput. 143, 539–541 (2003)
15. Ramos, J.I.: Linearization techniques for singular initial-value problems of ordinary differential equations. Appl. Math. Comput. 161, 525–542 (2005)
16. Marzban, H.R., Tabrizidooz, H.R., Razzaghi, M.: Hybrid functions for nonlinear initial-value problems with applications to Lane-Emden type equations. Phys. Let. A. 372, 5883–5886 (2008)
17. Costabile, F., Dellaccio, F., Gualtieri, M.I.: A new approach to Bernoulli polynomials. Rendiconti di Matematica, Serie VII 26, 1–12 (2006)
18. Arfken, G.: Mathematical Methods for Physicists, 3rd edn. Academic Press, San Diego (1985)
19. Kreyszig, E.: Introductory Functional Analysis with Applications. John Wiley and Sons Press, New York (1978)
20. Mashayekhi, S., Ordokhani, Y., Razzaghi, M.: Hybrid functions approach for nonlinear constrained optimal control problems. Commun. Nonli. Sci. Numer. Simul. 17, 1831–1843 (2012)
21. Davis, P.J., Rabinowitz, P.: Methods of Numerical Integration. Academic Press (1975)

Error Estimation in Energy Norms: Is It Necessary to Fit the Mesh to Boundary Layers

Hans-G. Roos and Martin Schopf

University of Technology Dresden
{Hans-Goerg.Roos,Martin.Schopf}@tu-dresden.de

Abstract. We demonstrate for two typical model problems that one observes uniform convergence of the Galerkin FEM on standard meshes with respect to the perturbation parameter in energy norms if the energy norm of the layers is small. Moreover, it is also possible only to resolve the strong layer using a layer adapted mesh but to do nothing concerning the weaker layer.

Keywords: boundary layer, energy norm, Galerkin FEM.

1 Reaction Diffusion

1.1 Introduction

Consider the singularly perturbed reaction-diffusion problem

$$\mathcal{L}u := -\varepsilon^2 \Delta u + cu = f \quad \text{in } \Omega \subset \mathbb{R}^2, \tag{1a}$$

$$u = 0 \quad \text{on } \partial\Omega, \tag{1b}$$

where Ω is polygonal and convex, $0 < \varepsilon \ll 1$ and $f \in L_2(\Omega)$, $c \in C(\overline{\Omega})$ with $c \geq c_0 > 0$.

Then (1) has a unique solution $u \in H_0^1(\Omega) \cap H^2(\Omega)$. Furthermore typically u exhibits sharp boundary layers near $\partial\Omega$.

For the numerical solution of (1) we introduce a shape-regular mesh $\mathcal{T}_h$ with discretisation parameter $h := \max_{T \in \mathcal{T}_h} \operatorname{diam} T$ and the space $V_h \subset H_0^1(\Omega)$ of linear or bilinear elements. The Galerkin approximation $u_h \in V_h$ satisfies

$$\varepsilon^2 (\nabla u_h, \nabla v_h) + (cu_h, v_h) = (f, v_h) \quad \text{for all } v_h \in V_h. \tag{2}$$

Here $(\cdot, \cdot)$ denotes the inner product in $L_2(\Omega)$. Moreover we use C to denote a generic constant that is independent of ε and the mesh. It is said that the family of approximations u_h converges uniformly in ε in the norm $\|\|\cdot\|\|$ of order μ if, with a constant C independent of ε, h,

$$\sup_{0 < \varepsilon \leq \varepsilon_0} \|\|u - u_h\|\| \leq Ch^\mu.$$

Schatz and Wahlbin [10] stated without a rigorous proof that the Galerkin finite element method on a shape-regular mesh cannot converge uniformly in ε in

I. Dimov, I. Faragó, and L. Vulkov (Eds.): NAA 2012, LNCS 8236, pp. 95–109, 2013.

the global $L_\infty(\Omega)$ norm. They proved interior uniform convergence in $L_\infty(\tilde\Omega)$, ($\tilde\Omega \subset \Omega$) but also the uniform result

$$\|u - u_h\|_0 \leq Ch^{1/2} \tag{3}$$

in the $L_2(\Omega)$ norm assuming $f \in H^{1/2,\infty}(\Omega)$ and $\nabla c \in L_\infty(\Omega)$. In several papers [1,3,4,13] the authors study the finite element method for (1) on special anisotropic meshes and prove error estimates in the energy norm

$$\|v\|_\varepsilon^2 := \varepsilon^2 |v|_1^2 + \|v\|_0^2. \tag{4}$$

A natural question is: can one prove a similar result as (3) also in the energy norm (4)? So far such an estimate is only known for LDG least-squares method, see [5]. Based on a priori estimates of u we will do that and support our theoretical findings by a detailed numerical study of the behaviour of $\|u - u_h\|_0$, $\|u - u_h\|_\varepsilon$ and $\varepsilon |u - u_h|_1$.

Our results confirm the observation in [6,9] that error estimates in the energy norm are not fitted to the behaviour of the solution and should be replaced by estimates in a stronger balanced norm.

1.2   Error Estimation in the Energy Norm

First we prove a priori estimates for u. Denote by $u_0 = f/c$ the solution of the "reduced" problem to (1).

Lemma 1. *Assume $u_0 \in H^1(\Omega)$. Then*

$$\varepsilon^3 |u|_2^2 + \varepsilon |u|_1^2 \leq C \left(\varepsilon |u_0|_1^2 + \|u_0\|_{0,\partial\Omega}^2 \right), \tag{5}$$

moreover,

$$\|u - u_0\|_0^2 \leq C \left(\varepsilon^2 |u_0|_1^2 + \varepsilon \|u_0\|_{0,\partial\Omega}^2 \right). \tag{6}$$

Proof. The equation

$$-\Delta u = \varepsilon^{-2}(f - cu) = \varepsilon^{-2}c(u_0 - u)$$

implies

$$|u|_2 \leq C\varepsilon^{-2}\|u_0 - u\|_0.$$

Thus from (6) follows the estimate for $\varepsilon^3 |u|_2^2$.

We have

$$\varepsilon^2 (\nabla u, \nabla v) + (c(u - u_0), v) = \varepsilon^2 \int_{\partial\Omega} \frac{\partial u}{\partial n} v \quad \text{for all } v \in H^1(\Omega).$$

Setting $v := u - u_0$ we get

$$\varepsilon^2 (\nabla u, \nabla(u - u_0)) + (c(u - u_0), u - u_0) = -\varepsilon^2 \int_{\partial\Omega} \frac{\partial u}{\partial n} u_0$$

or

$$\min\{1, c_0\} \left(\varepsilon^2 |u|_1^2 + \|u - u_0\|_0^2\right) \leq -\varepsilon^2 \int_{\partial\Omega} \frac{\partial u}{\partial n} u_0 + \varepsilon^2 (\nabla u, \nabla u_0).$$

Next we use several times

$$\left| \int h_1 h_2 \right| \leq \frac{\alpha}{2} \int h_1^2 + \frac{1}{2\alpha} \int h_2^2$$

and choose α adequately. Combined with a trace inequality we get, for instance,

$$\left| \varepsilon^2 \int_{\partial\Omega} \frac{\partial u}{\partial n} u_0 \right| \leq \frac{\alpha\varepsilon}{2} \int_{\partial\Omega} u_0^2 + \frac{C\varepsilon^3}{2\alpha} |u|_1 |u|_2$$

$$\leq \frac{\alpha\varepsilon}{2} \|u_0\|_{0,\partial\Omega}^2 + \frac{C}{2\alpha} \left(\frac{\beta}{2} \varepsilon^2 |u|_1^2 + \frac{1}{2\beta} \varepsilon^4 |u|_2^2 \right)$$

$$\leq \frac{\alpha\varepsilon}{2} \|u_0\|_{0,\partial\Omega}^2 + \frac{C\beta}{4\alpha} \varepsilon^2 |u|_1^2 + \frac{C}{4\alpha\beta} \|u - u_0\|^2.$$

Summarizing, we get

$$\varepsilon^2 |u|_1^2 + \|u - u_0\|_0^2 \leq C \left(\varepsilon^2 |u_0|_1^2 + \varepsilon \|u_0\|_{0,\partial\Omega}^2 \right),$$

which proves (5) and (6). □

Remark 1. Lin and Stynes [6] assume $c \in C(\overline{\Omega})$, $c \geq c_0$ and prove

$$\varepsilon^{3/2} \|\Delta u\|_0 + \varepsilon^{1/2} |u|_1 + \|u\|_0 \leq C \left(\|f\|_1 + \|f\|_\infty \right). \tag{7}$$

It turns out that it is possible to weaken the assumptions a bit and to replace (7) by (5) (the estimate $\|u\|_0 \leq C\|f\|_0$ is trivial). Lin and Stynes also present a nice example showing that

$$\varepsilon |u|_1^2 + \|u\|_0^2 \leq C\|f\|_0^2$$

cannot be true, in general.

For estimating the finite element error, we have only to estimate the projection error $\|u - \pi u\|_\varepsilon$ with $\pi u \in V_h$. Our technique requires the L_2 and H^1 stability of the projection: on a shape regular mesh we would use the L_2 projection or the Clément's interpolant.

We get

$$\|u - \pi u\|_0^2 \leq 2 \left(\|u_0 - \pi u_0\|_0^2 + \|u - u_0 - \pi(u - u_0)\|_0^2 \right)$$

$$\leq Ch^2 |u_0|_1^2 + Ch|u - u_0|_1 \|u - u_0\|_0$$

$$\leq Ch^2 + Ch\varepsilon^{-1/2}\varepsilon^{1/2} \leq Ch$$

if $|u_0|_1$ and $\|u_0\|_{0,\partial\Omega}$ are bounded. Similarly,

$$\varepsilon^2 |u - \pi u|_1^2 = \varepsilon^2 |u - \pi u|_1 |u - \pi u|_1$$

$$\leq C\varepsilon^2 h|u|_2 |u|_1 \leq C\varepsilon^2 h\varepsilon^{-3/2}\varepsilon^{-1/2} \leq Ch,$$

if $|u_0|_1$ and $\|u_0\|_{0,\partial\Omega}$ are bounded.

Theorem 1. *If $|u_0|_1$ amd $\|u_0\|_{0,\partial\Omega}$ are bounded, the error of the finite element method with linear or bilinear finite elements on a shape-regular mesh satisfies the uniform in ε estimate*

$$\|u - u_h\|_\varepsilon \le Ch^{1/2}. \tag{8}$$

Remark 2. As our numerical experiments show, the error in the energy norm is for $h \gg \varepsilon$ dominated by the $L_2(\Omega)$ error, the influence of $\varepsilon|u - u_h|_1$ is small. This is theoretically clear because

$$\varepsilon|u - u_h|_1 \le C\varepsilon^{1/2} + \varepsilon|u_h|_1 \le C\varepsilon^{1/2} + C\frac{\varepsilon}{h}.$$

In both, the $L_\infty(\Omega)$ and the discrete $L_\infty(\Omega)$ norm we observe stagnation of the error as long as $h > \varepsilon$. As for convection-diffusion problems one can prove (see [8, Remark I.2.85]), that uniform convergence in the discrete maximum norm requires a mesh width of order $\mathcal{O}(\varepsilon)$ in the layer region.

2   A Convection-Diffusion Problem: Characteristic Layers and Neumann Outflow Condition

Consider

$$-\varepsilon\Delta u - bu_x + cu = f \quad \text{in} \quad \Omega = (0,1)^2, \tag{9a}$$

$$\left.\frac{\partial u}{\partial x}\right|_{x=0} = 0, \quad u|_{x=1} = 0 \quad \text{and} \quad u|_{y=0} = u|_{y=1} = 0. \tag{9b}$$

We assume $b \in W^{1,\infty}(\Omega)$, $c \in L_\infty(\Omega)$, $b \ge \beta > 0$ with some constant β, $0 < \varepsilon \ll 1$ and

$$c + \frac{1}{2}b_x \ge \gamma > 0. \tag{10}$$

The problem is characterized by characteristic (or parabolic) boundary layers at $y = 0$ and $y = 1$, moreover due to the Neumann condition on the outflow boundary there exists a weak exponential layer.

We assume additionally: u can be decomposed as

$$u = v + w_1 + w_2 + w_{12}, \tag{11}$$

where we have for all $(x,y) \in \Omega$ and $0 \le i + j \le 2$ the pointwise estimates

$$
\begin{aligned}
&\left|\frac{\partial^{i+j}v}{\partial x^i \partial y^j}(x,y)\right| \le C, \quad \left|\frac{\partial^{i+j}w_1}{\partial x^i \partial y^j}(x,y)\right| \le C\varepsilon^{1-i}e^{-\beta x/\varepsilon}, \\
&\left|\frac{\partial^{i+j}w_2}{\partial x^i \partial y^j}(x,y)\right| \le C\varepsilon^{-j/2}\left(e^{-y/\sqrt{\varepsilon}}e^{-(1-y)/\sqrt{\varepsilon}}\right), \\
&\left|\frac{\partial^{i+j}w_{12}}{\partial x^i \partial y^j}(x,y)\right| \le C\varepsilon^{-(i+j/2-1)}e^{-\beta x/\varepsilon}\left(e^{-y/\sqrt{\varepsilon}} + e^{-(1-y)/\sqrt{\varepsilon}}\right)
\end{aligned} \tag{12}
$$

(see [7]). w_1 is the weak exponential layer, the other terms cover characteristic layers and corner layer.

The solution decomposition implies following estimates:

Corollary 1. *u and u − v satisfy the following a priori estimates:*

$$|u|_1 \leq C\varepsilon^{-1/4}, \qquad |u|_2 \leq C\varepsilon^{-3/4} \tag{13}$$

and

$$\|u - v\|_0 \leq C\varepsilon^{1/4}, \qquad |(u - v)_x|_1 \leq C\varepsilon^{-1/2}, \qquad \|(u - v)_x\|_0 \leq C\varepsilon^{1/4}. \tag{14}$$

Next we discretize (9) on a shape-regular mesh with linear or bilinear elements and denote as in Section 1 by $\pi u \in V_h$ some L_2 and H^1 stable projection.

The projection error satisfies as in Section 1

$$\|u - \pi u\|_\varepsilon \leq C h^{1/2}. \tag{15}$$

Note that the energy norm associated with problem (9) is now given by $\|v\|_\varepsilon^2 := \varepsilon |v|_1^2 + \|v\|_0^2$. To estimate the error, we have additionally to estimate $((u - \pi u)_x, v_h)$. But we can directly use Cauchy-Schwarz and

$$\begin{aligned}
\|(u - \pi u)_x\|_0^3 &\leq \frac{8}{3} \left(\|(v - \pi v)_x\|_0^3 + \|((u - v) - \pi(u - v))_x\|_0^3 \right) \\
&\leq C h^3 + \|(u - v)_x\|_0^2 \|((u - v) - \pi(u - v))_x\|_0 \\
&\leq C h^3 + C\varepsilon^{1/2} h |(u - v)_x|_1 \\
&\leq C h^3 + C h.
\end{aligned}$$

Summarizing we get

Theorem 2. *The error of the finite element method with linear or bilinear finite elements on a shape-regular mesh satisfies the uniform in ε estimate*

$$\|u - u_h\|_\varepsilon \leq C h^{1/3}.$$

3 Characteristic Layers But a Fitted Mesh only for the (Strong) Exponential Layer

3.1 Introduction

Consider

$$\begin{aligned}
-\varepsilon \Delta u - b u_x + c u &= f \quad \text{in } \Omega = (0,1)^2, \\
u|_{\partial\Omega} &= 0,
\end{aligned} \tag{16}$$

assuming $b \in W^{1,\infty}(\Omega)$, $c \in L_\infty(\Omega)$, $b \geq \beta > 0$ with some constant β, $0 < \varepsilon \ll 1$ and

$$c + \frac{1}{2} b_x \geq \gamma > 0. \tag{17}$$

The presence of the small parameter and the orientation of convection give rise to an exponential layer in the solution near the outflow boundary at $x = 0$ and two parabolic layers near the characteristic boundaries $y = 0$ and $y = 1$.

The error of the Galerkin finite element method using bilinear elements on appropriately constructed Shishkin meshes satisfies

$$\|u - u^N\|_\varepsilon \leq CN^{-1}\ln N, \tag{18}$$

here $\|v\|_\varepsilon^2 := \varepsilon|v|_1^2 + \|v\|_0^2$. Remark that the mesh used is fine as well at $x = 0$ as at $y = 0$, $y = 1$. Because the characteristic layers are weaker than the exponential layers (see the next Subsection for details) we are interested to use a fine mesh only at $x = 0$ and to ignore the characteristic layers. The question is: what error behaviour can we expect on such a mesh?

3.2   Error Estimation

Our error analysis is based on the asymptotic approximation

$$u_{\mathrm{as}} = v + w, \tag{19}$$

where v is the smooth solution of the reduced problem

$$-bv_x + cv = f \quad \text{in } \Omega, \qquad v|_{x=1} = 0$$

and w corresponds to a layer term at $x = 0$ such that $u_{\mathrm{as}}|_{x=0} = u_{\mathrm{as}}|_{x=1} = 0$ and

$$\left|\frac{\partial^{i+j}w}{\partial x^i \partial y^j}(x,y)\right| \leq C\varepsilon^{-i}e^{-\beta x/\varepsilon} \quad \text{for } (x,y) \in \Omega \text{ and } 0 \leq i+j \leq 2. \tag{20}$$

We do not use the solution decomposition which contains the characteristic layer term w_p with

$$\left|\frac{\partial^{i+j}w_p}{\partial x^i \partial y^j}(x,y)\right| \leq C\varepsilon^{-j/2}\left(e^{-y/\sqrt{\varepsilon}} + e^{-(1-y)/\sqrt{\varepsilon}}\right) \quad \text{for } (x,y) \in \Omega, 0 \leq i+j \leq 2 \tag{21}$$

and, additionally, a corner layer term (see [2]). Because, for instance, $\frac{\partial w_p}{\partial y} = \mathcal{O}(\varepsilon^{-1/2})$ but $\frac{\partial w}{\partial x} = \mathcal{O}(\varepsilon^{-1})$ we call the characteristic layers "weaker" than the exponential layer.

Schieweck [12, Remark 5.4.] proved the estimate

$$\|u - u_{\mathrm{as}}\|_\varepsilon \leq C\varepsilon^{1/4} \tag{22}$$

which reflects the fact that measured in the norm $\|\cdot\|_\varepsilon$ the characteristic layer terms are of order $\mathcal{O}(\varepsilon^{1/4})$ (in contrast to the exponential layers which are of order $\mathcal{O}(1)$).

We use a Shishkin mesh typically for exponential layers:

$$x_i = ih \qquad \text{for} \quad i = 0, 1, \ldots, \frac{N}{2} \qquad \text{with} \quad h = \frac{4\varepsilon}{\beta} N^{-1} \ln N,$$

$$x_i = \tau + \left(i - \frac{N}{2} \right) H \quad \text{for} \quad i = \frac{N}{2} + 1, \ldots, N \quad \text{with} \quad H = (1 - \tau)\frac{2}{N}$$

and $\tau = \frac{2\varepsilon}{\beta} \ln N$. In the y-direction the mesh is uniform with $y_j = jN^{-1}$, $j = 0, 1, \ldots, N$.

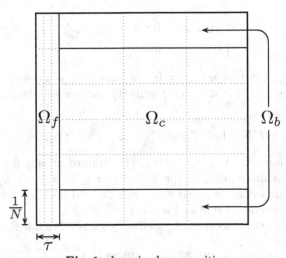

Fig. 1. domain decomposition

Let $V^N \subset H_0^1(\Omega)$ denote the space of bilinear finite elements on that mesh. Then the Galerkin approximation u^N of u satisfies

$$a(u^N, v^N) = (f, v^N) \quad \text{for all} \quad v^N \in V^N \tag{23}$$

with the bilinear form

$$a(w, v) := \varepsilon(\nabla u, \nabla w) - (bw_x, v) + (cw, v).$$

For estimating $u - u^N$, we use the splitting

$$u - u^N = u - u_{\text{as}} + \left(u_{\text{as}} - u^N \right) \tag{24}$$

and use some projection $\pi u_{\text{as}} \in V^N$:

$$u_{\text{as}} - u^N = (u_{\text{as}} - \pi u_{\text{as}}) + \left(\pi u_{\text{as}} - u^N \right). \tag{25}$$

The expression $\xi := \pi u_{as} - u^N \in V^N$ is estimated based on

$$\alpha \|\xi\|_\varepsilon^2 \le a(\xi, \xi) = a(\pi u_{as} - u, \xi) = a(\pi u_{as} - u_{as}, \xi) + a(u_{as} - u, \xi). \tag{26}$$

Because

$$|a(u_{as} - u, \xi)| \le C\varepsilon^{1/4} \|\xi\|_\varepsilon, \tag{27}$$

as also already shown by Schieweck [11] we have to estimate $u_{as} - \pi u_{as}$ and $a(u_{as} - \pi u_{as}, \xi)$.

We define the projections πv and πw by piecewise bilinear interpolation and, for instance,

$$(\pi v)(x_i, y_j) := \begin{cases} v(x_i, y_j) & \text{for } 0 \le i \le N, \, 1 \le j \le N-1 \\ 0 & \text{otherwise.} \end{cases}$$

Because $(v + w)|_{x=0} = (v + w)|_{x=1} = 0$, we have $\pi u_{as} = \pi v + \pi w \in V^N$.

On $\Omega_f := \{(x, y) \in \Omega : x < \tau\}$, cf. Figure 1, one can use standard arguments [8] to get

$$\|u_{as} - \pi u_{as}\|_{\varepsilon, \Omega_f} \le CN^{-1} \ln N \quad \text{and} \quad |a(u_{as} - \pi u_{as}, \xi)_{\Omega_f}| \le CN^{-1} \ln N \|\xi\|_\varepsilon.$$

Similarly, in Ω_c (here $x \ge \tau$, $y_1 \le y \le y_{N-1}$) standard arguments can be used.

It remains to estimate in $\Omega_b := \{(x, y) : \tau \le x \le 1, \, 0 \le y \le y_1 \text{ and } y_{N-1} \le y \le 1\}$, here the projection of v and w is nonstandard because the values at the boundaries are ignored and replaced by zero.

First we consider the smooth v. As well v as πv are bounded, consequently

$$\|v - \pi v\|_{0, \Omega_b} \le \|v - \pi v\|_{\infty, \Omega_b} (\text{meas } \Omega_b)^{1/2} \le CN^{-1/2}. \tag{28}$$

Next we estimate similarly

$$\|(v - \pi v)_x\|_{0, \Omega_b} \quad \text{and} \quad \|(v - \pi v)_y\|_{0, \Omega_b}$$

Because $(\pi v)_x$ is bounded but $(\pi v)_y = \mathcal{O}(H^{-1})$, we get

$$\|(v - \pi v)_x\|_{0, \Omega_b} \le CN^{-1/2} \quad \text{but} \quad \|(v - \pi v)_y\|_{0, \Omega_b} \le CN^{1/2}.$$

Therefore, we have

$$\|v - \pi v\|_{\varepsilon, \Omega_b} \le C \left(N^{-1/2} + (\varepsilon N)^{1/2} \right) \tag{29}$$

and, similarly,

$$|a(v - \pi v, \xi)_{\Omega_b}| \le C \left(N^{-1/2} + (\varepsilon N)^{1/2} \right) \|\xi\|_\varepsilon.$$

The w component is small for $x \ge \tau$ and of order (pointwise) $\mathcal{O}(N^{-2})$. Consequently

$$\|w - \pi w\|_{\varepsilon, \Omega_b} \le C \left(\varepsilon^{1/2} N^{-1} + N^{-2} \right).$$

Similarly,

$$|a(w - \pi w, \xi)_{\Omega_b}| \leq C \left(\varepsilon^{1/2} N^{-1} + N^{-1} \right) \|\xi\|_\varepsilon.$$

Summarizing we obtain

$$\|u - u^N\|_\varepsilon \leq C \left(\varepsilon^{1/4} + N^{-1/2} + (\varepsilon N)^{1/2} + N^{-1} \ln N \right)$$

or

$$\|u - u^N\|_\varepsilon \leq C \left(N^{-1/2} + (\varepsilon N)^{1/2} + N^{-1} \ln N \right) \tag{30}$$

because

$$\varepsilon^{1/4} = \left(N^{-1/2} (\varepsilon N)^{1/2} \right)^{1/2} \leq \frac{1}{2} \left(N^{-1/2} + (\varepsilon N)^{1/2} \right).$$

Theorem 3. *If a Shishkin mesh is used to fit the exponential layer but close to characteristic layers no layer adaption takes place, the error of the finite element method with bilinear elements satisfies*

$$\|u - u^N\|_\varepsilon \leq C \left(N^{-1/2} + (\varepsilon N)^{1/2} + N^{-1} \ln N \right).$$

Remark 3. If ε is small in comparison to N^{-1}, i.e. $\varepsilon \leq CN^{-2}$ theorem 3 states convergence of order $1/2$. Our numerical examples show that this result is in fact optimal. However a relation between ε and N appears to be not substantial and uniform convergence is observed.

Remark 4. Schieweck proved a similar result using exponential splines in Ω_f.

4   Numerical Experiments

4.1   Reaction Diffusion

Consider the test boundary value problem

$$-\varepsilon^2 \Delta u + u = f \quad \text{in } \Omega = (0,1)^2,$$
$$u = 0 \quad \text{on } \partial\Omega$$

where $\varepsilon \in (0, 1]$ and f is chosen in such a way that

$$u(x, y) = \hat{u}(x)\hat{u}(y), \quad \hat{u}(t) = -\frac{1 - e^{-1/\varepsilon}}{1 - e^{-2/\varepsilon}} \left(e^{-t/\varepsilon} + e^{-(1-t)/\varepsilon} \right) + 1$$

is the exact solution, which exhibits typical boundary layer behavior. Let Ω_N denote the uniform mesh that is generated as the tensor product of two uniform 1D meshes dissecting the interval $(0, 1)$ into N elements. Moreover we define V^N as the FE-space of bilinear functions on Ω_N. We denote by u_N the solution of the

Table 1. $\varepsilon = 1$

N	$\varepsilon\lvert u-u_N\rvert_1$ error	rate	$\lVert u-u_N\rVert_0$ error	rate	$\lVert u-u_N\rVert_\infty$ error	rate	$\lVert u^I-u_N\rVert_\infty$ error	rate
8	3.92e-3	1.00	1.22e-4	2.00	3.57e-3	2.08	1.66e-4	2.01
16	1.96e-3	1.00	3.05e-5	2.00	8.41e-4	2.02	4.12e-5	2.00
32	9.78e-4	1.00	7.63e-6	2.00	2.07e-4	2.01	1.03e-5	2.00
64	4.89e-4	1.00	1.91e-6	2.00	5.16e-5	2.00	2.57e-6	2.00
128	2.45e-4	1.00	4.77e-7	2.00	1.29e-5	2.00	6.42e-7	2.00
256	1.22e-4	1.00	1.19e-7	2.00	3.22e-6	2.00	1.60e-7	2.00
512	6.11e-5		2.98e-8		8.06e-7		4.01e-8	

Table 2. $\varepsilon = 10^{-1}$

N	$\varepsilon\lvert u-u_N\rvert_1$ error	rate	$\lVert u-u_N\rVert_0$ error	rate	$\lVert u-u_N\rVert_\infty$ error	rate	$\lVert u^I-u_N\rVert_\infty$ error	rate
8	1.27e-1	0.94	4.08e-2	1.92	9.54e-2	1.53	4.74e-2	2.24
16	6.65e-2	0.98	1.08e-2	1.98	3.29e-2	1.74	1.00e-2	2.05
32	3.36e-2	1.00	2.73e-3	2.00	9.87e-3	1.86	2.41e-3	2.01
64	1.69e-2	1.00	6.84e-4	2.00	2.72e-3	1.93	5.98e-4	2.00
128	8.44e-3	1.00	1.71e-4	2.00	7.15e-4	1.96	1.49e-4	2.00
256	4.22e-3	1.00	4.28e-5	2.00	1.83e-4	1.98	3.73e-5	2.00
512	2.11e-3		1.07e-5		4.64e-5		9.31e-6	

Table 3. $\varepsilon = 10^{-2}$

N	$\varepsilon\lvert u-u_N\rvert_1$ error	rate	$\lVert u-u_N\rVert_0$ error	rate	$\lVert u-u_N\rVert_\infty$ error	rate	$\lVert u^I-u_N\rVert_\infty$ error	rate
8	1.30e-1	0.14	2.96e-1	0.84	8.19e-1	0.42	5.65e-1	0.30
16	1.18e-1	0.37	1.65e-1	1.21	6.13e-1	0.87	4.57e-1	0.88
32	9.17e-2	0.69	7.11e-2	1.63	3.36e-1	1.35	2.48e-1	1.68
64	5.69e-2	0.90	2.30e-2	1.88	1.32e-1	1.46	7.70e-2	2.34
128	3.06e-2	0.97	6.24e-3	1.97	4.81e-2	1.69	1.53e-2	2.01
256	1.56e-2	0.99	1.60e-3	1.99	1.49e-2	1.83	3.78e-3	2.02
512	7.84e-3		4.01e-4		4.20e-3		9.30e-4	

Table 4. $\varepsilon = 10^{-4}$

N	$\varepsilon\|u - u_N\|_1$		$\|u - u_N\|_0$		$\|u - u_N\|_\infty$		$\|u^I - u_N\|_\infty$	
	error	rate	error	rate	error	rate	error	rate
8	1.41e-2		3.72e-1		9.99e-1		6.08e-1	
		0.00		0.49		0.00		0.00
16	1.41e-2		2.65e-1		9.99e-1		6.07e-1	
		0.00		0.50		0.00		0.00
32	1.41e-2		1.88e-1		9.97e-1		6.07e-1	
		0.00		0.51		0.00		0.00
64	1.41e-2		1.32e-1		9.94e-1		6.07e-1	
		0.01		0.52		0.01		0.00
128	1.40e-2		9.17e-2		9.86e-1		6.07e-1	
		0.02		0.55		0.04		0.01
256	1.38e-2		6.27e-2		9.60e-1		6.03e-1	
		0.03		0.60		0.10		0.03
512	1.35e-2		4.13e-2		8.97e-1		5.90e-1	

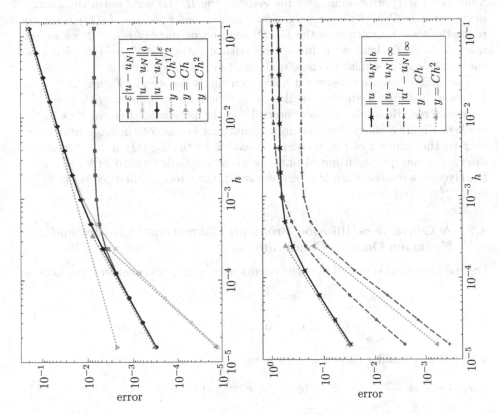

Fig. 2. 1D problem: energy norm errors for $\varepsilon = 10^{-4}$

Fig. 3. 1D problem: errors for $\varepsilon = 10^{-4}$

Galerkin FE-method. We use some adaptive quadrature algorithm to compute all integrals with a tolerance of 10^{-10}.

In Table 1 we see the performance of the method in the setting $\varepsilon = 1$. In agreement with classical analysis one observes convergence of second order in the $L_2(\Omega)$ norm as well as in the $L_\infty(\Omega)$ norm and its discrete version. However, the error of the method measured in the energy norm is $\mathcal{O}(N^{-1})$ because its $H^1(\Omega)$ semi norm component is only first oder convergent and dominates the $L_2(\Omega)$ error (even for small N). Remark that similar results will follow if for the element diameter $h(N)$ it holds $h(N) \leq \varepsilon$, see Table 2.

If ε is small in comparison to $h(N)$ — as is the case in Table 3 — one observes that the rates of converges decline. Furthermore the error in the energy norm is dominated by the $L_2(\Omega)$ norm component for small values of N where the $H^1(\Omega)$ semi norm component attains convergence rates smaller than $1/2$.

Finally Table 4 shows that if the ratio $\varepsilon/h(N)$ is further decreased one observes no error reduction in the $H^1(\Omega)$ semi norm or the $L_\infty(\Omega)$ norm. However the $L_2(\Omega)$ error seems to be $\mathcal{O}(N^{-1/2})$. In fact we have $\|u - u_N\|_0 \leq 1.06N^{-1/2}$. Since the $L_2(\Omega)$ error dominates the error in the $H^1(\Omega)$ semi norm the energy norm error is $\mathcal{O}(N^{-1/2})$, as well. Note that in order for the $L_2(\Omega)$ error to reach the same magnitude as the $H^1(\Omega)$ semi norm, namely $1.4e-2$, we would need $N > 5000$ elements reducing the critical ratio $\varepsilon/h(N)$ and resulting in convergence rates similar to those of Table 3 (with $N > 32$).

The different phases described above can nicely be seen in Figure 2. Here a 1D reaction-diffusion problem similar to the original test problem with $\varepsilon = 10^{-4}$ is considered. Hence, a broader interval for the parameter $h = h(N)$ can be studied. Figure 3 shows that the approximations u_N do not converge uniformly in ε to the solution of the problem u neither in the $L_\infty(\Omega)$ norm nor in its discrete counterpart. An initial phase of error stagnation (until $h(N) \leq \varepsilon$ and the layers are resolved) can also be observed in the stronger balanced norm $\|\cdot\|_b$, i.e. $\|v\|_b^2 := \varepsilon|v|_1^2 + \|v\|_0$.

4.2   A Convection-Diffusion Problem: Characteristic Layers and Neumann Outflow Condition

For the numerical verification of the results of Section 2 we consider the problem

$$-\varepsilon\Delta u - 2u_x + u = f \quad \text{in} \quad \Omega = (0,1)^2,$$

$$\left.\frac{\partial u}{\partial x}\right|_{x=0} = 0, \quad u|_{x=1} = 0 \quad \text{and} \quad u|_{y=0} = u|_{y=1} = 0, \tag{31}$$

with $0 < \varepsilon \leq 1$ and f chosen in such a way that

$$u(x,y) = \left(\cos\frac{\pi x}{2} - 2x + 2 - \varepsilon\big(e^{-2x/\varepsilon} - e^{-2/\varepsilon}\big)\right)\left(1 - e^{-y/\sqrt{\varepsilon}}\right)\left(1 - e^{-(1-y)/\sqrt{\varepsilon}}\right)$$

is the exact solution of (31). It exhibits typical boundary layer behavior for this kind of problem. We denote by u_N the finite element solution determined by

$$\varepsilon(\nabla u_N, \nabla v_N) + (u_N - 2u_{N,x}, v_N) = (f, v_N) \quad \text{for all } v_N \in V^N.$$

Here V^N is the FE-space of bilinear functions on a uniform mesh with N^2 elements like in the previous Subsection.

In Table 5 we see the error of the method measured in the $\sqrt{\varepsilon}$-weighted $H^1(\Omega)$ semi norm. For the last two columns in that table ($\varepsilon = 10^{-8}$ and $\varepsilon = 10^{-12}$) we observe no significant error reduction. In fact the error behaves almost like $\approx 1.8\varepsilon^{1/4}$ independently of N. Since u is of the same quality (according to (13): $|u|_1 \leq C\varepsilon^{-1/4}$) this indicates that for $\sqrt{\varepsilon} \ll h_N := N^{-1}$ bilinear functions are not able to yield good approximations for a sharp layer function. For a bigger value of ε, namely $\varepsilon = 10^{-4}$ we find that the rates of convergence increase significantly if some hundred elements are considered. Hence, the ratio $\sqrt{\varepsilon}/h_N$ appears to be significant. Remark that a parabolic boundary layer has a width $\mathcal{O}(\sqrt{\varepsilon})$. If $\sqrt{\varepsilon}/h_N \geq 1$ we observe first order convergence.

The $L_2(\Omega)$ errors and corresponding rates are depicted in Table 6. For $\sqrt{\varepsilon}/h_N \geq 1$ second order convergence can be observed. If that ratio is smaller than one the rates start to fall but $1/2$ appears to be a lower bound for all $L_2(\Omega)$ rates, independently of the quotient $\sqrt{\varepsilon}/h_N$. Remark also that for $\sqrt{\varepsilon}/h_N \ll 1$ the $L_2(\Omega)$ error dominates the corresponding $\sqrt{\varepsilon}$-weighted $H^1(\Omega)$ seminorm error. Hence, in this case the energy norm error is essentially given by its $L_2(\Omega)$ component.

Table 5. $\sqrt{\varepsilon}|u - u_N|_1$ for different values of N and ε

N	$\varepsilon = 1$		$\varepsilon = 10^{-2}$		$\varepsilon = 10^{-4}$		$\varepsilon = 10^{-8}$	$\varepsilon = 10^{-12}$
	error	rate	error	rate	error	rate	error	error
8	6.98e-2	1.00	2.04e-1	0.94	1.73e-1	0.13	1.86e-2	1.86e-3
16	3.49e-2	1.00	1.06e-1	0.98	1.58e-1	0.34	1.86e-2	1.86e-3
32	1.74e-2	1.00	5.37e-2	0.98	1.25e-1	0.68	1.85e-2	1.86e-3
64	8.72e-3	1.00	2.72e-2	0.99	7.78e-2	0.91	1.85e-2	1.86e-3
128	4.36e-3	1.00	1.37e-2	1.00	4.13e-2	0.98	1.84e-2	1.87e-3
256	2.18e-3		6.89e-3		2.09e-2		1.82e-2	1.87e-3

Table 6. $\|u - u_N\|_0$ for different values of N and ε

N	$\varepsilon = 1$		$\varepsilon = 10^{-2}$		$\varepsilon = 10^{-4}$		$\varepsilon = 10^{-8}$	
	error	rate	error	rate	error	rate	error	rate
8	2.23e-3	2.00	5.89e-2	1.99	4.00e-1	0.86	4.98e-1	0.50
16	5.56e-4	2.00	1.49e-2	1.99	2.20e-1	1.24	3.52e-1	0.51
32	1.39e-4	2.00	3.74e-3	1.98	9.34e-2	1.71	2.48e-1	0.51
64	3.47e-5	2.00	9.47e-4	1.98	2.85e-2	1.97	1.74e-1	0.52
128	8.68e-6	2.00	2.39e-4	1.99	7.30e-3	2.01	1.21e-1	0.55
256	2.17e-6		6.01e-5		1.82e-3		8.25e-2	

4.3 Characteristic Layers But a Fitted Mesh only for the (Strong) Exponential Layer

Table 7. $\sqrt{\varepsilon}|u - u^N|_1$ for different values of N and ε

N	$\varepsilon = 1$ error	rate	$\varepsilon = 10^{-2}$ error	rate	$\varepsilon = 10^{-4}$ error	rate	$\varepsilon = 10^{-8}$ error	rate	$\varepsilon = 10^{-12}$ error	rate
8	2.21e-2		2.59e-1		3.61e-1		3.90e-1		3.90e-1	
		1.00		0.60		0.62		0.52		0.52
16	1.10e-2		1.70e-1		2.35e-1		2.71e-1		2.72e-1	
		1.00		0.68		0.74		0.58		0.57
32	5.52e-3		1.06e-1		1.41e-1		1.82e-1		1.82e-1	
		1.00		0.74		0.80		0.60		0.59
64	2.76e-3		6.35e-2		8.10e-2		1.20e-1		1.21e-1	
		1.00		0.78		0.81		0.61		0.59
128	1.38e-3		3.70e-2		4.60e-2		7.86e-2		8.01e-2	
		1.00		0.81		0.83		0.61		0.58
256	6.90e-4		2.11e-2		2.59e-2		5.14e-2		5.36e-2	
		1.00		0.83		0.84		0.63		0.56
512	3.45e-4		1.18e-2		1.44e-2		3.31e-2		3.64e-2	

Table 8. $\|u - u^N\|_0$ for different values of N and ε

N	$\varepsilon = 1$ error	rate	$\varepsilon = 10^{-2}$ error	rate	$\varepsilon = 10^{-4}$ error	rate	$\varepsilon = 10^{-8}$ error	rate	$\varepsilon = 10^{-12}$ error	rate
8	6.55e-4		2.57e-2		1.53e-1		1.91e-1		1.91e-1	
		2.00		2.07		0.87		0.51		0.51
16	1.63e-4		6.12e-3		8.40e-2		1.34e-1		1.34e-1	
		2.00		2.00		1.24		0.51		0.50
32	4.09e-5		1.53e-3		3.56e-2		9.42e-2		9.50e-2	
		2.00		1.92		1.71		0.51		0.50
64	1.02e-5		4.05e-4		1.09e-2		6.60e-2		6.72e-2	
		2.00		1.87		1.96		0.52		0.50
128	2.55e-6		1.11e-4		2.80e-3		4.59e-2		4.75e-2	
		2.00		1.84		2.00		0.55		0.50
256	6.38e-7		3.11e-5		6.99e-4		3.14e-2		3.36e-2	
		2.00		1.81		2.00		0.60		0.50
512	1.60e-7		8.85e-6		1.75e-4		2.07e-2		2.37e-2	

In the problem of Section 3 the energy norm associated with (16) is strong enough to capture the exponential layer in contrast to the previous problems. The numerical results reflect this exceptional feature. Let u^N denote the solution of (23) on the tensor product mesh composed by a Shishkin mesh in x-direction and a uniform mesh in y-direction introduced in Section 3. As in the previous Subsection we choose $b = 2$ and $c = 1$. Moreover we determine f in such way that

$$u(x, y) = \left(\cos\frac{\pi x}{2} - \frac{e^{-2x/\varepsilon} - e^{-2\varepsilon}}{1 - e^{-2\varepsilon}}\right)\left(1 - e^{-y/\sqrt{\varepsilon}}\right)\left(1 - e^{-(1-y)/\sqrt{\varepsilon}}\right)$$

is the exact solution of (16).

In Table 7 and Table 8 the $\sqrt{\varepsilon}$-weighted $H^1(\Omega)$ semi norm errors and the errors in $L_2(\Omega)$ are presented, respectively. In contrast to previous numerical results both components are almost of the same magnitude if ε is very small. For the

$\sqrt{\varepsilon}$-weighted $H^1(\Omega)$ semi norm error we observe first order convergence in the non-perturbed case and $|u - u^N| \leq CN^{-1}\ln N$ if the singular perturbation is mild ($\varepsilon \geq 10^{-4}$). In the case of a very small perturbation parameter convergence with a rate slightly greater than $1/2$ can be observed uniformly in ε. For the $L_2(\Omega)$ error the rates of convergence are close to two for $\epsilon \geq 10^{-4}$ and $N \geq 64$. If $\sqrt{\varepsilon} \ll N^{-1}$ we observe uniform convergence of order $1/2$. Hence, the method is of order $1/2$ in the energy norm and our theoretical findings are sharp.

References

1. Apel, T., Lube, G.: Anisotropic mesh refinement for a singularly perturbed reaction diffusion model problem. Appl. Numer. Math. 26(4), 415–433 (1998), doi:10.1016/S0168-9274(97)00106-2
2. Kellogg, R.B., Stynes, M.: Sharpened bounds for corner singularities and boundary layers in a simple convection-diffusion problem. App. Math. Lett. 20(5), 539–544 (2007), doi:10.1016/j.aml.2006.08.001
3. Li, J., Navon, I.M.: Uniformly convergent finite element methods for singularly perturbed elliptic boundary value problems I: Reaction-diffusion Type. Comput. Math. Appl. 35(3), 57–70 (1998), doi:10.1016/S0898-1221(97)00279-4
4. Li, J.: Convergence and superconvergence analysis of finite element methods on highly nonuniform anisotropic meshes for singularly perturbed reaction-diffusion problems. Appl. Numer. Math. 36(2-3), 129–154 (2001)
5. Lin, R.: Discontinuous Discretization for Least-Squares Formulation of Singularly Perturbed Reaction-Diffusion Problems in One and Two Dimensions. SIAM J. Numer. Anal. 47, 89–108 (2008), doi:10.1137/070700267
6. Lin, R., Stynes, M.: A balanced finite element method for singularly perturbed reaction-diffusion problems. SIAM J. Numer. Anal. 50(5), 2729–2743 (2012), doi:10.1137/110837784
7. Naughton, A., Kellogg, R.B., Stynes, M.: Regularity and derivative bounds for a convection-diffusion problem with a Neumann outflow condition. J. Differential Equations 247(9), 2495–2516 (2009), doi:10.1016/j.jde.2009.07.030
8. Roos, H.-G., Stynes, M., Tobiska, L.: Robust numerical methods for singularly perturbed differential equations. Springer (2008)
9. Roos, H.-G., Schopf, M.: Convergence and stability in balanced norms of finite element methods on Shishkin meshes for reaction-diffusion problems (submitted)
10. Schatz, A.H., Wahlbin, L.B.: On the finite element method for singularly perturbed reaction-diffusion problems in two and one dimensions. Math. Comp. 40(161), 47–89 (1983)
11. Schieweck, F.: Eine asymptotisch angepate Finite-Element-Methode für singulär gestörte elliptische Randwertaufgaben. PhD thesis, Technical University Magdeburg (1986)
12. Schieweck, F.: On the role of boundary conditions for CIP stabilization of higher order finite elements. Electron. Trans. Numer. Anal. 32, 1–16 (2008)
13. Zhu, G., Chen, S.: Convergence and superconvergence analysis of an anisotropic nonconforming finite element methods for singularly perturbed reaction-diffusion problems. J. Comput. Appl. Math. 234(10), 3048–3063 (2010), doi: 10.1016/j.cam, 04.021

The Finite Element Method for Boundary Value Problems with Strong Singularity and Double Singularity

Viktor A. Rukavishnikov and Elena I. Rukavishnikova

Computing Center, Far-Eastern Branch, Russian Academy of Sciences,
Kim-U-Chena st. 65, 680000 Khabarovsk, Russian Federation
Far Eastern State Transport University,
ul. Serysheva 47, Khabarovsk, 680021 Russia
vark0102@mail.ru

Abstract. A boundary value problem is said to possess strong singularity if its solution u does not belong to the Sobolev space W_2^1 (H^1) or, in other words, the Dirichlet integral of the solution u diverges.

We consider the boundary value problems with strong singularity and with double singularity caused the discontinuity of coefficients in the equation on the domain with slot and presence of the corners equal 2π on boundary of this domain.

The schemes of the finite element method is constructed on the basis of the definition on R_ν-generalized solution to these problems, and the finite element space contains singular power functions. The rate of convergence of the approximate solution to the R_ν-generalized solution in the norm of the Sobolev weighted space is established and, finally, results of numerical experiments are presented.

Keywords: Finite element method, the R_ν-generalized solution, singularity of solution.

1 Introduction

Boundary value problems with strong singularity caused by the singularity in the initial data or by internal properties of the solution are found in the physics of plasma and gas discharge, electrodynamics, nuclear physics, nonlinear optics, and other branches of physics. In particular cases, numerical method for problems of electrodynamics and quantum mechanics with strong singularity, were constructed based on separation of singular and regular components, mesh refinement near singular points, multiplicative extraction of singularities, etc. (see, e.g., [1–5]).

In [6] it was proposed to define the solution to a boundary value problem with strong singularity as an R_ν-generalized solution in a weighted Sobolev space. Such a new concept of solution led to the distinction of two classes of boundary value problems: problems with coordinated and uncoordinated degeneracy of input data; it also made it possible to study the existence and uniqueness of solutions as well as its coercivity and differential properties in weighted Sobolev spaces (see [7–12]).

I. Dimov, I. Faragó, and L. Vulkov (Eds.): NAA 2012, LNCS 8236, pp. 110–121, 2013.
© Springer-Verlag Berlin Heidelberg 2013

The finite element method were constructed and investigated for different kinds of boundary value problems with strong singularity of solution (see [13–15] and references therein).

In this paper we consider the FEM for boundary value problems with strong singularity and double singularity.

2 Basic Notations

We denote the two-dimensional Euclidean space by R^2 with $x = (x_1, x_2)$ and $dx = dx_1 dx_2$. Let $\Omega \subset R^2$ be a bounded domain with piecewise smooth boundary $\partial \Omega$, and let $\bar{\Omega}$ be the closure of Ω, i.e. $\bar{\Omega} = \Omega \cup \partial \Omega$. We denote by $\bigcup_{i=1}^n \tau_i$ a set of points τ_i $(i = 1, \ldots, n)$ belonging to $\partial \Omega$, including the points of intersection of its smooth pieces.

Let O_i^δ be a disk of radius $\delta > 0$ with its center in τ_i $(i = 1, \ldots, n)$, i.e. $O_i^\delta = \{x \colon \|x - \tau_i\| \leq \delta\}$, and suppose that $O_i^\delta \cap O_j^\delta = \emptyset$, $i \neq j$. Let $\Omega' = \Omega \cap \bigcup_{i=1}^n O_i^\delta$.

Let $\rho(x)$ be a function that is infinitely differentiable, positive everywhere, except in $\bigcup_{i=1}^n \tau_i$, and satisfies the following conditions:

(a) $\rho(x) = \delta$ for $x \in \Omega \setminus \bigcup_{i=1}^n O_i^\delta$,
(b) $\rho(x) = ((x_1 - x_1^{(i)})^2 + (x_2 - x_2^{(i)})^2)^{1/2}$, $(x_1^{(i)}, x_2^{(i)}) = \tau_i$ for $x \in \Omega \cap O_i^{\delta/2}$,
(c) $\delta/2 \leq \rho(x) \leq \delta$ for $x \in \Omega \setminus O_i^{\delta/2}$ $(i = 1, \ldots, n)$.

Moreover, it is assumed that

$$\left| \frac{\partial \rho}{\partial x_i} \right| \leq \delta', \quad i = 1, 2 \tag{1}$$

where $\delta' > 0$ is a real number.

We introduce the weighted spaces $H_{2,\alpha}^k(\Omega)$ and $W_{2,\alpha}^k(\Omega)$ with norms:

$$\|u\|_{H_{2,\alpha}^k(\Omega)} = \left(\sum_{|\lambda| \leq k} \int_\Omega \rho^{2(\alpha + |\lambda| - k)} |D^\lambda u|^2 \, dx \right)^{1/2}, \tag{2}$$

$$\|u\|_{W_{2,\alpha}^k(\Omega)} = \left(\sum_{|\lambda| \leq k} \int_\Omega \rho^{2\alpha} |D^\lambda u|^2 \, dx \right)^{1/2}, \tag{3}$$

where $D^\lambda = \partial^{|\lambda|} / \partial x_1^{\lambda_1} \partial x_2^{\lambda_2}$, $\lambda = (\lambda_1, \lambda_2)$, $|\lambda| = \lambda_1 + \lambda_2$; λ_1, λ_2 are integer nonnegative numbers, α is some real nonnegative number, k is an integer nonnegative number. For $k = 0$ we use the notation $H_{2,\alpha}^0(\Omega) = W_{2,\alpha}^0(\Omega) = L_{2,\alpha}(\Omega)$.

By $W_{2,\alpha+l-1}^l(\Omega, \delta)$ for $l \geq 1$ we denote a set of functions satisfying the following conditions:

(a) $|D^k u(x)| \leq C_1 \gamma^k k! (\rho^{\alpha+k}(x))^{-1}$ for $x \in \Omega'$, where $k = 0, \ldots, l$, the constants $C_1, \gamma \geq 1$ do not depend on k;
(b) $\|u\|_{L_{2,\alpha}(\Omega \setminus \Omega')} \geq C_2$, $C_2 = $ const; with squared norm

$$\|u\|^2_{W^l_{2,\alpha+l-1}(\Omega,\delta)} = \sum_{|\lambda|\leq l} \|\rho^{\alpha+l-1}|D^\lambda u|\|^2_{L_2(\Omega)}. \tag{4}$$

The spaces $\mathring{H}^k_{2,\alpha}(\Omega) \subset H^k_{2,\alpha}(\Omega)$, $\mathring{W}^k_{2,\alpha}(\Omega) \subset W^k_{2,\alpha}(\Omega)$ and the set $\mathring{W}^k_{2,\alpha}(\Omega,\delta) \subset W^k_{2,\alpha}(\Omega,\delta)$ are defined as the closures of the set of infinitely differentiable and finite in Ω functions in norms (2)–(4), respectively.

Let $L_{\infty,-\alpha}(\Omega,C_3)$ and $H^k_{\infty,-\alpha}(\Omega,C_4)$ ($k \geq 0$, $\alpha \in R$) be the set of functions with the norms satisfying the inequalities

$$\|u\|_{L_{\infty,-\alpha}(\Omega,C_3)} = \operatorname*{ess\,sup}_{x\in\Omega} |\rho^{-\alpha}u| \leq C_3\,,$$

$$\|u\|_{H^k_{\infty,-\alpha}(\Omega,C_4)} = \max_{|\lambda|\leq k} \operatorname*{ess\,sup}_{x\in\Omega} |\rho^{-\alpha+|\lambda|}D^\lambda u| \leq C_4$$

with positive constants C_3, C_4 independent of u.

3   The Boundary Value Problems with Strong Singularity

3.1   The BVP with Coordinated Degeneration of the Input Data

In the domain Ω we consider the differential equation

$$-\sum_{l,s=1}^{2} a_{ls}(x)\frac{\partial^2 u}{\partial x_l \partial x_s} + \sum_{l=1}^{2} a_l(x)\frac{\partial u}{\partial x_l} + a(x)u(x) = f(x)\,, \quad x \in \Omega \tag{5}$$

with the boundary condition

$$u = 0\,, \quad x \in \partial\Omega. \tag{6}$$

Definition 1. *The boundary value problem (5)–(6) is called the Dirichlet problem with coordinated degeneration of the input data, or Problem A, if $a_{ls}(x) = a_{sl}(x)$ ($l,s = 1,2$) and for some real number β*

$$a_{ls} \in H^1_{\infty,-\beta}(\Omega,C_5)\,, \ a_l \in L_{\infty,-(\beta-1)}(\Omega,C_6)\,, \ a \in L_{\infty,-(\beta-2)}(\Omega,C_7)\,, \tag{7}$$

$$\sum_{l,s=1}^{2} a_{ls}(x)\xi_l\xi_s \geq C_8\rho^\beta(x)\sum_{s=1}^{2}\xi_s^2\,, \ a(x) \geq C_9\rho^{\beta-2}(x)$$

and right-hand side of (5) satisfies $f \in L_{2,\mu}(\Omega)$,

where C_i ($i = 5,\ldots,9$) are positive constants independent of x; ξ_1, ξ_2 are any real parameters; μ is some nonnegative real number.

Definition 2. *A function u_ν from the space $\mathring{H}^1_{2,\nu+\beta/2}(\Omega)$ is called an R_ν-generalized solution of the Dirichlet problem with coordinated degeneration of input data if $u_\nu = 0$ almost everywhere on $\partial\Omega$ and the identity*

$$a(u_\nu,v) = l(v) \quad \forall v \in \mathring{H}^1_{2,\nu+\beta/2}(\Omega)$$

holds, where ν is arbitrary but fixed and satisfies the inequality $\nu \geq \mu + \beta/2 - 1$.

In Definition 2

$$a(u,v) = \int\limits_{\Omega} \left[\sum_{l,s=1}^{2} a_{ls}\rho^{2\nu}\frac{\partial u}{\partial x_l}\frac{\partial v}{\partial x_s} + a_{ls}\frac{\partial \rho^{2\nu}}{\partial x_l}\frac{\partial u}{\partial x_s}v + \frac{\partial a_{ls}}{\partial x_l}\rho^{2\nu}\frac{\partial u}{\partial x_s}v + \right.$$

$$\left. + a_l\rho^{2\nu}\frac{\partial u}{\partial x_l}v + a\rho^{2\nu}uv \right]dx,$$

$$l(v) = \int\limits_{\Omega} \rho^{2\nu}fv\,dx$$

are the bilinear and linear forms respectively.

Remark 1. We say that a boundary value problem is strongly singular if it is impossible to determine a generalized solution for it, that is, a solution which does not belong to the Sobolev space W_2^1 (H^1), or, in other words, the Dirichlet integral of the solution diverges. In [6, 7] it was proposed to define the solution to a boundary value problems with strong singularity as a R_ν-generalized one in a weighted Sobolev space. The main idea of this approach is that we introduce into the integral identity of the generalized solution a weight function raised to some power. The weight function coincides with the distance to the singularity points in some neighborhoods of them. The role of this function is to eliminate the singularity appearing in the solution due to problem non-regularity and to achieve the convergence of the integrals in both parts of the integral identity. Taking into account the local character of the singularity, we define the weight function as the distance to each singularity point inside the disk of radius δ centered in that points, and outside of these disks the weight function equals δ. The power of the weight function in the definition of the R_ν-generalized solution, as well as the weighted space containing it, depend on the spaces which contain the initial data of the problem, on the geometrical singularities of the domain boundary (presence of the reentrant corners) and on the change of the type of the boundary conditions.

The existence and uniqueness, the coercitive and differential properties of the R_ν generalized solution for the differential equation (5) with the boundary conditions the first and third types and coordinated degeneration of initial data were established in [7, 8, 15, 16]. The finite element method for this boundary value problems were studied in [13, 15] and references therein.

3.2   The Finite Element Method for Boundary Value Problem with Uncoordinated Degeneration of Input Data

We consider the boundary value problem

$$-\sum_{l=1}^{2}\frac{\partial}{\partial x_l}\left(a_{ll}(x)\frac{\partial u}{\partial x_l}\right) + a(x)u(x) = f(x), \quad x \in \Omega, \tag{8}$$

$$u = 0, \quad x \in \partial\Omega. \tag{9}$$

Definition 3. *Boundary value problem* (8), (9) *is called the Dirichlet problem with uncoordinated degeneration of initial data, or Problem B, if for some real number* β

$$a_{ll} \in H^1_{\infty,-\beta}(\Omega, C_{10}), \quad a \in L_{\infty,-\beta}(\Omega, C_{11}),$$ (10)

$$\sum_{l=1}^{2} a_{ll}(x)\xi_l^2 \geq C_{12}\rho^\beta(x) \sum_{l=1}^{2} \xi_l^2, \quad a(x) \geq C_{13}\rho^\beta(x)$$ (11)

and the right-hand side of the equation satisfies the condition

$$f \in L_{2,\mu}(\Omega, \delta)$$ (12)

for some nonnegative real number μ. *Here* C_i $(i = 10, \ldots, 13)$ *are positive constants not depending on* x; ξ_1, ξ_2 *are arbitrary real parameters.*

Definition 4. *A function* u_ν *from the set* $\overset{\circ}{W}{}^1_{2,\nu+\beta/2}(\Omega, \delta)$ *is called an* R_ν-*generalized solution of Problem B if* $u_\nu = 0$ *almost everywhere on* $\partial\Omega$ *and for all* v *from* $\overset{\circ}{W}{}^1_{2,\nu+\beta/2}(\Omega, \delta)$ *the following integral identity holds:*

$$b(u_\nu, v) \equiv \int_\Omega \left[\sum_{l=1}^{2} a_{ll}\rho^{2\nu} \frac{\partial u_\nu}{\partial x_l} \frac{\partial v}{\partial x_l} + a_{ll} \frac{\partial \rho^{2\nu}}{\partial x_l} \frac{\partial u_\nu}{\partial x_l} v + a\rho^{2\nu} u_\nu v \right] dx = \int_\Omega \rho^{2\nu} fv \, dx \equiv l(v)$$

for any fixed value ν *satisfying the inequality*

$$\nu \geq \mu + \beta/2.$$ (13)

Remark 2. In Definition 2 of an R_ν-generalized solution to a Problem A all summands of the bilinear form $a(u_\nu, v)$ at the left-hand side of the integral identity are of the some order in each neighbourhood of the points of singularity. This is caused by the fulfillment of requirement (7) on the coefficients of the equation. According to conditions (10), the coefficients of the equation for Problem B are asymptotically equal and hence the summands of the bilinear form $b(u, v)$ have different orders near the points of singularity. Problems of this type call for a special approach to the study of properties of the R_ν-generalized solution. This is expressed in the fact that the solution is sought for in the set $W^1_{2,\nu+\beta/2}(\Omega, \delta)$ because in the space $W^1_{2,\nu+\beta/2}(\Omega)$ there exists a bundle of R_ν-generalized solutions in the neighbourhoods of singularity points (see [8]), and unique R_ν-generalized solution can be extracted from the bundle only by choosing the parameters ν and δ (see [9, 11]). In [11, 12] a coercitive and differential properties of the R_ν-generalized solution were investigated for Problem B.

Remark 3. The introduction of an R_ν-generalized solution for problems with uncoordinated degeneration of initial data allows us to construct efficient numerical methods no only for problems with strong singularity where the Dirichlet integral of the solution diverges, but also for boundary value problem with "bad" singularity of the solution caused by the presence of angular and conical points in the boundary of the domain.

We construct the scheme of the finite element method for determination of an R_ν-generalized solution to the Dirichlet problem with coordinated degeneracy of the input data. To do this, we perform a quasiuniform triangulation of the domain Ω and we introduce a special system of basic functions.

Inscribe in Ω a polygonal domain Ω_h. We triangulate Ω_h so that: (1) a polygonal domain $\Omega_h = \bigcup_{i=1}^{N} K_i$, where $\{K\} = \{K_1, K_2, \ldots, K_N\}$, is a set of closed triangles; here h is greatest of the side lengths of triangles K_i, $i = \overline{1,N}$, and $\partial\Omega_h$ is the boundary of the domain Ω_h; (2) triangles K_i, $i = \overline{1,N}$, may share only common sides or vertices; (3) all vertices in K_i, $i = \overline{1,N}$, sitting on $\partial\Omega$ belong also to $\partial\Omega_h$; points τ_i, $i = \overline{1,n}$, are a subset of the vertex set of triangles to $\partial\Omega_h$, and moreover, one of K_i, $i = \overline{1,N}$, contains almost one point from $\bigcup_{i=1}^{n} \tau_i$ (obviously, $h < \delta$ in this instance); (4) an angle that is minimal among all angles in triangles is strictly positive and does not depend on the triangulation; (5) all triangles K_i, $i = \overline{1,N}$, have areas of the same order; (6) a distance between points on $\partial\Omega_h$ and on $\partial\Omega$ does not exceed ηh^2, where $\eta > 0$, and does not depend on h.

The vertices $P_1, \ldots, P_{N_h}$ in K_i, $i = \overline{1,N}$, are called triangulation nodes. The number N_h is represented as the sum $N_h = \overline{N}_h + n + m$, where $\overline{N}_h$ is the number of triangulation nodes not belonging to the broken line $\partial\Omega_h$, and m is the number of nodes belonging to $\partial\Omega_h \setminus \left\{ \bigcup_{i=1}^{n} \tau_i \right\}$. We write Ω'' to denote a set of segments formed by pieces of the boundary $\partial\Omega$ and by intervals of the broken line $\partial\Omega_h$, i.e. $\Omega'' = \Omega \setminus \overline{\Omega}_h$. The triangulation properties imply that not more than one point τ_i, $i = \overline{1,n}$, belongs to every segment. To each node P_i, we assign a function of the form

$$\psi_i = \rho^{-(\nu+\beta/2+1)}\varphi_i, \quad i = 1, \ldots, \overline{N}_h,$$

where $\varphi_i(x)$ is linear over every triangle K_i, $i = \overline{1,N}$, equals one at the point P_i, $i = \overline{1,N_h}$ and zero at all the other nodes. We denote by $V^h(\Omega_h)$ the linear span $\{\psi_i\}_{i=1}^{\overline{N}_h}$. The functions $v^h \in V^h(\Omega_h)$ are extended to Ω'' so as to be identically equal to zero. In this way, we shall in fact construct the space $V^h(\Omega)$. Obviously, $V^h(\Omega)$ is subset $\overset{\circ}{W}_{2,\nu+\beta/2}^1(\Omega, \delta)$.

The approximate R_ν-generalized solution of the Problem B be the function $u_\nu^h = \sum_{i=1}^{\overline{N}_h} a_i \psi_i$ ($a_i = \rho^{\nu+\beta/2+1}(P_i)b_i$) satisfying the equality

$$b(u_\nu^h, v^h) = l(v^h) \quad \forall v^h \in V^h(\Omega).$$

Theorem 1 ([14]). *Let conditions (1), (10)–(13) hold and constant C_{13} be sufficiently large, then there exists a constant C_{14} which is independent of f and h and is such that*

$$\|u_\nu - u_\nu^h\|_{W_{2,\nu+\beta/2+1}^1(\Omega,\delta)} \leq C_{14}h\|f\|_{L_{2,\mu}(\Omega,\delta)}.$$

3.3   The Numerical Experiments

We have carried out a set numerical test for boundary value problems with singularity using our finite element method and GMRES-method for solving system of algebraic equation. The errors of numerical approximation to the R_ν-generalized solution in the norms of space $W^1_{2,\nu+\beta/2+1}(\Omega)$ was computed. Besides, for obtained approximate R_ν-generalized and generalized solutions we calculate the module between approximate and exact solutions $\delta(P_{ij})$ at each node of the grid, the quantity n_i, $i = 1, 2, \ldots$ and coordinates of nodes with errors exceeding given limit values $\bar{\delta}_i$, $i = 1, 2, \ldots$. Numerical experiments were realized on meshes with different step h.

Example. Let $\Omega = (-1, 1) \times (-1, 1) \setminus [0, 1] \times [-1, 0]$ be a L-shaped domain with one reentrant corner on its boundary $\partial\Omega$, τ_0 be the point with the coordinates $(0, 0)$.

We consider the boundary value problem (8), (9), where

$$a_{11} = a_{22} = 1, \quad a = 0,$$

$$f(x) = -\frac{2(27x_1^4 - 10x_1^2x_2^2 - 14x_1^2 + 27x_2^4 - 14x_2^2 - 16)x_1x_2}{9\left(x_1^2 + x_2^2\right)^{5/3}}.$$

The exact solution of this problem is

$$u(x) = \left(\sqrt{x_1^2 + x_2^2}\right)^{2/3} \sin\varphi\cos\varphi\left(1 - x_1^2\right)\left(1 - x_2^2\right), \quad \varphi = \operatorname{arctg}\frac{x_2}{x_1}.$$

We denote the set of the mesh points on the domain $\bar{\Omega}$ by

$$\{P_{ij}\} = \{P_{ij} : P_{ij} = ((-1 + ih), (-1 + jh)), \ h = \frac{2}{M}, \ i, j = \overline{0, M}, \ P_{ij} \in \bar{\Omega}\}.$$

Here M is a positive even integer. Let $\rho(x) = \min\{\delta, \operatorname{dist}(x, \tau_0)\}$.

Table 1. The influence of the mesh-size variations on the behaviour of the error in the norm of the weight space $W^1_{2,\nu+\beta/2+1}$ ($\nu = 0.2$, $\beta = 0$, $\gamma = 1$, $\delta = 2.03125 \cdot 10^{-2}$)

h	0.0625	0.03125	0.0156
M	32	64	128
$\|u_\nu - u_\nu^h\|_{W^1_{2,\nu+1}}$	$2.02 \cdot 10^{-3}$	$6.34 \cdot 10^{-4}$	$1.99 \cdot 10^{-4}$

Figure 1 (a, b) show the nodes of the mesh where the error exceeds some given limit value for the generalized (a) and R_ν-generalized (b) solutions ($h = 0.0156$, $\nu = 0.2$, $\beta = 0$, $\delta = 2.03125 \cdot 10^{-2}$). Here for limit values of the error we use following values: 10^{-3}, $2 \cdot 10^{-4}$, $2 \cdot 10^{-5}$.

The series of numerical experiments showed that:

(i) the rate of convergence of the approximate solution to the exact R_ν-generalized solution has first order in the norm of the weight space $W^1_{2,\nu+\beta/2+1}(\Omega)$;

(ii) if we choose parameters ν and δ near to optimal, the accuracy of the approximation in the case of the R_ν-generalized solution is in general two orders higher than for the generalized solution;

Table 2. The influence of the mesh-size variations on the behaviour of the error of the generalized ($\nu = 0$) and R_ν-generalized solutions for $\nu = 0.2$, $\beta = 0$, $\delta = 2.03125 \cdot 10^{-2}$, $\bar{\delta}_1 = 10^{-3}$, $\bar{\delta}_2 = 2 \cdot 10^{-4}$

h	0.0625		0.03125		0.0156	
M	32		64		128	
	Gen. sol.	R_ν-gen. sol.	Gen. sol.	R_ν-gen. sol.	Gen. sol.	R_ν-gen. sol.
n_1	183	18	78	12	30	12
n_2	309	132	885	66	444	9

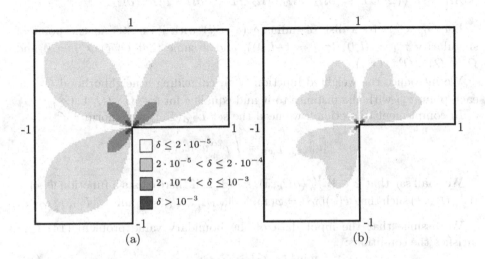

Fig. 1. Error of generalized (a) and R_ν-generalized (b) solutions

(iii) if the input data had strong singularity, it is impossible to find the generalized solution ($\nu = 0$), because the computation is interrupted by an exception, while the R_ν-generalized solution can be computed with high accuracy (see, for example [15]).

4 The Boundary Value Problem with Double Singularity

4.1 Problem Formulation: The R_ν-generalized Solution

Let the differential equation

$$-\Delta u + \lambda u(x) = f(x), \quad x \in \Omega \tag{14}$$

be given in the domain Ω with boundary conditions

$$u = \varphi(x),\ x \in \partial\Omega_1, \quad \frac{\partial u}{\partial \eta} = 0,\ x \in \partial\Omega_2 \tag{15}$$

and continuity conditions on the interface L

$$[u] = 0, \quad \left[\frac{\partial u}{\partial \eta}\right] = \psi(x), \quad x \in L, \tag{16}$$

where η is the unit vector of the outer normal and

$$\lambda = \lambda_i = const, \quad x \in \Omega_i, \quad i = 1,2, \quad \lambda_1 \neq \lambda_2. \tag{17}$$

Here $x = (x_1, x_2)$, $\Omega = \Omega_1 \cup \Omega_2$, were

$$\Omega_1 = \{x : x_1 \in (-s_1, s_1), x_2 \in (0, 2s_1)\}, \bar{\Omega}_1 = \{x : x_1 \in [-s_1, s_1], x_2 \in [0, 2s_1]\},$$
$$\Omega_2 = \{x : x_1 \in (-s_2, s_2), x_2 \in (-2s_2, 0)\}, \bar{\Omega}_2 = \{x : x_1 \in [-s_2, s_2], x_2 \in [-2s_2, 0]\},$$
$$L = \{x : x_1 \in (-l, l), x_2 = 0\}, \quad l < s_2 < s_1, \quad s_1 = 2s_2,$$
$$\partial\Omega_1 = \bar{\Omega}_1 \setminus (\Omega_1 \cup L), \quad \partial\Omega_2 = \bar{\Omega}_2 \setminus (\Omega_2 \cup L), \quad \partial\Omega = \partial\Omega_1 \cup \partial\Omega_2.$$

Let O_l^δ (O_{-l}^δ) be a disk of radius δ ($\delta < l$) with its center in the point of singularity $\tau_{+l} = (l, 0)$ ($\tau_{-l} = (-l, 0)$), and assume that $O_l^\delta \cap O_{-l}^\delta = \emptyset$ and $\Omega' = \Omega \cap (O_l^\delta \cup O_{-l}^\delta)$.

We introduce the weighted function $\rho(x)$, coinciding a neighborhood $O_{\pm l}^\delta$ of each point $\tau_{\pm l}$ with the distance to it and equals δ for $x \in \bar{\Omega} \setminus (O_l^\delta \cup O_{-l}^\delta)$.

In complement to section 2 we need the set $L_{2,\alpha}(L, \delta)$ with norm

$$\|u\|_{L_{2,\alpha}(L,\delta)} = \int_L \rho^{2\alpha} u^2 \, dx_1. \tag{18}$$

We shall say that $\varphi \in W_{2,\alpha}^{1/2}(\partial\Omega_k, \delta)$, $k = 1,2$, if there exists a function $\Phi(x) \in W_{2,\alpha}^1(\Omega_k, \delta)$ such that $\Phi(x)|_{\partial\Omega_k} = \varphi(x)$, $\|\varphi\|_{W_{2,\alpha}^{1/2}(\partial\Omega_k,\delta)} = \inf_{\Phi|_{\partial\Omega_k}=\varphi} \|\Phi\|_{W_{2,\alpha}^1(\Omega_k,\delta)}$.

We assume that the input data of the boundary value problem (14)–(17) satisfies the conditions:

$$\min\{\lambda_1, \lambda_2\} \geq C_{15} > 0, \tag{19}$$

$$f \in L_{2,\mu}(\Omega, \delta), \quad \varphi \in W_{2,\mu}^{1/2}(\partial\Omega_1, \delta), \quad \psi \in L_{2,\mu}(L, \delta), \tag{20}$$

where μ is a nonnegative real.

We introduce the bilinear and linear forms:

$$a(u,v) = \sum_{k=1}^{2} \int_{\Omega_k} \left(\sum_{i=1}^{2} \left(\rho^{2\nu} \frac{\partial u}{\partial x_i} \frac{\partial v}{\partial x_i} + \frac{\partial \rho^{2\nu}}{\partial x_i} \frac{\partial u}{\partial x_i} v \right) + \lambda_k \rho^{2\nu} uv \right) dx,$$

$$F(v) = \int_\Omega \rho^{2\nu} fv \, dx - \int_L \rho^{2\nu} \psi v \, dx.$$

Definition 5. *A function u_ν from the set $W_{2,\nu}^1(\Omega, \delta)$ is called an R_ν-generalized solution of the boundary value problem (14)–(17), (19), (20) if $u_\nu = \varphi$ almost everywhere on $\partial\Omega_1$, if for any v from $W_{2,\nu}^1(\Omega, \delta) \cap \mathring{W}_{2,\nu}^1(\Omega_1, \delta)$ the identity $a(u_\nu, v) = F(v)$ hold, where ν is arbitrary but fixed and satisfies the inequality $\nu \geq \mu$, and following weak continuity conditions on the interface L*

$$\int_L \rho^{2\nu}(u_{\nu 1}|_L - u_{\nu 2}|_L)v \, dx_1 = 0, \quad \int_L \rho^{2\nu} \left(\frac{u_{\nu 1}}{\partial \eta_1} - \frac{u_{\nu 2}}{\partial \eta_2} \right) v \, dx_1 = \int_L \rho^{2\nu} \psi v \, dx_1$$

are valid.

Here η_k is the unit vector of the outward normal to L about Ω_k, $k = 1, 2$.

4.2 The Scheme of the Finite Element Method: Numerical Experiments

We construct the scheme of the finite element method using Definition 5 of an R_ν-generalized solution. With this goal each subdomain Ω_1 and Ω_2 is decomposed by means of vertical and horizontal lines into elementary squares with the lengths of the sides h_1 and h_2 $(0 < h_2 < h_1 < 1)$ respectively. The vertices of the domain $\bar{\Omega}_1$ $(\bar{\Omega}_2)$ will be called the nodes of the grid $\bar{\Omega}_1^h$ $(\bar{\Omega}_2^h)$. We choose the step of the grid h $(h = h_2)$ so that the points of singularity $\tau_{\pm l}$ do not coincide with nodes. The nodes of the grid $\bar{\Omega}_1^h$ on interface L do not coincide with the nodes of the grid $\bar{\Omega}_2^h$, that is, the grids do not join on interface L.

We define the finite-dimensional space $V^h(\bar{\Omega}_k^h) \subset W_{2,\nu}^1(\Omega, \delta)$ $(k = 1, 2)$ of continuous functions v_k^h that are bilinear on each finite element.

Let $\{P_i^{(2)}\}_{i=0}^{i=n}$ be the set of nodes of the grid $\bar{\Omega}_2^h$ locating on the interface L and $I = \left[P_0^{(2)}, P_n^{(2)}\right]$, we introduce the space

$$V^h(L) = \left\{v^h(x_1, 0) \in C(L), v^h|_I \in S_3(I), v^h \in P_0(L \setminus I)\right\},$$

where $S_3(I)$ is the space of cubic splines, $P_0(L \setminus I)$ is the space of polynomials of degree 0. The norm in $V^h(L)$ is defined by means of the equality (18).

Definition 6. *A function u_ν^h in the space $V^h(\Omega_1^h) \cup V^h(\Omega_2^h)$ is called an approximate R_ν-generalized solution of the boundary value problem (14)–(17), (19), (20) by the finite element method if $u_{\nu 1}^h = \varphi$ at the nodes of the grid $\partial\Omega_1$ and the identity $a(u_\nu^h, v^h) = F(v^h)$ holds for all $v^h \in \mathring{V}^h(\Omega_1^h) \cup V^h(\Omega_2^h)$ and condition*

$$\|u_{\nu 1}^h|_L - u_{\nu 2}^h|_L\|_{L_{2,\nu}(L, \delta)} = O(h^n), \quad n \ge 2$$

is valid on the interface L.

Theorem 2 ([17]). *Let an R_ν-generalized solution u_ν belongs to the set $W_{2,\nu+1}^2(\Omega_k, \delta)$ for $k = 1, 2$. Then there exists a constant C_{16}, independent of u_ν, u_ν^h, f, φ, ψ and h, such that the convergence estimate*

$$\sum_{k=1}^{2} \|u_\nu - u_\nu^h\|_{W_{2,\nu}^1(\Omega_k, \delta)} + h^{-\frac{1}{2}}\|[u_\nu] - [u_\nu^h]\|_{L_{2,\nu}(L, \delta)} \le C_{16} h \sum_{k=1}^{2} \|u_\nu\|_{W_{2,\nu+1}^2(\Omega_k, \delta)}$$

holds for the constructed triangulation of the domain $\bar{\Omega}$.

We conducted on computer the numerical analysis of modeling boundary value problems (14)–(17) on domain Ω using our finite element method. For approximate solutions we calculated the following values: the error $\delta(x_h) = |v(x_h) - u^h(x_h)|$ at each node of the mesh $\bar{\Omega}_h$, the maximal error $\Delta = \max \delta(x_h)$, the number of nodes n_i, $i = 1, \ldots, m$, where the errors exceeds the given limit values $\bar{\delta}_i$, $i = 1, \ldots, m$.

Example. We consider the boundary value problem (14)–(17) with

$$u(x) = ((x_1 + l)^2 + x_2^2)^{0.2} + ((x_1 - l)^2 + x_2^2)^{0.2}, \quad \lambda_1 = 1, \quad \lambda_2 = 10,$$

$$f(x) = -0.16[((x_1 + l)^2 + x_2^2)^{-0.8} + ((x_1 - l)^2 + x_2^2)^{-0.8}]+$$
$$+\lambda_k[((x_1 + l)^2 + x_2^2)^{0.2} + ((x_1 - l)^2 + x_2^2)^{0.2}], \quad x \in \Omega_k, \quad k = 1, 2.$$

The results of computations see on Figure 2.

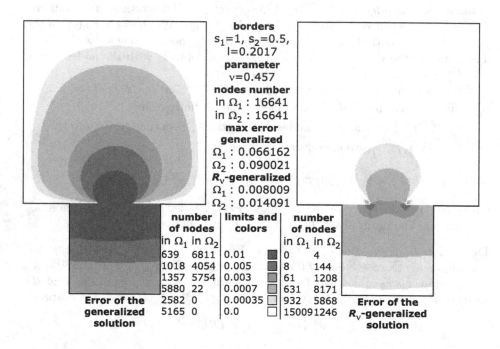

borders
$s_1=1$, $s_2=0.5$,
$l=0.2017$
parameter
$\nu=0.457$
nodes number
in Ω_1 : 16641
in Ω_2 : 16641
max error
generalized
Ω_1 : 0.066162
Ω_2 : 0.090021
R_ν-**generalized**
Ω_1 : 0.008009
Ω_2 : 0.014091

Error of the generalized solution

number of nodes in Ω_1	in Ω_2	limits and colors		number of nodes in Ω_1	in Ω_2
639	6811	0.01	■	0	4
1018	4054	0.005	■	8	144
1357	5754	0.003	■	61	1208
5880	22	0.0007	■	631	8171
2582	0	0.00035	□	932	5868
5165	0	0.0	□	15009	1246

Error of the R_ν-generalized solution

Fig. 2. Error of generalized and R_ν-generalized solutions

Series of calculations results have shown:

(1) the value of error is always decreases as points mesh removes from points of singularity;
(2) if we choose parameters ν and δ near to optimal, the error for founded approximate of the R_ν-generalized solution, as a rule, on two order better, than the error for founded approximate of the generalized solution.

Acknowledgements. The authors would like to acknowledge financial support from Russian Foundation of Basic Research (grant no. 10-01-00060, 11-01-98502_r_east_a), Far-Eastern Branch, Russian Academy of Sciences (12-I-P18-01, 12-III-A-01M-002).

References

1. Assous, F., Ciarlet Jr., P., Segré, J.: Numerical solution to the time-dependent Maxwell equations in two-dimentional singular domains: the singular complement method. J. Comp. Phys. 161, 218–249 (2000)
2. Costabel, M., Dauge, M., Schwab, C.: Exponential convergence of hp-FEM for Maxwell equations with weighted regularization in polygonal domains. Math. Models Meth. Appl. Sci. 15, 575–622 (2005)
3. Arroyo, D., Bespalov, A., Heuer, N.: On the finite element method for elliptic problems with degenerate and singular coefficients. Math. Comp. 76, 509–537 (2007)
4. Li, H., Nistor, V.: Analysis of a modified Schrödinger operator in 2D: Regularity, index, and FEM. J. Comp. Appl. Math. 224, 320–338 (2009)
5. Rukavishnikov, V.A., Mosolapov, A.O.: New numerical method for solving time-harmonic Maxwell equations with strong singularity. Journal of Computational Physics 231, 2438–2448 (2012)
6. Rukavishnikov, V.A.: On a weighted estimate of the rate of convergence of difference schemes. Sov. Math. Dokl. 22, 826–829 (1986)
7. Rukavishnikov, V.A.: On differentiability properties of an R_ν-generalized solution of Dirichlet problem. Sov. Math. Dokl. 40, 653–655 (1990)
8. Rukavishnikov, V.A.: On the Dirichlet problem for the second order elliptic equation with noncoordinated degeneration of input data. Differ. Equ. 32, 406–412 (1996)
9. Rukavishnikov, V.A.: On the uniqueness of R_ν-generalized solution for boundary value problem with non-coordinated degeneration of the input data. Dokl. Math 63, 68–70 (2001)
10. Rukavishnikov, V.A., Ereklintsev, A.G.: On the coercivity of the R_ν-generalized solution of the first boundary value problem with coordinated degeneration of the input data. Differ. Equ. 41, 1757–1767 (2005)
11. Rukavishnikov, V.A., Kuznetsova, E.V.: Coercive estimate for a boundary value problem with noncoordinated degeneration of the data. Differ. Equ. 43, 550–560 (2007)
12. Rukavishnikov, V.A., Kuznetsova, E.V.: The R_ν-generalized solution of a boundary value problem with a singularity belongs to the space $W_{2,\nu+\beta/2+k+1}^{k+2}(\Omega,\delta)$. Differ. Equ. 45, 913–917 (2009)
13. Rukavishnikov, V.A., Rukavishnikova, E.I.: The finite element method for the first boundary value problem with compatible degeneracy of the input data. Russ. Acad. Sci., Dokl. Math. 50, 335–339 (1995)
14. Rukavishnikov, V.A., Kuznetsova, E.V.: A finite element method scheme for boundary value problems with noncoordinated degeneration of input data. Numer. Anal. Appl. 2, 250–259 (2009)
15. Rukavishnikov, V.A., Rukavishnikova, H.I.: The finite element method for a boundary value problem with strong singularity. J. Comp. Appl. Math. 234, 2870–2882 (2010)
16. Rukavishnikov, V.A.: On differential properties R_ν-generalized solution of the Dirichlet problem with coordinated degeneration of the input data. ISRN Mathematical Analysis, Article ID 243724, 18 p. (2011), doi: 10.5402/2011/243724
17. Rukavishnikov, V.A., Rukavishnikova, H.I.: The numerical method for boundary value problem with double singularity. Inform. and Control Systems 17, 47–52 (2008)

Some New Approaches to Solving Navier-Stokes Equations for Viscous Heat-Conducting Gas

Vladimir V. Shaydurov[1,2], Galina I. Shchepanovskaya[1],
and Maxim V. Yakubovich[1]

[1] Institute of Computational Modeling of Siberian Branch
of Russian Academy of Science, Krasnoyarsk, Russia
{shidurov,gi,yakubovich}@icm.krasn.ru
[2] Beihang University, Beijing, China
shaidurov04@gmail.com

Abstract. The algorithm for numerical solving the Navier-Stokes equations for two-dimensional motion for viscous heat-conducting gas is proposed. The discretization of equations is performed by a combination of a special version of the trajectory method for a substantial derivative and the finite element method with piecewise bilinear basis functions for other terms. The results of numerical studies of the structure of a supersonic flow in a plane channel with an obstacle for a wide range of Mach numbers and Reynolds numbers are presented. Velocity and pressure fields are investigated, and the vortex structure of flow is studied in the circulation area of the obstacle.

Keywords: Navier-Stokes equations, viscous heat-conducting gas, numerical modeling, trajectories, finite element method.

1 Introduction

Gas flow in channels with obstacles is encountered in many technical devices and installations. Although many numerical algorithms and special software complexes have been developed (see [1–6] and references therein), the problem of development and application of efficient numerical methods and algorithms remains open.

In this paper, to approximate a total (substantial or Lagrangian) time derivative in each equation the method of trajectories is used. It consists in the approximation of this derivative by the backward difference time derivative along the particle trajectory. This method (called the method of characteristics) first has been proposed in [7] for the equation of mass transfer. Later, it was called the modified method of characteristics and has been repeatedly used to solve a parabolic equation (see [8] and references therein). Since a parabolic equation has no characteristic and in gas dynamics characteristics are refereed to other objects, we use a more appropriate name "method of trajectories". Space discretization of the remaining terms of the Navier-Stokes equations at each time step is performed by the finite element method with piecewise bilinear basis functions and the use of simple quadrature formulas.

I. Dimov, I. Faragó, and L. Vulkov (Eds.): NAA 2012, LNCS 8236, pp. 122–131, 2013.
© Springer-Verlag Berlin Heidelberg 2013

2   Formulation of the Problem and the Original Equations

Consider a two-dimensional laminar flow of gas in a plane channel with an obstacle under a supersonic flow velocity at the inlet. The shape of the computational domain is shown in Fig. 1. The point A is assumed to be the origin. Besides, h_c is the width of the channel, b and c are the width and the length of the obstacle at the center of the channel inlet, respectively. To describe the motion of gas, we use the time-dependent Navier-Stokes equations without any simplifying assumption. To introduce dimensionless variables, the width h_c of the channel is taken as the length scale unit; the density ρ_∞ of the incoming flow is taken as the density scale unit; flow velocity u_∞ at the channel inlet is taken as the velocity scale unit; h_c/u_∞ is taken as the time scale unit; the scale units of pressure, temperature and internal energy values are taken from the condition of a perfect gas.

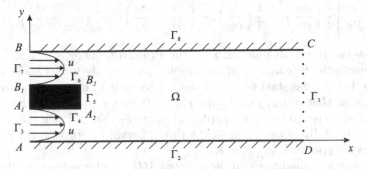

Fig. 1. The channel with an obstacle at the inlet

We write the differential equations of two-dimensional viscous heat-conducting gas in the form of dimensionless equations of continuity, momentum, and internal energy

$$\frac{d\rho}{dt} + \rho\frac{\partial u}{\partial x} + \rho\frac{\partial v}{\partial y} = 0 , \tag{1}$$

$$\rho\frac{du}{dt} = -\frac{\partial P}{\partial x} + \frac{\partial \tau_{xx}}{\partial x} + \frac{\partial \tau_{xy}}{\partial y} , \tag{2}$$

$$\rho\frac{dv}{dt} = -\frac{\partial P}{\partial y} + \frac{\partial \tau_{xy}}{\partial x} + \frac{\partial \tau_{yy}}{\partial y} , \tag{3}$$

$$\rho\frac{de}{dt} + P\left(\frac{\partial u}{\partial x} + \frac{\partial v}{\partial y}\right) = -\frac{\partial q_x}{\partial x} - \frac{\partial q_y}{\partial y} + \Phi . \tag{4}$$

Here $d(\cdot)/dt$ is a substantial or total derivative:

$$\frac{d\rho}{dt} = \frac{\partial \rho}{\partial t} + u\frac{\partial \rho}{\partial x} + v\frac{\partial \rho}{\partial y} ; \tag{5}$$

ρ is density, u and v are the projections of the velocity vector on the axes x and y respectively; $P = (\gamma - 1)\rho e$ is pressure; $\mu = (\gamma(\gamma - 1)M^2 e)^\omega$ is dynamic viscosity; e is internal energy. The components τ_{xx}, τ_{yy}, τ_{xy} of the stress tensor, projections q_x, q_y of the heat flux and the dissipation function Φ are expressed as follows:

$$\tau_{xx} = \frac{2}{3\mathrm{Re}}\mu\left(2\frac{\partial u}{\partial x} - \frac{\partial v}{\partial y}\right) \ , \quad \tau_{yy} = \frac{2}{3\mathrm{Re}}\mu\left(2\frac{\partial v}{\partial y} - \frac{\partial u}{\partial x}\right) \ ,$$

$$\tau_{xy} = \tau_{yx} = \frac{\mu}{\mathrm{Re}}\left(\frac{\partial u}{\partial y} + \frac{\partial v}{\partial x}\right) \ ,$$

$$q_x = -\frac{\gamma}{\mathrm{Pr}\,\mathrm{Re}}\mu\frac{\partial e}{\partial x} \ , \quad q_y = -\frac{\gamma}{\mathrm{Pr}\,\mathrm{Re}}\mu\frac{\partial e}{\partial y} \ , \tag{6}$$

$$\Phi = \frac{\mu}{\mathrm{Re}}\left[\frac{2}{3}\left(\frac{\partial u}{\partial x}\right)^2 + \frac{2}{3}\left(\frac{\partial v}{\partial y}\right)^2 + \left(\frac{\partial v}{\partial x} + \frac{\partial u}{\partial y}\right)^2 + \frac{2}{3}\left(\frac{\partial u}{\partial x} - \frac{\partial v}{\partial y}\right)^2\right] \ ,$$

where Re is the Reynolds number, Pr is the Prandtl number, $\gamma = 1.4$.

To complete the formulation of the problem, we specify initial and boundary conditions. Let the gas start to move from left to right from the rest state inside the domain, so that $\rho(0, x, y) = 1$, $u(0, x, y) = 0$, $v(0, x, y) = 0$. Internal energy of a perfect gas is set to be equal to $e(0, x, y) = (\gamma(\gamma-1)M^2)^{-1}$. On the boundary areas AA_1 and BB_1 of the input in the time interval $t \in (0, t_{\mathrm{fin}})$ the following parameters are specified: $\rho|_{AA_1} = 1$, $e|_{AA_1} = (\gamma(\gamma-1)M^2)^{-1}$, $v|_{AA_1} = 0$ as well as the symmetrical conditions on the segment BB_1. As for velocity u, its profile $u(t, 0, y)$ in the inlet section is set in the following way:

$$u(t, 0, y) = \begin{cases} (2a - y)y/a^2, & y \in (0, a), \\ 1, & y \in [a, d - a], \\ (d - 2a - y)(y - d)/a^2, & y \in (d - a, d) \ , \end{cases} \tag{7}$$

where $d = (h_c - b)/2$ is the length of the segment AA_1, and a is a free parameter, which in subsequent calculations is taken equal to $1/20$. In the area BB_1 the velocity profile $u(t, 0, y)$ is chosen symmetrically. The selected profile is designed to ensure continuity of $u(t, x, y)$ at the points A, A_1, B and B_1. Otherwise, not only there is no convergence, but also parasitic oscillations manifested due to the difference differentiation with respect to the space variables in the vicinity of these points are observed. As for the jump between zero initial conditions and the values in (7) for $t > 0$, the monotonic approximation of the time derivative being used leads to rapidly smoothing the gap with time.

On fixed solid walls, the slip conditions $u|_{\Gamma_s} = 0$ and $v|_{\Gamma_s} = 0$ and the condition of thermal insulation, i.e., the condition $\partial e/\partial n|_{\Gamma_s} = 0$ for the derivative of internal energy along the normal to a solid wall, where $\Gamma_s = \Gamma_2 \cup \Gamma_4 \cup \Gamma_5 \cup \Gamma_6 \cup \Gamma_8$ is solid boundary, are imposed. At the exit of the channel cross section CD, the functions u, v, e are assumed to satisfy the zero Neumann conditions.

3   The Reduction of the Original Equations

We rearrange the equations (1) and (4) to a new form. To do this, given non-negativity of density and internal energy, we introduce the functions

$$\rho = \sigma^2 \,, \quad e = \varepsilon^2 \,. \tag{8}$$

Substituting (8) in the equation of continuity (1) and dividing it by 2σ, we get

$$\frac{d\sigma}{dt} + \frac{1}{2}\sigma\left(\frac{\partial u}{\partial x} + \frac{\partial v}{\partial y}\right) = 0 \,. \tag{9}$$

Repeat the procedure for the internal energy equation (4). To do this, substitute (8) in (4) and divide it by 2ε. This gives

$$\rho\frac{d\varepsilon}{dt} + \frac{1}{2\varepsilon}\frac{\partial q_x}{\partial x} + \frac{1}{2\varepsilon}\frac{\partial q_y}{\partial y} = -\frac{P}{2\varepsilon}\left(\frac{\partial u}{\partial x} + \frac{\partial v}{\partial y}\right) + \frac{1}{2\varepsilon}\Phi \,. \tag{10}$$

We use (8) also in the expressions for the heat flux q_x, q_y from (6). Taking derivatives with respect to x and y, we arrive at

$$q_x = -\frac{2\gamma}{\Pr\text{Re}}\mu\varepsilon\frac{\partial\varepsilon}{\partial x} \,, \quad q_y = -\frac{2\gamma}{\Pr\text{Re}}\mu\varepsilon\frac{\partial\varepsilon}{\partial y} \,, \tag{11}$$

$$\frac{\partial q_x}{\partial x} = -\frac{2\gamma}{\Pr\text{Re}}\left(\mu\left(\frac{\partial\varepsilon}{\partial x}\right)^2 + \varepsilon\frac{\partial}{\partial x}\left(\mu\frac{\partial\varepsilon}{\partial x}\right)\right) \,,$$

$$\frac{\partial q_y}{\partial y} = -\frac{2\gamma}{\Pr\text{Re}}\left(\mu\left(\frac{\partial\varepsilon}{\partial y}\right)^2 + \varepsilon\frac{\partial}{\partial y}\left(\mu\frac{\partial\varepsilon}{\partial y}\right)\right) \,. \tag{12}$$

In view of (11), (12), and the expressions for the dissipation function Φ from (6), the equation (10) takes the form:

$$\rho\frac{d\varepsilon}{dt} - \frac{\gamma}{\Pr\text{Re}}\left(\frac{\mu}{\varepsilon}\left(\frac{\partial\varepsilon}{\partial x}\right)^2 + \frac{\partial}{\partial x}\left(\mu\frac{\partial\varepsilon}{\partial x}\right)\right) -$$

$$\frac{\gamma}{\Pr\text{Re}}\left(\frac{\mu}{\varepsilon}\left(\frac{\partial\varepsilon}{\partial y}\right)^2 + \frac{\partial}{\partial y}\left(\mu\frac{\partial\varepsilon}{\partial y}\right)\right) = -\frac{P}{2\varepsilon}\left(\frac{\partial u}{\partial x} + \frac{\partial v}{\partial y}\right) +$$

$$\frac{1}{2\text{Re}\,\varepsilon}\frac{\mu}{\varepsilon}\left[\frac{2}{3}\left(\frac{\partial u}{\partial x}\right)^2 + \frac{2}{3}\left(\frac{\partial v}{\partial y}\right)^2 + \left(\frac{\partial v}{\partial x} + \frac{\partial u}{\partial y}\right)^2 + \frac{2}{3}\left(\frac{\partial u}{\partial x} - \frac{\partial v}{\partial y}\right)^2\right] \,. \tag{13}$$

So, we shall solve the system of the equations (9), (2), (3), (13).

4   The Method of Trajectories

As the domain of the problem, we take the polygon Ω bounded by the broken line $AA_1A_2B_2B_1BCDA$ with the boundary Γ, consisting of eight segments:

$$\begin{aligned}
\Gamma_1 &= \{(x,y): \quad x = 5.0, \qquad y \in (0.0, 1.0)\}; \\
\Gamma_2 &= \{(x,y): \quad x \in [0.0, 5.0], \quad y = 0.0\}; \\
\Gamma_3 &= \{(x,y): \quad x = 0.0, \qquad y \in (0.0, 0.4)\}; \\
\Gamma_4 &= \{(x,y): \quad x \in [0.0, 0.6], \quad y = 0.4\}; \\
\Gamma_5 &= \{(x,y): \quad x = 0.6, \qquad y \in (0.4, 0.6)\}; \\
\Gamma_6 &= \{(x,y): \quad x \in [0.0, 0.6], \quad y = 0.6\}; \\
\Gamma_7 &= \{(x,y): \quad x = 0.0, \qquad y \in (0.6, 1.0)\}; \\
\Gamma_8 &= \{(x,y): \quad x \in [0.0, 5.0], \quad y = 1.0\}.
\end{aligned}$$

For simplicity sake we take a uniform square grid in space with coordinates $x_i = ih$, $y_j = jh$, $i = 0, 1, ..., n+1$, $j = 0, 1, ..., n_1 + 1$, with meshsize $h = 1/n_1$, entirely fitted in the horizontal and vertical directions of the polygon Ω. This grid subdivides the computational domain $\overline{\Omega}$ into square cells $\omega_{i,j} = (x_i, x_{i+1}) \times (y_j, y_{j+1})$. We denote the set of nodes of the grid in the rectangle $BCDA$ by $S_h = \{s_{i,j} = (x_i, y_j) : i = 0, 1, ..., n, \ j = 0, 1, ..., n_1\}$ and introduce the grid domain $\overline{\Omega}_h = S_h \cap \overline{\Omega}$. We denote the set of "computational nodes" by $\Omega_h = S_h \cap (\Omega \cup \Gamma_1)$, and the set of boundary nodes with 'known values' of the velocity components by $\Gamma_h^D = \overline{\Omega}_h \cap (\Gamma \setminus \Gamma_1)$. We also denote two portions of the grid boundary by $\Gamma_h^{out} = \overline{\Omega}_h \cap \Gamma_1$ and $\Gamma_h^{in} = \overline{\Omega}_h \cap (\Gamma_3 \cup \Gamma_7)$.

To approximate the substantial derivatives in each equation of the system (9), (2), (3), and (13), we use the method of trajectories which is in the approximation of this derivative with the backward difference time derivative along the trajectory defined by (1) [6]. For this purpose, we introduce a uniform grid with time step $\tau = t_{\text{fin}}/m$:

$$\overline{\omega}_\tau = \{t_k: \quad t_k = k\tau, \quad k = 0, ..., m\} \ .$$

For an arbitrary function $\varphi(t, x, y)$ we use the notations $\varphi^k(x, y) = \varphi(t_k, x, y)$ and $\varphi_{i,j}^k = \varphi(t_k, x_i, y_j)$.

Thus, the substantial derivative in the equation (9) is replaced by the difference derivative of first-order consistency:

$$\left. \frac{d\sigma_{i,j}}{dt} \right|_{t_{k+1}} \approx \frac{\sigma_{i,j}^{k+1} - \sigma^k(X_i^k, Y_j^k)}{\tau} \ ,$$

where $X_i^k = x(t_k)$, $Y_j^k = y(t_k)$ are the coordinates of the trajectory at the instant of time $t = t_k$, which passes through the node (x_i, y_j) for $t = t_{k+1}$. In principle, to determine (X_i^k, Y_j^k) it is required to solve the following problem on this trajectory on the interval $t \in [t_k, t_{k+1}]$ in backward time:

$$\begin{cases} \dfrac{dx}{dt} = u\left(t, x(t), y(t)\right), \\[2mm] \dfrac{dy}{dt} = v\left(t, x(t), y(t)\right), \end{cases} \qquad \begin{cases} x(t_{k+1}) = x_i, \\[2mm] y(t_{k+1}) = y_j. \end{cases}$$

Instead, we implement one time step of the explicit Euler scheme (being first-order consistency as well). As a result, we obtain the approximate values

$$X_i^k \approx \bar{X}_i^k = x_i - \tau u_{i,j}^k \text{ and } Y_j^k \approx \bar{Y}_j^k = y_j - \tau v_{i,j}^k .$$

It is clear that in general case the coordinates $\bar{X}_i^k$, $\bar{Y}_j^k$ do not fall at a node. Therefore, the value of $\sigma^k(\bar{X}_i^k, \bar{Y}_j^k)$ is defined by the linear interpolation:

$$\sigma^k(\bar{X}_i^k, \bar{Y}_j^k) = \sigma_{i,j}^k + \frac{\sigma^k(\bar{x}, y_j) - \sigma_{i,j}^k}{\bar{x} - x_i}(\bar{X}_i^k - x_i) + \frac{\sigma^k(x_i, \bar{y}) - \sigma_{i,j}^k}{\bar{y} - y_j}(\bar{Y}_j^k - y_j) =$$

$$\sigma_{i,j}^k - \tau u_{i,j}^k \frac{\sigma^k(\bar{x}, y_j) - \sigma_{i,j}^k}{\bar{x} - x_i} - \tau v_{i,j}^k \frac{\sigma^k(x_i, \bar{y}) - \sigma_{i,j}^k}{\bar{y} - y_j} . \tag{14}$$

Coordinates $\bar{x}$ and $\bar{y}$ are chosen to provide monotonicity of the difference approximation:

$$\bar{x} = \begin{cases} x_{i-1}, & \text{if} \quad u_{i,j}^k \geq 0, \\ x_{i+1} & \text{otherwise,} \end{cases} \qquad \bar{y} = \begin{cases} y_{j-1}, & \text{if} \quad v_{i,j}^k \geq 0, \\ y_{j+1} & \text{otherwise.} \end{cases}$$

As a result, monotonicity (non-positiveness of off-diagonal elements) is achieved under the condition

$$\tau \leq h/(|u_{i,j}^k| + |v_{i,j}^k|) \quad \text{for all nodes of } \overline{\Omega}_h = S_h \cap \overline{\Omega} . \tag{15}$$

Thus, the substantial derivatives in the equations (9), (2) can be approximated as follows:

$$\frac{d\sigma_{i,j}}{dt}\bigg|_{t_{k+1}} \approx \frac{\sigma_{i,j}^{k+1} - \sigma^k(\bar{X}_i^k, \bar{Y}_j^k)}{\tau} , \qquad \rho_{i,j}\frac{du_{i,j}}{dt}\bigg|_{t_{k+1}} \approx \rho_{i,j}^{k+1}\frac{u_{i,j}^{k+1} - u^k(\bar{X}_i^k, \bar{Y}_j^k)}{\tau} .$$

For $v_{i,j}^{k+1}$ and $\varepsilon_{i,j}^{k+1}$ we have the similar expressions. The values of $u^k(\bar{X}_i^k, \bar{Y}_j^k)$, $v^k(\bar{X}_i^k, \bar{Y}_j^k)$, and $\varepsilon^k(\bar{X}_i^k, \bar{Y}_j^k)$ are calculated by the linear interpolation formula similar to (14).

5   The Finite Element Method

In principle, after the approximation of a substantial derivative at each time step $t = t_{k+1}, k = 0, ..., m - 1$, in $\Omega \cup \Gamma_1$ we obtain the equations

$$\frac{\sigma}{\tau} + \frac{1}{2}\sigma\left(\frac{\partial u}{\partial x} + \frac{\partial v}{\partial y}\right) = f_1 , \tag{16}$$

$$\frac{\rho u}{\tau} = -\frac{\partial P}{\partial x} + \frac{\partial \tau_{xx}}{\partial x} + \frac{\partial \tau_{xy}}{\partial y} + f_2 , \tag{17}$$

$$\frac{\rho v}{\tau} = -\frac{\partial P}{\partial y} + \frac{\partial \tau_{xy}}{\partial x} + \frac{\partial \tau_{yy}}{\partial y} + f_3 , \tag{18}$$

$$\frac{\rho\varepsilon}{\tau} - \frac{\gamma}{\Pr\operatorname{Re}}\left(\frac{\mu}{\varepsilon}\left(\frac{\partial\varepsilon}{\partial x}\right)^2 + \frac{\partial}{\partial x}\left(\mu\frac{\partial\varepsilon}{\partial x}\right)\right) - \frac{\gamma}{\Pr\operatorname{Re}}\left(\frac{\mu}{\varepsilon}\left(\frac{\partial\varepsilon}{\partial y}\right)^2 + \frac{\partial}{\partial y}\left(\mu\frac{\partial\varepsilon}{\partial y}\right)\right) =$$

$$f_4 - \frac{P}{2\varepsilon}\left(\frac{\partial u}{\partial x} + \frac{\partial v}{\partial y}\right) + \frac{1}{2\operatorname{Re}}\frac{\mu}{\varepsilon}\left[\frac{2}{3}\left(\frac{\partial u}{\partial x}\right)^2 + \frac{2}{3}\left(\frac{\partial v}{\partial y}\right)^2 + \right.$$

$$\left.\left(\frac{\partial v}{\partial x} + \frac{\partial u}{\partial y}\right)^2 + \frac{2}{3}\left(\frac{\partial u}{\partial x} - \frac{\partial v}{\partial y}\right)^2\right]. \tag{19}$$

with right-hand sides f_1, f_2, f_3, f_4, which involve the terms known from the previous time layer.

Generally speaking, in the previous section the approximations in an explicit form are constructed for nodal points for the whole domain $\Omega \cup \Gamma_1$. In fact, in the finite element method after the use of quadrature formulas there is no need in approximations at other points.

For each node $s_{i,j} \in \overline{\Omega}_h$ we introduce the basis function

$$\varphi_{i,j}(x,y) = \begin{cases} (1 - |x_i - x|/h)\,(1 - |y_j - y|/h), \\[2mm] (x,y) \in ([x_{i-1}, x_{i+1}] \times [y_{j-1}, y_{j+1}]) \cap \overline{\Omega}, \\[2mm] 0 \quad \text{otherwise}, \end{cases}$$

which is equal to unity at $s_{i,j}$ and zero at all other nodes $\overline{\Omega}_h$. We seek an approximate solution as follows:

$$\sigma^h(t,x,y) = \sum_{s_{i,j}\in\overline{\Omega}_h} \sigma_{i,j}(t)\varphi_{i,j}(x,y)\,, \quad u^h(t,x,y) = \sum_{s_{i,j}\in\overline{\Omega}_h} u_{i,j}(t)\varphi_{i,j}(x,y)\,,$$

$$v^h(t,x,y) = \sum_{s_{i,j}\in\overline{\Omega}_h} v_{i,j}(t)\varphi_{i,j}(x,y)\,, \quad \varepsilon^h(t,x,y) = \sum_{s_{i,j}\in\overline{\Omega}_h} \varepsilon_{i,j}(t)\varphi_{i,j}(x,y)\,.$$

The values of σ^h, u^h, and v^h on Γ_h^D are known, and the values of e^h are known only on Γ_h^{in}. This follows from the fact that the (natural) Neumann boundary conditions in contrast to the (main) Dirichlet conditions do not eliminate the degrees of freedom at the corresponding nodes on the boundary.

After the standard application of the finite element method (Bubnov-Galerkin method), we apply the trapezoidal quadrature formula to calculate integrals on the segments and its Cartesian product for the integrals in the cells $\omega_{i,j} = (x_i, x_{i+1}) \times (y_j, y_{j+1})$. As a result, at the interior nodes of the computational domain $S_h \cap \Omega$ we have the following grid analogue of the continuity equation (from here on, for all the functions the superscript $k+1$ characterizing the time dependence is omitted):

$$\frac{\sigma_{i,j}}{\tau} + \frac{1}{4h}\sigma_{i,j}(u_{i+1,j} - u_{i-1,j}) + \frac{1}{4h}\sigma_{i,j}(v_{i,j+1} - v_{i,j-1}) = f_1 \;\forall\, s_{i,j} \in S_h \cap \Omega\,.$$

At some boundary nodes, the grid equations for u^h, v^h, ε^h are simplified due to boundary conditions or due to the smaller support of test functions, for example, on the border Γ_h^{out}. This results is a wide variety of such equations omitted here. We confine ourselves to the equations at interior nodes, which give an adequate idea of the form of the obtained difference equations. Therefore, for u we give the grid analogue of the momentum equation (14) also only at the interior nodes $S_h \cap \Omega$:

$$\frac{\rho_{i,j} u_{i,j}}{\tau} + \frac{2}{3h^2 \mathrm{Re}} \left((u_{i,j} - u_{i-1,j})(\mu_{i-1,j} + \mu_{i,j}) - (u_{i+1,j} - u_{i,j})(\mu_{i,j} + \mu_{i+1,j}) \right) +$$

$$\frac{1}{6h^2 \mathrm{Re}} \left((v_{i+1,j+1} - v_{i+1,j-1})\mu_{i+1,j} - (v_{i-1,j+1} - v_{i-1,j-1})\mu_{i-1,j} \right) +$$

$$\frac{1}{2h^2 \mathrm{Re}} \left((u_{i,j} - u_{i,j-1})(\mu_{i,j-1} + \mu_{i,j}) - (u_{i,j+1} - u_{i,j})(\mu_{i,j} + \mu_{i,j+1}) \right) +$$

$$\frac{1}{4h^2 \mathrm{Re}} \left((v_{i+1,j-1} - v_{i-1,j-1})\mu_{i,j-1} - (v_{i+1,j+1} - v_{i-1,j+1})\mu_{i,j+1} \right) =$$

$$f_2 - \frac{1}{2h} \left(P_{i+1,j} - P_{i-1,j} \right) \quad \forall s_{i,j} \in S_h \cap \Omega \ .$$

At the nodes of the grid boundary Γ_h^{out}, due to the Neumann condition, these equations are simplified, and the grid boundary conditions on Γ_h^D are derived from the boundary conditions of the original differential problem.

Grid analogues for the velocity v components and energy ε^h at the interior nodes $S_h \cap \Omega$ are obtained similarly.

Thus, we obtain the variational-difference scheme of first-order consistency, both in time and space. To solve systems of linear algebraic equations at each time step, we use the pointwise Jacobi method. The convergence of this method and of nonlinearity iteration is much faster when the quadratic extrapolation in time of the values from two time layers is used as an initial guess rather than a simple transfer of values from the previous layer. In view of the significant diagonal dominance, an average number of iteration steps required for convergence of the Jacobi method on a grid of 1001×201 units was not greater than 10.

6   Calculation of Flow in the Channel Bottom Beyond a Rectangular Obstacle

The algorithm is implemented for the above problem of a gas flow for supersonic velocity at the inlet. As the equations (5), we used in the calculations the Sutherland equation of state for a perfect gas:

$$T = \gamma(\gamma - 1)\mathrm{M}^2 e, \ \mu = T^\omega, \ \omega = 0.8 \ .$$

The calculations were performed on a grid consisting of 1001×201 nodes, the mesh size is $h = 0.005$, the time step is $\tau = 0.0005$. The Reynolds Re, Prandtl Pr and Mach M numbers were taken as follows: Re $= 2 \times 10^3, 10^4$, Pr $= 0.7$, M $= 2, 4$.

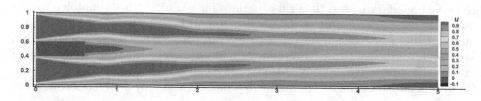

Fig. 2. The distribution of the longitudinal velocity component

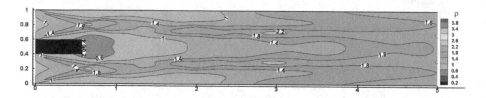

Fig. 3. Density distribution

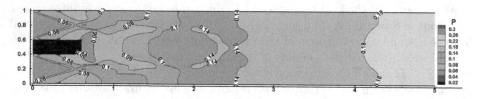

Fig. 4. Pressure distribution

In Fig. 2-4 the flow pattern in the channel behind a rectangular obstacle with the Mach number M = 4 and the Reynolds number Re = 2×10^3 is shown at the instant of time $t = 20.0$.

Figure 2 shows that the body generates stagnant zones with reverse currents and velocity values near zero. Dark color shows an area in which the velocity component u has values in the interval $[0, -0.1)$. As follows from the calculations, stagnant zone beyond the body decreases with time and takes a constant size.

Further, for the same parameters and the instant of time, Fig. 3-4 shows the density and pressure fields. The numerical results show that with time the obstacle generates regions with density less than that in the free stream. Thus, we have constructed a pattern of the bottom of the supersonic flow of a viscous heat-conducting gas with a rectangular obstacle in a channel. Beyond the body the stagnant zone is formed with smaller velocity as compared to the incident flow. The calculations show that an area with negative velocity is formed, which indicates the occurrence of a reverse flow.

7   Conclusion

It is useful to note that the combination of the method of trajectories and the finite element method does not require matching triangulations at the neighboring time layers. This greatly facilitates the dynamic coarsening or condensation of space triangulations for the optimization of computational work or improvements of the approximation of boundary layers and shock waves. For solving systems of algebraic equations due to considerable diagonal dominance the Jacobi method in combination with external nonlinearity iteration is used.

Acknowledgement. This work was supported by RFBR (grant N 11-01-00224) and the International Project TRISTAM.

References

1. ADIGMA - A European Initiative on the Development of Adaptive Higher-Order Variational Methods for Aerospace Applications. Notes on Numerical Fluid Mechanics and Multidisciplinary Design vol. 113. Springer (2010)
2. Oberkampf, W.L., Trucano, T.G.: Verification and validation in computational fluid dynamics. Progress in Aerospace Sciences 38, 209–272 (2002)
3. Shaidurov, V.V., Shchepanovskaya, G.I., Yakubovich, M.V.: Application of the trajectories and finite element method in modeling the motion of a viscous heat-conducting gas. Computational Methods and Programming 12, 275–281 (2011)
4. Bosnyakov, S., Kursakov, I., Lysenkov, A., Matyash, S., Mikhailov, S., Vlasenko, V., Quest, J.: Computational tools for supporting the testing of civil aircraft configurations in wind tunnels. Progress in Aerospace Sciences 44, 67–120 (2008)
5. Vos, J.B., Rizzi, A., Darracq, D., Hirschel, E.H.: Navier-Stokes solvers in European aircraft design. Progress in Aerospace Sciences 38, 601–697 (2002)
6. The Handbook of Fluid Dynamics. CRC Press LLC & Springer (1998)
7. Pironneau, O.: On the Transport-Diffusion Algorithm and Its Applications to the Navier-Stokes Equations. Numerische Mathematik 38, 309–332 (1982)
8. Chen, H., Lin, Q., Shaidurov, V.V., Zhou, J.: Error Estimates for Trangular and Tetrahedral Finite Elements in Combination with a Trajectory Approximation of the First Derivatives for Advection-Diffusion Equations. Numerical Analysis and Applications 4(4), 345–362 (2011)

Flux-Splitting Schemes for Parabolic Problems

Petr N. Vabishchevich

Nuclear Safety Institute, 52, B. Tulskaya, 115191 Moscow, Russia
North-Eastern Federal University, 58, Belinskogo, 677000 Yakutsk, Russia

Abstract. To solve numerically boundary value problems for parabolic
equations with mixed derivatives, the construction of difference schemes
with prescribed quality faces essential difficulties. In parabolic problems,
some possibilities are associated with the transition to a new formulation
of the problem, where the fluxes (derivatives with respect to a spatial
direction) are treated as unknown quantities. In this case, the original
problem is rewritten in the form of a boundary value problem for the system of equations in the fluxes. This work deals with studying schemes
with weights for parabolic equations written in the flux coordinates. Unconditionally stable flux locally one-dimensional schemes of the first and
second order of approximation in time are constructed for parabolic equations without mixed derivatives. A peculiarity of the system of equations
written in flux variables for equations with mixed derivatives is that there
do exist coupled terms with time derivatives.

1 Introduction

Investigating many applied problems, we can consider a second-order parabolic
equation with mixed derivatives as the basic equation. An example is diffusion
processes in anisotropic media. In desining various approximations for the corresponding boundary-value problems, we focus on the inheritance of the primary
properties of the differential problem during the construction of the discrete
problem.

Locally one-dimensional difference schemes are obtained in a simple enough
way for second-order parabolic equations without mixed derivatives [1, 2]. Mixed
derivatives complicate essentially the construction of unconditionally stable schemes of splitting with respect to the spatial variables for parabolic equations with
variable derivatives, even for two-dimensional problems.

In some problems, it is convenient to use the fluxes (derivatives with respect to
a spatial direction) as unknow quantities. This idea may be implemented in the
most simple manner for one-dimensional problems [3]. To introduce fluxes, mixed
and hybrid finite elements are applied [4,5]. The original parabolic equation with
mixed derivatives may be written as a system of equations for the fluxes. The
basic peculiarity of this system is that the time derivatives for the fluxes in
separate equations are interconnected to each other. For the problem in the flux
variables, unconditionally stable schemes with weights are developed. Locally
one-dimensional schemes are proposed for problems without mixed derivatives.

I. Dimov, I. Faragó, and L. Vulkov (Eds.): NAA 2012, LNCS 8236, pp. 132–142, 2013.

2   Differential Problem

In a bounded domain Ω, the unknown function $u(\boldsymbol{x}, t)$, $\boldsymbol{x} = (x_1, x_2, ..., x_m)$, satisfies the equation

$$\frac{\partial u}{\partial t} - \sum_{\alpha, \beta=1}^{m} \frac{\partial}{\partial x_\alpha} \left(k_{\alpha\beta}(\boldsymbol{x}) \frac{\partial u}{\partial x_\beta} \right) = f(\boldsymbol{x}, t), \quad \boldsymbol{x} \in \Omega, \quad 0 < t \le T. \tag{1}$$

Assume that the coefficients $k_{\alpha\beta}$, $\alpha, \beta = 1, 2, ..., m$ satisfy the conditions

$$\underline{k} \sum_{\alpha=1}^{m} \xi_\alpha^2(\boldsymbol{x}) \le \sum_{\alpha, \beta=1}^{m} k_{\alpha\beta}(\boldsymbol{x}) \xi_\alpha(\boldsymbol{x}) \xi_\beta(\boldsymbol{x}) \le \overline{k} \sum_{\alpha=1}^{m} \xi_\alpha^2(\boldsymbol{x}),$$

$$k_{\alpha\beta} = k_{\beta\alpha}, \quad \alpha, \beta = 1, 2, ..., m, \quad \boldsymbol{x} \in \Omega \tag{2}$$

for any $\xi_\alpha(\boldsymbol{x})$, $\alpha = 1, 2, ..., m$ with constant $\underline{k} > 0$. Consider the boundary value problem for equation (1) with homogeneous Dirichlet boundary conditions

$$u(\boldsymbol{x}, t) = 0, \quad \boldsymbol{x} \in \partial\Omega, \quad 0 < t \le T \tag{3}$$

and the initial conditions in the form

$$u(\boldsymbol{x}, 0) = u^0(\boldsymbol{x}), \quad \boldsymbol{x} \in \Omega. \tag{4}$$

We introduce a vector quantity $\boldsymbol{q} = (q_1, q_2, ..., q_m)^T$ (the index T denotes transposition) such that

$$\boldsymbol{q} = -\mathcal{K} \operatorname{grad} u, \tag{5}$$

where $\mathcal{K} = (k_{\alpha\beta})$ is a square matrix $m \times m$ ($\mathcal{K} \in \mathbb{R}^{mm}$) with elements $k_{\alpha\beta}(\boldsymbol{x})$, $\alpha, \beta = 1, 2, ..., m$. Using this notation, equation (1) may be written as

$$\frac{\partial u}{\partial t} + \operatorname{div} \boldsymbol{q} = f, \quad \boldsymbol{x} \in \Omega, \quad 0 < t \le T. \tag{6}$$

We can write the above problem (3)–(5) in the operator form. Scalar functions are considered in the Hilbert space $\mathcal{H} = L_2(\Omega)$ with the scalar product and norm defined by the rules

$$(u, v) = \int_\Omega u(\boldsymbol{x}) v(\boldsymbol{x}) d\boldsymbol{x}, \quad \|u\| = (u, u)^{1/2}.$$

For vector functions, we use the Hilbert space $\mathcal{V} = \mathbf{L}_2(\Omega)$, where

$$(\mathbf{q}, \mathbf{g}) = \sum_{\alpha=1}^{m} \int_\Omega q_\alpha(\boldsymbol{x}) g_\alpha(\boldsymbol{x}) d\boldsymbol{x}, \quad \|\mathbf{q}\| = (\mathbf{q}, \mathbf{q})^{1/2}.$$

Taking into account (2), we can treat the matrix $\mathcal{K}$ as a linear, bounded, self-adjoint, and positive definite operator in $\mathcal{V}$:

$$\mathcal{K} : \mathcal{V} \to \mathcal{V}, \quad \mathcal{K} = \mathcal{K}^*, \quad \underline{k}\mathcal{E} \le \mathcal{K} \le \overline{k}\mathcal{E}, \quad \underline{k} > 0, \tag{7}$$

where $\mathcal{E}$ is the identity operator in $\mathcal{V}$. Suppose $\mathcal{D}u = -\operatorname{grad} u$, i.e.,

$$\mathcal{D} : \mathcal{H} \to \mathcal{V}, \quad \mathcal{D} = \left(-\frac{\partial}{\partial x_1}, -\frac{\partial}{\partial x_2}, \ldots, -\frac{\partial}{\partial x_m} \right)^T. \tag{8}$$

On the set of functions that satisfy the boundary conditions (3), for the gradient and divergence operators, we have

$$\int_\Omega u \operatorname{div} \mathbf{q} \, d\boldsymbol{x} + \int_\Omega \mathbf{q} \operatorname{grad} u \, d\boldsymbol{x} = 0.$$

It follows from this that $\mathcal{D}^* \mathbf{q} = \operatorname{div} \mathbf{q}$, i.e.,

$$\mathcal{D}^* : \mathcal{V} \to \mathcal{H}, \quad \mathcal{D}^* = \left(\frac{\partial}{\partial x_1}, \frac{\partial}{\partial x_2}, \ldots, \frac{\partial}{\partial x_m} \right). \tag{9}$$

In the above notation (7)–(9), from (3)–(5), we obtain the Cauchy problem for the system of operator-differential equations

$$\frac{du}{dt} + \mathcal{D}^* \mathbf{q} = f(t), \quad 0 < t \le T, \tag{10}$$

$$\mathbf{q} = \mathcal{K}\mathcal{D}u, \tag{11}$$

$$u(0) = u^0. \tag{12}$$

For the problem (1)–(4), the following equation corresponds

$$\frac{du}{dt} + \mathcal{D}^* \mathcal{K}\mathcal{D}u = f(t), \quad 0 < t \le T, \tag{13}$$

which is supplemented by the initial condition (12). Taking into account that

$$\frac{d}{dt}\mathcal{K} - \mathcal{K}\frac{d}{dt} = 0,$$

it is possible to eliminate u from the system of equations (10), (11) that gives

$$\mathcal{C}\frac{d\mathbf{q}}{dt} + \mathcal{D}\mathcal{D}^* \mathbf{q} = \mathcal{D}f, \quad \mathcal{C} = \mathcal{K}^{-1}, \quad 0 < t \le T. \tag{14}$$

In view of (11) and (12), we put

$$\mathbf{q}(0) = \mathbf{q}^0 \equiv \mathcal{K}\mathcal{D}u^0. \tag{15}$$

In constructing locally one-dimensional schemes (schemes based on splitting with respect to spatial directions), we focus on the coordinatewise formulation of equations (10), (11), (14) and (14). Let

$$\mathcal{D} = (\mathcal{D}_1, \mathcal{D}_2, \ldots, \mathcal{D}_m)^T, \quad \mathcal{K} = (\mathcal{K}_{\alpha\beta}), \quad \mathcal{C} = (\mathcal{C}_{\alpha\beta}),$$

then the basic system of equations (10), (11) takes the form

$$\frac{du}{dt} + \sum_{\alpha=1}^{m} \mathcal{D}_{\alpha}^{*} q_{\alpha} = f(t), \quad 0 < t \le T, \tag{16}$$

$$q_{\alpha} = \sum_{\beta=1}^{m} \mathcal{K}_{\alpha\beta} \mathcal{D}_{\beta} u, \quad \alpha = 1, 2, ..., m. \tag{17}$$

The equation (13) for u is reduced to

$$\frac{du}{dt} + \sum_{\alpha,\beta=1}^{m} \mathcal{D}_{\alpha}^{*} \mathcal{K}_{\alpha\beta} \mathcal{D}_{\beta} u = f(t), \quad 0 < t \le T. \tag{18}$$

For the flux components (see (14)), we obtain

$$\sum_{\beta=1}^{m} \mathcal{C}_{\alpha\beta} \frac{dq_{\beta}}{dt} + \sum_{\beta=1}^{m} \mathcal{D}_{\alpha} \mathcal{D}_{\beta}^{*} q_{\beta} = \mathcal{D}_{\alpha} f, \quad 0 < t \le T. \tag{19}$$

The equations of the system (19) are connected with each other, and, moreover, the time derivatives are interconnected. The problem (12), (18) seems to be much easier — we have a single equation instead of the system of m equations. Nevertheless, some possibilities to design locally one-dimensional schemes for the system of equations are still there.

Here we present elementary a priori estimates for the solution of the above Cauchy problems for operator-differential equations, which will serve us as a checkpoint in the study of discrete problems. Multiplying equation (13) scalarly in $\mathcal{H}$ by u, we obtain

$$\|u\| \frac{d}{dt} \|u\| + (\mathcal{K} \mathcal{D} u, \mathcal{D} u) = (f, u).$$

Taking into account (7) and

$$(f, u) \le \|f\| \|u\|,$$

we arrive at

$$\frac{d}{dt} \|u\| \le \|f\|.$$

This inequality implies the estimate

$$\|u(t)\| \le \|u^{0}\| + \int_{0}^{t} \|f(\theta)\| d\theta \tag{20}$$

for the solution of the problem 12), (18).

Now we investigate the problem (14), (15). By the properties (7) of the operator $\mathcal{K}$, for $\mathcal{C}$, we have

$$\mathcal{C} : \mathcal{V} \to \mathcal{V}, \quad \mathcal{C} = \mathcal{C}^{*}, \quad \underline{c} \mathcal{E} \le \mathcal{C} \le \overline{c} \mathcal{E}, \quad \underline{c} = \overline{k}^{-1} > 0, \quad \overline{c} = \underline{k}^{-1}. \tag{21}$$

In view of (21), we define the Hilbert space $\mathcal{V}_\mathcal{C}$, where the scalar product and norm are

$$(\mathbf{q}, \mathbf{g})_\mathcal{C} = (\mathcal{C}\mathbf{q}, \mathbf{g}), \quad \|\mathbf{q}\|_\mathcal{C} = (\mathbf{q}, \mathbf{q})_\mathcal{C}^{1/2}.$$

Multiplying equation (15) scalarly in $\mathcal{V}$ by $\mathbf{q}$, we obtain

$$\|\mathbf{q}\|_\mathcal{C} \frac{d}{dt} \|\mathbf{q}\|_\mathcal{C} + (\mathcal{D}^* \mathbf{q}, \mathcal{D}^* \mathbf{q}) = (\mathcal{D}f, \mathbf{q}).$$

In view of

$$(\mathcal{D}f, \mathbf{q}) \le \|\mathcal{D}f\|_\mathcal{K} \|\mathbf{q}\|_\mathcal{C},$$

we arrive at a priori estimate

$$\|\mathbf{q}(t)\|_\mathcal{C} \le \|\mathbf{q}^0\|_\mathcal{C} + \int_0^t \|\mathcal{D}f(\theta)\|_\mathcal{K} \, d\theta \tag{22}$$

for the solution of the problem (14), (15).

3   Approximation in Space

We conduct a detailed analysis using a model two-dimensional parabolic problem in a rectangle

$$\Omega = \{\mathbf{x} \mid \mathbf{x} = (x_1, x_2), \quad 0 < x_\alpha < l_\alpha, \quad \alpha = 1, 2\}.$$

In Ω, we introduce a uniform rectangular grid

$$\overline{\omega} = \{\mathbf{x} \mid \mathbf{x} = (x_1, x_2), \quad x_\alpha = i_\alpha h_\alpha, \quad i_\alpha = 0, 1, ..., N_\alpha, \quad N_\alpha h_\alpha = l_\alpha\}$$

and let ω be the set of interior nodes ($\overline{\omega} = \omega \cup \partial\omega$). On this grid, scalar grid functions are given. For grid functions $y(\mathbf{x}) = 0$, $\mathbf{x} \in \partial\omega$, we define the Hilbert space $H = L_2(\omega)$ with the scalar product and norm

$$(y, w) \equiv \sum_{\mathbf{x} \in \omega} y(\mathbf{x}) w(\mathbf{x}) h_1 h_2, \quad \|y\| \equiv (y, y)^{1/2}.$$

To determine vector grid functions, we have two main possibilities. The first approach deals with specifying vector functions on the same grid as it used for scalar functions. The second possibility, which is traditionally widely used, e.g., in computational fluid dynamics, is based on the grid arrangement, where each individual component of a vector quantity is referred to its own mesh. Here we restrict ourselves to the use of the same grid for all quantities, in particular, for setting the coefficients $k_{\alpha\beta}(\mathbf{x})$, $\alpha, \beta = 1, 2, ..., m$.

Consider approximations for the differential operators

$$\mathcal{L}_{\alpha\beta} u = -\frac{\partial}{\partial x_\alpha} \left(k_{\alpha\beta}(\mathbf{x}) \frac{\partial u}{\partial x_\beta} \right), \quad \alpha, \beta = 1, 2, ..., m.$$

We apply the standard index-free notation from the theory of difference schemes [6] for the difference operators:

$$u_x = \frac{u(x+h) - u(x)}{h}, \quad u_{\bar{x}} = \frac{u(x) - u(x-h)}{h}.$$

If we set the coefficients of the elliptic operator at the grid points, then

$$L_{\alpha\alpha}y = -\frac{1}{2}(k_{\alpha\alpha}u_{x_\alpha})_{\bar{x}_\alpha} - \frac{1}{2}(k_{\alpha\alpha}u_{\bar{x}_\alpha})_{x_\alpha}, \quad \alpha = 1, 2. \tag{23}$$

More opportunities are available in approximation of operators with mixed derivatives. As the basic discretization [6], we emphasize

$$L_{\alpha\beta}^{(1)}y = -\frac{1}{2}(k_{\alpha\beta}u_{x_\alpha})_{\bar{x}_\beta} - \frac{1}{2}(k_{\alpha\beta}u_{\bar{x}_\alpha})_{x_\beta}, \tag{24}$$

$$L_{\alpha\beta}^{(2)}y = -\frac{1}{2}(k_{\alpha\beta}u_{x_\alpha})_{x_\beta} - \frac{1}{2}(k_{\alpha\beta}u_{\bar{x}_\alpha})_{\bar{x}_\beta}, \quad \alpha, \beta = 1, 2, \quad \alpha \neq \beta. \tag{25}$$

Instead of $L_{\alpha\beta}^{(1)}, L_{\alpha\beta}^{(2)}$, we can take their linear combination. In particular, it is possible [7] to put

$$L_{\alpha\beta}^{(3)} = \frac{1}{2}L_{\alpha\beta}^{(1)} + \frac{1}{2}L_{\alpha\beta}^{(2)}, \quad \alpha, \beta = 1, 2, \quad \alpha \neq \beta. \tag{26}$$

In the general case, we set

$$L_{\alpha\beta} = \chi L_{\alpha\beta}^{(1)} + (1 - \chi)L_{\alpha\beta}^{(2)}, \quad \alpha, \beta = 1, 2, \quad \alpha \neq \beta, \quad \chi = \text{const}. \tag{27}$$

The introduced discrete operators approximate the corresponding differential operators with the second order:

$$L_{\alpha\alpha}u = \mathcal{L}_{\alpha\alpha}u + \mathcal{O}(h_\alpha^2), \quad L_{\alpha\beta}u = \mathcal{L}_{\alpha\beta} + \mathcal{O}(h^2), \quad \beta \neq \alpha, \quad \alpha, \beta = 1, 2, \tag{28}$$

where $h^2 = h_1^2 + h_2^2$.

We define a grid subset $\bar{\omega}$, where the corresponding components of vector quantities are defined. Let

$$\omega_1^+ = \{x \mid x_1 = i_1, \ i_1 = 0, 1, ..., N_1 - 1, \ x_2 = i_2 h_2, \ i_2 = 1, 2, ..., N_2 - 1\},$$

$$\omega_1^- = \{x \mid x_1 = i_1, \ i_1 = 1, 2, ..., N_1, \ x_2 = i_2 h_2, \ i_2 = 1, 2, ..., N_2 - 1\},$$

$$\omega_2^+ = \{x \mid x_1 = i_1 h_1, \ i_1 = 1, 2, ..., N_1 - 1, \ x_2 = i_2 h_2, \ i_2 = 0, 1, ..., N_2 - 1\},$$

$$\omega_2^- = \{x \mid x_1 = i_1 h_1, \ i_1 = 1, 2, ..., N_1 - 1, \ x_2 = i_2 h_2, \ i_2 = 1, 2, ..., N_2\},$$

and

$$\tilde{\omega} = \omega_1^+ \cup \omega_1^- \cup \omega_2^+ \cup \omega_2^-.$$

For the grid vector variables, instead of two components, we will use four components, putting

$$\mathbf{q} = (q_1^+, q_1^-, q_2^+, q_2^-)^T, \quad q_\alpha^{\pm} = q_\alpha^{\pm}(x), \quad x \in \omega_\alpha^{\pm}, \quad \alpha = 1, 2.$$

For the grid functions defined on grids $\omega_\alpha^\pm$, $\alpha = 1, 2$, we define the Hilbert spaces $H_\alpha^\pm$, $\alpha = 1, 2$, where

$$(y, w)_\alpha^\pm \equiv \sum_{x \in \omega_\alpha^\pm} y(x) w(x) h_1 h_2, \quad \|y\|_\alpha^\pm \equiv ((y, y)_\alpha^\pm)^{1/2}, \quad \alpha = 1, 2.$$

For the grid vector functions in $V = H_1^+ \oplus H_1^- \oplus H_2^+ \oplus H_2^-$, we set

$$(\mathbf{q}, \mathbf{g}) = \sum_{\alpha=1}^{2} ((q_\alpha^+, g_\alpha^+)_\alpha^+ + (q_\alpha^-, g_\alpha^-)_\alpha^-), \quad \|\mathbf{q}\| = (\mathbf{q}, \mathbf{q})^{1/2}.$$

Now we construct the discrete analogs of differential operators $\mathcal{D}_\alpha$, $\mathcal{D}_\alpha^*$, $\alpha = 1, 2$ introduced according to (8), (9). Using the above difference derivatives in space, we set

$$D_\alpha^+ y = -y_{x_\alpha}, \quad x \in \omega_\alpha^+, \quad \alpha = 1, 2, \tag{29}$$

so that $D_\alpha^+ : H \to H_\alpha^+$, $\alpha = 1, 2$. Similarly, we define $D_\alpha^- : H \to H_\alpha^-$, $\alpha = 1, 2$, where

$$D_\alpha^- y = -y_{\bar{x}_\alpha}, \quad x \in \omega_\alpha^-, \quad \alpha = 1, 2. \tag{30}$$

Thus

$$D : H \to V, \quad D = (D_1^+, D_1^-, D_2^+, D_2^-)^T. \tag{31}$$

For the adjoint operator, we have

$$D^* : V \to H, \quad D^* = ((D_1^+)^*, (D_1^-)^*, (D_2^+)^*, (D_2^-)^*), \tag{32}$$

and

$$(D_\alpha^+)^* : H_\alpha^+ \to H, \quad (D_\alpha^+)q = q_{\bar{x}_\alpha}, \tag{33}$$

$$(D_\alpha^-)^* : H_\alpha^- \to H, \quad (D_\alpha^-)q = q_{x_\alpha}, \quad x \in \omega, \quad \alpha = 1, 2. \tag{34}$$

The above discrete operators approximate the corresponding differential operators with the first order:

$$D_\alpha^\pm u = \mathcal{D}_\alpha u + \mathcal{O}(h_\alpha), \quad (D_\alpha^\pm)^* u = \mathcal{D}_\alpha^* u + \mathcal{O}(h_\alpha), \quad \alpha = 1, 2. \tag{35}$$

For the operator-differential equation (13), we put into the correspondence the equation

$$\frac{dy}{dt} + D^* K D y = \varphi(t), \quad 0 < y \le T, \tag{36}$$

where, e.g., $\varphi(t) = f(x, t)$, $x \in \omega$. For equation (36), we consider the Cauchy problem

$$y(0) = u^0. \tag{37}$$

The construction of the operator K is associated with the approximations (23)–(27). The most important properties are self-adjointness and positive definiteness of the operator K. The equation (36) approximates the differential equation (13) with the second order.

The system of equations (10), (11) is attributed to the system

$$\frac{dy}{dt} + D^*\mathbf{g} = \varphi(t), \quad 0 < t \le T, \tag{38}$$

$$\mathbf{g} = KDy. \tag{39}$$

For the flux problem (14), (15), we put into the correspondence the problem

$$C\frac{d\mathbf{g}}{dt} + DD^*\mathbf{g} = D\varphi(t), \quad C = K^{-1}, \quad 0 < t \le T, \tag{40}$$

$$\mathbf{g}(0) = KDu^0. \tag{41}$$

Similarly to (20), we prove the following estimate for the solution of the problem (36), (37):

$$\|y(t)\| \le \|u^0\| + \int_0^t \|\varphi(\theta)\| d\theta. \tag{42}$$

For the estimate (22), we put into the correspondence the estimate

$$\|\mathbf{g}(t)\|_C \le \|Du^0\|_K + \int_0^t \|D\varphi(\theta)\|_K \, d\theta \tag{43}$$

for the solution of the problem (40), (41).

4   Operator-Difference Schemes

We introduce a uniform grid in time with a step τ and let $y^n = y(t^n)$, $t^n = n\tau$, $n = 0, 1, ..., N$, $N\tau = T$. For numerical solving the problem (36), (37), we apply the standard two-level scheme with weights, where equation (36) is approximated by the scheme

$$\frac{y^{n+1} - y^n}{\tau} + A(\sigma y^{n+1} + (1 - \sigma)y^n) = \varphi^n, \quad n = 0, 1, ..., N - 1, \tag{44}$$

where

$$A = D^*KD, \quad A = A^* > 0 \tag{45}$$

and, e.g., $\varphi^n = f(\sigma t^{n+1} + (1 - \sigma)t^n)$. Taking into account (37), the operator-difference equation (44) is supplemented with the initial condition

$$y^0 = u^0. \tag{46}$$

The truncation error of the difference scheme (44)–(46) is $\mathcal{O}(|h|^2 + \tau^2 + (\sigma - 0.5)\tau)$.

The study of the difference scheme is conducted using the general theory of stability (well-posedness) for operator-difference schemes [6, 8]. Let us formulate a typical result on stability of difference schemes with weights for an evolutionary equation of first order.

Theorem 1. *The scheme (44)–(46) is unconditionally stable for $\sigma \geq 0.5$, and the difference solution satisfies the levelwise estimate*

$$\|y^{n+1}\| \leq \|y^n\| + \tau\|\varphi^n\|, \quad n = 0, 1, ..., N - 1. \tag{47}$$

From (47), in the standard way, we get the desired stability estimate

$$\|y^{n+1}\| \leq \|u^0\| + \sum_{k=0}^{n} \tau\|\varphi^k\|,$$

which may be treated as a direct discrete analogue of the a priori estimate (20) for the solution of the differential problem (12), (18).

Schemes with weights for a system of semi-discrete equations (38), (39) are constructed in a similar way. We put

$$\frac{y^{n+1} - y^n}{\tau} + D^*(\sigma \mathbf{g}^{n+1} + (1 - \sigma)\mathbf{g}^n) = \varphi^n, \quad n = 0, 1, ..., N - 1, \tag{48}$$

$$\mathbf{g}^n = KDy^n, \quad n = 0, 1, ..., N. \tag{49}$$

The scheme (48), (48) is equivalent to the scheme (44). In view of Theorem 1, it is stable under the restriction $\sigma \geq 0.5$, and for the solution of difference problem (45), (48), (48), the a priori estimate (47) holds.

The special consideration should be given to the flux problem (40), (41). To solve it numerically, we apply the scheme

$$C\frac{\mathbf{g}^{n+1} - \mathbf{g}^n}{\tau} + DD^*(\sigma \mathbf{g}^{n+1} + (1 - \sigma)\mathbf{g}^n) = D\varphi^n, \quad n = 0, 1, ..., N - 1, \tag{50}$$

$$\mathbf{g}^0 = KDu^0. \tag{51}$$

Theorem 2. *The difference scheme (50), (51) is unconditionally stable for $\sigma \geq 0.5$, and the difference solution satisfies the estimate*

$$\|\mathbf{g}^{n+1}\|_C \leq \|\mathbf{g}^n\|_C + \tau\|D\varphi^n\|_K, \quad n = 0, 1, ..., N - 1. \tag{52}$$

From (52), it follows the estimate

$$\|\mathbf{g}^{n+1}\|_C \leq \|Du^0\|_K + \sum_{k=0}^{n} \tau\|D\varphi^k\|_K, \quad n = 0, 1, ..., N - 1,$$

which corresponds to the estimate (43) for the solution of the problem (40), (41).

The computational implementation of the unconditionally stable operator-difference schemes (44)–(46) for the parabolic equation (1) with mixed derivatives is based on solving discrete elliptic problems at every time step. For the problem (36), (37), it seems more convenient to employ additive schemes (operator-splitting schemes) that provide the transition to a new time level using simpler problems associated with the inversion of the individual operators

$D_\alpha^* D_\alpha$, $\alpha = 1, 2$ rather then their combinations. By the nature of the operators D_α^*, D_α, $\alpha = 1, 2$, in this case, we speak of locally one-dimensional schemes.

The issues of designing unconditionally stable locally one-dimensional schemes for a parabolic equation without mixed derivatives have been studied in detain. For parabolic equations with mixed derivatives, locally one-dimensional schemes were constructed in several papers (see, e.g., [9, 10]). Strong results on unconditional stability of operator-splitting schemes can be proved only in a uninteresting case with pairwise commutative operators (the equation with constant coefficients). For our problems (1)–(4), the construction of locally one-dimensional schemes requires separate consideration.

Let us investigate approaches to constructing locally one-dimensional schemes for the problem (40), (41). The computational implementation of the scheme with weights (50), (51), which is unconditionally stable for $\sigma \geq 0.5$, is associated with solving the system of difference equations for four components of the vector $\mathbf{g}^{n+1}$. The equations of this system are strongly coupled to each other, and this interconnection does exist not only for the spatial derivatives (operators $D_1^\pm D_2^{\pm *}$, $D_2^\pm D_1^{\pm *}$), but also for the time derivatives ($k_{12} = k_{21} \neq 0$). Thus, we need to resolve the problem of splitting for the operator at the time derivative, too.

The simplest case is splitting of the spatial operator without coupling the time derivatives. Such a technique is directly applicable for the construction of locally one-dimensional schemes for parabolic equations without mixed derivatives, where

$$k_{\alpha\beta}(\mathbf{x}) = k_{\beta\alpha}(\mathbf{x}) = 0, \quad \alpha \neq \beta = 1, 2, ..., m, \quad \mathbf{x} \in \Omega \tag{53}$$

in equation (1).

Assume that

$$R = DD^*, \quad Q = \operatorname{diag}(D_1^+ (D_1^+)^*, D_1^- (D_1^-)^*, D_2^+ (D_2^+)^*, D_2^- (D_2^-)^*),$$

i.e., Q is the diagonal part of R. For numerical solving the problem (40), (41), we employ the difference scheme, where only the diagonal part of R is shifted to the upper time level. In our notation, we set

$$C \frac{\mathbf{g}^{n+1} - \mathbf{g}^n}{\tau} + Q(\sigma \mathbf{g}^{n+1} + (1 - \sigma)\mathbf{g}^n) + (R - Q)\mathbf{g}^n = D\varphi^n, \quad n = 0, 1, ..., N - 1, \tag{54}$$

with the initial conditions according to (51).

Theorem 3. *The difference scheme (51), (54) is unconditionally stable for $\sigma \geq 2$, and the difference solution satisfies the estimate*

$$\|\mathbf{g}^{n+1}\|_B \leq \|\mathbf{g}^n\|_B + \tau \|D\varphi^n\|_{B^{-1}}, \quad n = 0, 1, ..., N - 1, \tag{55}$$

where

$$B = C + \sigma\tau P - \frac{\tau}{2}R.$$

The scheme (51), (54) has the first-order approximation in time. It seems more preferable, in terms of accuracy, to apply the scheme that is based on the triangular decomposition of the self-adjoint matrix operator $\boldsymbol{R}$:

$$\boldsymbol{R} = \boldsymbol{R}_1 + \boldsymbol{R}_2, \quad \boldsymbol{R}_1^* = \boldsymbol{R}_2. \tag{56}$$

For the problem (40), (41), we construct the additive scheme with the splitting (56), where

$$(C + \sigma\tau R_1)C^{-1}(C + \sigma\tau R_2)\frac{\mathbf{g}^{n+1} - \mathbf{g}^n}{\tau} + R\mathbf{g}^n = D\varphi^n, \quad n = 0, 1, ..., N-1. \tag{57}$$

The main result is formulated in the following statement.

Theorem 4. *The difference scheme (51), (56)–(57) is unconditionally stable for $\sigma \geq 0.5$, and the difference solution satisfies the estimate (55) with*

$$B = (C + \sigma\tau R_1)C^{-1}(C + \sigma\tau R_2) - \frac{\tau}{2}R.$$

The alternating triangle operator-difference scheme (51), (56)–(57) belongs to the class of schemes that are based on a pseudo-time evolution process — the solution of the steady-state problem is obtained as a limit of this pseudo-time evolution. It has the second-order accuracy in time if $\sigma = 0.5$, and ony the first order for other values of σ.

References

1. Marchuk, G.I.: Handbook of Numerical Analysis, Splitting and alternating direction methods, vol. I. Elsevier Science Publishers B.V., North-Holland (1990)
2. Samarskii, A.A., Vabishchevich, P.N.: Additive schemes for problems of mathematical physics. Nauka (1999) (in Russian)
3. Degtyarev, L.M., Favorskii, A.P.: A flow variant of the sweep method. USSR Comput. Math. Math. Phys. 8(3), 252–261 (1968)
4. Brezzi, F., Fortin, M.: Mixed and Hybrid Finite Element Methods. Springer, Heidelberg (1991)
5. Roberts, J.E., Thomas, J.M.: Handbook of Numerical Analysis, Mixed and hybrid methods. Amsterdam, vol. II. Elsevier Science Publishers B.V., North-Holland (1991)
6. Samarskii, A.A.: The theory of difference schemes. Marcel Dekker Inc., New York (2001)
7. Matus, P., Rybak, I.: Difference schemes for elliptic equations with mixed derivatives. Computational Methods in Applied Mathematics 4(4), 494–505 (2004)
8. Samarskii, A.A., Matus, P.P., Vabishchevich, P.N.: Difference schemes with operator factors. Kluwer Academic Pub. (2002)
9. McKee, S., Mitchell, A.R.: Alternating direction methods for parabolic equations in two space dimensions with a mixed derivative. The Computer Journal 13(1), 81–86 (1970)
10. Hout, K.J., Mishra, C.: Stability of the modified craig-sneyd scheme for twodimensional convection-diffusion equations with mixed derivative term. Mathematics and Computers in Simulation 81(11), 2540–2548 (2011)

A Fourth-Order Iterative Solver for the Singular Poisson Equation

Stéphane Abide, Xavier Chesneau, and Belkacem Zeghmati

Univ. Perpignan via Domitia,
LAboratoire de Mathématiques et PhySique, EA 4217,
F-66860, Perpignan, France
stephane.abide@univ-perp.fr

Abstract. A compact fourth-order finite difference scheme solver devoted to the singular-Poisson equation is proposed and verified. The solver is based on a mixed formulation: the Poisson equation is splitted into a system of partial differential equations of the first order. This system is then discretized using a fourth-order compact scheme. This leads to a sparse linear system but introduces new variables related to the gradient of an unknow function. The Schur factorization allows us to work on a linear sub-problem for which a conjugated-gradient preconditioned by an algebraic multigrid method is proposed.Numerical results show that the new proposed Poisson solver is efficient while retaining the fourth-order compact accuracy.

1 Introduction

We consider a high-order solution for the Poisson equation:

$$-(\partial_x\partial_x\phi + \partial_y\partial_y\phi + \partial_z\partial_z\phi) = s(x,y,z), \tag{1}$$

on a cubical domain Ω with Neumann's boundary conditions:

$$\partial_n\phi = g(x,y,z) \text{ on } \partial\Omega \tag{2}$$

where $\partial\Omega$ is the boundary of the domain Ω, n is the normal vector to the domain boundary, and the symbol ∂_n indicates the derivative normal to the boundary. This problem arises, in particular, at the solution of the incompressible Navier-Stokes equations. In this framework, the staggered approximation of derivatives is one of essential features of the Poisson equation. Literature on efficient methods of solution of this problem is abundant for second-order finite differences [1]. Several open source toolboxes have been developed. High-order spatial approximations are known for saving computational cost, despite a more complex algorithm than a standard second-order accurate one. Compact finite difference schemes [2] belong to this category, and was widely used to simulate incompressible flows [3-7]. But, the formationa and solution of the pressure-like, or the singular-Poisson, equation still remains a topical issue.

I. Dimov, I. Faragó, and L. Vulkov (Eds.): NAA 2012, LNCS 8236, pp. 143–150, 2013.
© Springer-Verlag Berlin Heidelberg 2013

High-order compact finite differences have been studied for a long time [8]. The abundant literature on this subject make difficult to resume this amount of work. However, several important tendencies could be brought out. First, a substantial proportion of results concerns with developments of compact schemes on regular multi-dimensional grids. The principle idea consists to utilize the governing partial differential equation leading to lower-derivative expressions equivalent to the higher-order truncation error terms [8]. These expressions can then be discretized using the standard centered finite-differences. This approach has shown a high accuracy on the convection/diffusion equations while leading to a solution of sparse linear systems. Extension to three dimensional problems [9], non-uniform mesh [10, 11], variable convection coefficients [12] have also been proposed. In particular, Zhuang and Sun [13] have proposed a high-order fast direct solver for the singular Poisson equation. The spatial discretization is based on the results of [8], and a singular value decomposition have been designed to remove singularity of the problem. The aforementioned works were built on collocated grids. Conversely, staggered grids have shown a better resolution and conservation properties when finite difference schemes are used to approximate the Navier-Stokes equations [6, 7]. On such a grid, the Poisson equation is defined as the combination of two staggered first-order derivatives. Thus, high-order compact finite difference schemes complicate the approximation of the Poisson equation since, by definition, such schemes are implicit [2]. Early works of Schiestel and Viazzo [6] relate an iterative solution method for the Poisson equation, while using a staggered grid and fourth-order compact schemes. This approach relies on a second order discretization of the pressure correction equation, which tends toward zero at the end of iterations. Recently, Knikker [7] has used a similar method. The diagonalization method, which is a direct solver, is retained by Vedy *et al.* [14] and extended by Abide and Viazzo [15] to complex geometries. Brüger *et al.* [4] have proposed an incompressible Navier-Stokes solver for which spatial derivatives are approximated using fourth-order compact schemes. The originality of their work is the iterative solution method for the underlying pressure-like, or the singular Poisson, equation. Due to a preconditioner build from ILU factorization of the second-order discretization of the Poisson equation, they have shown that the condition number behaves as $o(h^{-1})$ (h being the grid space). The present work concerns with the development of an iterative solution method for the Poisson equation using fourth-order compact approximations on a staggered grid. First, finite difference approximations and a formulation of the Poisson equation are presented. Then, the iterative method and the preconditioner are described. Finally, numerical test are perfomed to assess the accuracy and efficiency of the method proposed.

2   Formulation of the Problem

We consider the Poisson equation expressed in a mixed formulation, namely,

$$\begin{aligned} \sigma - \nabla\phi &= 0 \quad \text{in } \Omega, \\ -\nabla \cdot \sigma &= f \quad \text{in } \Omega, \end{aligned} \tag{3}$$

Then, new vectorial variable $\sigma = \nabla\phi$, consistent with a flux vector, is introduced. Thus the Poisson equation results in the application of a conservation principle $-\nabla \cdot \sigma = f$ [16]. In this work, only the Neumann boundary conditions are considered for the Poisson equation, which are written as $\sigma \cdot n = g$. Expressed in a three-dimensional cartesian coordinates system, the mixed formulation of the Poisson equation is

$$
\begin{cases}
\sigma^x - \partial_x\phi & = 0 \\
\sigma^y - \partial_y\phi & = 0 \\
\sigma^z - \partial_z\phi & = 0 \\
-\left(\partial_x\sigma^x + \partial_y\sigma^y + \partial_z\sigma^y\right) & = s
\end{cases}
\tag{4}
$$

The unknowns functions ϕ, σ^x, σ^y and σ^z are distributed on a staggered grid. The primary variable ϕ is defined on the node $(x_{i+1/2}, y_{j+1/2}, z_{k+1/2})$, where each component is defined by $\xi_{p+1/2} = (p + 1/2)\Delta\xi$ with $0 \le p \le n_\xi - 1$. This location corresponds to cell centers of the grid. The flux component σ^ξ is located at the half of a grid spacing away from cell centers and along the direction of ξ. For instance, the component σ^y is located at $(x_{i+1/2}, y_j, z_{k+1/2})$, where $y_j = j\Delta y$ with $0 \le j \le n_y$. This corresponds to the location of y-faces.

A fourth-order compact scheme discretization is used to approximate the variable $\sigma = \nabla\phi$. If we consider a one-dimensional discretization, saying the direction x, the evaluation of the staggered derivative cell-to-face is given by

$$
\frac{1}{24}\sigma_{i-1}^x + \frac{11}{12}\sigma_i^x + \frac{1}{24}\sigma_{i+1}^x = \frac{1}{\Delta x}\left(\phi_{i+1/2} - \phi_{i-1/2}\right),
\tag{5}
$$

For the sake of clarity, the indices of the second and third components are dropped out. Thus, σ_i^x and $\phi_{i+1/2}$ should be read respectively as $\sigma_{i,j+1/2,k+1/2}^x$ and $\phi_{i+1/2,j+1/2,k+1/2}$. Since the Neumann boundary conditions are prescribed, the boundary values σ_0^x and $\sigma_{n_x}^x$ occur for $i = 1$ and $i = n_x - 1$ in Eq. (5). So, the following linear system

$$
A_x^{cf}\sigma^x = B_x^{cf}\phi,
$$

holds for Eq. (5), with A_x^{cf} and B_x^{cf} being tridiagonal and bidiagonal respectively. Extension to a higher dimension of this finite difference scheme is made by tensor products, viz.

$$
\begin{cases}
\left(A_x^{cf} \otimes I_y \otimes I_z\right)\sigma^x - \left(B_x^{cf} \otimes I_y \otimes I_z\right)\phi = 0 \\
\left(I_x \otimes A_y^{cf} \otimes I_z\right)\sigma^y - \left(I_x \otimes B_y^{cf} \otimes I_z\right)\phi = 0 \\
\left(I_x \otimes I_y \otimes A_z^{cf}\right)\sigma^z - \left(I_x \otimes I_y \otimes B_z^{cf}\right)\phi = 0
\end{cases}
\tag{6}
$$

where I_ξ stands for the identity matrix following direction ξ. The next step of the formulation consists in the approximation of the conservation law $-\nabla \cdot \sigma = s$. For inner nodes, the relation is the shift version of Eq. (5)

$$
\frac{1}{24}\partial_x\sigma_{i-1/2}^x + \frac{11}{12}\partial_x\sigma_{i+1/2}^x + \frac{1}{24}\partial_x\sigma_{i+3/2}^x = \frac{1}{\Delta x}\left(\sigma_{i+1}^x - \sigma_i^x\right),
\tag{7}
$$

A special care is needed for the boundary nodes since a discrete form of global conservation is required [2, 17]. Thus, the following boundary relations [17] are considered:

$$\partial_x \sigma_{1/2}^x = \frac{1}{\Delta x} \left(-\frac{23}{24}\sigma_0^x + \frac{7}{8}\sigma_1^x + \frac{1}{8}\sigma_2^x - \frac{1}{24}\sigma_3^x \right) \tag{8}$$

and

$$\frac{1}{12}\partial_x \sigma_{1/2}^x + \frac{5}{6}\partial_x \sigma_{3/2}^x + \frac{1}{12}\partial_x \sigma_{5/2}^x = \frac{1}{\Delta x} \left(-\frac{1}{24}\sigma_0^x - \frac{7}{8}\sigma_1^x + \frac{7}{8}\sigma_2^x + \frac{1}{24}\sigma_3^x \right) \tag{9}$$

The other boundary relations are readily obtained by symmetry considerations. Once again, the extension to a higher dimension is given by

$$- \left((A_x^{fc})^{-1} B_x^{fc} \otimes I_y \otimes I_z \right) \sigma^x - \left(I_x \otimes (A_y^{fc})^{-1} B_y^{fc} \otimes I_z \right) \sigma^y$$
$$- \left(I_x \otimes I_y \otimes (A_z^{fc})^{-1} B_z^{fc} \right) \sigma^z = s \tag{10}$$

where each matrix associated to this linear sytem is full. A sparse form is obtained by mutlipliyng Eq. (10) with $\left(A_x^{fc} \otimes A_y^{fc} \otimes A_z^{fc} \right)$:

$$- \left(B_x^{fc} \otimes A_y^{fc} \otimes A_z^{fc} \right) \sigma^x - \left(A_x^{fc} \otimes B_y^{fc} \otimes A_z^{fc} \right) \sigma^y$$
$$- \left(A_x^{fc} \otimes A_y^{fc} \otimes B_z^{fc} \right) \sigma^z = \left(A_x^{fc} \otimes A_y^{fc} \otimes A_z^{fc} \right) s \tag{11}$$

Equations (6) and (11) form a block linear system for the unknown fluxes $(\sigma^x, \sigma^y, \sigma^z, \phi)^t$, which is given by:

$$\begin{pmatrix} F_x & 0 & 0 & G_x \\ 0 & F_y & 0 & G_y \\ 0 & 0 & F_z & G_z \\ D_x & D_y & D_z & 0 \end{pmatrix} \begin{pmatrix} \sigma^x \\ \sigma^y \\ \sigma^z \\ \phi \end{pmatrix} = \begin{pmatrix} F & G \\ D & 0 \end{pmatrix} \begin{pmatrix} \sigma \\ \phi \end{pmatrix} = \begin{pmatrix} 0 \\ \tilde{s} \end{pmatrix} \tag{12}$$

where the blocks are defined by

$$\begin{array}{lll} F_x = A_x^{cf} \otimes I_y \otimes I_z, & G_x = B_x^{cf} \otimes I_y \otimes I_z, & D_x = B_x^{fc} \otimes A_y^{fc} \otimes A_z^{fc} \\ F_y = I_x \otimes A_y^{cf} \otimes I_z, & G_y = I_x \otimes B_y^{cf} \otimes I_z, & D_y = A_x^{fc} \otimes B_y^{fc} \otimes A_z^{fc} \\ F_z = I_x \otimes I_y \otimes A_z^{cf}, & G_z = I_x \otimes I_y \otimes B_z^{cf}, & D_z = A_x^{fc} \otimes A_y^{fc} \otimes B_z^{fc} \end{array} \tag{13}$$

and the modified source term is $\tilde{s} = \left(A_x^{fc} \otimes A_y^{fc} \otimes A_z^{fc} \right) s$. The following formulation yields a sparse linear system. F_ξ are tridiagonal matrices, whereas G_ξ and D_ξ are bidiagonal matrices. A straightforward analysis using Taylor series shows that the underlying operator corresponding to F_ξ could be interpreted to a filter. In like manner, G_ξ and D_ξ are the staggered second-order derivatives. Theses remarks allow us to design the following preconditioned iterative methods.

3   Description of the Iteration Method

First, the number of unknowns in Eq. (12) is reduced. The first row is multiplied by DF^{-1} and the result subtracted from the last row. Hence, the variable ϕ satisfies the linear system:

$$-DF^{-1}G\phi = A\phi = \tilde{s} \tag{14}$$

The matrix F^{-1} is full, nevertheless, its action on an arbitrary vector could be performed with a linear complexity, since F is a tridiagonal system. Moreover, the operators D and G are also evaluated with a linear complexity. So, the action of A on the vector ϕ could be performed with a number of floating point operations, which increases linearly respect with to the number of mesh nodes. The conjugated gradient is used to solve this linear system because A is symmetric. However, without preconditioning, a poor convergence rate have been observed. An improvement of the convergence is achieved by using the preconditioned conjugated gradient solver. The preconditioner was chosen by noting that if F is approximated by the identity matrix, then $DF^{-1}G$ reduces to the second-order discretization of the Poisson equation, denoted $K = DG$. More precisely, the evaluation of K^{-1} is performed by a multigrid iteration of the AGMG solver [18].

The condition number $\kappa(K^{-1}A)$ is computed to give insights on the convergence rate. The periodic conditions hold on the three directions to simplify the derivation of the eigenvalues. In this case, for an arbitrary direction ξ, it can be proved that:

$$\lambda(D_\xi G_\xi) = -\frac{4}{\Delta\xi^2}\sin^2(\omega/2) \tag{15}$$

and:

$$\lambda(D_\xi F_\xi^{-1} G_\xi) = -\frac{1}{\Delta\xi^2}\frac{48\sin^2(\omega/2)}{11 + \cos(\omega)} \tag{16}$$

where $-\pi \leq \omega \leq \pi$. So, the condition number of $K^{-1}A$ is bounded by $1 \leq \kappa(K^{-1}A) \leq 6/5$. This result is also valid for the three-dimensional case, since the higher dimension is achieved by means of the tensor products. From a practical point of view, this result implies that the independence of the convergence rate against the problem size could be expected, at least, if the periodic directions are assumed.

4   Numerical Results

To test the efficiency of our formulation and its associated solution methods for the singular Poisson equation, four testing problems with solutions of different order of differentiability are retained. They are :

1. $\phi(x, y, z) = (xyz)^{3.5}[1 - \cos(xyz)]$, (five times differentiable);
2. $\phi(x, y, z) = x^{4.5} + y^{4.5} + z^{4.5}$, (four times differentiable);
3. $\phi(x, y, z) = (x + y + z)^{2.5}\sin(x)$, (three times differentiable);
4. $\phi(x, y, z) = (x + y + z)^{2.5}$, (two times differentiable);

The domain is a unit cube $[0, 1] \times [0, 1] \times [0, 1]$, and a uniform mesh size $h = 1/n$ is chosen in each direction, n being the number of grid cells in each all directions. All tests were performed on a Dell desktop station. In order to facilitate comparisons, a second-order discretization of the singular-Poisson equation has been implemented, and the AGMG multigrid has been used to solve it.

Table 1. Numerical error, iteration number and computational time for different grid sizes (fourth-order compact finite differences)

N	problem 1			problem 2			problem 3			problem 4		
	err.	it.	cpu (s)	err.	it.	cpu (s)	err.	it.	cpu (s)	err.	it.	cpu (s)
16	0.11e-04	42	0.18	1.40e-04	40	0.18	2.23e-05	43	0.20	4.03e-05	46	0.20
24	0.22e-05	41	0.63	3.28e-05	40	0.62	4.77e-06	43	0.65	1.53e-05	44	0.67
32	0.68e-06	39	0.14	1.17e-05	37	1.34	1.61e-06	41	1.45	7.63e-06	40	1.45
48	0.13e-06	38	0.48	2.76e-06	37	4.76	3.54e-07	41	5.22	2.84e-06	39	4.98
64	0.42e-07	38	0.12	9.90e-07	39	12.83	1.22e-07	38	12.60	1.40e-06	40	13.08
96	0.83e-08	36	0.44	2.34e-07	36	45.05	2.75e-08	39	48.46	5.15e-07	40	49.65

Table 2. Numerical error, iteration number and computational time for different grid sizes (second-order finite differences)

N	problem 1			problem 2			problem 3			problem 4		
	err.	it.	cpu (s)	err.	it.	cpu (s)	err.	it.	cpu (s)	err.	it.	cpu (s)
16	2.69e-04	20	0.02	5.19e-03	21	0.01	1.62e-03	23	0.02	6.38e-04	21	0.02
24	1.62e-04	21	0.05	2.39e-03	21	0.06	7.26e-04	22	0.05	2.93e-04	22	0.06
32	1.07e-04	20	0.12	1.37e-03	21	0.13	4.10e-04	22	0.14	1.68e-04	22	0.14
48	5.54e-05	21	0.46	6.19e-04	22	0.49	1.83e-04	22	0.48	7.63e-05	23	0.50
64	3.35e-05	22	1.23	3.51e-04	21	1.17	1.03e-04	22	1.22	4.34e-05	22	1.22
96	1.59e-05	22	4.44	1.57e-04	22	4.44	4.58e-05	22	4.43	1.96e-05	22	4.42

Tables 1 and 2 present results for the problems 1-4 and for the fourth and second-order formulations. For all formulations, the absolute numerical error, cpu time, and number of iterations to reach a residual of (10^{-12}) have been noted.

First, the accuracy of the present procedure is verified. The absolute numerical error is plotted versus the mesh size for all problems and for the both discretizations, in figure 1. Using a logarithmic scales, curves with -2 and -4 slopes are observed, depending on the used discretization. These expected values are the effective accuracy order of the scheme. However, it should be noted that for the problem 4, the fourth-order accuracy is not achieved. This is due to the fact that the solution itself is only two times differentiables, so that regularity assumptions [2] are not satisfied.

The second point of the discussion concerns with the computational cost of the present method. First, table 1 indicates a convergence of the iterative method after nearly 40 iterations for each mesh size. This behaviour was explained in section 3. For the second order finite difference, the convergence is achieved after 20 iterations, illustrating the mesh size independence of the AGMG multigrid solver. A rapid look into the computational cost by node, indicates that the second order scheme ($\simeq 5 \times 10^{-6}$sec/node) is ten times faster than the proposed fourth-order one. However, the higher-order discretization gives an accurate solution with a lower computational cost than the second-order one. This remark have to be qualified by considering the achieved accuracy. For instance, the nu-

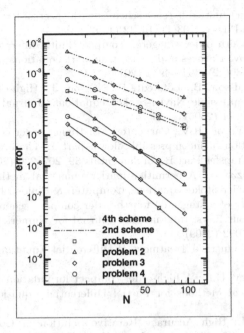

Fig. 1. Absolute numerical error versus the mesh size

merical error for the problem is 2.23×10^{-5} and the computational is about 0.2 seconds, by using our procedure. The second-order numerical solution of the problem 3 is obtained, with the same absolute error level (10^{-5}), takes 4.4 seconds. These observations indicate the superiority of our method with respect to the second-order solution methods, at least in the computational cost.

5   Conclusion

A fourth compact scheme formulation for the singular-Poisson equation has been proposed. This formulation is based on a mixed formulation and implemented on staggered grids. An iterative solution procedure has been also proposed. Several numerical examples have been treated, in comparison with the second-order discretization. It has been shown that using the proposed procedure, the iteration number is independent to the grid size. Moreover, despite a more tricky algorithm, the present fourth-order scheme remains faster than the second-order procedure to achieve a desired accuracy.

References

1. Ferziger, J.H., Perić, M.: Computational Methods for Fluid Dynamics. Springer-verlag edn. Springer (2002)

2. Lele, S.K.: Compact finite difference schemes with spectral-like resolution. Journal of Computational Physics 103(1), 16–42 (1992)
3. Boersma, B.J.: A 6th order staggered compact finite difference method for the incompressible Navier-Stokes and scalar transport equations. Journal of Computational Physics 230(12), 4940–4954 (2011)
4. Brüger, A., Gustafsson, B., Lötstedt, P., Nilsson, J.: High-order accurate solution of the incompressible Navier-Stokes equations. Journal of Computational Physics 203(1), 49–71 (2005)
5. Demuren, A.O., Wilson, R.V., Carpenter, M.: Higher order compact schemes for numerical simulation of incompressible flows, part 1: Theroretical development. Numerical Heat Transfer Part B Fundamentals 39, 207–230 (2001)
6. Schiestel, R., Viazzo, S.: A hermitian-fourier numerical method for solving the incompressible Navier-Stokes equations. Computers & Fluids 24(6), 739–752 (1995)
7. Knikker, R.: Study of a staggered fourth-order compact scheme for unsteady incompressible viscous flows. International Journal for Numerical Methods in Fluids 59(10), 1063–1092 (2009)
8. Collatz, L.: The Numerical Treatment of Differential Equations. Springer, Berlin (1960)
9. Spotz, W.F., Carey, G.F.: A high-order compact formulation for the 3D Poisson equation. Numerical Methods for Partial Differential Equations 12(2), 235–243 (1996)
10. Ge, L., Zhang, J.: High Accuracy Iterative Solution of Convection Diffusion Equation with Boundary Layers on Nonuniform Grids. Journal of Computational Physics 171(2), 560–578 (2001)
11. Spotz, W.F.: Formulation and experiments with high-order compact schemes for nonuniform grids. International Journal of Numerical Methods for Heat Fluid Flow 8(3), 288–303 (1998)
12. Gupta, M.M., Zhang, J.: High accuracy multigrid solution of the 3D convection diffusion equation. Applied Mathematics and Computation 113 (2000)
13. Zhuang, Y., Sun, X.H.: A High-Order Fast Direct Solver for Singular Poisson Equations. Journal of Computational Physics 171(1), 79–94 (2001)
14. Vedy, E., Viazzo, S., Schiestel, R.: A high-order finite difference method for incompressible fluid turbulence simulations. International Journal for Numerical Methods in Fluids 42(11), 1155–1188 (2003)
15. Abide, S., Viazzo, S.: A 2D compact fourth-order projection decomposition method. Journal of Computational Physics 206(1), 252–276 (2005)
16. Carey, G.F., Spotz, W.F.: Higher-order compact mixed methods. Communications in Numerical Methods in Engineering 13(7), 553–564 (1997)
17. Nagarajan, S., Lele, S.K., Ferziger, J.H.: A robust high-order compact method for large eddy simulation. Journal of Computational Physics 191(2), 392–419 (2003)
18. Notay, Y.: An aggregation-based algebraic multigrid method. Electronic Transactions on Numerical Analysis 37, 123–146 (2010)

Unconditionally Stable Schemes
for Non-stationary Convection-Diffusion
Equations

Nadezhda Afanasyeva[1], Petr N. Vabishchevich[2], and Maria Vasilyeva[1]

[1] North-Eastern Federal University, Belinskogo Str 58, 677000 Yakutsk, Russia
[2] Nuclear Safety Institute, B. Tulskaya Str 52, 115191 Moscow, Russia

Abstract. Convection-diffusion problem are the base for continuum mechanics. The main features of these problems are associated with an indefinite operator the problem. In this work we construct unconditionally stable scheme for non-stationary convection-diffusion equations, which are based on use of new variables. Also, we consider these equations in the form of convection-diffusion-reaction and construct unconditionally stable schemes when explicit-implicit approximations are used with splitting of the reaction operator.

Keywords: non-stationary convection-diffusion equations, two-layer weighted scheme, unconditionally stable schemes.

1 Introduction

Convection-diffusion equation are basic in the mathematical modelling of the problems of continuum mechanics. The main features of these problems are connected with the nonselfadjoint property of elliptic operator and domination of convective transport. When considering compressible media, an operator of convection-diffusion problem is indefinite. In this case, given process can be nondissipative, i.e. norm of the homogeneous problem solutions does not decrease with time. This behavior of the norm solutions need to pass on the discrete level in choosing of approximations in time.

In the numerical solution of non-stationary problems for convection-diffusion equations the most widely used two- and three-layer scheme. Investigation of the stability and convergence of approximate solutions can be performed using the general theory Samarskii A. A. of stability (correctness) of operator-difference schemes [3,4]. Must be kept in mind that for convection-diffusion problems direct application of the general stability criteria can be difficult due to non-selfadjoint operators. Note also that, in view of indefinite operator of problem we need to oriented ϱ-stable ($\varrho > 1$) operator-difference schemes. In the solution non-stationary problems of long periods of time preference should be given asymptotically stable schemes [5]. For these schemes ensures the correct behavior of the solutions with the release of the fundamental solutions for large time and damping of others.

In this paper, we construct unconditionally stable scheme for the approximate solution of non-stationary convection-diffusion problems. Such schemes can be

I. Dimov, I. Faragó, and L. Vulkov (Eds.): NAA 2012, LNCS 8236, pp. 151–157, 2013.

applied to other problems with an indefinite operator. The study conducted by the example of a model two-dimensional boundary-value problem in a rectangle. Used the simplest approximation of the operators of diffusive and convective transfer on a uniform rectangular grid. Constructed unconditionally ϱ-stable difference scheme based on the introduction of new variables and the explicit-implicit approximation.

2   The Convection-Diffusion Problem

We consider the Neumann problem in a rectangle for the non-stationary convection-diffusion equation. For simplicity, assume that the coefficient of diffusion transport is a constant (independent of time, but depends on the point of the computational domain). The coefficient of convective transport is natural to consider the variables both in space and time.

In the rectangle

$$\Omega = \{\boldsymbol{x} \mid \boldsymbol{x} = (x_1, x_2), \quad 0$$

We consider the non-stationary convection-diffusion equation with the convective transport in divergent form,

$$\frac{\partial u}{\partial t} + \sum_{\alpha=1}^{2} \frac{\partial}{\partial x_\alpha} \left(v_\alpha\left(\boldsymbol{x},t\right) u\right) - \sum_{\alpha=1}^{2} \frac{\partial}{\partial x_\alpha} \left(k(\boldsymbol{x}) \frac{\partial u}{\partial x_\alpha}\right) = f\left(\boldsymbol{x},t\right), \quad \boldsymbol{x} \in \Omega, \quad (1)$$

in the standard assumptions $k_1 \leq k\left(\boldsymbol{x}\right) \leq k_2$, $k_1 > 0$, $T > 0$. This equation is supplemented by Neumann boundary conditions

$$\frac{\partial u\left(\boldsymbol{x},t\right)}{\partial n} = 0, \quad \boldsymbol{x} \in \partial\Omega, \quad 0 < t \leq T. \tag{2}$$

For the unique solvability of the nonstationary problem the initial condition is given

$$u\left(\boldsymbol{x},0\right) = u^0(\boldsymbol{x}), \quad \boldsymbol{x} \in \Omega. \tag{3}$$

On the set of functions $u\left(\boldsymbol{x},t\right)$, which satisfy the boundary conditions (2), non-stationary convection-diffusion problem written in the form of differential-operator equation

$$\frac{du}{dt} + \mathcal{A}u = f(t), \quad \mathcal{A} = \mathcal{C}(t) + \mathcal{D}, \quad 0 < t \leq T. \tag{4}$$

The diffusion operator $\mathcal{D}$ is defined by

$$\mathcal{D}u = -\sum_{\alpha=1}^{2} \frac{\partial}{\partial x_\alpha} \left(k\left(\boldsymbol{x}\right) \frac{\partial u}{\partial x_\alpha}\right)$$

and convection operator $\mathcal{C}$

$$\mathcal{C}u = \sum_{\alpha=1}^{2} \frac{\partial}{\partial x_\alpha} \left(v_\alpha\left(\boldsymbol{x},t\right) u\right).$$

Cauchy problem is considered for the evolution equation (4):

$$u\,(0) = u^0. \tag{5}$$

For convection operator we have the following representation

$$\mathcal{C} = \mathcal{C}_0 + \frac{1}{2} v \mathcal{E}, \quad \mathcal{C}_0 u = \frac{1}{2} \sum_{\alpha=1}^{2} \left(v_\alpha\,(\boldsymbol{x}, t)\,\frac{\partial u}{\partial x_\alpha} + \frac{\partial}{\partial x_\alpha}\,(v_\alpha(\boldsymbol{x}, t)u) \right),$$

where $\mathcal{E}$ — the identity operator and $\mathcal{C}_0$ -the operator of convective transport in a symmetric form.

For arbitrary functions $u(\boldsymbol{x})$ $w(\boldsymbol{x})$, we define the Hilbert space $\mathcal{H} = L_2\,(\Omega)$ with inner product and norm

$$(u, w) = \int_\Omega u\,(\boldsymbol{x})\,w\,(\boldsymbol{x})\,dx, \quad ||u|| = (u, u)^{1/2}.$$

Diffusion operator $\mathcal{D}$ on the set of functions satisfying (2), is self-adjoint and positive define

$$\mathcal{D} = \mathcal{D}^* \geq 0. \tag{6}$$

The operator of convective transport is considered under the assumption that the normal component of the medium velocity $v = (v_1, v_2)$ on the boundary is zero:

$$v_n(x) = \boldsymbol{v}\,\boldsymbol{n} = 0, \quad x \in \partial\Omega, \tag{7}$$

where n — outward normal to $\partial\Omega$. In $\mathcal{H}$ the convection operators have the following properties:

$$\mathcal{C}_0 = -\mathcal{C}_0^*. \tag{8}$$

Also useful upper estimates for convective transport operator $\mathcal{C}$:

$$|(\mathcal{C}u, u)| \leq \gamma||u||^2, \quad \gamma = \frac{1}{2}||\,\mathrm{div}\,v||_{C(\Omega)}. \tag{9}$$

3   The Differential-Difference Problem

For an approximate solution of the non-stationary convection-diffusion problem we use a uniform grid in the area Ω:

$$\omega = \{\boldsymbol{x} \mid \boldsymbol{x} = (x_1, x_2),\quad x_\alpha = \left(i_\alpha + \frac{1}{2} \right) h_\alpha,$$

$$i_\alpha = 0, 1, ..., N_\alpha, \quad (N_\alpha + 1)h_\alpha = l_\alpha, \quad \alpha = 1, 2\}.$$

We define the Hilbert space $H = L_2\,(\omega)$ for grid functions, where the inner product and norm are defined as follows:

$$(y, w) \equiv \sum_{\boldsymbol{x} \in \omega} y\,(\boldsymbol{x})\,w\,(\boldsymbol{x})\,h_1 h_2, \quad ||y|| \equiv (y, y)^{1/2}.$$

For the difference operator of the diffusion transfer D is used additive representation

$$D = \sum_{\alpha=1}^{2} D^{(\alpha)}, \quad \alpha = 1, 2, \quad \boldsymbol{x} \in \omega, \tag{10}$$

here $D^{(\alpha)}$, $\alpha = 1, 2$ is associated with the corresponding differential operator in one direction.

The difference operator of diffusion transport (10) in H is self-adjoint and positive definite [3]

$$D = D^* \geq 0. \tag{11}$$

The convective terms are approximated with second-order, using the central difference derivatives and shifted grids to specify the velocity components. For the difference operator of convective transport are also using additive representation

$$C = \sum_{\alpha=1}^{2} C^{(\alpha)}. \tag{12}$$

Difference operator of convective transport in symmetric form have the following basic property:

$$C_0^* = -C_0. \tag{13}$$

We also [5] have the grid analogue of inequality (9):

$$|(Cy, y)| \leq \delta \|y\|^2 \tag{14}$$

with a constant

$$\delta = \frac{1}{2} \max_{\boldsymbol{x} \in \omega} \left| \frac{b^{(1)}(x_1 + 0.5h_1, x_2) - b^{(1)}(x_1 - 0.5h_1, x_2)}{h_1} \right.$$

$$\left. + \frac{b^{(2)}(x_1, x_2 + 0.5h_2) - b^{(2)}(x_1, x_2 - 0.5h_2)}{h_2} \right|.$$

In case of velocity is independent from time and sufficiently smooth velocity components and solutions of the differential problem, we can assume,

$$b^{(\alpha)}(\boldsymbol{x}) = v_\alpha(\boldsymbol{x}), \quad \boldsymbol{x} \in \Omega, \quad 0 < x_\alpha < l_\alpha,$$

$$b^{(\alpha)}(\boldsymbol{x}) = 0, \quad x_\alpha = 0, \quad x_\alpha = l_\alpha, \quad \alpha = 1, 2.$$

Therefore, from the equation (4) we arrive at the differential-operator equation

$$\frac{dy}{dt} + Ay = \phi(t), \quad A = A(t) = C + D, \quad 0 < t \leq T, \tag{15}$$

on the set of grid functions $y(t) \in H$ with the initial condition

$$y(0) = y^0. \tag{16}$$

Difference convection and diffusion operators in the differential-difference problem inherit the basic properties of differential operators.

4   Unconditionally Stable Schemes

For simplicity, we restrict ourselves to a uniform grid in time

$$\bar{\omega}_\tau = \omega_\tau \cup \{T\} = \{t^n = n\tau, \quad n = 0, 1, ..., N, \quad \tau N = T\}.$$

For an approximate solution of (15), (16) commonly used two-layer weighted scheme, which have a following restrictions on the time step $\tau < \tau_0 = \frac{1}{\sigma\delta}$.

To construct the unconditionally stable schemes for the solution of the differential problem (15), (16) with $A \geq -\delta E$, $\delta > 0$ we define a new function w:

$$y = \exp(\delta t)w. \tag{17}$$

Substitution of (17) in (15), (16) with homogeneous right-hand side gives the following problem for the w:

$$\frac{dw}{dt} + \tilde{A}w = 0, \quad \tilde{A} = A + \delta E, \quad 0 < t \leq T, \tag{18}$$

$$w(0) = y^0. \tag{19}$$

Under this transformation, operator $\tilde{A}$ is a nonnegative ($\tilde{A} \geq 0$).

To solve the problem (18), (19) we use a two-layer weighted difference scheme, which is unconditionally stable for standard restrictions $\sigma \geq 0.5$. We write the scheme for the grid function y^n

$$\frac{\exp(-\delta\tau)y^{n+1} - y^n}{\tau} + (A + \delta E)\left(\sigma \exp(-\delta\tau)y^{n+1} + (1 - \sigma)y^n\right) = 0, \tag{20}$$

$$y^0 = u^0, \quad t^n \in \omega_\tau,. \tag{21}$$

In contrast to the non-standard schemes considered in [6], the positive effect is achieved not only through the use of a new approximation of the time, but also by correcting the problem operator.

Theorem 1. *The difference scheme (20), (21) with $\sigma \geq 0.5$ unconditionally ϱ-stable in H with*

$$\varrho = \exp(\delta\tau), \tag{22}$$

with the a priori estimate for solutions

$$\|y^{n+1}\| \leq \varrho\|y^n\|. \tag{23}$$

Proof. We rewrite the scheme (20), (21) in form

$$\frac{\exp(-\delta\tau)y^{n+1} - y^n}{\tau} + \tilde{A}p^{n+1} = 0, \quad t^n \in \omega_\tau, \tag{24}$$

where

$$p^{n+1} = \sigma \exp(-\delta\tau)y^{n+1} + (1 - \sigma)y^n = \tau\left(\sigma - \frac{1}{2}\right)r^{n+1} + \frac{1}{2}\left(\exp(-\delta\tau)y^{n+1} - y^n\right),$$

$$r^{n+1} = \frac{\exp(-\delta\tau)y^{n+1} - y^n}{\tau}.$$

Multiplying the scalar equation (24) by p^{n+1}, we obtain the equality

$$\tau\left(\sigma - \frac{1}{2}\right)\left(r^{n+1}, r^{n+1}\right) + \tilde{A}\left(p^{n+1}, p^{n+1}\right)$$

$$+\frac{1}{2\tau}\left(\left(\exp(-\delta\tau)y^{n+1}, \exp(-\delta\tau)y^{n+1}\right) - (y^n, y^n)\right) = 0$$

From this equation, under the condition $\sigma \geq 0.5$ and $\tilde{A} \geq 0$, yields the estimate of stability (23),(22).   □

Equation (1) can be written in the form of convection-diffusion-reaction equation with the convective terms in the symmetric form

$$\frac{\partial u}{\partial t} + C_0 u + Du + Ru = f(x, t), \quad x \in \Omega, \quad t > 0. \tag{25}$$

where

$$Ru = r(x, t)y, \quad r(x, t) = \frac{1}{2}\operatorname{div} v.$$

For the reaction operator we have the estimate

$$R = R^*, \quad -\delta\mathcal{E} \leq R \leq \delta E. \tag{26}$$

To construct the unconditionally stable scheme without the assumption of nonnegativity operator of problem we will use the explicit-implicit approximation for the equation (25) [7]. The problem is generated by the reaction operator therefore we split it into two:

$$R = R_+ + R_-, \quad R_+ = R_+^*, \quad R_- = R_-^*, \quad 0 \leq R_+ \leq \delta E, \quad -\delta E \leq R_- < 0. \tag{27}$$

When using the two-layer explicit-implicit schemes, we can only count on first-order accuracy in time. Therefore it is natural oriented to purely implicit approximation of the basic terms of the operator and define following difference scheme

$$\frac{y^{n+1} - y^n}{\tau} + (C^n + D + R_+^n)y^{n+1} + R_-^n y^n = 0, \quad n = 0, 1, ..., N - 1. \tag{28}$$

Theorem 2. *Explicit-implicit difference scheme (28), (21) is unconditionally ϱ-stable in H with*

$$\varrho = 1 + \delta\tau \tag{29}$$

for the numerical solution we have the estimate:

$$\|y^{n+1}\| \leq \varrho\|y^n\|, \quad n = 0, 1, ..., N - 1. \tag{30}$$

It is important to note that, in contrast to the ordinary weighted scheme, stability is obtained without restrictions on the time step. The transition to a new time layer associated with the solution of the grid problem

$$(E + \sigma\tau(A + \delta E))y^{n+1} = \chi^n \tag{31}$$

for scheme (20) and

$$(E + \tau(C + D + R_+))y^{n+1} = r^n \tag{32}$$

for scheme (28).

Equation (31) and (32) is a system of linear algebraic equations with a positive definite nonselfadjoint matrix. The standard iterative methods can be used for numerical solution.

References

1. Morton, K.W., Kellogg, R.B.: Numerical solution of convection-diffusion problems. Chapman & Hall, London (1996)
2. Hundsdorfer, W.H., Verwer, J.G.: Numerical solution of time-dependent advection-diffusion-reaction equations. Springer, Berlin (2003)
3. Samarskii, A.A.: The theory of difference schemes. CRC Press (2001)
4. Samarskii, A.A., Matus, P.P., Vabishchevich, P.N.: Difference schemes with operator factors. Kluwer, Dordrecht (2002)
5. Samarskii, A.A., Vabishchevich, P.N.: Computational Heat Transfer. Wiley, Chichester (1995)
6. Mickens, R.E.: Nonstandard Finite Difference Schemes for Differential Equations. Journal of Difference Equations and Applications 8(9), 823–847 (2002)
7. Vabishchevich, P.N., Vasilyeva, M.V.: Explicit-implicit schemes for convection-diffusion-reaction problems. Numerical Analisis and Application 5(1), 297–306 (2012)

Nonconforming Rectangular Morley Finite Elements

A.B. Andreev[1] and M.R. Racheva[2]

[1] Department of Informatics,
Technical University of Gabrovo 5300 Gabrovo, Bulgaria
[2] Department of Mathematics,
Technical University of Gabrovo 5300 Gabrovo, Bulgaria

Abstract. We analyze some approximation properties of modified rectangular Morley elements applied to fourth-order problems. Degrees of freedom of integrals type are used which yields superclose property. Further asymptotic error estimates for biharmonic solutions are derived. Some interesting and new numerical results concerning plate vibration problems are also presented.

1 Introduction and Preliminaries

The Morley nonconforming element has been widely used in computational mechanics and structural engineering because of its simplicity in implementation. In many practical cases, it seems better than some conforming finite elements. This phenomenon causes the great interest of many mathematicians who study finite elements.

In this paper we consider the following model biharmonic equation which arises in fluid mechanics as well as in thin elastic plate problems:

$$\Delta^2 u = f \quad \text{in } \Omega, \tag{1}$$

where Ω is bounded domain in $\mathbf{R}^2$. We also consider homogeneous boundary conditions

$$u = \frac{\partial u}{\partial \nu} = 0 \quad \text{on } \partial\Omega \tag{2}$$

and $\dfrac{\partial}{\partial \nu}$ denotes outer normal derivative.

To formulate the variational equivalent to (1), (2), we introduce the variational space $V = H_0^2(\Omega)$. Here $H^k(\Omega)$ is the usual Sobolev space of order k and $(\cdot, \cdot)$ denotes the L_2-inner product.

The variational problem associated with (1), (2) is given by

$$a(u, v) = (f, v) \quad \forall v \in V, \tag{3}$$

where

$$a(u, v) = \int_\Omega \Delta u \Delta v \, dx \quad \forall u, v \in V.$$

I. Dimov, I. Faragó, and L. Vulkov (Eds.): NAA 2012, LNCS 8236, pp. 158–165, 2013.
© Springer-Verlag Berlin Heidelberg 2013

Obviously, the bilinear form $a(\cdot,\cdot)$ is symmetric and V–elliptic. We shall approximate the solutions of (3) by a nonconforming finite element method.

Consider a family of rectangulations $\tau_h = \bigcup_i K_i$ of $\overline{\Omega}$. Finite elements K_i fulfil standard assumptions (see [1], Chapter 3). Let $h = \max_i h_i$ be the finite element parameter corresponding to any partition τ_h.

With a partition τ_h we associate a finite dimensional space V_h by means of Morley rectangles (see Fig. 1).

Let τ_h consist of rectangles with edges parallel to the coordinate axes and $K \in \tau_h$ be a rectangle with vertices a_j and edges l_j, $j = 1, 2, 3, 4$ (see e.g. [2,3]).

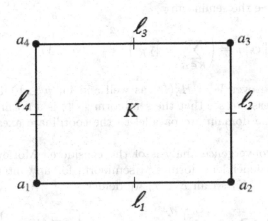

Fig. 1. The Morley rectangle

We choose the following set of degrees of freedom (v is a test function): $v(a_j)$ and $\int_{l_j} \dfrac{\partial v}{\partial \nu}\, dl,\ j = 1, 2, 3, 4.$

There are two variants for the polynomial space $\mathcal{P}_K$, namely (see [3,4]):

$$\mathcal{P}_K^{(1)} = P_2 + \operatorname{span}\{x_1^3, x_2^3\}$$

and

$$\mathcal{P}_K^{(2)} = P_2 + \operatorname{span}\{x_1^3 - 3x_1 x_2^2,\, x_2^3 - 3x_1^2 x_2\},$$

i.e. $\mathcal{P}_K$ consists of polynomials of degree ≤ 3 for which the only terms of third degree are x_1^3 and x_2^3 or $x_1^3 - 3x_1 x_2^2$ and $x_2^3 - 3x_1^2 x_2$, respectively.

Thus $P_2 \subset \mathcal{P}_K^{(i)}$, $i = 1, 2$. Obviously, the set of degrees of freedom is $\mathcal{P}_K-$ unisolvent.

Now, we consider the finite element space V_h using $\mathcal{P}_K^{(1)}$ or $\mathcal{P}_K^{(2)}$. Let us emphasize that the terms $x_1^3 - 3x_1 x_2^2$ and $x_2^3 - 3x_1^2 x_2$ are the unique polynomials of degree greater than or equal to 3 which are harmonic.

Then the approximate variational problem of (3) is: find $u_h \in V_h$ such that

$$a_h(u_h, v_h) = (f, v_h), \quad \forall v_h \in V_h, \tag{4}$$

where

$$a_h(u_h, v_h) = \sum_{K \in \tau_h} \int_K \Delta u_h \Delta v_h \, dx.$$

2   Main Results

First, on V_h we define the seminorm

$$\|v_h\|_h = \left(\sum_{K \in \tau_h} |v_h|_{2,K}^2 \right)^{1/2} \quad \forall v_h \in V_h.$$

It may be defined over $V = H_0^2(\Omega)$ as well and for $v \in V$ it is valid that $\|v\|_h = |v|_{2,\Omega}$. It is easy to see that the seminorm $\| \cdot \|_h$ is a norm on V_h in case when the sides of the domain are parallel to the coordinate axes (rectangular plates).

In order to get convergence analysis of the considered Morley element, we introduce the mesh-dependent norm and seminorm for any integer m. If $v \in L_2(\Omega)$ with $v_{|K} \in H^m(K)$ for all $K \in \tau_h$ we define:

$$\|v\|_{m,h} = \left\{ \sum_{K \in \tau_h} \|v\|_{m,K}^2 \right\}^{1/2}, \quad |v|_{m,h} = \left\{ \sum_{K \in \tau_h} |v|_{m,K}^2 \right\}^{1/2}.$$

The piecewise interpolation functions $i_h u \in \mathcal{P}_K$ are formulated as (Fig. 1):

$$i_h u(a_j) = u(a_j),$$

$$\int_{l_j} \frac{\partial i_h u}{\partial \nu} \, dl = \int_{l_j} \frac{\partial u}{\partial \nu} \, dl, \quad j = 1, 2, 3, 4,$$

where a_j and l_j are vertices and edges for any rectangle $K \in \tau_h$, respectively.

First, we shall prove the following property of the rectangular Morley element.

Lemma 1. *Let the finite element space V_h be engendered by the polynomials $\mathcal{P}_K^{(2)}$. Then for any function $u \in H_0^2(\Omega)$*

$$a_h(u - i_h u, v_h) = 0, \quad \forall v_h \in V_h. \tag{5}$$

Proof. Since $v_{h|K}$ contains cubic harmonic monomials then $\Delta v_{h|K} = \text{const}$. Now

$$a_h(u - i_h u, v_h) = \sum_{K \in \tau_h} \int_K \Delta(u - i_h u) \Delta v_h \, dx$$

$$= \sum_{K \in \tau_h} \Delta v_{h|K} \int_K \Delta(u - i_h u) \, dx.$$

Using Green's formula and by standard orientation arguments we obtain

$$a_h(u - i_h u, v_h) = \sum_{K \in \tau_h} \Delta v_{h|K} \oint_{\partial K} \frac{\partial(u - i_h u)}{\partial \nu} \, dl$$

$$= \sum_{K \in \tau_h} \Delta v_{h|K} \left(\int_{l_1} - \int_{l_3} \right) \frac{\partial(u - i_h u)}{\partial x_2} \, dx_1 + \left(\int_{l_2} - \int_{l_4} \right) \frac{\partial(u - i_h u)}{\partial x_1} \, dx_2 = 0.$$

The next theorem gives a main result for the convergence and precision of the rectangular Morley elements in case when $\mathcal{P}_K = \mathcal{P}_K^{(2)}$.

Theorem 1. *Let τ_h be regular partitions of Ω containing rectangular elements with sides parallel to the coordinate axes. Let also V_h be the space used in the previous lemma. If u and u_h are the solutions of (3) and (4) respectively and $u \in H^4(\Omega) \cap H_0^2(\Omega)$ then*

$$\|u - u_h\|_h \leq Ch\|u\|_{4,\Omega}.$$

Proof. The basic approach is to apply the Strang's theorem [1]. If $a_h(\cdot, \cdot)$ is V_h−elliptic and continuous, then there exists a constant C, independent of h, such that

$$\|u - u_h\|_h \leq C \left\{ \inf_{v_h \in V_h} \|u - v_h\|_h + \sup_{v_h \in V_h} \frac{|(f, v_h) - a_h(u, v_h)|}{\|v_h\|_h} \right\}. \tag{6}$$

Since $i_h u \in V_h$, we have for the first term of (6) that

$$\inf_{v_h \in V_h} \|u - v_h\|_h \leq |u - i_h u|_{2,h} = \left\{ \sum_{K \in \tau_h} |u - i_h u|_{2,K}^2 \right\}^{1/2}.$$

From (5) and using locally Poincare's inequality for L_2 (see [6]) we have:

$$|u - i_h u|_{2,K} \leq Ch|u|_{4,K}.$$

Thus

$$\inf_{v_h \in V_h} \|u - v_h\|_h \leq Ch|u|_{4,\Omega}. \tag{7}$$

In order to estimate the second term of (6) we define

$$E_h(u, v_h) = (f, v_h) - a_h(u, v_h)$$

for $u \in H^4(\Omega) \cap H_0^2(\Omega)$, $v_h \in V_h$.

For any $v_h \in V_h$ we take a sequence $\{v_{h,n}\}$ of smooth enough functions converging to v_h in $H_0^1(\Omega)$ and equal to zero at the degrees of freedom located on the boundary $\partial \Omega$. Hence, by Green's formula,

$$\int_\Omega f v_{h,n} \, dx = \int_\Omega \Delta u \Delta v_{h,n} \, dx = - \int_\Omega (\text{grad}\Delta u)(\text{grad} v_{h,n}) \, dx.$$

Since both sides of the above relation are continuous linear functionals on $L_2(\Omega)$ vanishing over $\partial\Omega$, we can pass to the limit in order to obtain for all $v_h \in V_h$

$$(f, v_h) = \int_\Omega f v_h \, dx = -\int_\Omega (\mathrm{grad}\Delta u)\,(\mathrm{grad} v_h)\,dx. \tag{8}$$

On the other hand, using again the Green's formula we have:

$$a_h(u, v_h) = \sum_{K\in\tau_h} \int_K \Delta u \Delta v_h \, dx$$

$$= \sum_{K\in\tau_h} \left[-\int_K \mathrm{grad}\Delta u \, \mathrm{grad} v_h \, dx + \oint_{\partial K} \Delta u \frac{\partial v_h}{\partial \nu}\, dl \right].$$

We substitute the last equality and (8) in $E_h(u, v_h)$ to get

$$E_h(u, v_h) = -\sum_{K\in\tau_h} \oint_{\partial K} \Delta u \frac{\partial v_h}{\partial \nu}\, dl. \tag{9}$$

We shall transform the equality (9). First, for any $K \in \tau_h$ we set $\varphi = -\Delta u$ and $w = \dfrac{\partial v_h}{\partial x_j}$, $j = 1, 2$.

Here $\varphi \in H^2(\Omega)$, since $u \in H^4(\Omega)$. Also

$$w = \frac{\partial v_h}{\partial x_j}, \quad v_h \in \mathcal{P},$$

where $\mathcal{P}$ is the restriction of $\mathcal{P}_K$ on any side of the rectangle K.

Let us introduce the bilinear forms (see Fig. 1)

$$\delta_{j,K}(\varphi, w) = \left(\int_{l_j} - \int_{l_{j+2}} \right) (\varphi w)\, dl, \quad j = 1, 2.$$

From now on the sign $\widehat{}$ refers to the reference finite element $\widehat{K}$ (for example unit square). Thus, by a simple (affine) change of variables

$$\delta_{j,K}(\varphi, w) = h_i \delta_{j,\widehat{K}}(\widehat\varphi, \widehat w), \tag{10}$$

where h_i, $i = 1, 2$, $i \neq j$ is the length of l_j (and l_{j+2}).

It is easy to see that $\delta_{j,\widehat{K}}(\widehat\varphi, \widehat w) = 0$ for $\widehat\varphi \in \mathcal{P}_0$, $\widehat w \in \widehat{\mathcal{P}}$.

Note that the bilinear form $\delta_{j,\widehat{K}}$ is continuous. Indeed ($j = 1, 2$),

$$\left| \delta_{j,\widehat{K}}(\widehat\varphi, \widehat w) \right| \leq C \|\widehat\varphi\|_{0,\partial\widehat{K}} \|\widehat w\|_{0,\partial\widehat{K}}.$$

By the trace theorem we get [1]:

$$\left| \delta_{j,\widehat{K}}(\widehat\varphi, \widehat w) \right| \leq C \|\widehat\varphi\|_{1,\widehat{K}} \|\widehat w\|_{1,\widehat{K}}.$$

Now we can apply the bilinear lemma [1] to the bilinear form $\delta_{j,\widehat{K}}$ to get

$$\left|\delta_{j,\widehat{K}}(\widehat{\varphi},\widehat{w})\right| \leq C \left|\widehat{\varphi}\right|_{1,\widehat{K}} \left|\widehat{w}\right|_{1,\widehat{K}}. \tag{11}$$

Having in mind that the transformation $\widehat{K} \to K$ is accomplished by (nonsingular) matrix B_K, we have the relations

$$\left|\widehat{\varphi}\right|_{1,\widehat{K}} \leq C\|B_K\| \ |\det B_K|^{-1/2} \ |\varphi|_{1,K},$$

$$\left|\widehat{w}\right|_{1,\widehat{K}} \leq C\|B_K\| \ |\det B_K|^{-1/2} \ |w|_{1,K}.$$

Here $\|B_K\| \leq C h_K$ and $|\det B_K| = \text{meas}(K)/\text{meas}(\widehat{K}) \geq C\rho_K^2$, where ρ_K is the radius of the inscribed circle in K such that $h_K/\rho_K \leq \text{const.}$ Also $h_j \leq h_K$, $j = 1, 2$.

Thus, from (10) and (11) we obtain

$$|\delta_{j,K}(\varphi,w)| \leq C h_K \|u\|_{4,K} \|v_h\|_{2,K}, \quad j = 1, 2.$$

These inequalities lead us to the estimate

$$|(f, v_h) - a_h(u, v_h)| \leq C h \|u\|_{4,\Omega} \|v_h\|_h. \tag{12}$$

To complete the proof of the theorem, it remains to replace the estimates (7) and (12) in (6).

Remark 1. For the triangular Morley element there is an analogous estimate by Lascaux and Lasaint (see [6], Theorem 3.1), while the convergence of Morley rectangle in case $\mathcal{P}_K = \mathcal{P}_K^{(1)}$ could be found in [3].

The variational eigenvalue problem corresponding to the biharmonic problem (2) is: find $(\lambda, u) \in \mathbf{R} \times H_0^2(\Omega)$ such that

$$a(u, v) = \lambda(u, v), \quad \forall v \in V. \tag{13}$$

We determine the approximate eigenpairs (λ_h, u_h) using nonconforming rectangular Morley finite elements. Then, the eigenvalue problem corresponding to (13) is: find $(\lambda_h, u_h) \in \mathbf{R} \times V_h$ such that

$$a_h(u_h, v_h) = \lambda_h(u_h, v_h), \quad \forall v_h \in V_h. \tag{14}$$

Theorem 1 is a basic tool to determine the order of convergence for the eigenpairs of (13) by means of the solution operator approach (see e.g. [7]):

Theorem 2. *Let $u \in H^4(\Omega) \cap H_0^2(\Omega)$ and $u_h \in V_h$ be the solutions of (13) and (14) respectively. If the conditions of Theorem 1 concerning the partitions of Ω by nonconforming Morley rectangles are fulfilled, then*

$$\|u - u_h\|_{2,h} \leq C h \|u\|_{4,\Omega},$$

$$|\lambda - \lambda_h| \leq C h^2 \|u\|_{4,\Omega}^2.$$

Remark 2. The numerical results using rectangular Morley elements for forth-order spectral problem give asymptotically lower bounds of the exact eigenvalues [8]. The proof of this fact as well as of Theorem 2 will be under consideration in separate investigation.

Remark 3. It is easy to see that an interpolation-equivalent choice for degrees of freedom is (see [2]): function values at the vertices a_j, $j = 1, 2, 3, 4$ of the rectangle $K \in \tau_h$ and the first derivatives in normal direction at the midside nodes of l_j, $j = 1, 2, 3, 4$ (Fig.1) for $v \in C^1(K), K \in \tau_h$.

3   Numerical Results

To illustrate our theoretical results we shall refer to the example on related eigenvalue problem (13) when Ω is the unit square. For this problem, numerical results concerning the first eigenvalue by means of nonconforming Adini element and triangular Morley element could be seen in [7]. Let us note that numerical experiments for fourth-order eigenvalue problems could not be easily find in the bibliography, especially by means of nonconforming FEMs.

The main purpose of this section is to compare the results obtained using rectangular Morley elements and those obtained using Adini element for which it is well known that it approximates eigenvalues from below [9]. On this base it is illustrated that the rectangular Morley elements give lower bounds for the exact eigenvalues. In addition to that, the numerical experiment by means of Morley rectangle is implemented for both versions of degrees of freedom – (i)

Table 1. First three eigenvalues computed by Adini FE; Morley rectangle (i); Morley rectangle (ii)

n	FE	$\lambda_{1,h}$	$\lambda_{2,h}$	$\lambda_{3,h}$
	Adini	1185.550861	4944.317694	4994.393228
4	Morley (i)	1102.196140	4902.648339	4902.648339
	Morley (ii)	1003.056682	4107.340189	4107.340190
	Adini	1254.152526	5164.790407	5219.985096
8	Morley (i)	1225.392738	5055.592884	5055.592884
	Morley (ii)	1187.878467	4770.160362	4770.160362
	Adini	1274.357984	5258.865533	5308.864358
12	Morley (i)	1261.666381	5213.480371	5213.480371
	Morley (ii)	1243.088531	5068.724011	5068.724011
	Adini	1283.199186	5307.705444	5342.189148
16	Morley (i)	1275.718722	5283.435248	5283.435248
	Morley (ii)	1264.829904	5197.493919	5197.493919

using the integral values of the derivatives in normal direction on the sides of the elements and, (ii) using the values of the derivatives in normal direction at the midside nodes of the elements.

The domain Ω is uniformly divided into n^2 rectangles, where $n = 4; 8; 12; 16$, respectively. Our numerical results show that both variants of Morley rectangle give eigenvalues less than these obtained by Adini element. The eigenvalues obtained using the Morley rectangle (ii) are less than those obtained using the Morley rectangle (i). In particular, this fact illustrates the advantage of the degrees of freedom of integral type. It is seen from the table above, the difference between eigenvalues obtained by means of the considered three different type of finite elements is significant in case of course mesh and decreases when we refine the mesh.

Let us recall that under uniform mesh, Adini element gives an improved accuracy:

$$|\lambda - \lambda_h| \le Ch^4 \|u\|_{4,\Omega}^2,$$

thus it approximates more accurately any eigenvalues λ_j than Morley rectangle [6]. Consequently, nonconforming Morley rectangular finite elements give also lower bounds for the eigenvalues.

Acknowledgement. This work has been supported by the Bulgarian National Science Fund under grant DFNI-I01/5.

References

1. Ciarlet, P.G.: Basic error estimates for elliptic problems. In: Ciarlet, P.G., Lions, J.L. (eds.) Finite Element Methods (Part 1), Handbook of Numerical Analysis, vol. 2, pp. 21–343. Elsevier Science Publishers, North-Holland (1991)
2. Wang, L., Xie, X.: Uniformly stable rectangular elements for fourth order elliptic singular perturbation problems. eprint arXiv:1101.1218 (arXiv1101.1218W) (2011)
3. Zhang, H., Wang, M.: The Mathematical Theory of Finite Elements. Science Press, Beijing (1991)
4. Nicaise, S.: A posteriori error estimations of some cell-centered finite volume methods. SIAM J. Numer. Anal. 43(04), 1481–1503 (2005)
5. Brenner, S., Scott, L.R.: The Mathematical Theory for Finite Element Methods. Springer-Verlag, New York (1994)
6. Lascaux, P., Lesaint, P.: Some nonconforming finite elements for the plate bending problem. Rev. Française Automat. Informat. Recherche Operationnelle Sér. Rouge Anal. Numer. R-1, 9–53 (1975)
7. Rannacher, D.: Nonconforming finite element method for eigenvalue problems in linear plate theory. Numer. Math. 3, 23–42 (1979)
8. Racheva M.R.: Approximation from below of the exact eigenvalues by means of nonconforming FEMs. Mathematica Balkanica (2012) (to appear)
9. Yang, Y.D.: A posteriori error estimates in Adini finite element for eigenvalue problems. J. Comput. Math. 18, 413–418 (2000)

Symplectic Numerical Schemes for Stochastic Systems Preserving Hamiltonian Functions

Cristina Anton[1], Yau Shu Wong[2], and Jian Deng[2]

[1] Grant MacEwan University, Edmonton, AB T5J 4S2, Canada
popescuc@macewan.ca
[2] University of Alberta, Edmonton, AB T6G 2G1, Canada

Abstract. We present high-order symplectic schemes for stochastic Hamiltonian systems preserving Hamiltonian functions. The approach is based on the generating function method, and we show that for the stochastic Hamiltonian systems, the coefficients of the generating function are invariant under permutations. As a consequence, the high-order symplectic schemes have a simpler form than the explicit Taylor expansion schemes with the same order. Moreover, we demonstrate numerically that the symplectic schemes are effective for long time simulations.

Keywords: stochastic Hamiltonian systems, symplectic integration, mean-square convergence, high-order numerical schemes.

1 Introduction

We consider the autonomous stochastic differential equations (SDEs) in the sense of Stratonovich:

$$
dP_i = -\frac{\partial H^{(0)}(P,Q)}{\partial Q_i}dt - \sum_{r=1}^{m}\frac{\partial H^{(r)}(P,Q)}{\partial Q_i}\circ dw_t^r, \quad P(t_0) = p
$$

$$
dQ_i = \frac{\partial H^{(0)}(P,Q)}{\partial P_i}dt + \sum_{r=1}^{m}\frac{\partial H^{(r)}(P,Q)}{\partial P_i}\circ dw_t^r, \quad Q(t_0) = q, \tag{1}
$$

where P, Q, p, q are n-dimensional vectors with the components P^i, Q^i, p^i, q^i, $i = 1, \ldots, n$, and $w_t^r, r = 1, \ldots, m$ are independent standard Wiener processes. The SDEs (1) are called the Stochastic Hamiltonian System (SHS) ([6]).

The stochastic flow $(p,q) \longrightarrow (P,Q)$ of the SHS (1) preserves the symplectic structure (Theorem 2.1 in [6]) as follows:

$$
dP \wedge dQ = dp \wedge dq, \tag{2}
$$

i.e. the sum over the oriented areas of its projections onto the two dimensional plane (p^i, q^i) is invariant. Here, we consider the differential 2-form

$$
dp \wedge dq = dp^1 \wedge dq^1 + \cdots + dp^n \wedge dq^n, \tag{3}
$$

I. Dimov, I. Faragó, and L. Vulkov (Eds.): NAA 2012, LNCS 8236, pp. 166–173, 2013.
© Springer-Verlag Berlin Heidelberg 2013

and differentiation in (1) and (2) have different meanings: in (1) p, q are fixed parameters and differentiation is done with respect to time t, while in (2) differentiation is carried out with respect to the initial data p, q. We say that a method based on the one step approximation $\bar{P} = \bar{P}(t+h; t, p, q)$, $\bar{Q} = \bar{Q}(t+h; t, p, q)$ preserves symplectic structure if $d\bar{P} \wedge d\bar{Q} = dp \wedge dq$.

Milstein et al. [6] [7] introduced the symplectic numerical schemes for SHS, and they demonstrated the superiority of the symplectic methods for long time computation. Recently, Wang et al. [4],[8] proposed generating function methods to construct symplectic schemes for SHS. In the present study, we focus on SHS that preserve the Hamiltonian function (i.e. SHS for which $dH^{(r)} = 0$, $r = 0, \ldots, m$). We propose higher order symplectic schemes that are computationally efficient for this special type of SHS .

2   The Generating Function Method and Symplectic Schemes

Similar with the deterministic case [3], we have the following result [4] relating the solutions of the Hamilton-Jacobi partial differential equation (HJ PDE) and the solutions of the SHS (1):

Theorem 1. *If $S^1_\omega(P, q)$ is a solution of the HJ PDE*

$$dS^1_\omega = H^{(0)}(P, q + \frac{\partial S^1_\omega}{\partial P})dt + \sum_{r=1}^{m} H^{(r)}(P, q + \frac{\partial S^1_\omega}{\partial P}) \circ dw^r_t, \quad S^1_\omega|_{t=t_0} = 0, \quad (4)$$

and if the matrix $(\frac{\partial^2 S_\omega}{\partial P_i \partial q_j})$ is invertible, then the map $(p, q) \to (P(t, \omega), Q(t, \omega))$ defined by

$$P = p - \frac{\partial S^1_\omega}{\partial q}(P, q), \quad Q = q + \frac{\partial S^1_\omega}{\partial P}(P, q), \quad (5)$$

is the flow of the SHS (1).

The key idea for deriving high order symplectic schemes via generating functions is to obtain an approximation of the solution of HJ PDE, and then to construct the symplectic numerical scheme through the relations (5). It is reasonable to assume that the generating function can be expressed by the following expansion locally [4]

$$S^1_\omega(P, q, t) = G_{(0)}(P, q)J_{(0)} + G_{(1)}(P, q)J_{(1)} + G_{(0,1)}(P, q)J_{(0,1)} + \cdots = \sum_\alpha G_\alpha J_\alpha, \quad (6)$$

where $\alpha = (j_1, j_2, \ldots, j_l), j_i \in \{0, 1, \ldots, m\}$, $i = 1, \ldots, l$ is a multi-index of length $l(\alpha) = l$, and, with $dw^0_s := ds$, J_α is the multiple Stratonovich integral

$$J_\alpha = \int_0^t \int_0^{s_l} \cdots \int_0^{s_2} \circ dw^{j_1}_{s_1} \cdots \circ dw^{j_{l-1}}_{s_{l-1}} \circ dw^{j_l}_{s_l}. \quad (7)$$

If the multi-index $\alpha = (j_1, j_2, \ldots, j_l)$ with $l > 1$, then $\alpha- = (j_1, j_2, \ldots, j_{l-1})$. For any two multi-indexes $\alpha = (j_1, j_2, \ldots, j_l)$ and $\alpha' = (j'_1, j'_2, \ldots, j'_{l'})$, we define the concatenation operation $'*'$ as $\alpha * \alpha' = (j_1, j_2, \ldots, j_l, j'_1, j'_2, \ldots, j'_{l'})$. The concatenation of a collection Λ of multi-indexes with the multi-index α gives the collection $\Lambda * \alpha = \{\alpha' * \alpha\}_{\alpha' \in \Lambda}$.

For any multi-index $\alpha = (j_1, j_2, \ldots, j_l)$ with no duplicated elements (i.e., $j_m \neq j_n$ if $m \neq n$, $1 \leq m, n \leq l$), we define the set $R(\alpha)$ to be the empty set $R(\alpha) = \Phi$ if $l = 1$ and $R(\alpha) = \{(j_m, j_n) | m < n, 1 \leq m, n \leq l\}$ if $l \geq 2$. $R(\alpha)$ defines a partial order on the set formed with the numbers included in the multi-index α, defined by $i \prec j$ if and only if $(i, j) \in R(\alpha)$. We suppose that there are no duplicated elements in or between the multi-indexes $\alpha = (j_1, j_2, \ldots, j_l)$ and $\alpha' = (j'_1, j'_2, \ldots, j'_{l'})$, and we define

$$\Lambda_{\alpha,\alpha'} = \{\beta \in \mathcal{M} | R(\alpha) \cup R(\alpha') \subseteq R(\beta) \text{ and } \beta \text{ has no duplicates}\} \qquad (8)$$

where $\mathcal{M} = \{(\hat{j}_1, \hat{j}_2, \ldots, \hat{j}_{l+l'}) | \hat{j}_i \in \{j_1, j_2, \ldots, j_l, j'_1, j'_2, \ldots, j'_{l'}\}, i = 1, \ldots, l+l'\}$. Analogously if there are no duplicated elements in or between any of the multi-indexes $\alpha = (j_1^{(1)}, j_2^{(1)}, \ldots, j_{l_1}^{(1)})$, $\ldots$, $\alpha_n = (j_1^{(n)}, j_2^{(n)}, \ldots, j_{l_n}^{(n)})$, then we define

$$\Lambda_{\alpha_1,\ldots,\alpha_n} = \{\beta \in \mathcal{M} | \cup_{k=1}^{n} R(\alpha_k) \subseteq R(\beta) \text{ and } \beta \text{ has no duplicates}\}, \qquad (9)$$

where $\mathcal{M} = \{(\hat{j}_1, \hat{j}_2, \ldots, \hat{j}_{\hat{l}}) | \hat{j}_i \in \{j_1^{(1)}, j_2^{(1)}, \ldots, j_{l_1}^{(1)}, \ldots, j_1^{(n)}, j_2^{(n)}, \ldots, j_{l_n}^{(n)}\}, i = 1, \ldots, \hat{l}, \hat{l} = l_1 + \cdots + l_n\}$. For multi-indexes with duplicated elements, we extend the previous definitions by assigning a different subscript to each duplicated element, for example, $\Lambda_{(2,0),(0,1)} = \Lambda_{(2,0_1),(0_2,1)} = \{(2, 0_2, 1, 0_1), (0_2, 2, 1, 0_1), (0_2, 1, 2, 0_1), (0_2, 2, 0_1, 1), (2, 0_1, 0_2, 1), (2, 0_2, 0_1, 1)\} = \{(2, 0, 1, 0), (0, 2, 1, 0), (0, 1, 2, 0), (0, 2, 0, 1), (2, 0, 0, 1), (2, 0, 0, 1)\}$.

We can easily verify that $\Lambda_{\alpha,\alpha'} = \Lambda_{\alpha',\alpha}$, and the length of the multi indexes $\beta \in \Lambda_{\alpha,\alpha'}$, is $l(\beta) = l(\alpha) + l(\alpha')$.

It can be proved [2] that the multiplication of a finite sequence of multiple-indexes can be expressed by the following summation:

$$\prod_{i=1}^{n} J_{\alpha_i} = \sum_{\beta \in \Lambda_{\alpha_1,\ldots,\alpha_n}} J_\beta. \qquad (10)$$

Inserting (6) into the HJ PDE (4), and using the previous equation, we get

$$S_\omega^1 = \int_0^t H^{(0)}(P, q + \sum_\alpha \frac{\partial G_\alpha}{\partial P} J_\alpha) ds + \sum_{r=1}^{m} \int_0^t H^{(r)}(P, q + \sum_\alpha \frac{\partial G_\alpha}{\partial P} J_\alpha) \circ dw_s^r$$

$$= \sum_{r=0}^{m} \sum_{i=0}^{\infty} \sum_{k_1,\ldots,k_i=1}^{n} \sum_{\alpha_1,\ldots,\alpha_i} \sum_{\beta \in \Lambda_{\alpha_1,\ldots,\alpha_i}} \frac{1}{i!} \frac{\partial^i H^{(r)}}{\partial q_{k_1} \ldots \partial q_{k_i}} \frac{\partial G_{\alpha_1}}{\partial P_{k_1}} \cdots \frac{\partial G_{\alpha_i}}{\partial P_{k_i}} J_{\beta*(r)} \qquad (11)$$

where $(\sum_\alpha \frac{\partial G_{\alpha_i}}{\partial P})_{k_i}$ is the k_i-th component of the column vector $\sum_\alpha \frac{\partial G_{\alpha_i}}{\partial P}$. Equating the coefficients of J_α in (6) and (11), we obtain a recurrence formula for determining G_α.

If $\alpha = (r)$, $r = 0, \ldots, m$ then $G_\alpha = H^{(r)}$. If $\alpha = (i_1, \ldots, i_{l-1}, r)$, $l > 1$, $i_1, \ldots, i_{l-1}, r = 0, \ldots, m$ has no duplicates then

$$G_\alpha = \sum_{i=1}^{l(\alpha)-1} \frac{1}{i!} \sum_{k_1, \ldots, k_i = 1}^{n} \frac{\partial^i H^{(r)}}{\partial q_{k_1} \ldots \partial q_{k_i}} \sum_{\substack{l(\alpha_1) + \cdots + l(\alpha_i) = l(\alpha)-1 \\ \alpha - \in \Lambda_{\alpha_1, \ldots, \alpha_i}}} \frac{\partial G_{\alpha_1}}{\partial P_{k_1}} \cdots \frac{\partial G_{\alpha_i}}{\partial P_{k_i}}. \quad (12)$$

If the multi-index α contains any duplicates, then we apply formula (12) after we associate different subscripts to the repeating numbers.

In [2], we prove that the symplectic schemes based on truncations of S_ω^1 for multi-indexes $\alpha \in \mathcal{A}_k = \{\alpha : l(\alpha) + n(\alpha) \leq 2k\}$ have mean square order k, for $k = 1, 1.5, 2$. Here $n(\alpha)$ is the number of components equal with 0 in the multi-index α. For example, using the following truncation of S_ω^1 based on $\mathcal{A}_1$, we can get a scheme with mean square order 1:

$$S_\omega^1 \approx G_{(0)} J_{(0)} + \sum_{r=1}^{m} \left(G_{(r)} J_{(r)} + G_{(r,r)} J_{(r,r)} \right) + \sum_{i,j=1, i\neq j}^{m} G_{(i,j)} J_{(i,j)}. \quad (13)$$

3   Symplectic Schemes for SHS Preserving the Hamiltonian Functions

Unlike the deterministic cases, in general the SHS (1) no longer preserves the Hamiltonian functions $H_i, i = 0, \ldots, n$ with respect to time.

Proposition 1. *The Hamiltonian functions $H^{(i)}, i = 0, \ldots, m$ are invariant for the flow of the system (1), if and only if $\{H^{(i)}, H^{(j)}\} = 0$ for $i, j = 0, \ldots, m$, where the Poisson bracket is defined as $\{H^{(i)}, H^{(j)}\} = \sum_{k=1}^{n} \left(\frac{\partial H^{(j)}}{\partial Q_k} \frac{\partial H^{(i)}}{\partial P_k} - \frac{\partial H^{(i)}}{\partial Q_k} \frac{\partial H^{(j)}}{\partial P_k} \right).$*

Proof. By the chain rule of the Stratonovich stochastic integration, the Hamiltonian functions $H^{(i)}, i = 0, \ldots, m$ are invariant for the system (1), if and only if for every $i = 0, \ldots, m$

$$dH^{(i)} = \sum_{k=1}^{n} \left(\frac{\partial H^{(i)}}{\partial P_k} dP_k + \frac{\partial H^{(i)}}{\partial Q_k} dQ_k \right) = \sum_{k=1}^{n} \left(-\frac{\partial H^{(i)}}{\partial P_k} \frac{\partial H^{(0)}}{\partial Q_k} \right.$$
$$\left. + \frac{\partial H^{(i)}}{\partial Q_k} \frac{\partial H^{(0)}}{\partial P_k} \right) dt + \sum_{r=1}^{m} \sum_{k=1}^{n} \left(-\frac{\partial H^{(i)}}{\partial P_k} \frac{\partial H^{(r)}}{\partial Q_k} + \frac{\partial H^{(i)}}{\partial Q_k} \frac{\partial H^{(r)}}{\partial P_k} \right) \circ dw_t^r = 0. \quad (14)$$

For any permutation on $\{1, \ldots, l\}$, $l \geq 1$ (i.e. for any bijective function $\pi : \{1, \ldots, l\} \to \{1, \ldots, l\}$), and for any multi-index $\alpha = (i_1, \ldots, i_l)$ with $l(\alpha) = l$, let denote by $\pi(\alpha)$ the multi-index defined as $\pi(\alpha) := (i_{\pi(1)}, \ldots, i_{\pi(l)})$. For systems preserving the Hamiltonian functions, the coefficients G_α of S_ω^1 are invariant under the permutations on α, when $l(\alpha) = 2$ because for any $r_1, r_2 = 0, \ldots, m$, we have

$$G_{(r_1,r_2)} = \sum_{k=1}^{n} \frac{\partial H^{(r_2)}}{\partial q_k} \frac{\partial H^{(r_1)}}{\partial P_k} = \sum_{k=1}^{n} \frac{\partial H^{(r_1)}}{\partial q_k} \frac{\partial H^{(r_2)}}{\partial P_k} = G_{(r_2,r_1)}. \tag{15}$$

A simple calculation verifies that G_α are invariant under the permutations on α when $l(\alpha) = 3$. By induction we can prove that this invariance also holds in the general case ([1]).

Proposition 2. *For SHS preserving the Hamiltonian functions, the coefficients G_α are invariants to permutations, i.e $G_\alpha = G_{\pi(\alpha)}$.*

The invariance under permutations of G_α makes higher order symplectic schemes computationally attractive for systems preserving the Hamiltonian functions. For example, for the system (1) with $m = 1$, since $J_{(0,1)} + J_{(1,0)} = J_{(1)}J_{(0)}$ and $J_{(0,1,1)} + J_{(1,0,1)} + J_{(1,1,0)} = J_{(1,1)}J_{(0)}$ (see (10)), we get the following generating function based on the set $\mathcal{A}_2$

$$S_\omega^1 \approx G_{(0)}h + G_{(1)}\sqrt{h}\xi_h + \frac{G_{(0,0)}}{2}h^2 + \frac{G_{(1,1)}}{2}h\xi_h^2 + G_{(1,0)}\xi_h h^{\frac{3}{2}}$$
$$+ \frac{G_{(1,1,1)}}{6}h^{\frac{3}{2}}\xi_h^3 + \frac{G_{(1,1,0)}}{2}\xi_h^2 h^2 + \frac{G_{(1,1,1,1)}}{24}h^2\xi_h^4. \tag{16}$$

Here, we proceed as reported in [7] to construct an implicit scheme based on S_ω^1 and ensuring it is well-defined. If the time step $h < 1$, then when simulating the stochastic integrals J_1, J_{11}, J_{110}, J_{111} and J_{1111}, we replace the random variable $\xi \sim N(0,1)$ with the bounded random variable ξ_h:

$$\xi_h = \begin{cases} -A_h(2) & \text{if } \xi < -A_h(2) \\ \xi & \text{if } |\xi| \leq A_h(2) \\ A_h(2) & \text{if } \xi > A_h(2), \end{cases} \tag{17}$$

where $A_h(2) = 2\sqrt{2|\ln h|}$. Using (5) and (16) we construct the following symplectic scheme:

$$P_i(k+1) = P_i(k) - \left(\frac{\partial G_{(0)}}{\partial Q_i}h + \frac{\partial G_{(1)}}{\partial Q_i}\sqrt{h}\xi_h + \frac{\partial G_{(0,0)}}{\partial Q_i}\frac{h^2}{2} + \frac{\partial G_{(1,1)}}{\partial Q_i}\frac{h\xi_h^2}{2} \right.$$
$$\left. + 2\frac{\partial G_{(1,0)}}{\partial Q_i}\xi_h h^{\frac{3}{2}} + \frac{\partial G_{(1,1,1)}}{\partial Q_i}\frac{h^{\frac{3}{2}}\xi_h^3}{6} + \frac{\partial G_{(1,1,0)}}{\partial Q_i}\frac{3\xi_h^2 h^2}{2} + \frac{\partial G_{(1,1,1,1)}}{\partial Q_i}\frac{h^2\xi_h^4}{24} \right)$$
$$Q_i(k+1) = Q_i(k) + \left(\frac{\partial G_{(0)}}{\partial P_i}h + \frac{\partial G_{(1)}}{\partial P_i}\sqrt{h}\xi_h + \frac{\partial G_{(0,0)}}{\partial P_i}\frac{h^2}{2} + \frac{\partial G_{(1,1)}}{\partial P_i}\frac{h\xi_h^2}{2} \right. \tag{18}$$
$$\left. + 2\frac{\partial G_{(1,0)}}{\partial P_i}\xi_h h^{\frac{3}{2}} + \frac{\partial G_{(1,1,1)}}{\partial P_i}\frac{h^{\frac{3}{2}}\xi_h^3}{6} + \frac{\partial G_{(1,1,0)}}{\partial P_i}\frac{3\xi_h^2 h^2}{2} + \frac{\partial G_{(1,1,1,1)}}{\partial P_i}\frac{h^2\xi_h^4}{24} \right),$$

where everywhere the arguments are $(P(k+1), Q(k))$. From [7] we know that $E(\xi - \xi_h)^2 \leq h^4$ and $0 \leq E(\xi^2 - \xi_h^2) \leq 7h^{7/2}$, so proceeding as in [2] we can prove that (18) is a mean square second-order scheme.

Based on (5) and the truncation (13) we can build the symplectic mean square first-order scheme:

$$P_i(k+1) = P_i(k) - \left(\frac{\partial G_{(0)}}{\partial Q_i} h + \frac{\partial G_{(1)}}{\partial Q_i} \sqrt{h}\zeta_h + \frac{\partial G_{(1,1)}}{\partial Q_i} \frac{h\zeta_h^2}{2} \right)$$

$$Q_i(k+1) = Q_i(k) + \left(\frac{\partial G_{(0)}}{\partial P_i} h + \frac{\partial G_{(1)}}{\partial P_i} \sqrt{h}\zeta_h + \frac{\partial G_{(1,1)}}{\partial P_i} \frac{h\zeta_h^2}{2} \right),$$

(19)

where everywhere the arguments are $(P(k+1), Q(k))$ and ζ_h is defined as in (17), but with $A_h(2)$ replaced by $A_h(1) = 2\sqrt{|\ln h|}$.

4   Numerical Simulations and Conclusions

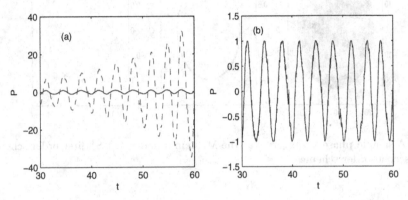

Fig. 1. Sample path for (20): (-) exact solution , (- -) numerical solution from (a) the explicit Milstein scheme and (b) the mean square second-order symplectic scheme

The mathematical model for the Kubo oscillator is given by

$$dP = -aQdt - \sigma Q \circ dw_t^1, \quad P(0) = p_0,$$

$$dQ = aPdt + \sigma P \circ dw_t^2, \quad Q(0) = q_0,$$

(20)

where a and σ are constants. This example has been studied in [7] to demonstrate the performance of the stochastic symplectic scheme for long time computation. The linear system with constant coefficients (20) can be solved analytically (see chapter 4 in [5]) , so we can easily simulate trajectories of the exact solution. The Hamiltonian functions are $H^{(0)}(P(t), Q(t)) = a\frac{P(t)^2 + Q(t)^2}{2}$ and $H^{(1)}(P(t), Q(t)) = \sigma\frac{P(t)^2 + Q(t)^2}{2}$, and it is easy to verify that they are preserved under the phase flow of the systems. As a consequence, the phase trajectory of (20) lies on the circle with the center at the origin and the radius $\sqrt{p_0^2 + q_0^2}$.

Replacing in (12), we obtain the following coefficients $G_\alpha(P, q)$ of $S_\omega^1(P, q)$:

$$G_{(0)} = \frac{a}{2}(P^2 + q^2), \quad G_{(1)} = \frac{\sigma}{2}(P^2 + q^2), \quad G_{(0,0)} = a^2 Pq, \quad G_{(1,1)} = \sigma^2 Pq,$$

$$G_{(1,0)} = G_{(0,1)} = a\sigma Pq, \quad G_{(0,0,0)} = a^3(P^2 + q^2), \quad G_{(1,1,1)} = \sigma^3(P^2 + q^2),$$

$$G_{(1,1,0)} = G_{(1,0,1)} = G_{(0,1,1)} = a\sigma^2(P^2 + q^2), \quad G_{(1,1,1,1)} = 5\sigma^4 Pq. \tag{21}$$

Here, we consider the mean square first-order scheme (19), and the mean square

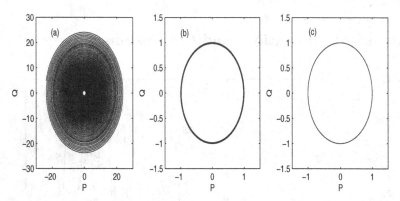

Fig. 2. A sample phase trajectory: (a) the Milstein scheme; (b) S_ω^1 first-order scheme; (c) S_ω^1 second-order scheme

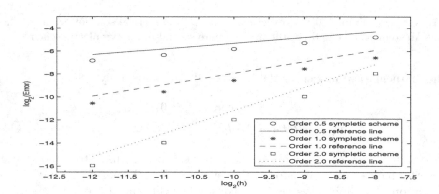

Fig. 3. Convergence rate of different order S_ω^1 symplectic schemes

second-order scheme given in (18). Fig. 1 displays sample paths computed using the scheme (18) and the explicit mean square order one Milstein scheme ([5]) for $a = 2$, $\sigma = 0.3$, $p_0 = 1$ and $q_0 = 0$. Comparing with the exact solution we notice that the explicit scheme gives a divergent solution (see Fig. 1 a), while the symplectic scheme (18) produce accurate results (see Fig. 1 b).

Moreover, to validate the performance of symplectic schemes for long term simulations, in Fig. 2, we display sample phase trajectories of (20) computed using the explicit order one Milstein scheme given in [5] and the mean square order one and two symplectic schemes proposed in this paper. The time interval is $0 \le T \le 200$ and the time step $h = 2^{-8}$. It is clear that the phase trajectory of the Milstein non-symplectic scheme deviates from the circle $P(t)^2 + Q(t)^2 = 1$, while the proposed symplectic schemes produce accurate numerical solutions.

In [7], a mean square order 0.5 symplectic scheme is presented. Fig. 3 confirms the expected convergence rate for the symplectic schemes with the mean square orders 0.5, 1, 2, where the error is the maximum error of (P, Q) at $T = 100$.

4.1 Conclusions

We construct high-order symplectic schemes based on the generating functions for stochastic Hamiltonian systems preserving Hamiltonian functions. Since the coefficients of the generating function are invariant under permutations, the high-order implicit symplectic schemes have simpler forms and require less multiple stochastic integrals than the explicit Taylor expansion schemes. Based on the numerical simulations presented in this study, we conclude that the symplectic schemes are very effective for long term computations.

Acknowledgement. The authors are grateful for the support provided by the Natural Sciences and Engineering Research Council of Canada.

References

1. Anton, C., Deng, J., Wong, Y.S.: Symplectic schemes for stochastic Hamiltonian systems preserving Hamiltonian functions. Int. J. Num. Anal. Mod. (submitted)
2. Deng, J., Anton, C., Wong, Y.S.: High-order symplectic schemes for stochastic Hamiltonian systems. Comm. Comp. Physics (submitted)
3. Hairer, E.: Geometric numerical integration: structure-preserving algorithms for ordinary differential equations. Springer, Berlin (2006)
4. Hong, J., Wang, L., Scherer, R.: Simulation of stochastic Hamiltonian systems via generating functions. In: Proceedings IEEE 2011 4th ICCSIT (2011)
5. Kloeden, P., Platen, E.: Numerical solutions of stochastic differential equations. Springer, Berlin (1992)
6. Milstein, G.N., Tretyakov, M.V., Repin, Y.M.: Symplectic integration of Hamiltonian systems with additive noise. SIAM J. Num. Anal. 39, 2066–2088 (2002)
7. Milstein, G.N., Tretyakov, M.V., Repin, Y.M.: Numerical methods for stochastic systems preserving symplectic structure. SIAM J. Num. Anal. 40, 1583–1604 (2002)
8. Wang, L.: Variational integrators and generating functions for stochastic Hamiltonian systems. Dissertation, University of Karlsruhe, Germany, KIT Scientific Publishing (2007), http://www.ksp.kit.edu

High Order Accurate Difference Schemes
for Hyperbolic IBVP

Allaberen Ashyralyev and Ozgur Yildirim

Fatih University, Department of Mathematics,
Buyukcekmece, 34500 Istanbul, Turkey
Yildiz Technical University, Department of Mathematics,
Davutpasa, 34210 Istanbul, Turkey
aashyr@fatih.edu.tr,
ozgury@yildiz.edu.tr

Abstract. In the present paper the initial-boundary value problem for multidimensional hyperbolic equation with Dirichlet condition is considered. The third and fourth orders of accuracy difference schemes for the approximate solution of this problem are presented and the stability estimates for the solutions of these difference schemes are obtained. Some results of numerical experiments are presented in order to support theoretical statements.

Keywords: Hyperbolic equation, Stability, Initial boundary value problem.

1 Introduction

Partial differential equations of the hyperbolic type play an important role in many branches of science and engineering. The study of this type of problems is driven not only by a theoretical interest but also by the fact that several phenomena in engineering and physics can be modeled in this way. For example, acoustics, electromagnetics, hydrodynamics, elasticity, fluid mechanics, and other areas of physics lead to partial differential equations of the hyperbolic type (see, e.g., [1]- [7] and the references given therein). In the development of numerical techniques for solving these equations, the stability has been an important research topic (see [8]-[27]).

A large cycle of works on difference schemes for hyperbolic partial differential equations, in which stability was established under the assumption that the magnitude of the grid steps τ and h with respect to time and space variables are connected (see, e.g., [23]- [27] and the references given therein). We are interested in studying the high order of accuracy difference schemes for hyperbolic PDEs, in which stability is established without any assumption in respect of grid steps τ and h.

Many of the scientists have studied unconditionally stable difference schemes for approximately solving partial differential equations of the hyperbolic type (see, [8]-[22]). The high order of accuracy two-step difference schemes generated

I. Dimov, I. Faragó, and L. Vulkov (Eds.): NAA 2012, LNCS 8236, pp. 174–181, 2013.

by an exact difference scheme or by the Taylor decomposition on the three points for the numerical solution of the same problem were presented and the stability estimates for approximate solution of paper these difference schemes were obtained in [8]. However, the difference methods of [8] are difficult for a realization.

In the present paper, the following initial value problem

$$
\begin{cases}
\frac{\partial^2 u(t,x)}{\partial t^2} - \sum_{r=1}^{m}(a_r(x)u_{x_r})_{x_r} = f(t,x), \\
x = (x_1,\ldots,x_m) \in \Omega, \ 0 < t < 1, \\
u(0,x) = \varphi(x), u_t(0,x) = \psi(x), x \in \overline{\Omega}, \\
u(t,x) = 0, x \in S
\end{cases}
\tag{1}
$$

for the multidimensional hyperbolic equation with Dirichlet condition is considered. Let Ω be a unit open cube in the m-dimensional Euclidean space $\mathbb{R}^m \{x = (x_1,\cdots,x_m) : 0 < x_j < 1, 1 \leq j \leq m\}$ with boundary S, $\overline{\Omega} = \Omega \cup S$. The third and fourth orders of accuracy difference schemes for the approximate solution of (1) are constructed using integer powers of the space operator generated by problem (1), and the stability estimates for the solution of these difference schemes are presented.

Note that boundary value problems for parabolic equations, elliptic equations and equations of mixed types have been studied extensively by many scientists (see, e.g., [16]-[30] and the references given therein).

2 Third Order of Accuracy Difference Scheme

In the first step, let us define the grid sets

$$
\widetilde{\Omega}_h = \{x = x_r = (h_1 r_1, \cdots, h_m r_m), r = (r_1,\cdots,r_m),
\tag{2}
$$

$$
0 \leq r_j \leq N_j, h_j N_j = 1, j = 1,\cdots,m\}, \Omega_h = \widetilde{\Omega}_h \cap \Omega, S_h = \widetilde{\Omega}_h \cap S.
\tag{3}
$$

We introduce the Banach space $L_{2h} = L_2(\widetilde{\Omega}_h)$, $W_{2h}^1 = W_{2h}^1\left(\widetilde{\Omega}_h\right)$ and $W_{2h}^2 = W_{2h}^2\left(\widetilde{\Omega}_h\right)$ of the grid functions $\varphi^h(x) = \{\varphi(h_1 r_1,\cdots,h_m r_m)\}$ defined on $\widetilde{\Omega}_h$, equipped with norms

$$
\|\varphi^h\|_{L_2(\widetilde{\Omega}_h)} = \left(\sum_{x \in \overline{\Omega}_h} |\varphi^h(x)|^2 h_1 \cdots h_m \right)^{1/2},
\tag{4}
$$

$$
\|\varphi^h\|_{W_{2h}^1} = \|\varphi^h\|_{L_{2h}} + \left(\sum_{x \in \overline{\Omega}_h} \sum_{r=1}^{m} |(\varphi^h)_{\overline{x}_r, j_r}|^2 h_1 \cdots h_m \right)^{1/2},
\tag{5}
$$

and

$$\|\varphi^h\|_{W_{2h}^2} = \|\varphi^h\|_{W_{2h}^1} + \left(\sum_{x\in\overline{\Omega}_h}\sum_{r=1}^m \left|(\varphi^h)_{x_r\overline{x}_r,j_r}\right|^2 h_1\cdots h_m\right)^{1/2}, \quad (6)$$

respectively. To the differential operator A generated by problem (1), we assign the difference operator A_h^x by the formula

$$A_h^x u^h = -\sum_{r=1}^m \left(a_r(x)u_{\overline{x}_r}^h\right)_{x_r,j_r} \quad (7)$$

acting in the space of grid functions $u^h(x)$, satisfying the conditions $u^h(x) = 0$ for all $x \in S_h$.

It is known that A_h^x is a self-adjoint positive definite operator in $L_2(\widetilde{\Omega}_h)$. With the help of A_h^x we arrive at the Cauchy problem

$$\begin{cases} \dfrac{d^2 v^h(t,x)}{dt^2} + A_h^x v^h(t,x) = f^h(t,x), 0 < t < 1, \ x \in \Omega_h, \\ v^h(0,x) = \varphi^h(x), x \in \widetilde{\Omega}_h, \\ \dfrac{dv^h(0,x)}{dt} = \psi^h(x), x \in \widetilde{\Omega}_h \end{cases} \quad (8)$$

for an infinite system of ordinary differential equations.

In the second step, we replace problem (8) by the following difference scheme

$$\begin{cases} \dfrac{u_{k+1}^h(x)-2u_k^h(x)+u_{k-1}^h(x)}{\tau^2} + \frac{2}{3}A_h^x u_k^h(x) + \frac{1}{6}A_h^x\left(u_{k+1}^h(x)+u_{k-1}^h(x)\right) \\ +\frac{1}{12}\tau^2\left(A_h^x\right)^2 u_{k+1}^h(x) = f_k^h(x), \\ f_k^h(x) = \frac{2}{3}f^h(t_k,x) + \frac{1}{6}\left(f^h(t_{k+1},x) + f^h(t_{k-1},x)\right) \\ -\frac{1}{12}\tau^2\left(-Af^h(t_{k+1},x) + f_{tt}^h(t_{k+1},x)\right), x \in \Omega_h, \\ t_k = k\tau, 1 \le k \le N-1, N\tau = 1, \\ u_0^h(x) = \varphi^h(x), x \in \widetilde{\Omega}_h, \\ \left(I + \frac{\tau^2}{12}\left(A_h^x\right) + \frac{\tau^4}{144}\left(A_h^x\right)^2\right)\tau^{-1}\left(u_1^h(x) - u_0^h(x)\right) \\ = \left(-\left(\frac{\tau}{2}\left(A_h^x\right)\right)\varphi^h(x) + \left(I - \frac{\tau^2}{12}\left(A_h^x\right)\right)\psi^h(x) + \tau f_{1,1}^h(x)\right), x \in \widetilde{\Omega}_h, \\ f_{1,1}^h(x) = \frac{1}{2}f^h(0,x) + \frac{\tau}{6}f_t^h(0,x), x \in \widetilde{\Omega} \end{cases} \quad (9)$$

Theorem 1. *Let τ and $|h|$ be sufficiently small numbers. Then, the solution of difference scheme (9) satisfies the following stability estimates:*

$$\max_{0 \leq k \leq N} \left\| u_k^h \right\|_{L_{2h}} + \max_{0 \leq k \leq N} \left\| u_k^h \right\|_{W_{2h}^1} + \max_{1 \leq k \leq N} \left\| \tau^{-1} \left(u_k^h - u_{k-1}^h \right) \right\|_{L_{2h}}$$

$$\leq M_1 \left[\max_{1 \leq k \leq N-1} \left\| f_k^h \right\|_{L_{2h}} + \left\| \psi^h \right\|_{L_{2h}} + \left\| \varphi^h \right\|_{W_{2h}^1} + \tau \left\| f_{1,1}^h \right\|_{L_{2h}} \right],$$

$$\max_{1 \leq k \leq N-1} \left\| \tau^{-2} \left(u_{k+1}^h - 2u_k^h + u_{k-1}^h \right) \right\|_{L_{2h}} + \max_{0 \leq k \leq N} \left\| u_k^h \right\|_{W_{2h}^2}$$

$$+ \max_{1 \leq k \leq N} \left\| \tau^{-1} \left(u_k^h - u_{k-1}^h \right) \right\|_{W_{2h}^1} \leq M_1 \left[\left\| f_1^h \right\|_{L_{2h}} + \max_{2 \leq k \leq N-1} \left\| \tau^{-1} \left(f_k^h - f_{k-1}^h \right) \right\|_{L_{2h}} \right.$$

$$\left. + \left\| \psi^h \right\|_{W_{2h}^1} + \left\| \varphi^h \right\|_{W_{2h}^2} + \tau \left\| f_{1,1}^h \right\|_{W_{2h}^1} \right].$$

Here M_1 does not depend on τ, h, $\varphi^h(x)$, $\psi^h(x)$, $f_{1,1}^h(x)$ and $f_k^h(x)$, $1 \leq k < N$.

The proof of Theorem 1 is based on symmetry property of the operator A_h^x defined by formula (7) and the following theorem on the coercivity inequality for solution of elliptic difference problem in L_{2h}.

Theorem 2. *For the solutions of the elliptic difference problem*

$$A_h^x u^h(x) = \omega^h(x), x \in \Omega_h, \tag{10}$$

$$u^h(x) = 0, x \in S_h \tag{11}$$

the following coercivity inequality holds (see, [20]) :

$$\sum_{r=1}^{m} \left\| u_{x_r \bar{x}_r, j_r}^h \right\|_{L_{2h}} \leq M \| \omega^h \|_{L_{2h}}. \tag{12}$$

Here M does not depend on h, $\omega^h(x)$.

3 Fourth Order of Accuracy Difference Scheme

Now, let us consider the fourth order of accuracy difference scheme for approximately solving initial value problem (1). In a similar way as described in the previous section the discretization of problem (1) is carried out in two steps. The first step is exactly the same with the step previously presented. In the second step, we replace problem (8) by the following difference scheme

$$\begin{cases} \frac{u_{k+1}^h(x)-2u_k^h(x)+u_{k-1}^h(x)}{\tau^2} + \frac{5}{6}A_h^x u_k^h(x) + \frac{1}{12}A_h^x\left(u_{k+1}^h(x)+u_{k-1}^h(x)\right) \\[2mm] -\frac{1}{72}\tau^2\left(A_h^x\right)^2 u_{k+1}^h(x) + \frac{1}{144}\tau^2\left(A_h^x\right)^2\left(u_{k+1}^h(x)+u_{k-1}^h(x)\right) = f_k^h(x), \\[2mm] f_k^h(x) = \frac{5}{6}f^h(t_k,x) + \frac{1}{12}\left(f^h(t_{k+1},x)+f^h(t_{k-1},x)\right) \\[2mm] +\frac{1}{72}\tau^2\left(-Af^h(t_k,x)+f_{tt}^h(t_k,x)\right) \\[2mm] -\frac{1}{144}\tau^2\left(-A\left(f^h(t_{k+1},x)+f^h(t_{k-1},x)\right)\right. \\[2mm] \left.+f_{tt}^h(t_{k+1},x)+f_{tt}^h(t_{k-1},x)\right), x\in\Omega_h, \\[2mm] t_k = k\tau, N\tau = 1, 1\le k\le N-1, \\[2mm] u_0^h(x) = \varphi^h(x), x\in\widetilde{\Omega}_h, \\[2mm] \left(I + \frac{\tau^2}{12}\left(A_h^x\right) + \frac{\tau^4}{144}\left(A_h^x\right)^2\right)\tau^{-1}\left(u_1^h(x)-u_0^h(x)\right) \\[2mm] = \left(-\left(\frac{\tau}{2}\left(A_h^x\right)\right)\varphi^h(x) + \left(I-\frac{\tau^2}{12}\left(A_h^x\right)\right)\psi^h(x) + \tau f_{2,2}^h(x)\right), \\[2mm] f_{2,2}^h(x) = \frac{1}{2}f^h(0,x) + \frac{\tau}{6}f_t^h(0,x) + \frac{\tau^2}{24}f_{tt}^h(0,x) \; x\in\widetilde{\Omega}_h. \end{cases} \tag{13}$$

Theorem 3. *Let τ and $|h|$ be sufficiently small numbers. Then, the solution of difference scheme (13) satisfies the following stability estimates:*

$$\max_{0\le k\le N}\left\|\frac{u_k^h+u_{k-1}^h}{2}\right\|_{L_{2h}} + \max_{0\le k\le N}\left\|\frac{u_k^h+u_{k-1}^h}{2}\right\|_{W_{2h}^1}$$

$$+ \max_{1\le k\le N-1}\left\|\frac{u_{k+1}^h-u_{k-1}^h}{2\tau}\right\|_{L_{2h}}$$

$$\le M_1\left[\max_{1\le k\le N-1}\left\|f_k^h\right\|_{L_{2h}} + \left\|\psi^h\right\|_{L_{2h}} + \left\|\varphi^h\right\|_{W_{2h}^1} + \tau\left\|f_{2,2}\right\|_{L_{2h}}\right],$$

$$\max_{1\le k\le N-1}\left\|\frac{u_{k+1}^h-u_{k-1}^h}{2\tau}\right\|_{W_{2h}^1} + \max_{0\le k\le N}\left\|\frac{u_k^h+u_{k-1}^h}{2}\right\|_{W_{2h}^2}$$

$$\le M_1\left[\left\|f_1^h\right\|_{L_{2h}} + \max_{2\le k\le N-1}\left\|\tau^{-1}\left(f_k^h-f_{k-1}^h\right)\right\|_{L_{2h}}\right.$$

$$\left.+ \left\|\psi^h\right\|_{W_{2h}^1} + \left\|\varphi^h\right\|_{W_{2h}^2} + \tau\left\|f_{2,2}^h\right\|_{W_{2h}^1}\right].$$

Here M_1 does not depend on τ, h, $\varphi^h(x)$, $\psi^h(x)$, $f_{2,2}^h(x)$ and $f_k^h(x), 1\le k < N$.

The proof of Theorem 3 is based on the symmetry property of the operator A_h^x defined by formula (7) and Theorem 2, on the coercivity inequality for the solution of the elliptic difference problem in L_{2h}.

4   Numerical Results

In the last section, some results of numerical experiments in order to support our theoretical statements are presented. The following problem

$$\begin{cases} \frac{\partial^2 u(t,x)}{\partial t^2} - \frac{\partial^2 u(t,x)}{\partial x^2} = 2e^{-t}\sin x, 0 < t < 1, 0 < x < \pi, \\ u(0,x) = \sin x, u_t(0,x) = -\sin x, 0 \le x \le \pi, \\ u(t,0) = u(t,\pi) = 0, 0 \le t \le 1 \end{cases} \tag{14}$$

for one dimensional hyperbolic equation is considered for the numerical results. The exact solution of this problem is

$$u(t,x) = e^{-t}\sin x. \tag{15}$$

For the approximate solution of problem (14), the third and fourth orders of accuracy difference schemes are used respectively and a system of linear equations with matrix coefficients is obtained. Solving this system a procedure of modified Gauss elimination method with respect to n with matrix coefficients is applied. The implementations of numerical experiments are carried out by Matlab. Errors are computed by the following formula

$$E_M^N = \max_{1 \le k \le N-1, 1 \le k \le M-1} \left| u(t_k, x_n) - u_n^k \right| \tag{16}$$

Here, $u(t_k, x_n)$ represents the exact solution of problem (14) and u_n^k represents the numerical solution of problem (14) at (t_k, x_n).

Table 1. Errors for the approximate solution of problem (14)

Method		
	$N = 10, M = 30$	$N = 20, M = 50$
Difference scheme (9)	$0{,}7107.10^{-4}$	$0{,}7645.10^{-5}$
	$N = 10, M = 100$	$N = 20, M = 400$
Difference scheme (13)	$0{,}2459.10^{-4}$	$0{,}1628.10^{-5}$

The results are presented for different step numbers M and N which represent time and space variables respectively.

Acknowledgments. The authors would like to thank Prof. P. E. Sobolevskii for his helpful suggestions to the improvement of this paper.

References

1. Lamb, H.: Hydrodynamics. Cambridge University Press, Cambridge (1993)
2. Lighthill, J.: Waves in Fluids. Cambridge University Press, Cambridge (1978)
3. Hudson, J.A.: The Excitation and Propagation of Elastic Waves. Cambridge University Press, Cambridge (1980)
4. Jones, D.S.: Acoustic and Electromagnetic Waves. The Clarendon Press/Oxford University Press, New York (1986)
5. Taflove, A.: Computational Electrodynamics: The Finite-Difference Time-Domain Method. Artech House, Boston (1995)
6. Ciment, M., Leventhal, S.H.: A note on the operator compact implicit method for the wave equation. Mathematics of Computation 32(141), 143–147 (1978)
7. Twizell, E.H.: An explicit difference method for the wave equation with extended stability range. BIT 19(3), 378–383 (1979)
8. Ashyralyev, A., Sobolevskii, P.E.: Two new approaches for construction of the high order of accuracy difference schemes for hyperbolic differential equations. Discrete Dynamics Nature and Society 2(1), 183–213 (2005)
9. Ashyralyev, A., Sobolevskii, P.E.: New Difference Schemes for Partial Differential Equations, Operator Theory: Advances and Applications. Birkhäuser, Basel (2004)
10. Ashyralyev, A., Sobolevskii, P.E.: A note on the difference schemes for hyperbolic equations. Abstract and Applied Analysis 6(2), 63–70 (2001)
11. Ashyralyev, A., Koksal, M.E., Agarwal, R.P.: A difference scheme for Cauchy problem for the hyperbolic equation with self-adjoint operator. Mathematical and Computer Modelling 52(1-2), 409–424 (2010)
12. Ashyralyev, A., Aggez, N.: A note on the difference schemes of the nonlocal boundary value problems for hyperbolic equations. Numerical Functional Analysis and Optimization 25(5-6), 1–24 (2004)
13. Ashyralyev, A., Yildirim, O.: Second order of accuracy stable difference schemes for hyperbolic problems subject to nonlocal conditions with self-adjoint operator. In: 9th International Conference of Numerical Analysis and Applied Mathematics, pp. 597–600. AIP Conference Proceedings, New York (2011)
14. Ashyralyev, A., Yildirim, O.: On multipoint nonlocal boundary value problems for hyperbolic differential and difference equations. Taiwanese Journal of Mathematics 14(1), 165–194 (2010)
15. Sobolevskii, P.E., Chebotaryeva, L.M.: Approximate solution by method of lines of the Cauchy problem for an abstract hyperbolic equations. Izv. Vyssh. Uchebn. Zav. Matematika 5(1), 103–116 (1977)
16. Samarskii, A.A., Lazarov, R.D., Makarov, V.L.: Difference Schemes for Differential Equations with Generalized Solutions. Vysshaya Shkola, Moscow (1987) (in Russian)
17. Ashyralyev, A., Ozdemir, Y.: On stable implicit difference scheme for hyperbolic-parabolic equations in a Hilbert space. Numerical Methods for Partial Differential Equations 25(1), 1110–1118 (2009)
18. Ashyralyev, A., Gercek, O.: On second order of accuracy difference scheme of the approximate solution of nonlocal elliptic-parabolic problems. Abstract and Applied Analysis 2010 (ID 705172), 17 p. (2010)
19. Ashyralyev, A., Judakova, G., Sobolevskii, P.E.: A note on the difference schemes for hyperbolic-elliptic equations. Abstract and Applied Analysis 2006, 01–13 (2006)
20. Sobolevskii, P.E.: Difference Methods for the Approximate Solution of Differential Equations. Izdat. Gosud. Univ: Voronezh (1975)

21. Gordezani, D., Meladze, H., Avalishvili, G.: On one class of nonlocal in time problems for first-order evolution equations. Zh. Obchysl. Prykl. Mat. 88(1), 66–78 (2003)
22. Sapagovas, M.: On stability of the finite difference schemes for a parabolic equations with nonlocal condition. Journal of Computer Applied Mathematics 1(1), 89–98 (2003)
23. Mitchell, A.R., Griffiths, D.F.: The Finite Difference Methods in Partial Differential Equations. Wiley, New York (1980)
24. Lax, P.D., Wendroff, B.: Difference schemes for hyperbolic equations with high order of accuracy. Comm. Pure Appl. Math. 17, 381–398 (1964)
25. Gustafson, B., Kreiss, H.O., Oliger, J.: Time dependent problems and difference methods. Wiley, New York (1995)
26. Martín-Vaquero, J., Queiruga-Dios, A., Encinas, A.H.: Numerical algorithms for diffusion-reaction problems with non-classical conditions. Applied Mathematics and Computation 218(9), 5487–5495 (2012)
27. Simos, T.E.: A finite-difference method for the numerical solution of the Schrodinger equation. Journal of Computational and Applied Mathematics 79(4), 189–205 (1997)
28. Fattorini, H.O.: Second Order Linear Differential Equations in Banach Space. Notas de Matematica, North-Holland (1985)
29. Krein, S.G.: Linear Differential Equations in a Banach Space. Nauka, Moscow (1966)
30. Piskarev, S., Shaw, Y.: On certain operator families related to cosine operator function. Taiwanese Journal of Mathematics 1(4), 3585–3592 (1997)

Modified Crank-Nicholson Difference Schemes for Ultra Parabolic Equations with Neumann Condition

Allaberen Ashyralyev[1,2] and Serhat Yilmaz[1]

[1] Department of Mathematics, Fatih University, Istanbul, Turkey
[2] Department of Mathematics, ITTU, Ashgabat, Turkmenistan

Abstract. In this paper, our interest is studying the stability of difference schemes for the approximate solution of the initial boundary value problem for ultra-parabolic equations. For approximately solving the given problem, the second-order of accuracy modified Crank-Nicholson difference schemes are presented. Theorem on almost coercive stability of these difference schemes is established. Numerical example is given to illustrate the applicability and efficiency of our method.

Keywords: Ultra parabolic equations, difference schemes, stability estimates, matlab implementation, numerical solutions.

1 Introduction

We refer to [1-13] and the references therein for a series of papers dealing with the initial boundary value problem for ultra parabolic equations

$$
\begin{cases}
\frac{\partial u(t,s)}{\partial t} + \frac{\partial u(t,s)}{\partial s} + Au(t,s) = f(t,s), \ \ 0 < t, s < T, \\[2mm]
u(0,s) = \psi(s), \ \ 0 \le s \le T, \\[2mm]
u(t,0) = \varphi(t), \ \ 0 \le t \le T
\end{cases}
\tag{1}
$$

which arise in many natural phenomenons. In paper [13], for approximately solving problem (1) r-modified Crank-Nicolson difference schemes of the second-order of accuracy

$$
\begin{cases}
\frac{u_{k,m} - u_{k-1,m-1}}{\tau} + Au_{k,m} = f_{k,m} \ , \ \ 1 \le k, m \le r, \\[2mm]
\frac{u_{k,m} - u_{k-1,m-1}}{\tau} + \frac{A}{2}(u_{k,m} + u_{k-1,m-1}) = f_{k,m} \ , \ \ r+1 \le k, m \le N, \\[2mm]
f_{k,m} = f(t_k - \frac{\tau}{2}, s_m - \frac{\tau}{2}), \, t_k = k\tau, \ s_m = m\tau, \ 1 \le k, m \le N, \ N\tau = 1, \\[2mm]
u_{0,m} = \psi_m, \psi_m = \psi(s_m), \ 0 \le m \le N, \\[2mm]
u_{k,0} = \varphi_k, \varphi_k = \varphi(t_k), \ 0 \le k \le N
\end{cases}
\tag{2}
$$

I. Dimov, I. Faragó, and L. Vulkov (Eds.): NAA 2012, LNCS 8236, pp. 182–189, 2013.
© Springer-Verlag Berlin Heidelberg 2013

were presented. The following theorem on stability estimates for the solution of difference schemes (2) was established.

Theorem 1. *For the solution of problem (2), we have the following stability inequality*

$$\max_{1\leq k,m\leq N}\|u_{k,m}\|_E \leq C\left(\max_{0\leq m\leq N}\|\psi_m\|_E + \max_{0\leq k\leq N}\|\varphi_k\|_E + \max_{1\leq k,m\leq N}\|f_{k,m}\|_E\right),$$

where C is independent of τ, ψ_m, φ_k, and $f_{k,m}$.

The main result of the present paper is study of almost coercivity of difference schemes. Theorem on almost coercivity of (2) is established. In application the almost coercive stability estimates in maximum norm for the solution of difference schemes for multidimensional ultra parabolic equations with Neumann condition are obtained. The theoretical statements are supported by a numerical example.

Theorem 2. *For the solution of problem (2), we have the almost coercivity inequality*

$$\max_{1\leq k,m\leq N}\left\|\frac{u_{k,m}-u_{k-1,m-1}}{\tau}\right\|_E$$

$$+ \max_{1\leq k,m\leq r}\|Au_{k,m}\|_E + \max_{r+1\leq k,m\leq N}\left\|\frac{A}{2}(u_{k,m}+u_{k-1,m-1})\right\|_E$$

$$\leq C\left(\max_{0\leq m\leq N}\|A\psi_m\|_E + \max_{0\leq k\leq N}\|A\varphi_k\|_E\right.$$

$$\left. + \min\left\{\ln\frac{1}{\tau},\ 1+|\ln\|A\|_{E\to E}|\right\}\max_{1\leq k,m\leq N}\|f_{k,m}\|\right),$$

where C is independent of τ, ψ_m, φ_k, and $f_{k,m}$.

Proof. Using the identities(see [13])

$$u_{k,m} = R^k\psi_{m-k} + \sum_{j=1}^{k}\tau R^{k-j+1}f_{j,m-k+j}, k\leq m$$

$$u_{k,m} = R^m\phi_{k-m} + \sum_{j=1}^{m}\tau R^{m-j+1}f_{k-m+j,j}, m\leq k \tag{3}$$

and estimates(see [14])

$$\|R^k\|_{E\to E}\leq M$$

$$\|AR^k\|_{E\to E}\leq \frac{M}{k\tau} \tag{4}$$

we obtain

$$\|Au_{k,m}\|_E \leq \sum_{j=1}^{k} \tau \|AR^{k-j+1}\|_{E \to E} \|f_{j,m-k+j}\|_E$$

$$\leq M \sum_{j=1}^{k} \frac{1}{k+1-j} \max_{1 \leq j,m \leq N} \|f_{j,m}\|_E.$$

Since

$$\sum_{j=1}^{k} \frac{1}{k+1-j} \leq \int_{1}^{k} \frac{ds}{k+1-s} = lnk,$$

we have

$$\|Au_{k,m}\|_E \leq Mlnk \max_{1 \leq j,m \leq N} \|f_{j,m}\|_E,$$

hence

$$\max_{1 \leq k,m \leq r} \|Au_{k,m}\|_E \leq Mln\frac{1}{\tau} \max_{1 \leq j,m \leq N} \|f_{j,m}\|_E. \tag{5}$$

Further, using identities (3) we obtain

$$\|Au_{k,m}\|_E \leq \sum_{j=1}^{k} \tau \|AR^{k-j+1}\|_{E \to E} \|f_{j,m-k+j}\|_E$$

$$\leq \sum_{j=1}^{k} \tau \|AR^{k-j+1}\|_{E \to E} \max_{1 \leq j,m \leq N} \|f_{j,m}\|_E.$$

It remains to estimate the quantity

$$J_k = \sum_{j=1}^{k} \tau \|AR^{k-j+1}\|_{E \to E} = \sum_{j=1}^{k} \tau \|AR^s\|_{E \to E}.$$

From the last identity it is clear that it suffices to estimate J_N. Using estimates (4), we obtain

$$\|AR^s\|_{E \to E} \leq Mmin\{\tfrac{1}{s\tau}, \|A\|_{E \to E}\}.$$

If $\|A\|_{E \to E} > N$, then

$$J_N \leq M \sum_{j=1}^{N} \frac{\tau}{s\tau} \leq M \int_{\|A\|_{E \to E}}^{1} \frac{ds}{s} \leq M \mid ln\|A\|_{E \to E} \mid .$$

If $\|A\|_{E \to E} \leq 1$, then

$$J_N \leq M \sum_{j=1}^{N} \|A\|_{E \to E} \tau \leq M\|A\|_{E \to E} \leq M.$$

Finally, if $\|A\|_{E \to E} \leq N$, then

$$J_N \leq M \left(\sum_{s=1}^{[N\|A\|_{E \to E}^{-1}]} \|A\|_{E \to E} \tau + \sum_{[N\|A\|_{E \to E}^{-1}]+1}^{N} \frac{\tau}{s\tau} \right)$$

$$\leq M(1 + \int_{1}^{\|A\|_{E \to E}^{-1}} \frac{ds}{s}) = M(1 + ln\|A\|_{E \to E}).$$

Thus, in all three cases we have the estimate

$$J_N \leq M(1 + ln\|A\|_{E \to E}),$$

which yields

$$\max_{1 \leq k,m \leq r} \|Au_{k,m}\|_E \leq M\left[1 + ln\|A\|_{E \to E}\right] \max_{1 \leq j,m \leq N} \|f_{j,m}\|_E. \tag{6}$$

From the estimates (5) and (6) we obtain the estimate

$$\max_{1 \leq k,m \leq r} \|Au_{k,m}\|_E$$

$$\leq C\left(\max_{0 \leq m \leq N} \|A\psi_m\|_E + \max_{0 \leq k \leq N} \|A\varphi_k\|_E \right.$$

$$\left. + \min\left\{\ln\tfrac{1}{\tau},\ 1 + |\ln\|A\|_{E \to E}|\right\} \max_{1 \leq k,m \leq N} \|f_{k,m}\|\right).$$

In a similar manner one can show that

$$\max_{r+1 \leq k,m \leq N} \|\tfrac{A}{2}(u_{k,m} + u_{k-1,m-1})\|_E$$

$$\leq C\left(\max_{0 \leq m \leq N} \|A\psi_m\|_E + \max_{0 \leq k \leq N} \|A\varphi_k\|_E \right.$$

$$\left. + \min\left\{\ln\tfrac{1}{\tau},\ 1 + |\ln\|A\|_{E \to E}|\right\} \max_{1 \leq k,m \leq N} \|f_{k,m}\|\right).$$

Theorem 2 is proved.

2 Application

For application, let Ω be the unit open cube in the n-dimensional Euclidean space $\mathbb{R}^n$ $(0 < x_k < 1, 1 \leq k \leq n)$ with boundary S, $\overline{\Omega} = \Omega \cup S$. In $[0,1] \times [0,1] \times \overline{\Omega}$ we consider the boundary-value problem for the multidimensional ultra-parabolic equation

$$
\begin{cases}
\dfrac{\partial u(t,s,x)}{\partial t} + \dfrac{\partial u(t,s,x)}{\partial s} - \displaystyle\sum_{r=1}^{n} \alpha_r(x)\dfrac{\partial^2 u(t,s,x)}{\partial x_r^2} + \delta u(t,s,x) = f(t,s,x), \\[2mm]
x = (x_1, \cdots, x_n) \in \Omega, 0 < t, s < 1, \\[2mm]
u(0,s,x) = \psi(s,x), s \in [0,1]\ u(t,0,x) = \varphi(t,x), t \in [0,1], x \in \overline{\Omega}, \\[2mm]
\dfrac{\partial u(t,s,x)}{\partial \overrightarrow{n}} = 0,\ t, s \in [0,1], x \in S,
\end{cases} \tag{7}
$$

where $\alpha_r(x) > a > 0(x \in \Omega)$ and $f(t, s, x)$ $(t, s \in (0, 1)$, $x \in \Omega)$ are given smooth functions and $\delta > 0$ is a sufficiently large number, $\overrightarrow{n}$ is the normal vector to Ω.

The discretization of problem (7) is carried out in two steps. In the first step, let us define the grid sets

$$\widetilde{\Omega}_h = \{x = x_m = (h_1 m_1, \cdots, h_n m_n), m = (m_1, \cdots, m_n),$$
$$0 \leq m_r \leq N_r, h_r N_r = L, r = 1, \cdots, n\},$$
$$\Omega_h = \widetilde{\Omega}_h \cap \Omega, S_h = \widetilde{\Omega}_h \cap S.$$

We introduce the Banach space $C_h = C_h(\widetilde{\Omega}_h)$ of grid functions $\varphi^h(x) = \{\varphi(h_1 m_1, \cdots, h_n m_n)\}$ defined on $\widetilde{\Omega}_h$, equipped with the norm

$$\| \varphi^h \|_{C(\overline{\Omega}_h)} = \max_{x \in \overline{\Omega}_h} |\varphi^h(x)|.$$

To the differential operator A generated by problem (7) we assign the difference operator A_h^x by the formula

$$A_h^x u^h = -\sum_{r=1}^{n} a_r(x)(u_{\underline{x_r}}^h)_{x_r, j_r} + \delta u^h$$

acting in the space of grid functions $u^h(x)$, satisfying the condition $D^h u^h(x) = 0$ for all $x \in S_h$. Here D^h is the difference operator, it is the first order of approximation of $\frac{\partial \cdot}{\partial \overrightarrow{n}}$.

With the help of A_h^x we arrive at the initial boundary-value problem

$$\begin{cases} \frac{\partial u^h(t,s,x)}{\partial t} + \frac{\partial u^h(t,s,x)}{\partial s} + A_h^x u^h(t, s, x) = f^h(t, s, x), \ 0 < t, s < 1, \ x \in \Omega_h \\ \\ u^h(0, s, x) = \psi^h(s, x), \ 0 \leq s \leq 1, \ x \in \widetilde{\Omega}_h, \\ \\ u^h(t, 0, x) = \varphi^h(t, x), 0 \leq t \leq 1, \ x \in \widetilde{\Omega}_h \end{cases} \quad (8)$$

for an infinite system of ordinary differential equations. In the second step, we replace problem (8) by difference scheme (2)

$$\begin{cases} \frac{u_{k,m}^h - u_{k-1,m-1}^h}{\tau} + A_h^x u_{k,m}^h = f_{k,m}^h(x), x \in \Omega_h, 1 \leq k, m \leq r \\ \\ \frac{u_{k,m}^h - u_{k-1,m-1}^h}{\tau} + \frac{1}{2}A_h^x(u_{k,m} + u_{k-1,m-1}) = f_{k,m}^h(x), x \in \Omega_h, \\ \\ r + 1 \leq k, m \leq N, \\ \\ f_{k,m}^h(x) = f^h(t_k - \frac{\tau}{2}, s_m - \frac{\tau}{2}, x), t_k = k\tau, s_m = m\tau, \\ \\ 1 \leq k, m \leq N, \ x \in \widetilde{\Omega}_h, \\ \\ u_{0,m}^h = \psi_m^h, 0 \leq m \leq N, u_{k,0}^h = \varphi_k^h, 0 \leq k \leq N \end{cases} \quad (9)$$

It is known that A_h^x is a positive operator in $C(\widetilde{\Omega}_h)$ (see [15]). Let us give a number of corollary of the Theorems 1 and 2.

Theorem 3. *For the solution of problem (2), we have the following stability inequality*

$$\max_{1 \leq k,m \leq N} \|u_{k,m}^h\|_{C(\widetilde{\Omega}_h)}$$

$$\leq C_1 \left(\max_{0 \leq m \leq N} \|\psi_m^h\|_{C(\widetilde{\Omega}_h)} + \max_{0 \leq k \leq N} \|\varphi_k^h\|_{C(\widetilde{\Omega}_h)} + \max_{1 \leq k,m \leq N} \|f_{k,m}^h\|_{C(\widetilde{\Omega}_h)} \right),$$

where C is independent of τ, ψ_m^h, φ_k^h, and $f_{k,m}^h$.

Theorem 4. *For the solution of problem (2), we have the following almost coercivity inequality*

$$\max_{1 \leq k,m \leq N} \left\| \frac{u_{k,m}^h - u_{k-1,m-1}^h}{\tau} \right\|_{C(\widetilde{\Omega}_h)} + \max_{1 \leq k,m \leq r} \sum_{r=1}^{n} \|u^h_{(\bar{x}_r x_r,\; j_r)\; k,m}\|_{C(\widetilde{\Omega}_h)}$$

$$+\frac{1}{2} \max_{1 \leq k,m \leq r+1} \sum_{r=1}^{n} \|u^h_{(\bar{x}_r x_r,\; j_r)\; k,m} + u^h_{(\bar{x}_r x_r,\; j_r)\; k-1,m-1}\|_{C(\widetilde{\Omega}_h)}$$

$$\leq C \ln \frac{1}{|h|} \left(\max_{1 \leq m \leq N} \sum_{r=1}^{n} \|\psi^h_{(\bar{x}_r x_r,\; j_r)\; m}\|_{C(\widetilde{\Omega}_h)} + \max_{0 \leq k \leq N} \sum_{r=1}^{n} \|\varphi^h_{(\bar{x}_r x_r,\; j_r)\; k}\|_{C(\widetilde{\Omega}_h)} \right)$$

$$+C \ln \frac{1}{\tau+|h|} \max_{1 \leq k,\; m \leq N} \|f^{\;h}_{k,m}\|_{C(\widetilde{\Omega}_h)}.$$

where C is independent of τ, ψ_m^h, φ_k^h, and $f_{k,m}^h$.

3   Numerical Analysis

In this section, the initial boundary value problem

$$\begin{cases} \dfrac{\partial u(s,t,x)}{\partial t} + \dfrac{\partial u(s,t,x)}{\partial s} - \dfrac{\partial^2 u(s,t,x)}{\partial x^2} + 2u(t,s,x) = f(t,s,x), \\[2mm] f(t,s,x) = e^{-(t+s)} \cos \pi x, \; 0 < s,t < 1, \; 0 < x < 1, \\[2mm] u(0,s,x) = e^{-s} \cos \pi x, \; 0 \leq t \leq 1, \; 0 \leq x \leq 1, \\[2mm] u(t,0,x) = e^{-t} \cos \pi x, \; 0 \leq s \leq 1, \; 0 \leq x \leq 1, \\[2mm] u_x(s,t,0) = u_x(s,t,\pi) = 0, \quad 0 \leq s,t \leq 1 \end{cases} \qquad (10)$$

for one dimensional ultra parabolic equations is considered.
The exact solution of problem (10) is

$$u(t,s,x) = e^{-(t+s)} \cos \pi x.$$

Using the second order of accuracy in t and s implicit difference scheme (9), we obtain the difference scheme second order of accuracy in t and s and second order of accuracy in x

$$
\begin{cases}
\dfrac{u_n^{k,m} - u_n^{k-1,m-1}}{\tau} - \dfrac{u_{n+1}^{k,m} - 2u_n^{k,m} + u_{n-1}^{k,m}}{h^2} + 2u_n^{k,m} = f_{k,m}^h, \ 1 \le k, m \le r, \\[3mm]
\dfrac{u_n^{k,m} - u_n^{k-1,m-1}}{\tau} - \dfrac{u_{n+1}^{k,m} - 2u_n^{k,m} + u_{n-1}^{k,m}}{h^2} + 2u_n^{k,m} \\[3mm]
\quad - \dfrac{u_{n+1}^{k-1,m-1} - 2u_n^{k-1,m-1} + u_{n-1}^{k-1,m-1}}{h^2} + 2u_n^{k-1,m-1} = f_{k,m}^h, \\[3mm]
r+1 \le k, m \le N, \\[3mm]
f_{k,m}^h = f(t_k - \tfrac{\tau}{2}, s_m - \tfrac{\tau}{2}, x_n) = e^{-(t_k + s_m - \tau)} \sin x_n, \\[3mm]
1 \le k, m \le N, \ 1 \le n \le M - 1, \\[3mm]
u_n^{0,m} = e^{-s_m} \cos x_n, \ 0 \le m \le N, \ 0 \le n \le M, \\[3mm]
u_n^{k,0} = e^{-t_k} \cos x_n, \ 0 \le k \le N, \ 0 \le n \le M, \\[3mm]
u_0^{k,m} = u_1^{k,m}, u_M^{k,m} = u_{M-1}^{k,m}, \ 0 \le k, m \le N, \\[3mm]
t_k = k\tau, s_m = m\tau, \ 1 \le k, m \le N, \ N\tau = 1, \\[3mm]
x_n = nh, \ 1 \le n \le M, \ Mh = \pi,
\end{cases}
\tag{11}
$$

for approximate solutions of initial boundary value problem (10). For the solution of (11), we will use the modified Gauss elimination method(see [16]).

Errors computed by

$$
E_N^{K,M} = \max_{1 \le k, m \le N, 1 \le n \le M-1} \left| u(t_k, s_m, x_n) - u_n^{k,m} \right|
$$

of the numerical solutions, where $u(t_k, s_m, x_n)$ represents the exact solution and $u_n^{k,m}$ represents the numerical solution at (t_k, s_m, x_n) and the results for $r = 1$ and $r = 2$ are given in following table.

Table 1. Difference Schemes(7) for different N=M values

	$N = 10$	$N = 15$	$N = 20$	$N = 30$
r=1	0.0432	0.0223	0.0109	0.0031
r=2	0.0574	0.0286	0.0197	0.0091

References

1. Dyson, J., Sanches, E., Villella-Bressan, R., Weeb, G.F.: An age and spatially structured model of tumor invasion with haptotaxis. Discrete Continuous Dynam. Systems B. 8, 45–60 (2007)
2. Kunisch, K., Schappacher, W., Weeb, G.F.: Nonlinear age-dependent population dynamics with random diffusion. Comput. Math. Appl. 11, 155–173 (1985)
3. Kolmogorov, A.N.: Zur Theorie der stetigen zufälligen prozesse. Math. Ann. 108, 149–160 (1933)
4. Kolmogorov, A.N.: Zufällige bewegungen. Ann. of Math. 35, 116–117 (1934)
5. Genčev, T.G.: Ultraparabolic equations. Dokl. Akad. Nauk SSSR 151, 265–268 (1963)
6. Deng, Q., Hallam, T.G.: An age structured population model in a spatially heterogeneousenvironment: Existence and uniqueness theory. Nonlinear Anal. 65, 379–394 (2006)
7. Di Blasio, G., Lamberti, L.: An initial boundary value problem for age-dependent population diffusion. SIAM J. Appl. Math. 35, 593–615 (1978)
8. Di Blasio, G.: Nonlinear age-dependent diffusion. UJ. Math. Biol. 8, 265–284 (1979)
9. Tersenov, S.A.: On boundary value problems for a class of ultraparabolic equations and their applications. Matem. Sbornik. 175, 529–544 (1987)
10. Ashyralyev, A., Yilmaz, S.: Second order of accuracy difference schemes for ultra parabolic equations. In: AIP Conference Proceedings, vol. 1389, pp. 601–604 (2011)
11. Ashyralyev, A., Yilmaz, S.: An Approximation of ultra-parabolic equations. Abstr. Appl. Anal, Article ID 840621, 14 pages (2012)
12. Ashyralyev, A., Yilmaz, S.: On the numerical solution of ultra-parabolic equations with the Neumann Condition. In: AIP Conference Proceedings, vol. 1470, pp. 240–243 (2012)
13. Ashyralyev, A., Yilmaz, S.: Modified Crank-Nicholson difference schemes for ultra-parabolic equations. Comput. Math. Appl. 64, 2756–2764 (2012)
14. Ashyralyev, A., Sobolevskii, P.E.: Well-Posedness of Parabolic Difference Equations. Operator Theory Advances and Applications, vol. 69. Birkhäuser Verlag, Basel (1994)
15. Alibekov, K.A., Sobolevskii, P.E.: Stability and convergence of difference schemes of a high order for parabolic partial differential equations. Ukrain. Math. Zh. 32, 291–300 (1980)
16. Samarskii, K.A., Nikolaev, E.S.: Numerical Methods for Grid Equations. Iterative Methods, vol. 2. Birkhäuser, Basel (1989)

Bifurcations in Long Josephson Junctions with Second Harmonic in the Current-Phase Relation: Numerical Study

Pavlina Atanasova[1] and Elena Zemlyanaya[2]

[1] University of Plovdiv, FMI, Plovdiv 4003, Bulgaria
atanasova@uni-plovdiv.bg
[2] Laboratory of Information Technologies, Joint Institute for Nuclear Research
141980 Dubna, Moscow Region, Russia
elena@jinr.ru

Abstract. Critical regimes in the long Josephson junction (LJJ) are studied within the frame of a model accounting the second harmonic in the current-phase relation (CPR). Numerical approach is shown to provide a good agreement with analytic results. Numerical results are presented to demonstrate the availabilities and advantages of the numerical scheme for investigation of bifurcations and properties of the magnetic flux distributions in dependence on the sign and value of the second harmonic in CPR.

Keywords: Long Josephson junction, double sine-Gordon equation, continuous analogue of Newton's method, numerical continuation, stability, bifurcations.

1 Introduction

Physical properties of magnetic flux in Josephson junctions (JJs) play important role in the modern nanoelectronics. Generally, the current-phase relation (CPR) in the JJ is taken as the Fourier decomposition of sinuses [12]. For JJs of the "superconductor–insulator–superconductor" type, the CPR is close to a sinusoidal function of phase while another terms in the CPR Fourier decomposition are negligible. In those cases, the magnetic flux distributions are described by the sine-Gordon (SG) equation.

However, in a number of JJs models, the second harmonic contribution of the CPR Fourier expansion should be accounted, see for example, [14,9,10]. In the frame of corresponding models the magnetic flux distributions satisfy the following double sine-Gordon equation:

$$\varphi'' - \ddot{\varphi} - \alpha\dot{\varphi} = a_1 \sin\varphi + a_2 \sin 2\varphi - \gamma, \quad t > 0, \quad x \in (-l, l). \quad (1)$$

Here and below the prime means a derivative with respect to the coordinate x and the dot – with respect to the time t. The case of the overlap-contact of a finite length yields the following form of boundary conditions

$$\varphi'(\pm l, t) = h_e. \quad (2)$$

I. Dimov, I. Faragó, and L. Vulkov (Eds.): NAA 2012, LNCS 8236, pp. 190–197, 2013.

In (1),(2), φ is a magnetic flux distribution, h_e – an external magnetic field, γ – the external current, $\alpha \geq 0$ – the dissipation coefficient, l is the semilength of the junction, a_1 and a_2 are parameters corresponding the contribution of 1st and 2nd harmonic in CPR, respectively. The sign of a_2 can be positive or negative in the frame of different physical applications.

Static regimes of the magnetic flux distributions are described by the nonlinear boundary problem [12,11,7]:

$$-\varphi'' + a_1 \sin\varphi + a_2 \sin 2\varphi - \gamma = 0, \quad x \in (-l, l), \quad \varphi'(\pm l) = h_e \qquad (3)$$

that follows from a necessary condition for the full energy functional extremum. Bifurcations of φ correspond the transitions of junction from superconducting to resistive state where the voltage measurement changes from zero to nonzero value. Stable distributions correspond to superconductive state where the voltage is equal to zero.

The stability analysis of solution $\varphi(x, p)$ where $p = (l, a_1, a_2, h_e, \gamma)$ is a vector of parameters) can be reduced to the numerical solution of the corresponding Sturm-Liouville problem (SLP) [8,13]:

$$-\psi'' + q(x)\psi = \lambda\psi, \quad \psi'(\pm l) = 0, \quad q(x) = a_1 \cos\varphi + 2a_2 \cos 2\varphi. \qquad (4)$$

The case of positive minimal eigenvalue $\lambda_0(p) > 0$ corresponds the minimum of the distribution energy and, hence, the stable solution φ. In case $\lambda_0(p) < 0$ solution $\varphi(x, p)$ is unstable. The case $\lambda_0(p) = 0$ indicates the bifurcation (the transition of junction from superconducting to resistive state) with respect to one of parameters p.

Our numerical approach is based on the consideration of Eqs.(3),(4) as unique problem with respect to functions φ, ψ, and one of the parameters p. In comparison with a standard direct numerical simulation of Eq.(1), this method simplifies an obtaining of the dependence of critical current γcr on external magnetic field h_e – an important physical observable measured in experiments. This idea was successfully applied to reproduce critical states in different models described by SG (i.e. with $a_2 = 0$), see for example [13,1] and references there. We extend this technique for the case of nonzero a_2 and investigate the effect of the second harmonic accounting on the critical magnetic flux distributions.

Beside, our numerical scheme is furnished with a continuation algorithm [17,5] providing analysis of interconnection between coexisting (stable and unstable) distributions. We study the critical curves and the stability areas of the corresponding solutions under the influence of the second harmonic in CPR. In Sect. 2 we describe details of our numerical approach. Results of numerical study are discussed in Sect. 3. Concluding remarks are given in Sect. 4.

2   Numerical Approach

Supplying the system (3),(4) with a normalization condition for the eigenfunction $\psi(x)$ of SLP (4)

$$\int_{-l}^{+l} \psi^2(x)\, dx = 1. \qquad (5)$$

we proceed the nonlinear problem at $x \in (-l; l)$. The system (3-5) is considered as the unified nonlinear functional equation for the functions $\varphi(x)$, $\psi(x)$, and one of the five parameters p (which we denote by ξ). The other four parameters $\bar{p} \in p$ are assumed to be known. Thus, the system (3-5) can be rewritten in a vector form as follows:

$$\mathcal{F}(y, \bar{p}) = 0\,, \qquad (6)$$

where $y = (\varphi, \psi, \xi)$ is unknown element in Banach space and $\bar{p}$ is the set of parameters.

In the simplest case, when all the elements of the vector p are defined, the problem is split into independent subsystems (3) and (4-5). The eigenvalue λ is to be found, which corresponds to the solution $\varphi(x)$ of (3). This approach was applied in our papers [2,6,3].

In case the eigenvalue of SLP is assumed to be fixed $\lambda = 0$, we obtain the closed system (3-5) with respect to the unknown Banach element y. Considering the case $\xi = \gamma$ and solving the problem (6) with respect to $y = (\varphi, \psi, \gamma)$ yields the value $\gamma = \gamma_{cr}$ that corresponds the bifurcation magnetic flux distribution φ. The bifurcation solution y is path-followed in the continuation parameter h_e and the critical dependence $\gamma_{cr}(h_e)$ is determined.

In the case $\xi = \gamma$ or $\xi = h_e$ the continuation of the bifurcation solution in the parameter a_2 produces the dependence of stability region on external current γ or on external magnetic field h_e.

Below, we present the Newtonian iteration scheme for numerical solution of the problem (6) for the case of $\xi = \gamma$.

The continuous analog of Newton's method (CANM) [13] reduces the problem (6) to the auxiliary linear problem:

$$\frac{\partial \mathcal{F}}{\partial y} w + \mathcal{F}(y) = 0\,, \qquad (7)$$

where w denotes the iteration increment of y: $w = (u, v, \Gamma)$, and $\partial \mathcal{F}/\partial y$ is the Frechet derivative. For each element $\mathcal{F}_i$ of the vector-function $\mathcal{F}$ we have

$$\frac{\partial \mathcal{F}_1}{\partial y} w = -u'' + qu - \Gamma\,, \qquad (8)$$

$$\frac{\partial \mathcal{F}_2}{\partial y} w = u'(-l)\,, \qquad \frac{\partial \mathcal{F}_3}{\partial y} w = u'(l)\,, \qquad (9)$$

$$\frac{\partial \mathcal{F}_4}{\partial y} w = -v'' + ru\psi + qv - \lambda v\,, \qquad r = -a_1 \sin \varphi - 4a_2 \sin 2\varphi \qquad (10)$$

$$\frac{\partial \mathcal{F}_5}{\partial y} w = v'(-l)\,, \qquad \frac{\partial \mathcal{F}_6}{\partial y} w = v'(l)\,, \qquad (11)$$

$$\frac{\partial \mathcal{F}_7}{\partial y} w = 2 \int_{-l}^{l} \psi v \, dx\,. \qquad (12)$$

Finally, we obtain the linear system with respect to the unknowns $w = (u, v, \Gamma)$:

$$-u'' + qu - \Gamma - \varphi'' + f = 0, \tag{13}$$

$$u'(\pm l) + \varphi'(\pm l) - h_e = 0, \tag{14}$$

$$-v'' + ru\psi + qv - \lambda v - \psi'' + q\psi - \lambda\psi = 0, \tag{15}$$

$$v'(\pm l) + \psi'(\pm l) = 0, \tag{16}$$

$$2\int_{-l}^{l} \psi v \, dx + \int_{-l}^{l} \psi^2 \, dx - 1 = 0. \tag{17}$$

Solutions $u(x)$ and $v(x)$ of the system (13-17) are decomposed in a form of linear combination

$$u(x) = u_1(x) + \Gamma u_2(x), \qquad v(x) = v_1(x) + \Gamma v_2(x), \tag{18}$$

where $x \in (-l; l)$ and u_1, u_2, v_1, v_2 are solutions, respectively, of the following boundary-value problems:

$$-u_1'' + qu_1 - \varphi'' + f = 0, \quad u_1'(\pm l) + \varphi'(\pm l) - h_e = 0, \tag{19}$$

$$-u_2'' + qu_2 - 1 = 0, \quad u_2'(\pm l) = 0, \tag{20}$$

$$-v_1'' + ru\psi + (q - \lambda)v_1 - \psi'' + (q - \lambda)\psi = 0, \quad v_1'(\pm l) + \psi'(\pm l) = 0, \tag{21}$$

$$-v_2'' + ru\psi + (q - \lambda)v_2 = 0, \quad v_2'(\pm l) = 0, \tag{22}$$

Quantity Γ can be determined from the following expression:

$$2\int_{-l}^{l} \psi v_1 \, dx + 2\Gamma \int_{-l}^{l} \psi v_2 dx + \int_{-l}^{l} \psi^2 \, dx - 1 = 0. \tag{23}$$

The above formulae (18-23) define the following sequence of calculations at each newtonian iteration for the fixed value of the continuation parameter h_e. Let us assume, at n-th iteration we have n-th approximation of solution $\varphi^n(x)$, $\psi^n(x)$ and γ^n. At $(n+1)$-th iteration:

1. We calculate the functions $u_1^n(x)$ and $u_2^n(x)$ from the linear boundary-value problems (19) and (20).
2. Then we solve linear boundary-value problems (21) and (22)) with obtained functions $u_1^n(x)$ and $u_2^n(x)$ and determine solutions $v_1^n(x)$ and $v_2^n(x)$.
3. Quantity Γ^n is calculated using Eq.(23).
4. The $(n + 1)$-th approximation of $\varphi^{n+1}(x)$, $\psi^{n+1}(x)$ and γ^{n+1} is defined by means of the formulas

$$\varphi^{n+1}(x) = \varphi^n(x) + \tau_n[u_1^n(x) + \Gamma^n u_2^n(x)], \tag{24}$$

$$\psi^{n+1}(x) = \psi^n(x) + \tau_n[v_1^n(x) + \Gamma^n v_2^n(x)], \tag{25}$$

$$\gamma^{n+1} = \gamma^n + \tau_n \Gamma^n, \tag{26}$$

where the iteration parameter τ_n is calculated as follows [15]:

$$\tau_n = \max\left(\tau_{min}, \frac{\delta_n(0)}{\delta_n(0) + \delta_n(1)}\right). \tag{27}$$

Here τ_{min} is a fixed minimal value of parameter τ_n $(0 < \tau_{min} \leq 1)$;

$$\delta_n(0) \equiv \|\mathcal{F}(\varphi^n, \psi^n, \gamma^n)\|, \quad \delta_n(1) \equiv \|\mathcal{F}(\bar{\varphi}^{n+1}, \bar{\psi}^{n+1}, \bar{\gamma}^{n+1})\|$$

where $\| \cdot \|$ means the standard C-norm and $\bar{\varphi}^{n+1}, \bar{\psi}^{n+1}, \bar{\gamma}^{n+1}$ are obtained by means Eqs.(24-26) with $\tau_n = 1$.

5. The iterations are finished when the inequality $\| \mathcal{F} \| \leq \epsilon$ holds true, where $\epsilon > 0$ is a small number chosen beforehand.

Convergence of the CANM-based iteration process is proved in [16].

For numerical solution of the linear boundary-value problems (19-22) at each Newtonian iteration, we apply Numerov's finite-difference approximation of the 4th order accuracy [4].

3   Numerical Results

Together with the well-known distributions (standardly called M_π and M_0) the nonzero a_2 in Eq.(3) gives a rise another uniform state (called $M_{\pm ac}$ in [6]). Stability and bifurcations of $M_{\pm ac}$ have been investigated in [6,2].

New fluxon solution inspired by the second harmonic contribution was observed in direct numerical simulation [9] and denoted "small fluxon" in contradiction to the standard "large fluxon" solution Φ^1. Later, the "small fluxon" was reproduced in [2] in the frame of CANM-based numerical approach.

"Large fluxon" Φ_1 and "small fluxon" are characterized, respectively, with $N = 1$ and $N = 0$ where the quantity N (denoted "number of fluxons" in [6]) is determined as follows

$$N = \frac{1}{2l\pi} \int\limits_{-l}^{l} \varphi(x)\,dx. \tag{28}$$

One more a_2-inspired one-fluxon solution Φ^{1*} (existing at $a_2 < -0.5$) was obtained in [4]. As the standard solution Φ^1, the Φ^{1*} fluxon characterized by the "number of fluxons" $N = 1$

Figure 1 exhibits the full stability chart of two fluxons: Φ^{1*} and Φ^1 on the (h_e, a_2)-plane for the case $a_1 = 1$. Stability domains of Φ^1 and Φ^{1*} are bounded, respectively, by solid and dashed curves. In the region where two stability domains overlap, the different stable fluxon distributions coexist.

Our numerical study shows that, beside of bound states of two, three and more identical fluxons Φ^1, Eq.(3) holds the stable the mixed bound states of different types of fluxons. Stable mixed solutions φ_1 and φ_2 are shown in figs 2,3 for $h_e = 0.6$, $a_1 = 1$, $a_2 = -1$. They are characterized by $(N[\varphi_1] + N[\varphi_2])/2 = 1$ while quantities $N[\varphi_1]$ and $N[\varphi_2]$ are fractional. Note that stability of the mixed

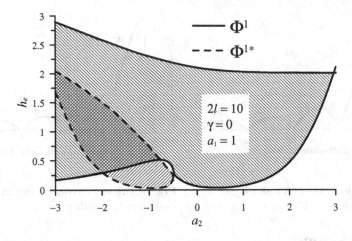

Fig. 1. Bifurcation diagram of two fluxons Φ^1 and Φ^{1*} at the plane of parameters a_2 and h_e. Here $a_1 = 1$, $2l = 10$, $\gamma = 0$.

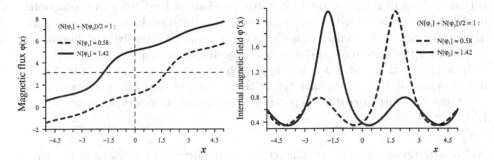

Fig. 2. The magnetic flux distribution $\varphi(x)$ for the stable mixed bound state of two fluxons with $h_e = 0.6$, $a_1 = 1$, $a_2 = -1$, $2l = 10$, $\gamma = 0$

Fig. 3. The internal magnetic field distribution $\varphi'(x)$ for the stable mixed bound state of two fluxons with $h_e = 0.6$, $a_1 = 1$, $a_2 = -1$, $2l = 10$, $\gamma = 0$

bound states depends on a_2 and h_e. For growing a_2 this solution stabilizes for sufficiently small h_e.

Figures 4 and 5 demonstrate the critical current curves $\gamma_{cr}(h_e)$ for $a_2 = 0$, $a_2 = 0.3$, and $a_2 = 1$. Maximal value of the critical current corresponds the M_0 distribution for $h_e = 0$. Another portions of critical current curve (for growing h_e) correspond, sequentially, one-fluxon solution Φ^1, two-fluxon solution Φ^2, three-fluxon solution Φ^3, etc. Here, we present the case of relatively short junctions $2l = 1/\pi$ in order to show that our results are in a good agreement with results of direct numerical simulation of Eq. (1) [9]. Note, for longer l and for growing a_2 a complexity of the $\gamma_{cr}(h_e)$ dependence increases, see fig.5.

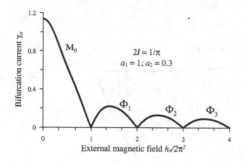 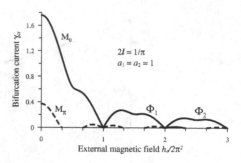

Fig. 4. Critical curves for short junction $2l = 1/\pi$ with $a_1 = 1$ and $a_2 = 0.3$

Fig. 5. Critical curves for the short junction $2l = 1/\pi$ with $a_1 = a_2 = 1$

4   Conclusions

The aim of this contribution is to present the numerical scheme for the bifurcation analysis of static magnetic flux distributions in long Josephson junctions described by double sine-Gordon equation (1). Instead of direct numerical simulation of partial differential equation (1) we numerically solve the boundary problem for the system of nonlinear ordinary differential equations.

It is shown that our numerical approach allows one to obtain new fluxon distributions of Eq. (1), to investigate their stability and bifurcations and to construct the critical current dependence, important physical characteristic of LJJs.

Acknowledgement. The work is partially supported in the frame of the Program for collaboration of JINR-Dubna and Bulgarian scientific centers "JINR – Bulgaria". E.V.Z. was partially supported by RFFI under grant 12-01-00396a. P.Kh.A. is partially supported by the project NI11-FMI-004.

References

1. Atanasova, P., Kh., B.T.L., Dimova, S.N.: Numerical Modeling of Critical Dependence for Symmetric Two-Layer Josephson Junctions. Comput. Maths. and Math. Phys. 46(4), 666–679 (2006)
2. Atanasova, P.K., Boyadjiev, T.L., Shukrinov, Y.M., Zemlyanaya, E.V., Seidel, P.: Influence of Josephson current second harmonic on stability of magnetic flux in long junctions. J. Phys. Conf. Ser. 248, 012044 (2010), arXiv:1007.4778v1
3. Atanasova, P.K., Boyadjiev, T.L., Zemlyanaya, E.V., Shukrinov, Y.M.: Numerical study of magnetic flux in the LJJ model with double sine-gordon equation. In: Dimov, I., Dimova, S., Kolkovska, N. (eds.) NMA 2010. LNCS, vol. 6046, pp. 347–352. Springer, Heidelberg (2011)
4. Atanasova, P.K., Boyadjiev, T.L., Shukrinov, Y.M., Zemlyanaya, E.V.: Numerical investigation of the second harmonic effects in the LJJ. In: Koleva, M., Vulkov, L. (eds.) Proc. the FDM 2010 Conf., Lozenetz, Bulgaria, pp. 1–8 (July 2010), arXiv:1005.5691v1

5. Atanasova, P.K., Zemlyanaya, E., Shukrinov, Y.: Numerical study of fluxon solutions of sine-gordon equation under the influence of the boundary conditions. In: Adam, G., Buša, J., Hnatič, M. (eds.) MMCP 2011. LNCS, vol. 7125, pp. 201–206. Springer, Heidelberg (2012)

6. Atanasova, P.K., Zemlyanaya, E.V., Boyadjiev, T.L., Shukrinov, Y.M.: Numerical modeling of long Josephson junctions in the frame of double sine-Gordon equation. Mathematical Models and Computer Simulations 3(3), 388–397 (2011)

7. Buzdin, A., Koshelev, A.E.: Periodic alternating 0-and π-junction structures as realization of φ-Josephson junctions. Phys. Rev. B. 67, 220504(R) (2003)

8. Galpern, Y.S., Filippov, A.T.: Joint solution states in inhomogeneous Josephson junctions. Sov. Phys. JETP 59, 894 (1984) (Russian)

9. Goldobin, E., Koelle, D., Kleiner, R., Buzdin, A.: Josephson junctions with second harmonic in the current-phase relation: Properties of junctions. Phys. Rev. B. 76, 224523 (2007)

10. Goldobin, E., Koelle, D., Kleiner, R., Mints, R.G.: Josephson junction with magnetic-field tunable ground state. Phys. Rev. Lett. 107, 227001 (2011)

11. Hatakenaka, N., Takayanag, H., Kasai, Y., Tanda, S.: Double sine-Gordon fluxons in isolated long Josephson junction. Physica B. 284-288, 563–564 (2000)

12. Likharev, K.K.: Introduction in Josephson junction dynamics, M. Nauka, GRFML (1985) (Russian)

13. Puzynin, I.V., et al.: Methods of Computational Physics for Investigation of Models of Complex Physical Systems. Physics of Particles and Nuclei 38(1), 70–116 (2007)

14. Ryazanov, V.V., et al.: Coupling of two superconductors through a ferromagnet: evidence for a pi junction. Phys. Rev. Lett. 36, 2427–2430 (2001)

15. Ermakov, V.V., Kalitkin, N.N.: The optimal step and regularisation for Newton's method, Comput. Maths. and Math. Phys. 21(2), 235–242 (1981)

16. Zhanlav, T., Puzynin, I.V.: On iterations convergence on the base of continuous analog of Newton's method. Comput. Maths. and Math. Phys. 32(6), 729–737 (1992)

17. Zemlyanaya, E.V., Barashenkov, I.V.: Numerical study of the multisoliton complexes in the damped-driven NLS. Math. Modelling 16(3), 3–14 (2004) (Russian)

On the Unsymmetrical Buckling
of the Nonuniform Orthotropic Circular Plates

Svetlana M. Bauer and Eva B. Voronkova

St.Petersburg State University,
Department of Theoretical and Applied Mechanics
Universitetckii pr. 28, 198504, St. Petersburg, Russia
s_bauer@mail.ru, voronkova.eva@math.spbu.ru
http://www.spbu.ru

Abstract. This work is concerned with the numerical study of unsymmetrical buckling of clamped orthotropic plates under uniform pressure. The effect of material heterogeneity on the buckling load is examined. The refined 2D shell theory is employed to obtain the governing equations for buckling of a clamped circular shell.

The unsymmetric part of the solution is sought in terms of multiples of the harmonics of the angular coordinate. A numerical method is employed to obtain the lowest load value, which leads to the appearance of waves in the circumferential direction. It is shown that if the elasticity modulus decreases away from the center of a plate, the critical pressure for unsymmetric buckling is sufficiently lower than for a plate with constant mechanical properties.

Keywords: Circular plate, unsymmetrical buckling, inhomogeneity.

1   Introduction

The present paper is devoted to study the unsymmetrical buckling of clamped heterogeneous circular plate subjected to uniform surface load.

For the first time, unsymmetrical buckling of the thin clamped circular isotropic plates under normal pressure was analyzed by Panov and Feodosev in 1948 [1]. The authors assumed that under sufficiently large load an unsymmetric state branched from the axisymmetric one and waves developed near the edge of the plates. Nonaxisymmetric displacement was represented in the form $W = (1 - r^2)^2 (A + Br^4 \cos n\theta)$ and the bending problem was studied by Galerkin procedure [1]. In this approach the pre-buckling axisymmetric state was approximated by function with only one unknown parameter. Later, in 1963, Feodos'ev showed that under large deformations the elastic surfaces of plates or shells were exposed to strong changes, and the stress-strain state of these structures could not be described by one or two unknown parameters in approximating functions [2].

Still later, Cheo and Reiss examined the same problem on the unsymmetric wrinkling of clamped circular plates subjected to surface load [3]. The critical

I. Dimov, I. Faragó, and L. Vulkov (Eds.): NAA 2012, LNCS 8236, pp. 198–205, 2013.

buckling load and the corresponding wave number obtained in [3] differs significantly from the results in [1]. Cheo and Reiss suspected that Panov and Feodos'ev had found unstable unsymmetric state, and underlined the approximation function with two unknown parameters was "too inaccurate to adequately describe the wrinkling of the plate".

In this paper we study buckling of a circular plate with varying mechanical characteristics. Such a plate can be used as the simplest model of Lamina Cribrosa (LC) in the human eye [4]. The buckling of the LC in the nonaxisymmetric state in the neighborhood of the edge could cause edamas and folds at the periphery of the LC and loss of sight.

2   Problem Statement

Let us consider a circular thin plate of radius R, thickness h $(h/R \ll 1)$ subjected to uniform normal load p. The plate displays cylindrical orthotropy and Ambartsumyan's theory of anisotropic plates is employed to study large deformations of the plate [5]. We assume radial inhomogeneity for the plate, i.e. the elastic moduli continuously vary from point to point in the radial direction.

In general case, the fundamental equations can be written in the form

$$
\begin{aligned}
&\phi' + \frac{\phi}{r} + \frac{\dot{\psi}}{r} = -L(w, F) - p \\
&g_1 L_1(w) + g_1' \mathcal{L}_1^+(w) - L_{1\phi}(\phi) - L_{1\psi}(\psi) = -\phi, \\
&g_1 L_2(w) + g_1' \mathcal{L}_2(w) - L_{2\phi}(\phi) - L_{2\psi}(\psi) = -\psi, \\
&g_2 L_3(F) + g_2' \mathcal{L}_3^-(F) + g_2'' \mathcal{L}_1^-(F) = -\frac{\lambda^2}{2} L(w, w), \\
&(\,)' = \frac{\partial(\,)}{\partial r}, \quad (\dot{\,}) = \frac{\partial(\,)}{\partial \theta}
\end{aligned}
\tag{1}
$$

with non-linear operator

$$
L(x, y) = x'' \left(y'/r + \ddot{y}/r^2 \right) + y'' \left(x'/r + \ddot{x}/r^2 \right) - 2 \left(\dot{x}/r \right)' \left(\dot{y}/r \right)'
$$

and linear operators L_i, $\mathcal{L}_i$, $L_{j\phi}$, $L_{j\psi}$, $i = 1, 2, 3$, $j = 1, 2$ are as defined in Appendix.

System (1) is obtained from determining relations of the nonlinear theory of anisotropic plates [5]. Here $w(r, \theta)$, $F(r, \theta)$, $\phi(r, \theta)$, $\psi(r, \theta)$ are the nondimensional out-of-plane deflection, the Airy stress function and the force functions, respectively. Dimensionless quantities are related with those with dimensions by the expressions

$$
r = \frac{r^*}{R}, \; w = \frac{\beta w^*}{h}, \; p = \frac{\beta^3 R^4}{E_r^{av} h^4} p^*, \; F = \frac{\beta^2 F^*}{E_r^{av} h^3},
$$
$$
\{\phi, \psi\} = \frac{R^3 \beta^3}{12 E_r^{av} h} \{\phi^*, \psi^*\}, \; \beta^2 = 12(1 - \nu_r \nu_\theta).
$$

E_r^{av} is an average value of the elastic modulus in the radial direction

$$E_r^{av} = \frac{1}{\pi R^2} \int_0^{2\pi} \int_0^R E_r(r) r \, dr \, d\theta, \quad E_r(r) = E_r^0 f(r) \tag{2}$$

and $g_1(r) = E_r^0 f(r)/E_r^{av}$, $g_2(r) = 1/g_1(r)$.

We suppose that the edge of the plate is clamped but moves freely in the plate's plane. This results in the following set of conditions at $r = 1$ [5]

$$w = 0, \quad w' = 5\mu_r \phi/2, \quad F'/r + \ddot{F}/r^2 = -\left(\dot{F}/r\right)' = 0. \tag{3}$$

In addition, all sought-for functions must fulfil the boundedness condition at the center of the plate.

In case of an isotropic plate Eqs. (1), (3) can be transformed into corresponding equations of the classical plate theory

$$\begin{aligned}
g_1 \Delta \Delta w + g_1' \mathcal{L}_3^+(w) + g_2'' \mathcal{L}_1^+(w) &= p + L(w, F), \\
g_2 \Delta \Delta F + g_2' \mathcal{L}_3^-(F) + g_2'' \mathcal{L}_1^-(F) &= -L(w, w)/2, \\
w = w' = 0, \quad F'/r + \ddot{F}/r^2 &= -\left(\dot{F}/r\right)' = 0 \quad \text{at } r = 1,
\end{aligned} \tag{4}$$

where $\Delta = (\ )'' + (\ )'/r + (\ddot{\ })/r^2$ is the Laplacian in polar coordinates.

3   Numerical Solution

We seek for solutions of equations (1) in the form

$$\begin{Bmatrix} \phi(r, \theta) \\ w(r, \theta) \\ F(r, \theta) \end{Bmatrix} = \begin{Bmatrix} \phi_s(r) \\ w_s(r) \\ F_s(r) \end{Bmatrix} + \varepsilon \begin{Bmatrix} \phi_n(r) \\ w_n(r) \\ F_n(r) \end{Bmatrix} \cos n\theta, \quad \psi(r, \theta) = \psi_n(r) \sin n\theta, \tag{5}$$

where $\phi_s(r)$, $w_s(r)$, $F_s(r)$ describe prebuckling axisymmetric state, ε is a small parameter, n is a mode number and $\phi_n(r), \psi_n(r), w_n(r), F_n(r)$ are the non-symmetrical components.

For the symmetrical problem Eqs. (1) can be reduced to

$$\begin{aligned}
\phi_s &= -\frac{pr}{2} - \frac{W_0 \Phi_0}{r}, \\
g_1 \left(W_0'' + \frac{W_0'}{r} - \lambda^2 \frac{W_0}{r^2} \right) + g_1' \left(W_0' + \frac{\nu_\theta}{r} W_0 \right) &= \phi_s \left(\frac{\mu_\theta}{r^2} - 1 \right) - \frac{\mu_r}{r} (r\phi_s')', \\
g_2 \left(\Phi_0'' + \frac{\Phi_0'}{r} - \frac{\Phi_0}{r^2} \right) + g_2' \left(\Phi_0' - \frac{\nu_\theta}{r} \Phi_0 \right) &= -\frac{W_0^2}{2r}, \\
W_0 = w_s', \quad \Phi_0 &= F_s'
\end{aligned} \tag{6}$$

with boundary conditions

$$W_0(1) = 5\mu_r \phi_s/2, \quad \Phi_0(1) = 0. \tag{7}$$

Deriving system (6) we use the limiting conditions for the function $\phi_s(r)$ at the plate center $(r = 0)$.

Substituting (5) in (1), (3), using Eqs. (6), (7) we obtain after linearization with respect to ε

$$\phi_n' + \frac{\phi_n}{r} + \frac{n\psi_n}{r} = \frac{w_n''}{r}\Phi_0 + \frac{F_n''}{r}W_0 - W_0'\left(\frac{F_n'}{r} - \frac{n^2}{r^2}F_n\right) - \Phi_0'\left(\frac{w_n'}{r} - \frac{n^2}{r^2}w_n\right)$$

$$g_1 L_1^n(w_n) + g_1'\mathcal{L}_1^{n+} + (w_n) - L_{1\phi_n}^n(\phi) - L_{1\psi}^n(\psi_n) = -\phi_n,$$

$$g_1 L_2^n(w_n) + g_1'\mathcal{L}_2^n(w_n) - L_{2\phi}^n(\phi_n) - L_{2\psi}^n(\psi_n) = -\psi_n,$$

$$g_2 L_3(F_n) + g_2'\mathcal{L}_3^{n-}(F_n) + g_2''\mathcal{L}_1^{n-}(F_n) = -W_0'\left(\frac{w_n'}{r} - \frac{n^2}{r^2}w_n\right) - \frac{w_n''}{r}W_0$$

$$\tag{8}$$

with the constraints

$$w_n(1) = 0, \quad w_n'(1) = 5\mu_r\phi_n/2, \quad F_n'(1) = F_n(1) = 0. \tag{9}$$

Expressions for L_i^n, $\mathcal{L}_i^n$, $L_{j\phi}^n$, $L_{j\psi}^n$, $i = 1,2,3$, $j = 1,2$ are listed in Appendix.

Buckling equations (8)-(9) constitute an eigenvalue problem, in which the parameter p is implicit and appears in the equations through the functions W_0 and Φ_0.

We use the shooting method to solve nonlinear axisymmetric problem (7) together with (8). To determine the value of p, for which (8)-(9) have nontrivial solution, the finite difference method is employed [3]. We refer to the smallest of these eigenvalues as the buckling load. The step of the difference grid is chosen so that by reducing the step by 2 the value of the critical load varied less than 1%.

4   Results and Discussion

The physical mechanism that initiates the buckling about the axisymmetric state into an unsymmetric equilibrium state is proposed in [1], [3]. A ring of large circumferential compressive stress develops near the edge of the plate and indicates possibility of wrinkling near the edge. For non-uniform plate dimensionless axisymmetrical circumferential stress for increasing values of load parameter p has been plotted in Fig. 1. As one can see, the compressive stress intensity increases with the load, but the width of the compressive ring decreases. Thus for sufficiently large load, the plate may buckle unsymmetrically into circumferential waves near the boundary.

In order to study the effect of the varying rate of inhomogeneity on the critical load and buckling mode, we solved the corresponding problem for two different law of material inhomogeneity: $E = \bar{E}_0 \exp^{-q_1 r}$ and $E = \widehat{E}_0 \exp^{-q_2 r^2}$. The buckling load for unsymmetrical buckling was calculated numerically over a large range of parameters $\bar{E}_0$, $\widehat{E}_0$, $q_{1,2}$, but for constant average value of the in-plane elastic modulus (2). The results are summarized in Table 1 and Fig. 2. The parameter value $q = 0$ corresponds to uniform plate with constant Young's modulus.

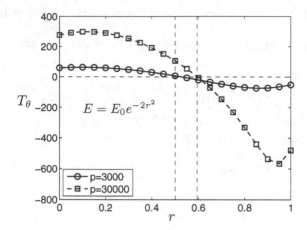

Fig. 1. Dimensionless axisymmetrical circumferensial stress T_θ for increasing values of load p. Law of material inhomogeneity is taken as $f(r) = \exp^{-2r^2}$.

Table 1. Buckling load and corresponding wave numbers for the nonuniform plate

		$q = 0$	$q = 0.5$	$q = 1$	$q = 3$	$q = 5$
$E = \bar{E}_0 e^{-qr}$	p_{cr}	64522	56841	49207	26324	12123
	n	14	14	14	15	17
$E = \widehat{E}_0 e^{-qr^2}$	p_{cr}	64522	53287	44137	20843	9103
	n	14	14	14	16	20

As the rate of inhomogeneity q increases the buckling mode shows more and more waves in the circumferential directions (Tab. 1). The buckling load of nonuniform circular plate is approximately 6 times less than the buckling load of uniform plate, see Fig. 2.

For the consecutive wave number we noted closely adjacent values of the critical load, e.g. for the uniform plate the critical loads differ between each other by less than 1% ($p_{cr} = 64522$ for $n = 14$ and $p_{cr} = 64929$ for $n = 13$). The heterogeneous plate (with the rate function $f(r) = \exp^{-4r}$) wrinkles at $p_{cr} = 18355$ and the buckling mode has 16 waves, while for 15 waves the critical load is 18416. Thus, the considered plate is sensitive to initial imperfections of form or to initial stresses.

Buckling of axisymmetric equilibrium states of isotropic homogeneous circular plates was studied in [3]. The authors took the grid step to be $\delta = 0.02$ and obtained the critical load value of 62800. We used the mesh size $\delta = 0.01$ in most calculations. For a twice larger mesh our results coincide with those found in [3].

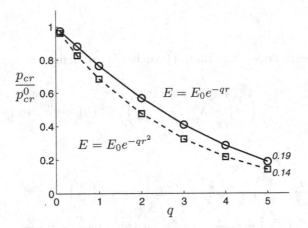

Fig. 2. Buckling load of heterogeneous plate. p_{cr}^0 denotes the buckling pressure for uniform plate.

Pre-buckling stress-state in a narrow zone near the plates's edge makes a major contribution to the unsymmetrical buckling mode and the value of the critical load. Change of elastic moduli ratio E_r/E_θ affects the deformed shape in the same manner as varying the in-plane Young's modulus in the radial direction. Thus, decreasing the ratio E_r/E_θ causes the smaller critical load and the large mode number in the circumferential directions.

For numerical examples the plate properties are taken from literature as material properties of the Lamina Cribrosa (LC) of the human eye [6]: the in-plane modulus and Poisson's ratio are assumed to be E_r^{av} =0.3 MPa, ν = 0.45, $h/R = 0.1$. By taking inhomogeneity parameter $q = 5$ we find that the non-axisymmetric buckling occurs under pressure about 60 mm Hg. So, from mechanical point of view, folds at the periphery of LC could be explained by the bucking of the axisymmetric state of the LC in the nonaxisymmetric state.

5   Conclusion

The critical pressure for unsymmetric buckling is significantly lower than for a plate with constant elastic moduli, if the elasticity moduli decrease away from the center of a plate. Number of waves in the circumferential direction increases with degree of nonuniformity. The folders in the narrow zone at the periphery of the Lamina Cribrosa (LC) of the human eye could be explained by the buckling of the axisymmetric state of LC in the nonaxisymmetric state.

Appendix

The linear differential operators that appear in (1) are define by

$$L_1(y) = y''' + \frac{y''}{r} + \frac{\lambda_{r\theta}^2}{r}\left(\frac{\ddot{y}}{r}\right)' - \lambda^2\left(\frac{y'}{r^2} + \frac{\ddot{y}}{r^3}\right), \quad \mathcal{L}_1^{\pm}(y) = y'' \pm \nu_\theta\left(\frac{\ddot{y}}{r^2} + \frac{y'}{r}\right),$$

$$L_2(y) = \left(\lambda^2\left(\frac{y'}{r^2} + \frac{\ddot{y}}{r^3}\right) + \lambda_{r\theta}^2\frac{y''}{r}\right)^{\cdot}, \quad \mathcal{L}_2(y) = \lambda_k^2\left(\frac{\dot{y}}{r}\right)',$$

$$L_{1\phi}(y) = \mu_r\left(y'' + \frac{y'}{r}\right) - \mu_\theta\frac{y}{r^2} + \mu_k\frac{\ddot{y}}{r^2}, \quad L_{2\phi}(y) = \mu_\theta\frac{\dot{y}}{r^2} + \mu_{r\theta}\frac{\dot{y}'}{r} - \mu_k\left(\frac{\dot{y}}{r}\right)',$$

$$L_{1\psi}(y) = \eta_{r\theta}\frac{\dot{y}'}{r} - \eta_\theta\frac{\dot{y}}{r^2} - \eta_k\left(\frac{\dot{y}}{r^2} + \frac{\dot{y}'}{r}\right), \quad L_{2\psi}(y) = \eta_\theta\frac{\ddot{y}}{r^2} + \eta_k\left(y'' + \frac{y'}{r}\right)',$$

$$L_3(y) = y'''' + \frac{2y'''}{r} + \frac{\kappa^2}{r^2}\left(y'' - \frac{y'}{r} + \frac{y}{r^2}\right)^{\cdot\cdot} + \lambda^2\left(\frac{y'}{r^3} - \frac{y''}{r^2} + \left(2\frac{y}{r^4} + \frac{\ddot{y}}{r^4}\right)^{\cdot\cdot}\right),$$

$$\mathcal{L}_3^{\pm}(y) = 2y''' + \frac{2 \pm \nu_\theta}{r}y'' - \lambda^2\left(\frac{y'}{r^2} + \frac{\ddot{y}}{r^2}\right) - \kappa^2\frac{1}{r}\left(\frac{\ddot{y}}{r}\right)',$$

with the short-hand notations

$$\lambda^2 = \frac{D_\theta}{D_r}, \quad \lambda_{r\theta}^2 = \frac{D_{r\theta}}{D_r}, \quad \lambda_k^2 = \frac{D_k}{D_r}, \quad \kappa^2 = \frac{E_\theta}{G_{r\theta}} - 2\nu_\theta,$$

$$\mu_i = \frac{6}{5hR^2}\frac{D_i}{G_{rz}}, \quad i = \{r, r\theta, \theta, k\}, \quad \eta_j = \frac{6}{5hR^2}\frac{D_j}{G_{\theta z}}, \quad j = \{r\theta, \theta, k\},$$

$$D_i = \frac{h^3 E_i}{12(1 - \nu_r\nu_\theta)}, \ (i = r, \theta), \quad D_k = \frac{h^3}{12}G_{r\theta}, \quad D_{r\theta} = D_r\nu_\theta + 2D_k.$$

Here D_r, D_θ, D_k are the bending stiffnesses, E_r, E_θ are the Young moduli in the radial and the circumferential directions, respectively; ν_r, ν_θ are corresponding Poisson's ratios; $G_{r\theta}, G_{rz}, G_{\theta z}$ are shear moduli characterising changes of angle between the r- and θ-, the r- and z-, the θ- and z-directions, respectively. The equality $E_r/E_\theta = \nu_r/\nu_\theta$ must be satisfied due of the symmetry.

For an isotropic plate $E_r = E_\theta = E$, $\nu_r = \nu_\theta = \nu$, $G_{r\theta} = E/2(1+\nu)$, $D_r = D_\theta = D_{r\theta}$ so $\lambda = \lambda_{r\theta} = \lambda_k = 1$, $\kappa^2 = 2$.

After separation of variables the linear differential operators take the form

$$L_1^n(y) = y''' + \frac{y''}{r} - (\lambda^2 + n^2\lambda_{r\theta}^2)\frac{y'}{r^2} + n^2\frac{\lambda^2 + \lambda_{r\theta}^2}{r^3},$$

$$\mathcal{L}_1^{n\pm}(y) = y'' \pm \nu_\theta\left(\frac{y'}{r} - \frac{n^2 y}{r^2}\right), \quad L_{1\phi}^n(y) = \mu_r\left(y'' + \frac{y'}{r}\right) - (\mu_\theta + n^2\mu_k)\frac{y}{r^2},$$

$$L_{1\psi}^n(y) = n\left(y'\frac{\eta_{r\theta} - \eta_k}{r} - y\frac{\eta_\theta + \eta_k}{r^2}\right),$$

$$L_2^n(y) = -\lambda^2\frac{ny'}{r^2} - n\lambda^2\left(\frac{y'}{r^2} - \frac{n^2 y}{r^3}\right), \quad \mathcal{L}_2^n(y) = -n\lambda_k^2\left(\frac{y'}{r} - \frac{y}{r^2}\right),$$

$$L_{2\phi}^n(y) = n\left(y'\frac{\mu_{r\theta} + \mu_k}{r} - y\frac{\mu_\theta + \mu_k}{r^2}\right), \quad L_{2\psi}^n(y) = \eta_k\left(y'' + \frac{y'}{r}\right)' - \eta_\theta\frac{n^2 y}{r^2},$$

$$L_3^n(y) = y'''' + \frac{2}{r}y''' - \frac{\lambda^2 + n^2\kappa^2}{r^2}y'' + \frac{\lambda^2 + n^2\kappa^2}{r^3}y' + n^2(n^2\lambda^2 - (2\lambda^2 + \kappa^2))\frac{y}{r^4},$$

$$\mathcal{L}_3^{\pm}(y) = 2y''' + \frac{2 \pm \nu_\theta}{r}y'' - \lambda^2\left(\frac{y'}{r^2} - \frac{n^2 y}{r^2}\right) + \kappa^2\frac{n^2}{r^2}\left(y' - \frac{y}{r}\right),$$

References

1. Panov, D.Y., Feodos'ev, V.I.: Equilibrum and Loss of Stability of Shallow Shells with Large Deflections. P.M.M. 12, 389–406 (1948) (in Russian)
2. Feodos'ev, V.I.: On a method of solution of the nonlinear problems of stability of deformable systems. J. App. Math. Mech. 27, 392–404 (1963)
3. Cheo, L.S., Reiss, E.L.: Unsymmetric Wrinkling of Circular Plates. Quart. Appl. Math. 31, 75–91 (1973)
4. Bauer, S.M.: Mechanical Models of the Development of Glaucoma. In: Advances in Mechanics of Solids In Memory of Prof. E. M. Haseganu, pp. 153–178. World Scientific Publishing, Singapore (2006)
5. Ambartsumyan, S.A.: Theory of Anisotropic Plates. Technomic Publishing, Stamford (1970)
6. Edwards, M.E., Good, T.A.: Use of a mathematical model to estimate stress and strain during elevated pressure induced lamina cribrosa deformation. Curr. Eye Res. 3, 215–225 (2001)

A Positivity-Preserving Splitting Method for 2D Black-Scholes Equations in Stochastic Volatility Models

Tatiana P. Chernogorova and Radoslav L. Valkov

Sofia University, Faculty of Mathematics and Informatics
{chernogorova,rvalkov}@fmi.uni-sofia.bg

Abstract. In this paper we present a locally one-dimensional (LOD) splitting method to the two-dimensional Black-Scholes equation, arising in the Hull & White model for pricing European options with stochastic volatility. The parabolic equation degenerates on the boundary $x = 0$ and we apply to the one-dimensional subproblems the fitted finite-volume difference scheme, proposed in [8], in order to resolve the degeneration. Discrete maximum principle is proved and therefore our method is positivity-preserving. Numerical experiments are discussed.

1 Introduction

In 1987 Hull & White proposed a model for valuing an option with a stochastic volatility of the price of the underlying stock [7]. The Hull & White PDE constitutes an important two-dimenstional extension to the celebrated, one-dimensional, Black-Scholes PDE. Over the years, several numerical methods have been developed to solve different two-dimensional problems [4,6,5].

This paper deals with the numerical solution of the Black-Scholes equation in stochastic volatility models. The features of this time-dependant, two-dimensional convection-reaction-diffusion problem is the presence of a *mixed spatial derivative term*, stemming from the correlation between the two underlying stochastic processes for the asset price and it's variance, and *degeneration* of the parabolic equation on the part of the domain boundary. Existence of solutions to the degenerate parabolic PDEs such as the Hull & White model does not follow from classical theory and additional analysis is needed [5].

We formulate the differential problem and present a brief analysis for existence and uniqueness of a weak solution in Section 2. Section 3 contains the full description of the splitting method. In Section 4 we perform numerical experiments with the splitting scheme, where we analyze global errors in the strong norm and the L_2-norm.

2 The Differential Problem

Consider a European option with stochastic volatility $\sqrt{y}$ and an expiry date T. It has been shown in [7] that it's price, u, satisfies the following second-order

I. Dimov, I. Faragó, and L. Vulkov (Eds.): NAA 2012, LNCS 8236, pp. 206–213, 2013.

differential equation

$$-\frac{\partial u}{\partial t} - \frac{1}{2}\left[x^2 y \frac{\partial^2 u}{\partial x^2} + 2\rho\xi xy^{3/2}\frac{\partial^2 u}{\partial x \partial y} + \xi^2 y^2 \frac{\partial^2 u}{\partial y^2}\right] - rx\frac{\partial u}{\partial x} - \mu y\frac{\partial u}{\partial y} + ru = 0, \quad (1)$$

for $(x,y,t) \in (0,X) \times (\zeta, Y) \times (0,T) := \Omega \times (0,T)$ with appropriate final (pay-off) and Dirichlet boundary conditions of the form

$$u(x,y,T) = u_T(x,y), \ (x,y) \in \Omega, \tag{2}$$
$$u(x,y,t) = u_D(x,y,t), \ (x,y,t) \in \partial\Omega \times (0,T), \tag{3}$$

where x denotes the price of the underlying stock, ξ and μ are constants from the stochastic process, governing the variance y, ρ is the instantaneous correlation between x and y and ζ, X, Y and T are positive constants, defining the solution domain. In (3) $\partial\Omega$ denotes the boundary of Ω and $u_T(x)$ and $u_D(x,t)$ are given functions. For the choices of these functions we refer to [5]. As mentioned in [7], ρ can not take negative value, $\rho \in [0,1)$. In this paper we assume that $\rho \in [0,1)$ is a constant. Also, the independent variable y satisfies, in general, $y \geq 0$. However the case that $y = 0$ is trivial because it means that the volatility of the stock is zero in the market. This stock then becomes deterministic, which is impossible unless the stock is a risk-less asset. In this case the price of the option is deterministic. Therefore it is reasonable to assume that $y \geq \zeta$ for a (small) positive constant ζ.

Introducing a new variable $\tilde{u} = \exp(\beta t)u$ and coming back to the previous notation, (1) is rewritten in the following general equation after a change in the time variable $\tilde{t} = T - t$

$$\frac{\partial u}{\partial t} - \nabla \cdot (k(u)) + cu = g, \tag{4}$$

$k(u) = A\nabla u + \mathbf{b}u$ is the flux, $\mathbf{b} = \left(rx - \frac{3}{4}\rho y^{1/2}\xi x - yx, \mu y - \frac{1}{2}\rho y^{1/2}\xi - \xi^2 y\right)^T$,

$$A = \begin{pmatrix} a_{11} & a_{12} \\ a_{21} & a_{22} \end{pmatrix} = \begin{pmatrix} \frac{1}{2}yx^2 & \frac{1}{2}\rho y^{3/2}\xi x \\ \frac{1}{2}\rho y^{3/2}\xi x & \frac{1}{2}\xi^2 y^2 \end{pmatrix},$$
$$c = \beta + 2r - \frac{3}{4}\rho y^{1/2}\xi - y + \mu - \frac{3}{4}\rho y^{1/2}\xi - \xi^2. \tag{5}$$

We consider the ramp payoff terminal condition, given by

$$u_T(x,y) = max(0, x - E), \ (x,y) \in \bar{I}_x \times \bar{I}_y, \ I_x = (0,X) \text{ and } I_y = (0,Y), \quad (6)$$

where $E < X$ denotes the exercise price of the option.

The solution domain of the above problem contains four boundary surfaces, defined by $x = 0$, $x = X$, $y = \zeta$ and $y = Y$. The boundary conditions at $x = 0$ and $x = X$ are simply taken to be the extension of the terminal conditions at the points, i.e.

$$u_D(0,y,t) = u_T(0,y) = 0, \text{ and } u_D(X,y,t) = u_T(X,y). \tag{7}$$

To determine the boundary conditions at $y = \zeta$ and $y = Y$ we need to solve the standard one-dimensional Black-Scholes equation, obtained by taking $\xi = \mu = 0$ in (1) for two particular values $\sigma = \sqrt{\zeta}$ and $\sigma = \sqrt{Y}$ with the boundary and terminal conditions defined above.

3 LOD Additive Splitting and Full Discretization

In this section we present the numerical method.

3.1 The Splitting Method

We start with rewriting our equation in a conservative form

$$
\frac{\partial u}{\partial t} = \frac{\partial}{\partial x}\left(a_{11}\frac{\partial u}{\partial x} + \left(b_1 - \frac{\partial a_{12}}{\partial y}\right)u\right) - c_1 u
$$
$$
+ \frac{\partial}{\partial y}\left(a_{22}\frac{\partial u}{\partial y} + \left(b_2 + \frac{\partial a_{21}}{\partial x}\right)u\right) - c_2 u + \frac{\partial}{\partial y}\left((a_{12} + a_{21})\frac{\partial u}{\partial x}\right),
$$

where a_{11}, a_{22}, $a_{12} = a_{21}$ and b_1, b_2 are as given in (5) and $c_1 + c_2 = c$. Our flux-based finite volume spatial discretization benefits from the following representation

$$
\frac{\partial u}{\partial t} = \frac{\partial}{\partial x}\left(xw(x,y,u)\right) - qu + \frac{\partial}{\partial y}\left(y\hat{w}(y,u)\right) - \hat{q}u + \frac{\partial}{\partial y}\left(k(x,y)\frac{\partial u}{\partial x}\right),
$$

$$
w(x,y,u) = 0.5xy\frac{\partial u}{\partial x} + \left(r - y - 1.5\rho\xi y^{1/2}\right)u, q(y) = 1.5r - y - 1.5\rho\xi y^{1/2},
$$
$$
\hat{w}(x,y,u) = 0.5\xi^2 y\frac{\partial u}{\partial y} + \left(\mu - \xi^2\right)u, \hat{q}(y) = 0.5r + \mu - \xi^2, k(x,y) = \rho\xi xy^{3/2}.
$$
$$\tag{8}$$

Let us introduce the notations:

$$
L_1 u = \frac{\partial}{\partial x}\left(xw(x,y,u)\right) - qu,
$$

$$
L_2 u = \frac{\partial}{\partial y}\left(y\hat{w}(y,u)\right) - \hat{q}u + \frac{\partial}{\partial y}\left(k(x,y)\frac{\partial u}{\partial x}\right).
$$

An equidistant truncation of $[0,T]$ $\{t_k = k\tau, k = 0,1,\ldots,K, \tau = \frac{T}{K}\}$, a non-uniform mesh $\overline{w} = \overline{w}_x \times \overline{w}_y$ by space steps for x and y are h_i^x, $i = 0,\ldots,N-1$ and $h_j^y, j = 0,\ldots,M-1$ respectively and a secondary mesh $x_{i\pm1/2} = 0.5(x_{i\pm1} + x_i)$, $y_{j\pm1/2} = 0.5(y_{j\pm1} + y_j)$, $x_{-1/2} \equiv x_0 = 0$, $x_{N+1/2} \equiv x_N = X$, $y_{-1/2} \equiv y_0 = \zeta$, $y_{M+1/2} \equiv y_M = Y$ allow us to consider the LOD additive scheme:

$$
u_{(1)}\begin{cases}
\frac{\partial u_{(1)}}{\partial t} = L_1 u_{(1)}, & t_k < t \le t_{k+1}, \\
u_{(1)}(x,y,0) = u_T(x), & (x,y) \in [0,X] \times [\zeta,Y], \\
u_{(1)}(0,y,t) = u_D(0,y,t), & (y,t) \in [\zeta,Y] \times [0,T], \\
u_{(1)}(X,y,t) = u_D(X,y,t), & (y,t) \in [\zeta,Y] \times [0,T],
\end{cases}
\tag{9}
$$

$$u_{(2)} \begin{cases} \frac{\partial u_{(2)}}{\partial t} = L_2 u_{(2)}, \quad t_k < t \leq t_{k+1}, k = 1, 2, \ldots, K, \\ u_{(2)}(x, y, t_k) = u_{(1)}(x, y, t_{k+1}), \ (x, y) \in [0, X) \times [\zeta, Y], \\ u_{(2)}(x, \zeta, t) = u_D(x, \zeta, t), \ (x, t) \in (0, X] \times (0, T], \\ u_{(2)}(x, Y, t) = u_D(x, Y, t), \ (x, t) \in (0, X] \times (0, T]. \end{cases} \quad (10)$$

3.2 Analysis of Time Semi-discretization

In order to proceed with the analysis of the time semi-discretization we assume that the following condition is fulfilled:

Assumption 1. $u, L_1 u, L_2 u, L_1^2 u, L_2^2 u, L_1 L_2 u, L_2 L_1 u \subset C(\Omega \times [0, T])$

as well as the boundary and initial conditions are compatible and smooth enough.

We define the global error $E_\tau = \sup_{k \leq \frac{T}{\tau}} \|u(t_k) - u^k\|_{\infty, \Omega}$ for semi-discretization, where u^k is the solution of the semi-discrete system, resulting from (9) and (10). By similar considerations as in [2], one can prove the following theorem:

Theorem 1. *If the solution of the continuous problem (1)-(3) satisfies the restrictions in Assumption 1 then $E_\tau \leq C\tau$, where C is a constant, independent of τ.*

3.3 The Finite Volume Method and Full Discretization

We now proceed to the derivation of the full discretization of problem (1)-(3). By (8) we have

$$w(u) = 0.5xy \frac{\partial u}{\partial x} + \left(r - y - 1.5\rho\xi y^{1/2} \right) u =: \bar{a}(y)x \frac{\partial u}{\partial x} + \bar{b}(y)u,$$

where $\bar{a}(y) = 0.5y$ and $\bar{b}(y) = r - y - 1.5\rho\xi y^{1/2}$ are notations, used in the next considerations.

Let y is fixed. After *time discretization* we integrate the first equation in (9) w.r.t. x in the interval $(x_{i-1/2}, x_{i+1/2})$, $i = 1, 2, \ldots, N - 1$ and applying the mid-point quadrature rule to the integrals in the equation we arrive at

$$\frac{u_i^{k+1/2} - u_i^k}{\tau} \hbar_i^x = \left(x_{i+1/2} \, w(u^{k+1/2}) \Big|_{x_{i+1/2}} \right.$$
$$\left. - x_{i-1/2} \, w(u^{k+1/2}) \Big|_{x_{i-1/2}} \right) - c_1(y) u_i^{k+1/2} \hbar_i^x, \quad (11)$$

where $\hbar_i^x = x_{i+1/2} - x_{i-1/2}$, $u_i = u(x_i, y, t)$ and $c_1(y) = 1.5r - y - 1.5\rho\xi y^{1/2}$. Following [8], see also [1], in order to obtain an approximation for the flux in the node $x_{i+1/2}$, we consider the following BVP

$$\left(\bar{a}_{i+1/2}(y)xv' + \bar{b}_{i+1/2}(y)v \right)'_x = 0, \ x \in I_i, \ v(x_i) = u_i, \ v(x_{i+1}) = u_{i+1}.$$

The solution of that problem is

$$w_{i+1/2}(u) = \bar{b}(y)\frac{x_{i+1}^{\bar{\alpha}_i(y)}u_{i+1} - x_i^{\bar{\alpha}_i(y)}u_i}{x_{i+1}^{\bar{\alpha}_i(y)} - x_i^{\bar{\alpha}_i(y)}}, \quad \bar{\alpha}_i = \frac{\bar{b}_{i+1/2}}{\bar{a}_{i+1/2}}.$$

When deriving the approximation of the flux at $x_{1/2}$, because of the degeneration, we consider the BVP with an extra degree of freedom [8]

$$\left(\bar{a}(y)xv' + \bar{b}(y)v\right)_x' = C_1, x \in I_0, \ v(0) = u_0, \ v(x_1) = u_1$$

and the approximation for $w_{1/2}(u)$ is $\frac{1}{2}\left[\left(\bar{a}(y) + \bar{b}(y)\right)u_1 - \left(\bar{a}(y) - \bar{b}(y)\right)u_0\right]$.

It was first mentioned in [3] that the boundary conditions deteriorate the accuracy of the splitting method if the discrete equations on the boundaries differ from the equations for the inner nodes of the mesh. We adopt the following correction technique [9]:

$$\bar{u}_{0,j} = u_D(0, y_j, t^{n+1}), \bar{u}_{X,j} = u_D(X, y_j, t^{n+1}), \ j = 0, \dots, M,$$
$$\bar{\Lambda}_1\bar{u}_{i,0} - \Lambda_1 u_{i,0}^n = 0, \bar{\Lambda}_1\bar{u}_{i,Y} - \Lambda_1 u_{i,Y}^n = 0, \ i = 1, \dots, N-1,$$

where the discrete operators $\bar{\Lambda}_1$ and Λ_1 match the presented discretization in the x-direction and $\bar{u}$ is the numerical solution, corresponding to $u^{k+1/2}$.

Next, after substituting the obtained approximations for the flux in (11), considering the boundary conditions, we arrive at the scalar form of the discrete problem for $\bar{u}$,

$$B_0\bar{u}_{0,j} + C_0\bar{u}_{1,j} = F_0,$$
$$A_1\bar{u}_{0,j} + B_1\bar{u}_{1,j} + C_1\bar{u}_{2,j} = F_1,$$
$$\dots\dots\dots\dots\dots\dots\dots\dots\dots\dots\dots\dots$$
$$A_i\bar{u}_{i-1,j} + B_i\bar{u}_{i,j} + C_i\bar{u}_{1,j} = F_i,$$
$$A_1\bar{u}_{N-1,j} + B_N\bar{u}_{N,j} = F_N,$$

for $i = 2, \dots, N-1$, where

$$B_0 = 1, \ C_0 = 0, \ F_0 = u_D(0, y, t^{k+1}), \ A_N = 0, \ B_N = 1, \ F(N) = u_D(X, y, t^{k+1}),$$

$$A_1 = -\frac{x_{1/2}}{2}\left(\bar{a}(y_j) - \bar{b}(y_j)\right), \ C_1 = -\frac{x_{3/2}\bar{b}(y_j)x_2^{\alpha(y_j)}}{x_2^{\alpha(y_j)} - x_1^{\alpha(y_j)}},$$

$$B_1 = \frac{\hbar_2^x}{\tau} + \frac{x_{3/2}\bar{b}(y_j)x_1^{\alpha(y_j)}}{x_2^{\alpha(y_j)} - x_1^{\alpha(y_j)}} + \frac{x_{1/2}}{2}\left(\bar{a}(y_j) + \bar{b}(y_j)\right) + \hbar_2^x c_1(y_j), \ F_1 = \frac{\hbar_2^x}{\tau},$$

$$A_i = -\frac{x_{i-1/2}\bar{b}(y_j)x_{i-1}^{\alpha(y_j)}}{x_i^{\alpha(y_j)} - x_{i-1}^{\alpha(y_j)}}, \ C_i = -\frac{x_{i+1/2}\bar{b}(y_j)x_{i+1}^{\alpha(y_j)}}{x_{i+1}^{\alpha(y_j)} - x_i^{\alpha(y_j)}},$$

$$B_i = \frac{\hbar_i^x}{\tau} + \frac{x_{i+1/2}\bar{b}(y_j)x_i^{\alpha(y_j)}}{x_{i+1}^{\alpha(y_j)} - x_i^{\alpha(y_j)}} + \frac{x_{i-1/2}\bar{b}(y_j)x_i^{\alpha(y_j)}}{x_i^{\alpha(y_j)} - x_{i-1}^{\alpha(y_j)}} + \hbar_i^x c_1(y_j), \ F_i = \frac{\hbar_i^x}{\tau}.$$

The discretization of the problem (10) is obtained by similar considerations. Introducing $\hat{\alpha}_j = \frac{\bar{b}_{j+1/2}}{\bar{a}_{j+1/2}}$, we obtain

$$\hat{u}_{i,0} = u_D(x_i, \zeta, t_{k+1}),$$

$$y_{j-1/2} \frac{\hat{b}_{j-1/2} y_{j-1}^{\hat{\alpha}_{j-1}}}{y_j^{\hat{\alpha}_{j-1}} - y_{j-1}^{\hat{\alpha}_{j-1}}} \hat{u}_{i,j-1} - \left[\frac{\hbar_j^y}{\tau} + y_{j+1/2} \frac{\hat{b}_{j+1/2} y_j^{\hat{\alpha}_j}}{y_{j+1}^{\hat{\alpha}_j} - y_j^{\hat{\alpha}_j}} + y_{j-1/2} \frac{\hat{b}_{j-1/2} y_j^{\hat{\alpha}_{j-1}}}{y_j^{\hat{\alpha}_{j-1}} - y_{j-1}^{\hat{\alpha}_{j-1}}} \right.$$

$$\left. - \hbar_j^y \hat{c} \right] \hat{u}_{i,j} + y_{j+1/2} \frac{\hat{b}_{j+1/2} y_{j+1}^{\hat{\alpha}_j}}{y_{j+1}^{\hat{\alpha}_j} - y_j^{\hat{\alpha}_j}} \hat{u}_{i,j+1} = -\frac{\hbar_j^y \bar{u}_{i,j}}{\tau} - 0.25 k(x_i, y_{j+1/2})$$

$$\times \left(\frac{\bar{u}_{i+1,j+1} - \bar{u}_{i,j+1} + \bar{u}_{i+1,j} - \bar{u}_{i,j}}{h_i^x} + \frac{\bar{u}_{i,j+1} - \bar{u}_{i-1,j+1} + \bar{u}_{i,j} - \bar{u}_{i-1,j}}{h_{i-1}^x} \right)$$

$$+ 0.25 k(x_i, y_{j-1/2}) \left(\frac{\bar{u}_{i+1,j} - \bar{u}_{i,j} + \bar{u}_{i+1,j-1} - \bar{u}_{i,j-1}}{h_i^x} \right.$$

$$\left. + \frac{\bar{u}_{i,j} - \bar{u}_{i-1,j} + \bar{u}_{i,j-1} - \bar{u}_{i-1,j-1}}{h_{i-1}^x} \right).$$

$$\hat{u}_{i,M} = u_D(x_i, Y, t_{k+1}),$$

By similar considerations as given in [8] we prove the following lemma and theorem:

Lemma 2. *The system matrices for both $\bar{u}$ and $\hat{u}$ are (can be reduced to) M-matrices.*

Theorem 3. *For a non-negative functions $g(x)$ and $\phi(x, y, t)$ the numerical solution $\hat{u}$, generated by the our splitting method, is also non-negative.*

One can see that the consistency of the space discretization above relies on the consistency of the flux $w(u)$ with respect to x (similar is the case for y). As it was shown in [8] this discretization scheme admits a finite volume formulation with special trial space S_h. Therefore the following analogue of the estimate in Lemma 4.2 in [8] holds

$$\|w(v) - w_h(v_I)\|_{\infty, I_i} \leq C \|w(v)\|_{\infty, I_i}, i = 1, 2, \dots, N,$$

where v is the sufficiently smooth function and v_I is the S_h interpolant.

Theorem 4. *Let u be the exact solution of (1)-(3) and u^n is the numerical solution. Then, there exists a positive constant C, independent of N, M and τ such that the global error satisfies*

$$\left\| u(t_n)|_{\Omega_h} - u^n \right\|_\infty \leq C(\tau + h_x + h_y),$$

where $u(t_n)|_{\Omega_h}$ *is the restriction of the exact solution on the product of the meshes with respect to x and y.*

4   Numerical Experiments

Numerical experiments, presented in this section, illustrate the properties of the constructed method. We solve numerically the following Test Problem:

(TP). *Call option* with final condition (6). Parameters: $X = 100$, $Y = 1$, $T = 1$, $\zeta = 0.01$, $r = 0.1$, $\rho = 0.9$, $\xi = 1$, $\mu = 0$ and $E = 57$.

Table 1 shows the temporal convergence of the numerical solution to the chosen exact solution, $u = x\exp(-yt)$, of (1). We use the parameters $X = Y = T = 1$, $\xi = 1$ and $\zeta = 0.01$. The size of the spatial mesh is now fixed to 512×512 as the time step varies. The obtained results show that our numerical method profits from the boundary corrections since it is able to sustain the first order of temporal convergence.

Table 1.

K	$\rho = 0.5, r = 0, \mu = 0$				$\rho = 0.9, r = 0.1, \mu = 0.1$			
	E_∞^N	RC	E_2^N	RC	E_∞^N	RC	E_2^N	RC
16	2.000e-2	-	7.138e-3	-	3.235e-2	-	1.157e-2	-
32	9.859e-3	*1.02*	3.585e-3	*0.99*	1.562e-2	*1.05*	5.753e-3	*1.01*
64	4.848e-3	*1.02*	1.796e-3	*1.00*	7.549e-3	*1.05*	2.864e-3	*1.01*
128	2.398e-3	*1.02*	8.980e-4	*1.00*	3.721e-3	*1.02*	1.427e-3	*1.01*
256	1.197e-3	*1.00*	4.477e-4	*1.00*	1.862e-3	*1.00*	7.099e-4	*1.01*

We now solve numerically the original problem TP, characterized by non-smoothness of the terminal (initial) condition (6) on an uniform spatial mesh sized $N \times N$ with $2N$ time layers. In the following Table 2 the mesh C-norm and $RMSE$ (*root mean square error*)-norm are computed w.r.t. the numerical solution on a very fine mesh sized $512 \times 512 \times 1024$. The boundary conditions in direction x are derived by the terminal condition (7). The boundary conditions in direction y are obtained as explained in Section 2, see Figures 1, 2. The RMSE is computed on the region $[0.9E, 1.1E] \times [\zeta, Y]$. The numerical solution of $TP1$ is visualized on Figure 3.

Table 2.

N	8	16	32	64	128	256
E_∞	4.0877	2.0678	0.9911	0.4559	0.1944	0.0649
		(0.983)	*(1.061)*	*(1.120)*	*(1.230)*	*(1.584)*
E_{RMSE}	0.7641	0.2649	0.1197	0.0551	0.0236	0.0079
		(1.528)	*(1.146)*	*(1.119)*	*(1.223)*	*(1.571)*

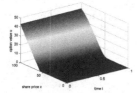
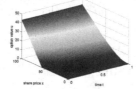
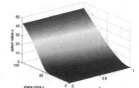

Fig. 1. BC $y = 0.01$ **Fig. 2.** BC $y = Y$ **Fig. 3.** Option Value TP

5 Conclusion

In this paper we develop a locally one-dimensional splitting scheme, where the spatial discretization is performed as a flux-based finite volume method [8]. We prove the non-negativity of the numerical solution and first order convergence in both space and time is shown by numerical experiments.

Acknowledgement. The authors are supported by the Sofia University Foundation under Grant No 181/2012. The first author is also supported by the European Social Fund through the Human Resource Development Operational Programme under contract BG05 1PO001-3.3.06-0052 (2012/2014), while the second author is supported by the Bulgarian National Fund under Project DID 02/37/09.

References

1. Chernogorova, T., Valkov, R.: Finite volume difference scheme for a degenerate parabolic equation in the zero-coupon bond pricing. Math. and Comp. Modeling 54, 2659–2671 (2011)
2. Clavero, C., Jorge, J.C., Lisbona, F.: Uniformly convergent schemes for singular perturbation problems combining alternating directions and exponential fitting techniques. In: Miller, J.J.H. (ed.) Applications of Advanced Computational Methods for Boundary and Interior Layers, pp. 33–52. Boole Press, Dublin (1993)
3. D'Yakonov, E.G.: Difference schemes with splitting operator for multidimensional non-stationary problem. Zh. Vychisl. Mat. i Mat. Fiz. 2, 549–568 (1962)
4. int' Hout, K.J., Foulon, S.: ADI finite difference schemes for option pricing in the Heston model with correlation. Int. J. Numer. Anal. Mod. 7, 303–320 (2010)
5. Huang, C.-S., Hung, C.-H., Wang, S.: A fitted finite volume method for the valuation of options on assets with stochastic volatilities. Computing 77, 297–320 (2006)
6. Hundsdorfer, W., Verwer, J.: Numerical Solution of Time-Dependent Advection-Diffusion-Reaction Equations. Springer, Heidelberg (2003)
7. Hull, J., White, A.: The pricing of options on assets with stochastic volatilities. J. Financ. 42, 281–300 (1987)
8. Wang, S.: A novel fitted finite volume method for Black-Sholes equation governing option pricing. IMA J. of Numer. Anal. 24, 699–720 (2004)
9. Yanenko, N.N.: The Method of Fractional Steps. Springer, Berlin (1971)

Two Splitting Methods for a Fixed Strike Asian Option

Tatiana P. Chernogorova[1] and Lubin G. Vulkov[2]

[1] FMI, University of Sofia, 5, J. Bourchier Blvd., 1164 Sofia, Bulgaria
`chernogorova@fmi.uni-sofia.bg`
[2] FNSE, Uiversity of Rousse, Studentska 8 Str. 7017 Rousse , Bulgaria
`lvalkov@uni-ruse.bg`

Abstract. The valuation of Asian Options can often be reduced to the study of initial boundary problems for ultra-parabolic equations. Two splitting methods are used to transform the whole time-dependent problem of a fixed strike Asian option into two unsteady subproblems of a smaller complexity. The first subproblem is a time-dependent convection-diffusion and the finite volume difference method of S. Wang [6] is applied for its discretization. The second one is a transport problem and is approximated by monotone weighted difference schemes. The positivity property of the numerical methods is established. Numerical experiments are discussed.

1 Introduction

Asian options are exotic financial derivative products whose price must be calculated by numerical evaluation. We consider here the financial instrument denoted a *fixed strike Asian option*, which is basically a contract written at time $t = 0$ between a buyer and a seller, giving the buyer the right to receive an a priori unknown, non-negative sum of money, the *terminal payoff*, at a predefined time T sometimes in the future. T is called the *time of expiration* of the option. The terminal payoff is related to the value of something called the *underlying risky assets*; anything whose value $S(t)$ at time t is determined by a certain stochastic process will do, e.g. a stock. For an *Asian option* the terminal payoff must depend on the value $S(t)$ at time T of the risky asset. The terminal value at time T is written $\max\{k(S(T), A(T)), 0\}$ for some continuous function k to be negotiated at time 0 between the buyer and the seller of the option, as for example the average value put option $k = K - A(T)$ (see [2,5] for other cases) for some predetermined *strike price K*. Following the so called axiom of no arbitrage of our financial model (see [2,3,7]) the seller will want to get some money up from to do you the service of selling you the option. The question is how much money? Since no cheating is allowed, the seller will ask you exactly the *fair price V* which is the amount of money that needs to go out in the market and by copper enough to exactly pay for the loss he will have then. It turns out that

I. Dimov, I. Faragó, and L. Vulkov (Eds.): NAA 2012, LNCS 8236, pp. 214–221, 2013.
© Springer-Verlag Berlin Heidelberg 2013

fair price $V(S, A, t)$ is tractable and can be described as the solution to the value problem

$$\frac{\partial V}{\partial \tau} = \frac{\sigma^2(t)}{2} S^2 \frac{\partial^2 V}{\partial S^2} + (r(t) - \gamma(t)) S \frac{\partial V}{\partial S} + S \frac{\partial V}{\partial A} - r(t) V, \quad (S, x, \tau) \in \Omega_\infty,$$

where $\Omega_\infty = (0, \infty) \times (0, \infty) \times (0, T)$ and $V(S, A, T) = \max\{k(S, A), 0\}$ in $\overline{\Omega}_\infty|_{t=T}$. Here $\gamma, \sigma > 0$ and r are the *dividend yield* and the *volatility* (on the risky asset) and the *market interest* rate (on the risk free assets). Financially, the solution on the entire domain Ω_∞ is not of much interest. Generally, it is possible to insert artificial positive cut off values S_{max} and A_{max} so that the interesting computational domain becomes the box $\{(S, A, t) \in (0, S_{max}) \times (0, A_{max}) \times (0, T)\}$. Then changing the variables $x = A_{max} - A$ we will solve the initial-boundary value problem in $\Omega = (0, S_{max}) \times (0, x_{max}) \times (0, T)$,

$$\frac{\partial V}{\partial \tau} = \frac{\sigma^2(t)}{2} S^2 \frac{\partial^2 V}{\partial S^2} + (r(t) - \gamma(t)) S \frac{\partial V}{\partial S} - S \frac{\partial V}{\partial x} - r(t) V, \quad (S, x, \tau) \in \Omega, \quad (1)$$

$$V(S, x, 0) = \max\left\{\frac{x_{max} - x}{T} - K, \, 0\right\}, \quad (2)$$

$$V(0, x, \tau) = e^{-r\tau} \max\left\{\frac{x_{max} - x}{T} - K, \, 0\right\}, \quad (3)$$

$$V(S_{max}, x, \tau) = \max\left\{e^{-r\tau}\left(\frac{x_{max} - x}{T} - K\right) + \frac{S_{max}}{rT}\left(1 - e^{-r\tau}\right), \, 0\right\}, \quad (4)$$

$$V(S, 0, \tau) = \left(\frac{x_{max}}{T} - K\right) e^{-r\tau} + \frac{S}{rT}\left(1 - e^{-r\tau}\right). \quad (5)$$

In recent years several numerical methods have been introduced for valuation of Asian options [3-6], while the FEM for population balance equations proposed in [1] could be implemented to our problem.

The rest of the paper is organized as follows. We construct two splittings for the problem (1)-(4) in Section 2. In Section 3 we describe finite difference approximations of the $1D$ subproblems. In Section 4 we provide numerical experiments to demonstrate the performance of these splittings.

2   Two Splittings

We will describe two splittings of problem (1) - (5) into two subproblems: the first with respect to (S, τ) and second one - with respect to (x, τ). Let introduce the non-uniform mesh in time: $0 = \tau_1 < \tau_2 < \cdots < \tau_n < \tau_{n+1} < \ldots \tau_{P+1} = T$, $\Delta \tau_n = \tau_{n+1} - \tau_n$.

2.1 Splitting 1

Problems 1.1. For given $V(S, x, \tau_n)$, find the function $u(S, x, \tau)$ that satisfies the problem

$$\frac{1}{2}\frac{\partial u}{\partial \tau} = \frac{\sigma^2(t)}{2}S^2\frac{\partial^2 u}{\partial S^2} + (r(t) - \gamma(t))S\frac{\partial u}{\partial S} - r(t)u, \tag{6}$$
$$(S, x, \tau) \in (0, S_{\max}) \times (0, x_{\max}) \times (\tau_n, \tau_{n+\frac{1}{2}}],$$

$$u(S, x, \tau_n) = V(S, x, \tau_n), \tag{7}$$

$$u(0, x, \tau_{n+\frac{1}{2}}) = e^{-r\tau_{n+\frac{1}{2}}} \max\left\{\frac{A_{\max} - x}{T} - K, 0\right\}, \tag{8}$$

$$u(S_{\max}, x, \tau_{n+\frac{1}{2}}) = \max\left\{e^{-r\tau_{n+\frac{1}{2}}}\left(\frac{A_{\max} - x}{T} - K\right) + \frac{S_{\max}}{rT}\left(1 - e^{-r\tau_{n+\frac{1}{2}}}\right), 0\right\} \tag{9}$$

Problem 1.2. For given $u(S, x, \tau_{n+\frac{1}{2}})$, find the function $V(S, x, \tau)$, that satisfies

$$\frac{1}{2}\frac{\partial V}{\partial \tau} + S\frac{\partial V}{\partial x} = 0, \quad (S, x, \tau) \in (0, S_{\max}) \times (0, x_{\max}) \times (\tau_{n+\frac{1}{2}}, \tau_{n+1}], \tag{10}$$

$$V(S, 0, \tau_{n+1}) = \left(\frac{x_{\max}}{T} - K\right)e^{-r\tau_{n+1}} + \frac{S}{rT}\left(1 - e^{-r\tau_{n+1}}\right), \tag{11}$$

$$V(S, x, \tau_{n+\frac{1}{2}}) = u(S, x, \tau_{n+\frac{1}{2}}). \tag{12}$$

2.2 Splitting 2

We start from the initial condition (2) with given $V(S, x, 0)$ and solve consequently two problems on each of the sub-intervals $(\tau_n, \tau_{n+1}]$, $n = 1, 2, \ldots, P$.

Problems 2.1. For given $V(S, x, \tau_n)$ find $u(S, x, \tau)$ that satisfies the problem

$$\frac{\partial u}{\partial \tau} = \frac{\sigma^2(t)}{2}S^2\frac{\partial^2 u}{\partial S^2} + (r(t) - \gamma(t))S\frac{\partial u}{\partial S} - r(t)u, \tag{13}$$
$$(S, x, \tau) \in (0, S_{\max}) \times (0, x_{\max}) \times (\tau_n, \tau_{n+1}],$$
$$u(S, x, \tau_n) = V(S, x, \tau_n), \tag{14}$$

$$u(0, x, \tau_{n+1}) = e^{-r\tau_{n+1}} \max\left\{\frac{A_{\max} - x}{T} - K, 0\right\}, \tag{15}$$

$$u(S_{\max}, x, \tau_{n+1}) = \max\left\{e^{-r\tau_{n+1}}\left(\frac{A_{\max} - x}{T} - K\right) + \frac{S_{\max}}{rT}\left(1 - e^{-r\tau_{n+1}}\right), 0\right\}. \tag{16}$$

Problems 2.2. For given $u(S, x, \tau_{n+1})$ find $V(S, x, \tau)$ that satisfies the problem

$$\frac{\partial V}{\partial \tau} + S\frac{\partial V}{\partial x} = 0, \quad (S, x, \tau) \in (0, S_{\max}) \times (0, x_{\max}) \times (\tau_n, \tau_{n+1}], \tag{17}$$

$$V(S, 0, \tau_{n+1}) = \left(\frac{x_{\max}}{T} - K\right)e^{-r\tau_{n+1}} + \frac{S}{rT}\left(1 - e^{-r\tau_{n+1}}\right), \tag{18}$$

$$V(S, x, \tau_n) = u(S, x, \tau_{n+1}). \tag{19}$$

3 Approximation of Problems 1.1, 1.2, 2.1, 2.2

We will implement the S. Wang difference scheme [6]. For simplicity we will present the formulas for the case of constant with respect t coefficients in (1).

3.1 Approximation of the Problems 1.1, 2.1

Let introduce on $(0, S_{max})$ the mesh $w_h^s = \{0 = S_1 < S_2 < \cdots < S_{N+1} = S_{max}, h_i = S_{i+1} - S_i, i = 1, 2, \ldots, N\}$ and auxiliary mesh $\widetilde{w}_h^s = \{S_{i+1/2} = 0.5(S_i + S_{i+1}), i = 1, 2, \ldots, N\}$. First we consider the *problem 1.1*. Let us rewrite the equation (6) in conservative form.

$$\frac{1}{2}\frac{\partial u}{\partial \tau} = \frac{\partial}{\partial S}\left(aS^2\frac{\partial u}{\partial S} + bSu\right) - cu, \quad a = \frac{\sigma^2}{2}, \quad b = r - \sigma^2, \quad c = 2r - \sigma^2. \quad (20)$$

We integrate equation (20) on the interval $(S_{i-1/2}, S_{i+1/2})$ and using the central rectangles formula to obtain

$$\frac{1}{2}\left.\frac{\partial u}{\partial \tau}\right|_{S_i} \hbar_i = \left[S_{i+1/2}\,\rho(u)|_{S_{i+1/2}} - S_{i-1/2}\,\rho(u)|_{S_{i-1/2}}\right] - c\hbar_i u_i, \quad (21)$$

where $\hbar_i = S_{i+1/2} - S_{i-1/2}$, $u_i = u(S_i, x, \tau)$, $\rho(u) = aS\frac{\partial u}{\partial S} + bu$. In order to obtain an approximation of the flux $\rho(u)$ in the points $S_{i+1/2}$, $i = 2, 3, \ldots, N$ at fixed x and τ, we consider the boundary value problem:

$$(aSw' + bw)' = 0, \quad S \in I_i, \quad w(S_i) = u_i, \quad w(S_{i+1}) = u_{i+1}.$$

After suitable calculations we find

$$\rho(u)|_{S_{i+1/2}} = \rho_i(u) = b\frac{S_{i+1}^\alpha u_{i+1} - S_i^\alpha u_i}{S_{i+1}^\alpha - S_i^\alpha}, \quad \rho_{i-1}(u) = b\frac{S_i^\alpha u_i - S_{i-1}^\alpha u_{i-1}}{S_i^\alpha - S_{i-1}^\alpha}. \quad (22)$$

This approach is not applicable to the flux on the intervals $(S_1, S_2) = (0, S_2)$ because (20) degenerates. After tedious calculations, we obtain

$$\rho_1(u) = 0.5\left[(a + b)\,u_2 - (a - b)\,u_1\right]. \quad (23)$$

Now, using (22), (23), we define a global piecewise constant approximation to $\rho(u)$ by satisfying $\rho_h(u) = \rho_i(u)$ for $x \in I_i$. Finally, placing $\rho_h(u)$ in (21) at fixed x, we obtain the system of ODEs

$$\frac{1}{2}\left.\frac{\partial u}{\partial \tau}\right|_{S=S_2} \hbar_2 = \left[S_{5/2}b\frac{S_3^\alpha u_3 - S_2^\alpha u_2}{S_3^\alpha - S_2^\alpha} - \frac{h_1}{2}\cdot\frac{1}{2}\left[(a+b)\,u_2 - (a-b)\,u_1\right]\right] - \hbar_2 c u_2,$$

$$\frac{1}{2}\left.\frac{\partial u}{\partial \tau}\right|_{S=S_i} \hbar_i = \left[S_{i+1/2}b\frac{S_{i+1}^\alpha u_{i+1} - S_i^\alpha u_i}{S_{i+1}^\alpha - S_i^\alpha} - S_{i-1/2}b\frac{S_i^\alpha u_i - S_{i-1}^\alpha u_{i-1}}{S_i^\alpha - S_{i-1}^\alpha}\right] - \hbar_i c u_i,$$

$$i = \overline{3, N}. \quad (24)$$

We approximate (24) in time t by weighting scheme adding the initial and boundary conditions and taking into account the variable x. To this end we introduce the mesh $0 = x_1 < x_2 < \ldots < x_j < x_{j+1} < \ldots < x_{M+1} = x_{\max}$, $h_j^x = x_{j+1} - x_j$. Then we get on the new time level for the unknowns $\bar{u}_{i,j} \approx V(S_i, x_j, \tau_n)$, $i = 1, 2, \ldots, N+1$, $j = 1, 2, \ldots, M+1$, $n = 1, 2, \ldots P$, the system of algebraic equation (at fixed j):

$$
\left|
\begin{array}{l}
L_h \bar{u}_1 = -C_1 \bar{u}_1 + B_1 \bar{u}_2 \;\;\;\;\;\;\;\;\;\;\;\;\; = F_1, \\
L_h \bar{u}_i = A_i \bar{u}_{i-1} - C_i \bar{u}_i + B_i \bar{u}_{i+1} = F_i, \;\;\; i = 2, 3, \ldots, N, \\
L_h \bar{u}_N = A_{N+1} \bar{u}_N - C_{N+1} \bar{u}_{N+1} = F_{N+1},
\end{array}
\right.
\tag{25}
$$

$$
F_1 = -e^{-r\tau_{n+\frac{1}{2}}} \max\left\{\frac{A_{\max} - x_j}{T} - K, 0\right\}, \; j = 2, 3, \ldots, M, \; n = 1, 2, \ldots, P,
$$

$$
C_1 = 1, \; B_1 = 0, \; A_2 = \frac{\theta h_1}{4}(a - b), \; B_2 = \frac{\theta S_{5/2} b S_3^\alpha}{S_3^\alpha - S_2^\alpha},
$$

$$
C_2 = \frac{\hbar_2}{\Delta \tau_n} + \frac{\theta h_1}{4}(a + b) + \frac{\theta S_{5/2} b S_2^\alpha}{S_3^\alpha - S_2^\alpha} + \theta \hbar_2 c,
$$

$$
F_2 = -\frac{\hbar_2}{\Delta \tau_n} u_{2,j} - (1 - \theta)\left[S_{\frac{5}{2}} b \frac{S_3^\alpha u_{3,j} - S_2^\alpha u_{2,j}}{S_3^\alpha - S_2^\alpha} - \frac{h_1}{4}\left[(a + b)u_{2,j} - (a - b)u_{1,j}\right]\right]
$$

$$
+(1 - \theta)\hbar_2 c u_{2,j}, \;\;\; j = 2, 3, \ldots, M, \; A_i = \frac{\theta S_{i-1/2} b S_{i-1}^\alpha}{S_i^\alpha - S_{i-1}^\alpha}, \;\;\; B_i = \frac{\theta S_{i+1/2} b S_{i+1}^\alpha}{S_{i+1}^\alpha - S_i^\alpha},
$$

$$
C_i = \frac{\hbar_i}{\Delta \tau_n} + \frac{\theta S_{i-1/2} b S_i^\alpha}{S_i^\alpha - S_{i-1}^\alpha} + \frac{\theta S_{i+1/2} b S_i^\alpha}{S_{i+1}^\alpha - S_i^\alpha} + \theta \hbar_i c,
$$

$$
F_i = -\frac{\hbar_i}{\Delta \tau_n} u_{i,j} - (1 - \theta)\left[S_{i+1/2} b \frac{S_{i+1}^\alpha u_{i+1,j} - S_i^\alpha u_{i,j}}{S_{i+1}^\alpha - S_i^\alpha} - S_{i-1/2} b \frac{S_i^\alpha u_{i,j} - S_{i-1}^\alpha u_{i-1,j}}{S_i^\alpha - S_{i-1}^\alpha}\right]
$$

$$
+(1 - \theta)\hbar_i c u_{i,j}, \;\;\; i = 3, 4, \ldots, N, \;\;\; j = 2, 3, \ldots, M+1, \;\;\; A_{N+1} = 0, \;\;\; C_{N+1} = 1,
$$

$$
F_{N+1} = -\max\left\{e^{-r\tau_{n+1/2}}\left(\frac{A_{\max} - x_j}{T} - K\right) + \frac{S_{\max}}{rT}\left(1 - e^{-r\tau_{n+1/2}}\right), 0\right\},
$$

$$
j = 1, 2, \ldots, M, \;\;\; n = 1, 2, \ldots, P.
$$

Lemma 1. *Suppose that $u_{i,j} \geq 0$, $i = \overline{1, N+1}$, $j = \overline{1, M+1}$. Then for sufficiently small $\Delta \tau_n$ we have $\bar{u}_{i,j} \geq 0$, $i = \overline{1, N+1}$, $j = \overline{1, M+1}$*

Scetch of the Proof. Let us first investigate the off-diagonal entries of the system matrix. For A_i at $i = 3, 4, \ldots, N$ we have

$$
A_i = \frac{\theta S_{i-1/2} b S_{i-1}^\alpha}{S_i^\alpha - S_{i-1}^\alpha} = \frac{\theta S_{i-1/2} a \alpha S_{i-1}^\alpha}{S_i^\alpha - S_{i-1}^\alpha} = \theta S_{i-1/2} a \frac{\alpha}{\bar{S}_{i-1} - 1} > 0,
$$

because when $\alpha < 0$ $\;\;0 < \bar{S}_{i-1} = \left(\frac{S_i}{S_{i-1}}\right)^\alpha < 1$, and when $\alpha > 0$ $\;\;\bar{S}_{i-1} > 1$. Analogously $B_i > 0$, for $i = 3, 4, \ldots, N$. For sufficiently small $\Delta \tau_n$ the coefficients

$C_i > 0$, $i = 3, 4, \ldots, N$. Different is the situation for $i = 2$ and $i = 3$. From the second and the third equation of (25) we get

$$\bar{u}_2 = \frac{B_2}{C_2}\bar{u}_3 - \frac{F_2 + A_2 F_1}{C_2}, \quad -\tilde{C}_3 \bar{u}_3 + B_3 \bar{u}_4 = \tilde{F}_3,$$

$$\tilde{C}_3 = C_3 - \frac{A_3 B_2}{C_2}, \quad \tilde{F}_3 = F_3 + A_3 \frac{F_2 + A_2 F_1}{C_2}.$$

When $\Delta \tau_n$ is sufficiently small $C_2 = O\left(\frac{1}{\Delta \tau_n}\right)$, $C_3 = O\left(\frac{1}{\Delta \tau_n}\right)$. Therefore for sufficiently small $\Delta \tau_n$ $\tilde{C}_3 > 0$, $\tilde{C}_3 = O\left(\frac{1}{\Delta \tau_n}\right)$. Now, the discrete maximum principle provides us that for sufficiently small time-step $\Delta \tau_n$ the numerical solution $\{\bar{u}_{i,j}\}$ is non-negative.

Let us consider the *second problem*. For the boundary condition (11) we have:

$$\hat{V}_{i,1} = \left(\frac{x_{\max}}{T} - K\right) e^{-r\tau_{n+1}} + \frac{S_i}{rT}\left(1 - e^{-r\tau_{n+1}}\right), i = 2, 3, \ldots, N, n = 2, 3, \ldots, P. \tag{26}$$

The initial condition is as follows

$$V(S_i, x_j, \tau_{n+1/2}) = u(S_i, x_j, \tau_{n+1/2}). \tag{27}$$

We approximate equation (10) by the θ_1 - scheme:

$$\frac{\hat{V}_{i,j} - \bar{u}_{i,j}}{\Delta \tau_n} + \theta_1 S_i \frac{\hat{V}_{i,j} - \hat{V}_{i,j-1}}{h_{j-1}^x} + (1 - \theta_1) S_i \frac{\bar{u}_{i,j} - \bar{u}_{i,j-1}}{h_{j-1}^x} = 0, \tag{28}$$

$$i = \overline{2, N}, \quad j = \overline{2, M+1}.$$

The following local approximation error $\theta_1 \neq \frac{1}{2}$ for $O(\Delta \tau + h)$, $\Delta \tau = \max \Delta \tau_n$, $h = \max h_j$ and $\theta_1 = \frac{1}{2}(O(\tau^2 + h)$. If $\theta_1 = 0$ scheme (25)-(27) is monotone at $\Delta \tau \leq \frac{\min\limits_{1 \leq j \leq M} h_j^x}{S_{\max}}$, for $\theta_1 = 1$ is unconditionally monotone and for $\theta_1 = 0.5$ it is monotone if $\Delta \tau \leq \frac{2 \min\limits_{1 \leq j \leq M} h_j^x}{S_{\max}}$. >From (28) we get for $j = \overline{2, M+1}$

$$\hat{V}_{i,j} = \frac{\frac{\theta_1 S_i \Delta \tau_n}{h_{j-1}^x}\hat{V}_{i,j-1} + \bar{u}_{i,j} - (1 - \theta_1)S_i \Delta \tau_n \frac{\bar{u}_{i,j} - \bar{u}_{i,j-1}}{h_{j-1}^x} + \Delta \tau_n f}{1 + \frac{\theta_1 S_i \Delta \tau_n}{h_{j-1}^x}}, \quad i = \overline{2, N}. \tag{29}$$

Theorem 1. *For sufficiently small time step $\Delta \tau$ the numerical solutions obtained by the Splittings 1,2 are non-negative.*

From the discretization above, one can see that the consistency of the Splittings 1,2 relies on the consistency of the flux $\rho(u)$ approximation. As it was shown in [6] the discretization scheme in 3.1 admits a finite element formulation with special trial space S_h. Then in our case the following analog of the estimate in Lemma 4.2 in [6] holds:

$$\|\rho(w) - \rho_h(w)\|_{\infty,I_i} \le C\|\rho'(w)\|_{\infty,I}, \quad i = 1, 2, \dots, N.$$

where w is a sufficiently smooth function and w_I is the S_h interpolant. Now, using technique similar to those in Section u of [7], one can obtain the following convergence result.

Theorem 2. *Let V be exact solution of (1)-(4) and $\{V^n\}$ the numerical solution of the splitting 1(2). Then, there exists a positive constant C, independent of N, M and $\Delta\tau$, such that the global error satisfies*

$$\|V(t_n)|_{\Omega_h} - V^n\|_\infty \le C(\Delta\tau + h),$$

where $u(t_n)|_{\overline{\Omega}_h}$ is the restriction of the exact solution on the product of the meshes with respect to S and x.

4   Numerical Experiments

In order to observe the behaviour of the error for the two splittings we use the analytical solution

$$V_a(S, x, \tau) = (2x_{\max} - x)\,(S/S_{\max})^2\,e^{-r\tau}.$$

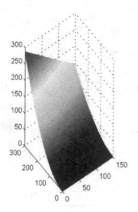
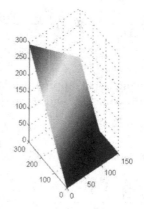

Fig. 1. Analytical solution V_a at $t = T$.

Fig. 2. Numerical solution at $t = T$.

Numerical experiments were performed for the different problems derived with: the two splittings; consequently (S, τ), (x, τ) and on the back (x, τ), (S, τ); the three schemes for the transport equation. For every one of the experiments the time-step decreases to establishment of the first four digits of the relative C-norm of the error at the last level $\tau = T$, which is computed by by the formula

$$\left\|V_a^{P+1} - V^{P+1}\right\|_C = \max_{i,j}\left|(V_a)_{i,j}^{P+1} - V_{i,j}^{P+1}\right| / \max_{i,j}\left|(V_a)_{i,j}^{P+1}\right|.$$

The most effective is the scheme for Splitting 1 performed in order to $(S, \tau), (x, \tau)$ combined with implicit scheme $(\theta_1 = 1)$ for the transport equation. Results from computation are presented in Table 1. The rate of convergence (RC) is calculated using double mesh principle

$$RC = \log_2(ER^K/ER^{2K}), \quad ER^K = \left\| V_a^{P+1} - V^{P+1} \right\|_C, \quad K = (N, M).$$

Table 1. Exact solution test

N	M	P	Relative C−norm of the error	RC
30	15	700	7.906 E-5	-
60	30	1200	2.077 E-5	1.93
120	60	6000	5.353 E-6	1.96
240	120	26000	1.358 E-6	1.98
480	240	130000	3.428 E-7	1.99

5   Conclusion

We derive and implement two splitting algorithms for a fixed strike Asian option model. We use the fitted finite volume method for the spatial discretization of the one-dimensional parabolic subproblems in combination with monotone weighted difference schemes for the hyperbolic subproblems. The positivity preserving in time of the numerical solution is shown. Numerical experiments suggest second-order of convergence.

Acknowledgment. The first author is supported by the Sofia University Foundation under Grant No 181/2012 and the second author, by Bulgarian National Fund of Sciences under project DID 02/37/09 and by the European Union under Grant Agreement number 304617 (FP7 Marie Curie Action Project Multi-ITN STRIKE-Novel Methods in Computational Finance).

References

1. Ahmed, N., Matthias, G., Tobiska, L.: Finite element methods of an operator splitting applied to population balance equations. J. Comput. Appl. Math. 236, 1604–1621 (2011)
2. Hugger, J.: Wellposedness of the boundary value formulation of a fixed strike Asian option. J. Comp. Appl. Math. 105, 460–481 (2006)
3. Marcozzi, M.D.: An addaptive extrapolation discontinuous Galerkin method for the valuation of Asian options. J. Comp. Math. 235, 3632–3645 (2011)
4. Meyer, G.H.: On pricing American and Asian options with PDE methods. Acta Math. Univ. Comenianne, LXX 1, 153–165 (2000)
5. Oosterlee, C.W., Frish, J.C., Gaspar, F.J.: TVD, WENO and blended BDF discretizations for Asian options. Comp. and Visul. in Sci. 6, 131–138 (2004)
6. Wang, S.: A novel fitted finite volume method for the Black-Sholes equation governing option pricing. IMAJ of Numer. Anal. 24, 699–720 (2004)
7. Wilmott, P., Dewynne, J., Howison, S.: Option Pricing. Oxford Financial Press (1993)

New Family of Iterative Methods with High Order of Convergence for Solving Nonlinear Systems

Alicia Cordero[1], Juan R. Torregrosa[1], and María P. Vassileva[2]

[1] Instituto de Matemática Multidisciplinar,
Universitat Politècnica de València,
Camino de Vera, s/n 40022 Valencia, España
{acordero,jrtorre}@mat.upv.es
[2] Instituto Tecnológico de Santo Domingo (INTEC),
av. Los Próceres, Galá, Santo Domingo,
República Dominicana
maria.penkova@intec.edu.do

Abstract. In this paper we present and analyze a set of predictor-corrector iterative methods with increasing order of convergence, for solving systems of nonlinear equations. Our aim is to achieve high order of convergence with few Jacobian and/or functional evaluations. On the other hand, by applying the pseudocomposition technique on each proposed scheme we get to increase their order of convergence, obtaining new high-order and efficient methods. We use the classical efficiency index in order to compare the obtained schemes and make some numerical test.

Keywords: Nonlinear systems, Iterative methods, Jacobian matrix, Convergence order, Efficiency index.

1 Introduction

Many relationships in nature are inherently nonlinear, which according to these effects are not in direct proportion to their cause. In fact, a large number of such real-world applications are reduce to solve nonlinear systems numerically. Approximating a solution ξ of a nonlinear system $F(x) = 0$, is a classical problem that appears in different branches of science and engineering.

Recently, for $n = 1$, many robust and efficient methods have been proposed with high convergence order, but in most of cases the method cannot be extended for several variables. Few papers for the multidimensional case introduce methods with high order of convergence. The authors design in [1] a modified Newton-Jarrat scheme of sixth-order; in [6] a third-order method is presented for computing real and complex roots of nonlinear systems; Shin et al. compare in [8] Newton-Krylov methods and Newton-like schemes for solving big-sized nonlinear systems; in [2] a general procedure to design high-order methods for problems in several variables is presented.

I. Dimov, I. Faragó, and L. Vulkov (Eds.): NAA 2012, LNCS 8236, pp. 222–230, 2013.

The pseudocomposition technique (see [5]) consists of the following: we consider a method of order of convergence p as a predictor, whose penultimate step is of order q, and then we use a corrector step based on the Gaussian quadrature. So, we obtain a family of iterative schemes whose order of convergence is $min\{q+p, 3q\}$. This is a general procedure to improve the order of convergence of known methods.

To analyze and compare the efficiency of the proposed methods we use the classic efficiency index $I = p^{1/d}$ due to Ostrowski [7], where p is the order of convergence, d is the number of functional evaluations, per iteration.

In this paper, we present three new Newton-like schemes, of order of convergence four, six and eight, respectively. After the analysis of convergence of the new methods, we apply the pseudocomposition technique in order to get higher order procedures.

The convergence theorem in Section 2 can be demonstrated by means of the n-dimensional Taylor expansion of the functions involved. Let $F : D \subseteq R^n \longrightarrow R^n$ be sufficiently Frechet differentiable in D. By using the notation introduced in [1], the qth derivative of F at $u \in R^n$, $q \geq 1$, is the q-linear function $F^{(q)}(u)$: $R^n \times \cdots \times R^n \longrightarrow R^n$ such that $F^{(q)}(u)(v_1, \ldots, v_q) \in R^n$. It is easy to observe that

1. $F^{(q)}(u)(v_1, \ldots, v_{q-1}, \cdot) \in \mathcal{L}(R^n)$,
2. $F^{(q)}(u)(v_{\sigma(1)}, \ldots, v_{\sigma(q)}) = F^{(q)}(u)(v_1, \ldots, v_q)$, for all permutation σ of $\{1, 2 \ldots, q\}$.

So, in the following we will denote:

(a) $F^{(q)}(u)(v_1, \ldots, v_q) = F^{(q)}(u)v_1 \ldots v_q$,
(b) $F^{(q)}(u)v^{q-1}F^{(p)}v^p = F^{(q)}(u)F^{(p)}(u)v^{q+p-1}$.

It is well known that, for $\xi + h \in R^n$ lying in a neighborhood of a solution ξ of the nonlinear system $F(x) = 0$, Taylor's expansion can be applied (assuming that the Jacobian matrix $F'(\xi)$ is nonsingular), and

$$F(\xi + h) = F'(\xi)\left[h + \sum_{q=2}^{p-1} C_q h^q\right] + O[h^p], \tag{1}$$

where $C_q = (1/q!)[F'(\xi)]^{-1}F^{(q)}(\xi)$, $q \geq 2$. We observe that $C_q h^q \in R^n$ since $F^{(q)}(\xi) \in \mathcal{L}(R^n \times \cdots \times R^n, R^n)$ and $[F'(\xi)]^{-1} \in \mathcal{L}(R^n)$.

In addition, we can express the Jacobian matrix of F, F', as

$$F'(\xi + h) = F'(\xi)\left[I + \sum_{q=2}^{p-1} qC_q h^{q-1}\right] + O[h^p], \tag{2}$$

where I is the identity matrix. Therefore, $qC_q h^{q-1} \in \mathcal{L}(R^n)$. From (2), we obtain

$$[F'(\xi + h)]^{-1} = [I + X_2 h + X_3 h^2 + X_4 h^3 + \cdots][F'(\xi)]^{-1} + O[h^p], \tag{3}$$

where $X_2 = -2C_2$, $X_3 = 4C_2^2 - 3C_3, \ldots$

We denote $e_k = x^{(k)} - \xi$ the error in the kth iteration. The equation $e_{(k+1)} = Le_k{}^p + O[e_k{}^{p+1}]$, where L is a p-linear function $L \in \mathcal{L}(R^n \times \cdots \times R^n, R^n)$, is called the *error equation* and p is the *order of convergence*.

The rest of the paper is organized as follows: in the next section, we present the new methods of order four, six and eight, respectively. Moreover, the convergence order is increased when the pseudocomposition technique is applied. Section 3 is devoted to the comparison of the different methods by means of several numerical tests.

2   Proposed High-Order Methods

Let us introduce now a new Jarratt-type scheme of five steps which we will denote as M8. It can be proved that its first three steps are a fourth-order scheme, denoted by M4, and its four first steps become a sixth-order method that will be denoted by M6. The coefficients involved have been obtained optimizing the order the convergence and the whole scheme requires three functional evaluations of F and two of F' to attain eighth-order of convergence. Let us also note that the linear systems to be solved in first, second and last step have the same matrix and also have the third and fourth steps, so the number of operations involved is not as high as it can seem.

Theorem 1. *Let $F : \Omega \subseteq R^n \to R^n$ be sufficiently differentiable in a neighborhood of $\xi \in \Omega$ which is a solution of the nonlinear system $F(x) = 0$. We suppose that $F'(x)$ is continuous and nonsingular at ξ and $x^{(0)}$ close enough to the solution. Then, the sequence $\{x^{(k)}\}_{k \geq 0}$ obtained by*

$$
\begin{aligned}
y^{(k)} &= x^{(k)} - \frac{2}{3} \left[F'\left(x^{(k)}\right) \right]^{-1} F\left(x^{(k)}\right), \\
z^{(k)} &= y^{(k)} + \frac{1}{6} \left[F'\left(x^{(k)}\right) \right]^{-1} F\left(x^{(k)}\right), \\
u^{(k)} &= z^{(k)} + \left[F'\left(x^{(k)}\right) - 3F'\left(y^{(k)}\right) \right]^{-1} F\left(x^{(k)}\right), \\
v^{(k)} &= z^{(k)} + \left[F'\left(x^{(k)}\right) - 3F'\left(y^{(k)}\right) \right]^{-1} \left[F\left(x^{(k)}\right) + 2F\left(u^{(k)}\right) \right], \\
x^{(k+1)} &= v^{(k)} - \frac{1}{2} \left[F'\left(x^{(k)}\right) \right]^{-1} \left[5F'\left(x^{(k)}\right) - 3F'\left(y^{(k)}\right) \right] \left[F'\left(x^{(k)}\right) \right]^{-1} F\left(v^{(k)}\right),
\end{aligned}
\tag{4}
$$

converges to ξ with order of convergence eight. The error equation is:

$$
e_{k+1} = \left(C_2^2 - \frac{1}{2}C_3 \right) \left(2C_2^3 + 2C_3C_2 - 2C_2C_3 - \frac{20}{9}C_4 \right) e_k^8 + O[e_k^9].
$$

By applying the next result, it is known (see [5]) that, the pseudocomposition technique allows us to design methods with higher order of convergence.

Theorem 2. *[5] Let $F : \Omega \subseteq R^n \to R^n$ be differentiable enough in Ω and $\xi \in \Omega$ a solution of the nonlinear system $F(x) = 0$. We suppose that $F'(x)$ is* *continuous and nonsingular at* ξ. *Let* $y^{(k)}$ *and* $z^{(k)}$ *be the penultimate and final steps of orders* q *and* p, *respectively, of a certain iterative method. Taking this scheme as a predictor we get a new approximation* $x^{(k+1)}$ *of* ξ *given by*

$$x^{(k+1)} = y^{(k)} - 2 \left[\sum_{i=1}^{m} \omega_i F'(\eta_i^{(k)}) \right]^{-1} F(y^{(k)}),$$

where $\eta_i^{(k)} = \dfrac{1}{2}\left[(1 + \tau_i)z^{(k)} + (1 - \tau_i)y^{(k)}\right]$ *and* τ_i, ω_i $i = 1, \ldots, m$ *are the nodes and weights of the orthogonal polynomial corresponding to the Gaussian quadrature used. Then,*

1. *the obtained set of families will have an order of convergence at least* q;
2. *if* $\sigma = 2$ *is satisfied, then the order of convergence will be at least* $2q$;
3. *if, also,* $\sigma_1 = 0$ *the order of convergence will be* $min\{p + q, 3q\}$;

where $\displaystyle\sum_{i=1}^{n} \omega_i = \sigma$ *and* $\displaystyle\sum_{i=1}^{n} \dfrac{\omega_i \tau_i^j}{\sigma} = \sigma_j$ *with* $j = 1, 2$.

Each of the families obtained will consist of subfamilies that are determined by the orthogonal polynomial corresponding to the Gaussian quadrature used. Furthermore, in these subfamilies it can be obtained methods using different number of nodes corresponding to the orthogonal polynomial used (see Table 1). According to the proof of Theorem 2 the order of convergence of the obtained methods does not depend on the number of nodes used.

Table 1. Quadratures used

Number of nodes	Quadratures							
	Chebyshev		Legendre		Lobatto		Radau	
	σ	σ_1	σ	σ_1	σ	σ_1	σ	σ_1
1	π	0	2	0	2	0	2	-1
2	π	0	2	0	2	0	2	0
3	π	0	2	0	2	0	2	0

Let us note that these methods, obtained by means of Gaussian quadratures, seem to be known interpolation quadrature schemes such as midpoint, trapezoidal or Simpson's method (see [4]). It is only a similitude, as they are not applied on the last iteration $x^{(k)}$, and the last step of the predictor $z^{(k)}$, but on the two last steps of the predictor. In the following, we will use a midpoint-like as a corrector step, which corresponds to a Gauss-Legendre quadrature with one node; for this scheme the order of convergence will be at least $min\{q + p, 3q\}$, by applying Theorem 2.

The pseudocomposition can be applied on the proposed scheme M8 with iterative expression (4), but also on M6. By pseudocomposing on M6 and M8 there can be obtained two procedures of order of convergence 10 and 14 (denoted by PsM10 and PsM14), respectively. Let us note that it is also possible to pseudocompose on M4, but the resulting scheme would be of third order of convergence, which is worst than the original M4, so it will not be considered.

Following the notation used in (4), the last step of PsM10 is

$$x^{(k+1)} = u^{(k)} - 2 \left[F' \left(\frac{v^{(k)} + u^{(k)}}{2} \right) \right]^{-1} F(u^{(k)}), \tag{5}$$

and the last three steps of PsM14 can be expressed as

$$v^{(k)} = z^{(k)} + \left[F' \left(x^{(k)} \right) - 3F' \left(y^{(k)} \right) \right]^{-1} \left[F \left(x^{(k)} \right) + 2F \left(u^{(k)} \right) \right],$$
$$w^{(k+1)} = v^{(k)} - \frac{1}{2} \left[F' \left(x^{(k)} \right) \right]^{-1} \left[5F' \left(x^{(k)} \right) - 3F' \left(y^{(k)} \right) \right] \left[F' \left(x^{(k)} \right) \right]^{-1} F \left(v^{(k)} \right), \tag{6}$$
$$x^{(k+1)} = v^{(k)} - 2 \left[F' \left(\frac{w^{(k)} + v^{(k)}}{2} \right) \right]^{-1} F(v^{(k)}).$$

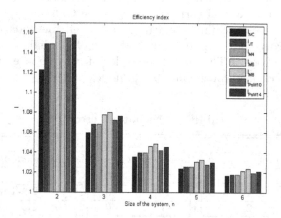

Fig. 1. Efficiency index of the different methods for different sizes of the system

If we analyze the efficiency indices (see Figure 1), we deduce the following conclusions: the new methods M4, M6 and M8 (and also the pseudocomposed PsM10 and PsM14) improve Newton and Jarratt's schemes (in fact, the indices of M4 and Jarratt's are equal). Indeed, for $n \geq 3$ the best index is that of M8. Nevertheless, none of the pseudocomposed methods improve the efficiency index of their original partners.

3   Numerical Results

In this section, we test the developed methods to illustrate its effectiveness compared with other methods. Numerical computations have been performed in MATLAB R2011a by using variable-precision arithmetic, which uses floating-point representation of 2000 decimal digits of mantissa. The computer specifications are: Intel(R) Core(TM) i5-2500 CPU @ 3.30GHz with 16.00GB of RAM. Each iteration is obtained from the former by means of an iterative expression $x^{(k+1)} = x^{(k)} - A^{-1}b$, where $x^{(k)} \in \mathbb{R}^n$, A is a real matrix $n \times n$ and $b \in \mathbb{R}^n$. The matrix A and vector b are different according to the method used, but in any case, we use to calculate inverse $-A^{-1}b$ the solution of the linear system $Ay = b$, with Gaussian elimination with partial pivoting. The stopping criterion used is $||x^{(k+1)} - x^{(k)}|| < 10^{-200}$ or $||F(x^{(k)})|| < 10^{-200}$.

Firstly, let us consider the following nonlinear systems of different sizes:

1. $F_1 = (f_1(x), f_2(x), \ldots, f_n(x))$, where $x = (x_1, x_2, \ldots, x_n)^T$ and $f_i : \mathbb{R}^n \to \mathbb{R}, i = 1, 2, \ldots, n$, such that

$$f_i(x) = x_i x_{i+1} - 1, \; i = 1, 2, \ldots, n - 1$$
$$f_n(x) = x_n x_1 - 1.$$

   When n is odd, the exact zeros of $F_1(x)$ are: $\xi_1 = (1, 1, \ldots, 1)^T$ and $\xi_2 = (-1, -1, \ldots, -1)^T$.

2. $F_2(x_1, x_2) = (x_1^2 - x_1 - x_2^2 - 1, -\sin(x_1) + x_2)$ and the solutions are $\xi_1 \approx (-0.845257, -0.748141)^T$ and $\xi_2 \approx (1.952913, 0.927877)^T$.

3. $F_3(x_1, x_2) = (x_1^2 + x_2^2 - 4, -\exp(x_1) + x_2 - 1)$, being the solutions $\xi_1 \approx (1.004168, -1.729637)^T$ and $\xi_2 \approx (-1.816264, 0.837368)^T$.

4. $F_4(x_1, x_2, x_3) = (x_1^2 + x_2^2 + x_3^2 - 9, x_1 x_2 x_3 - 1, x_1 + x_2 - x_3^2)$ with three roots $\xi_1 \approx (2.14025, -2.09029, -0.223525)^T$, $\xi_2 \approx (2.491376, 0.242746, 1.653518)^T$ and $\xi_1 \approx (0.242746, 2.491376, 1.653518)^T$.

Table 2 presents results showing the following information: the different iterative methods employed (Newton (NC), Jarratt (JT), the new methods M4, M6 and M8 and the pseudocomposed PsM10 and PsM14), the number of iterations $Iter$ needed to converge to the solution Sol, the value of the stopping factors at the last step and the computational order of convergence ρ (see [3]) approximated by the formula:

$$\rho \approx \frac{ln(||x^{(k+1)} - x^{(k)}||)/(||x^{(k)} - x^{(k-1)}||)}{ln(||x^{(k)} - x^{(k-1)}||)/(||x^{(k-1)} - x^{(k-2)}||)}. \tag{7}$$

The value of ρ which appears in Table 2 is the last coordinate of the vector ρ when the variation between their coordinates is small. Also the elapsed time, in seconds, appears in Table 2, being the mean execution time for 100 performances of the method (the command cputime of Matlab has been used).

We observe from Table 2 that, not only the order of convergence and the number of new functional evaluations and operations is important in order to

Table 2. Numerical results for functions F_1 to F_4

Function	Method	Iter	Sol	$\|x^{(k)} - x^{(k-1)}\|$	$\|F(x^{(k)})\|$	ρ	e-time (sec)
F_1	NC	8	ξ_1	1.43e-121	2.06e-243	2.0000	8.6407
$x^{(0)} = (0.8, \ldots, 0.8)$ $n = 99$	JT	4	ξ_1	1.69e-60	2.06e-243	4.0000	3.9347
	M4	4	ξ_1	1.69e-60	2.06e-243	4.0000	3.7813
	M6	4	ξ_1	6.94e-193	4.33e-1160	6.0000	5.3911
	M8	3	ξ_1	9.40e-50	3.51e-4011	8.0913	5.0065
	PsM10	3	ξ_1	1.28e-91	9.54e-921	10.0545	4.9061
	PsM14	3	ξ_1	4.65e-164	0	14.0702	6.1018
F_1	NC	17	ξ_1	3.37e-340	1.14e-340	-	9.2128
$x^{(0)} = (0.0015, \ldots, 0.0015)$ $n = 99$	JT	9	ξ_1	8.18e-085	1.14e-340	4.0000	10.1416
	M4	9	ξ_1	8.18e-085	1.14e-340	4.0000	10.9104
	M6	7	ξ_1	1.40e-035	9.46e-216	-	12.3266
	M8	19	ξ_1	9.50e-030	1.29e-240	-	59.4832
	PsM10	6	ξ_1	3.02e-102	5.23e-1027	-	17.9957
	PsM14	5	ξ_1	1.84e-162	0	-	22.6130
F_2	NC	9	ξ_1	2.45e-181	5.92e-362	2.0148	0.2395
$x^{(0)} = (-0.5, -0.5)$	JT	5	ξ_1	9.48e-189	8.13e-754	4.0279	0.3250
	M4	5	ξ_1	9.48e-189	8.13e-754	4.0279	0.1841
	M6	4	ξ_1	1.34e-146	2.14e-878	5.9048	0.2744
	M8	3	ξ_1	1.90e-038	1.23e-302	7.8530	0.3718
	PsM10	3	ξ_1	6.72e-72	2.68e-714	9.9092	0.4674
	PsM14	3	ξ_1	2.13e-122	1.95e-1706	13.9829	0.3187
F_2	NC	13	ξ_1	2.20e-182	2.73e-374	1.9917	0.3713
$x^{(0)} = (-5, -3)$	JT	7	ξ_1	2.10e-179	4.51e-716	3.9925	0.4001
	M4	7	ξ_1	2.10e-179	4.51e-716	3.9925	0.7535
	M6	8	ξ_1	2.55e-036	5.81e-216	-	0.9382
	M8	> 5000					
	PsM10	4	ξ_1	2.59e-021	3.51e-208	-	0.4363
	PsM14	29	ξ_2	9.45e-020	5.05e-273	-	7.8090
F_3	NC	10	ξ_1	1.65e-190	4.61e-380	2.0000	1.4675
$x^{(0)} = (2, -3)$	JT	5	ξ_1	8.03e-113	7.59e-450	3.9995	0.3151
	M4	5	ξ_1	8.03e-113	7.59e-450	3.9995	0.3034
	M6	4	ξ_1	1.25e-082	2.83e-493	6.0015	0.3696
	M8	4	ξ_1	1.54e-162	3.16e-1296	7.9993	0.4463
	PsM10	3	ξ_1	5.59e-044	1.40e-436	9.4708	0.4682
	PsM14	3	ξ_1	3.46e-068	3.45e-948	13.1659	0.5925
F_3	NC	35	ξ_1	3.71e-177	2.33e-253	-	1.4828
$x^{(0)} = (0.2, 0.1)$	JT	11	ξ_1	3.29e-143	1.67e-574	-	0.7781
	M4	11	ξ_1	3.29e-143	1.67e-574	-	0.7535
	M6	9	ξ_1	1.31e-064	3.61e-385	-	0.8001
	M8	n.c.	ξ_1				
	PsM10	5	ξ_1	6.85e-156	1.06e-1555	-	0.6352
	PsM14	8	ξ_2	7.87e-155	0	-	1.1870
F_4	NC	10	ξ_1	1.03e-135	1.55e-270	1.9995	2.3263
$x^{(0)} = (1, -1.5, -0.5)$	JT	5	ξ_1	9.94e-073	2.09e-289	4.0066	0.5296
	M4	5	ξ_1	9.94e-073	2.09e-289	4.0066	0.6340
	M6	4	ξ_1	9.31e-057	4.86e-338	5.9750	0.7443
	M8	4	ξ_1	4.43e-046	1.08e-364	-	0.8282
	PsM10	3	ξ_1	1.43e-031	1.04e-311	9.6674	0.8100
	PsM14	3	ξ_1	1.91e-033	4.05e-462	13.9954	1.0465
F_4	NC	12	ξ_1	1.08e-192	1.55e-384	1.9996	2.7271
$x^{(0)} = (7, -5, -5)$	JT	6	ξ_1	2.31e-103	7.97e-412	4.0090	0.7761
	M4	6	ξ_1	2.31e-103	7.97e-412	4.0090	1.0301
	M6	5	ξ_1	2.99e-086	4.69e-515	-	1.0090
	M8	15	ξ_3	1.77e-071	1.48e-568	-	3.4007
	PsM10	4	ξ_1	6.86e-067	1.25e-666	-	1.0245
	PsM14	7	ξ_2	1.09e-130	9.15e-1825	-	1.8179

obtain new efficient iterative methods to solve nonlinear systems of equations. A key factor is the range of applicability of the methods. Although they are slower than the original methods when the initial estimation is quite good, when we are far from the solution or inside a region of instability, the original schemes do not converge or do it more slowly, the corresponding pseudocomposed procedures usually still converge or do it faster.

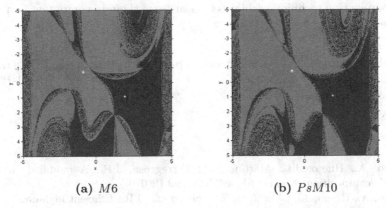

(a) $M6$ (b) $PsM10$

Fig. 2. Real dynamical planes for system (b) and methods M6 and PsM10

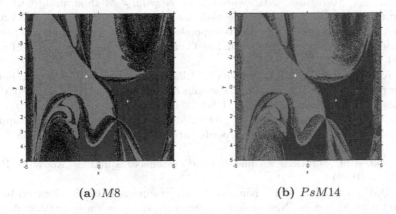

(a) $M8$ (b) $PsM14$

Fig. 3. Real dynamical planes for system (b) and methods M8 and PsM14

The advantage of pseudocomposition can be observed in Figures 2a, 2b (methods M6 and PsM10) and 3a, 3b (methods M8 and PsM14) where the dynamical plane on R^2 is represented: let us consider a system of two equations and two unknowns (the case $F_2(x) = 0$ is showed), for any initial estimation in $\mathbb{R}^2$ represented by its position in the plane, a different color (blue or orange, as there exist only two solutions) is used for the different solutions found (marked by a white point in the figure). Black color represents an initial point in which the method converges to infinity, and the green one means that no convergence is found (usually because any linear system cannot be solved). It is clear that when many initial estimations tend to infinity (see Figure 3a), the pseudocomposition "cleans" the dynamical plane, making the method more stable as it can find one of the solutions by using starting points that do not allow convergence with the original scheme (see Figure 2b).

We conclude that the presented schemes M4, M6 and M8 show to be excellent, in terms of order of convergence and efficiency, but also that the pseudocomposition technique achieves to transform them in competent and more robust new schemes.

Acknowledgement. This research was supported by Ministerio de Ciencia y Tecnología MTM2011-28636-C02-02 and by FONDOCYT 2011-1-B1-33, República Dominicana.

References

1. Cordero, A., Hueso, J.L., Martínez, E., Torregrosa, J.R.: A modified Newton-Jarratt's composition. Numer. Algor. 55, 87–99 (2010)
2. Cordero, A., Hueso, J.L., Martínez, E., Torregrosa, J.R.: Efficient high-order methods based on golden ratio for nonlinear systems. Applied Mathematics and Computation 217(9), 4548–4556 (2011)
3. Cordero, A., Torregrosa, J.R.: Variants of Newton's Method using fifth-order quadrature formulas. Applied Mathematics and Computation 190, 686–698 (2007)
4. Cordero, A., Torregrosa, J.R.: On interpolation variants of Newton's method for functions of several variables. Journal of Computational and Applied Mathematics 234, 34–43 (2010)
5. Cordero, A., Torregrosa, J.R., Vassileva, M.P.: Pseudocomposition: a technique to design predictor-corrector methods for systms of nonlinear equtaions. Applied Mathematics and Computation 218(23), 11496–11504 (2012)
6. Nikkhah-Bahrami, M., Oftadeh, R.: An effective iterative method for computing real and complex roots of systems of nonlinear equations. Applied Mathematics and Computation 215, 1813–1820 (2009)
7. Ostrowski, A.M.: Solutions of equations and systems of equations. Academic Press, New York (1966)
8. Shin, B.-C., Darvishi, M.T., Kim, C.-H.: A comparison of the Newton-Krylov method with high order Newton-like methods to solve nonlinear systems. Applied Mathematics and Computation 217, 3190–3198 (2010)

A Finite Difference Approach
for the Time-Fractional Diffusion Equation with
Concentrated Capacity

Aleksandra Delić

University of Belgrade, Faculty of Mathematics
Studentski trg 16, 11000 Belgrade, Serbia
adelic@matf.bg.ac.rs

Abstract. In this paper we consider finite-difference scheme for the
time-fractional diffusion equation with Caputo fractional derivative of
order $\alpha \in (0, 1)$ with the coefficient at the time derivative containing
Dirac delta distribution.

Keywords: Fractional derivative, Boundary value problem, Dirac distribution, Finite difference method.

1 Introduction

In recent years fractional differential equations have attracted many researchers
due to its demonstrated applications in engineering, physics, chemistry, and other
sciences (see e.g. [7,8]). In many cases fractional-order models are more adequate
than integer-order models, because fractional derivatives and integrals enable
the description of the memory properties of various materials and processes.
The analytical solutions of most fractional differential equations cannot be obtained, and as consequence, approximate and numerical techniques are playing
important role in identifying the solutions behavior of such fractional equations.
Using the energy inequality method, a priori estimates for the solution of the
first and third boundary value problems for the diffusion-wave equation with
Caputo fractional derivative have been obtained in [2]. Lin and Xu at [6] examined a finite difference/Legandre spectral method to solve the initial-boundary
value time-fractional diffusion problem on a finite domain and they obtained
estimates of $(2 - \alpha)$-order convergence in time and exponential convergence in
space. Some other methods for solving fractional differential equations can be
seen in [5], [11], [12]. In the present paper, we consider finite-difference scheme for
the time-fractional diffusion equation with Caputo fractional derivative of order
$\alpha \in (0, 1)$ with the coefficient at the time derivative containing Dirac delta distribution. Analogous problem for integer order diffusion equation is considered
in [3,4].

I. Dimov, I. Faragó, and L. Vulkov (Eds.): NAA 2012, LNCS 8236, pp. 231–238, 2013.

2 Problem Formulation

Let $\Lambda = (0,1)$, $I = (0,T)$, be space and time domain respectively and $Q = \Lambda \times I$. Consider the model boundary value problem with concentrated capacity at an interior point $x = \xi$:

$$\partial_{0t}^{\alpha} u - \frac{\partial^2 u}{\partial x^2} = f(x,t), \quad x \in (0,\xi) \cup (\xi,1), \quad t \in I, \tag{1}$$

with homogeneous Dirichlet boundary conditions

$$u(0,t) = 0, \quad u(1,t) = 0, \quad t \in I, \tag{2}$$

and initial condition

$$u(x,0) = 0, \quad x \in \Lambda, \tag{3}$$

where

$$\partial_{0t}^{\alpha} u(x,t) = \frac{1}{\Gamma(1-\alpha)} \int_0^t \frac{u_{\tau}(x,\tau)}{(t-\tau)^{\alpha}} d\tau, \tag{4}$$

is the Caputo fractional derivative of order α, $0 < \alpha < 1$. As a result the conjugation conditions are fulfilled:

$$[u]_{x=\xi} = u(\xi + 0, t) - u(\xi - 0, t) = 0, \quad t \in I, \tag{5}$$

and

$$\left[\frac{\partial u}{\partial x}\right]_{x=\xi} = K\partial_{0t}^{\alpha} u(\xi, t), \quad t \in I, \quad K = const. > 0, \tag{6}$$

Using the theory of generalized functions [14], the equation (1) and conditions (5) and (6) can be rewritten as follows:

$$(1 + K\delta(x-\xi))\partial_{0t}^{\alpha} u - \frac{\partial^2 u}{\partial x^2} = f(x,t), \quad (x,t) \in Q, \quad K = const. > 0, \tag{7}$$

where $\delta(x)$ is the Dirac's delta generalized function.
There is another option for computing fractional derivatives [10]; the Riemann-Liouville fractional derivative:

$$^R\partial_{0t}^{\alpha} u(x,t) = \frac{1}{\Gamma(1-\alpha)} \frac{\partial}{\partial t} \int_0^t \frac{u(x,\tau)}{(t-\tau)^{\alpha}} d\tau, \quad 0 < \alpha < 1,$$

and it is connected to Caputo fractional derivative by relation:

$$^R\partial_{0t}^{\alpha} u(x,t) = \frac{u(x,0)}{\Gamma(1-\alpha)t^{\alpha}} + \partial_{0t}^{\alpha} u(x,t). \tag{8}$$

In contrast to the Caputo fractional derivative, when solving differential equations using Riemann-Liouville's definition, it is necessary to define the fractional order initial conditions. It is worthwhile to note, by virtue of (8), that for homogeneous condition considered here, the Riemann-Liouville definition coincides with Caputo version.

Let H be a real separable Hilbert space endowed with inner product $(\cdot, \cdot)$ and norm $|| \cdot ||$. We also define the Sobolev spaces $H^s((a,b); H) = W_2^s((a,b); H)$, $H^0((a,b); H) = L_2((a,b); H)$ of the function $u = u(t)$ mapping interval $(a,b) \subset R$ into H.

We define space:

$$B^{1,\frac{\alpha}{2}}(Q) = L_2(I; H_0^1(\Lambda)) \cap B^{\frac{\alpha}{2}}(I; \tilde{L}_{2,K}(\Lambda)), \tag{9}$$

equipped with the norm

$$||u||_{B^{1,\frac{\alpha}{2}}(Q)} = \left(||u||^2_{L_2(I; H_0^1(\Lambda))} + ||u||^2_{B^{\frac{\alpha}{2}}(I; \tilde{L}_{2,K}(\Lambda))} \right)^{\frac{1}{2}}, \tag{10}$$

where

$$||u||^2_{B^{\frac{\alpha}{2}}(I; \tilde{L}_{2,K}(\Lambda))} = \frac{1}{\Gamma(1-\alpha)} \int_0^T (T-t)^{-\alpha} ||u||^2_{\tilde{L}_{2,K}(\Lambda)} dt, \tag{11}$$

and

$$||u||^2_{\tilde{L}_{2,K}(\Lambda)} = ||u||^2_{L_2(\Lambda)} + K u^2(\xi). \tag{12}$$

Lemma 1. [2] For any function $v(t) \in H^1(0,T)$, one has the inequality

$$v(t)\partial_{0t}^\alpha v(t) \geq \frac{1}{2}\partial_{0t}^\alpha v^2(t), \quad 0 < \alpha < 1. \tag{13}$$

Theorem 1. The solution of the problem (1)-(6) satisfies the a priori estimate:

$$||u||_{B^{1,\frac{\alpha}{2}}(Q)} \leq ||f||_{L_2(Q)}. \tag{14}$$

Proof. Multiplying equation (7) by $u(x,t)$ and integrating by parts with respect to x from 0 to 1 we obtain the identity:

$$\int_\Lambda u\partial_{0t}^\alpha u dx + Ku(\xi,t)\partial_{0t}^\alpha u(\xi,t) + \int_\Lambda \left|\frac{\partial u}{\partial x}\right|^2 dx = \int_\Lambda f u dx.$$

From virtue of inequality $(u,f) \leq \varepsilon||u||^2 + \frac{1}{4\varepsilon}||f||^2$, Poincaré-Friedrics inequality $||u||^2_{L_2(\Lambda)} \leq \frac{1}{2}||u_x||^2_{L_2(\Lambda)}$ and lemma 1 we further obtain

$$\partial_{0t}^\alpha ||u||^2_{\tilde{L}_{2,K}(\Lambda)} + ||u||^2_{H_0^1(\Lambda)} \leq ||f||^2_{L_2(\Lambda)}. \tag{15}$$

Integrating (15) with respect to t from 0 to T and using property that Caputo and Rimman-Louville fractional derivative are the same when $u(x,0) = 0$ we finally obtain:

$$\frac{1}{\Gamma(1-\alpha)} \int_0^T (T-t)^{-\alpha}||u||^2_{\tilde{L}_{2,K}(\Lambda)} dt + ||u||^2_{L_2(I; H_0^1(\Lambda))} \leq ||f||^2_{L_2(Q)}. \tag{16}$$

$\square$

It follows from the a priori estimate (14) that the solution of problem (7)-(3) is unique and continuously depends on the input data.

3   The Finite Difference Scheme

In order to construct a finite difference scheme to the boundary problem (1)-(6) we construct a mesh on the rectangle $\overline{Q} = [0,1] \times [0,T]$. Let $h = 1/N$ be the mesh-size in the x-direction and $\tau = T/M$ the mesh-size in the t-direction, for some positive integers N and M. We define the uniform mesh $Q_{h\tau}$ on Q by

$$\overline{Q}_{h\tau} = \overline{\omega}_h \times \overline{\omega}_\tau = \{(x_j, t_m) : x_j = jh, 0 \le j \le N, t_m = m\tau, 0 \le m \le M\},$$

where

$$\overline{\omega}_h = \{x_j = jh : 0 \le j \le N\} = \omega_h \cup \{0,1\},$$

and

$$\overline{\omega}_\tau = \{t_m = m\tau : 0 \le m \le M\} = \omega_\tau \cup \{0,T\}.$$

We will use standard notation from the theory of finite difference schemes [9]:

$$v = v(x,t), \quad v_j^m = v(x_j, t_m),$$

$$v_x = \frac{v(x+h,t) - v(x,t)}{h} = v_{\bar{x}}(x+h,t), \quad v_t = \frac{v(x,t+\tau) - v(x,t)}{h} = v_{\bar{t}}(x,t+\tau)$$

Suppose for the simplicity that ξ is rational number. Then we can choose the step h, so that $\xi \in \omega_h$.

If $f(x,t) \in C(\overline{Q})$ then the problem (1)-(6) can be approximated on the mesh $\overline{Q}_{h\tau}$ by the implicit difference scheme:

$$(1 + K\delta_h(x - \xi))D_\tau^\alpha v^m - v_{x\bar{x}}^m = f^m, \quad x \in \omega_h, \quad m = 1, 2, ..., M, \tag{17}$$

$$v(0,t) = 0, \, v(1,t) = 0, \quad t \in \omega_\tau \cup \{T\} \tag{18}$$

$$v(x,0) = 0, \quad x \in \overline{\omega}_h, \tag{19}$$

where

$$\delta_h(x - \xi) = \begin{cases} 0, & x \in \omega_h \setminus \{\xi\} \\ 1/h, & x = \xi \end{cases}$$

is Dirac's mesh function. Also, the Caputo fractional derivative is approximated by [13]:

$$D_\tau^\alpha v^m = \frac{1}{\Gamma(2-\alpha)\tau^\alpha} \sum_{k=0}^{m-1} a_{m-k} v_{\bar{\tau}}^k,$$

where $a_{m-k} = (m-k)^{1-\alpha} - (m-k-1)^{1-\alpha}$.

Finite difference scheme (17)-(19) requires, at each time step, to solve a tridiagonal system of linear equations where the right-hand side utilizes all the history of the computed solution up to that time.

Let H_h denotes the set of functions defined on the mesh $\bar{\omega}_h$ and equal to zero at $x = 0$ and $x = 1$. We define the inner product

$$(u, w)_h = h \sum_{x \in \omega_h} v(x)w(x),$$

discrete L_2 and $\tilde{L}_{2,K}$-norms

$$||v||_h^2 = (v, v)_h, \quad ||v||_{\tilde{h}}^2 = h \sum_{x \in \bar{\omega}_h \setminus \{0\}} v^2(x), \quad ||v||_{h,K}^2 = (v, v)_h + Kv^2(\xi),$$

and the discrete Sobolev norm

$$||v||_{1,h}^2 = ||v||_h^2 + ||v_{\bar{x}}]|_{\tilde{h}}^2.$$

We also define the norm

$$||v||_{B_h^{1, \frac{\alpha}{2}}(Q_{h\tau})}^2 = \frac{1}{\Gamma(2 - \alpha)} \sum_{t \in \omega_\tau} \left((T-t)^{1-\alpha} - (T-(t+\tau))^{1-\alpha}\right) ||v||_{h,K}^2 + \sum_{t \in \omega_\tau} \tau ||v||_{1,h}^2.$$

Lemma 2. *[1] For any function $v(t)$ defined on the grid $\bar{\omega}_\tau$, the following inequality is valid:*

$$v^m D_\tau^\alpha v^m \geq \frac{1}{2} D_\tau^\alpha (v^m)^2.$$

Theorem 2. *The difference scheme (17)-(19) is absolutely stable and for its solution the following a priori estimate is valid:*

$$||v||_{B_h^{1, \frac{\alpha}{2}}(Q_{h\tau})} \leq C||f||_{L_2(Q_{h\tau})}. \tag{20}$$

Proof. If we proceed in the same way as in the proof of theorem 1, using lemma 2, we get:

$$(1, D_\tau^\alpha (v^m)^2)_h + K D_\tau^\alpha v^2(\xi, t_m)) + ||v^m||_{h,1}^2 \leq C||f^m||_h^2. \tag{21}$$

Multiplying inequality (21) by τ and summing over m from 1 to $M - 1$, we obtain a priori estimate (20). The a priori estimate (20) implies the stability and convergence of the difference scheme (17)-(19). □

4 Numerical Experiment

To check the stability and convergence properties of the numerical method we solved the problem (1)-(6) for $K = 4\pi$, $\xi = \frac{1}{2}$, $T = 1$ and

$$f(x, t) = \sin(\pi x) \left[\frac{2t^{2-\alpha}}{\Gamma(2 - \alpha)} + \pi^2 t^2 \right] + |\sin(2\pi x)| \left[\frac{2t^{2-2\alpha}}{\Gamma(3 - \alpha)} + 4\pi^2 \right].$$

The exact solution of the above problem is $u(x,t) = \sin(\pi x)t^2 + |\sin(2\pi x)|\frac{2t^{2-\alpha}}{\Gamma(2-\alpha)}$. Denote the maximum error

$$\|e(h,\tau)\|_\infty = \|u - v\|_\infty = \max_{\bar\omega_\tau}\left(\max_{\bar\omega_h}|u - v|\right).$$

and error in $B_h^{1,\alpha/2}(Q_{h\tau})$ norm

$$\|e(h,\tau)\|_B = \|u - v\|_{B_h^{1,\alpha/2}(Q_{h\tau})} \qquad (22)$$

Table 1 lists the computational results with different time step sizes τ when space step size is fixed as $h = 2^{-13}$. From the table, we can draw the conclusion that the order of convergence in time direction is $2 - \alpha$.

Table 2 gives numerical results for small and fixed $\tau = 2^{-14}$ with different h. The reason why we have used a very small τ is to make sure that the dominated error is from space discretization. From the table, we can see that the order of convergence in space direction is two.

The results for T=1 with $h = \tau = 1/60$ are displayed in Fig.1, where the exact solution is also depicted for comparison. From the results, we can see that the numerical solution obtained by finite difference (17)-(19) is in good agreement with exact solution.

Table 1. The experimental error results and convergence order in time direction (the last column) with $h = 2^{-13}$

α	τ	$\|e(h,\tau)\|_\infty$	$\|e(h,\tau)\|_B$	$\log_2 \frac{\|e(h,\tau)\|_B}{\|e(h,\tau/2)\|_B}$
0.5	2^{-5}	2.205968e-3	6.190359e-3	1.47
	2^{-6}	7.936232e-4	2.222931e-3	1.48
	2^{-7}	2.839064e-4	7.995761e-4	1.48
	2^{-8}	1.011713e-4	2.857433e-5	1.49
	2^{-9}	3.594643e-5	1.018082e-5	no data
0.7	2^{-5}	6.428188e-3	1.448575e-2	1.29
	2^{-6}	2.636782e-3	5.916592e-3	1.29
	2^{-7}	1.076979e-3	2.416312e-3	1.29
	2^{-8}	4.389248e-4	9.859266e-4	1.29
	2^{-9}	1.786251e-5	4.018979e-4	no data
0.9	2^{-5}	1.704275e-2	2.768540e-2	1.11
	2^{-6}	7.995080e-3	1.285650e-2	1.10
	2^{-7}	3.742127e-3	5.998412e-3	1.10
	2^{-8}	1.749317e-3	2.803601e-3	1.10
	2^{-9}	8.171687e-4	1.311121e-3	no data
1	2^{-5}	2.690193e-2	3.435696e-2	1.01
	2^{-6}	1.347861e-2	1.703289e-2	1.01
	2^{-7}	6.746233e-3	8.479882e-3	1.00
	2^{-8}	3.337484e-3	4.230770e-3	1.00
	2^{-9}	1.687831e-3	2.113081e-3	no data

Table 2. The experimental error results and convergence order in space direction (the last column) with $\tau = 2^{-14}$

α	h	$\|e(h,\tau)\|_\infty$	$\|e(h,\tau)\|_B$	$\log_2 \frac{\|e(h,\tau)\|_B}{\|e(h/2,\tau)\|_B}$
0.5	2^{-4}	1.449066e-2	4.464869e-2	1.99
	2^{-5}	3.660336e-3	1.120727e-2	1.98
	2^{-6}	9.136818e-4	2.848326e-3	1.98
	2^{-7}	2.285259e-4	7.218827e-4	no data
0.7	2^{-4}	1.691375e-2	5.220505e-2	1.99
	2^{-5}	4.257733e-3	1.313580e-2	1.99
	2^{-6}	1.063423e-3	3.314222e-3	1.99
	2^{-7}	2.668165e-4	8.351539e-4	no data
0.9	2^{-4}	1.929084e-2	5.879114e-2	1.97
	2^{-5}	4.847840e-3	1.502163e-2	1.98
	2^{-6}	1.217797e-3	3.815850e-3	1.99
	2^{-7}	3.109223e-4	9.617699e-4	no data
1	2^{-4}	2.042462e-2	6.185711e-2	1.96
	2^{-5}	5.137748e-3	1.595421e-2	1.97
	2^{-6}	1.303556e-3	4.073481e-3	1.98
	2^{-7}	3.447313e-4	1.029635e-3	no data

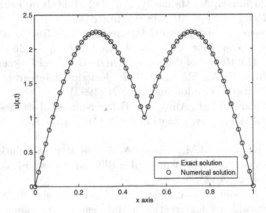

Fig. 1. Solution behavior at t=1 with h=τ=1/60 and α=0.5

5   Conclusion

In this paper, a finite difference-scheme (17)-(19) on the uniform meshes for the model problem (1)-(6) is derived. The stability of the difference scheme is proved. Numerical example illustrate the efficiency of the proposed method. Consideration of the other numerical methods and eventual selection of the optimal method will be the subject of my future work.

Acknowledgement. The research was supported by Ministry of Education and Science of Republic of Serbia under project 174015.

References

1. Alikhanov, A.A.: Boundary value problems for the diffusion equation of the variable order in the differential and difference settings. arXiv:1105.2033v1 [Math. N.A.] (May 10, 2011)
2. Alikhanov, A.A.: A priori estimates for solutions of boundary value problems for fractional-order equations. Differ. Equ. 46(5), 660–666 (2010)
3. Jovanović, B.S., Vulkov, L.G.: On the convergence of finite difference schemes for the heat equation with concentrated capacity. Numer. Math. 89(4), 715–734 (2001)
4. Jovanović, B.S., Vulkov, L.G.: Operator's approach to the problems with concentrated factors. In: Vulkov, L.G., Waśniewski, J., Yalamov, P. (eds.) NAA 2000. LNCS, vol. 1988, pp. 439–450. Springer, Heidelberg (2001)
5. Khader, M.M.: On the numerical solutions for the fractional diffusion equation. Commun. Nonlinear Sci. Numer. Simul. 16(6), 2535–2542 (2011)
6. Lin, Y., Xu, C.: Finite difference/spectral approximations for the time-fractional diffusion equation. J. Comput. Phys. 225, 1533–1552 (2007)
7. Mainardy, F.: Fractional calculus: Some basic problems in continuum and statistical mechanics. In: Carpinteri, A., Mainardy, F. (eds.) Fractals and Fractional Calculus in Continuum Mechanics, pp. 291–348. Springer (1997)
8. Podlubny, I.: Geometric and physical interpretation of fractional integration and fractional differentiation. Frac. Calc. Appl. Anal. 5(4), 367–386 (2002)
9. Samarskii, A.A.: The theory of difference schemes. Marcel Dekker (2001)
10. Samko, S.G., Kilbas, A.A., Marichev, O.I.: Fractional integrals and derivatives: Theory and applications. Gordon and Breach (1993)
11. Sweilam, N.H., Khader, M.M., Almwarm, H.M.: Numerical studies for the variable-order nonlinear fractional wave equation. Frac. Calc. Appl. Anal. 235(9), 2832–2841 (2011)
12. Sweilam, N.H., Khader, M.M., Nagy, A.M.: Numerical solution of two-sided space-fractional wave equation using finite difference method. J. Comput. Appl. Math. 15(4), 669–683 (2012)
13. Taukenova, F.I., Shkhanukov-Lafishev, M.K.: Difference methods for solving boundary value problems for fractional differential equations. Comput. Math. Math. Phys. 46(10), 1785–1795 (2006)
14. Vladimirov, V.S.: Equations of mathematical physics. Nauka, Moscow (1988) (in Russian)

Model-Based Biological Control
of the Chemostat

Neli S. Dimitrova[1] and Mikhail I. Krastanov[1,2]

[1] Institute of Mathematics and Informatics
Bulgarian Academy of Sciences
Acad. G. Bonchev Str. Bl. 8, 1113 Sofia, Bulgaria
{nelid,krast}@math.bas.bg
[2] Faculty of of Mathematics and Informatics, Sofia University
5 James Bourchier Blvd., 1164 Sofia, Bulgaria
krast@fmi.uni-sofia.bg

Abstract. In this paper we investigate a known competition model between two species in a chemostat with general (nonmonotone) response functions and distinct removal rates. Based on the competitive exclusion principle A. Rappaport and J. Harmand (2008) proposed the concept of the so called biological control. Here we present a generalization of this result.

1 Introduction

The competitive exclusion principle (CEP) is a well known concept in microbial ecology. CEP means that when two or more microbial species grow on a single resource in a chemostat, at most one species eventually survives – this is the species that possesses the best affinity to the substrate. Consider the following model of the chemostat

$$\dot{s} = (s_0 - s)D - \sum_{i=1}^{n} \mu_i(s)x_i$$

$$\dot{x}_i = (\mu_i(s) - D_i)x_i \tag{1}$$

$$i = 1, 2, \ldots, n; \quad s(0) \ge 0, \quad x_i(0) > 0,$$

where s_0 is the input concentration of the nutrient in the chemostat, D is the dilution rate of the chemostat, x_i are the concentrations of the microorganisms with response (growth rate) functions $\mu_i(s)$ and removal rates D_i.

Let the following assumptions be fulfilled.

Assumption A1. For $i = 1, 2, \ldots, n$, the functions $\mu_i(s)$ are nonnegative with $\mu_i(0) = 0$, and Lipschitz continuous.

Assumption A2. There exist unique, positive real numbers α_i and β_i with $\alpha_i < \beta_i$ (β_i possibly equal to $+\infty$) such that

$$\mu_i(s) \begin{cases} < D_i, & \text{if } s \notin [\alpha_i, \beta_i] \\ > D_i & \text{if } s \in (\alpha_i, \beta_i) \end{cases}, \quad i = 1, 2.$$

I. Dimov, I. Faragó, and L. Vulkov (Eds.): NAA 2012, LNCS 8236, pp. 239–246, 2013.

The numbers α_i and β_i are the steady state components of (1) with respect to s; they are also called break-even concentrations. Obviously, if $\mu_i(s)$ is a monotone increasing function (like the Monod law), then $\beta_i = +\infty$.

The equilibrium solutions $(s, x_1, x_2, \ldots, x_n)$ of the model are of the form

$$E_0 = (s_0, 0, 0, \ldots, 0)$$
$$E_i = \left(\alpha_i, 0, \ldots, 0, \frac{D(s_0 - \alpha_i)}{D_i}, 0, \ldots, 0 \right)$$
$$F_i = \left(\beta_i, 0, \ldots, 0, \frac{D(s_0 - \beta_i)}{D_i}, 0, \ldots, 0 \right), \quad i = 1, 2, \ldots, n,$$

that is all components of E_i (F_i) are equal to zero except for the first and the $(i+1)$-th, which are $s = \alpha_i$ ($s = \beta_i$) and $x_i = \dfrac{D(s_0 - \alpha_i)}{D_i}$ $\left(x_i = \dfrac{D(s_0 - \beta_i)}{D_i} \right)$.
The equilibrium E_i (F_i) exists for all $i = 1, 2, \ldots, n$, such that $\alpha_i < s_0$ ($\beta_i < s_0$). If $\mu_i(s)$ is monotone increasing then the equilibrium F_i does not exist. Also, each equilibrium F_i is not stable if it exists (cf. [10]).

There are lot of papers devoted to the stability analysis of the model (1). A survey of results is given in [10], see also [1], [6], [7], [11], [12] and the references therein. The main objective is to give sufficient conditions for global asymptotic stability of the equilibrium points.

The main result of G. J. Butler and G. S. K. Wolkovicz [1] is that in the case $D = D_1 = D_2 = \cdots = D_n$, every solution converges to one of the above equilibrium points. In particular, since at most one population has a nonzero component at equilibrium, no more than one population can survive. If $s_0 < \alpha_1$, then E_0 is the global attractor.

A more general result (for the case of different removal rates $D_i \neq D_j$, $i \neq j$ and $D \neq D_i$) is given by B. Li in [7]. Different removal rates typically appear in chemostats with output membranes that remove the biomass selectively, depending on the size of the microorganisms. The usual assumption is $D_i < D$. When however the mortality of a species is predominant, one may consider $D_i > D$. Denote for convenience

$$D_{\max} = \max\{D, D_1, D_2\}, \quad D_{\min} = \min\{D, D_1, D_2\},$$
$$s_0^{\min} = \frac{s_0 D}{D_{\max}}, \quad s_0^{\max} = \frac{s_0 D}{D_{\min}}. \tag{2}$$

Theorem 1. *(cf. [7]). Assume that $\alpha_1 < \alpha_2 \leq \cdots \leq \alpha_n$. If $\alpha_1 < s_0 < \beta_1$ and $s_0^{\max} - s_0^{\min} < \alpha_2 - \alpha_1$ then all solutions of (1) satisfy*

$$\lim_{t \to +\infty} (s(t), x_1(t), \ldots, x_n(t)) = E_1.$$

Biological control of the chemostat. Based on CEP, the original concept of the so called biological control of the chemostat has been recently developed by A. Rapaport and J. Harmand in [9]. The main idea consists in adding particular species in the chemostat to globally stabilize the biological process to a desired outcome. More precisely, consider (1) for $n = 1$:

$$\dot{s} = (s_0 - s)D - \mu_1(s)x_1$$
$$\dot{x}_1 = (\mu_1(s) - D_1)x_1 \tag{3}$$
$$s(0) \geq 0, \quad x_1(0) > 0.$$

Assume that the uptake function $\mu_1(s)$ is not monotone (such as the Haldane law). Let α_1 and β_1 be defined as in Assumption A2 with $\beta_1 < s_0$. Then it is well known that the dynamics (3) possesses two locally stable equilibrium points, the wash-out steady state $E_0 = (s_0, 0)$ and the positive steady state $E_1 = \left(\alpha_1, \dfrac{D(s_0 - \alpha_1)}{D_1}\right)$ (see e. g. [1], [10], [12]): from some initial conditions. the dilution rate can lead to wash-out (extinction) of the biomass and breaking-down of the process. Different approaches are known from the literature (like feedback control, cf. [2], [3], [8]) aimed to make the positive equilibrium E_1 a globally asymptotically stable point of the closed-loop system.

A. Rapaport and J. Harmand proposed in [9] a new way of controlling a biosystem through the use of an additional species with particular characteristics to globally asymptotically stabilize the process towards the equilibrium E_1. Below we present the main result of [9] (cf. also [5]).

Consider the model (1) with two populations

$$\dot{s} = (s_0 - s)D - \mu_1(s)x_1 - \mu_2(s)x_2$$
$$\dot{x}_1 = (\mu_1(s) - D_1)x_1 \tag{4}$$
$$\dot{x}_2 = (\mu_2(s) - D_2)x_2$$
$$s(0) \geq 0, \quad x_1(0) > 0, \quad x_2(0) > 0.$$

With $d \in \{D, D_1, D_2\}$ define the sets $(\alpha_i(d), \beta_i(d)) = \{s \geq 0 : \mu_i(s) > d\}$; in particular, denote $\alpha_i = \alpha_i(D_i)$, $\beta_i = \beta_i(D_i)$, $i = 1, 2$.

Assumption A3. The sets (α_i, β_i), $(\alpha_i(D), \beta_i(D))$, $i = 1, 2$, are intervals, where β_i and/or $\beta_i(D)$ is possibly equal to $+\infty$.

Denote further $\overline{D}_i = \max(D, D_i)$, $i = 1, 2$.

Assumption A4. Let the following inequalities be fulfilled:

$$\beta_1 \leq s_0^{\min}, \quad \alpha_1(\overline{D}_1) < \alpha_2(\overline{D}_2) < \beta_1(\overline{D}_1) \quad \text{and} \quad s_0 < \beta_2(\overline{D}_2).$$

Define the point $\bar{s} = \min\left\{s \in (\alpha_2(\overline{D}_2), \beta_1(\overline{D}_1)) : \mu_1(s) - \overline{D}_1 = \mu_2(s) - \overline{D}_2\right\}$.

Assumption A5. Let the following inequality be fulfilled:

$$\mu_1(\bar{s}) - \overline{D}_1 = \mu_2(\bar{s}) - \overline{D}_2 > \frac{s_0 - s_0^{\min}}{s_0^{\min} - \bar{s}} \cdot D.$$

Theorem 2. *(cf. [9]). Let the assumptions A1 to A5 be fulfilled. If $s_0^{\max} - s_0^{\min} < \alpha_2 - \alpha_1$, then any solution of (4) converges asymptotically towards* $E^* = \left(\alpha_1, \dfrac{D(s_0 - \alpha_1)}{D_1}, 0\right).$

2   Generalization of CEP

Consider the model (4) including two populations with concentrations x_1 and x_2, which compete for a single substrate s. Let the Assumption A1 holds true for $i = 1, 2$. Instead of Assumption A2 we use the following one

Assumption B2. There exist unique positive real numbers α_i and β_i with $\alpha_i < \beta_i$ (β_2 possibly equal to $+\infty$) such that

$$\mu_i(s) \begin{cases} < D_i, & \text{if } s \notin [\alpha_i, \beta_i] \\ > D_i & \text{if } s \in (\alpha_i, \beta_i) \end{cases} , \quad i = 1, 2;$$

let $\alpha_1 < \alpha_2 < \beta_1 < s_0 < \beta_2$ be satisfied.

Define the function

$$H(s) = (s_0 - s)D - \min\{\mu_1(s), \mu_2(s)\} \cdot (s_0^{\min} - s). \tag{5}$$

Assumption B3. There exists points s_1, s_2 with $s_1 < s_2$ and $[s_1, s_2] \in (\alpha_2, \beta_1)$ such that $H(s) < 0$ for all $s \in (s_1, s_2)$.

Assumption B4. Let the inequality $s_0^{\max} - s_0^{\min} < \alpha_2 - \alpha_1$ hods true.

The following Lemmata will be used further.

Barbălat's Lemma *(cf. [4]). If $f : (0, \infty) \to R$ is Riemann integrable and uniformly continuous, then $\lim\limits_{t \to \infty} f(t) = 0$.*

Lemma 1. *(cf. [12]) Let the Assumption A1 be satisfied. Then for any $\varepsilon > 0$, the solutions $s(t)$, $x_1(t)$, $x_2(t)$ of (4) satisfy $s_0^{\min} - \varepsilon < s(t) + x_1(t) + x_2(t) < s_0^{\max} + \varepsilon$ for all sufficiently large $t > 0$.*

Lemma 2. *(cf. [11]) Let the Assumptions A1 and B2 be satisfied. Then $s(t) < s_0$ for all sufficiently large $t > 0$.*

The main result of the paper is contained in the next Theorem 3.

Theorem 3. *Let Assumptions A1 (with $n = 2$), B2, B3 and B4 be fulfilled. Then any solution of (4) converges asymptotically towards $E^* = (\alpha_1, x_1^*, 0)$, with $x_1^* = \dfrac{D(s_0 - \alpha_1)}{D_1}$.*

Proof. Lemma 1 implies that there exists a sufficiently large $T_2 > 0$, so that for each $t \geq T_2$ the following inequality holds true $s_0^{\min} - s(t) - x_1(t) - x_2(t) < \varepsilon$.

Assume that $s(t) \geq \alpha_2$ for all $t \geq T_2$. The derivative $\dot{x}_2(t)$ of $x_2(t)$ is uniformly continuous (because $\mu_2(\cdot)$ and $x_2(\cdot)$ are Lipschitz continuous) and Riemann integrable. Clearly, $\dot{x}_2(\cdot) \geq 0$ whenever $s \in [\alpha_2, s_0]$, and so $x_2(t) \geq x_2(T_2) > 0$ for $t \geq T_2$. Applying Barbălat's Lemma, we obtain $\lim_{t \to \infty} \mu_2(s(t)) = D_2$. According to Assumption B2, α_2 is the unique point from the interval $[\alpha_2, s_0]$ such that $\mu_2(s) = D_2$; therefore $\lim_{t \to \infty} s(t) = \alpha_2$. This means that for each positive integer n there exists $T_n > 0$ such that $s(t) \in [\alpha_2, \alpha_2 + 1/n]$ for each $t \geq T_n$.

Since $\mu_1(\alpha_2) - D_1 = \eta > 0$ there exists a positive integer n such that $\mu_1(s) - D_1 \geq \eta/2$ for each $s \in [\alpha_2, \alpha_2 + 1/n]$. Therefore, for each $t \geq T_n$ we have

$$x_1(t) = x_1(T_n) + \int_{T_n}^t \dot{x}_1(\tau)\, d\tau = x_1(T_n) + \int_{T_n}^t (\mu_1(s(\tau)) - D_1)\, d\tau$$

$$\geq x_1(T_n) + \int_{T_n}^t \frac{\eta}{2}\, d\tau \to \infty \quad \text{as } t \to \infty.$$

But this is impossible because $x_1(\cdot)$ is bounded. Hence, there exists $T_3 > T_2$ such that $s(T_3) < \alpha_2$.

Consider the function $H(s)$ from (5). Let us fix a sufficiently small $\varepsilon > 0$ and choose a point $\tilde{s} \in (s_1, s_2)$, so that $(s_0 - \tilde{s})D < (s_0^{\min} - \tilde{s} - \varepsilon)\min\{\mu_1(\tilde{s}), \mu_2(\tilde{s})\}$. Let us assume that there exists $\tilde{t} \geq T_3$ such that $s(\tilde{t}) = \tilde{s}$; set $\tilde{x}_1 = x_1(\tilde{t})$ and $\tilde{x}_2 = x_2(\tilde{t})$. Then

$$\dot{s}(\tilde{t}) = D(s_0 - s(\tilde{t})) - \mu_1(s(\tilde{t}))x_1(\tilde{t}) - \mu_2(s(\tilde{t}))x_2(\tilde{t})$$

$$= D(s_0^{\min} - \tilde{s} - \tilde{x}_1 - \tilde{x}_2) + D(s_0 - s_0^{\min}) - (\mu_1(\tilde{s}) - D)\tilde{x}_1 - (\mu_2(\tilde{s}) - D)\tilde{x}_2$$

$$< D\varepsilon + D(s_0 - s_0^{\min}) - (\tilde{x}_1 + \tilde{x}_2)\left(\frac{\tilde{x}_1}{\tilde{x}_1 + \tilde{x}_2}(\mu_1(\tilde{s}) - D) + \frac{\tilde{x}_2}{\tilde{x}_1 + \tilde{x}_2}(\mu_2(\tilde{s}) - D)\right)$$

$$\leq D\varepsilon + D(s_0 - s_0^{\min}) - (\tilde{x}_1 + \tilde{x}_2)\min\{(\mu_1(\tilde{s}) - D), (\mu_2(\tilde{s}) - D)\}$$

$$< (s_0 - \tilde{s})D - (s_0^{\min} - \tilde{s} - \varepsilon)\min\{\mu_1(\tilde{s}), \mu_2(\tilde{s})\} < 0.$$

The last inequality implies that $s(t) \leq \tilde{s}$ for each $t \geq T_3$. The proof further follows the same idea as in the proof of Theorem 1 (cf. [7]). We sketch it for completeness.

We define the function $g(s) = \dfrac{\mu_2(s)(\mu_1(s) - D_1)(s_0 - \alpha_1)}{D_1(s_0 - s)(\mu_2(s) - D_2)}$ and the constant $M > \max_{s \in (0, \alpha_1)} g(s)$. Following [7], we choose $d_{\max} > D_{\max}$ and $0 < d_{\min} < D_{\min}$; then Lemma 1 implies that for each $t \geq T_3$,

$$\frac{Ds_0}{d_{\max}} < s(t) + x_1(t) + x_2(t) < \frac{Ds_0}{d_{\min}} \tag{6}$$

Taking into account the assumption B4, we define a continuously differentiable function $F(u)$ with nonnegative values whose derivative satisfies the relations:

$$F'(u) := \begin{cases} 0, & \text{if } u \geq \dfrac{Ds_0}{d_{\max}} - \alpha_1; \\[2ex] -M + M\dfrac{\left(u - \dfrac{Ds_0}{d_{\min}} + \alpha_2\right)}{\left(\dfrac{Ds_0}{d_{\max}} - \alpha_1 - \dfrac{Ds_0}{d_{\min}} + \alpha_2\right)}, & \text{if } \dfrac{Ds_0}{d_{\min}} - \alpha_2 \leq u \leq \dfrac{Ds_0}{d_{\max}} - \alpha_1; \\[2ex] -M, & \text{if } 0 \leq u \leq \dfrac{Ds_0}{d_{\min}} - \alpha_2. \end{cases}$$

Taking into account (6), we obtain that

$$x_1 + x_2 \left\{ < \frac{Ds_0}{d_{\min}} - \alpha_2 \text{ if } s \geq \alpha_2; \ > \frac{Ds_0}{d_{\max}} - \alpha_1 \text{ if } s \leq \alpha_1 \right\}.$$

Following [6] and [7], we set

$$V(s, x_1, x_2) := \int_{\alpha_1}^{s} \frac{(\mu_1(\xi) - D_1)x_1^*}{D(s_0 - \xi)} \, d\xi + \int_{x_1^*}^{x_1} \frac{\zeta - x_1^*}{\zeta} \, d\zeta + Mx_2 + F(x_1 + x_2).$$

One can directly check that the Lie derivative of V with respect to the trajectories of the system (4) is

$$\dot{V}(s, x_1, x_2) = (\mu_1(s) - D_1)x_1 \left(1 - \frac{\mu_1(s)(s_0 - \alpha_1)}{D_1(s_0 - s)} + F'(x_1 + x_2) \right) \quad (7)$$

$$+ (\mu_2(s) - D_2)x_2 \left(M + F'(x_1 + x_2) - g(s) \right). \quad (8)$$

Since

$$\frac{\mu_1(s)(s_0 - \alpha_1)}{D_1(s_0 - s)} \begin{cases} \leq 1, & \text{if } 0 < s \leq \alpha_1 \\ > 1, & \text{if } \alpha_1 < s < \beta_1, \end{cases} \qquad F'(x_1 + x_2) \begin{cases} = 0, & \text{if } 0 < s \leq \alpha_1 \\ < 0, & \text{if } \alpha_1 < s < \beta_1, \end{cases}$$

we obtain that the first term (7) in the expression for $\dot{V}$ is always nonpositive. Further, if $0 < s \leq \alpha_1$, then $(\mu_2(s) - D_2)(M + F'(x_1 + x_2) - g(s)) = (\mu_2(s) - D_2)(M - g(s)) < 0$, because $\mu_2(s) - D_2 < 0$ and according to the choice of M. If $\alpha_1 < s < \alpha_2$, then $M + F'(x_1 + x_2) > 0$ and $g(s) < 0$. If $s = \alpha_2$, then one can verify that $\dot{V} < 0$. If $\alpha_2 < s \leq \beta_1$, then $M + F'(x_1 + x_2) = 0$ and hence $(\mu_2(s) - D_2)(M + F'(x_1 + x_2) - g(s)) = -(\mu_2(s) - D_2)g(s) \leq 0$. This means that the second term (8) in the expression of $\dot{V}$ is also nonpositive. Therefore, $\dot{V}(s, x_1, x_2) \leq 0$ for all $s \in (0, \beta_1]$. According to the LaSalle invariance principle, every solution $(s(t), x_1(t), x_2(t))$, $t \geq T_3$, approaches the largest invariant set $\mathcal{L}_\infty$ contained in the set $\mathcal{L} = \{(s, x_1, x_2) : \dot{V}(s, x_1, x_2) = 0, s \in [0, \beta_1], x_1 \geq 0, x_2 \geq 0\}$. One can directly check that $\mathcal{L} = \{(s, x_1, x_2) : (\mu_1(s) - D_1)x_1 = 0, x_2 = 0, s \in [0, \beta_1], x_1 \geq 0\}$. If $x_1 = 0$, then (because $x_2 = 0$) we have $\dot{s}(t) \geq D(s_0 - \tilde{s}) > 0$, and so $\lim_{t \to \infty} s(t) = +\infty$, which is impossible. Hence $x_1 > 0$ and $\mu_1(s) = D_1$. Further, $s(t) \leq \tilde{s} < s_2 < \beta_1$ for sufficiently large $t > 0$, thus $\mu_1(s) - D_1$ cannot vanish at $s = \beta_1$. This shows that $\mathcal{L}_\infty = \{(\alpha_1, x_1^*, 0)\}$.

3   Numerical Example

We demonstrate the theoretical results from Theorem 3 on a numerical example. To make a comparison with the results of Rapaport and Harmand (Theorem 2), we consider the following response functions

$$\mu_1(s) = \frac{m_1 s}{a_1 + s + \gamma_1 s^2}, \quad \mu_2(s) = \frac{m_2 s}{a_2 + s}$$

with coefficient values $a_1 = 20$, $m_1 = 200$, $\gamma_1 = 1/90$, $a_2 = 265$, $m_2 = 620$; let $s_0 = 180$, $D = 88$, $D_1 = 80$, $D_2 = 96.45$. Here, $D_{\max} = D_2$; we have $\alpha_1 = 15$, $\alpha_2 = 48.82$, $\beta_1 = 120$ and $s_0 < \beta_2 = +\infty$; hence Assumption B2 is satisfied. The response functions are visualized on Figure 1, left plot. The right plot in Figure 1

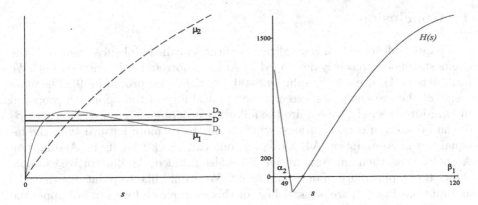

Fig. 1. The graphs of $\mu_1(s)$ and $\mu_2(s)$ (left). The graph of $H(s)$ (right); the solid circles on the horizontal axis correspond to the points s_1 and s_2 in Assumption B3.

visualizes the graph of the function $H(s) = (s_0 - s) - \min\{\mu_1(s), \mu_2(s)\}(s_0^{\min} - s)$; obviously, there exist the points $s_1 = 51.13$ and $s_2 = 56.61$ (marked by solid circles on the horizontal axis), such that $H(s) < 0$ for all $s \in (s_1, s_2)$; therefore Assumption B3 is satisfied. Since $s_0^{\max} - s_0^{\min} - (\alpha_2 - \alpha_1) \approx -0.0493 < 0$, Assumption B4 is also fulfilled.

In this example Assumption A5 is however not satisfied. Indeed, the point $\bar{s}$ satisfying the equality $\mu_1(s) - \bar{D}_1 = \mu_2(s) - \bar{D}_2$ is $\bar{s} \approx 56.72$; obviously, $\bar{s}$ belongs to the interval $(\alpha_2(\bar{D}_2), \beta_1(\bar{D}_1)) = (48.82, 95.75)$ but $\mu_1(\bar{s}) - \bar{D}_1 - \dfrac{s_0 - s_0^{\min}}{s_0^{\min} - \bar{s}} D \approx -0.0439 < 0$. Nevertheless, the equilibrium point E_1^* is globally asymptotically stable according to our Theorem 3.

Figure 2 presents the solutions $x_1(t)$ and $x_2(t)$ of the model with initial point $(s(0), x_1(0), x_2(0)) = (180, 0.01, 0.01)$. The horizontal dash line on the left plot passes through the equilibrium point x_1^*.

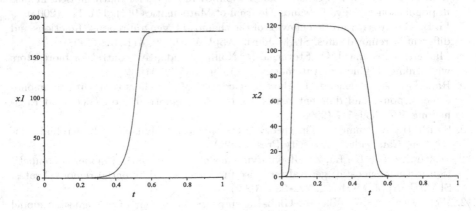

Fig. 2. The solution $x_1(t)$ (left) and $x_2(t)$ (right)

4 Conclusion

The paper is devoted to a generalization of the so-called biological control of the chemostat model, recently developed by A. Rapaport and J. Harmand in [9]. We formulate in Theorem 2 the sufficient stability condition proved in [9]. The main result of the present paper extends the applicability of the approach proposed in [9]. More precisely, we require the following ordering $\alpha_1 < \alpha_2 < \beta_1 < s_0 < \beta_2$ of the break-even concentrations, which seems to be more natural than the inequalities in Assumption A4. Moreover, one can check that if the Assumption A5 holds true, then our Assumption B3 is also fulfilled. An illustrative example shows the applicability of our main result. We would like to point out that the assumptions from [9] are not satisfied for this example and we can not apply the result of A. Rapaport and J. Harmand.

Acknowledgement. The work of the second author has been partially supported by the Sofia University under contract No. 126/ 09.05.2012.

References

1. Butler, G.J., Wolkowicz, G.S.K.: A mathematical model of the chemostat with a general class of functions describing nutrient uptake. SIAM Journ. Appl. Math. 45, 138–151 (1985)
2. De Leenheer, P., Smith, H.: Feedback control for chemostat models. J. Math. Biol. 46, 48–70 (2003)
3. Dimitrova, N., Krastanov, M.: Nonlinear adaptive control of a model of an uncertain fermentation process. Int. J. Robust Nonlinear Control 20, 1001–1009 (2010)
4. Golpalsamy, K.: Stability and oscillations in delay differential equations of population dynamics. Kluwer Academic Publishers, Dordrect (1992)
5. Harmand, J., Rapaport, A., Dochain, D., Lobry, C.: Microbial ecology and bioprocess control: opportunities and challehges. J. Proc. Control 18, 865–875 (2008)
6. Hsu, S.-B.: A survey of construction Lyapunov functions for mathrmatical models in population biology. Taiwanese Journal of Mathematics 9(2), 151–173 (2005)
7. Li, B.: Global asymptotic behavior of the chemostat: general reponse functions and differential removal rates. SIAM Journ. Appl. Math. 59, 411–422 (1998)
8. Maillert, L., Bernard, O., Steyer, J.-P.: Nonlinear adaptive control for bioreactors with unknown kinetics. Automatica 40, 1379–1385 (2004)
9. Rapaport, A., Harmand, J.: Biological control of the chemostat with nonmonotonic response and different removal rates. Mathematical Biosciences and Engineering 5(3), 539–547 (2008)
10. Smith, H., Waltman, P.: The theory of the chemostat, dynamics of microbial competition. Cambridge University Press (1995)
11. Wolkowicz, G.S.K., Lu, Z.: Global dynamics of a mathematical model of competition in the chemostat: general response functions and differential removuval rates. SIAM J. Appl. Math. 52, 222–233 (1992)
12. Wolkowicz, G.S.K., Xia, H.: Global asymptotic behaviour of a chemostat model with discrete delays. SIAM J. Appl. Math. 57, 1281–1310 (1997)

Variance-Based Sensitivity Analysis of the Unified Danish Eulerian Model According to Variations of Chemical Rates

Ivan Dimov[1], Rayna Georgieva[1], Tzvetan Ostromsky[1], and Zahari Zlatev[2]

[1] Institute of Information and Communication Technologies, Bulgarian Academy of Sciences, Acad. G. Bonchev 25A, 1113 Sofia, Bulgaria
ivdimov@bas.bg, {rayna,ceco}@parallel.bas.bg
[2] Department of Environmental Science - Atmospheric Environment, Aarhus University, Frederiksborgvej 399, Building ATMI, 4000, Roskilde, Denmark
zz@dmu.dk

Abstract. A special computational technology for sensitivity analysis of ozone concentrations according to variations of rates of chemical reactions is developed. It allows us to study a larger number of reactions than we have considered in our previous study. The reactions are taken from the standardized scheme for air-pollution chemistry CBM-IV. A number of numerical experiments with a large-scale air pollution model (Unified Danish Eulerian Model, UNI-DEM) have been carried out to compute Sobol sensitivity measures. The sensitivity study has been done for the areas of four European cities (Genova, Milan, Manchester, and Edinburgh) with different geographical locations.

1 Introduction

Environmental security is rapidly becoming a significant topic of present interest all over the world, and environmental modelling has a very high priority in various scientific fields, respectively. Such complicated multidisciplinary problem requires joined research and collaboration between experts in the area of environmental modeling, numerical analysis and scientific computing.

Our current study aims an exploration of model output sensitivity and/or a model improvement in the air pollution transport. This is done for a large-scale mathematical model describing the remote transport of air pollutants (**Unified Danish Eulerian Model, UNI-DEM**, [13, 14]). The motivation to choose UNI-DEM is that it is one of the models of atmospheric chemistry in which the chemical processes are taken into account in a very accurate way.

Perturbations of chemical reaction rates as well as emission levels and boundary conditions (of the system of partial differential equations describing the mathematical model mentioned above) are crucial issues for the model output variability. Our main objective here is to provide a reliable global sensitivity analysis of this mathematical model and, in particular, to analyze the influence of variations of the rates of some chemical reactions on the model results.

I. Dimov, I. Faragó, and L. Vulkov (Eds.): NAA 2012, LNCS 8236, pp. 247–254, 2013.

Here a sensitivity study of ozone concentrations according to variations of rates of a larger number of chemical reactions than the set of chemical reactions during the previous study [1–3] has been performed. It has been done in order to increase the reliability of the results produced by the model, and to identify processes that must be studied more carefully, as well as to find input parameters that need to be measured with a higher precision.

Sensitivity analysis is a powerful tool to study of how uncertainty in the output of a model (numerical or otherwise) can be apportioned to different sources of uncertainty in the model input [8]. The general procedure for sensitivity analysis can be described in the following steps: (i) definition of probability distributions for the parameters under study, (ii) generation of samples according to the defined probability distributions using a sampling strategy, (iii) sensitivity analysis of the output variance in relation to the variation of the inputs.

2   Background

The input data for sensitivity analysis has been obtained during runs of the mathematical model for remote transport of air pollutants - UNI-DEM. Here we apply the Sobol approach [10], as a reliable variance-based tool to provide global sensitivity analysis. A brief description of the mathematical model under consideration and the Sobol approach applied during the current study is given in the next two subsections.

2.1   Description of the Unified Danish Eulerian Model and SA-DEM

The model gives the possibility to study concentration variations in time of a high number of air pollutants and other species over a large geographical region (4800 × 4800 km), covering the whole of Europe, the Mediterranean and some parts of Asia and Africa which is important for environmental protection, agriculture, health care. It takes into account the main physical, chemical and photochemical processes between the studied species, the emissions, the quickly changing meteorological conditions. Both non-linearity and stiffness of the equations are mainly introduced by the chemistry [14]. The chemical scheme used in the model is the well-known condensed CBM-IV (Carbon Bond Mechanism). This chemical scheme is one of the most accurate, but computationally expensive as well.

The development and improvements of UNI-DEM throughout the years has lead to a variety of different versions with respect to the grid-size/resolution, vertical layering (2D or 3D model respectively) and the number of species in the chemical scheme. A coarse-grain parallelization strategy based on partitioning of the spatial domain in strips or blocks is mainly used in UNI-DEM. For the purpose of sensitivity analysis studies a specialized version of this package, called SA-DEM, was recently developed. More details about its parallel implementation properties can be found in [4–6].

Our main aim here is to study the sensitivity of the ozone concentration according to the rate variation of some chemical reactions: ## $1, 3, 7, 22$ (time-dependent) and $27, 28$ (time independent) reactions of the condensed CBM-IV scheme ([13]). The simplified chemical equations of these reactions are as follows:

[#1] $NO_2 + h\nu \Longrightarrow NO + O$ [#22] $HO_2 + NO \Longrightarrow OH + NO_2$

[#3] $O_3 + NO \Longrightarrow NO_2$ [#27] $HO_2 + HO_2 \Longrightarrow H_2O_2$

[#7] $NO_2 + O_3 \Longrightarrow NO_3$ [#28] $OH + CO \Longrightarrow HO_2$

Note, that the ozone does not necessarily participate in all these reactions. Important precursors of ozone participate instead.

The most commonly used output of UNI-DEM are the mean monthly concentrations of a set of dangerous chemical species (or groups of species) in dependence with the particular chemical scheme, calculated in the grid points of the computational domain. We consider the chemical raction rates to be input parameters and the concentrations of pollutants to be output parameters.

The first stage of computations consists of generation of input data necessary for the particular sensitivity analysis study. In our case this means to perform a number of experiments with UNI-DEM by doing certain perturbations in the data for chemical reaction rates.

The specialized version SA-DEM is used here to perform the necessary computations for a set of different values of the vector $\alpha = (\alpha_1, \ldots, \alpha_6)$ in the domain under consideration (in our case - the 6-dimensional hypercubic domain $[0.6, 1.4]^6$). The values of α are selected to lie on the edges from the vertex $(1,1,1,1,1,1)$ (representing the "basic scenario" with the true emissions for the corresponding year) to all the other vertices of the above hypercube. Along each edge the samples of α are distributed regularly by decreasing all its variable coordinates with a fixed step h ($h = 0.1$ in our experiments).

This require a huge computational effort and a powerful computer system. To meet these challenges, we developed an advanced highly parallel modification of UNI-DEM, specially adjusted to be used in various sensitivity analysis studies. This code, called SA-DEM, was recently developed and implemented on the most powerful supercomputer in Bulgaria - IBM BlueGene/P. It was used successfully in sensitivity analysis of UNI-DEM with respect to the rate coefficients of several chemical reactions [2, 5]. In our particular sensitivity analysis study regular perturbations have to be done on the input data of the chemical rate coefficients (in the chemical submodel). This leads to generation of multiple data-independent tasks, appropriate for parallel execution. Thus an additional opportunity for a coarse-grain parallelism appear in SA-DEM, which is efficiently exploited on the highly parallel IBM BlueGene/P. This is the highest level of parallelism in SA-DEM on the top of the grid-partitioning level, the basis for distributed-memory MPI parallelization in UNI-DEM. Moreover, our target machine offers a limited amount of shared memory parallelism. It is exploited on the lowest (finer-grain) level of parallelism in our algorithm by using OpenMP standard directives.

As a result of parallel computations with the use of SA-DEM and the IBM BlueGene/P supercomputer, on this stage we obtain the needed mesh function defined on $[0.6, 1.4]^6$.

2.2   Sobol Global Sensitivity Analysis

Variance-based sensitivity methods are an useful tool for an advanced study of relations between the input parameters of a model, output results and internal mechanisms regulating the system under consideration. They deliver global, quantitative and model-independent sensitivity measures and are efficient of the computational point of view. Its computational cost for estimating all first-order and total sensitivity measures is proportional to the sample size and the number of input parameters. In Sobol approach the variance of the square integrable model function is decomposed into terms of increasing dimension. The sensitivity of model output to each parameter or parameter interaction is measured by its contribution to the total variance. An important advantage of this approach is that it allows to compute also higher-order interaction effects in a way similar to the computation of the main effects. The total effect of a fixed parameter can be calculated with just one Monte Carlo integral per factor.

Consider a scalar model output u $= f(\mathbf{x})$ corresponding to a number of non-correlated model parameters $\mathbf{x} = (x_1, x_2, \ldots, x_d)$ with a joint probability density function $p(\mathbf{x}) = p(x_1, \ldots, x_d)$ in the d-dimensional unit cube $U^d = [0; 1]^d$. In Sobol approach [10] the parameter importance is studied via numerical integration in the terms of **analysis of variance** (ANOVA) model representation [8, 10]:

$$f(\mathbf{x}) = f_0 + \sum_{\nu=1}^{d} \sum_{l_1 < \ldots < l_\nu} f_{l_1 \ldots l_\nu}(x_{l_1}, x_{l_2}, \ldots, x_{l_\nu}), \quad l_1, \ldots, l_\nu \in \{1, \ldots, d\},$$

where $f_0 = \int_{U^d} f(\mathbf{x}) d\mathbf{x} = const$, $f(\mathbf{x})$ is a square integrable model function, $f_{l_1 \ldots l_\nu}(x_{l_1}, x_{l_2}, \ldots, x_{l_\nu})$ are the terms of increasing dimension in the ANOVA representation of $f(\mathbf{x})$ satisfying the following condition $\int_0^1 f_{l_1 \ldots l_\nu}(x_{l_1}, \ldots, x_{l_\nu}) dx_{l_k}$ $= 0$, $1 \leq k \leq \nu$, $\nu = 1, \ldots, d$. The quantities $\mathbf{D} = \int_{U^d} f^2(\mathbf{x}) d\mathbf{x} - f_0^2$, $\mathbf{D}_{l_1 \ldots l_\nu} = \int f_{l_1 \ldots l_\nu}^2 dx_{l_1} \ldots dx_{l_\nu}$ are called variances (total and partial variances, respectively), where $f(\mathbf{x})$ is a square integrable function. Based on the above assumptions about the model function and the output variance, the following quantities

$$S_{l_1 \ldots l_\nu} = \frac{\mathbf{D}_{l_1 \ldots l_\nu}}{\mathbf{D}}, \quad \nu \in \{1, \ldots, d\} \tag{1}$$

are called Sobol global sensitivity indices [9, 10]. The main sensitivity measures introduced in the Sobol approach represent ratios between the corresponding partial variances and total variance (see (1), [10]). The basic assumption underlying the so called High Dimensional Model Representation is that the major features of the model functions describing typical real-live problems can be shown by low-order subsets of inputs - constants, terms of first and second order. This means that one can use low-order indices only, but should be able to control the contribution of higher order terms.

The mathematical treatment of the problem of providing global sensitivity analysis consists in evaluating total sensitivity indices and in particular Sobol global sensitivity indices (1) of corresponding order. It leads to computing multidimensional integrals (from the mathematical representation of variances)

$I = \int_\Omega g(\mathbf{x})p(\mathbf{x})\,d\mathbf{x}$, $\Omega \subset \mathbf{R}^d$, where $g(\mathbf{x})$ is a square integrable function in Ω and $p(\mathbf{x}) \geq 0$ is a p.d.f., such that $\int_\Omega p(\mathbf{x})\,d\mathbf{x} = 1$.

A more detailed description of the Sobol approach is also given in [2].

Several approaches for evaluating *small* sensitivity indices (to avoid loss of accuracy because the analyzed database comes under this case) have been applied: standard (initial) Sobol approach, reducing of the mean value (proposed by I.M. Sobol', 1990), *correlated sampling* technique (proposed in [7, 11]), and a combined approach between second and third ones [12]. The partial and the total variance estimations referred to *correlated sampling* approach are presented in the following way using two independent samples $\mathbf{x} = (\mathbf{y}, \mathbf{z})$ and $\mathbf{x}' = (\mathbf{y}', \mathbf{z}')$:

$$D_y = \int_\Omega f(\mathbf{x})\,[f(\mathbf{y}, \mathbf{z}') - f(\mathbf{x}')]d\mathbf{x}d\mathbf{x}', \quad D = \int_\Omega f(\mathbf{x})[f(\mathbf{x}) - f(\mathbf{x}')]\,d\mathbf{x}d\mathbf{x}', \quad \Omega \equiv U^{2d}.$$

3   Analysis of Numerical Results and Discussion

The second stage of computations consists of the following steps: (i) Approximation, and (ii) Computing of Sobol global sensitivity indices.

We use polynomials of second degree as an approximation tool, where $p_s^{(k)}(\mathbf{x})$ is the polynomial that approximates the mesh function given in the table that corresponds to the s-th chemical species:

$$p_s^{(k)}(\mathbf{x}) = \sum_{j=0}^{k} \sum_{(\nu_1, \nu_2, \ldots, \nu_d) \in N_j^k} a_{\nu_1 \ldots \nu_d}\, x_1^{\nu_1} x_2^{\nu_2} \ldots x_d^{\nu_d}, \qquad k = 2,$$

where $\quad N_j^k = \left\{ (\nu_1, \nu_2, \ldots, \nu_d) \mid \nu_i = 0, 1, \ldots, k, \sum_{i=1}^{d} \nu_i = j \right\}.$

To estimate the accuracy of the approximation the squared 2-vector norm is used. It is defined as $\| p_s - r_s \|_2^2 = \sum_{l=1}^{n} [p_s(\mathbf{x}_l) - r_s(\mathbf{x}_l)]^2$, where $\mathbf{x}_l \in [0.6; 1.4]^6$, and $r_s(\mathbf{x}_l), l = 1, \ldots, n$ are the corresponding table values obtained as a result of runs of SA-DEM. In the case of a polynomial of 2-nd degree in six variables the squared 2-vector norm derived during numerical experiments is presented as follows: (i) Genova - 0.00478; (ii) Milan - 0.00460; (iii) Manchester - 0.00423; (iv) Edinburgh - 0.00504. A number of preliminary numerical experiments on the approximation step with polynomials of different degree has been done. Second degree polynomials (28 unknown coefficients) and third degree polynomials (84 unknown coefficients) are used for data approximation. Unfortunately, it has been impossible to obtain reliable approximation results with forth degree polynomials (210 unknown coefficients) since the number of model values in the database is less than the number of unknowns. On this stage, one can see that second degree polynomials fully satisfy the requirements for accuracy. It is studied numerically (just one model input varies, the others are fixed to 1.0) how various chemical rate reactions influence air pollution concentrations. An example is shown on Figure 1. Analyzing presented results for reactions under consideration (see Section 2.1) of CBM-IV scheme one can conclude that the

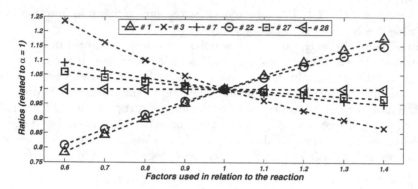

Fig. 1. Sensitivity of ozone concentrations (Genova, July 1998)

Table 1. First-order sensitivity indices of input parameters obtained using different approaches for sensitivity analysis (Genova, July 1998)

approach / estimated quantity	Initial Sobol approach	Approaches for small indices correlated sampling
integrand $g(x)$	$f(x)$	$f(x)$
g_0	0.26588	0.26588
D	0.20849	0.00249
S_1	0.09013	0.35858
S_2	0.08918	0.29485
S_3	0.08546	0.04652
S_4	0.08872	0.26462
S_5	0.08477	4.3e-07
S_6	0.08505	0.01904

influence of rates of the reactions ## 1, 3 and 22 on ozone concentrations is very important. The impact of the rates of the reactions ## 7 and 27 is smaller, but significant. At the same time the influence of the rate of the reaction # 28 can be neglected. The numerical experiments caried out to compute sensitivity indices via Mathematica [15] show that *correlated sampling* approach leads to similar results to the combined approach. In such a way, *correlated sampling* approach has been applied since the original model function has been used as an integrand. The results presented in Table 1 demonstrate significant difference for the values of sensitivity indices obtained using standard Sobol approach and *correlated sampling* approach. The numerical results referred to the second one follow strictly the influence behaviour of the rates of the chosen chemical reactions towards ozone concentrations. Table 2 contains first-, second-order and total sensitivity indices of model inputs under consideration. One can expect that the values of higher-order sensitivity indices are comparatively small and close to zero taking into account that the values of both first-order and total sensitivity indices are close to each other. It means that the mathematical model is additive

Table 2. Sensitivity indices of input parameters (for ozone concentrations)

town	Genova	Milan	Manchester	Edinburgh
f_0	0.26588	0.26566	0.26526	0.26616
D	0.00249	0.00256	0.00245	0.00136
S_1	0.35858	0.36281	0.37165	0.33487
S_2	0.29485	0.29936	0.26509	0.23399
S_3	0.04652	0.04129	0.00997	0.05559
S_4	0.26462	0.26276	0.32358	0.30133
S_5	4.34e-07	1.8e-07	0.00023	0.00009
S_6	0.01904	0.01703	0.00857	0.04653
$\sum_{i=1}^{6} S_i$	0.98361	0.98325	0.97909	0.97241
S_{12}	0.00556	0.00574	0.00568	0.00457
S_{13}	0.00048	0.00049	0.00024	0.00106
S_{14}	0.00516	0.00563	0.00809	0.00837
S_{16}	0.00031	0.00025	0.00018	0.00104
S_{23}	0.00038	0.00033	0.00005	0.00075
S_{24}	0.00349	0.00343	0.00516	0.00457
S_{34}	0.00045	0.00040	0.00015	0.00068
S_{36}	0.00016	0.00014	0.00039	0.00435
$\sum_{i=1}^{6} S_i$	0.01639	0.01675	0.02092	0.02759
S_1^{tot}	0.37009	0.37493	0.38599	0.34993
S_2^{tot}	0.30442	0.30897	0.27625	0.24471
S_3^{tot}	0.04799	0.04267	0.01098	0.06274
S_4^{tot}	0.27391	0.27239	0.33719	0.31559
S_5^{tot}	0.00015	0.00013	0.00089	0.00091
S_6^{tot}	0.01983	0.01766	0.00963	0.05371

according to the chosen input parameters. The numerical results obtained here confirm the conclusions about the importance of some of the model inputs made on the base of our previous study [3]. At the same time one can observe that now a new important input parameter (the rate of the time-dependent chemical reaction # 1) appears.

4 Conclusions

In this work the results from the sensitivity analysis provided for ozone concentrations according to variations of rates of a larger number of chemical reactions than in our previous study are described. A number of numerical experiments with a large-scale air pollution model (Unified Danish Eulerian Model, UNI-DEM) have been carried out to compute Sobol sensitivity measures.

The sensitivity study has been done for the areas of four European cities (Genova, Milan, Manchester, and Edinburgh). The results are similar and show unsignificant correlation between geographical location and sensitivity study conclusions. The main conclusions are: (i) the mathematical model UNI-DEM under consideration is additive according to the chosen input parameters - rates of chemical reactions; (ii) the results obtained during current study are fully consistent with the conclusions about importance of model inputs obtained in the previous study [1–3]; (iii) a new important input parameter (the rate of the time-dependent chemical reaction # 1) is identified.

Acknowledgment. The research reported in this paper is partly supported by the Bulgarian NSF Grants DTK 02-44/2009, DMU 03/61/2011, and DCVP 02/1.

References

1. Dimov, I., Georgieva, R.: Monte Carlo Adaptive Technique for Sensitivity Analysis of a Large-Scale Air Pollution Model. In: Lirkov, I., Margenov, S., Waśniewski, J. (eds.) LSSC 2009. LNCS, vol. 5910, pp. 387–394. Springer, Heidelberg (2010)
2. Dimov, I.T., Georgieva, R., Ivanovska, I., Ostromsky, T., Zlatev, Z.: Studying the Sensitivity of Pollutants' Concentrations Caused by Variations of Chemical Rates. J. Comput. Appl. Math. 235, 391–402 (2010)
3. Dimov, I.T., Georgieva, R., Ostromsky, T.: Monte Carlo Sensitivity Analysis of an Eulerian Large-scale Air Pollution Model. Reliability Engineering & System Safety 107, 23–28 (2012), doi:10.1016/j.ress.2011.06.007.
4. Ostromsky, T., Dimov, I.T., Georgieva, R., Zlatev, Z.: Air pollution modelling, sensitivity analysis and parallel implementation. International Journal of Environment and Pollution 46(1/2), 83–96 (2011)
5. Ostromsky, T., Dimov, I., Georgieva, R., Zlatev, Z.: Parallel Computation of Sensitivity Analysis Data for the Danish Eulerian Model. In: Lirkov, I., Margenov, S., Waśniewski, J. (eds.) LSSC 2011. LNCS, vol. 7116, pp. 307–315. Springer, Heidelberg (2012)
6. Ostromsky, T., Dimov, I.T., Marinov, P., Georgieva, R., Zlatev, Z.: Advanced sensitivity analysis of the Danish Eulerian Model in parallel and grid environment. In: Proc. Third International Conference AMiTaNS 2011, AIP Conf. Proceedings, Albena Bulgaria, June 20-25, vol. 1404, pp. 225–232 (2011), AIP Conf. Proceedings
7. Saltelli, S.: Making best use of model valuations to compute sensitivity indices. Computer Physics Communications 145, 280–297 (2002)
8. Saltelli, A., Ratto, M., Andres, T., Campolongo, F., Cariboni, J., Gatelli, D., Saisana, M., Tarantola, S.: Global Sensitivity Analysis. The Primer. John Wiley & Sons Ltd. (2008) ISBN: 978-0-470-05997-5
9. Sobol, I.M.: Sensitivity estimates for nonlinear mathematical models. Mathematical Modeling and Computational Experiment 1, 407–414 (1993)
10. Sobol, I.M.: Global sensitivity indices for nonlinear mathematical models and their Monte Carlo estimates. Mathematics and Computers in Simulation 55(1-3), 271–280 (2001)
11. Sobol, I.M., Tarantola, S., Gatelli, D., Kucherenko, S., Mauntz, W.: Estimating the approximation error when fixing unessential factors in global sensitivity analysis. Reliability Engineering and System Safety 92, 957–960 (2007)
12. Sobol, I., Myshetskaya, E.: Monte Carlo Estimators for Small Sensitivity Indices. Monte Carlo Methods and Applications 13(5-6), 455–465 (2007)
13. Zlatev, Z.: Computer Treatment of Large Air Pollution Models. KLUWER Academic Publishers, Dorsrecht (1995)
14. Zlatev, Z., Dimov, I.: Computational and Numerical Challenges in Environmental Modelling. Elsevier, Amsterdam (2006)
15. http://www.wolfram.com/mathematica/

Comparison of Two Numerical Approaches to Boussinesq Paradigm Equation

Milena Dimova and Daniela Vasileva

Institute of Mathematics and Informatics, Bulgarian Acad. Sci.,
Acad. G. Bonchev str., bl.8, 1113 Sofia, Bulgaria
{mkoleva,vasileva}@math.bas.bg

Abstract. In order to study the time behavior and structural stability of the solutions of Boussinesq Paradigm Equation, two different numerical approaches are designed. The first one (A1) is based on splitting the fourth order equation to a system of a hyperbolic and an elliptic equation. The corresponding implicit difference scheme is solved with an iterative solver. The second approach (A2) consists in devising of a finite difference factorization scheme. This scheme is split into a sequence of three simpler ones that lead to five-diagonal systems of linear algebraic equations. The schemes, corresponding to both approaches A1 and A2, have second order truncation error in space and time. The results obtained by both approaches are in good agreement with each other.

1 Introduction

The aim of this paper is the numerical study of time dependent solutions of the two-dimensional Boussinesq Paradigm Equation (BPE) [1]:

$$u_{tt} = \Delta \left[u - F(u) + \beta_1 u_{tt} - \beta_2 \Delta u \right], \tag{1}$$

$$u(x,y,0) = u_0(x,y), \quad u_t(x,y,0) = u_1(x,y), \tag{2}$$

where $F(u) := \alpha u^2$ or $F(u) := \alpha(u^3 - \sigma u^5)$, $u(x,y,t)$ is the surface elevation, β_1, $\beta_2 > 0$ are two dispersion coefficients, and $\alpha > 0$ is an amplitude parameter. The main difference between (1) and the original Boussinesq equation [2] is the presence of one more term for $\beta_1 \neq 0$ called "rotational inertia".

It has been recently shown that the 2D BPE with quadratic or qubic-quintic nonlinearity admits localized solutions that propagate stationary with a prescribed phase velocity. These solutions can be constructed using either a perturbation [3, 4], finite differences [5] or Galerkin spectral method [6]. It is of utmost importance to answer the question about the structural stability of these solutions when used as initial conditions for (1). The first results on the problem with quadratic nonlinearity are reported in the pioneering work [7]. In order to investigate further the time evolution of the localized solutions, alternative techniques for (1) have to be developed. In this study we consider two different numerical approaches A1 and A2. Some results for quadratic nonlinearity using A1 have been already described in [8, 9]. Here we present another numerical approach A2 and compare the numerical results obtained by both approaches for quadratic nonlinearity as well as for qubic-quintic nonlinearity.

I. Dimov, I. Faragó, and L. Vulkov (Eds.): NAA 2012, LNCS 8236, pp. 255–262, 2013.
© Springer-Verlag Berlin Heidelberg 2013

2 Numerical Methods for Solving BPE

Approach A1 is presented in [8, 9]. Here we describe it briefly. We set

$$v(x, y, t) := u - \beta_1 \Delta u. \tag{3a}$$

Upon substituting it in (1) we get the following equation for v

$$v_{tt} = \frac{\beta_2}{\beta_1} \Delta v + \frac{\beta_1 - \beta_2}{\beta_1^2}(u - v) - \Delta F(u). \tag{3b}$$

The following implicit time stepping is designed for the system (3)

$$\frac{v_{ij}^{n+1} - 2v_{ij}^n + v_{ij}^{n-1}}{\tau^2} = \frac{\beta_2}{2\beta_1} \Lambda[v_{ij}^{n+1} + v_{ij}^{n-1}] - \Lambda G(u_{ij}^{n+1}, u_{ij}^n, u_{ij}^{n-1})$$

$$+ \frac{\beta_1 - \beta_2}{2\beta_1^2}[u_{ij}^{n+1} - v_{ij}^{n+1} + u_{ij}^{n-1} - v_{ij}^{n-1}], \tag{4a}$$

$$u_{ij}^{n+1} - \beta_1 \Lambda u_{ij}^{n+1} = v_{ij}^{n+1}, \quad i = 1, \dots, N_x, \quad j = 1, \dots, N_y. \tag{4b}$$

By u_{ij}^n and v_{ij}^n we denote a discrete approximation to u and v at (x_i, y_j, t_n), where $t_n = \tau n$ and τ is a time increment, $\Lambda = \Lambda^{xx} + \Lambda^{yy}$ stands for the difference approximation of the Laplace operator Δ on a uniform or non-uniform grid and $G(u_{ij}^{n+1}, u_{ij}^n, u_{ij}^{n-1})$ is an approximation to the nonlinear term $F(u)$. There are different possibilities to treat the nonlinear term [8, 9, 12–14]. Here we use

$$G(u_{ij}^{n+1}, u_{ij}^n, u_{ij}^{n-1}) = 2[g((u_{ij}^{n+1} + u_{ij}^n)/2) - g((u_{ij}^n + u_{ij}^{n-1})/2)]/(u_{ij}^{n+1} - u_{ij}^{n-1}), \tag{5}$$

where $g(u) = \int_0^u F(s)\,ds$. The values of the sought functions at the $(n-1)$-st and n-th time stages are considered as known when computing the $(n+1)$-st stage. The nonlinear term G is linearized using Picard method for nonlinear PDE [10], i.e., we perform successive iterations for u and v on the $(n+1)$-st stage, starting with initial condition from the already computed n-th stage. Usually 5-10 nonlinear iterations are sufficient for convergence with tolerance 10^{-16}. An energy conserving numerical approach for Boussinesq equation based on this kind of linearization was proposed and investigated in [11].

The unconditional stability of the scheme, the convergence and the conservation of the energy are shown in [15, 12, 14].

Thus, we have two *coupled* equations for the two unknown grid functions $u_{ij}^{n+1}, v_{ij}^{n+1}$. Two different grids on the computational domain $\Omega_h = [-L_1, L_1] \times [-L_2, L_2]$, uniform and non-uniform ones, are used:

$$\begin{aligned} x_i = -L_1 + ih_x, \ i = 0, \dots, N_x + 1, \ h_x = 2L_1/(N_x + 1), \\ y_j = -L_2 + jh_y, \ j = 0 \dots, N_y + 1, \ h_y = 2L_2/(N_y + 1), \ \text{or} \end{aligned} \tag{6}$$

$$x_i = \sinh[\hat{h}_x(i - n_x)], \ x_{N_x + 1 - i} = -x_i, \ i = n_x + 1, \dots, N_x + 1, \ x_{n_x} = 0,$$

$$y_j = \sinh[\hat{h}_y(j - n_y)], \ y_{N_y + 1 - j} = -y_j, \ j = n_y + 1, \dots, N_y + 1, \ y_{n_y} = 0,$$

where N_x, N_y are odd numbers, $n_x = (N_x + 1)/2$, $n_y = (N_y + 1)/2$, $\hat{h}_x = 2D_x/(N_x + 1)$, $\hat{h}_y = 2D_y(/N_y + 1)$, $D_x = \mathrm{arcsinh}(L_1)$, $D_y = \mathrm{arcsinh}(L_2)$.

The boundary conditions can be set equal to zero because of the localization of the wave profile. This forms the first set of b.c.'s used in this approach. In order to avoid the influence of the boundary, the second set of b.c.'s used here consists of the asymptotic boundary conditions formulated in [5]

$$x\frac{\partial u}{\partial x} + y\frac{\partial u}{\partial y} \approx -2u, \quad x\frac{\partial v}{\partial x} + y\frac{\partial v}{\partial y} \approx -2v, \quad \sqrt{x^2 + y^2} \gg 1. \tag{7}$$

We chose the following approximation for $(7)_1$ at the numerical infinities:

$$u_{i,N_y+1}^{n+1} = u_{i,N_y-1}^{n+1} + \frac{h_{N_y}^y + h_{N_y-1}^y}{y_{N_y}}\left[-2u_{i,N_y}^{n+1} - \frac{x_i}{h_i^x + h_{i-1}^x}(u_{i+1,N_y}^{n+1} - u_{i-1,N_y}^{n+1})\right],$$

$$u_{N_x+1,j}^{n+1} = u_{N_x-1,j}^{n+1} + \frac{h_{N_x}^x + h_{N_x-1}^x}{x_{N_x}}\left[-2u_{N_x,j}^{n+1} - \frac{y_j}{h_j^y + h_{j-1}^y}(u_{N_x,j+1}^{n+1} - u_{N_x,j-1}^{n+1})\right],$$

$i = 0, \ldots, N_x$, $j = 0, \ldots, N_y$. The implementation of $(7)_2$ is the same.

The first initial condition in (2) is approximated by $u_{ij}^0 = u_0(x_i, y_j)$. The approximations

$$(u_{ij}^1 - u_{ij}^{-1})/(2\tau) = u_1(x_i, y_j), \quad (v_{ij}^1 - v_{ij}^{-1})/(2\tau) = u_1(x_i, y_j) - \beta_1 \Delta u_1(x_i, y_j)$$

to the second initial condition are used and (4a) is modified for $n = 0$.

The coupled system of equations (4) is solved by the Bi-Conjugate Gradient Stabilized Method with ILU preconditioner [16].

Approach A2. For the discretization of (1) we use the uniform mesh (6). The suggested numerical approach is based on the weighted finite difference scheme proposed in [13, 14]

$$B\left(\frac{u_{ij}^{n+1} - 2u_{ij}^n + u_{ij}^{n-1}}{\tau^2}\right) = \Lambda u_{ij}^n - \beta_2\Lambda^2 u_{ij}^n + \Lambda G(u_{ij}^{n+1}, u_{ij}^n, u_{ij}^{n-1}). \tag{8}$$

Here $B = I - (\beta_1 + \theta\tau^2)\Lambda + \theta\tau^2\beta_2\Lambda^2$, I is the identity operator, $\Lambda^2 = (\Lambda^{xx} + \Lambda^{yy})^2$ is the discrete biLaplacian, G is defined by (5), $\theta \in \mathbb{R}$ is a parameter.

An $O(|h|^2 + \tau^2)$ approximation to the second initial condition in (2) is given by

$$u_{i,j}^1 = u_0(x_i, y_j) + \tau u_1(x_i, y_j) + \frac{\tau^2}{2(I - \beta_1\Lambda)}\left(\Lambda u_0 - \beta_2\Lambda^2 u_0 - \Lambda F(u_0)\right)(x_i, y_j).$$

In this approach we use the following boundary conditions $u_{ij}^{n+1} = 0$, $\Lambda u_{ij}^{n+1} = 0$ for $i = 0, N_x + 1$ or $j = 0, N_y + 1$. A second order of convergence in space and time and a preservation of the discrete energy for the above scheme (8) are proved in [14]. These theoretical results are confirmed numerically in the 1D case [13, 14].

The main idea of the numerical approach A2 consists in replacing the operator B in (8) by the factorized operator $\tilde{B}$, i.e. $\tilde{B} = B_1 B_2 B_3$, where

$$B_1 = (I - \theta\tau^2\Lambda^{xx} + \theta\tau^2\beta_2\Lambda^{xxxx}), B_2 = (I - \theta\tau^2\Lambda^{yy} + \theta\tau^2\beta_2\Lambda^{yyyy}), B_3 = (I - \beta_1\Lambda).$$

The factorization is based on the regularization method [17]. The form of the operator $\tilde{B}$ is proposed in [13]. In such a way we get a second order in space and time conservative factorized scheme

$$B_1 B_2 B_3 \left(\frac{u_{ij}^{n+1} - 2u_{ij}^n + u_{ij}^{n-1}}{\tau^2} \right) = \Lambda u_{ij}^n - \beta_2 \Lambda^2 u_{ij}^n + \Lambda G(u_{ij}^{n+1}, u_{ij}^n, u_{ij}^{n-1}). \quad (9)$$

It can be shown that for $\theta \geq 1/2$ the factorized scheme is unconditionally stable. The main advantage of the above factorized scheme is that it can be split into a sequence of three simpler schemes. As in approach A1 we apply Picard method for the linearization of (9). Thus at the time stage $(n + 1)$ we perform successive iterations starting with already computed solution at stage n as initial approximation. Each iteration consists of the following four steps:

- Step 1: Solve the problem for the unknown $w_{ij}^{(1)}$:

$$\begin{aligned} B_1 w_{ij}^{(1)} &= \Lambda u_{ij}^n - \beta_2 \Lambda^2 u_{ij}^n + \alpha \Lambda G(u_{ij}^{n+1}, u_{ij}^n, u_{ij}^{n-1}), & i \neq 0, N_x + 1, \\ w_{ij}^{(1)} &= 0, \quad \Lambda^{xx} w_{ij}^{(1)} = 0, & i = 0, N_x + 1. \end{aligned} \quad (10)$$

- Step 2: Define the unknown $w_{ij}^{(2)}$ as a solution of the following problem:

$$\begin{aligned} B_2 w_{ij}^{(2)} &= w_{ij}^{(1)}, & j \neq 0, N_y + 1, \\ w_{ij}^{(2)} &= 0, \Lambda^{yy} w_{ij}^{(2)} = 0, & j = 0, N_y + 1. \end{aligned} \quad (11)$$

- Step 3: Compute $w_{i,j}^{(3)}$ by solving

$$\begin{aligned} B_3 w_{ij}^{(3)} &= w_{ij}^{(2)}, & i \neq 0, N_x + 1, \quad j \neq 0, N_y + 1, \\ w_{ij}^{(3)} &= 0, & i = 0, N_x + 1 \text{ or } j = 0, N_y + 1. \end{aligned} \quad (12)$$

- Step 4: Finally, compute the solution of (9): $u_{ij}^{n+1} = 2u_{ij}^n - u_{ij}^{n-1} + \tau^2 w_{ij}^{(3)}$.

Let us emphasize that the discrete operators B_1 and B_2 are one-dimensional operators, since they depend on one spatial variable only. In such a way the solution of the first problem (10) is reduced to a sequence of 1D problems on the rows of the domain Ω_h, while for problem (11) we have a sequence of 1D problems on the columns of Ω_h. For both problems the resulting systems of linear algebraic equations are five-diagonal with constant matrix coefficients. For solving these systems we apply a special kind of nonmonotonic Gaussian elimination with pivoting [18, 19]. The third problem (12) is solved by a Conjugate Gradient type Method specially designed for the discrete Laplacian equation [20].

3   Numerical Experiments

We denote by $u^s(x, y; c)$ the best-fit approximation to the stationary translating with velocity c solution of (1), obtained in [3, 4]

$$u^s(x, y; c) = f^s(x, y) + c^2 g^s(x, y; \beta_1) + c^2 h^s(x, y; \beta_1) \cos[2 \arctan(y/x)]. \quad (13)$$

The formulas for the functions f^s, g^s, h^s can be found in [3] and [4] for the case of quadratic and qubic-quintic nonlinearity respectively. In [4] the approximations (13) are computed for two particular values of the parameter σ: $\sigma = 3/16$ and $\sigma = 0.95$. We consider equation (1) subject to the initial conditions (2)

$$u_0(x,y) := u^s(x,y;c), \qquad u_1(x,y) := -cu^s_y(x,y;c),$$

that correspond to a solution moving along the y-axis with the velocity c.

The numerical experiments are performed for $\beta_1 = 3$, $\beta_2 = 1$, $\alpha = 1$. The solutions are computed by both approaches on two different uniform grids in the domain $x, y \in [-25, 25]^2$ with 500^2, and 1000^2 grid points respectively. On the coarse grid the time increment is $\tau = 0.1$, and on the fine grid $\tau = 0.05$. In approach A2 we set the parameter $\theta = 1/2$. The solutions are also computed by A1 on a nonuniform grid in the region $[-250, 250]^2$ with 500^2 grid points and $\tau = 0.1$, as well as on the uniform grid in $[-25, 25]^2$ with 500^2 grid points and $\tau = 0.1$, using the asymptotic boundary conditions (7).

(i) The case of quadratic nonlinearity, $F(u) = \alpha u^2$. The numerical results in [7–9] show that the behaviour of the solution significantly changes when the velocity $c \in [0.2, 0.3]$. That is why we are focusing on these values of c.

Example 1. First, we present results for the case $c = 0.2$. As it is seen in Fig. 1, for $t > 8$ the solution cannot keep its form, and transforms into a propagating wave. The values of the maximum of the solution and its trajectory as function of time are also shown in Fig. 1. The notation y_{max} is used for the y-coordinate of the maximum of the solution. For $t < 8$, the solution not only moves with a velocity, close to $c = 0.2$, but also behaves like a soliton, i.e., preserves its shape, albeit its maximum decreases slightly. For larger times, the solution transforms into a diverging propagating wave with a front deformed in the direction of propagation.

The behaviour of the solution is the same on all grids and for all times steps, and does not depend on the type of the boundary conditions used (the trivial one or (7)). The approach A2 produces slightly different results for the maximum of the solution and its position on the coarse grid, but on the fine grid the results are very close to those obtained by A1.

Example 2. In Fig. 2 results for $c = 0.26$ are presented. For $t < 10$ the solution moves with a velocity, very close to $c = 0.26$, and behaves like a soliton. For larger times the solution transforms into a diverging propagating wave, except in the case of A2 on the coarser grid, where the soliton keeps its form till $t < 20$. But on the finer grid A2 leads to a solution, very close to those, produced by A1 on all grids and with both boundary conditions.

Here we do not present the results computed for $c = 0.27$ and $c = 0.28$ (some may be found in [9]). All A1 solutions and the A2 solution on the finer grid have similar behaviour – the solutions keep their form and move with the prescribed velocity till $t \approx 10$. After that they transform into diverging waves for $c = 0.27$ or blow-up for $c = 0.28$. On the coarser grid the A2 solution blows-up for $c = 0.27$.

The results from these experiments confirm once again that a mechanism for having a balance between the nonlinearity and dispersion is present, but the

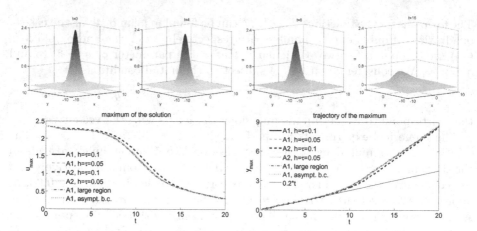

Fig. 1. Evolution of the solution for $c = 0.2$, the maximum $u(0, y_{max})$, and the trajectory of the maximum

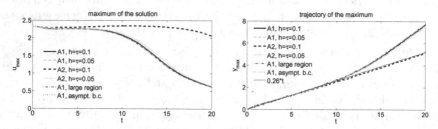

Fig. 2. $c = 0.26$, the maximum $u(0, y_{max})$, and the trajectory of the maximum

solution is not robust (even when it is stable as a time stepping process) and takes the path to the attractor presented by the propagating wave for $c \leq 0.27$ or blows-up for $c \geq 0.28$.

These results were a motivation for investigating BPE with a different non-linear term.

(ii) The case of qubic-quintic nonlinearity, $F(u) = \alpha(u^3 - \sigma u^5)$.

Example 3. Results for the case $\sigma = 3/16$ and $c = 0.3$ are presented in Fig. 3. The solution cannot keep its form even for small times, and transforms into a propagating wave, which is almost concentric for $t > 8$. The maximum of the solution moves with a velocity, much faster than $c = 0.3$. The behaviour of the solution is the same on all grids and for all times steps, and does not depend on the type of the boundary conditions used.

Example 4. The next results are for $\sigma = 3/16$ and $c = 0.6$. As it is seen in Fig. 4 they are very similar to those in the previous example, i.e., the solution cannot keep its form and transform into a diverging wave. Slightly different results are obtained for the maximum of the A2 solution on the coarse grid, but on the fine grid the results are closer to those for A1.

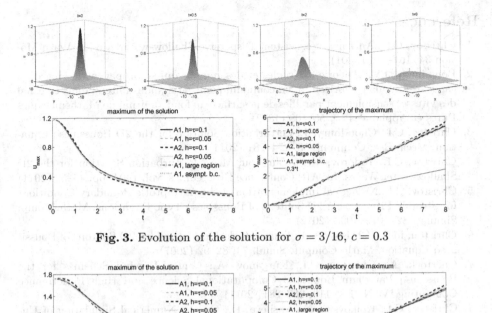

Fig. 3. Evolution of the solution for $\sigma = 3/16$, $c = 0.3$

Fig. 4. $\sigma = 3/16$, $c = 0.6$, the maximum $u(0, y_{max})$, and the trajectory of the maximum

We do not show here results for $\sigma = 0.95$, as they are very similar to the already presented results for $\sigma = 3/16$. Let us note that the investigated here 2D solutions of BPE with qubic-quintic nonlinearity do not blow-up even for larger values of c, but unfortunately they seem to be less structurally stable in comparison with the 2D solutions of BPE with quadratic nonlinearity.

4 Conclusion

We have compared the results obtained by approach A1 with these obtained by A2 for quadratic and qubic-quintic nonlinearity and have shown that they are in good agreement with each other. This fact is a good criteria for the reliability of the proposed approaches A1 and A2. In the case of quadratic nonlinearity we have confirmed the results from [7–9] – the solution preserves its shape for small times, but for larger times it either disperses in the form of decaying ring wave or blows-up. The threshold for the appearance of blow-up seems to be near $c \approx 0.28$. For qubic-quintic nonlinearity the solution does not blow-up even for relatively large values of c, but is much less stable and transforms into a diverging propagating wave.

Acknowledgment. This work has been supported by Grant DDVU02/71 from the Bulgarian National Science Fund. The article is dedicated to the memory of Professor Christo I. Christov, who initiated our collaborative research for BPE.

References

1. Christov, C.I.: An Energy-consistent Dispersive Shallow-water Model. Wave Motion 34, 161–174 (2001)
2. Boussinesq, J.V.: Théorie des ondes et des remous qui se propagent le long d'un canal rectangulaire horizontal, en communiquant au liquide contenu dans ce canal des vitesses sensiblement pareilles de la surface au fond. Journal de Mathématiques Pures et Appliquées 17, 55–108 (1872)
3. Christov, C.I., Choudhury, J.: Perturbation Solution for the 2D Boussinesq Equation. Mech. Res. Commun. 38, 274–281 (2011)
4. Christov, C.I., Todorov, M.T., Christou, M.A.: Perturbation Solution for the 2D Shallow-water Waves. In: AIP Conference Proceedings, vol. 1404, pp. 49–56 (2011)
5. Christov, C.I.: Numerical Implementation of the Asymptotic Boundary Conditions for Steadily Propagating 2D Solitons of Boussinesq Type Equations. Math. Comp. Simulat. 82, 1079–1092 (2012)
6. Christou, M.A., Christov, C.I.: Fourier-Galerkin Method for 2D Solitons of Boussinesq Equation. Math. Comput. Simul. 74, 82–92 (2007)
7. Chertock, A., Christov, C.I., Kurganov, A.: Central-Upwind Schemes for the Boussinesq Paradigm Equation. Computational Science and High Performance Computing IV, NNFM 113, 267–281 (2011)
8. Christov, C.I., Kolkovska, N., Vasileva, D.: On the Numerical Simulation of Unsteady Solutions for the 2D Boussinesq Paradigm Equation. In: Dimov, I., Dimova, S., Kolkovska, N. (eds.) NMA 2010. LNCS, vol. 6046, pp. 386–394. Springer, Heidelberg (2011)
9. Christov, C.I., Kolkovska, N., Vasileva, D.: Numerical Investigation of Unsteady Solutions for the 2D Boussinesq Paradigm Equation. 5th Annual Meeting of the Bulgarian Section of SIAM. In: BGSIAM 2010 Proceedings, pp. 11–16 (2011)
10. Ames, W.F.: Nonlinear Partial Differential Equations in Engineering. Academic Press (1965)
11. Christov, C.I., Velarde, M.G.: Inelastic interaction of Boussinesq solitons. J. Bifurcation & Chaos 4, 1095–1112 (1994)
12. Kolkovska, N.: Convergence of Finite Difference Schemes for a Multidimensional Boussinesq Equation. In: Dimov, I., Dimova, S., Kolkovska, N. (eds.) NMA 2010. LNCS, vol. 6046, pp. 469–476. Springer, Heidelberg (2011)
13. Dimova, M., Kolkovska, N.: Comparison of Some Finite Difference Schemes for Boussinesq Paradigm Equation. In: Adam, G., Buša, J., Hnatič, M. (eds.) MMCP 2011. LNCS, vol. 7125, pp. 215–220. Springer, Heidelberg (2012)
14. Kolkovska, N., Dimova, M.: A New Conservative Finite Difference Scheme for Boussinesq Paradigm Equation. Cent. Eur. J. Math. 10(3), 1159–1171 (2012)
15. Kolkovska, N.: Two Families of Finite Difference Schemes for Multidimensional Boussinesq Equation. In: AIP Conference Series, vol. 1301, pp. 395–403 (2010)
16. van der Vorst, H.: Iterative Krylov Methods for Large Linear Systems. Cambridge Monographs on Appl. and Comp. Math. 13 (2009)
17. Samarskii, A.: The Theory of Difference Schemes. Marcel Dekker Inc. (2001)
18. Samarskii, A.A., Nikolaev, E.: Numerical Methods for Grid Equations. Birkhäuser Verlag (1989)
19. Christov, C.I.: Gaussian Elimination with Pivoting for Multidiagonal Systems. Internal Report, University of Reading 4 (1994)
20. Samarskii, A.A., Vabishchevich, P.N.: Numerical Methods for Solving Inverse Problems of Mathematical Physics. Walter de Gruyter (2007)

Modal Properties of Vertical Cavity Surface Emitting Laser Arrays under the Influence of Thermal Lensing

N.N. Elkin, A.P. Napartovich, and D.V. Vysotsky

State Science Center Troitsk Institute for Innovation and Fusion Research(TRINITI),
142190, Troitsk, Moscow Region, Russia
elkin@triniti.ru

Abstract. Modal behavior of an 5 × 5 array of vertical cavity surface emitting lasers (VCSEL) was studied numerically. Thermal lensing was simulated by the temperature profile set as a quadratic function of a polar radius. Mathematical formulation of the problem consists of self-consistent solution of the 3D Helmholtz wave equation and 2D non-linear diffusion equation as the material equation of the active laser medium. Complete formulation of the problem contains boundary conditions and an eigenvalue to be determined. Bidirectional beam propagation method was taken as a basis for numerical algorithms. Above-threshold operation of laser array was simulated using round-trip iterations similar to the Fox-Li method. In addition, the Arnoldi algorithm was implemented to find several high-order optical modes in a VCSEL array with gain and index distributions established by the oscillating mode.

1 Introduction

Coupled Vertical Cavity Surface-Emitting Laser (VCSEL) arrays are attractive means to increase the coherent output power of VCSELs. Thermal lensing is a serious obstacle for achievement of high-power single-mode laser output. Numerical study gives the opportunity to find optimal parameter of a laser device under thermal lensing effect. Modeling of VCSEL arrays represents a very difficult computational problem because of complicated geometry and non-linear partial differential equations containing eigenvalues. The traditional for optical resonators Fox-Li iteration method [1] is inapplicable owing to dispersion effects. We propose an efficient computational algorithm using moderate computational resources. The algorithm is based on the bidirectional beam propagation method (BiBPM) [2] and the modified round-trip operator technique [3].

2 Basic Equations and Boundary Conditions

The 5 × 5 VCSEL array [4] is schematically shown in Fig. 1. The device consists of a one-wave cavity placed between top and bottom distributed Bragg reflectors (p-DBR and n-DBR in Fig. 1). DBRs consist of several pairs of quarter-wave

I. Dimov, I. Faragó, and L. Vulkov (Eds.): NAA 2012, LNCS 8236, pp. 263–270, 2013.

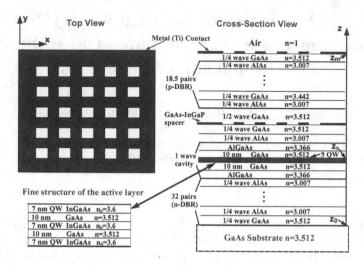

Fig. 1. Scheme of the VCSEL array in two projections

layers of GaAs and AlAs. The active layer shown as a black strip consists of $N_q = 3$ InGaAs quantum wells (QW) separated by GaAs layers. The QW has thickness 7 nm. The VCSEL array is mounted on the GaAs substrate which is unbounded from the point of view of mathematical modeling. Similarly, an unbounded layer of air must be accounted on top of a VCSEL. There exist some additional layers shown in Fig. 1. Geometrical thicknesses h_k of the layers presented in Fig. 1 are calculated by the formula: $h_k n = p\lambda_0$, where n is the refractive index, p is the thickness as a fraction of wavelength. The approximate value of wavelength $\lambda_0 = 980$ nm. This value is named the reference wavelength. The index and absorption are constant in each layer except the QWs. The metal (Ti) plate and GaAs-InGaP spacer have the 5×5 array of windows. Directing z-axis perpendicularly to the substrate surface we represent the VCSEL array as a pile of plane layers: $\{[z_{k-1}, z_k], k = 1, \ldots, m\}$, where m is the total number of layers, $\{z_k, k = 0, \ldots, m\}$ is an ascending sequence of coordinates of layer interfaces, $h_k = z_k - z_{k-1}$ is thickness of the k-th layer.

We start from Maxwell equations and assume that the polarization effects can be neglected and the scalar diffraction theory is applicable. Laser modes have a time dependence of the form $E(x, y, z, t) = U(x, y, z) \exp(-i\Omega t)$, $\Omega = \omega_0 + \Delta\omega - i\delta$, where ω_0 is the reference frequency, $\Delta\omega = \omega - \omega_0$ is the frequency shift and δ is the attenuation factor. The reference frequency, wavenumber and wavelength obey the standard relations: $\omega_0 = k_0 c$, $k_0 = 2\pi/\lambda_0$. Introducing new variables $\mu = 2\delta/c$, $\Delta k = \Delta\omega/c$, $\beta = \mu + i2\Delta k$, 3D Helmholtz wave equation reads

$$\frac{\partial^2 U}{\partial z^2} + Q^2 U = 0, \qquad Q = \sqrt{k_0^2 n^2 - i k_0 n g - i k_0 n^2 \beta + \frac{\partial^2}{\partial x^2} + \frac{\partial^2}{\partial y^2}}, \qquad (1)$$

with a complex eigenvalue β. Here n and g are index and gain respectively, Q is the operator of longitudinal wavenumber.

There is need to define the boundary conditions. We use condition of continuity for the wave field U and its normal derivative at the interfaces between adjoint layers. We use the Fresnel formulae for reflection and transmission of a plane wave from an interface as boundary conditions at the top of the device ($z = z_m$). The Dirichlet boundary condition $U \equiv 0$ were used at the substrate boundary $z = z_0$ because of large number of layers in the bottom DBR. Two options of conditions on lateral boundaries were implemented: 1) non-reflecting boundaries, 2) fully reflecting boundaries. In the first case absorbing boundary conditions [5] were used. The Dirichlet boundary conditions $U \equiv 0$ for the wave field $U(x, y, z)$ were used in the second case.

The 2D non-linear diffusion equation [6]:

$$\frac{\partial^2 Y}{\partial x^2} + \frac{\partial^2 Y}{\partial y^2} - \frac{Y}{D\tau_{nr}} - \frac{B}{D} N_{tr} Y^2 - \frac{|U|^2 \ln(\chi(Y))}{D\tau_{nr}} = -\frac{J}{N_q e D d N_{tr}}, \quad (2)$$

is correct for normalized carrier density $Y = N/N_{tr}$ at the QW. Here N is the charge carrier density, D is the diffusion coefficient, τ_{nr} is the recombination time, B is the spontaneous emission coefficient, d is thickness of the QW, e is the elementary charge, $N_{tr} = \left(-1/\tau_{nr} + \sqrt{1/\tau_{nr}^2 + 4BJ_{tr}/(ed)}\right)/(2B)$ is the carrier density for conditions of transparency, J_{tr} is the drive current density at transparency conditions, $|U|^2 = I/I_s$, I is the light intensity, $I_s = (\hbar c k_0 N_{tr})/(g_0 \tau_{nr})$ is the intensity of saturation. The drive current density is specified by the formula $J = \kappa J_{tr} f(x, y)$, where κ is the pump level, $f(x, y) = \exp\left(-(2x/R)^{16}\right) \exp\left(-(2y/R)^{16}\right)$ is the pump profile function, R is the transverse size of the current channel. Zero boundary conditions for $Y(x, y)$ are set at the lateral boundaries of the active layer. The function $\chi(Y)$, gain and index at the active layers are approximated by the formulae [7]. The temperature profile is specified by the quadratic function of the polar radius. As a result we have the formulae:

$$\chi(Y) = \begin{cases} \alpha + (1 - \alpha)Y^{1/(1-\alpha)}, & Y < 1 \\ Y, & Y \geq 1 \end{cases}, \quad g = g_0 \ln(\chi(Y)), \quad (3)$$

$$n = n_0 - \frac{F(g - g_{\min})}{2k_0} + \nu T, \quad T = T_0 \left(1 - (x^2 + y^2)/2R_0^2\right),$$

where $\alpha = \exp(g_{\min}/g_0)$, g_0 is the gain parameter, n_0 is the refractive index in the absence of carriers, F is the line enhancement factor, $g_{\min}$ is the minimum gain, ν is the thermo-optic coefficient, T is the increment of temperature, R_0 is the curvature radius of the temperature profile. We suppose that T_0 is proportional to the pump level, $T_0 = \theta\kappa$. The equation (2) is true for each of QWs because the wave field intensity $|U|^2$ is approximately the same in all of QWs.

3   Statement of the Problem and Numerical Algorithms

The problem consists of self-consistent solving of the wave field equation and material equations in order to find the magnitude and spatial profile of a laser

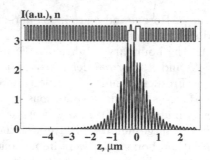

Fig. 2. Longitudinal profiles of the lasing mode and index

electromagnetic field and its frequency in steady-state mode of operation. The equation (1) jointly with the equations (2) and (3) supplemented with corresponding boundary conditions form the non-linear eigenvalue problem. The supplementary condition $\delta = 0$ ($\mathrm{Re}(\beta) = 0$) is required for steady-state operation.

We reformulate the problem using the bi-directional beam propagation method (BiBPM) [2]. The wave field can be represented in each z- invariant segment as the sum of the upward and downward propagating waves:

$$U = V^+ + V^-$$ (4)

The BiBPM is based on the principle that the wave fields in two arbitrary planes $z = \text{const}$ designated by symbols t and b are coupled by a transfer equation:

$$\left(V_t^+, \; V_t^-\right) = \mathbf{M} \left(V_b^+, \; V_b^-\right)$$ (5)

where $\mathbf{M}$ is a 2×2 transfer operational matrix.

The transfer matrix for a set of uniform layers is a product of the elementary interface and propagation matrices:

$$T_k = \frac{1}{2} \begin{pmatrix} 1 + Q_{k+1}^{-1}Q_k & 1 - Q_{k+1}^{-1}Q_k \\ 1 - Q_{k+1}^{-1}Q_k & 1 + Q_{k+1}^{-1}Q_k \end{pmatrix}, \quad P_k = \begin{pmatrix} \exp(iQ_kh_k) & 0 \\ 0 & \exp(-iQ_kh_k) \end{pmatrix},$$

where h_k is thickness of the k-th layer, Q_k is the operator of longitudinal wavenumber in the k-th layer.

We can replace the Helmholtz equation (1) with the equivalent system of transfer equations in the form of (5) covering all the layers. Then we choose the reference plane $z = z_r$, e.g. the plane of the active layer (see Fig. 1). Let $u(x, y) = V^+$ is the upward propagating wave outgoing from the reference plane. We can make a virtual round trip of the VCSEL using the transfer equations (5). Firstly, starting with the wave u we calculate the downward propagating wave $v(x, y) = V^-$ falling to the reference plane using the system of equations (5) connecting the reference plane and the plane under output windows ($z = z_m$) and the Fresnel boundary conditions. Then, we substitute the wave $v(x, y)$ into the system of equations (5) connecting the reference plane and the plane under bottom DBR ($z = z_0$). Solving this system with the boundary condition $U \equiv 0$ at $z = z_0$ we find the upward propagating wave $\tilde{u}$ at the reference plane. The wave field is reproduced after the round trip, if it satisfies the Helmholtz equation. This round-trip condition has a following form:

$$\mathbf{P}(g, n, \beta)u = \gamma u \qquad (6)$$

where $\gamma = 1$, $\mathbf{P}(g, n, \beta)u = \tilde{u}$ by definition of the round-trip operator.

Our approach consists in solution of the problem in the form of (6) for a function u and eigenvalue γ to be found provided the value of β is specified. The value β is adjusted until $\gamma = 1$ within a certain tolerance. As was shown formerly [3] for the one-wave cavity containing sufficiently thin active layer the solution of more simple problem

$$\mathbf{P}(g, n, 0)u = \gamma u \qquad (7)$$

gives good approach to (6) with $\gamma = \exp(-\beta L_e)$, where L_e is the effective length.

Equation (7) represents the eigenvalue problem for a non-linear operator because gain g and index n are determined by equations (2), (3) and depend on u. This problem is solved by the Fox-Li iteration method [1] which is schematically represented by the following diagram:

$$\begin{bmatrix} u \\ g, n \end{bmatrix} \Longrightarrow \begin{bmatrix} \tilde{u} = \mathbf{P}(g, n, 0)u \\ U \end{bmatrix} \Rightarrow \begin{bmatrix} \psi = \arg\left(\tilde{u}(x_0, y_0)\right) \end{bmatrix} \Rightarrow \begin{bmatrix} u = \exp(-i\psi)\tilde{u} \\ g, n \end{bmatrix}$$

where (x_0, y_0) is an arbitrary location in the field location area. The wave amplitude U is calculated as sum of the counterpropagating waves according to (4). If convergence is achieved we have obtained the eigen-pair: $\{u(x, y), \gamma = \exp(i\psi)\}$.

To analyze stability of the lasing mode we consider also the linear eigenvalue problem (7) when gain g and index n are established by lasing mode and "frozen". One of the solutions is the operating mode having $|\gamma| = 1$ ($\mathrm{Re}(\beta) = 0$). If all other solutions have eigenvalues satisfying the condition $|\gamma| < 1$ ($\mathrm{Re}(\beta) > 0$) then the single-mode operation is stable because all other modes decay in time. If at least one of the eigen-modes satisfies the condition $|\gamma| > 1$ ($\mathrm{Re}(\beta) < 0$), then single-mode lasing is instable. We have the linear non-hermitian eigenvalue problem (7) in this case. Only several eigenpairs (u, γ) are required, i.e. we have the partial eigenvalue problem.

Doing calculations for the uniform layers, the wave field was projected on the uniform mesh in (x, y)-plane and transformed by the fast Fourier transform (FFT) algorithm (sine-FFT in case of reflecting boundaries). In this case, the operators T_k and P_k turn into numerical matrices in the wavenumber space. To calculate the wave field propagation within non-uniform layers the inverse FFT was made at the boundaries of each layer. Then approach of locally uniform wave field was used, i.e. the spatial derivatives in the operators Q_k were neglected. This approach is permissible since thickness of the non-uniform layer is far less than the wavelength. The non-linear diffusion equation (2) was solved by the second-order finite-difference scheme using an iterative procedure.

In conclusion, we note that in such approach the numerical mesh over z-variable is not required. Usage of the FFT algorithm and a transfer-matrix approach for spectral components within plane layers structure give us the very fast algorithm for round-trip evaluation. The partial eigenvalue problem (7) for case of "frozen" medium can be solved efficiently by the Arnoldi method [8]. Calculation of all elements of matrix of $\mathbf{P}(g, n, \beta)$ is not required. It is necessary to calculate elements of vector $\mathbf{P}(g, n, \beta)u$ only.

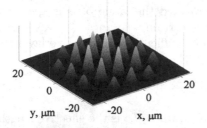

Fig. 3. Transverse profile of the lasing mode. Absorbing lateral boundaries.

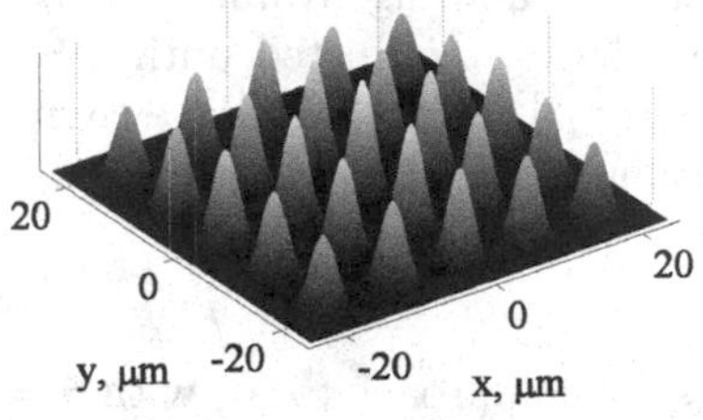

Fig. 4. Transverse profile of the lasing mode. Reflecting lateral boundaries.

4   Results and Discussion

The sizes of output windows are $5\,\mu$m, the sizes of spacer elements are $6\,\mu$m, the distances between neighbor elements are $3\,\mu$m. The absorption of spacer layer are specified so as product of absorbing coefficient and thickness was 0.0224. The top Bragg reflector consist of 16.5 pairs of layers for absorbing lateral boundaries and of 18.5 pairs of layers for reflecting lateral boundaries. These numbers of layers were chosen in order to reach maximum output power. The other parameters were given as follows: $D = 50\,\mathrm{cm^2s^{-1}}$, $\tau_{nr} = 10^{-9}\,$s, $B = 1.0 \times 10^{-10}\,\mathrm{cm^3s^{-1}}$, $J_{tr} = 50\,\mathrm{A\,cm^{-2}}$, $R = 44\,\mu$m, $F = 1$, $\theta = 0.083\,$K, $R_0 = 21.2\,\mu$m, $g_0 = 4400\,\mathrm{cm^{-1}}$, $n_0 = 3.6$, $g_{min} = -4400\,\mathrm{cm^{-1}}$, $\nu = 3 \times 10^{-4}\mathrm{K^{-1}}$. In case of reflection lateral boundaries their positions were set by formulae: $x = \pm L/2$, $y = \pm L/2$, where $L = 48\,\mu$m.

One more important approach was done in our numerical model. Thermal lensing zone is determined by overlapping of the heated region and location of the wave field intensity. We compress the thermal lensing zone into the active layer with renormalization of the thermo-optic coefficient: $\nu' = \nu h_l/h_a$ where $h_a = 41\,$nm is the thickness of the active layer, h_l is the effective thickness of the thermal lens to be estimated. Typical longitudinal profile of intensity and refractive index (step-wise function) are shown in Fig. 2. Usually, the heated region covers the field intensity location. The effective thickness of the thermal lens was set $h_l = 2\,\mu$m.

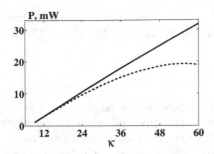

Fig. 5. Output power vs pump level κ. Absorbing lateral boundaries. Solid line - no thermal effect, dashed line - taking into account thermal effect.

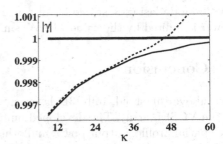

Fig. 6. Modulus of eigenvalues. Thick line - lasing mode ($|\gamma| \equiv 1$), solid thin line - competing mode (no thermal effect), dashed line - competing mode (taking into account thermal effect).

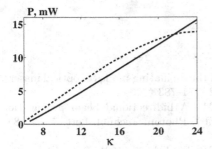

Fig. 7. Output power vs pump level κ. Reflecting lateral boundaries. Solid line - no thermal effect, dashed line - taking into account thermal effect.

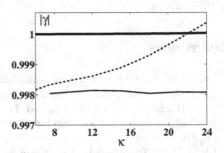

Fig. 8. Modulus of eigenvalues. Thick line - lasing mode ($|\gamma| \equiv 1$), solid thin line - competing mode (no thermal effect), dashed line - competing mode (taking into account thermal effect).

Typically, calculations were made using 252×252 mesh nodes in the (x, y)-plane. To estimate the error of discretization we have verified some calculations using 504×504 mesh. The relative change of results was approximately 10^{-2}.

The wave field intensities at the active layer under a weak pump current are presented in Figs. 3-4. Reflecting boundaries are more preferable for high output power because of uniform distribution of peaks of the light intensity.

Aggregate results for case of absorbing boundaries are presented in Figs. 5-6. In the absence of thermal effect the output power grows linearly with pump level, rate of growth $P'(\kappa) = 0.6\,\text{mW}$. Single-mode lasing is stable in all range of κ. On the contrary, under thermal effect the output power is limited by the value $19\,\text{mW}$, single-mode lasing is stable at $\kappa < 46$ and unstable at $\kappa > 46$.

Analogous results for case of reflecting boundaries are presented in Figs. 7-8. In the absence of thermal effect the output power grows linearly, rate of growth $P'(\kappa) = 0.88\,\text{mW}$ what considerably exceeds the rate for previous case.

Single-mode lasing is stable in all range of κ. Under thermal effect the output power is limited by the value 13 mW, single-mode lasing is stable at $\kappa < 22$ only.

5   Conclusion

The above-threshold, fully 3D laser-simulation program is employed for study of the VCSEL array. The developed numerical algorithms allows us to calculate the spatial profile, output power and other characteristics of an oscillating mode. To examine a single-mode operation, the numerical code was developed, which calculates a set of competing modes using gain and index variations produced by the oscillating mode. Strong influence of thermal effect on the output power and stability of single-mode lasing was found by means of numerical modeling.

Acknowledgments. Work is partially supported by the RFBR project No. 11-02-00298-a.

References

1. Fox, A.G., Li, T.: Effect of gain saturation on the oscillating modes of optical masers. IEEE Journal of Quantum Electronics QE-2, 774–783 (1966)
2. Rao, H., Scarmozzino, R., Osgood Jr., R.M.: A bidirectional beam propagation method for multiple dielectric interfaces. IEEE Photon. Technol. Lett. 7, 830–832 (1999)
3. Elkin, N.N., Napartovich, A.P., Vysotsky, D.V., Troshchieva, V.N.: Round-trip operator technique applied for optical resonators with dispersion elements. In: Boyanov, T., Dimova, S., Georgiev, K., Nikolov, G. (eds.) NMA 2006. LNCS, vol. 4310, pp. 542–549. Springer, Heidelberg (2007)
4. Bao, L., Kim, N.-H., Mawst, L.J., et al.: Modeling, Fabrication, and Characterization of Large Aperture Two-Dimensional Antiguided Vertical-Cavity Surface-Emitting Laser Arrays. IEEE Journal of Selected Topics in Quantum Electronics 11, 968–981 (2005)
5. Kosloff, R., Kosloff, D.: Absorbing boundaries for wave propagation problem. Journal of Computational Physics 63, 363–376 (1986)
6. Hadley, G.R.: Modeling of diode laser arrays. In: Botez, D., Scifres, D.R. (eds.) Diode Laser Arrays, ch. 4, pp. 1–72. Cambridge Univ. Press, Cambridge (1994)
7. Elkin, N.N., Napartovich, A.P., Sukharev, A.G., Vysotsky, D.V.: 3D modelling of diode laser active cavity. In: Li, Z., Vulkov, L.G., Waśniewski, J. (eds.) NAA 2004. LNCS, vol. 3401, pp. 272–279. Springer, Heidelberg (2005)
8. Saad, Y.: Numerical Methods for Large Eigenvalue Problem. Manchester University Press, Manchester (1992)

Stability of Implicit Difference Scheme for Solving the Identification Problem of a Parabolic Equation

Abdullah Said Erdogan and Allaberen Ashyralyev

Department of Mathematics, Fatih University, Istanbul, Turkey
{aserdogan,aashyr}@fatih.edu.tr

Abstract. We consider the inverse problem of reconstructing the right side of a parabolic equation with an unknown time dependent source function. Numerical solution and well-posedness of this type problem with local boundary conditions considered previously by A.A. Samarskii, P.N. Vabishchevich and V.T. Borukhov. In this paper, we focus on studying the stability of the problem with nonlocal conditions. A stable algorithm for the approximate solution of the problem is presented.

Keywords: Parabolic equations, identification problem, stability analysis, implicit difference scheme.

1 Introduction

Inverse problems take an important place in many branches of science and engineering and have been studied by different authors [1–9]. In this article, we deal with an inverse problem of reconstructing the right hand side (RHS) of a parabolic equation arising in heat transfer. The reader is referred to [5] and the references therein for a short discussion on RHS identification problems. The importance of well-posedness in the field of partial differential equations is well known (see [10–16]). Moreover, the well-posedness of the RHS identification problem for a parabolic equation where the unknown function p is in space variable is also well-investigated [17–19].

The inverse problem of reconstructing a RHS of a parabolic equation with local boundary conditions is investigated in [5] and [20]. The numerical solution of the identification problem and well-posedness of the algorithm are presented. For reconstructing the right hand side function $f(t, x) = p(t) q(x)$ where $p(t)$ is the unknown function, the solution is observed in the form of $u(t, x) = \eta(t) q(x) + w(t, x)$ where $\eta(t) = \int_0^t p(s)\, ds$. Then, an approximation is given for $w(t, x)$ via fully implicit difference scheme.

In this paper, we consider the inverse problem of reconstructing the right side of a parabolic equation with nonlocal conditions

I. Dimov, I. Faragó, and L. Vulkov (Eds.): NAA 2012, LNCS 8236, pp. 271–278, 2013.

$$\begin{cases} \dfrac{\partial u\left(t,x\right)}{\partial t} = a(x)\dfrac{\partial^2 u\left(t,x\right)}{\partial x^2} - \sigma u\left(t,x\right) + p\left(t\right)q\left(x\right) + f\left(t,x\right), \\ 0 < x < l,\ 0 < t \leq T, \\ u\left(t,0\right) = u\left(t,l\right), u_x\left(t,0\right) = u_x\left(t,l\right), 0 \leq t \leq T, \\ u\left(0,x\right) = \varphi\left(x\right),\ 0 \leq x \leq l, \\ u\left(t,x^*\right) = \rho\left(t\right),\ 0 \leq x^* \leq l, 0 \leq t \leq T, \end{cases} \qquad (1)$$

where $u\left(t,x\right)$ and $p\left(t\right)$ are unknown functions, $a\left(x\right) \geq \delta > 0$ and $\sigma > 0$ is a sufficiently large number with assuming that $q\left(x\right)$ is a sufficiently smooth function, $q\left(0\right) = q\left(l\right)$ and $q'\left(0\right) = q'\left(l\right)$ and $q\left(x^*\right) \neq 0$.

2   Difference Scheme

For the approximate solution of the problem (1), the purely implicit difference scheme

$$\begin{cases} \dfrac{u_n^k - u_n^{k-1}}{\tau} = a(x_n)\dfrac{u_{n+1}^k - 2u_n^k + u_{n-1}^k}{h^2} - \sigma u_n^k + p^k q_n + f\left(t_k, x_n\right), \\ p^k = p\left(t_k\right), q_n = q\left(x_n\right), x_n = nh,\ t_k = k\tau, \\ 1 \leq k \leq N, 1 \leq n \leq M-1, Mh = l, N\tau = T, \\ u_0^k = u_M^k, -3u_0^k + 4u_1^k - u_2^k = u_{M-2}^k - 4u_{M-1}^k + 3u_M^k, 0 \leq k \leq N, \\ u_n^0 = \varphi\left(x_n\right),\ 0 \leq n \leq M, \\ u^k\Big|_{\left\lfloor \frac{x^*}{h} \right\rfloor} = u_s^k = \rho\left(t_k\right), 0 \leq k \leq N, 0 \leq s \leq M \end{cases} \qquad (2)$$

is constructed. Here, $q_s \neq 0, q_0 = q_M$ and $-3q_0 + 4q_1 - q_2 = q_{M-2} - 4q_{M-1} + 3q_M$ are assumed. $\left\lfloor \frac{x^*}{h} \right\rfloor$ denotes the greatest integer less than or equal $\frac{x^*}{h}$.

Let A be a strongly positive operator. With the help of A, we introduce the fractional space $E'_\alpha\left(E, A\right), 0 < \alpha < 1$, consisting of all $v \in E$ for which the following norm is finite:

$$\|v\|_{E'_\alpha} = \sup_{\lambda > 0} \left\| \lambda^\alpha A\left(\lambda + A\right)^{-1} v \right\|_E.$$

To formulate our results, we introduce the Banach space $C_h^\alpha = C^\alpha\left[0,l\right]_h, \alpha \in \left(0,1\right)$, of all grid functions $\phi^h = \{\phi_n\}_{n=1}^{M-1}$ defined on

$$\left[0,l\right]_h = \{x_n = nh, 0 \leq n \leq M, Mh = l\}$$

with $\phi_0 = \phi_M$ equipped with the norm

$$\|\phi_h\|_{C_h^\alpha} = \|\phi_h\|_{C_h} + \sup_{1 \leq n < n+r \leq M} |\phi_{n+r} - \phi_n| \left(rh\right)^{-\alpha},$$

$$\|\phi_h\|_{C_h} = \max_{1 \leq n \leq M} |\phi_n|.$$

Moreover, $C_\tau(E) = C([0,T]_\tau, E)$ is the Banach space of all grid functions $\phi^\tau = \{\phi(t_k)\}_{k=1}^{N-1}$ defined on $[0,T]_\tau = \{t_k = k\tau, 0 \le k \le N, Nh = T\}$ with values in E equipped with the norm

$$\|\phi^\tau\|_{C_\tau(E)} = \max_{1 \le k \le N} \|\phi(t_k)\|_E .$$

Then, the following theorem on well-posedness of problem (2) is established.

Theorem 1. *For the solution of problem (2), coercive stability estimates*

$$\left\|\left\{\frac{u_k^h - u_{k-1}^h}{\tau}\right\}_{k=1}^N\right\|_{C_\tau(C_h^{2\alpha})} + \left\|\{D_h^2 u_k^h\}_{k=1}^N\right\|_{C_\tau(C_h^{2\alpha})}$$

$$\le M(q,s) \left\|\left\{\frac{\rho(t_k) - \rho(t_{k-1})}{\tau}\right\}_{k=1}^N\right\|_{C[0,T]_\tau}$$

$$+ M(\tilde{a}, \phi, \alpha, T)\left(\|D_h^2\varphi^h\|_{C_h^{2\alpha}} + \left\|\{f^h(t_k)\}_{k=1}^N\right\|_{C_\tau(C_h^{2\alpha})} + \|\rho^\tau\|_{C[0,T]_\tau}\right),$$

$$\|p^\tau\|_{C[0,T]_\tau} \le M(q,s) \left\|\left\{\frac{\rho(t_k) - \rho(t_{k-1})}{\tau}\right\}_{k=1}^N\right\|_{C[0,T]_\tau}$$

$$+ M(\tilde{a}, \phi, \alpha, T)\left[\|D_h^2\varphi^h\|_{C_h^{2\alpha}} + \left\|\{f^h(t_k)\}_{k=1}^N\right\|_{C_\tau(C_h^{2\alpha})} + \|\rho^\tau\|_{C[0,T]_\tau}\right]$$

hold. Here,

$$f^h(t_k) = \{f(t_k, x_n)\}_{n=1}^{M-1}, \varphi^h = \{\varphi(x_n)\}_{n=1}^{M-1}, \rho^\tau = \{\rho(t_k)\}_{k=0}^N,$$

$$D_h^2 u^h = \left\{\frac{u_{n+1} - 2u_n + u_{n-1}}{h^2}\right\}_{n=1}^{M-1}, \tilde{a} = \frac{1}{q_s}\left(aD_h^2 q^h - \sigma q^h\right).$$

Proof. We search the solution of problem (2) in the form

$$u_n^k = \eta^k q_n + w_n^k, \tag{3}$$

where

$$\eta^k = \sum_{i=1}^k p^i \tau, 1 \le k \le N, \eta^0 = 0. \tag{4}$$

From equation (3) taking difference derivatives, we get

$$\frac{u_n^k - u_n^{k-1}}{\tau} = \frac{\eta^k - \eta^{k-1}}{\tau} q_n + \frac{w_n^k - w_n^{k-1}}{\tau} = p^k q_n + \frac{w_n^k - w_n^{k-1}}{\tau}$$

and

$$\frac{u_{n+1}^k - 2u_n^k + u_{n-1}^k}{h^2} = \eta^k \frac{q_{n+1} - 2q_n + q_{n-1}}{h^2} + \frac{w_{n+1}^k - 2w_n^k + w_{n-1}^k}{h^2}$$

for any $n, 1 \leq n \leq M - 1$. Moreover for the interior grid point u_s^k, we have that

$$u_s^k = \eta^k q_s + w_s^k = \rho(t_k)$$

and

$$\eta^k = \frac{\rho(t_k) - w_s^k}{q_s}. \tag{5}$$

From the last equality, taking the difference derivative it follows that

$$p^k = \frac{1}{q_s} \left(\frac{\rho(t_k) - \rho(t_{k-1})}{\tau} - \frac{w_s^k - w_s^{k-1}}{\tau} \right). \tag{6}$$

Using the triangle inequality, we get

$$|p^k| \leq M(q, s) \left(\left| \frac{\rho(t_k) - \rho(t_{k-1})}{\tau} \right| + \left| \frac{w_s^k - w_s^{k-1}}{\tau} \right| \right)$$

$$\leq M(q, s) \left(\max_{1 \leq k \leq N} \left| \frac{\rho(t_k) - \rho(t_{k-1})}{\tau} \right| + \max_{1 \leq k \leq N} \max_{0 \leq s \leq M} \left| \frac{w_s^k - w_s^{k-1}}{\tau} \right| \right)$$

$$\leq M(q, s) \left(\max_{1 \leq k \leq N} \left| \frac{\rho(t_k) - \rho(t_{k-1})}{\tau} \right| + \max_{1 \leq k \leq N} \left\| \frac{w_k^h - w_{k-1}^h}{\tau} \right\|_{C_h^{2\alpha}} \right) \tag{7}$$

for any $k, 1 \leq k \leq N$.

Here, $\{w_k^h\}_{k=0}^N$ is the solution of the following difference scheme

$$\begin{cases} \dfrac{w_n^k - w_n^{k-1}}{\tau} = a(x_n) \dfrac{w_{n+1}^k - 2w_n^k + w_{n-1}^k}{h^2} \\[2mm] +a(x_n) \dfrac{\rho(t_k) - w_s^k}{q_s} \dfrac{q_{n+1} - 2q_n + q_{n-1}}{h^2} \\[2mm] -\sigma \dfrac{\rho(t_k) - w_s^k}{q_s} q_n - \sigma w_n^k + f(t_k, x_n), x_n = nh, \ t_k = k\tau, \\[2mm] 1 \leq k \leq N, 1 \leq n \leq M - 1, Mh = l, N\tau = T, \\[2mm] w_0^k = w_M^k, -3w_0^k + 4w_1^k - w_2^k = w_{M-2}^k - 4w_{M-1}^k + 3w_M^k, 0 \leq k \leq N, \\[2mm] w_n^0 = \varphi(x_n), \ 0 \leq n \leq M. \end{cases} \tag{8}$$

Therefore, the end of proof of Theorem 1 is based on inequality (7) and the following theorem.

Theorem 2. *For the solution of problem (8), the coercive stability estimate*

$$\left\| \left\{ \frac{w_k^h - w_{k-1}^h}{\tau} \right\}_{k=1}^N \right\|_{C_\tau(C_h^{2\alpha})} \leq M(\tilde{a}, \phi, \alpha, T)$$

$$\times \left(\left\| \varphi^h \right\|_{C_h^{2\alpha}} + \left\| \{ f^h (t_k) \}_{k=1}^N \right\|_{C_\tau \left(C_h^{2\alpha} \right)} + \left\| \rho^\tau \right\|_{C[0,T]_\tau} \right)$$

holds.

Proof. We can rewrite difference scheme (8) in the abstract form

$$\begin{cases} \dfrac{w_k^h - w_{k-1}^h}{\tau} + A_h^x w_k^h = \left(a \dfrac{q_{n+1} - 2q_n + q_{n-1}}{h^2} - \sigma q \right) \dfrac{\rho(t_k) - w_s^k}{q_s} \\ + f^h (t_k), t_k = k\tau, 1 \leq k \leq N, N\tau = T, \\ w_0^h = \varphi^h \end{cases} \tag{9}$$

in a Banach space $E = C[0,l]_h$ with the positive operator A_h^x defined by

$$A_h^x u^h = \left\{ -a(x_n) \frac{u_{n+1} - 2u_n + u_{n-1}}{h^2} + \sigma u \right\}_{n=1}^{M-1} \tag{10}$$

acting on grid functions u^h such that satisfies the condition

$$u_0 = u_M, -3u_0 + 4u_1 - u_2 = u_{M-2} - 4u_{M-1} + 3u_M.$$

Let denote $R = (I + \tau A_h^x)^{-1}$. In (9), we have that

$$w_k^h = R w_{k-1}^h + R\tau \left(\left(a \frac{q_{n+1} - 2q_n + q_{n-1}}{h^2} - \sigma q \right) \frac{\rho(t_k) - w_s^k}{q_s} + f^h (t_k) \right),$$

$\forall k, 1 \leq k \leq N$. By recurrence relations, we get

$$w_k^h = R^k \varphi^h + \sum_{m=1}^k R^{k-m+1} \frac{\tau}{q_s} \left(a \frac{q_{n+1} - 2q_n + q_{n-1}}{h^2} - \sigma q \right) \rho(t_m)$$

$$- \sum_{m=1}^k R^{k-m+1} \frac{\tau}{q_s} \left(a \frac{q_{n+1} - 2q_n + q_{n-1}}{h^2} - \sigma q \right) w_s^m + \sum_{m=1}^k R^{k-m+1} \tau f^h (t_m).$$

Taking the difference derivative of both sides, we obtain that

$$\frac{w_k^h - w_{k-1}^h}{\tau} = \frac{R^k - R^{k-1}}{\tau} \varphi^h + \frac{1}{q_s} \left(a \frac{q_{n+1} - 2q_n + q_{n-1}}{h^2} - \sigma q \right) \rho(t_k)$$

$$+ \sum_{m=1}^k \left(R^{k-m+1} - R^{k-m} \right) \frac{1}{q_s} \left(a \frac{q_{n+1} - 2q_n + q_{n-1}}{h^2} - \sigma q \right) \rho(t_m)$$

$$- \frac{1}{q_s} \left(a \frac{q_{n+1} - 2q_n + q_{n-1}}{h^2} - \sigma q \right) w_s^k$$

$$- \sum_{m=1}^k \left(R^{k-m+1} - R^{k-m} \right) \frac{1}{q_s} \left(a \frac{q_{n+1} - 2q_n + q_{n-1}}{h^2} - \sigma q \right) w_s^m$$

$$+f^h(t_k) + \sum_{m=1}^{k}\left(R^{k-m+1} - R^{k-m}\right) f^h(t_m).$$

Applying the formula

$$\sum_{m=1}^{k}\left(R^{k-m+1} - R^{k-m}\right) w_s^m = \sum_{m=1}^{k}\left(R^{k-m+1} - R^{k-m}\right)\varphi(x_s)$$

$$+\sum_{m=1}^{k}\left(R^{k-m+1} - R^{k-m}\right)\sum_{j=1}^{m}\frac{w_s^j - w_s^{j-1}}{\tau}\tau$$

and changing the order of summation, we obtain that

$$\sum_{m=1}^{k}\left(R^{k-m+1} - R^{k-m}\right) w_s^m = \sum_{m=1}^{k}\left(R^{k-m+1} - R^{k-m}\right)\varphi(x_s)$$

$$+\sum_{j=1}^{k}\sum_{m=j}^{k}\left(R^{k-m+1} - R^{k-m}\right)\frac{w_s^j - w_s^{j-1}}{\tau}\tau. \tag{11}$$

Then, the presentation of the solution of (8)

$$\frac{w_k^h - w_{k-1}^h}{\tau} = \frac{R^k - R^{k-1}}{\tau}\varphi^h + \frac{1}{q_s}\left(a\frac{q_{n+1} - 2q_n + q_{n-1}}{h^2} - \sigma q\right)\rho(t_k)$$

$$+\sum_{m=1}^{k}\left(R^{k-m+1} - R^{k-m}\right)\frac{1}{q_s}\left(a\frac{q_{n+1} - 2q_n + q_{n-1}}{h^2} - \sigma q\right)\rho(t_m)$$

$$-\frac{1}{q_s}\left(a\frac{q_{n+1} - 2q_n + q_{n-1}}{h^2} - \sigma q\right) w_s^k$$

$$-\sum_{m=1}^{k}\left(R^{k-m+1} - R^{k-m}\right)\frac{1}{q_s}\left(a\frac{q_{n+1} - 2q_n + q_{n-1}}{h^2} - \sigma q\right)\varphi(x_s)$$

$$-\sum_{j=1}^{k}\sum_{m=j}^{k}\left(R^{k-m+1} - R^{k-m}\right)\frac{1}{q_s}\left(a\frac{q_{n+1} - 2q_n + q_{n-1}}{h^2} - \sigma q\right)\frac{w_s^j - w_s^{j-1}}{\tau}\tau$$

$$+f^h(t_k) + \sum_{m=1}^{k}\left(R^{k-m+1} - R^{k-m}\right) f^h(t_m)$$

is obtained. Applying the definition of norm of the spaces E'_α and methods of monograph [21], we can show that,

$$\left\|\frac{w_k^h - w_{k-1}^h}{\tau}\right\|_{E'_\alpha} \le \left(1 - (1 + M_8(\phi, \alpha, T))\|\tilde{a}\|_{E'_\alpha}\tau\right)^{-1}\left[M_1\|A_h^x\varphi^h\|_{E'_\alpha}\right.$$

$$+ \max_{1 \le m \le N} |\rho(t_m)| M_8(\phi, \alpha, T) \|\widetilde{a}\|_{E'_\alpha} + M_9(\phi, \alpha, T) \|\widetilde{a}\|_{E'_\alpha} \|A_h^x \varphi\|_{E'_\alpha}$$

$$+ (M_{10}(\phi, \alpha, T) + 1) \left\| \{f^h(t_k)\}_{k=1}^N \right\|_{C_\tau(E'_\alpha)}$$

$$+ \|\widetilde{a}\|_{E'_\alpha} \left(\max_{1 \le k \le N} |\rho(t_k)| + M_{11} \|A_h^x \varphi^h\|_{E'_\alpha} \right)$$

$$+ \|\widetilde{a}\|_{E'_\alpha} (1 + M_8(\phi, \alpha, T)) \sum_{j=1}^{k-1} \left\| \frac{w_j^h - w_{j-1}^h}{\tau} \right\|_{E'_\alpha} \tau \Bigg].$$

Using the discrete analogue of Gronwall's inequality and the last inequality, we get

$$\left\| \frac{w_k^h - w_{k-1}^h}{\tau} \right\|_{E'_\alpha} \le e^{M_{12}(\widetilde{a}, \phi, \alpha, T)} \left[M_{13}(\widetilde{a}, \phi, \alpha, T) \|A_h^x \varphi^h\|_{E'_\alpha} \right.$$

$$\left. + M_{13}(\widetilde{a}, \phi, \alpha, T) \|\rho^\tau\|_{C[0,T]_\tau} + M_{14}(\widetilde{a}, \phi, \alpha, T) \left\| \{f^h(t_k)\}_{k=1}^N \right\|_{C_\tau(E'_\alpha)} \right]$$

for every k, $1 \le k \le N$. Then, we have that

$$\left\| \left\{ \frac{w_k^h - w_{k-1}^h}{\tau} \right\}_{k=1}^N \right\|_{C_\tau(E'_\alpha)} \le M_{15}(\widetilde{a}, \phi, \alpha, T) \left(\|A_h^x \varphi^h\|_{E'_\alpha} \right.$$

$$\left. + \left\| \{f^h(t_k)\}_{k=1}^N \right\|_{C_\tau(E'_\alpha)} + \|\rho^\tau\|_{C[0,T]_\tau} \right).$$

The following theorem finishes the proof of Theorem 2.

Theorem 3. [22] *For* $0 < \alpha < \dfrac{1}{2}$ *the norms of the spaces* $E'_\alpha \left(C[0,l]_h, A_h^x \right)$ *and* $C^{2\alpha}[0,l]_h$ *are equivalent.*

References

1. Hasanov, A.: Identification of unknown diffusion and convection coefficients in ion transport problems from flux data: an analytical approach. J. Math. Chem. 48(2), 413–423 (2010)
2. Shestopalov, Y., Smirnov, Y.: Existence and uniqueness of a solution to the inverse problem of the complex permittivity reconstruction of a dielectric body in a waveguide. Inverse Probl. 26(10), 105002 (2010)
3. Serov, V.S., Paivarinta, L.: Inverse scattering problem for two-dimensional Schrödinger operator. J. Inv. Ill-Pose. Probl. 14(3), 295–305 (2006)
4. Cannon, J.R., Lin, Y.L., Xu, S.: Numerical procedures for the determination of an unknown coefficient in semi-linear parabolic differential equations. Inverse Probl. 10, 227–243 (1994)

5. Borukhov, V.T., Vabishchevich, P.N.: Numerical solution of the inverse problem of reconstructing a distributed right-hand side of a parabolic equation. Comput. Phys. Commun. 126, 32–36 (2000)

6. Cannon, J.R., Yin, H.-M.: Numerical solutions of some parabolic inverse problems. Numer. Meth. Part. D. E. 2, 177–191 (1990)

7. Prilepko, A.I., Orlovsky, D.G., Vasin, I.A.: Methods for Solving Inverse Problems in Mathematical Physics. Marcel Dekker, New York (2000)

8. Isakov, V.: Inverse Problems for Partial Differential Equations. Applied Mathematical Sciences, vol. 127. Springer, New York (2006)

9. Ivanchov, N.I.: On the determination of unknown source in the heat equation with nonlocal boundary conditions. Ukr. Math. J. 47(10), 1647–1652 (1995)

10. Orlovsky, D., Piskarev, S.: On approximation of inverse problems for abstract elliptic problems. J. Inverse Ill-Pose. P. 17(8), 765–782 (2009)

11. Wang, Y., Zheng, S.: The existence and behavior of solutions for nonlocal boundary problems. Value Probl. 2009, Article ID 484879 (2009)

12. Agarwal, R.P., Shakhmurov, V.B.: Multipoint problems for degenerate abstract differential equations. Acta Math. Hung. 123(1-2), 65–89 (2009)

13. Di Blasio, G.: Maximal L^p regularity for nonautonomous parabolic equations in extrapolation spaces. J. Evol. Equ. 6(2), 229–245 (2006)

14. Eidelman, Y.S.: The boundary value problem for differential equations with a parameter. Differents. Uravneniya 14, 1335–1337 (1978)

15. Ashyralyev, A.: On the problem of determining the parameter of a parabolic equation. Ukr. Math. J. 62(9), 1200–1210 (2010)

16. Prilepko, A.I., Tikhonov, I.V.: Uniqueness of the solution of an inverse problem for an evolution equation and applications to the transfer equation. Mat. Zametki. 51(2), 77–87 (1992)

17. Choulli, M., Yamamoto, M.: Generic well-posedness of a linear inverse parabolic problem with diffusion parameter. J. Inverse Ill-Pose P. 7(3), 241–254 (1999)

18. Choulli, M., Yamamoto, M.: Generic well-posedness of an inverse parabolic problem-the Hölder-space approach. Inverse Probl. 12, 195–205 (1996)

19. Saitoh, S., Tuan, V.K., Yamamoto, M.: Reverse convolution inequalities and applications to inverse heat source problems. Journal of Inequalities in Pure and Applied Mathematics 3(5), Article 80 (2002) (electronic)

20. Samarskii, A.A., Vabishchevich, P.N.: Numerical Methods for Solving Inverse Problems of Mathematical Physics. Inverse and Ill-posed Problems Series. Walter de Gruyter, Berlin (2007)

21. Ashyralyev, A.: A note on fractional derivatives and fractional powers of operators. J. Math. Anal. Appl. 357, 232–236 (2009)

22. Ashyralyev, A., Sobolevskii, P.E.: Well-Posedness of Parabolic Difference Equations. Birkhäuser, Basel (1994)

Numerical and Analytical Modeling of the Stability of the Cylindrical Shell under the Axial Compression with the Use of the Non-classical Theories of Shells

Andrei M. Ermakov

Saint Petersburg State University, Department of Theoretical and Applied Mechanics,
Universitetsky Prospekt, 28, 198504 Peterhof, Sankt Petersburg, Russian Federation
khopesh_ra@mail.ru
http://www.math.spbu.ru/

Abstract. The problem of the buckling of a transveral-isotropic cylindrical shell under axial compression by means of new non-classical shell theories is studied. The local approach is used to solve the systems of differential equations. According to this approach the buckling deflection is sought in the form of a doubly periodic function of curvilinear coordinates. The well-known solutions obtained by classical shell theories are compared with the results of non-classical shell theories. For the non-classical theories of anisotropic shell of moderate thickness the buckling equations are constructed by the linearization of nonlinear equilibrium equations. Analytical and numerical results obtained with the use of 3D theory by the FEM code ANSYS 13 are also compared.

Keywords: Cylindrical Shell, Buckling, Non-Classical Theories of Shells, Numerical and Analytical Modeling.

1 Introduction

In this paper the problem of the buckling of the transversal-isotropic cylindrical shell under the axial compression by means of different nonclassical shell theories is studied. The following non-classical theories are considered: Ambartsumian (AMB) [1] theory of anisotropic shells, Paliy-Spiro (PS) theory of moderate-thickness shells and Rodionova-Titaev-Chernykh (RTCH) [2] iteration theory. The developed buckling equations for the shell theories of PS and RTCH are constructed by linearization of nonlinear equilibrium equations. The comparison of new solutions obtained by non-classical shell theories with well-known results of classical theories - Kirchhoff-Love (KL) and Timoshenko-Reissner (TR) [3] is done. In conclusion the comparison of the analytical results of shell theories with numerical results of three–dimensional theory by the FEM code Ansys 13 is given. The main focus is on the case of small cross-section shear modulus. Also we study the influence of relative thickness and length of the shell on the value of critical load. Let us denote a polar angle by α, and the length coordinate by β.

I. Dimov, I. Faragó, and L. Vulkov (Eds.): NAA 2012, LNCS 8236, pp. 279–286, 2013.

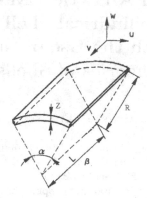

Fig. 1. An element of circular cylindrical shell

The radius of middle surface of the shell is R, the thickness is h, it's length is L, Young's modulus is E, Poisson ratio is ν and tangential shear modulus is G'. Lame coefficient and curvature coefficient which determine the geometry of cylindrical shell: $A_1 = R$, $A_2 = 1$, $k_1 = 1/R$, $k_2 = 0$.

We consider buckling equations of the shell which are constructed by the linearization of non-linear equilibrium equations. This method is very convenient in estimating of upper critical loading. It is enough to define the condition under which generalized stiffness of the construction is equal to zero. Using the method of linearization the solution of the problem is sought by summing of consequently calculated parameters of strain-stress state of the construction while loads are gradually increased. Thus at the each stage of the loading the linear shell problem is being solved.

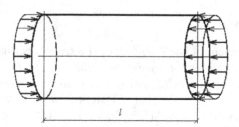

Fig. 2. The load applied to cylindrical shell

General equations are written down for increments in the components of inner forces, displacements and deformation parameters at this stage of loading. The components of the loading include parameters which describe the stress-strain state of the shell at the previous stage. If the change of the components is known and fixed, it is connected with the change of one scalar parameter. The initial state will be an implicit function of this parameter and there appears the eigenvalue problem.

In the given problem we use the classical hypothesis [3] holding that the basic stress-strain state of the shell before the loss of stability is membrane. Then one can take the well-known solution of the membrane shell theory as some function defining the distribution of force in a membrane shell:

$$Z = -T_2{}^0 \partial_{\beta,\beta} w \tag{1}$$

As a result, in this problem $T_2{}^0$ magnitude becomes the only one indefinite scalar parameter the eigenvalue of which should be found. In solving the systems of differential equations a local approach is used [4], according to which the bucking deflection is sought in the form of a doubly periodic function of curvilinear coordinates. The non-zero system solution is sought in the form of:

$$
\begin{aligned}
w(\alpha,\beta) &= w^0 \cos(n\alpha)\sin(m\beta), & \Phi(\alpha,\beta) &= \Phi^0 \cos(n\alpha)\sin(m\beta) \\
u(\alpha,\beta) &= u^0 \sin(n\alpha)\sin(m\beta), & \gamma_1(\alpha,\beta) &= \gamma_1{}^0 \sin(n\alpha)\sin(m\beta) \\
v(\alpha,\beta) &= v^0 \cos(n\alpha)\cos(m\beta), & \gamma_2(\alpha,\beta) &= \gamma_2{}^0 \cos(n\alpha)\cos(m\beta)
\end{aligned}
\tag{2}
$$

where u, v, w — displacement vector components of a point of mid-surface of shell, γ_1 and γ_2 —angles of normal turn in the planes $(\alpha,z), (\beta,z)$ respectively, $\Phi(\alpha,\beta)$ — force function.

2   Kirchhoff-Love Model Solution

Let us consider the well-known solution which is obtained by the classical theory of shells which is based on the following hypothesis:

1) the straight lines normal to the mid-surface remain straight and normal to the mid-surface after deformation;

2) the thickness of the shell does not change during a deformation.

Two-dimensional equation system of the shallow shell theory [3] has a form:

$$- D\Delta\Delta w + T_2^0 \partial_{\beta,\beta} w + \frac{1}{R}\partial_{\beta,\beta}\Phi = 0, \quad \frac{1}{Eh}\Delta\Delta\Phi + \frac{1}{R}\partial_{\beta,\beta} w = 0 \tag{3}$$

where Δ - Laplace operator; $D = Eh^3/(12(1-\nu^2))$ - cylindrical stiffness; T_2^0 - desired axial force.

If we substitute the expression (2) into this system (3) for force T_2^0, we will obtain:

$$- T_2^0 = f(m,n) = \frac{D(n^2+m^2)^2}{R^2 m^2} + \frac{Ehm^2}{(n^2+m^2)^2} \tag{4}$$

The critical load value is obtained as a result of minimization by the wave parameters m and n of the function $f(m,n)$.

$$T_2^0 = \sigma_0 h, \quad \sigma_0 = -\frac{E}{\sqrt{3(1-\nu^2)}}\frac{h}{R} = \sigma_{cl} \tag{5}$$

3   Ambartsumian Model Solution

The solution (5) being constructed by KL model does not allow taking into account the effect of stiffness on cross-section shear. Let us consider Ambartsumian [1] theory which is based on the following hypothesis:

1) displacement which is normal to the shell mid-surface does not depend on the normal coordinate;

2) shear stresses or the corresponding deformations change according to a quadratic law with respect to the plane thickness;

Let us write down the equations of Ambartsumian model which takes the influence of cross-section shear into account for a transversally-isotropic shell as:

$$-\Delta\Delta u - \frac{\nu}{R}\frac{\partial^3 w}{\partial\beta^3} + \frac{1}{R}\frac{\partial^3 w}{\partial\alpha^2\partial\beta} = 0 \quad -\Delta\Delta v - \frac{2+\nu}{R}\frac{\partial^3 w}{\partial\alpha\partial\beta^2} - \frac{1}{R}\frac{\partial^3 w}{\partial\alpha^3} = 0 \quad (6)$$

$$-D\Delta^4 w + \frac{Eh}{R^2}(1 - h_z\Delta)\frac{\partial^4 w}{\partial\beta^4} - T_2^0(1 - h_z\Delta)\Delta^2\frac{\partial^2 w}{\partial\beta^2} = 0 \quad h_z = \frac{Eh^2}{10(1-\nu^2)G'}$$

The simplified system of differential equations of the shell buckling which is used in Ambartsumian theory was obtained basing on the equations of the shallow shell theory.

Using the local approach (2) for solving this system for T_2^0 we obtain:

$$-T_2^0 = f(m,n) = \frac{D(n^2+m^2)^2}{R^2m^2(1+h_z(n^2+m^2))} + \frac{Ehm^2}{(n^2+m^2)^2} \quad (7)$$

The obtained value for critical load —

$$\sigma_0 = -\frac{E}{\sqrt{3(1-\nu^2)}}\frac{h}{R} + \frac{E^2}{10G'(1-\nu^2)}\left(\frac{h}{R}\right)^2 = \sigma_{cl}\left(1 - \frac{\sqrt{3}}{10\sqrt{(1-\nu^2)}}\frac{E}{G'}\frac{h}{R}\right) \quad (8)$$

agrees completely with the one being obtained by the theory of Timoshenko-Reissner [3]. It is known [4] that for an isotropic shells and plates the TR theory being asymptotically inconsistent refines the deflection of a body. But for bodies, which are made of transversal isotropic material "in case when material stiffness in tangential directions is much larger than its stiffness in the transversal direction" the TR theory makes the KL theory more precise and gives next asymptotical approximation of the three-dimensional theory. The bodies "with moderately small transverse shear stiffness" are thin-walled bodies for which small parameter g = G'/E (where E is the Young's modulus in the tangential direction, G' is the shear modulus for plane normal to the surface of isotropy) satisfies expression $(h/R)^2 \ll g \ll 1$.

4   Paliy-Spiro Model Solution

The situation is quite different when the buckling problems are considered with the use of improved theories. In this case the old representations are not always acceptable as there appear problems related to taking into account the change of the length and the turn of the normal to mid-surface.

The Paliy-Spiro [2] theory of moderate-thickness shells accepts the following hypothesis:

1) straight fibers of the shell which are perpendicular to its mid-surface before deformation remain also straight after deformation;

2) cosine of the slope angle of these fibers to the mid-surface of the deformed shell is equal to the averaged angle of transverse shear.

The mathematical formulation of the accepted hypotheses gives following equations:

$$
\begin{aligned}
&u_1 = u + \phi \cdot z, \qquad u_2 = v + \psi \cdot z, \\
&u_3 = w + F(\alpha, \beta, z), \\
&\phi = \gamma_1 + \phi_0, \qquad \psi = \gamma_2 + \psi_0, \\
&\phi_0 = -\frac{1}{A_1}\frac{\partial w}{\partial \alpha} + k_1 u, \quad \psi_0 = -\frac{1}{A_2}\frac{\partial w}{\partial \alpha} + k_2 v,
\end{aligned} \tag{9}
$$

where ϕ and ψ are the angles of rotation of the normal in the planes (α, z) and (β, z); $\phi_0, \psi_0, \gamma_1\ \gamma_2$ — the angles of rotation of the normal to the medial surface and the angles of displacement in the same planes. The function $F(\alpha, \beta, z)$ characterizes the variation of the length of normal to the middle surface.

The shell deformations $\varepsilon_1, \varepsilon_2, \varepsilon_{13}, \eta_1, \eta_2$ are expressed by displacement components with the following formulas:

$$
\varepsilon_1 = \frac{\partial_\alpha u}{R} + \frac{w}{R}, \quad \varepsilon_2 = \partial_\beta v, \quad \eta_1 = \frac{\partial_\alpha \phi}{R} \tag{10}
$$

$$
\eta_2 = \partial_\beta \psi, \quad w = \frac{\partial_\alpha v}{R} + \partial_\beta u, \quad \tau = \frac{\partial_\alpha \psi}{R} + \partial_\beta \phi
$$

Substituting the mentioned dependencies (10) into the constitutive relations (11), one can obtain the equations of relation between the components of displacement and forces and moments.

$$
\epsilon_1 = \frac{T_1 - \nu T_2}{Eh}, \quad \epsilon_2 = \frac{T_2 - \nu T_1}{Eh}, \quad \omega = \frac{S}{G'h}, \quad \eta_1 = \frac{12(M_1 - \nu M_2)}{Eh^3}, \tag{11}
$$

$$
\eta_2 = \frac{12(M_2 - \nu M_1)}{Eh^3}, \quad \tau = \frac{12H}{G'h^3}, \quad \gamma_1 = \frac{N_1}{G'h}, \quad \gamma_2 = \frac{N_2 + T_2{}^0 * \psi_0}{G'h}
$$

As one can see (11), PS theory includes the characteristical parameter $T_2{}^0$ in the equation of relation between normal slope γ_2 and shear force N_2.

The obtained equations of relation between components of displacement and forces and moments are substituted into equilibrium equations:

$$
\frac{\partial_\alpha T_1}{R} + \partial_\beta S + \frac{N_1}{R} = 0, \quad \frac{\partial_\alpha S}{R} + \partial_\beta T_2 = 0, \quad \frac{\partial_\alpha N_1}{R} + \partial_\beta N_2 - \frac{T_1}{R} = 0 \tag{12}
$$

$$
\frac{\partial_\alpha M_1}{R} + \partial_\beta H - N_1 = 0, \quad \frac{\partial_\alpha H}{R} + \partial_\beta M_2 - N_2 - T_2{}^0 * \psi_0 = 0.
$$

Resolution matrix in this case will be of [5*5] dimension. Nevertheless the value of critical load obtained is similar to the value of Ambartsumian theory (8). This is the factor of the second coefficient of asymptotical expansion by small parameter h/R that makes the results different. One can see that this numerical factor reduces the influence of transversal shear.

$$\sigma_0 = -\frac{E}{\sqrt{3(1-\nu^2)}}\frac{h}{R} + \frac{E^2}{12G'(1-\nu^2)}\left(\frac{h}{R}\right)^2 = \sigma_{cl}\left(1 - \frac{\sqrt{3}}{12\sqrt{(1-\nu^2)}}\frac{E}{G'}\frac{h}{R}\right) \tag{13}$$

5 Rodionova-Titaev-Chernykh Model Solution

The use of RTCH shell theory yields more interesting results [2]. This is a linear theory of non-homogeneous anisotropic shells which takes into account low transversal shear compliance and deformation towards the normal to the middle surface. It also takes into account transversal normal strains and supposes non-linear distribution of displacement vector component over shell thickness.

1) transverse tangential and normal stresses are distributed along the shell thickness according to quadratic and cubic laws respectively;

2) tangential and normal components of the displacement vector are distributed along the shell thickness according to quadratic and cubic laws;

The functions which describe shell displacement $u_1(\alpha, \beta, z), u_2(\alpha, \beta, z), u_3(\alpha, \beta, z)$ according to RTCH theory are supposed to be sought in the form of Legendre polynomial series P_0, P_1, P_2, P_3 from normal coordinate $z \in \left[-\frac{h}{2}, \frac{h}{2}\right]$.

$$u_1(\alpha, z) = u(\alpha, \beta)^* P_0(z) + \gamma_1(\alpha, \beta)^* P_1(z) + \theta_1(\alpha, \beta)^* P_2(z) + \varphi_1(\alpha, \beta)^* P_3(z),$$

$$u_2(\alpha, z) = v(\alpha, \beta)^* P_0(z) + \gamma_2(\alpha, \beta)^* P_1(z) + \theta_2(\alpha, \beta)^* P_2(z) + \varphi_2(\alpha, \beta)^* P_3(z),$$

$$u_3(\alpha, z) = w(\alpha, \beta)^* P_0(z) + \gamma_3(\alpha, \beta)^* P_1(z) + \theta_3(\alpha, \beta)^* P_2(z)$$

$$\tag{14}$$

$$P_0(z) = 1, \quad P_1(z) = \frac{2z}{h}, \quad P_2(z) = \frac{6z^2}{h^2} - \frac{1}{2}, \quad P_3(z) = \frac{20z^3}{h^3} - \frac{3z}{h} \tag{15}$$

where γ_3 and θ_3 characterize normal length variation to this surface, magnitudes θ_1 and φ_1, describe normal curvature in the plane (α, z) of a fiber, θ_2 φ_2, describe normal curvature in the plane (β, z) which before the deformation were perpendicular to the medial surface of the shell.

The shell deformations $\varepsilon_1, \varepsilon_2, \varepsilon_{13}$, η_1, η_2 are expressed by displacement components:

$$\epsilon_1 = \frac{\partial_\alpha u}{R} + \frac{w}{R}, \quad \epsilon_2 = \partial_\beta v, \quad \eta_1 = \frac{\partial_\alpha \gamma_1}{R} + \frac{\gamma_3}{R}, \quad \eta_2 = \partial_\beta \gamma_2, \quad \vartheta_0 = \partial_{\alpha,\alpha} w \tag{16}$$

$$\omega = \frac{\partial_\alpha v}{R} + \partial_\beta u, \quad \tau = \frac{\partial_\alpha \gamma_2}{R} + \partial_\beta \gamma_1, \quad \epsilon_{13} = \frac{\partial_\alpha w}{R} - \frac{u}{R} + \frac{2\gamma_1}{h}, \quad \epsilon_{23} = \partial_\beta w + \frac{2\gamma_2}{h}$$

The characteristical parameter $T_2{}^0$ is also included into the constitutive relations.

$$T_1 = \frac{Eh}{1-\nu^2}(\epsilon_1 + \nu\epsilon_2), \quad T_2 = \frac{Eh}{1-\nu^2}(\nu\epsilon_1 + \epsilon_2), \quad M_1 = \frac{Eh^2}{6(1-\nu^2)}(\eta_1 + \nu\eta_2),$$

$$M_2 = \frac{Eh^2}{6(1-\nu^2)}(\nu\eta_1 + \eta_2), \quad S = G'h\omega, \quad H = G'\frac{h^2}{6}\tau, \quad N_1 = \frac{5hG'}{6}\epsilon_{13}, \quad (17)$$

$$N_2 = \frac{5hG'}{6}\epsilon_{23} + \frac{T_2^0\vartheta_0}{6}, \quad \gamma_3 = -\frac{\nu}{1-\nu}\frac{h}{2}(\epsilon_1 + \epsilon_2),$$

The equations of relation forces and moments and displacement components were substituted into equilibrium equations (12).

In minimizing the determinant the following solution was obtained:

$$\sigma_0 = -\frac{E}{\sqrt{3(1-\nu^2)}}\sqrt{1 - \frac{E^2}{60G'^2(1-\nu^2)}\left(\frac{h}{R}\right)^2\left(\frac{h}{R}\right) + \frac{E^2}{15G'(1-\nu^2)}\left(\frac{h}{R}\right)^2}$$

(18)

This expansion into a series by a small parameter yields succeedent terms of expansion:

$$\sigma_0 = -\frac{E}{\sqrt{3(1-\nu^2)}}\frac{h}{R} + \frac{E^2}{15G'(1-\nu^2)}\left(\frac{h}{R}\right)^2 + O\left[\frac{h}{R}\right]^3 \quad (19)$$

6 The Comparison with Numerical Results

Let us compare the results which are obtained with the use of developed analytical formulae of shell theory and numerical results for three-dimensional theory. Unfortunately, the formulae for the critical load of the shell theory do not take into account the tube length. Being applied to the buckling problems the obtained solutions well agree with medium-length shells. For example, a three-dimensional model of steel tube under the influence of axial compression was studied under the following parameters $h/R = 2/15, \nu = 0.3$. The cross-section shear modulus is equal to $G' = E/(2(1+\nu))$. For modeling the three-dimensional problem in package Ansys 13 the finite element Solid186 was used. This is a higher order 3-D 20-node solid element that exhibits quadratic displacement behavior. The element is defined by 20 nodes having three degrees of freedom per node: translations in the nodal x, y, and z directions. The element supports plasticity, hyperelasticity, creep, stress stiffening, large deflection, and large strain capabilities. It also has mixed formulation capability for simulating deformations of nearly incompressible elastoplastic materials, and fully incompressible hyperelastic materials. [5] During mesh construction the tube thickness was split for five

Table 1. The comparison of critical load values

h/R	0.025	0.05	0.1	0.133	0.162
KL	0.01532	0.03103	0.0637	0.08069	0.09814
Amb	0.01513	0.03028	0.06054	0.07561	0.09063
PS	0.01516	0.03041	0.06107	0.07646	0.09188
RTCH	0.01519	0.03053	0.06155	0.07722	0.09297
Ansys	0.01445	0.02875	0.055	0.0595	0.0635

elements. Thus the value of the critical load according to the three-dimensional theory is smaller than shell theories. The value of a critical load for the considered tubes with length ranged from 1.5 to 3 diameters of mid-surface does not change considerably. Table 1 shows dimensionless values of critical load σ_0/E for different ratios of tube thickness to the radius of its middle surface.

7 Conclusions

As one can see in the table, as shell thickness increases, the values of critical load obtained by shell theories are not consistent with the results of three-dimensional theory. It can be noticed that error increases as the thickness grows. It is possible to claim that in spite of improvements of non-classical hypotheses reliable results can be obtained only for thin shells. However, as it was shown in [2] similar hypotheses suit well for defining stress-strain state of a shell.

References

1. Ambarcumian, S.A.: Theory of anisotropic plates. strength, stability, and vibrations. Technomic, Stamford, Conn. (1970)
2. Bauer, S.M., Ermakov, A.M., Kashtanova, S.V., Morozov, N.F.: Evaluation of the Mechanical Parameters of Multilayered Nanotubes by means of Nonclassical Theories of Shells. In: EUROMECH Colloquium 527 Shell-like Structures - Nonclassical Theories and Applications, Part 6, vol. 15, pp. 519–530. Springer, Heidelberg (2011)
3. Elishakoff, I.: Probabilistic resolution of the twentieth century conundrum in elastic stability. J. Thin-Walled Structures 59, 35–57 (2012)
4. Tovstik, P.E.: Stability of a transversally isotropic cylindrical shell under axial compression. J. Mechanics of Solids 44, 552–564 (2009)
5. ANSYS Mechanical APDL Element Reference. Release 13.0. ANSYS, Inc. (2010)

Ant Colony Optimization Start Strategies Performance According Some of the Parameters

Stefka Fidanova and Pencho Marinov

Institute of Information and Communication Technologies, Bulgarian Academy of Sciences, Acad. G. Bonchev str. bl25A, 1113 Sofia, Bulgaria
{stefka,pencho}@parallel.bas.bg

Abstract. Ant Colony Optimization (ACO) is a stochastic search method that mimic the social behavior of real ants colonies, which manage to establish the shortest rout to feeding sources and back. Such algorithms have been developed to arrive at near-optimal solutions to large-scale optimization problems, for which traditional mathematical techniques may fail. On this paper is proposed an ant algorithm with semi-random start. Several start strategies are prepared at the basis of the start nodes estimation. There are several parameters which manage the starting strategies. In this work we focus on influence on the quality of the achieved solutions of the parameters which shows the percentage of the solutions classified as good and as bad respectively. This new technique is tested on Multiple Knapsack Problem (MKP).

1 Introduction

Metaheuristic methods are general tools for solving hard (from computational point of view) optimization problems. Most of them use ideas coming from nature. Ant Colony Optimization is one of the most successive metaheuristic method. The main idea comes from collective intelligence of real ant when they look for a food. The problem is solved collectively by the whole colony. This ability is explained by the fact that ants communicate in an indirect way by laying trails of pheromone. The higher the pheromone trail within a particular direction, the higher the probability of choosing this direction.

The ACO algorithm uses a colony of artificial ants that behave as cooperative agents in a mathematical space where they are allowed to search and reinforce pathways (solutions) in order to find the optimal ones. The problem is represented by graph and the ants walk on the graph to construct solutions. The solutions are represented by paths in the graph. After the initialization of the pheromone trails, the ants construct feasible solutions, starting from random nodes, and then the pheromone trails are updated. At each step the ants compute a set of feasible moves and select the best one (according to some probabilistic rules) to continue the rest of the tour. The transition probability $p_{i,j}$, to choose the node j when the current node is i, is based on the heuristic information $\eta_{i,j}$ and the pheromone trail level $\tau_{i,j}$ of the move, where $i,j = 1,\ldots,n$.

I. Dimov, I. Faragó, and L. Vulkov (Eds.): NAA 2012, LNCS 8236, pp. 287–294, 2013.

$$p_{i,j} = \frac{\tau_{i,j}^a \eta_{i,j}^b}{\sum\limits_{k \in Unused} \tau_{i,k}^a \eta_{i,k}^b},$$

where $Unused$ is the set of unused nodes of the graph. The higher the value of the pheromone and the heuristic information, the more profitable it is to select this move and resume the search. In the beginning, the initial pheromone level is set to a small positive constant value τ_0; later, the ants update this value after completing the construction stage. ACO algorithms adopt different criteria to update the pheromone level.

The pheromone trail update rule is given by:

$$\tau_{i,j} \leftarrow \rho\tau_{i,j} + \Delta\tau_{i,j},$$

where ρ models evaporation in the nature and $\Delta\tau_{i,j}$ is the new added pheromone which is proportional to the quality of the solution.

As other metaheuristics, ACO algorithm is applied on hard (NP) combinatorial optimization problems coming from real life and industry. It is unpractical to apply exact methods or traditional numerical methods on this kind of problems, because they need huge amount of computational resources, time and memory. Examples of optimization problems are Traveling Salesman Problem [12], Vehicle Routing [13], Minimum Spanning Tree [11], Multiple Knapsack Problem [3], etc.

The ACO is a constructive methods. The ants construct solutions starting from random points. In serie of papers we propose and learn semirandom start of the ants with estimation of possible starting points. There are several parameters controlling the ant start. One of them is how many of the solution will be estimated as good (parameter A) and how many - as bad (parameter B). The aim of this paper is analysis of the influence of the parameters A and B on the ACO algorithm behavior. The success of the algorithm depends of the right values of it parameters.

The rest of the paper is organized as follows: in section 2 estimation of start node is introduced and several start strategies are proposed; in section 3 the strategies are applied on MKP and sensitivity analysis of the algorithm according parameters A and B is made. At the end some conclusions and directions for future work are done.

2   Subset Estimations

The essential part of ACO algorithm is starting from random node when ants create solutions. It is a kind of diversification of the search and leads to using small number of ants, which means less computational resources. For some optimization problems, especially subset problems, it is important from which node the search process starts. For example: if an ant starts from node which does not belong to the good solution, probability to construct it is zero. Therefore we divide the set of nodes of the graph of the problem to subsets. We estimate every subset, how good and how bad is to start from it, after we offer several start strategies keeping in some extent the random start.

Let the graph of the problem has m nodes. We divide the set of nodes on N subsets. There are different ways for dividing. Normally, the nodes of the graph are randomly enumerated. An example for creating of the nodes subsets, without loss of generality, is: the node number one is in the first subset, the node number two is in the second subset, etc. the node number N is in the $N - th$ subset, the node number $N + 1$ is in the first subset, etc. Thus the number of the nodes in the subsets are almost equal. We introduce estimations $D_j(i)$ and $E_j(i)$ of the node subsets, where $i \geq 2$ is the number of the current iteration. $D_j(i)$ shows how good is to start from node which belong to the jth subset and $E_j(i)$ shows how bad is to start from node which belong to the jth subset. $D_j(i)$ and $E_j(i)$ are weight coefficients of $j - th$ node subset $(1 \leq j \leq N)$.

$$D_j(i) = \varphi.D_j(i - 1) + (\psi - \varphi).F_j(i), \tag{1}$$

$$E_j(i) = \varphi.E_j(i - 1) + (\psi - \varphi).G_j(i), \tag{2}$$

where $i \geq 1$ is the current process iteration and for each j $(1 \leq j \leq N)$:

$$F_j(i) = \begin{cases} \frac{f_{j,A}}{n_j} & \text{if } n_j \neq 0 \\ F_j(i - 1) & \text{otherwise} \end{cases}, \tag{3}$$

$$G_j(i) = \begin{cases} \frac{g_{j,B}}{n_j} & \text{if } n_j \neq 0 \\ G_j(i - 1) & \text{otherwise} \end{cases}, \tag{4}$$

$f_{j,A}$ is the number of the solutions among the best $A\%$, $g_{j,B}$ is the number of the solutions among the worst $B\%$, where $A + B \leq 100$, $i \geq 2$ and

$$\sum_{j=1}^{N} n_j = n, \tag{5}$$

where n_j $(1 \leq j \leq N)$ is the number of solutions obtained by ants starting from nodes subset j, n is the number of ants. Initial values of the weight coefficients are: $D_j(1) = 1$ and $E_j(1) = 0$. With this estimation we take in to account the information from previous iterations as well as the information from current iteration. The information from previous iterations have less influence in the estimation because we divide to the number of iteration. The balance between the influence of the previous iterations and the last is important. At the beginning when the current best solution is far from the optimal one, some of the node subsets can be estimated as good. If the influence of the last iteration is too high then information for good and bad solutions from previous iterations is ignored, which can distort estimation too. We try to use the experience of the ants from previous iterations when they choose the better starting node. Other authors use this experience only by the pheromone, when the ants construct the solutions [2]. Let us fix threshold E for $E_j(i)$ and D for $D_j(i)$, than we construct several strategies to choose start node for every ant, the threshold E increases every iteration with $1/i$ where i is the number of the current iteration:

1 If $E_j(i)/D_j(i) > E$ then the subset j is forbidden for current iteration and we choose the starting node randomly from $\{j \,|\, j \text{ is not forbidden}\}$.

2 If $E_j(i)/D_j(i) > E$ then the subset j is forbidden for current simulation and we choose the starting node randomly from $\{j \,|\, j \text{ is not forbidden}\}$.

3 If $E_j(i)/D_j(i) > E$ then the subset j is forbidden for K_1 consecutive iterations and we choose the starting node randomly from $\{j \,|\, j \text{ is not forbidden}\}$.

4 Let $r_1 \in [R, 1)$ is a random number. Let $r_2 \in [0, 1]$ is a random number. If $r_2 > r_1$ we randomly choose node from subset $\{j \,|\, D_j(i) > D\}$, otherwise we randomly chose a node from the not forbidden subsets, r_1 is chosen and fixed at the beginning.

5 Let $r_1 \in [R, 1)$ is a random number. Let $r_2 \in [0, 1]$ is a random number. If $r_2 > r_1$ we randomly choose node from subset $\{j \,|\, D_j(i) > D\}$, otherwise we randomly chose a node from the not forbidden subsets, r_1 is chosen at the beginning and increase with r_3 every iteration.

Where $0 \leq K_1 \leq$ "number of iterations" is a parameter. If $K_1 = 0$, than strategy 3 is equal to the random choose of the start node. If $K_1 = 1$, than strategy 3 is equal to the strategy 1. If $K_1 =$ "maximal number of iterations", than strategy 3 is equal to the strategy 2.

A is a parameter which shows how many of the achieved solutions will be treated as good. B is a parameter which shows how many of the achieved solutions will be treated as bad. If B is a big number, many subsets will be estimated as bad and will become forbidden. If both A and B are small numbers, a lot of subsets will be treated neither good nor bad.

We can use more than one strategy for choosing the start node, but there are strategies which can not be combined. We distribute the strategies into two sets: $St1 = \{strategy1,\ strategy2,\ strategy3\}$ and $St2 = \{strategy4,\ strategy5\}$. The strategies from same set can not be used at once. Thus we can use strategy from one set or combine it with strategies from the other set. Exemplary combinations are $(strategy1)$, $(strategy2;\ strategy5)$, $(strategy3;\ strategy4)$. When we combine strategies from $St1$ and $St2$, first we apply the strategy from $St1$ and according it some of the regions (node subsets) become forbidden, and after that we choose the starting node from not forbidden subsets according the strategy from $St2$. For example if we combine $strategy2$ and $strategy5$. The $strategy2$ is applied first. Some of the node subsets become forbidden after it application. Next is apply $strategy5$. It is applied only on not forbidden subsets. The starting node is chosen from not forbidden subsets in a random way and the nodes from not forbidden subsets with good estimation will be with higher probability to be chosen.

3   Experimental Results

The influence of parameters A and B is analyzed in this section. Influence of other parameters is analyzed in previous authors works [5–9]. Like test is used

Multiple Knapsack Problem (MKP) because it is a good representative of subset problems. The Multiple Knapsack Problem has numerous applications in theory as well as in practice. It also arise as a subproblem in several algorithms for more complex problems and these algorithms will benefit from any improvement in the field of MKP. The following major applications can be mentioned: problems in cargo loading, cutting stock, bin-packing, budget control and financial management may be formulated as MKP. Other applications are industrial management, naval, aerospace, computational complexity theory.

The MKP can be thought as a resource allocation problem, where there are m resources (the knapsacks) and n objects and every object j has a profit p_j. Each resource has its own budget c_j (knapsack capacity) and consumption r_{ij} of resource i by object j. The aim is maximizing the sum of the profits, while working with a limited budget.

The MKP can be formulated as follows:

$$\max \sum_{j=1}^{n} p_j x_j$$

$$\text{subject to } \sum_{j=1}^{n} r_{ij} x_j \leq c_i \quad i = 1, \ldots, m \tag{6}$$

$$x_j \in \{0, 1\} \quad j = 1, \ldots, n$$

x_j is 1 if the object j is chosen and 0 otherwise.

There are m constraints in this problem, so MKP is also called m-dimensional knapsack problem. Let $I = \{1, \ldots, m\}$ and $J = \{1, \ldots, n\}$, with $c_i \geq 0$ for all $i \in I$. A well-stated MKP assumes that $p_j > 0$ and $r_{ij} \leq c_i \leq \sum_{j=1}^{n} r_{ij}$ for all $i \in I$ and $j \in J$. Note that the $[r_{ij}]_{m \times n}$ matrix and $[c_i]_m$ vector are both non-negative.

In the MKP one is not interested in solutions giving a particular order. Therefore a partial solution is represented by $S = \{i_1, i_2, \ldots, i_j\}$ and the most recent elements incorporated to S, i_j need not be involved in the process for selecting the next element. Moreover, solutions for ordering problems have a fixed length as one search for a permutation of a known number of elements. Solutions for MKP, however, do not have a fixed length. The graph of the problem is defined as follows: the nodes correspond to the items, the arcs fully connect nodes. Fully connected graph means that after the object i one can chooses the object j for every i and j if there are enough resources and object j is not chosen yet.

The computational experience of the ACO algorithm is shown using 10 MKP instances from "OR-Library" available within WWW access at http://people.brunel.ac.uk/~mastjjb/jeb/orlib, with 100 objects and 10 constraints. To provide a fair comparison for the above implemented ACO algorithm, a predefined number of iterations, $k = 100$, is fixed for all the runs. The developed technique has been coded in C++ language and implemented on a Pentium 4 (2.8 Ghz). The parameters are fixed as follows: $\rho = 0.5$, $a = 1$, $b = 1$, number of used ants is 20, $D = 1.5$, $E = 0.5$, $K_1 = 5$, $R = 0.5$, $r_3 = 0.01$. The values of ACO parameters (ρ, a, b) are from [4] and experimentally is found that they are best for MKP. The tests are run with 1, 2, 4, 5 and 10 nodes within the nodes subsets. Parameters φ and ψ have values $(0.45, 0.75)$ and they are from [10]. We test the algorithm on following values for parameters $\{A, B\} \in \{10, 20, 30, 40, 50, 60, 70, 80, 90\}$, and $A + B \leq 100$. For every experiment, the results are obtained by performing 30 independent runs, then averaging the fitness values obtained in order to ensure statistical confidence of the observed difference. The computational time which takes start strategies is negligible with respect to running time of the algorithm. Tests with all combinations of strategies and with random start (12 combinations), 5 node devisions and 45 combinations of the values for A and B are run. Thus we perform 81 000 tests.

Table 1. Estimation of strategies and rate of fuzziness

A	B	Best strategies	best estim.	A	B	Best strategies	best estim.
10	10	3 3-5 3-6	89	30	60	2 2-5 2-6	89
10	20	3	92	30	70	1 2-5 2-6	85
10	30	3	92	40	10	1 3	91
10	40	3 3-5 3-6	89	40	20	3-5 3-6	88
10	50	1	88	40	40	3	85
10	60	3	89	40	50	3-5 3-6	89
10	70	1	87	40	60	2	89
10	80	1-5 1-6	88	50	10	3-5 3-6	88
10	90	2	81	50	20	3	90
20	10	3 3-5 3-6	87	50	30	1	88
20	20	3-5 3-6	91	50	40	3-5 3-6	87
20	30	1-5 1-6	92	50	50	2-5 2-6	86
20	40	1	88	60	10	3	90
20	50	1	90	60	20	3	90
20	60	2	90	60	30	3-5 3-6	89
20	70	3-5 3-6	87	60	40	3-5 3-6	86
20	80	3-5 3-6	85	70	10	3-5 3-6	92
30	10	3-5 3-6	90	70	20	3	91
30	20	3	91	70	30	3	89
30	30	1-5 1-6 3-5 3-6	87	80	10	3	91
30	40	1	89	80	20	1 1-5 1-6 3	89
30	50	1-5 1-6	89	90	10	3-5 3-6	**93**

Average achieved result by some strategy, is better than without any strategy, for every test problem. For fair comparison, the difference d between the worst and best average result for every problem is divided to 10. If the average result for some strategy is between the worst average result and worst average plus $d/10$ it is appreciated with 1. If it is between the worst average plus $d/10$ and worst average plus $2d/10$ it is appreciated with 2 and so on. If it is between the best average minus $d/10$ and the best average, it is appreciated with 10. Thus for a test problem the achieved results for every strategy, every nodes devision and values for A and B is appreciated from 1 to 10. After that is summed the rate of all test problems for every strategy, every nodes devision and values for parameters A and B. So theirs rate becomes between 10 and 100 (see Table 1). It is like percentage of successes. We have applied ANOVA test to ensure the confidence of the work. Difference with one units is significant in our estimation system.

The best achieved results for all strategies are when there is only one node in the node subsets. Therefore in Table 1 we report only this case. With bold is a best found result.

The best found result is when $(A, B) = (90, 10)$ and the worst found result is when $(A, B) = (10, 90)$. We observe that we achieve good results when the value of parameter A is much greater then the value of the parameter B or A and B are small and have similar values. When the value of the parameter B is too big, the achieved result is bad. Big value of B means a lot of node subsets to be estimated like bad and to be forbidden. Thus the diversification of the search decrease. We can conclude that to achieve good results small number of node subsets to become forbidden or the value of the parameter B to be small.

4   Conclusion

In this paper we address on influence analysis of the parameters A and B of start strategies on the ACO algorithm applied on MKP. We vary the value of the parameters A and B in the interval $[10, 90]$. We found that small values of parameter B achieve better results, thus the best values are $(A, B) = (90, 10)$. Application of start strategies do not take significant time according running time of hall algorithm. The aim of this an some of the previous works of the authors is to find the best values of the algorithm parameters. Thus when someone prepare a software which include ant algorithm, the best algorithm parameters to be known. In a future we will apply start strategies on other problems different than subset problems.

Acknowledgments. This work has been partially supported by the Bulgarian National Scientific Fund under the grants DID 02/29 and DTK 02/44.

References

1. Dorigo, M., Gambardella, L.M.: Ant Colony System: A Cooperative Learning Approach to the Traveling Salesman Problem. IEEE Transactions on Evolutionary Computation 1, 53–66 (1997)
2. Dorigo, M., Stutzle, T.: Ant Colony Optimization. MIT Press (2004)
3. Fidanova, S.: Evolutionary Algorithm for Multiple Knapsack Problem. In: Int. Conference Parallel Problems Solving from Nature, Real World Optimization Using Evolutionary Computing, Granada, Spain (2002) ISBN No 0-9543481-0-9
4. Fidanova, S.: Ant colony optimization and multiple knapsack problem. In: Renard, J.P. (ed.) Handbook of Research on Nature Inspired Computing for Economics ad Management, pp. 498–509. Idea Grup Inc. (2006) ISBN 1-59140-984-5
5. Fidanova, S., Atanassov, K., Marinov, P., Parvathi, R.: Ant Colony Optimization for Multiple Knapsack Problems with Controlled Starts. Int. J. Bioautomation 13(4), 271–280
6. Fidanova, S., Marinov, P.: Intuitionistic Fuzzy Estimation of the Ant Methodology. J. of Cybernetics and Information Technologies 9(2), 79–88 (2009) ISSN 1311-9702
7. Fiodanova, S., Marinov, P., Atanassov, K.: Generalized Net Models of the Process of Anmt Colony Optimization with Different Strategies and Intuitionistic Fuzzy Estimations. In: Proc. Jangjeon Math., vol. 13(1), pp. 1–12 (2010) ISSN 1598-7264, Soc.
8. Fidanova, S., Atanassov, K., Marinov, P.: Start Strategies of ACO Applied on Subset Problems. In: Dimov, I., Dimova, S., Kolkovska, N. (eds.) NMA 2010. LNCS, vol. 6046, pp. 248–255. Springer, Heidelberg (2011)
9. Fidanova, S., Marinov, P., Atanassov, K.: Sensitivity Analysis of ACO Start Strategies for Subset Problems. In: Dimov, I., Dimova, S., Kolkovska, N. (eds.) NMA 2010. LNCS, vol. 6046, pp. 256–263. Springer, Heidelberg (2011)
10. Fidanova, S., Atanassov, K., Marinov, P.: Intuitionistic Fuzzy Estimation of the Ant Colony Optimization Starting Points. In: Lirkov, I., Margenov, S., Waśniewski, J. (eds.) LSSC 2011. LNCS, vol. 7116, pp. 222–229. Springer, Heidelberg (2012)
11. Reiman, M., Laumanns, M.: A Hybrid ACO algorithm for the Capacitated Minimum Spanning Tree Problem. In: Proc. of First Int. Workshop on Hybrid Metahuristics, Valencia, Spain, pp. 1–10 (2004)
12. Stutzle, T., Dorigo, M.: ACO Algorithm for the Traveling Salesman Problem. In: Miettinen, K., Makela, M., Neittaanmaki, P., Periaux, J. (eds.) Evolutionary Algorithms in Engineering and Computer Science, pp. 163–183. Wiley (1999)
13. Zhang, T., Wang, S., Tian, W., Zhang, Y.: ACO-VRPTWRV: A New Algorithm for the Vehicle Routing Problems with Time Windows and Re-used Vehicles based on Ant Colony Optimization. In: Sixth International Conference on Intelligent Systems Design and Applications, pp. 390–395. IEEE Press (2006)

Numerical and Asymptotic Modeling of Annular Plate Vibrations

Sergei Filippov and Mikhail Kolyada

St. Petersburg State University, Mathematics and Mechanics Faculty
Universitetsky Prospekt, 28, 198504 St. Petersburg, Russia
http://www.math.spbu.ru

Abstract. Free axisymmetric flexural vibrations of an annular elastic thin plate are studied. Numerical solutions of eigenvalue problem for various boundary conditions are obtained. The plate can be used as a model of the supporting frame of a shell. In this connection the boundary conditions corresponding the attaching of the plate to a cylindrical shell are also considered. The plate is called narrow if the ratio of its width to the radius of the inner edge is small. For the vibrations analysis of a narrow plate new asymptotic methods are elaborated. Comparison asymptotic and numerical results shows, that the error of the approximate formulae quickly decreases with reduction of the plate width.

Keywords: Free vibrations, Annular thin plate, Eigenvalue problems, Numerical solution, Asymptotic approach.

1 Introduction

The problem of free vibrations of annular plate is not too complex. Nevertheless, till now appear papers devoted to this theme and containing new results. For example, in [1] an exact solution for the free vibration problem of annular plates with parabolically varying rigidity was received.

2 Analytical Solution

We consider the radius r_0 of inner plate edge as the characteristic size. Then the non-dimensional equations describing the axisymmetric transverse flexural vibrations of the annular plate have the form

$$(sQ)' + \lambda s w = 0, \quad sQ = (sM_1)' - M_2,$$

$$M_1 = \frac{h^2}{12}(\kappa_1 + \nu\kappa_2), \quad M_2 = \frac{h^2}{12}(\kappa_2 + \nu\kappa_1), \tag{1}$$

$$\kappa_1 = \vartheta', \quad \kappa_2 = \vartheta/s, \quad \vartheta = -w', \quad \lambda = \frac{\sigma\rho r_0^2\omega^2}{E},$$

where $(')$ denotes the derivative with respect to the radial coordinate, $s \in [1, s_b]$, $s_b = 1 + b$, b is the dimensionless plate width, w is the deflection, Q_1, M_1, M_2 are

I. Dimov, I. Faragó, and L. Vulkov (Eds.): NAA 2012, LNCS 8236, pp. 295–302, 2013.

the dimensionless stress-resultant and stress-couples, h is the dimensionless plate thickness, ϑ is the angle of rotation, λ is the frequency parameter, $\sigma = 1 - \nu^2$, ν is Poisson's ratio, E is Young's modulus, ρ is the mass density, ω is the vibration frequency.

System (1) can be reduced to the equation

$$\Delta^2 w - \gamma^4 w = 0, \quad \Delta w = \frac{1}{s}(sw')', \quad \gamma^4 = \frac{12\lambda}{h^2}. \tag{2}$$

The exact solution of equation (2) has the form

$$w = C_1 J_0(\gamma s) + C_2 Y_0(\gamma s) + C_3 I_0(\gamma s) + C_4 K_0(\gamma s), \tag{3}$$

where C_j $(j = 1, 2, 3, 4)$ are the arbitrary constants, J_0 and Y_0 are the Bessel functions satisfying the equation

$$\Delta w - \gamma^2 w = 0,$$

I_0 and K_0 are the modified Bessel functions satisfying the equation

$$\Delta w + \gamma^2 w = 0.$$

The substitution of the solution (3) into boundary conditions provides the following linear algebraic equations:

$$\sum_{k=1}^{4} a_{jk} C_k = 0, \quad j = 1, 2, 3, 4. \tag{4}$$

The formulae for coefficients a_{jk} contain values of functions I_0, J_0, K_0, Y_0 and its derivatives at $s = 1$, $s = s_b$ and depend on the boundary conditions. System (4) has nontrivial solutions if its determinant vanishes:

$$\det \mathbf{A}(\gamma) = 0, \quad \mathbf{A} = \{a_{jk}\}. \tag{5}$$

We can calculate the frequency parameter, λ, using the following formula

$$\lambda = h^2 \gamma^4 / 12, \tag{6}$$

where γ is the root of equation (5).

3   Numerical Solution

One can solve equations (1) numerically, using the initial-value or shooting procedure [2]. After the introducing new variables

$$y_1 = w, \quad y_2 = \vartheta, \quad y_3 = 12 M_1 / h^2, \quad y_4 = 12 Q / h^2$$

equations (1) take the form

$$\begin{aligned} y_1' = -y_2, \quad y_2' = -\nu y_2 / s + y_3, \\ y_3' = \sigma y_2 / s^2 - (1 - \nu) y_3 + y_4, \quad y_4' = -\gamma^4 y_1 - y_4 / s. \end{aligned} \tag{7}$$

Consider for example the annular plate with clamped edges. In this case boundary conditions for equations (7) are

$$y_1(1) = y_2(1) = y_1(s_b) = y_2(s_b) = 0.$$

Using Runge-Kutta method we obtain solutions

$$y^{(1)} = (y_1^{(1)}, y_2^{(1)}, y_3^{(1)}, y_4^{(1)}), \quad y^{(2)} = (y_1^{(2)}, y_2^{(2)}, y_3^{(2)}, y_4^{(2)}), \quad s \in [1, s_b]$$

of two Cauchy problems with initial conditions

$$
\begin{aligned}
y_1^{(1)} &= 0, \quad y_2^{(1)} = 0, \quad y_3^{(1)} = 1, \quad y_4^{(1)} = 0, \\
y_1^{(2)} &= 0, \quad y_2^{(2)} = 0, \quad y_3^{(2)} = 0, \quad y_4^{(1)} = 0, \quad s = 1
\end{aligned}
$$

A linear combination of these two solutions $C_1 y^{(1)} + C_2 y^{(2)}$ satisfies the boundary conditions at $s = 1$. This combination satisfies also conditions at $s = s_b$ if

$$C_1 y_1^{(1)} + C_2 y_1^{(2)} = 0, \quad C_1 y_2^{(1)} + C_2 y_2^{(2)} = 0. \tag{8}$$

System of equations (8) has a nontrivial solution in the case

$$y_1^{(1)}(s_b) y_2^{(2)}(s_b) - y_1^{(2)}(s_b) y_2^{(1)}(s_b) = 0. \tag{9}$$

To obtain values of γ one have to find the roots of equation (9).

Compare algorithms base on analytical and numerical approaches. To find roots of equation (5) we must obtain formulae for coefficients a_{jk}, calculate the values of the Bessel functions and its derivatives and solve numerically equation (5). This procedure we have to do for each version of boundary conditions. If we use numerical algorithm based on formulae (7)–(9), then for the change of boundary conditions we need only change indexes into initial conditions and equation (9). It is obvious, that the numerical algorithm is more convenient for calculation of vibrations frequencies of a plate under various boundary conditions. With its help the results presented in Table 1 are obtained.

Table 1. The minimal positive root γ_1 of Eq. (9) vs. the dimensionless plate wide b for various boundary conditions

b	Root γ_1 of Eq. (9)		
	CL & CL	FS & FS	CL & FR
0.1	47.36	31.45	18.57
0.2	23.68	15.73	9.219
0.3	15.78	10.49	6.112
0.4	11.84	7.874	4.564
0.6	7.888	5.258	3.024
0.8	5.914	3.951	2.260
1.0	4.729	3.167	1.804

In the first column the values of the dimensionless plate wide b are given. The next columns contain the values of the minimal positive root γ_1 of equation (9) for various boundary conditions. The following designations are used: CL & CL — the plate with the clamped edges,

$$w = \vartheta = 0, \quad s = 1, \quad s = s_b, \tag{10}$$

FS & FS — the plate with the free supported edges,

$$w = M_1 = 0, \quad s = 1, \quad s = s_b, \tag{11}$$

CL & FR — the plate with the clamped inner edge $s = 1$ and free outer edge $s = s_b$

$$w = \vartheta = 0, \quad s = 1, \quad M_1 = Q = 0, \quad s = s_b, \tag{12}$$

The root γ_1 corresponds to the fundamental vibrations frequency of the plate.

4 Narrow Plate

Assume that the plate is narrow, i.e. $b \ll 1$. Replacing variable

$$s = 1 + bx \tag{13}$$

in equation (2) and neglecting small terms leads to the approximate equation

$$\frac{d^4 w}{dx^4} - \beta^4 w = 0, \quad x \in [0, 1] \tag{14}$$

where $\beta = b\gamma$. Equation (14) also describes the vibrations of a beam. Its solution has the form

$$w = C_1 \sin \beta x + C_2 \cos \beta x + C_3 \sinh \beta x + C_4 \cosh \beta. \tag{15}$$

After the replacing variable (13) instead boundary conditions (10)–(12) we obtain approximate boundary conditions

$$w = \frac{dw}{dx} = 0, \quad x = 0, \quad x = 1, \tag{16}$$

$$w = \frac{d^2 w}{dx^2} = 0, \quad x = 0, \quad x = 1, \tag{17}$$

$$w = \frac{dw}{dx} = 0, \quad x = 0, \quad \frac{d^2 w}{dx^2} = \frac{d^3 w}{dx^3} = 0, \quad x = 1, \tag{18}$$

corresponding vibrations of beam with clamped edges (16), free supported edges (17) and vibrations of cantilever beam (18).

Substituting (15) into boundary conditions (16)–(18) we obtain four linear homogeneous algebraic equations for unknowns C_1, C_2, C_3 and C_4. This system of equations have nontrivial solutions if its characteristic determinant

$$D(\beta) = 0. \tag{19}$$

Let β_k, $k = 1, 2, \ldots$ are positive roots of the equation (19). Then

$$\gamma_k = \frac{\beta_k}{b}, \quad \omega_k = \frac{h\beta_k^2}{b^2 r_0}\sqrt{\frac{E}{12\sigma\rho}}. \tag{20}$$

Therefore for narrow annular plate the vibrations frequencies are proportional to $1/b^2$.

So, the exact analytical solution (3) is practically useless. For calculation of frequencies it is more convenient to use a numerical method. The dependence of frequencies on parameters of a plate and boundary conditions can be found only by means of the approximate analytical solution.

For conditions (16), (17) and (18) the numbers β_k are the roots of equations

$$\cosh\beta \cos\beta - 1 = 0, \quad \sin\beta = 0, \quad \cosh\beta\cos\beta + 1 = 0.$$

The minimal positive roots of the last equations are 4.730, 3.142, 1.875. Using these roots and first formula (20) we can obtain the approximate values of γ_1 for boundary conditions (10)–(12) and various values of plate wide b. For example in the case when plate edges are clamped and $b = 0.1$ we get $\gamma_1 \simeq 4.730/0.1 = 47.30$.

Comparison approximate values of γ_1 and it exact values from Table 1 for $b \in [0.1, 1.0]$ shows that in the cases CL & CL, FS & FS and CL & FR the relative error is less than 0.1%, 1% and 3.9% correspondingly. Hence, the approximate formula (20) received for a narrow plate, gives good results even for enough great values of plate width.

5   Plate Joined with Shell

Consider free axisymmetric vibration of the thin annular plate joined with thin cylindrical shell (Fig. 1). This problem is of great importance for the theory

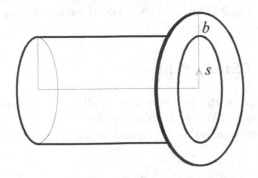

Fig. 1. Annular plate joined with cylindrical shell

of ring-stiffened cylindrical shells [3]. Asymptotic analysis [3] got the following boundary conditions on the inner plate edge

$$w = 0, \quad M_1 = c\vartheta, \quad s = 1, \quad c = 2(3\sigma)^{1/4} h_s^{5/2} h^{-3}, \tag{21}$$

where dimensionless shell thickness h_s and plate thickness h are small parameters. In the case $h_s = h$ the stiffness $c = 2(3\sigma)^{1/4}/h \gg 1$ and instead conditions (21) one can use the approximate boundary conditions

$$w(1) = \vartheta(1) = 0,$$

corresponding to the clamped edge. If $c \ll 1$ then approximate boundary conditions

$$w(1) = M_1(1) = 0,$$

correspond to the free supported edge. The outer plate edge if free, i.e.

$$M_1(s_b) = Q(s_b) = 0. \tag{22}$$

Assume that $b \ll 1$. After the replacing variable (13) we obtain eigenvalue problem for equation (14) with the boundary conditions

$$w = 0, \quad \frac{d^2 w}{dx^2} = c_* \frac{dw}{dx}, \quad x = 0, \quad \frac{d^2 w}{dx^2} = \frac{d^3 w}{dx^3} = 0, \quad x = 1, \tag{23}$$

where $c_* = bc$. The substitution of solution (15) into conditions (23) get system of linear algebraic equations. Equating to zero its characteristic determinant, we receive the equation

$$\beta(\sinh \beta \cos \beta - \cosh \beta \sin \beta) + c_*(\cosh \beta \cos \beta + 1) = 0. \tag{24}$$

We can find the positive roots β_k of equation (24) by means of a numerical method. Then for calculation of vibrations frequencies we use formula (20).

The minimal positive root of equation (24) $\beta_1 \to 0$ if $c_* \to 0$. Since approximate value of $\gamma_1 = \beta_1/b$ is very small for small c_*. From the other side numerical calculations show that for $c_* = 0$ exact value of $\gamma_1 \neq 0$. In particular, if $c_* = 0$ then $\gamma_1 = 3.94$ for $b = 0.1$. So, we need special asymptotic approach to obtain approximation for γ_1 in the case of small c_*.

6   The Case of Small Stiffness

Consider equation (2) with the boundary conditions (21) and (22). Replacing the variable according (13) and keeping in the equation and in the boundary conditions small terms including the terms of the order b^2 we obtain following eigenvalue problem

$$\frac{d^4 w}{dx^4} + 2b \frac{d^3 w}{dx^3} - b^2 \left(2x \frac{d^3 w}{dx^3} + \frac{d^2 w}{dx^2} \right) - \alpha^4 b^2 w = 0, \tag{25}$$

$$w = 0, \quad \frac{d^2 w}{dx^2} + \nu b \frac{dw}{dx} = c_0 b^2 \frac{dw}{dx}, \quad x = 0,$$

$$\frac{d^2 w}{dx^2} + \nu b(1-b)\frac{dw}{dx} = 0, \quad \frac{d^3 w}{dx^3} + b(1-b)\frac{d^2 w}{dx^2} - b^2 \frac{dw}{dx} = 0, \quad x = 1, \tag{26}$$

where $\alpha = \gamma_1 \sqrt{b}$, $c_0 = c/b = c_*/b^2$.

We seek solution of problem (25), (26) in the form

$$w = w_0 + b w_1 + b^2 w_2. \tag{27}$$

Substituting (27) into (25) and (26) and solving equations of the zeroth and first approximations we obtain

$$w_0 = x, \quad w_1 = -\frac{\nu}{2x} + C_1 x,$$

where C_1 is the arbitrary constant. The multiplication the equations of the second approximation

$$\frac{d^4 w_2}{dx^4} - \alpha^4 w_0 = 0$$

on w_0 and the integration by parts over the interval $[0,1]$ gives the relation

$$\alpha^4 \int_0^1 x^2 \, dx = \left[\frac{d^3 w_2}{dx^3} w_0 - \frac{d^2 w_2}{dx^2} \frac{dw_0}{dx} \right]_0^1. \tag{28}$$

Taken into account boundary conditions (26) from (28) we obtain

$$\alpha = \sqrt[4]{3(\sigma + c_0)}, \quad \gamma_1 = \alpha/\sqrt{b}. \tag{29}$$

The constant C_1 can be found from the third approximation.

Table 2 lists the values of the parameter γ_1 for $b = 0.1$ and various values of stiffness c presented in the first row. The results in the second row are obtained numerically. The approximate values of γ_1 in the third row are calculated by means of the solution of equation (24). The fourth row contain results found from asymptotic formula (29).

Table 2. The parameter γ_1 calculated by means of various methods vs. the dimensionless stiffness c for $b = 0.1$

c	0	0.1	0.2	0.5	1.0	3.0	10	30	∞
Num.	3.94	4.78	5.33	6.37	7.42	9.56	12.4	15.0	18.6
(24)		4.16	4.94	6.21	7,36	9.53	12.5	15.1	18.8
(29)	4.06	4.89	5.43	6.49	7.56	9.81			

If the stiffness c is not small the asymptotic results based on the solution of equation (24) are in the good agreement with the numerical ones. For small values of c one may use approximate formula (29). In the case $b = 0.1$ the relative error of asymptotic results is no more than 3%.

7   Conclusions

The three approach to the solution of eigenvalue problem describing free vibrations of annular plate are analyzed: the exact solution of equations, the numerical method and the asymptotic integration. The algorithm base on exact solution is more complicated then numerical one. Except for that the exact solution does not allow to study dependence of vibrations frequencies on parameters of the plate. Such dependence give the simple approximate formulae obtained for narrow plate by means of asymptotic methods. Comparison asymptotic and numerical results shows, that one can use approximate formulae for the enough wide plate.

The same roles play exact, numerical and asymptotic methods at the solutions of many other problems, in particular, at the analysis of cylindrical shells vibrations.

Acknowledgements. This work was supported by RFBR (grant 10-01-00244) which is gratefully acknowledged.

References

1. Wang, C.Y., Wang, C.M., Chen, W.Q.: Exact closed form solutions for free vibration of non-uniform annular plates. The IES Journal Part A: Civil & Structural Engineering 24, 50–55 (2012)
2. Ascher, U.M., Mattheij, R., Russell, R.: Numerical Solution of Boundary Value Problems for Ordinary Differential Equations, 2nd edn. SIAM, Philadelphia (1995)
3. Filippov, S.B.: Optimal design of stiffened cylindrical shells based on an asymptotic approach. Technische Mechanik 24, 221–230 (2004)

A Singularly Perturbed Reaction-Diffusion Problem with Incompatible Boundary-Initial Data

J.L. Gracia and E. O'Riordan

IUMA. Department of Applied Mathematics, University of Zaragoza, Spain
School of Mathematical Sciences, Dublin City University, Ireland

Abstract. A singularly perturbed reaction-diffusion parabolic problem with an incompatibility between the initial and boundary conditions is examined. A finite difference scheme is considered which utilizes a special finite difference operator and a piecewise uniform Shishkin mesh. Numerical results are presented for both nodal and global pointwise convergence, using bilinear interpolation and, also, an interpolation method based on the error function. These results show that the method is not globally convergent when bilinear interpolation is used but they indicate that, for the test problem considered, it is globally convergent using the second type of interpolation.

1 Introduction

The solution of a singularly perturbed problem typically has large derivatives in narrow subregions of the domain, called layer regions. Due to the presence of these large derivatives, classical numerical methods are not appropriate to numerically solve singularly perturbed problems [2].

In this paper we consider a singularly perturbed parabolic problem of reaction-diffusion type with an incompatibility between the initial and boundary conditions. This is a bi-singular problem where a classical singularity (due to the incompatibility in the data) is entwined with the singular nature of the differential operator when the diffusion parameter takes arbitrary small values. Parabolic problems with incompatible data appear in heat transfer at fluid/solid interfaces, geophysical fluid mechanics and in chemistry (see relevant references within [1]).

A special finite difference operator on a special fitted mesh of Shishkin type [2] which concentrates mesh points in the layer regions is used to approximate the solution of this kind of problems. The fitting coefficient of the method is defined as in Hemker et al. [4,5] where a classical and a singularly perturbed parabolic problem of reaction-diffusion type with a discontinuity in the initial condition was examined. In essence, this is an exact finite difference scheme on a uniform mesh for the parameter-dependent error function associated with the discontinuity in the initial condition.

I. Dimov, I. Faragó, and L. Vulkov (Eds.): NAA 2012, LNCS 8236, pp. 303–310, 2013.

In [3], the effect of regularizing the discontinuity in the data was examined and accurate approximations were obtained outside of a neighbourhood of the point of incompatibility. In this paper the nodal and global convergence of the proposed scheme are numerically studied in the whole domain. Nodal convergence is observed for a specific test problem, but the global convergence requires a more careful investigation. The bilinear interpolant of the numerical solution generated by the fitted operator method on the fitted mesh is examined, and we conclude that the numerical scheme is not globally accurate in the maximum pointwise norm. However, using an interpolation method based on the error function, global convergence is observed in the numerical output.

The class of singularly perturbed reaction-diffusion problems considered in this paper is given by

$$-\varepsilon u_{xx} + a(t)u + u_t = f(x,t), \ G := (x,t) \in (0,1) \times (0,1], \qquad (1.1a)$$

$$u(x,0) = \phi_B(x), \ u(0,t) = \phi_L(t), \ u(1,t) = \phi_R(t), \qquad (1.1b)$$

$$a \geq 0, \quad \phi_R(0) = \phi_B(1), \quad \phi_L(0) \neq \phi_B(0). \qquad (1.1c)$$

The singularity associated with $\phi_L(0) \neq \phi_B(0)$ is related to the following function (see [5])

$$(\phi_B(0) - \phi_L(0))w(x,t)e^{-\int_0^t a(s)ds},$$

where

$$w(x,t) := \frac{1}{2}\mathrm{erf}(\frac{x}{2\sqrt{\varepsilon t}}), \qquad \mathrm{erf}(\zeta) := \frac{2}{\sqrt{\pi}} \int_0^\zeta \exp(-\alpha^2)\,d\alpha. \qquad (1.2)$$

For $t = 0$, in $x = 0$, the error function is defined by continuous extension. In Figure 1 we display the computed generated "exact" error function and its contour lines for $\varepsilon = 2^{-10}$, where we observe the singular behavior of the solution and that it is constant along curves where $x/\sqrt{\varepsilon t} = $ constant.

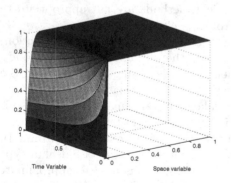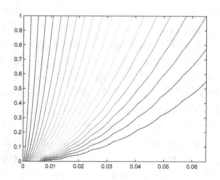

Fig. 1. Computed generated "exact" error function $w(x,t)$ for $\varepsilon = 2^{-10}$ in $\bar{G}$ and its associated contour plot in the vicinity of the edge $x = 0$

2   Numerical Scheme

To approximate the solution of problems of the form (1.1), we first examine a fitted operator (given in [5]) method defined on a special piecewise uniform mesh of Shishkin type [2], denoted by $\overline{G}^{N,M}$, where N and M are two positive integers corresponding to the space and time discretization parameters, respectively. The mesh is uniform in time and it splits the space domain $[0,1]$ into three subintervals

$$[0,\sigma],\ [\sigma,1-\sigma],\ [1-\sigma,1]; \quad \text{where } \sigma := \min\{\tfrac{1}{4}, 2\sqrt{\varepsilon}\ln N\}. \quad (2.1)$$

The grid points are uniformly distributed within each subinterval such that

$$x_0 = 0,\ x_{N/4} = \sigma,\ x_{3N/4} = 1-\sigma,\ x_N = 1.$$

Note that the mesh condenses near $x = 0$ and $x = 1$. The finite difference scheme is given by

$$-\varepsilon\kappa(x,t)\delta_x^2 U + a(t)U + D_t^- U = f(x,t), \quad (x,t) \in G^{N,M}, \quad (2.2\text{a})$$

$$U = u, \quad (x,t) \in \overline{\varGamma}^{N,M}, \quad (2.2\text{b})$$

where $G^{N,M} = G \cap \overline{G}^{N,M}$, $\varGamma^{N,M} = \overline{G}\backslash G^{N,M}$; D_t^- denotes the backward finite difference operator in time and δ_x^2 is the standard finite difference replacement for the second derivative in space. The fitting coefficient κ is defined in the subdomains $(0,\sigma)$ and $(\sigma, 1-\sigma)$, and $(1-\sigma, 1)$ (i.e., where the mesh is uniform) by

$$\kappa(x,t) := \frac{D_t^- w(x,t) + D_t^- u_0(x,t)}{\varepsilon\delta_x^2 w(x,t) + \varepsilon\delta_x^2 u_0(x,t)}, \quad (x,t) \in G^{N,M}, \quad (2.2\text{c})$$

where $w(x,t)$ is defined in (1.2) and $u_0(x,t) = -x^3 - 6\varepsilon x t$ (see [5] for more details on this choice of u_0). Note that the fitting factor $\kappa(x,t)$ is independent of the coefficient $a(t)$ in problem (1.1).

At the transition points σ and $1 - \sigma$, the fitting coefficient $\kappa(x,t)$ is specially defined since the Shishkin mesh is, in general, highly anisotropic at these points. The coefficient $\kappa := \kappa(x_{i_0}, t_j)$ at the points (σ, t_j) and $(1 - \sigma, t_j)$ is computed using simple linear interpolation based on the values $\kappa(x_{i_0-1}, t_j)$ and $\kappa(x_{i_0+1}, t_j)$.

3   Test Problem

We consider the following test problem

$$u_t - \varepsilon u_{xx} + (1+t^2)u = -(1-x)(2-x), \quad (x,t) \in (0,1) \times (0,1], \quad (3.1)$$

where the initial and boundary conditions are taken to be

$$u(x,0) = 1,\ x \in (0,1), \quad u(0,t) = 0,\ t \in [0,1],\ u(1,t) = 1,\ t \in [0,1].$$

The exact solution of problem (3.1) is not available, and the orders of convergence are estimated using the double–mesh principle [2, Chapter 8]: we define the double–mesh differences $D_\varepsilon^{N,M}$ for each value of $\varepsilon \in S_\varepsilon$ with $S_\varepsilon = \{2^0, 2^{-2}, \ldots, 2^{-20}\}$ and the uniform double–mesh differences $D^{N,M}$ by

$$D_\varepsilon^{N,M} := \left\| U^{N,M} - \bar{U}^{2N,2M} \right\|_{\bar{G}^{N,M}}, \quad D^{N,M} := \max_{\varepsilon \in S_\varepsilon} D_\varepsilon^{N,M},$$

where $U^{N,M}$, $U^{2N,2M}$ denote the discrete functions defined on the meshes $\bar{G}^{N,M}$ and $\bar{G}^{2N,2M}$, respectively; $\bar{U}^{2N,2M}$ is the bilinear interpolant of the numerical solution $U^{2N,2M}$. The computed order of convergence $Q_\varepsilon^{N,M}$ and the computed order of uniform convergence $Q^{N,M}$ are defined by

$$Q_\varepsilon^{N,M} := \log_2\left(\frac{D_\varepsilon^{N,M}}{D_\varepsilon^{2N,2M}} \right), \quad Q^{N,M} := \log_2\left(\frac{D^{N,M}}{D^{2N,2M}} \right).$$

Nodal double–mesh differences using scheme (2.2) to approximate the solution of problem (3.1) are given in Table 1, and we observe that the method is nodally convergent. It is worth remarking that this nodal accuracy is being observed on a mesh with mesh points concentrated near the corner $(0,0)$.

Nevertheless, a global approximation is more appropriate for problems with different scales in the solution. The double–mesh global differences

$$d_\varepsilon^{N,M} := \left\| U_I^{N,M} - U_I^{2N,2M} \right\|_{\bar{G}^{N,M} \cup \bar{G}^{2N,2M}}, \quad d^{N,M} := \max_{\varepsilon \in S_\varepsilon} d_\varepsilon^{N,M},$$

Table 1. The maximum and uniform double–mesh nodal differences $D_\varepsilon^{N,M}, D^{N,M}$ and their computed orders of convergence $Q_\varepsilon^{N,M}, Q^{N,M}$ for problem (3.1) using the numerical method (2.2)

	N=M=8	N=M=16	N=M=32	N=M=64	N=M=128	N=M=256	N=M=512
$\varepsilon = 2^0$	6.241E-002	3.318E-002	1.704E-002	8.625E-003	4.317E-003	2.186E-003	1.303E-003
	0.911	0.961	0.983	0.999	0.982	0.747	
$\varepsilon = 2^{-2}$	5.482E-002	2.770E-002	1.405E-002	7.094E-003	3.565E-003	1.786E-003	8.930E-004
	0.985	0.979	0.986	0.993	0.997	1.000	
$\varepsilon = 2^{-4}$	9.829E-002	2.862E-002	1.357E-002	6.662E-003	3.304E-003	1.646E-003	8.216E-004
	1.780	1.076	1.027	1.012	1.005	1.002	
$\varepsilon = 2^{-6}$	2.714E-001	9.473E-002	4.009E-002	2.563E-002	1.270E-002	6.549E-003	3.366E-003
	1.519	1.241	0.646	1.012	0.956	0.960	
$\varepsilon = 2^{-8}$	5.472E-001	2.827E-001	1.852E-001	9.267E-002	4.592E-002	2.858E-002	1.380E-002
	0.953	0.610	0.999	1.013	0.684	1.050	
$\varepsilon = 2^{-10}$	4.295E-001	3.372E-001	3.432E-001	3.293E-001	2.004E-001	9.753E-002	4.733E-002
	0.349	-0.025	0.060	0.716	1.039	1.043	
$\varepsilon = 2^{-12}$	4.288E-001	3.370E-001	3.295E-001	3.171E-001	2.654E-001	1.931E-001	1.233E-001
	0.348	0.033	0.055	0.257	0.459	0.647	
$\varepsilon = 2^{-14}$	4.286E-001	3.369E-001	3.295E-001	3.171E-001	2.654E-001	1.931E-001	1.233E-001
	0.347	0.032	0.055	0.257	0.459	0.647	
$\varepsilon = 2^{-16}$	4.284E-001	3.369E-001	3.295E-001	3.171E-001	2.654E-001	1.931E-001	1.233E-001
	0.347	0.032	0.055	0.257	0.459	0.647	
$\varepsilon = 2^{-18}$	4.284E-001	3.369E-001	3.295E-001	3.171E-001	2.654E-001	1.931E-001	1.233E-001
	0.347	0.032	0.055	0.257	0.459	0.647	
$\varepsilon = 2^{-20}$	4.284E-001	3.369E-001	3.295E-001	3.171E-001	2.654E-001	1.931E-001	1.233E-001
	0.347	0.032	0.055	0.257	0.459	0.647	
$D^{N,M}$	5.472E-001	3.372E-001	3.432E-001	3.293E-001	2.654E-001	1.931E-001	1.233E-001
$Q^{N,M}$	0.699	-0.025	0.060	0.311	0.459	0.647	

where $U^{N,M}, U^{2N,2M}$ denote mesh functions defined, respectively, on the meshes $\bar{G}^{N,M}$ and $\bar{G}^{2N,2M}$; and $U_I^{N,M}, U_I^{2N,2M}$ denote interpolated global functions. The choice of interpolation is discussed below. From these interpolated values we calculate computed orders of global convergence $q_\varepsilon^{N,M}$ and uniform computed orders of global convergence $q^{N,M}$ using, respectively,

$$q_\varepsilon^{N,M} := \log_2\left(\frac{d_\varepsilon^{N,M}}{d_\varepsilon^{2N,2M}}\right), \quad q^{N,M} := \log_2\left(\frac{d^{N,M}}{d^{2N,2M}}\right).$$

4   Bilinear Interpolation

First, we consider the case where the double–mesh global differences $d_\varepsilon^{N,M}$ are computed using bilinear interpolation. In Table 2 we show the numerical results using scheme (2.2) for test problem (3.1). From these numerical results, we observe a lack of global convergence of the numerical method when bilinear interpolation is used to generate a global approximation to the continuous solution.

Table 2. The maximum and uniform double–mesh global differences $d_\varepsilon^{N,M}, d^{N,M}$ and their computed orders of convergence $q_\varepsilon^{N,M}, q^{N,M}$ for problem (3.1) using the numerical method (2.2) and bilinear interpolation

	N=M=8	N=M=16	N=M=32	N=M=64	N=M=128	N=M=256	N=M=512
$\varepsilon = 2^0$	3.386E-001	3.788E-001	4.116E-001	4.365E-001	4.547E-001	4.679E-001	4.772E-001
	-0.162	-0.120	-0.085	-0.059	-0.041	-0.029	
$\varepsilon = 2^{-2}$	2.004E-001	2.655E-001	3.259E-001	3.740E-001	4.099E-001	4.359E-001	4.545E-001
	-0.406	-0.296	-0.198	-0.132	-0.089	-0.060	
$\varepsilon = 2^{-4}$	9.829E-002	9.344E-002	1.772E-001	2.576E-001	3.232E-001	3.730E-001	4.095E-001
	0.073	-0.923	-0.540	-0.328	-0.207	-0.135	
$\varepsilon = 2^{-6}$	2.893E-001	1.997E-001	9.225E-002	8.134E-002	1.725E-001	2.558E-001	3.226E-001
	0.535	1.114	0.182	-1.084	-0.569	-0.335	
$\varepsilon = 2^{-8}$	5.472E-001	4.139E-001	3.382E-001	2.165E-001	9.795E-002	7.850E-002	1.713E-001
	0.403	0.291	0.643	1.144	0.319	-1.126	
$\varepsilon = 2^{-10}$	4.295E-001	3.388E-001	3.987E-001	4.453E-001	3.501E-001	2.206E-001	9.936E-002
	0.342	-0.235	-0.159	0.347	0.666	1.151	
$\varepsilon = 2^{-12}$	4.288E-001	3.388E-001	3.782E-001	3.930E-001	3.786E-001	3.308E-001	2.562E-001
	0.340	-0.159	-0.055	0.054	0.195	0.368	
$\varepsilon = 2^{-14}$	4.286E-001	3.388E-001	3.782E-001	3.930E-001	3.786E-001	3.308E-001	2.562E-001
	0.339	-0.159	-0.055	0.054	0.195	0.368	
$\varepsilon = 2^{-16}$	4.284E-001	3.388E-001	3.782E-001	3.930E-001	3.786E-001	3.308E-001	2.562E-001
	0.339	-0.159	-0.055	0.054	0.195	0.368	
$\varepsilon = 2^{-18}$	4.284E-001	3.388E-001	3.782E-001	3.930E-001	3.786E-001	3.308E-001	2.562E-001
	0.338	-0.159	-0.055	0.054	0.195	0.368	
$\varepsilon = 2^{-20}$	4.284E-001	3.388E-001	3.782E-001	3.930E-001	3.786E-001	3.308E-001	2.562E-001
	0.338	-0.159	-0.055	0.054	0.195	0.368	
$d^{N,M}$ $q^{N,M}$	5.472E-001	4.139E-001	4.116E-001	4.453E-001	4.547E-001	4.679E-001	4.772E-001
	0.403	0.008	-0.114	-0.030	-0.041	-0.029	

This is illustrated in Figure 2 where the values of the parameters are $\varepsilon = 2^{-20}$, $N = 32$ and $N = 64$. Note that it is a zoom of the double–mesh global differences in the vicinity of the edge $x = 0$ and we can observe that the maximum double–mesh differences do not decrease as the number of grid points in time and space are doubled.

These observations can be supported by examining bounds on the interpolation error in approximating the solution u with its bilinear interpolant $\bar{u}$ over

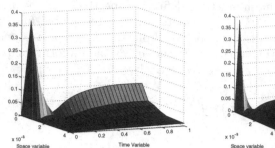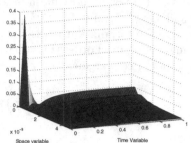

Fig. 2. Global double–mesh differences using a bilinear interpolant for problem (3.1) with $\varepsilon = 2^{-20}$, taking $M = N = 32$ (left figure), and $M = N = 64$ (right figure)

the computational cell $R_{i,j} := (x_i, x_{i+1}) \times (t_j, t_{j+1})$. We cite [5] for the following bounds on the partial derivatives of the solution

$$\left| \frac{\partial^{i+j} u}{\partial x^i \partial t^j} \right| \leq C(1 + \varepsilon^{-i/2} t^{-(i/2+j)} e^{-\frac{x}{2\sqrt{\varepsilon t}}}), \tag{4.1}$$

and we assume that the transition parameter $\sigma < 0.25$. From [6], we have the following bounds on the bilinear interpolation error

$$\|u - \bar{u}\|_{R_{i,j}} \leq C \min\left\{ (x_{i+1} - x_i)^2 \|u_{xx}\|_{R_{i,j}}, \max_{t_j \leq t \leq t_{j+1}} \int_{x_i}^{x_{i+1}} |u_x(s,t)| ds \right\}$$

$$+ C \min\left\{ (t_{j+1} - t_j)^2 \|u_{tt}\|_{R_{i,j}}, \max_{x_i \leq x \leq x_{i+1}} \int_{t_j}^{t_{j+1}} |u_t(x,s)| ds \right\}.$$

Combining these interpolation error estimates with (4.1), we obtain that

$$\|u - \bar{u}\|_{R_{i,j}} \leq C e^{-\frac{x_i}{2\sqrt{\varepsilon t_j}}} \min\left\{ (x_{i+1} - x_i)^2 \varepsilon^{-1} t_j^{-1}, (1 - e^{-\frac{(x_{i+1} - x_i)}{2\sqrt{\varepsilon t_j}}}) \right\}$$

$$+ C \min\left\{ e^{-\frac{x_i}{2\sqrt{\varepsilon t_j}}} (t_{j+1} - t_j)^2 t_j^{-2}, \int_{t_j}^{t_{j+1}} \frac{1}{s} e^{-\frac{x_i}{2\sqrt{\varepsilon s}}} ds \right\}.$$

Observe that if $x_i \geq 4\sqrt{\varepsilon}$, then

$$\int_{t_j}^{t_{j+1}} \frac{1}{s} e^{-\frac{x_i}{2\sqrt{\varepsilon s}}} ds \leq C(t_{j+1} - t_j) e^{-\frac{x_i}{2\sqrt{\varepsilon}}},$$

using the fact that $s^{-n} e^{-\delta/s} \leq e^{-\delta}$, if $n \leq \delta$. In the fine mesh region, where $x_{i+1} - x_i = C\sqrt{\varepsilon} N^{-1} \ln N$, if $t_j \geq \delta$, then we have that

$$\|u - \bar{u}\|_{R_{i,j}} \leq C(N^{-1} \ln N)^2 \delta^{-1} + CM^{-2} \delta^{-2}, \quad x_i < 2\sqrt{\varepsilon} \ln N, \ t_j \geq \delta;$$

and in the coarse mesh region, i.e., when $x_i \geq 2\sqrt{\varepsilon} \ln N$, it holds $e^{-\frac{x_i}{2\sqrt{\varepsilon t_j}}} \leq N^{-1}$, and therefore

$$\|u - \bar{u}\|_{R_{i,j}} \leq C N^{-1} + CM^{-1} N^{-1}; \quad x_i \geq 2\sqrt{\varepsilon} \ln N.$$

These bounds indicate that for a numerical method that is nodally parameter-uniform at all mesh points, global parameter-uniform convergence accuracy is obtained outside of the corner region $(x,t) \in (0,\sigma) \times (0,\delta), \delta > 0$, if one uses the fitted piecewise-uniform Shishkin mesh and bilinear interpolation. However, these interpolation bounds are of little value in the vicinity of the corner $(0,0)$.

5   Interpolation Based on the Error Function

Motivated by the previous numerical results, in this section we employ the following interpolant of the numerical solution $U^{N,M}$ to approximate the solution u of problem (1.1). For all $(x,t) \in [x_i, x_{i+1}] \times [t_j, t_{j+1}], 0 \le x_i < 1, t_j \ge 0$, we define

$$U_I^{N,M}(x,t) := \sum_{l,m=0}^{1} U_{i+l,j+m} T(t; x_{i+l}, t_{j+m}) S(x,t; x_{i+l}), \qquad (5.1)$$

where

$$T(t; x_{i+l}, t_{j+m}) := \frac{w(x_{i+l}, t) - w(x_{i+l}, t_{j+1-m})}{w(x_{i+l}, t_{j+m}) - w(x_{i+l}, t_{j+1-m})},$$

$$S(x,t; x_{i+l}) := \frac{w(x,t) - w(x_{i+1-l}, t)}{w(x_{i+l}, t) - w(x_{i+1-l}, t)}.$$

Note that this form of interpolation is globally exact for both the constant function 1 and for the error function $w(x,t)$. It is a two dimensional version of the interpolation discussed in §2 of [7]. In Table 3 we present the numerical results

Table 3. The maximum and uniform double–mesh global differences $d_\varepsilon^{N,M}, d^{N,M}$ and their computed orders of convergence $q_\varepsilon^{N,M}, q^{N,M}$ for problem (3.1) using the numerical method (2.2) and interpolant (5.1)

	N=M=8	N=M=16	N=M=32	N=M=64	N=M=128	N=M=256	N=M=512
$\varepsilon = 2^0$	1.021E-001	7.899E-002	4.957E-002	2.691E-002	1.400E-002	7.140E-003	3.573E-003
	0.371	0.672	0.881	0.943	0.971	0.999	
$\varepsilon = 2^{-2}$	1.742E-001	9.761E-002	5.004E-002	2.517E-002	1.257E-002	6.314E-003	3.159E-003
	0.835	0.964	0.991	1.001	0.994	0.999	
$\varepsilon = 2^{-4}$	2.574E-001	1.021E-001	6.199E-002	3.390E-002	1.424E-002	8.947E-003	4.527E-003
	1.333	0.721	0.871	1.251	0.670	0.983	
$\varepsilon = 2^{-6}$	7.721E-001	3.964E-001	2.288E-001	1.242E-001	6.394E-002	3.297E-002	1.676E-002
	0.962	0.793	0.881	0.958	0.956	0.976	
$\varepsilon = 2^{-8}$	1.856E-001	1.249E-001	4.851E-002	2.381E-002	1.190E-002	5.967E-003	3.017E-003
	0.572	1.364	1.027	1.001	0.996	0.984	
$\varepsilon = 2^{-10}$	1.813E-001	1.359E-001	9.781E-002	4.698E-002	1.415E-002	5.885E-003	2.963E-003
	0.416	0.475	1.058	1.731	1.266	0.990	
$\varepsilon = 2^{-12}$	1.791E-001	1.363E-001	9.531E-002	4.782E-002	1.943E-002	6.970E-003	2.933E-003
	0.394	0.516	0.995	1.299	1.479	1.249	
$\varepsilon = 2^{-14}$	1.779E-001	1.365E-001	9.537E-002	4.783E-002	1.943E-002	6.971E-003	2.930E-003
	0.382	0.517	0.996	1.299	1.479	1.251	
$\varepsilon = 2^{-16}$	1.773E-001	1.366E-001	9.541E-002	4.784E-002	1.944E-002	6.971E-003	2.928E-003
	0.376	0.518	0.996	1.300	1.479	1.251	
$\varepsilon = 2^{-18}$	1.770E-001	1.367E-001	9.542E-002	4.784E-002	1.944E-002	6.972E-003	2.928E-003
	0.373	0.518	0.996	1.300	1.479	1.252	
$\varepsilon = 2^{-20}$	1.768E-001	1.367E-001	9.543E-002	4.785E-002	1.944E-002	6.972E-003	2.927E-003
	0.371	0.519	0.996	1.300	1.479	1.252	
$d^{N,M}$	7.721E-001	3.964E-001	2.288E-001	1.242E-001	6.394E-002	3.297E-002	1.676E-002
$q^{N,M}$	0.962	0.793	0.881	0.958	0.956	0.976	

using scheme (2.2) for test problem (3.1) and we observe that the numerical method is globally convergent for this example when the specially designed interpolation (5.1) is used to generate a global approximation to the solution.

Conclusions

A fitted operator method on a piecewise-uniform Shishkin mesh has been designed to approximate the solution of problems from the class (1.1). Nodal pointwise accuracy was observed on this layer-adapted mesh. Global pointwise accuracy, as opposed to nodal convergence, was not observed numerically, when one combines a fitted finite difference operator, a rectangular layer–adapted mesh and bilinear interpolation. However, by replacing the bilinear interpolant by an interpolant based on the error function, pointwise accurate global approximations have been observed for the test problem considered in this paper. Theoretical analysis of the parameter uniform global convergence of the numerical method examined numerically in this paper remains open to future investigation.

Acknowledgement. This research was partially supported by the project MEC /FEDER MTM 2010-16917 and the Diputacion General de Aragon.

References

1. Chen, Q., Qin, Z., Temam, R.: Treatment of incompatible initial and boundary data for parabolic equations in higher dimensions. Math. Comp. 80(276), 2071–2096 (2011)
2. Farrell, P.A., Hegarty, A.F., Miller, J.J.H., O'Riordan, E., Shishkin, G.I.: Robust computational techniques for boundary layers. CRC Press (2000)
3. Gracia, J.L., O'Riordan, E.: A singularly perturbed parabolic problem with a layer in the initial condition, Appl. Math. Comp. 219, 498–510 (2012)
4. Hemker, P.W., Shishkin, G.I.: Approximation of parabolic PDEs with a discontinuous initial condition. East-West J. Numer. Math. 1, 287–302 (1993)
5. Hemker, P.W., Shishkin, G.I.: Discrete approximation of singularly perturbed parabolic PDEs with a discontinuous initial condition. Comp. Fluid Dynamics J. 2, 375–392 (1994)
6. Stynes, M., O'Riordan, E.: A uniformly convergent Galerkin method on a Shishkin mesh for a convection-diffusion problem. J. Math. Anal. Appl. 214, 36–54 (1997)
7. Zadorin, A.I., Zadorin, N.A.: Quadrature formulas for functions with a boundary layer component. Zh. Vychisl. Mat. Mat. Fiz. 51, 1952–1962 (2011) (Russian) ; Translation in Comput. Math. Math. Phys. 51, 1837–1846 (2011)

Method SMIF for Incompressible Fluid Flows Modeling

Valentin Gushchin and Pavel Matyushin

Institute for Computer Aided Design of the Russian Academy of Sciences,
Moscow, Russia
gushchin@icad.org.ru, pmatyushin@mail.ru

Abstract. For solving of the Navier-Stokes equations describing 3D incompressible viscous fluid flows the Splitting on physical factors Method for Incompressible Fluid flows (SMIF) with hybrid explicit finite difference scheme (second-order accuracy in space, minimum scheme viscosity and dispersion, capable for work in the wide range of Reynolds (Re) and internal Froude (Fr) numbers and monotonous) based on the Modified Central Difference Scheme and the Modified Upwind Difference Scheme with a special switch condition depending on the velocity sign and the signs of the first and second differences of the transferred functions has been developed and successfully applied. At the present paper the description of the numerical method SMIF and its application for simulation of the 3D separated homogeneous and density stratified fluid flows around a sphere are demonstrated.

Keywords: direct numerical simulation, viscous fluid, visualization of vortex structures, flow regime, formation mechanisms of vortices, sphere.

1 Introduction

Unsteady 3D separated and undulatory fluid flows around a moving blunt body are very wide spread phenomena in the nature. The understanding of such flows is very important both from theoretical and from practical points of view. In the experiments [1-2] the 2D internal waves structure in the vertical plane and the 3D vortex structure of the wake are observed. Using direct numerical simulation (DNS) the full 3D vortex structures of the flow (the 3D internal waves and the 3D wake) can be observed. Besides the numerical studies of the non-homogeneous (stratified) fluids are very rare. In this connection at the present paper the stratified viscous fluid flows around a sphere are investigated by means of DNS on the basis of the Navier-Stokes equations in the Boussinesq approximation on the supercomputers at the wide ranges of the main flow parameters.

2 Numerical Method SMIF

Let $\rho(x, y, z) = \rho_0(1 - x/(2A) + S(x, y, z))$ is the density of the linearly stratified fluid where x, y, z are the Cartesian coordinates; z, x, y are the streamwise, lift

I. Dimov, I. Faragó, and L. Vulkov (Eds.): NAA 2012, LNCS 8236, pp. 311–318, 2013.

and lateral directions (x, y, z have been non-dimensionalized by $d/2$, d is a sphere diameter); $A = \Lambda/d$ is the scale ratio, Λ is the buoyancy scale, which is related to the buoyancy frequency N and period $T_b(N = 2\pi/T_b, N^2 = g/\Lambda)$; g is the scalar of the gravitational acceleration; S is a dimensionless perturbation of salinity. The density stratified viscous fluid flows have been simulated on the basis of the Navier-Stokes equations in the Boussinesq approximation (1)–(3) (including the diffusion equation (1) for the stratified component (salt)) with four dimensionless parameters: $Fr = U/(N \cdot d)$, $Re = U \cdot d/\nu$, $A \gg 1$, $Sc = \nu/\kappa = 709.22$, where U is the scalar of the sphere velocity, ν is the kinematical viscosity, κ is the salt diffusion coefficient.

$$\frac{\partial S}{\partial t} + (\mathbf{v} \cdot \nabla) S = \frac{2}{Sc \cdot Re} \Delta S + \frac{v_x}{2A} \tag{1}$$

$$\frac{\partial \mathbf{v}}{\partial t} + (\mathbf{v} \cdot \nabla) \mathbf{v} = -\nabla p + \frac{2}{Re} \Delta \mathbf{v} + \frac{A}{2Fr^2} S \frac{\mathbf{g}}{g} \tag{2}$$

$$\nabla \cdot \mathbf{v} = 0 \tag{3}$$

In (1)-(3) $\mathbf{v} = (v_x, v_y, v_z)$ is the velocity vector (non-dimensionalized by U), p is a perturbation of pressure (non-dimensionalized by $\rho_0 U^2$). The spherical coordinate system $R, \theta, \phi(x = R \cdot sin\theta \cdot cos\phi, y = R \cdot sin\theta \cdot sin\phi, z = R \cdot cos\theta, \mathbf{v} = (v_R, v_\theta, v_\phi))$ and O-type grid are used. On the sphere surface the following boundary conditions have been used:

$$v_R = v_\theta = v_\phi = 0, \quad \frac{\partial p}{\partial R}\bigg|_{R=d/2} = \left(\frac{\partial S}{\partial R} - \frac{1}{2A}\frac{\partial x}{\partial R}\right)\bigg|_{R=d/2} = 0 \tag{4}$$

On the external boundary of the O-type grid the following boundary conditions have been used: 1) for $z < 0$: $v_R = cos\theta$, $v_\theta = -sin\theta$, $v_\phi = 0$, $S = 0$; 2) for $z \geq 0$: $v_R = cos\theta$, $v_\theta = -sin\theta$, $\frac{\partial v_\phi}{\partial R} = 0$, $\frac{\partial S}{\partial z} = 0$.

For solving of the Navier-Stokes equations (1)-(3) the Splitting on physical factors Method for Incompressible Fluid flows (SMIF) has been used [3-4].

Let the velocity, the perturbation of pressure and the perturbation of salinity are known at some moment $t_n = n \cdot \tau$, where τ is time step, and n is the number of time-steps. Then the calculation of the unknown functions at the next time level $t_{n+1} = (n + 1) \cdot \tau$ for equations (1)-(3) can be presented in the following four-step form:

$$\frac{S^{n+1} - S^n}{\tau} = -(\mathbf{v}^n \cdot \nabla) S^n + \frac{2}{Sc \cdot Re} \Delta S^n + \frac{v_x^n}{2A} \tag{5}$$

$$\frac{\tilde{\mathbf{v}} - \mathbf{v}^n}{\tau} = -(\mathbf{v}^n \cdot \nabla) \mathbf{v}^n + \frac{2}{Re} \Delta \mathbf{v}^n + \frac{A}{2Fr^2} S^{n+1} \frac{\mathbf{g}}{g} \tag{6}$$

$$\tau \Delta p = \nabla \cdot \tilde{\mathbf{v}} \tag{7}$$

$$\frac{\mathbf{v}^{n+1} - \tilde{\mathbf{v}}}{\tau} = -\nabla p \tag{8}$$

The Poisson equation (7) for the pressure has been solved by the diagonal Preconditioned Conjugate Gradients Method.

In order to understand the finite-difference scheme for the convective terms of the equations (1)–(2) let us consider the linear model equation:

$$f_t + u f_x = 0, \quad u = const. \tag{9}$$

Let

$$\frac{f_i^{n+1} - f_i^n}{\tau} + u \frac{f_{i+1/2}^n - f_{i-1/2}^n}{h} = 0 \tag{10}$$

be a finite-difference approximation of equation (9).

Let us investigate the class of the difference scheme which can be written in the form of the two-parameter family which depends on the parameters α and β in the following manner:

$$\begin{aligned} f_{i+1/2}^n &= \alpha f_{i-1}^n + (1 - \alpha - \beta) f_i^n + \beta f_{i+1}^n, \quad if \ \ u \geq 0, \\ f_{i+1/2}^n &= \alpha f_{i+2}^n + (1 - \alpha - \beta) f_{i+1}^n + \beta f_i^n, \quad if \ \ u < 0. \end{aligned} \tag{11}$$

In this case the first differential approximation for equation (10) has the form

$$f_t + u f_x = \left[\frac{h}{2} |u| (1 + 2\alpha - 2\beta) - \frac{\tau u^2}{2} \right] f_{xx} \tag{12}$$

If we put $\alpha = \beta = 0$ in (11) we'll obtain usual first order monotonic scheme which is stable when

$$0 < C = \frac{\tau |u|}{h} \leq 1, \quad where \ C \ is \ the \ Courant \ number. \tag{13}$$

If $\alpha = 0, \beta = 0.5$ we'll obtain the usual central difference scheme, and for $\alpha = -0.5, \beta = 0$ – the usual upwind scheme. Both last two schemes have second order of accuracy in space variable and are non-monotonic.

It is known that it is impossible to construct a homogeneous monotonic difference scheme of higher order than the first order of the approximation for equation (9). A monotonic scheme of higher order can therefore only be constructed either on the basis of second-order homogeneous scheme using smoothing operators, or on the basis of the hybrid schemes using different switch conditions from one scheme to another (depending on the nature of the solution), possibly with the use of smoothing. Here we are going to consider a hybrid monotonic difference scheme.

Let us investigate schemes with upwind differences, i.e. $\beta = 0$. The requirement that the scheme viscosity should be a minimum, as can readily be seen from equation (12), impose the following condition on α :

$$\alpha = -0.5 \cdot (1 - C) \tag{14}$$

For schemes with $\alpha = 0$, the analogous condition is

$$\beta = 0.5 \cdot (1 - C) \tag{15}$$

Since an explicit finite difference scheme considered, we shall restrict the subsequent analysis to the necessary condition for stability in the case of the explicit schemes (13).

Let us assume that there is a monotonic net function f_i^n , for example, $\Delta f_{i+1/2}^n \equiv f_{i+1}^n - f_i^n \geq 0$ at any i.

The function f_i^{n+1} will also be monotonic when the following conditions are satisfied:

(a) for a scheme with $\beta = 0$ and α from relationship (14), under the condition $\Delta f_{i+1/2}^n \geq \zeta(C) \cdot \Delta f_{i-1/2}^n$, where $\zeta(C) = 0.5 \cdot (1 - C)/(2 - C)$;

(b) for a scheme with $\alpha = 0$ and β from relationship (15), under the condition $\Delta f_{i+1/2}^n \leq \sigma(C) \cdot \Delta f_{i-1/2}^n$, where $\sigma(C) = 2 \cdot (1 + C)/C$.

It can be seen from this that the domains of monotonicity of the homogeneous schemes being considered have a non-empty intersection. Hence, a whole class of hybrid schemes is distinguished by the condition of switching from one homogeneous scheme to another. The general form of this condition is as follows:

$$\Delta f_{i+1/2}^n = \delta \cdot \Delta f_{i-1/2}^n, \text{ where } \zeta(C) \leq \delta \leq \sigma(C).$$

The choice of $\delta = 1$ corresponds to the points of the interchange of the sign of the second difference f_i^n and makes it possible to obtain the estimation $f_{xx} = O(h)$ for the required function f at the intersection points, by means of which a second-order approximation is retained with respect to the spatial variables of smooth solutions. We used the following switching condition:

if $(u \cdot \Delta f \Delta^2 f)_{i+1/2}^n \geq 0$, then the scheme with $\beta = 0$ (MUDS) is used;

if $(u \cdot \Delta f \Delta^2 f)_{i+1/2}^n < 0$, then the scheme with $\alpha = 0$ (MCDS) is used;

where $\Delta^2 f_{i+1/2}^n = \Delta f_{i+1}^n - \Delta f_i^n$.

On smooth solutions this scheme has a second order of approximation with respect to the time and spatial variables. It is stable when the Courant criterion (13) is satisfied and monotonic. More over it was shown that this hybrid scheme comes nearest to the third order schemes.

The generalization of the considered finite-difference scheme for 2D and 3D problems is easily performed for convective terms in (1) – (2). For the approximation of other space derivatives in equations (1)-(3) the central differences are used.

The efficiency of the method SMIF and the greater power of supercomputers make it possible adequately to model the 3D separated incompressible viscous flows past a sphere (Fig.1) and a circular cylinder at moderate Reynolds numbers [5-13] and the air, heat and mass transfer in the clean rooms [14].

3   The Visualization Techniques and the Basic Formation Mechanisms of Vortices (FMV)

For the visualization of the 3D vortex structures in the fluid flows the isosurfaces of β and λ_2 have been drawing, where β is *the imaginary part of the complex-conjugate eigen-values* of the velocity gradient tensor **G**[15] (Fig. 2b–d, 3), λ_2 is *the second*

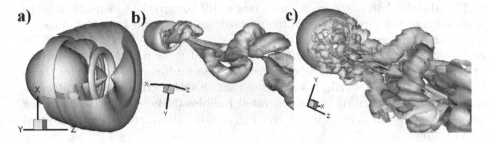

Fig. 1. Vortex structures of the sphere wake at $Fr > 10$: a–c – $Re = 200, 350, 5 \cdot 10^5$; $\lambda_2 = -10^{-6}$ and -0.16; $-2 \cdot 10^{-5}$; -10^{-4}

eigen-value of the $S^2 + \Omega^2$ tensor, where S and Ω are the symmetric and antisymmetric parts of $\mathbf{G}$[16] (Fig. 1, 2a).

The unsteady periodical flows at the moderate Re can be described through the chain of the basic formation mechanisms of vortices (FMV). For example for 2D circular cylinder (at $\rho = \rho_0$) the λ_2-visualization technique demonstrates the generation of a small vortex in the recirculation zone (RZ) (at the intersection of lines at Fig. 2a) due to Kelvin-Helmholtz instability (it is the basic FMV **1k**), connection of this vortex with vortex sheet (VSh) (surrounding RZ) and stretching the VSh downstream (**2s**), folding and separation of the top or bottom ends of VSh (**2t/b**) (formation of next wake vortex). Thus the detailed FMV during a period at $40 < Re < 191$ ($Fr = \infty$) can be described as **1k-2s-2t** – **1k-2s-2b** . At $Re < 40$ the formation of the steady symmetric RZ can be described as **1k**.

4   The Classification of Fluid Flow Regimes around a Sphere

For $Fr > 10$ the homogeneous viscous fluid flow regimes are observed in the sphere wake. The following classification of flows [5-7] has been obtained by SMIF at $Re < 5 \cdot 10^5$: 1) $Re \leq 20.5$ - without separation; 2) $20.5 < Re \leq 200$ - a steady axisymmetrical wake with a vortex ring in RZ and VSh surrounding RZ (**1k**, Fig.1a); 3) $200 < Re \leq 270$ - a steady double-thread wake with a deformed vortex ring in RZ (**1k-2s-1f**); 4) $270 < Re \leq 400$ – a periodical generation of the vortex loops (**VLs**) (facing upwards); the periodical separation of the one edge of VSh (**1k-2s-1k-3b** – **1f-3t-2t**) at $270 < Re \leq 290$; **1k-2s-1k-1k** – **2s-1f-3t-2t** at $290 < Re \leq 320$; **1k-2s-1k-1k** – **2s-1d-1f-3t-2t** at $320 < Re \leq 400$ (Fig.1b)); 5) $400 < Re \leq 3000$ – the periodical separation of the opposite edges of the irregularly rotating vortex sheet (**2s-1d-1k-1f-3b-2b** – **2s-1d-1k-1f-3t-2t**); 6) $3000 < Re \leq 4 \cdot 10^4$ - a turbulent wake (subcritical regime); 7) $4 \cdot 10^4 < Re < 5 \cdot 10^5$ - a laminar-turbulent transition (**LTT**) in the boundary layer (**BL**) on the sphere surface (critical regime); where **1f** is the formation of two filaments (threads) in RZ connected with the vortex half-ring generated due to **1k**; **1d** is the drift of the main vortex ring in RZ toward the sphere; **3t/b** is the generation of the forward part of the upward- and downward-oriented VLs [7]. Thus a new subregime has been discovered at $290 < Re \leq 320$[7] due to SMIF and β-visualization technique. At $5 \cdot 10^4 < Re < 4 \cdot 10^5$ the monotonous reduction of the time-averaged total drag coefficient has been observed (from value 0.455 to 0.155) due to LTT in BL. It was shown that this drag crisis manifests itself to us through the formation of the separated bubbles (**SBs**) within BL (near the primary separation line). SBs are growing, drifting downstream and converted to VLs [13].

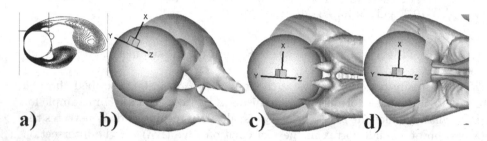

a)   b)   c)   d)

Fig. 2. a) 2D circular cylinder wake at $Re = 140, Fr = \infty, t = 310.1$ (isolines of $\lambda_2 < 0$ with step 0.04). b–d) Vortex structure of RZ of the sphere wake at $Re = 100, Fr = 2, 1, 0.6$ – isosurfaces of $\beta = 0.15, 0.1, 0.087$.

For $Fr < 10$ the code for DNS of the 3D separated density stratified viscous fluid flows around a sphere has been tested in the case of a resting sphere in a continuously stratified fluid. It was shown (for the first time) that the interruption of the molecular flow (by the resting sphere) not only generates the axisymmetrical flow on the sphere surface (from the equator to the poles) but also creates the short unsteady internal waves (**IWs**)[8-9]. At time more than $37 \cdot T_b$ the sizes and arrangement of IWs are stabilized and after some time the high gradient sheets of density with a thickness 2.2 mm are observed near the poles of the resting sphere.

The following classification of flow regimes around the sphere at $Re < 500$ [13] has been obtained by SMIF: I) $Fr > 10$ - the homogeneous case; II) $1.5 \leq Fr \leq 10$ - the quasi-homogeneous case (with four additional threads connected with VSh surrounding the sphere, Fig.2b); III) $0.9 < Fr < 1.5$ – the non-axisymmetric attached vortex in RZ (Fig.2c, 3a); IV) $0.6 < Fr \leq 0.9$ - the two symmetric vortex loops in RZ; V) $0.4 \leq Fr \leq 0.6$ - the absence of RZ (Fig.2d, 3c); VI) $0.25 < Fr < 0.4$ - a new RZ; VII) $Fr \leq 0.25$ – the two vertical vortices in new RZ (bounded by IWs) (Fig.3d-e). At $Fr \leq 0.3, Re > 120$ a periodical generation of the vortex loops (facing left or right) has been observed. The corresponding Strouhal numbers $0.19 < St = fd/U < 0.24$ (where f is the frequency of shedding) and horizontal and vertical separation angles are in a good agreement with the experiment[1]. The drag coefficients also correspond to experimental values. The interesting transformation of the four main threads

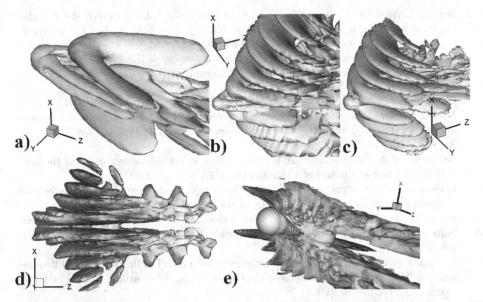

Fig. 3. The vortex structures in the stratified fluid around a moving sphere at $Re = 100$: a-e – $Fr = 1, 0.8, 0.5, 0.2, 0.08$; a-e – the isosurfaces of $\beta = 0.02; 0.005; 0.02; 0.02; 0.005$

(at $Fr = 1$, Fig.3a) into the high gradient sheets of density near the sphere poles (before the sphere) (at $Fr = 0.08$, Fig.3e) is shown at Fig.3.

5   Conclusions

The brief description of the numerical method SMIF and its application for simulation of the 3D separated homogeneous and density stratified fluid flows around a sphere have been demonstrated. The continuous changing of the complex 3D sphere wake vortex structure of the stratified viscous fluid with decreasing of Fr from 100 to 0.02 has been investigated at $Re < 500$ owing to the mathematical modeling on the supercomputers and the β–visualization of the 3D vortex structures.

Acknowledgments. This work has been supported by Russian Foundation for Basic Research (grants $10 - 01 - 92654, 11 - 01 - 00764$), by the grants from the Presidium of RAS and the Department of Mathematical Sciences of RAS.

References

1. Lin, Q., Lindberg, W.R., Boyer, D.L., Fernando, H.J.S.: Stratified flow past a sphere. J. Fluid Mech. 240, 315–354 (1992)
2. Chomaz, J.M., Bonneton, P., Hopfinger, E.J.: The structure of the near wake of a sphere moving horizontally in a stratified fluid. J. Fluid Mechanics 254, 1–21 (1993)

3. Belotserkovskii, O.M., Gushchin, V.A., Konshin, V.N.: Splitting method for studying stratified fluid flows with free surfaces. Zh. Vychisl. Mat. i Mat. Fiz (Computational Mathematics and Mathematical Physics) 27, 594–609 (1987)
4. Gushchin, V.A., Konshin, V.N.: Computational aspects of the splitting method for incompressible flow with a free surface. J. Computers and Fluids 21(3), 345–353 (1992)
5. Gushchin, V.A., Kostomarov, A.V., Matyushin, P.V., Pavlyukova, E.R.: Direct Numerical Simulation of the Transitional Separated Fluid Flows Around a Sphere and a Circular Cylinder. J. of Wind Engineering and Industrial Aerodynamics 90(4-5), 341–358 (2002)
6. Gushchin, V.A., Kostomarov, A.V., Matyushin, P.V.: 3D visualization of the separated fluid flows. Journal of Visualization 7(2), 143–150 (2004)
7. Gushchin, V.A., Matyushin, P.V.: Vortex formation mechanisms in the wake behind a sphere for $200 < Re < 380$. Fluid Dynamics 41(5), 795–809 (2006)
8. Baydulov, V.G., Matyushin, P.V., Chashechkin, Y.D.: Structure of a diffusion-induced flow near a sphere in a continuously stratified fluid. Doklady Physics 50(4), 195–199 (2005)
9. Baydulov, V.G., Matyushin, P.V., Chashechkin, Y.D.: Evolution of the diffusion-induced flow over a sphere submerged in a continuously stratified fluid. Fluid Dynamics 42(2), 255–267 (2007)
10. Gushchin, V.A., Mitkin, V.V., Rozhdestvenskaya, T.I., Chashechkin, Y.D.: Numerical and experimental study of the fine structure of a stratified fluid flow over a circular cylinder. Journal of Applied Mechanics and Technical Physics 48(1), 34–43 (2007)
11. Gushchin, V.A., Matyushin, P.V.: Numerical Simulation and Visualization of Vortical Structure Transformation in the Flow past a Sphere at an Increasing Degree of Stratification. Comput. Math. and Math. Physics 51(2), 251–263 (2011)
12. Gushchin, V.A., Rozhdestvenskaya, T.I.: Numerical study of the effects occurring near a circular cylinder in stratified fluid flows with short buoyancy periods. Journal of Applied Mechanics and Technical Physics 52(6), 905–911 (2011)
13. Matyushin, P.V., Gushchin, V.A.: Transformation of Vortex Structures in the wake of a sphere moving in the stratified fluid with decreasing of internal Froude Number. J. Phys.: Conf. Ser. 318, 62017 (2011)
14. Gushchin, V.A., Narayanan, P.S., Chafle, G.: Parallel computing of industrial aerodynamics problems: clean rooms. In: Schiano, P., Ecer, A., Periaux, J., Satofuka, N. (eds.) Parallel CFD 1997. Elsevier Science B.V. (1997)
15. Chong, M.S., Perry, A.E., Cantwell, B.J.: A general classification of three-dimensional flow field. Phys. Fluids A2(5), 765–777 (1990)
16. Jeong, J., Hussain, F.: On the identification of a vortex. J. Fluid Mech. 285, 69–94 (1995)

Finite Element Simulation of Nanoindentation Process

Roumen Iankov[1], Maria Datcheva[1], Sabina Cherneva[1], and Dimiter Stoychev[2]

[1] Institute of Mechanics, Acad G. Bontchev bl.4, BG-1113 Sofia, Bulgaria
[2] Institute of Physical Chemistry, Acad G. Bontchev bl.11, BG-1113 Sofia, Bulgaria

Abstract. Recently, nanoindentation technique is gaining importance in determination of the mechanical parameters of thin films and coatings. Most commonly, the instrumented indentation data are used to obtain two material characteristics of bulk materials: indentation modulus and indentation hardness. In this paper the authors discuss the possibility by means of numerical simulations of nanoindentation tests to obtained the force-displacement curve employing various constitutive models for both the substrate and the coating. Examples are given to demonstrate the influence of some features of the numerical model and the model assumptions on the quality of the simulation results. The main steps in creation of the numerical model and performing the numerical simulation of nanoindentation testing process are systematically studied and explained and the conclusions are drawn.

Keywords: finite element method, nanoindentation, thin films.

1 Introduction

In last years there is an increasing interest in the design of new materials whose microstructure can be controlled, e.g. nanostructured materials, functionally graded materials, thin layers, compositionally graded coatings, etc. Because an optimal material microstructure is the designer's target, the assessment of the mechanical properties in a local area using small volumes becomes an essential part of the designer process. Additionally, in many cases the material properties can not be determined via conventional macro experiments due to, e.g. the geometry, the size or the build-up of the samples. Instead, the application of instrumented indentation technique to characterise materials and device elements in a local area and using small volumes grew vast. Moreover, depth-sensing indentation or nanoindentation, where an indenter is pressed normal onto the surface of the specimen and penetration depth is continuously measured in nanometres, had been proven within the last decade to be a promising technique to investigate the mechanical properties of small volumes and thin films, [4,5]. During such tests the global variables load and displacement are continuously monitored and the mechanical properties such as indentation modulus and hardness are estimated based on these data and proper assumptions. However, the indentation

I. Dimov, I. Faragó, and L. Vulkov (Eds.): NAA 2012, LNCS 8236, pp. 319–326, 2013.

modulus and indentation hardness are not sufficient to characterise the material behaviour in case of numerical simulation of constructions and details whose components are fabricated using materials coated with this films. The reason is that in most practical cases the constitutive models representing properly the mechanical behaviour of the film-substrate system contain a number of other parameters that do not correlate with the indentation modulus and the indentation hardness. To overcome this shortage a technique for material model parameter identification using instrumented indentation test data together with a finite element simulation was intensively developed in recent years, [6,5,1,3]. In this paper a finite element model of nanoindentation on a system of thin film and substrate employing nonlinear material model is presented. The goal is to build the most suitable, robust and correct finite element model in order in the next step to be able to perform back analysis of real experiment and to achieve best fit of the experimental data varying selected parameters involved into the considered nonlinear material model. Furthermore, the numerical model and the obtained numerical solution are discussed in details and the simulated force-displacement curve is compared with data from real nanoindentaion tests. Conclusions are done on the applicability of the used numerical and material models, as well as the finite element technique for local material characterisation.

2   Experimental Setup

Nanoindentation testing requires a proper choice of indenter geometry and load. The decision depends on the structure of the tested material and particularly in the case of thin films and coatings it is important to formulate a testing programme accounting for the thickness of the film/coating, the expected ratio between the mechanical properties of the film and the substrate and the surface roughness of the tested sample probe. Nanoindentation test programme includes setting of several variables such as number of data points, thermal drift allowance, the percent and the rate of unloading, maximum load (penetration depth), and the way in which the load is applied. In this study, the experimental indentation procedure is realised by the use of Agilent Nano Indenter G200 equipped with a Berkovich three-sided diamond pyramid tip with a rounding of 20 nm and centreline-to-face angle 65.3°, [7]. The tested sample is a 100 μm thick copper foil with chemically deposited cobalt film with a thickness of 2.76 μm. In order to have better statistics the practice is to realise series of at least 9 indentations to collect reliable indentation test data. For the purpose of this study we performed a series of 25 indentations on each sample probe. We used several methods of indentation, e.g. with fixed maximum displacement and cyclic loading with a prescribed number of cycles and up to maximum load. The maximum load and displacement are chosen such that to guarantee that the indenter penetrates at a depth sufficiently far from the film-substrate contact surface. Two loading programmes are run: one cycle of loading up to maximal displacement of 200 nm and unloading up to 90% of the maximal load and 10 cycles up to a load of 50 gf, where each cycle introduced loading up to a progressively increasing

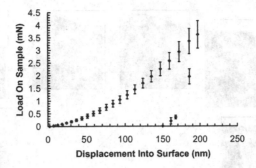

Fig. 1. Experimental load-displacement data for cobalt film on copper foil. The thickness of the film is $2.76 \mu m$.

maximal force and unloading up to 90% of this load. As a result we obtained the experimental load-displacement curve for the film-substrate system, and Fig. 1 illustrates such experimental data.

3   Numerical Modelling of Nanoindentation Process

The finite-element method (FEM) is a powerful technique in modelling and solving elastic-plastic contact problems and recently it is extensively used for material characterisation based on simulation of nanoindantation testing. One of the benefits of FEM–models is their ability to analyse the loading and unloading material response due to different material models and sample composition. Particularly, the application of FEM-models to nanoindenatation gives the possibility to assess not only the load-indentation depth curve, but also to analyse the distribution of the stress and strain fields, as well as the profile of the indentation imprint. An example of numerically obtained distribution of the stress field during both the loading and unloading phases of the indenentation process is given in Fig. 2. However, the most important is that the numerical simulation can be used for back analysis of the nanoindentation testing data and this way to obtain the material model parameters in cases where conventional experimental techniques can not be applied, [2]. While the estimation of constitutive parameters of bulk materials may be done by means of semi-empirical formulae, such approach is not applicable for determination of the mechanical properties of thin films and coatings, where inverse modelling of nanoindentation testing, often based on finite element (FE) simulation of the test, seems to be the only acceptable method.

The numerical model of nanoindentation is built depending on the indenter geometry and material isotropy. In the general case the problem is formulated in the 3D space. However, in the case of isotropic materials the indenter geometry can be introduced by an equivalent conical form whose geometric characteristics are known for the conventional indenter tips. This way the real and more complex 3D model is reduced to an axisymmetric one allowing to solve numerically

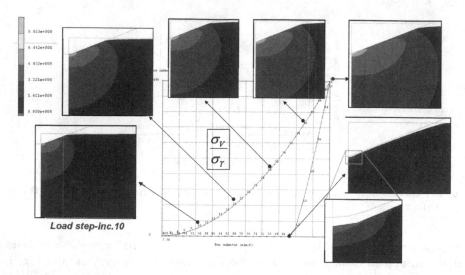

Fig. 2. Distribution during indentation of the equivalent von Mises stress normalised with respect to the one dimensional yield stress

a 2D problem and thus simplifying the numerical analysis and accelerating the process of nanoindentation back analysis procedure, [8]. For the Berkovich indenter used in the nanoindentation tests discussed in this paper, the equivalent conical indenter has a semi-angle of 70.3° ([9]) and the cone tip has a rounding of $R=20\,nm$. In our case it is important to account for the tip rounding because we analyse indentation tests on thin films, may be with thickness below $500\,nm$, and this may suggest very shallow indentation depths where the sharpness of the cone has an influence.

The boundary value problem for the nanoindentation experiment referred to in this article and explained in section 2 is specified as follows. All calculations were carried out using the axisymmetric cylindrical domain characterised by a radius $L=20\,\mu m$ and height of $102.76\,\mu m$ and representing the specimen. The copper substrate occupies a rectangular domain with dimensions $20\,\mu m \times 100\,\mu m$, Fig. 3. The material model for the copper substrate used in the present work is elastic-plastic model with linear hardening and von Mises yield surface [10]. The thin film takes up the upper rectangular part of the model with dimensions $20\,\mu m \times 2.76\,\mu m$.

The material response of the film is assumed to be isotropic elasticplastic with exponential hardening following a law, known as Hollomon's power hardening law obeying von Mises plastic criterion (for more detail description see [10]).

Normally, indenter material is very hard (usually it is a material with elastic modulus $\approx 10^3 GPa$). For that reason, it is accepted that the indenter can be modelled as a rigid body. The process of indenter penetration and separation from the specimen is simulated as a deformable-rigid contact problem via direct constraint procedure. The contact bodies are defined as a deformable (the specimen) and a rigid body (the indenter). Furthermore, the following assumptions are done:

- the indentation process is quasi-static;
- there is full adhesion between the film and the substrate, the thin film and the substrate are perfectly bonded and there is no delaminating or slippage at the interface;
- friction forces in the contact area are neglected;
- no stress-strain prehistory is taken into account.

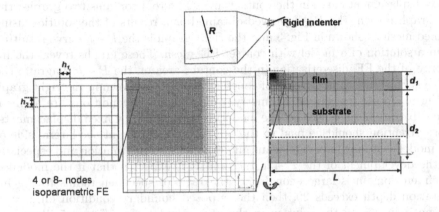

Fig. 3. Model geometry and finite element discretization: the whole model and detailed image near the indenter tip

The computational procedure for the elastic–plastic analysis employs the total Lagrangian approach, that means we use small strain elasto-plasticity (the discrete equations are formulated with respect to the reference configuration $t=0$), mean normal return mapping and additive decomposition of strain rates. In the total Lagrangian approach, the equilibrium can be expressed by the principle of virtual work as:

$$\int_{V_0} S_{ij}\,\delta E_{ij}\,dV = \int_{V_0} b_i^0\,\delta\eta_i dV + \int_{V_0} t_i^0\,\delta\eta_i dV \tag{1}$$

Here S_{ij} is the symmetric second Piola-Kirchhoff stress tensor, E_{ij} , is the Green-Lagrange strain, b_i^0 is the body force in the reference configuration, t_i^0 is the traction vector in the reference configuration, and η_i is the virtual displacements. Integrations are carried out in the original configuration at $t=0$. For small strain anaysis used in the examples given in this paper, the material law is formulated in true Cauchy stresses, σ, and true strains, ε, and in incremental form it reads:

$$d\sigma = L^{ep} : d\varepsilon, \quad L^{ep} = C - \frac{(C : \nabla\bar\sigma) \otimes (C : \nabla\bar\sigma)}{H + \bar\sigma : C : \bar\sigma} \tag{2}$$

Here C is the elastic stiffness matrix, H is the hardening coefficient, $\bar{\sigma}$ is the von Mises equivalent stress. The elastic stiffness and the hardening coefficient for the substrate and the film are introduced according to the assumed elasto-plastic models.

The specimen (substrate and thin film) is discretized with displacement formulated isoparametric quadrilateral finite elements (FE type 10, MSC.MARC [11]). The discretization is shown in Fig. 3, where it is seen that around the contact area and near the tip of the indenter we use finer discretization. The finite element mesh is done continuously coarser with getting away from the tip, as shown in Fig. 3. The characteristic size of the finite element in the contact area is $h_1 = h_2 = 5\,nm$. The FE–discretization was chosen in order to obtain a smooth force-displacement curve in the contact area. Figure 4 contains two graphs: the left graph present the $P - h$ curved obtained as a results of the solution using refined mesh as shown in Fig. 3 and the right graph is the $P - h$ curve resulting from a solution of a model with coarser FE–mesh. These graphs reveal the influence of the FE–discretization in the contact area on the $P - h$ diagram: the curve is non-smooth in the case of courser FE-mesh as shown in the right graph in Fig. 4. Along the axis of symmetry, roller boundary conditions are applied. The substrate base (bottom of the model) is constrained by fixed displacements. More attention should be paid to the boundary condition at the lateral side of the model as it may have significant influence on the final solution and especially on the development of the $P - h$ diagram. It is expected that if the modelled specimen domain is large enough, i.e. the ratio of specimen radius to the indentation depth exceeds 20, then the imposed boundary condition on $x = L$ has no influence on the solution in the near–contact zone. Figure 5 illustrates the influence of the lateral boundary condition on the load-displacement curve representing the resultant force from the contact between the specimen and the indenter at the maximum displacement of the indenter during cyclic nanoindentation. It can be concluded that up to a penetration depth of $1000\,nm$ there is no significant influence of the lateral boundary condition on the $P - h$ diagram.

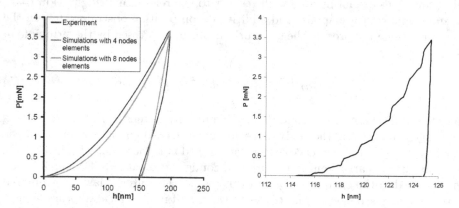

Fig. 4. Force-displacement curve obtained from nanoindentation test and from numerical simulation; simulated $P - h$ curve with course FE-mesh in the contact area

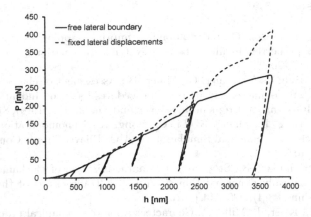

Fig. 5. Simulated $P - h$ diagram for different lateral boundary conditions

The movement of the indenter as a rigid body can be controlled by prescribed force, displacement or velocity. In the presented here example simulations this control was applied as a ramped function of time simulating the loading history during the experiment. The size of the loading sub-steps depends on the FE-discretization of the contact zone.

Figure 4 presents an example of simulation by means of the model given in Fig. 3 of a nanoindentation test on cobalt film with maximum depth of penetration 200 nm. The results in this figure also shows a comparison of solutions of the problem using 4 and 8 node isoparametric quadrilateral finite elements.

4   Closing Remarks

It is demonstrates that the numerical simulation of nanoindentation experiment may help in better understanding the material response to the indenter penetration and to picture the stress–strain behaviour in both the film and the substrate as well as to explain the apparent response of the whole system.

Several important features of numerical simulation of nanoindentation testing via FEM have been pointed out and discussed, namely: the FE–discretization, the boundary conditions and the motivation for simplification of the geometry and the behaviour at the contact.

A conclusion may be done that it is important to build the most suitable, robust and correct finite element model in order in the next step to be able to perform back analysis of real experiment and to achieve best fit of the experimental data varying selected parameters involved into the considered nonlinear material model.

Acknowledgments. Authors gratefully acknowledge for financial support of Bulgarian National Science Fund under grant No. TK01/0185.

References

1. Bocciarelli, M., Bolzon, G.: Indentation and imprint mapping for the identification of interface properties in film-substrate systems. International Journal of Fracture 155, 1–17 (2009)
2. Bolzon, G., Buljaka, V., Maier, G., Miller, B.: Assessment of elastic–plastic material parameters comparatively by three procedures based on indentation test and inverse analysis. Inverse Problems in Science and Engineering 19(6), 815–837 (2011)
3. Zhang, L., Yang, P., Shang, S., Li, C., Song, X.: Nanoindentation Experimental Approach and Numerical Simulation of AlCr Bilayer Films. Composite Interfaces 18(7), 615–626 (2011)
4. Cherneva, S., Iankov, R., Stoychev, D.: Characterisation of mechanical properties of electrochemically deposited thin silver layers Transactions of the Institute of Metal Finishing 88(4), 209–214 (2010)
5. Pelletier, H., Krier, J., Mille, P.: Characterization of mechanical properties of thin films using nanoindentation test. Mechanics of Materials 38, 1182–1198 (2006)
6. Cherneva, S., Iankov, R., Stoychev, D.: Determination of Mechanical Properties of Electrochemically Deposited Thin Gold Films. Journal of Theoretical and Applied Mechanics, Sofia 39(4), 65–72 (2009)
7. U9820A Agilent Nano Indenter G200,
   http://www.agilent.com/find/nanoindenter
8. Qin, J., Huang, Y., Xiao, J., Hwang, K.C.: The equivalence of axisymmetric indentation model for three-dimensional indentation hardness. Journal of Materials Research 24, 776–783 (2009)
9. Lichinchi, M., Lenardi, C., Haupt, J., Vitali, R.: Simulation of Berkovich nanoindentation experiments on thin films using finite element method. Thin Solid Films 312(1-2), 240 (1998)
10. Iankov, R., Cherneva, S., Datcheva, M., Stoychev, D.: Mechanical Characterization of Layers and Thin Films via Nanoindentation and Numerical Simulations, chapter in Series in Applied Mathematics and Mechanics. Mechanics of Nanomaterials and Nanotechnology, 261–286 (2012)
11. MSC Software Corporation, MSC.MARC Volume A: Theory and User Information (2005)

Runge-Kutta Methods with Equation Dependent Coefficients

L.Gr. Ixaru

"Horia Hulubei" National Institute of Physics and Nuclear Engineering,
Department of Theoretical Physics,
P.O. Box MG-6, Bucharest, Romania
and
Academy of Romanian Scientists,
54 Splaiul Independenţei, 050094, Bucharest, Romania
ixaru@theory.nipne.ro

Abstract. The simplest two and three stage explicit Runge-Kutta methods are examined by a conveniently adapted form of the exponential fitting approach. The unusual feature is that the coefficients of the new versions are no longer constant, as in standard versions, but depend on the equation to be solved. Some valuable properties emerge from this. Thus, in the case of three-stage versions, although in general the order is three, that is the same as for the standard method, this is easily increased to four by a suitable choice of the position of the stage abscissas. Also, the stability properties are massively enhanced. Two particular versions of order four are A-stable, a fact which is quite unusual for explicit methods. This recommends them as efficient tools for solving stiff differential equations.

Keywords: Runge-Kutta method, Exponential fitting, A-stability.

1 Introduction

There are different ways of approaching Runge-Kutta methods. The standard way is as in [1] but in this contribution we adopt a different one. We report on some recent results ([2],[3]) where a procedure inspired by the exponential fitting (EF) technique is used for the derivation of the coefficients. As a matter of fact, the literature on the latter is quite vast, see, e.g., [4]–[19] and references therein.

In the context of Runge-Kutta methods the EF-based procedure is not as general as the standard one. It can be used only for simpler forms of the method but, whenever it can be applied, the resulting versions and the methods derived by the standard procedure look quite different: the coefficients are no longer constants but equation dependent.

This has a number of attractive consequences for practice, and perhaps the most salient of these is the massive improvement in stability. Specifically, in [2],[3] two forms have been derived in this frame, the diagonally explicit two and three stage methods, and a surprizing result was that the new explicit three-stage method has A-stable versions of order four; this contrasts the commonly

I. Dimov, I. Faragó, and L. Vulkov (Eds.): NAA 2012, LNCS 8236, pp. 327–336, 2013.

accepted view that explicit methods of higher order cannot be A-stable. For implicit A-stable Runge-Kutta methods see [20].

2   Derivation of the Algorithm for the Scalar Problem

The one-step problem to solve is

$$y' = f(x, y), \quad x \in [x_n, x_{n+1} = x_n + h], \quad y(x_n) = y_n, \tag{1}$$

and the general algorithm of an s-stage RK method is (see [1], [21])

$$y_{n+1} = y_n + h \sum_{i=1}^{s} b_i f(x_n + c_i h, Y_i),$$

where

$$Y_i = y_n + h \sum_{j=1}^{s} a_{ij} f(x_n + c_j h, Y_j), \quad i = 1, 2, ..., s.$$

This allows computing y_{n+1} in terms of the input y_n by the formula written in the first line (so called final stage), in which the values of Y_i are as resulting from the set of formulae in the second line (internal stages). The coefficients are usually collected in the Butcher array

$$
\begin{array}{c|cccc}
c_1 & a_{11} & a_{12} & \cdots & a_{1s} \\
c_2 & a_{21} & a_{22} & \cdots & a_{2s} \\
\cdot & \cdot & \cdot & \cdots & \cdot \\
c_s & a_{s1} & a_{s2} & \cdots & a_{ss} \\
\hline
 & b_1 & b_2 & \cdots & b_s
\end{array}
$$

As a rule the value $c_1 = 0$ is used, and when $a_{ij} = 0$ for all $j \geq i$, $i = 1, 2, ..., s$, the method is called explicit. If the only nonvanishing elements are $a_{i,i-1}$, $i = 2, 3, ..., s$ then the method is called diagonally explicit. In [2] and [3] the cases $s = 2, 3$ of the diagonally explicit form were investigated by the EF-based technique.

2.1   Two-Stage Version

Its Butcher array is:

$$
\begin{array}{c|cc}
0 & 0 & 0 \\
c_2 & a_{21} & 0 \\
\hline
 & b_1 & b_2
\end{array}
$$

With $c_2 \in (0, 1]$ as a free parameter the coefficients are:

1. Standard version:

$$a_{21} = c_2, \quad b_1 = 1 - 1/2c_2, \quad b_2 = 1 - b_1. \tag{2}$$

2. New version:

$$a_{21} = c_2, \quad b_1 = 1 - 1/[2c_2(1 - c_2 M_2/2)], \quad b_2 = 1 - b_1, \qquad (3)$$

where $M_2 = hf_y(x_n + c_2 h, Y_2)$. Coefficients b_1 and b_2 are therefore equation dependent.

To get some idea on how the new version has been derived by means of the EF procedure, we present briefly the steps followed in [2].

Stage 1. This is simply $Y_1 = y_n$, where the localizing assumption $y_n = y(x_n)$ of (1) is permanently accepted hereinafter. No extra investigation is needed.
Stage 2. This consists in computing

$$Y_2 = y_n + ha_{21} f(x_n, Y_1), \qquad (4)$$

and, since it is tacitly assumed that Y_2 is an approximation to $y(x_n + c_2 h)$, $c_2 \neq 0$, it makes sense to search for parameter a_{21} such that this approximation is the closest. With this aim the function

$$\mathcal{Y}_2(x) = y(x) + ha_{21} y'(x),$$

is introduced, whose value at $x = x_n$ is Y_2. As announced, the EF procedure is used (for a detailed description of this procedure see [7]). Thus, a linear operator $\mathcal{L}_2$ acting on function space is defined in the following way:

$$\mathcal{L}_2 y(x) = y(x + c_2 h) - \mathcal{Y}_2(x), \qquad (5)$$

which measures the error of (4). Then $\mathcal{L}_2 y(x)$ is evaluated for $y(x) = x^k$, $k = 0, 1, 2, \ldots$ to obtain expressions which depend on x, h, c_2 and a_{21} but, as explained in [7] or [22], only the expressions corresponding to $x = 0$, called moments and denoted L_{k_2}, are of concern. More general, the moments allow writing the action of any linear operator $\mathcal{L}$ on a function $y(x)$ as a Taylor-like expansion

$$\mathcal{L}y(x) = \sum_{k=0}^{\infty} \frac{1}{k!} L_k y^{(k)}(x), \qquad (6)$$

where L_k are its moments. The moments of $\mathcal{L}_2$ are

$$L_{2_0} = 0, \quad L_{2_1} = h(-a_{21} + c_2), \quad L_{2_k} = (hc_2)^k, \quad k = 2, 3, \ldots \qquad (7)$$

Notice that $L_{2_k} \sim h^k$ and therefore it makes sense to accept that, once c_2 was fixed, the best value of parameter a_{21} is the one which results by imposing the condition that as many successive moments as possible are vanishing. The first moment is zero but the second is vanishing only if we take $a_{21} = c_2$; the third and subsequent moments are always nonvanishing because $c_2 \neq 0$. It follows that we can write $y(x + c_2 h) = \mathcal{Y}_2(x) + err_2(x)$ where

$$err_2(x) = \sum_{k=2}^{\infty} \frac{1}{k!} L_k y^{(k)}(x) = \frac{1}{2}(hc_2)^2 y''(x) + \frac{1}{6}(hc_2)^3 y^{(3)}(x) + \mathcal{O}(h^4). \qquad (8)$$

As a matter of fact, evaluations within $\mathcal{O}(h^4)$ are sufficient for our purpose. In particular, in the following we will not use the whole err_2 but its truncated form $terr_2$ which contains only the terms explicitly written; we therefore have $err_2 = terr_2 + \mathcal{O}(h^4)$.

External Stage. This is

$$y_{n+1} = y_n + h[b_1 f(x_n, y_n) + b_2 f(x_n + c_2 h, Y_2)], \tag{9}$$

and it allows propagating the solution from x_n to $x_{n+1} = x_n + h$. We introduce the operator $\mathcal{L}_{prop}$ as follows:

$$\mathcal{L}_{prop} y(x) = y(x + h) - y(x) - h[b_1 y'(x) + b_2 f(x + c_2 h, \mathcal{Y}_2(x))], \tag{10}$$

but, in contrast to $\mathcal{L}_2$, this can be approached only after linearization. We have successively (notice that $err_2 \sim \mathcal{O}(h^2)$)

$$y'(x + c_2 h) = f(x + c_2 h, y(x + c_2 h)) = f(x + c_2 h, \mathcal{Y}_2(x) + err_2(x)) = f(x + c_2 h, \mathcal{Y}_2(x)) +$$

$$J_2(x)\, err_2(x) + \mathcal{O}(err^2) = f(x + c_2 h, \mathcal{Y}_2(x)) + J_2(x)\, terr_2(x) + \mathcal{O}(h^4),$$

where $J_2(x) = f_y(x + c_2 h, y)|_{y = \mathcal{Y}_2(x)}$ is the usual Jacobian function at the quoted arguments. It follows that, within this level of approximation, we can safely use

$$f(x + c_2 h, \mathcal{Y}_2(x)) = y'(x + c_2 h) - J_2(x)\, terr_2(x), \tag{11}$$

in (10), and in this way the desired linear approximation

$$\mathcal{L}_{prop} y(x) = y(x + h) - y(x) - h[b_1 y'(x) + b_2(y'(x + c_2 h) - J_2(x)\, terr_2(x))] \tag{12}$$

is obtained. Its moments are

$$L_{prop\,0} = 0, \; L_{prop\,1} = h(1 - b_1 - b_2),$$
$$L_{prop\,2} = h^2 \left[1 - b_2 c_2 (2 - c_2 M_2)\right]\},$$
$$L_{prop\,3} = h^3 \left[1 - b_2 c_2^2 (3 - c_2 M_2)\right]\}, \ldots$$

and, upon solving the linear algebraic system

$$L_{prop\,1} = L_{prop\,2} = 0 \tag{13}$$

for b_1 and b_2 we get the coefficients (3).
The local truncation error formula of this algorithm is

$$LTE = \frac{1}{3!} L_{prop\,3} y^{(3)}(x_n) + \mathcal{O}(h^4) = h^3 \frac{2 - 3c_2 + c_2(c_2 - 1)M_2}{6(2 - c_2 M_2)} y^{(3)}(x_n) + \mathcal{O}(h^4). \tag{14}$$

The order of the method is then 2 but if $c_2 = 2/3$ it becomes 3. To see this we examine the factor which multiplyes h^3. Parameter M_2 behaves as h and then it can be disregarded. That factor thus becomes $(2 - 3c_2)/12$, and this vanishes when $c_2 = 2/3$.

2.2   Three-Stage Version

Its Butcher array is

$$
\begin{array}{c|ccc}
0 & 0 & 0 & 0 \\
c_2 & a_{21} & 0 & 0 \\
c_3 & 0 & a_{32} & 0 \\
\hline
& b_1 & b_2 & b_3
\end{array}
$$

c_2 and c_3 are taken as free parameters and all other depend on these. These are:

1. Standard version:

$$
a_{21} = c_2, \ a_{32} = c_3, \ b_3 = -\frac{1}{18c_2^2(c_2 - 1)}, \ b_2 = \frac{3c_2 - 1}{6c_2^2}, \ b_1 = 1 - b_2 - b_3,
$$

   where c_2 and c_3 are correlated, $c_3 = 3c_2(1 - c_2)$, see [21]. Each method in this family is of order 3.

2. New version, obtained by exponential fitting (see [3]):

$$
a_{21} = c_2, \ a_{32} = c_3, \ b_i = b_i^{num}/b_i^{den}, \ i = 2, 3, \ \text{and} \ b_1 = 1 - b_2 - b_3, \quad (15)
$$

   where

$$
\begin{aligned}
b_2^{num} &= 2 - 3c_3 + [2c_2 - 3c_2^2 + (c_3 - 1)c_3]M_3 + (c_2 - 1)c_2^2 M_2 M_3, \\
b_3^{num} &= -2 + 3c_2 - (c_2 - 1)c_2 M_2, \ b_2^{den} = c_2 B, \ b_3^{den} = c_3 B,
\end{aligned}
$$

   with $M_i = hf_y(x_n + c_i h, Y_i)$, $i = 2, 3$ and

$$
B = 6(c_2 - c_3) + c_2(3c_3 - 2c_2)M_2 + c_3(2c_3 - 3c_2)M_3 + c_2 c_3(c_2 - c_3)M_2 M_3. \quad (16)
$$

   Each of these methods is of order 3 but when c_2 and c_3 are correlated,

$$
c_3 = \frac{3 - 4c_2}{4 - 6c_2}. \quad (17)
$$

   the order becomes 4.

Remark. For systems of N equations, y and f are column vectors with N elements, and f_y is the $N \times N$ Jacobian matrix. Due to that, the b_i coefficients are no longer scalar functions but $N \times N$ matrices.

3   Linear Stability

The linear stability analysis consists in taking the scalar problem $y' = \lambda y$, $x \geq 0$, $y(0) = y_0 \neq 0$, on which the behaviour of the numerical solution is investigated when x is increased, for different values of complex λ with $\mathrm{Re}\,\lambda < 0$, see e.g. [21] for a rigorous presentation of these things. In essence, since the exact solution $y(x) = y_0 e^{\lambda x}$ tends to 0 when $x \to \infty$, the same behaviour is expected also for the numerical solution but for most methods this does not happen for all λ and

h. The product $\nu = \lambda h$ is playing a central role in this context.
For all discussed methods a function $R_f(\nu)$ exists such that $y_n = [R_f(\nu)]^n y_0$ and therefore the desired behaviour that $y_n \to 0$ when n increases holds true only if

$$|R_f(\nu)| < 1. \tag{18}$$

Function $R_f(\nu)$ and the stated condition are called stability function and stability condition, respectively, and the region in the complex ν plane where that condition is satisfied is called stability region. Of course, we are mainly interested in the left-half part $\mathrm{Re}\,\nu < 0$ of this plane, and the larger its extension the better is the method for stability. The best in this respect is when the whole left-half plane belongs to the stability region, and in such a case the method is called A-stable.

All discussed methods have R of form

$$R_f(\nu) = 1 + \nu + (b_2 c_2 + b_3 c_3)\nu^2 + b_3 c_2 c_3 \nu^3.$$

3.1   Two-Stage Method

1. Standard version: We have $b_2 = 1/2c_2$ and $b_3 = c_3 = 0$. The stability function is a polynomial, $R_f(\nu) = 1 + \nu + \nu^2/2$, and then the stability region has limited extension.
2. New version: For the test equation $y' = \lambda y$ we have $M_2 = \nu$ and then $b_2 = 1/2c_2(1 - c_2\nu)$, $b_3 = c_3 = 0$, such that

$$R_f(\nu) = \frac{2 + (2 - c_2)\nu + (1 - c_2)\nu^2}{2 - c_2\nu}.$$

When $c_2 = 1$ this becomes

$$R_f(\nu) = \frac{2 + \nu}{2 - \nu}$$

and therefore this version is A-stable.

Notice the difference: the stability function is a polynomial in ν for the standard version but a rational function for the new version. At the origin of this difference is the fact that all coefficients are constants in the former method while some of them are rational functions in the latter.

3.2   Three-Stage Method

The same difference holds true also in this case. Of particular importance are two methods with c_2 and c_3 correlated by (17). These are:
1. $c_2 = 1/2$, $c_3 = 1$, $b^{den} = 12 - 4M_2 - 2M_3 + M_2 M_3$ and

$$b_1 = \frac{2 - 3M_2}{b^{den}}, \; b_2 = \frac{8 - 2M_3 + M_2 M_3}{b^{den}}, \; b_3 = \frac{2 - M_2}{b^{den}}; \tag{19}$$

2. $c_2 = 1$, $c_3 = 1/2$, $b^{den} = 12 - 2M_2 - 4M_3 + M_2M_3$ and

$$b_1 = \frac{2 - 2M_2 + M_3 + M_2M_3}{b^{den}}, \ b_2 = \frac{2 - 5M_3}{b^{den}}, \ b_3 = \frac{8}{b^{den}}. \tag{20}$$

In fact, these two methods of order 4 have one and the same stability function,

$$R_f(\nu) = \frac{12 + 6\nu + \nu^2}{12 - 6\nu + \nu^2}, \tag{21}$$

a form which guarantees that they are A-stable.

3.3   Link between A-stability and Capacity of Solving Stiff Problems

There is generally accepted that an A-stable method gives good results for stiff problems but in our case a special provision must be added. This is because the standard methods have constant coefficients and therefore they are fully defined. For contrast, in the new methods the coefficients are updated at each step and therefore such a method can be coined as defined only when these coefficients are computed accurately. Thus, the new A-stable methods work well only if the coefficients are computed accurately, and this represents some limitation. The following test case illustrates this:

$$y^{1'} = (10\lambda + 9)y^1 - 10(\lambda + 1)y^2, \ \ y^{2'} = -9(\lambda + 1)y^1 - (9\lambda + 10)y^2, \tag{22}$$

$$x \in [x_{min} = 0, x_{max} = 5], \ y^1(0) = y_0^1, \ y^2(0) = y_0^2.$$

In matrix notations this has the form $y' = Jy$, $y(0) = y_0$ where

$$J = \begin{bmatrix} 10\lambda + 9 & -10(\lambda + 1) \\ -9(\lambda + 1) & -(9\lambda + 10) \end{bmatrix}, \ y(x) = \begin{bmatrix} y^1(x) \\ y^2(x) \end{bmatrix}, \ y_0 = \begin{bmatrix} y_0^1 \\ y_0^2 \end{bmatrix}. \tag{23}$$

The exact solution is

$$y^1(x) = 10(y_0^1 - y_0^2)e^{\lambda x} + (-9y_0^1 + 10y_0^2)e^{-x},$$

$$y^2(x) = 9(y_0^1 - y_0^2)e^{\lambda x} + (-9y_0^1 + 10y_0^2)e^{-x}.$$

We take $y_0^1 = y_0^2 = 1$. The exact solution is now independent of λ: $y^1(x) = y^2(x) = e^{-x}$, and if stability were not an issue then the results at $h = 1/2$ or $1/4$ must be sufficiently accurate irrespective of the value of λ.

We compare three methods: standard fourth order Runge-Kutta method (RK4) and two new A-stable methods: two-stage (that is, with $c_2 = 1$), which is of second order, and three-stage (with $c_2 = 1/2$), of fourth order. We take a large sample of negative λ and run the methods with $h = 1/2^m$, $m = 1, ..., 6$.

To compute the b_i coefficients of the new methods (they are 2×2 matrices), one matrix inversion is involved, of $S_2 = I - hJ/2$ for the two stage method, and of $S_3 = 12I - 6hJ + h^2J^2$ for the three-stage one; I is the unit matrix. Now, the eigenvalues of J are -1 and λ, and therefore those of S_2 and S_3 are

Table 1. Dependence on λ of the relative errors at $x = 5$ for new A-stable explicit two-stage Runge-Kutta method. Notation $a(b)$ means $a \times 10^b$.

			λ			
h	-1	-10^2	-10^4	-10^6	-10^8	-10^{10}
1/ 2	1.03(-01)	1.03(-01)	1.03(-01)	1.03(-01)	1.03(-01)	-7.97(+01)
1/ 4	2.59(-02)	2.59(-02)	2.59(-02)	2.59(-02)	2.59(-02)	2.78(24)
1/ 8	6.50(-03)	6.50(-03)	6.50(-03)	6.50(-03)	6.50(-03)	1.45(50)
1/ 16	1.63(-03)	1.63(-03)	1.63(-03)	1.63(-03)	1.63(-03)	-7.81(105)
1/ 32	4.07(-04)	4.07(-04)	4.07(-04)	4.07(-04)	4.07(-04)	-5.15(103)
1/ 64	1.02(-04)	1.02(-04)	1.02(-04)	1.02(-04)	1.02(-04)	2.25(14)

$1 + h/2$, $1 - \nu/2$, and $12 + 6h + h^2$, $12 - 6\nu + \nu^2$, respectively, where $\nu = \lambda h$. It is well known that the accuracy of the computation of the inverse of a matrix depends on the ratio between its largest and smallest eigenvalues: the bigger the ratio the worse the accuracy. In our case the ratio asymptotically increases as $-\nu/2$ for the two-stage method but as $\nu^2/12$ for the three-stage method and therefore the first method is expected to work well for much more negative values of λ than the second.

These behaviours are nicely illustrated on data presented in Tables 1 and 2; all computations were carried out in double precision arithmetics. Table 1 collects data from the two-stage A-stable method. It is seen that the relative errors are similar for all λ up to -10^8, and they also confirm that this version is of second order: the error is smaller by a factor 4 at each halving of h. As for the drastic alteration when $\lambda = -10^{10}$, this is because the matrix inversion is now inaccurate.

On table 2 we compare RK4 and the new three-stage version. For RK4 we present data only from $\lambda = -10$ and -1000 but this is enough to see that this classical method works well only for $\lambda = -10$ which is consistent with the

Table 2. Dependence on λ of the relative errors at $x = 5$ for classical RK4 and new A-stable three-stage explicit Runge Kutta method

	RK4		New three $-$ stage method			
h	-10	-10^3	-10	-10^3	-10^5	-10^7
1/ 2	-1.28(-02)	-6.96(72)	-4.41(-04)	-4.41(-04)	-4.41(-04)	1.49(39)
1/ 4	-2.01(-04)	6.44(108)	-2.72(-05)	-2.72(-05)	-2.73(-05)	4.57(74)
1/ 8	-1.13(-05)	-7.69(260)	-1.70(-06)	-1.70(-06)	-1.67(-06)	7.07(102)
1/ 16	-6.70(-07)	-NaN	-1.06(-07)	-1.06(-07)	-1.09(-07)	3.70(159)
1/ 32	-4.08(-08)	-NaN	-6.62(-09)	-6.63(-09)	-6.62(-09)	2.41(152)
1/ 64	-2.52(-09)	-NaN	-4.14(-10)	-4.15(-10)	-5.43(-10)	2.71(165)

known fact that the method is stable only when $\nu > -2.8$, see,e.g., [21]. The new method works much better, viz. for all λ up to -10^5. When $\lambda = -10^7$ the results are inacceptable, again due to the inaccuracy in the matrix inversion which, as expected, now appears earlier than in the two-stage method.

In conclusion, the new methods are massively better than the classical ones for stability. In spite of being explicit they are A-stable, but in practice much attention should be given to the accurate computation of the coefficients, especially for the three-stage version. Finding highly accurate numerical procedures represents a real challenge although this will remove only partially the limitations for stiff systems. Much more promising seems the idea of concentrating our attention on the form of the algorithm. For example, it would be interesting to see if explicit three-stage versions can be derived under the condition that the matrix to be inverted is *linear* in the jacobian matrix, not *quadratic*, as in the present version.

Acknowledgement. This work was supported by the project PN-II-ID-PCE-2011-3-0092 of the Romanian Ministry of Education and Research.

References

1. Butcher, J.C.: Numerical Methods for Ordinary Differential Equations, 2nd edn. Wiley, Chichester (2008)
2. D'Ambrosio, R., Ixaru, L.G., Paternoster, B.: Construction of the ef-based Runge-Kutta methods revisited. Comput. Phys. Commun. 182, 322–329 (2011)
3. Ixaru, L.G.: Runge-Kutta method with equation dependent coefficients. Comput. Phys. Commun. 183, 63–69 (2012)
4. Calvo, M., Franco, J.M., Montijano, J.I., et al.: Sixth-order symmetric and symplectic exponentially fitted Runge-Kutta methods of the Gauss type. J. Comput. Appl. Math. 223, 387–398 (2009)
5. Franco, J.M.: An embedded pair of exponentially fitted explicit Runge-Kutta methods. J. Comput. Appl. Math. 149, 407–414 (2002)
6. Franco, J.M.: Exponentially fitted explicit Runge-Kutta-Nystrom methods. J. Comput. Appl. Math. 167, 1–19 (2004)
7. Ixaru, L.G., Vanden Berghe, G.: Exponential Fitting. Kluwer Academic Publishers (2004)
8. Martin-Vaquero, J., Janssen, B.: Second-order stabilized explicit Runge-Kutta methods for stiff problems. Comput. Phys. Commun. 180, 1802–1810 (2009)
9. Ozawa, K., Japan, J.: Indust. Appl. Math. 18, 107 (2001)
10. Paternoster, B.: Runge-Kutta(-Nystrom) methods for ODEs with periodic solutions based on trigonometric polynomials. Appl. Numer. Math. 28, 401–412 (1998)
11. Simos, T.: An exponentially-fitted Runge-Kutta method for the numerical integration of initial-value problems with periodic or oscillating solutions. Comput. Phys. Commun. 115, 1–8 (1998)
12. Vanden Berghe, G., De Meyer, H., Van Daele, M., et al.: Exponentially-fitted explicit Runge-Kutta methods. Comput. Phys. Commun. 123, 7–15 (1999)
13. Vanden Berghe, G., De Meyer, H., Van Daele, M., et al.: Exponentially fitted Runge-Kutta methods. J. Comput. Appl. Math. 125, 107–115 (2000)

14. Vanden Berghe, G., Ixaru, L.G., De Meyer, H.: Frequency determination and step-length control for exponentially-fitted Runge-Kutta methods. J. Comput. Appl. Math. 132, 95–105 (2001)

15. Ixaru, L.G., Vanden Berghe, G., De Meyer, H.: Exponentially fitted variable two-step BDF algorithm for first order ODEs. Comput. Phys. Commun. 150, 116–128 (2003)

16. Vanden Berghe, G., Van Daele, M., Vyver, H.V.: Exponential fitted Runge-Kutta methods of collocation type: fixed or variable knot points? J. Comput. Appl. Math. 159, 217–239 (2003)

17. Van De Vyver, H.: Frequency evaluation for exponentially fitted Runge-Kutta methods. J. Comput. Appl. Math. 184, 442–463 (2005)

18. Van de Vyver, H.: On the generation of P-stable exponentially fitted Runge-Kutta-Nystrom methods by exponentially fitted Runge-Kutta methods. J. Comput. Appl. Math. 188, 309–318 (2006)

19. Vigo-Aguiar, J., Martin-Vaquero, J., Ramos, H.: Exponential fitting BDF-Runge-Kutta algorithms. Comput. Phys. Commun. 178, 15–34 (2008)

20. Jesus, V.-A., Higinio, R.: A family of A-stable Runge-Kutta collocation methods of higher order for initial-value problems. IMA J. Numerical Analysis 27, 798–817 (2007)

21. Lambert, J.D.: Numerical methods for ordinary differential systems: The initial value problem. Wiley, Chichester (1991)

22. Ixaru, L.G.: Operations on oscillatory functions. Comput. Phys. Commun. 105, 1–19 (1997)

Method of Lines and Finite Difference Schemes with Exact Spectrum for Solving Some Linear Problems of Mathematical Physics

Harijs Kalis[1,2], Sergejs Rogovs[1], and Aigars Gedroics[1]

[1] University of Latvia: Department of Physics and Mathematics
[2] Institute of Mathematics and Informatics, Zellu iela 8, Rīga LV-1002, Latvija
kalis@lanet.lv, {rogovs.sergejs,aigors}@inbox.lv

Abstract. In this paper linear initial-boundary-value problems of mathematical physics with different type boundary conditions (BCs) and periodic boundary conditions (PBCs) are studied. The finite difference scheme (FDS) and the finite difference scheme with exact spectrum (FDSES) are used for the space discretization. The solution in the time is obtained analytically and numerically, using the method of lines and continuous and discrete Fourier methods.

1 Introduction

We consider here linear parabolic and hyperbolic type problems, which are used for modelling different problem of mathematical pysics. For numerical investigations we consider the linear parabolic and hyperbolic heat conduction problems in the following form:

$$
\begin{cases}
\alpha_2 \frac{\partial^2 T(x,t)}{\partial t^2} + \alpha_1 \frac{\partial T(x,t)}{\partial t} = \frac{\partial}{\partial x}(\bar{k}\frac{\partial T(x,t)}{\partial x}) + f(x,t), x \in (0,l), t \in (0,t_f), \\
\frac{\partial T(0,t)}{\partial x} - \sigma_1(T(0,t) - T_l(t)) = g_1(t), \\
\frac{\partial T(l,t)}{\partial x} + \sigma_2(T(l,t) - T_r(t)) = g_2(t), t \in (0,t_f), \\
T(x,0) = T_0(x), \frac{\partial T(x,0)}{\partial t} = \bar{T}_0(x), x \in (0,l),
\end{cases}
\tag{1}
$$

where $\bar{k} > 0, \alpha_1 \geq 0, \alpha_2 \geq 0, \sigma_1 \geq 0, \sigma_2 \geq 0 (\sigma_1^2 + \sigma_2^2 \neq 0)$, are the constant parameters, t_f is the final time, $T_l(t), T_r(t), T_0(x), \bar{T}_0(x)$ are given functions. We consider also a problem with periodic boundary conditions (in the following form: $T(0,t) = T(l,t), \frac{\partial T(0,t)}{\partial x} = \frac{\partial T(l,t)}{\partial x}, t \in (0,t_f)$. In this case the functions $T_0(x), \bar{T}_0(x)$ are also periodic.

Solutions of these problems are obtained analytically and numerically, using the finite second order difference scheme (FDS) and the finite difference scheme with exact spectrum (FDSES) [4].

We obtain new transcendental equation and algorithms for solving the last two eigenvalues and eigenvectors of the finite difference scheme, using the spectral method for BCs of the third kind [2], [3]. We define the FDSES method, using the finite difference matrix A in the form $A = WDW^T$ (W, D are the matrices,

I. Dimov, I. Faragó, and L. Vulkov (Eds.): NAA 2012, LNCS 8236, pp. 337–344, 2013.

containing finite difference eigenvectors and eigenvalues), where the elements of diagonal matrix D are replaced with the first eigenvalues of the second order differential operator. Also we consider the method of lines and Fourier methods for solving the problems with homogenous BCs and periodic BCs.

2   Finite Difference Approximations and Spectral Problem with Homogenous BCs of the Third Kind

We consider the uniform grid $x_j = jh, j = \overline{0,N}, Nh = l$. Using the finite differences of second order approximation $(O(h^2))$ for partial derivatives of the second order with respect to x, for (1) we obtain the initial value problem for a system of ordinary differential equations (ODEs):

$$\begin{cases} \alpha_2 \ddot{U}(t) + \alpha_1 \dot{U}(t) + \bar{k}AU(t) = F(t), \\ U(0) = U_0, \dot{U}(0) = \bar{U}_0, \end{cases} \tag{2}$$

where A is a 3-diagonal matrix of $N+1$ order [5], $U(t), \dot{U}(t), \ddot{U}(t), U_0, \bar{U}_0, F(t)$ are the column-vectors of $N+1$ order with elements $u_j(t) \approx T(x_j,t))$, $\dot{u}_j(t) \approx \frac{\partial T(x_j,t)}{\partial t}, \ddot{u}_j(t) \approx \frac{\partial^2 T(x_j,t)}{\partial t^2}$, $u_j(0) = T_0(x_j), v_j(0) = \bar{T}_0(x_j), f_j(t) = \bar{f}(x_j,t), j = \overline{0,N}$.

For the 3-diagonal matrix A the solution of the spectral problem $Ay^n = \mu_n y^n, n = \overline{1,N+1}$ is in following form [5]: the elements of eigenvectors $y_j^n = C_n^{-1}(\frac{\sin(p_n h)}{h}\cos(p_n x_j) + \sigma_1 \sin(p_n x_j))$ and the eigenvalues $\mu_n = \frac{4}{h^2}\sin^2(p_n h/2)$, where p_n are the positive roots of the following transcendental equation

$$\cot(p_n l) = \frac{\sin^2(p_n h) - h^2 \sigma_1 \sigma_2}{h(\sigma_1 + \sigma_2)\sin(p_n h)}, n = \overline{1,N+1} \tag{3}$$

Since the scalar product $[y^k, y^m] = h(\sum_{j=1}^{N-1} y_j^k y_j^m + 0.5(y_0^k y_0^m + y_N^k y_N^m)) = 0, k \neq m$, the eigenvectors are orthogonal, $C_n^2 = [y^n, y^n]$ [2]. Hence we have the orthonormal eigenvectors y^n, y^m with $[y^n, y^m] = \delta_{n,m}$. The experimental calculations with MATLAB show [2], that the last two roots p_N, p_{N+1} can not be obtained from (3). Depending on the parameter $Q = \frac{l\sigma_1 \sigma_2}{\sigma_1 + \sigma_2}$ one $(Q < 1)$ or two $(Q \geq 1)$ roots from the following new transcendental equations can be obtained:

$$\coth(p_n l) = \frac{\sinh^2(p_n h) + h^2 \sigma_1 \sigma_2}{h(\sigma_1 + \sigma_2)\sinh(p_n h)}, n = \overline{N,N+1} \tag{4}$$

and the eigenvalues and eigenvectors are in the form:
$y_j^n = C_n^{-1}(-1)^j(\frac{\sinh(p_n h)}{h}\cosh(p_n x_j) - \sigma_1 \sinh(p_n x_j)), j = \overline{0,N}$,
$\mu_n = \frac{4}{h^2}\cosh^2(p_n h/2), n \geq N, C_n^2[y^n, y^n]$ [2].

Therefore the matrix A can be represented in the form $A = WDW^T(AW = WD)$, where the column of the matrix W and the diagonal matrix D contains $M = N+1$ orthonormed eigenvectors y^n and eigenvalues $\mu_n, n = \overline{1,M}$. From $W^T W = E$ follows that $W^{-1} = W^T$.

The solution of the spectral problem $w''(x) + \lambda^2 w(x) = 0, x \in (0, l), w'(0) - \sigma_1 w(0) = 0, w'(l) + \sigma_2 w(l) = 0$, is

$$w_n(x) = C_n^{-1}(\lambda_n \cos(\lambda_n x) + \sigma_1 \sin(\lambda_n x)), \ C_n^2 = 0.5(l(\lambda_n^2 + \sigma_1^2) + \frac{\sigma_2(\lambda_n^2 + \sigma_1^2)}{\lambda_n^2 + \sigma_2^2} + \sigma_1),$$

where $(w_n, w_m) = \int_0^1 w_n(x) w_m(x) dx = \delta_{n,m}$ and λ_n are positive roots of the following transcendental equation:

$$\cot(\lambda_n l) = \frac{\lambda_n}{\sigma_1 + \sigma_2} - \frac{\sigma_1 \sigma_2}{\lambda_n(\sigma_1 + \sigma_2)}, n = \overline{1, N+1}.$$

For the finite difference scheme with the exact spectrum (FDSES) the matrix A is represented in the form form $A = WDW^T$, where the diagonal matrix D contains the eigenvalues $d_k = \lambda_k^2, \ k=\overline{1, N+1}$ of the differential operator $(-\frac{\partial^2}{\partial x^2})$. For the boundary conditions of the first kind $W = W^T = W^{-1}$ is a symmetric orthogonal matrix with the elements $w_{i,j} = \sqrt{\frac{2}{N}} \sin \frac{\pi i j}{N}, i, j = \overline{1, N-1}$,

$d_k = (k\pi/l)^2, \mu_k = \frac{4}{h^2} \sin^2 \frac{k\pi}{2N}, d_k > \mu_k, \ k=\overline{1, N-1}$.

3   Finite Difference Approximations and the Spectral Problem with Periodic BCs

In the case of periodic boundary conditions we consider different order of approximation.

3.1   Approximation Order $O(h^2)$

The finite difference 3-diagonal circulant matrix A of $M = N$ order is in the form: $A = \frac{1}{h^2}[2 \ -1 \ 0 \ ... \ 0 \ 0 \ -1]$ (this matrix is given only with first row [1]). The solution of the corresponding spectral problem $Ay^k = \mu_k y^k$ is

$$\mu_k = \frac{4}{h^2} \sin^2(k\pi/N), \ y_j^k = \sqrt{\frac{1}{N}} \exp(2\pi i k j/N), k, j = \overline{1, N}$$

and $(y^k, y^m) = \sum_{j=1}^N y_j^k y_{*,j}^m = \delta_{k,m,,}$ where $y_{*,j}^m = \sqrt{\frac{1}{N}} \exp(-2\pi i m j/N), m, j = \overline{1, N}, i = \sqrt{-1}$ [1].

The eigenvalues μ_k are symmetric: $\mu_{N/2+m} = \mu_{N/2-m}, m = \overline{1, N/2}$, where N is even number.

Using the orthonormal eigenvectors - matrices W, W_* with the elements y_j^k, y_{*j}^k we obtain the matrix representation:

$AW = WD, W^{-1} = W_*, A = WDW_*$, where the elements of the diagonal matrix D is $d_k = \mu_k, k = \overline{1, N}$.

3.2   Approximation Order $O(h^p), p = 2n, n \geq 1$.

We consider the finite difference approximation for second order derivative $u''(x_j)$, using the uniform grid $x_j = jh$ with $p+1, p = 2n$ points stencil $(x_{j-n}, \cdots, x_{j-1}, x_j, x_{j+1}, \cdots, x_{j-n})$.

Used the method of unknown coefficients C_k, E_p we obtain the approximation of the $O(h^p)$ order in following form :

$$u''(x_j) = \frac{1}{h^2} \sum_{k=-n}^n C_k u(x_{j-k}) + E_p \frac{h^p u^{(p+2)}(\xi)}{(p+2)!}, x_{j-n} < \xi < x_{j+n}, \text{ where } C_k = C_{-k}, C_0 = -2 \sum_{m=1}^n C_m. \text{ For the others coefficients } (m > 0) \text{ we get the system}$$

of linear algebraic equations with the Vandermonde matrix of the n order . Having solved the system, we have

$E_p = -2\sum_{m=1}^{n} C_m m^{2n+2}, C_m = \frac{2(n!)^2(-1)^{m-1}}{m^2(n-m)!(n+m)!}$

and for some particular cases we have:

1) $p = 2 : C_1 = 1, C_0 = -2, E_2 = -2,$

2) $p = 4 : C_1 = \frac{4}{3}, C_2 = -\frac{1}{12}, C_0 = -\frac{5}{2}, E_4 == 8,$

3) $p = 6 : C_1 = \frac{3}{2}, C_2 = -\frac{3}{20}, C_3 = \frac{1}{90}, C_0 = -\frac{49}{18}, E_4 = -72,$

4) $p = 8 : C_1 = \frac{8}{5}, C_2 = -\frac{1}{5}, C_3 = \frac{8}{315}, C_4 = -\frac{1}{560}, C_0 = -\frac{205}{72}, E_8 = 1152.$

The circulant matrix A is in the form:

$A = \frac{1}{h^2}[C_0, C_1, \cdots, C_{p/2}, 0, \cdots, 0, C_{p/2}, C_{p/2-1}, \cdots, C_2, C_1]$

and has following eigenvalues:

$\mu_k = \frac{4}{h^2}\sum_{m=1}^{n} C_m \sin^2\frac{\pi km}{N}, k = \overline{1, N}.$

Therefore for some particular cases we have the following eigenvalues ($k = \overline{1, N}$):

1) $p = 2 : \mu_k = \frac{4}{h^2}\sin^2(\pi k/N),$

2) $p = 4 : \mu_k = \frac{4}{h^2}(\sin^2(\pi k/N) + \frac{1}{3}\sin^4(\pi k/N)),$

3) $p = 6 : \mu_k = \frac{4}{h^2}(\sin^2(\pi k/N) + \frac{1}{3}\sin^4(\pi k/N) + \frac{8}{45}\sin^6(\pi k/N)),$

4) $p = 8 : \mu_k = \frac{4}{h^2}(\sin^2(\pi k/N) + \frac{1}{3}\sin^4(\pi k/N) + \frac{8}{45}\sin^6(\pi k/N) + \frac{4}{35}\sin^8(\pi k/N)).$

The orthonormal complex eigenvectors w^k, w_*^k of matrix A remain the same.

3.3   The Continuous and Discrete Fourier Series

For the continuous spectral problem $-w''(x) = \lambda w(x), x \in (0, l), w(0) = w(l),$ $w'(0) = w'(l)$

we obtain $\lambda_k = \frac{2\pi k}{l}, w^k(x) = \sqrt{\frac{1}{l}}\exp\frac{2\pi ikx}{l},$

$w_*^k(x) = \sqrt{\frac{1}{l}}\exp(-\frac{2\pi ikx}{l}), (w^k, w_*^m) = \int_0^l w^k(x)w_*^m(x)dx = \delta_{k,m}, k, m = \overline{-\infty, +\infty}.$

For a periodic function $g(x)$ with period l we have the complex Fourier series $g(x) = \sum_{k=-\infty}^{\infty} a_k w^k(x)$, where $a_k = (g, w_*^k) = \int_0^l g(x)w_*^k(x)dx$ or in the real Fourier series form:

$g(x) = \sum_{k=1}^{\infty}(a_k^{(1)}\cos\frac{2\pi kx}{l} + a_k^{(2)}\sin\frac{2\pi kx}{l}) + \frac{a_0^{(1)}}{2},$

where $a_k^{(1)} = \frac{2}{l}\int_0^l g(x)\cos\frac{2\pi kx}{l}dx, a_k^{(2)} = \frac{2}{l}\int_0^l g(x)\sin\frac{2\pi kx}{l}dx.$

Similarly expressions we can consider also for the discrete cases. For a N-order vector g with elements $g_j, j = \overline{1, N}$ we have the complex expression $g = \sum_{k=1}^{N} a_k y^k$, where $a_k = (g, y_*^k) = \sum_{j=1}^{N} g_j y_{*j}^k$ or $g = \sum_{k=1}^{*\bar{N}}(a_k y^k + a_{N-k}y^{N-k}) + \frac{a_N}{\sqrt{N}},$

$y^{N-k} = y_*^k, y_*^{N-k} = y^k, \mu_{N-k} = \mu_k, \sum_{k=1}^{*\bar{N}} a_k = \sum_{k=1}^{\bar{N}-1} a_k + a_{\bar{N}}/2, \bar{N} = N/2.$

Using the expressions $a_k = \frac{a_k + a_{N-k}}{2} + \frac{a_k - a_{N-k}}{2},$

$a_{N-k} = \frac{a_k + a_{N-k}}{2} - \frac{a_k - a_{N-k}}{2}, a_k^{(1)} = \frac{a_k + a_{N-k}}{\sqrt{N}}, a_k^{(2)} = \frac{i(a_k + a_{N-k})}{\sqrt{N}},$

we have real discrete Fourier series in the following form:

$g_j = \sum_{k=1}^{*\bar{N}}(a_k^{(1)}\cos\frac{2\pi kj}{N} + a_k^{(2)}\sin\frac{2\pi kj}{N}) + \frac{a_0^{(1)}}{2}, a_N^1 = a_0^1,$

where $a_k^{(1)} = \frac{1}{\sqrt{N}}(g, w_*^k + w_*^{N-k}) = \frac{2}{N}\sum_{j=1}^{N} g_j \cos\frac{2\pi kj}{N},$

$a_k^{(2)} = \frac{i}{\sqrt{N}}(g, w_*^k - w_*^{N-k}) = \frac{2}{N}\sum_{j=1}^{N} g_j \sin\frac{2\pi kj}{N}.$

For the FDSES the matrix A is represented in the form $A = WDW_*$ and the diagonal matrix D contains the first N eigenvalues $d_k = \lambda_k$, $k=\overline{1,N}$ from the differential operator $(-\frac{\partial^2}{\partial x^2})$ in following way:

1)$d_k = \lambda_k^2$ for $k = \overline{1, N_2}$, where $N_2 = N/2$.

2)$d_k = \lambda_{N-k}^2$ for $k = \overline{N_2, N-1}$, $d_N = 0$.

4   Analytical Solutions

We can consider the analytical solutions of (2) using the spectral representation of matrix A.

4.1   The Solution for BCs of the Third Kind

Using the transformation $V = W^T U$ follows the seperate system of ODEs

$$\begin{cases} \alpha_2 \ddot{V}(t) + \alpha_1 \dot{V}(t) + \bar{k}DV(t) = G(t), \\ V(0) = W^T U_0, \dot{V}(0) = W^T \bar{U}_0, \end{cases} \tag{5}$$

where $V(t), \dot{V}(t), \ddot{V}(t), V(0), \dot{V}(0), G(t) = W^T F(t)$ are the column-vectors of M order with elements $v_k(t), \dot{v}_k(t), \ddot{v}_k(t), v_k(0), \dot{v}_k(0), g_k(t), k = \overline{1, M}$.

The solution of this system is $(\alpha_2 \neq 0)$ [2]

$$\begin{aligned} v_k(t) = &exp(-0.5\alpha_1 t/\alpha_2)(C_k \sinh(\kappa_k t) + B_k \cosh(\kappa_k t)) \\ &+ \frac{1}{\kappa_k \alpha_2} \int_0^t exp(-0.5\frac{\alpha_1}{\alpha_2}(t - \tau)) \sinh(\kappa(t - \tau))g_k(\tau)d\tau, \end{aligned} \tag{6}$$

where $\kappa_k = \sqrt{0.25\alpha_1^2/\alpha_2^2 - \bar{k}d_k/\alpha_2}$, $B_k = v_k(0)$, $C_k = \frac{1}{\kappa}(\dot{v}_k(0) + \frac{\alpha_1}{2\alpha_2} v_k(0))$, $d_k = \mu_k$. If $4\bar{k}d_k\alpha_2/\alpha_1^2 > 1$, then the hyperbolic functions need to be replaced with the trigonometric. For FDSES $d_k = \lambda_k^2$.

We can also use the Fourier method for solving (1) in the form $T(x,t) = \sum_{k=1}^{\infty} v_k(t)w_k(x)$, where $w_k(x)$ are the orthonormed eigenvectors, $v_k(t)$ is the solution (6), with $g_k(t) = (f, w_k)$, $v_k(0) = (T_0, w_k)$, $\dot{v}_k(0) = (\bar{T}_0, w_k)$, $d_k = \lambda_k$.

For hyperbolic type equation $\alpha_1 = 0, \alpha_2 = 1$ the solution of (5) is $v_k(t) = \frac{\dot{v}_k(0)}{\kappa_k} \sin(\kappa_k t) + v_k(0) \cos(\kappa_k t) + \frac{1}{\kappa_k} \int_0^t \sin(\kappa_k(t - \tau))g_k(\tau)d\tau$, where $\kappa_k = \sqrt{\bar{k}d_k}, d_k = \mu_k$.

For parabolic type equation $\alpha_1 = 1, \alpha_2 = 0$, the solution is $v_k(t) = v_k(0) \exp(-\kappa_k t) + \int_0^t exp(-\kappa_k(t - \tau))g_k(\tau)d\tau$, where $\kappa_k = \bar{k}d_k$.

4.2   The Solution for Periodic BCs

The solution of (1) with PBCs, using the Fourier method is:
$T(x,t) = \sum_{k=-\infty}^{\infty} v_k(t)w^k(x)$, where $w^k(x)$ are orthonormal eigenvectors, $v_k(t)$ is the solution (6), with $d_k = \lambda_k, g_k(t) = (f, w_*^k), v_k(0) = (T_0, w_*^k), \dot{v}_k(0) = (\bar{T}_0, w_*^k)$.

The solution we can be also obtained in real Fourier series:

$T(x,t) = \sum_{k=1}^{\infty}(b_k(t)\cos\frac{2\pi kx}{l} + c_k(t)\sin\frac{2\pi kx}{l}) + \frac{b_0(t)}{2}$,

$f(x,t) = \sum_{k=1}^{\infty}(\bar{b}_k(t)\cos\frac{2\pi kx}{l} + \bar{c}_k(t)\sin\frac{2\pi kx}{l}) + \frac{\bar{b}_0(t)}{2}$,

$\bar{b}_k(t) = \frac{2}{l}\int_0^l f(\xi,t)\cos\frac{2\pi k\xi}{l}d\xi, \bar{c}_k(t) = \frac{2}{l}\int_0^l f(\xi,t)\sin\frac{2\pi k\xi}{l}d\xi$,

where $b_k(t), c_k(t)$ are the corresponding solutions of (6) by

$b_k(0) = \frac{2}{l}\int_0^l T_0(\xi)\cos\frac{2\pi k\xi}{l}d\xi, c_k(0) = \frac{2}{l}\int_0^l T_0(\xi)\sin\frac{2\pi k\xi}{l}d\xi$,

$g_k(t) = \bar{b}_k(t)$ or $\bar{c}_k(t)$.

We can also consider the analytical solution of (2) using the spectral matrix representation $A = WDW_*$. From transformation $V = W_*U (U = WV)$ follows the seperate system of ODEs (5), where the column-vectors are of the N- order. The solution is in the form (6).

Using the discrete Fourier transformation $U(t) = \sum_{k=1}^{N}v_k(t)y^k, F(t) = \sum_{k=1}^{N}g_k(t)y^k$,

we can obtained the functions $v_k(t)$ from (6), where $g_k(t) = (F(t), y_*^k), v_k(0) = (U_0, y_*^k), \dot{v}_k(0) = (\bar{U}_0, y_*^k)$

We can obtain the solution of the discrete problem also in the following real form:

$u_j(t) = \sum_{k=1}^{\bar{N}}(b_k(t)\cos\frac{2\pi kj}{N} + c_k(t)\sin\frac{2\pi kj}{N}) + \frac{b_0}{2}$,

$f_j(t) = \sum_{k=1}^{\bar{N}}(\bar{b}_k(t)\cos\frac{2\pi kj}{N} + \bar{c}_k(t)\sin\frac{2\pi kj}{N}) + \frac{b_0(t)}{2}$,

$\bar{b}_k(t) = \frac{2}{N}\sum_1^N f_j(t)\cos\frac{2\pi kj}{N}, \bar{c}_k(t) = \frac{2}{N}\sum_1^N f_j(t)\sin\frac{2\pi kj}{N}$,

where $b_k(t), c_k(t)$ are the corresponding solutions of (6) by

$b_k(0) = \frac{2}{N}\sum_{j=1}^N T_0(x_j)\cos\frac{2\pi kj}{N}, c_k(0) = \frac{2}{N}\sum_{j=1}^N T_0(x_j)\sin\frac{2\pi kj}{N}$,

$g_k(t) = \bar{b}_k(t)$ or $\bar{c}_k(t), d_k = \mu_k, \sum_{k=1}^{\bar{N}}a_k = \sum_{k=1}^{N/2-1}a_k + \frac{a_{N/2}}{2}$,

(for FDSES $d_k = \lambda_k$).

5   Some Examples and Numerical Results

In order to show, how do the methods work, we shall consider 2 simple boundary- value problems for ODEs and 3 initial- boundary- value problems for wave, heat transfer and hyperbolic heat conduction equations.

5.1   Example for ODEs with BCs of the Third Type

For finite value σ_1, σ_2 we consider following problem:

$-u''(x) = f(x), x \in (0, l), u'(0) - \sigma_1 u(0) = 0, u'(l) + \sigma_2 u(l) = 0$,

where $f(x) = 12x^2 C_0 + \sigma_1 \sin(x), C_0 = \frac{\sigma_1\cos(l)+\sigma_1\sigma_2\sin l+\sigma_2}{4l^3+\sigma_2 l^4}$.

The exact solution of the continuous problem is $u(x) = -x^4 C_0 + 1 + \sigma_1 \sin(x)$.

The solution of the problem is obtained in the form $y = A^{-1}F$ or $y = WD^{-1}W^T$, where D^{-1} is a diagonal matrix with elements $\mu_k^{-1}, k = \overline{1, N+1}$. If $\sigma_1, \sigma_2, N = 15, l = 11$ we have the maximal errors $\delta(FDS) = 0.25, \delta(FDSES) = 0.08$ and $p_N = 0.037, \mu_N = 7.44, p_{N+1} = 1.604, \mu_{N+1} = 10.32$.

5.2   Example for ODEs with Periodic BCs

We consider boundary-value problem with PBCs:
$-u''(x) = f(x), x \in (0, l), u(0) = u(l), u'(0) = u'(l)$.

This problem has unique solutions by $\int_0^l f(x)dx = 0, u(x_0) = u_0$, where $x_0 \in [0, l], u_0$ is fixed constant.

The exact solution of the problem with $f(x) = \cos(2\pi x/L)\exp(\sin(2\pi x/L))$ can be obtained use the Matlab operator "quad". The calculations by $l = 10, N = 10$ for FDS give following maximal errors:
$0.2097(O(h^2)), 0.0131(O(h^4)), 0.0049(O(h^6)), 0.0023(O(h^8))$, but for FDSES: 3.210^{-5}. Using the FDSES, we can obtain exact solution, if the function $f(x)$ is a linear combination of $\sin(2\pi p_1 x/l), \cos(2\pi p_2 x/l)$ and $N \geq 2 * \max(p_1, p_2)$.

5.3   Example for a Wave Equation

For a wave equation with BCs of the first kind (1), if $\alpha_1 = 0, \alpha_2 = 1, l = 1$, $T_0 = \sin(\pi x), \bar{T}_0 = 0, T(x, t) = \sin(\pi x)cos(\pi t)$, we obtain following maximal errors ($N = 40, t_f = 1$): 0.00748 (FDS), 10^{-16} (FDSES).

For numerical calculation of a wave equation with PBCs and $f = 0, T_0 = \sin(2\pi x), \bar{T}_0 = 0, T(x, t) = \sin(2\pi x)\cos(2\pi t)$ we obtain following maximal errors : FDS-0.0049 ($O(h^2)$), 0.000016 ($O(h^4)$), 10^{-7} ($O(h^6)$),2.10^{-10} ($O(h^8)$), FDSES- 10^{-14}.

5.4   Example for Heat Transfer Equation

For heat transfer equation with homogenous BC of first kind (the problem with discontinuous initial and boundary data) (1), if $\alpha_1 = 1, \alpha_2 = 0, l = 1, \bar{k} = 1, f = 0, T_0 = 1$. the maximal error by $t_f = 0.02, N = 10$ is 0.089 (FDS) and 0.0102(FDSES).

The results obtained with Fourier series have oscillations on $x = 0, x = l$ (Gibbs phenomenon). For the FDSES method these oscillations disappear. The maximal error by $t_f = 0.9, N = 10$ is 0.0000118 (FDS) and 0.0000015 (FDSES).

If the functions $f(x, t), T_0(x)$ are proportional to the eigenvector $w_p(x)$ and $p \leq N - 1$, then the FDSES is exact method.

For PBCs if the functions $f(x, t), T_0(x)$ are proportional to the functions $f_1(x) = \sin(2\pi p_1 x/l), f_2(x) = \cos(2\pi p_2 x/l)$, we have the exact solution for $\max(p_1, p_2) \leq N/2$, using the Fourier and FDSES methods.

5.5   Example for Hyperbolic Heat Conduction Equation

The numerical experiment for a hyperbolic heat conduction equation (1), if $l = 1, \sigma_1 = \sigma_2 = \infty, \bar{k} = 1, \alpha_2 = 0.1, \alpha_1 = 1, t_f = 0.2, f = 0, T_l = 1, T_r = 0, T_0 = 0, \bar{T}_0 = 0, N = 200$ (the initial and boundary conditions are discontinuous [2]) the numerical results are represented in Figs.1,2.

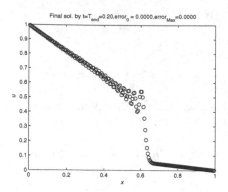
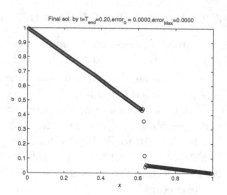

Fig. 1. Solution of finite-difference by $N = 200$, $t = 0.2$

Fig. 2. Solution of FDSES by $N = 200$, $t = 0.2$

6　Conclusions

The hyperbolic and parabolic type problems with BCs of the third kind and PBCs are solved with the method of lines in the time and with the finite difference scheme. The algorithms for discrete Fourier methods, which depend on the special parameter Q, are developed in different ways. For the last two eigenvalues and eigenvectors for finite difference operator new expressions are to work out, containing the hyperbolic functions.

The advantages of the FDSES in the case of BCs of the first kind and periodic BCs are demonstrated via several numerical examples in comparision with well-known finite difference scheme methods.

Acknowledgement. This work is partially supported by the projects 2009/0223/1DP/1.1.1.2.0/09/APIA/VIAA/008 of the European Social Fund and the grant 09.1572 of the Latvian Council of Science

References

1. Zvang, F.: Matrix theory. Basic results and techniques. Springer (1999)
2. Kalis, H., Buikis, A.: Method of lines and finite difference schemes with the exact spectrum for solution the hyperbolic heat conduction equation. Mathematical Modelling and Analysis 16(2), 220–232 (2011)
3. Kalis, H., Rogovs, S.: Finite differehce schemes with exact spectrum for solving differential equations with boundary conditions of the first kind. Int. Journ. of Pure and Applied Mathematics -IJPAM 71(1), 159–172 (2011)
4. Makarov, V.L., Gavrilyuk, I.P.: On constructing the best net circuits with the exact spectrum Dopov. Akad. Nauk Ukr. RSR, Ser. A, 1077–1080 (1975) (in Ukrainian)
5. Samarskij, A.A.: Theory of finite difference schemes Moscow. Nauka (1977) (in Russian)
6. Kalis, H., Rogovs, S., Gedroics, A.: On the mathematical modelling of the diffusion equation with piecewise constant coefficients in a multi-layered domain. International Journal of Pure and Applied Mathematics 81(4), 555–575 (2012)

The Numerical Solution of the Boundary Function Inverse Problem for the Tidal Models

Evgeniya Karepova and Ekaterina Dementyeva

Institute of Computational Modelling of SB RAS
660036 Akademgorodok, Krasnoyarsk, Russia
e.d.karepova@icm.krasn.ru
http://icm.krasn.ru

Abstract. The problem of propagation of long waves in a domain of an arbitrary form with the sufficiently smooth boundary on a sphere is considered. The boundary consists of "solid" parts passing along the coastline and "open liquid" parts passing through the water area. In general case the influence of the ocean through an open boundary is unknown and must be found together with components of a velocity vector and free surface elevation. For this purpose we use observation data of free surface elevation given only on a part of an "open liquid" boundary. We solve our ill-posed inverse problem by an approach based on the optimal control methods and adjoint equations theory.

Keywords: data assimilation problem, finite elements method and high performance computation.

Introduction

Researches [1–3] are devoted to the different aspects of mathematical and numerical modelling of tidal and free surface waves in large water areas taking into account the Earth's sphericity and the Coriolis acceleration based on vertically averaged equations of motion and continuity. One of the possible initial-boundary value problems for this equations is described in [4]. Useful a priori estimates providing stable and unique existence of solution are obtained ibid. In [5] for the same problem the finite element method is constructed and corresponding a priori estimates are obtained. Besides, numerical results on the special model grids and on the non-structured grids for the water areas of the Sea of Okhotsk and the World Ocean are presented.

Since in general case influence of an ocean on a water area through an open boundary is unknown then we consider the inverse problem on recovery of the boundary function which describes this influence. In [6] the iterative algorithm for recovery of the boundary function for every time-step, using the observation data of free surface elevation on the whole "open liquid" boundary, is considered. In the present work we discuss the opportunity of recovery of the influence of an ocean using an observation data of free surface elevation given only on a part of an "open liquid" boundary.

I. Dimov, I. Faragó, and L. Vulkov (Eds.): NAA 2012, LNCS 8236, pp. 345–354, 2013.

1   The Differential Formulation of a Problem

We consider the following problem. Let (r, λ, φ) be spherical coordinates with the origin at the terrestrial globe, $0 \leq \lambda \leq 2\pi$, $0 \leq \varphi < \pi$. Here λ means geographic longitude and instead of geographic latitude θ we use angle $\varphi = \pi/2 + \theta$. We put $r = R_E$, where R_E is the radius of the Earth which is assumed to be constant.

We formulate the problem on propagation of long waves in a water area as follows. Let $O_{\lambda\varphi}$ be image of (λ, φ) mapped onto a sphere $S^2_{R_E}$ with radius R_E, $\Omega = \{O_{\lambda\varphi}$, and let $(\lambda, \varphi) \in \Omega'\}$ be an open submanifold obtained by mapping Ω' onto a sphere of radius R_E with the piecewise smooth Lipchitz boundary $\Gamma = \Gamma_1 \cup \Gamma_2$ of the class $C^{(2)}$, where Γ_1 is a part of the boundary passing along a coastline and $\Gamma_2 = \Gamma \setminus \Gamma_1$ is a part of the boundary rounded a water area. Let denote characteristic functions of these parts of boundary by χ_1 and χ_2, respectively. Without loss of generality we may assume that the points $\varphi = 0$ and $\varphi = \pi$ (poles) are not involved in Ω. For the unknown functions $u = u(t, \lambda, \varphi)$, $v = v(t, \lambda, \varphi)$ and $\xi = \xi(t, \lambda, \varphi)$ in $\Omega \times (0, T)$ we write the vertically averaged equations of motion and continuity [1, 4] as following:

$$\frac{\partial u}{\partial t} = lv + mg \frac{\partial \xi}{\partial \lambda} - R_f u + f_1,$$

$$\frac{\partial v}{\partial t} = -lu + ng \frac{\partial \xi}{\partial \varphi} - R_f v + f_2, \tag{1}$$

$$\frac{\partial \xi}{\partial t} = m \left(\frac{\partial}{\partial \lambda} (Hu) + \frac{\partial}{\partial \varphi} \left(\frac{n}{m} Hv \right) \right) + f_3,$$

where u and v are components of the velocity vector $\mathbf{U}$ in λ and φ directions, respectively; ξ is a deviation of a free surface from the nonperturbed level; $H(\lambda, \varphi) > 0$ is depth of a water area at a point (λ, φ); the function $R_f = r_* |\mathbf{U}|/H$ takes into account the base friction force, r_* is the friction coefficient; $l = -2\omega \cos\varphi$ is the Coriolis parameter; $m = 1/(R_E \sin\varphi)$; $n = 1/R_E$; g is the acceleration of gravity; $f_1 = f_1(t, \lambda, \varphi)$, $f_2 = f_2(t, \lambda, \varphi)$ and $f_3 = f_3(t, \lambda, \varphi)$ are given functions of external forces.

We consider boundary conditions in the following form:

$$HU_n + \beta \chi_2 \sqrt{gH}\xi = \chi_2 \sqrt{gH} d \quad \text{on} \quad \Gamma \times (0, T), \tag{2}$$

where $U_n = \mathbf{U} \cdot \mathbf{n}$, $\mathbf{n} = (n_1, \frac{n}{m} n_2)$ is the vector of an outer normal to the boundary in the spherical coordinates; $\beta \in [0, 1]$ is a given parameter, $d = d(t, \lambda, \varphi)$ is a function defined on the boundary Γ_2. We extend d by zero on the boundary Γ_1.

We also impose initial conditions as following:

$$u(0, \lambda, \varphi) = u_0(\lambda, \varphi), \ v(0, \lambda, \varphi) = v_0(\lambda, \varphi), \ \xi(0, \lambda, \varphi) = \xi_0(\lambda, \varphi). \tag{3}$$

For time discretization we subdivide the segment $[0, T]$ into K subintervals by points: $0 = t_0 < t_1 < \cdots < t_K = T$ with the step $\tau = T/K$. We approximate time derivatives with backward differences and consider the system (1) – (2) on the interval (t_k, t_{k+1}) as:

$$\left(\frac{1}{\tau} + R_f\right) u - lv - mg\frac{\partial\xi}{\partial\lambda} = f_1 + \frac{1}{\tau}u^k \quad \text{in} \quad \Omega,$$

$$\left(\frac{1}{\tau} + R_f\right) v + lu - ng\frac{\partial\xi}{\partial\varphi} = f_2 + \frac{1}{\tau}v^k \quad \text{in} \quad \Omega, \tag{4}$$

$$\frac{1}{\tau}\xi - m\left(\frac{\partial}{\partial\lambda}(Hu) + \frac{\partial}{\partial\varphi}\left(\frac{n}{m}Hv\right)\right) = f_3 + \frac{1}{\tau}\xi^k \quad \text{in} \quad \Omega,$$

$$HU_n + \beta\chi_2\sqrt{gH}\xi = \chi_2\sqrt{gH}d \quad \text{on} \quad \Gamma, \quad k = 0, 1, \dots, K-1, \tag{5}$$

where for an arbitrary function $f(t, \lambda, \phi)$ we use $f^k = f(t_k, \lambda, \phi)$, $f = f(t_{k+1}, \lambda, \phi) = f^{k+1}$. Further the index $(k+1)$ in difference expressions is omitted if there is no ambiguity. Base friction $R_f = r_*|U^k|/H$ is taken from the previous time level.

System (4) – (5) is the subject of our investigation. In the direct problem (4) – (5) in time instant t_{k+1}, $k = 0, 1, ..., K-1$, we should find u, v, ξ when functions H, f_1, f_2, f_3, d are given. But in general case, the function d is unknown. So we formulate the inverse problem of finding the function d in the problem (4) – (5). In this case, to close the problem (4) – (5), we consider the following closed condition:

$$\xi = \xi_{obs} \quad \text{on} \quad \Gamma_0, \tag{6}$$

where $\xi_{obs} \in L_2(\Gamma_0)$ is a given function (for example, from observation data) on the some part of the boundary $\Gamma_0 \subset \Gamma$.

Thus, for the time instant t_{k+1}, $k = 0, 1, ..., K-1$, the differential problem (4) – (6) can be formulated as the problem on observation data assimilation in the following way [4].

Problem 1 (Inverse problem). *Assume that at the time instant t_{k+1}, $k = 0, 1, ..., K-1$, the function ξ_{obs} is defined on Γ_0, the function d is unknown on Γ_2 and vanishes on Γ_1. At a time instant t_{k+1} find u, v, ξ, d, satisfying the system (4), the boundary condition (5) and the closure condition (6).*

2   A Problem of Optimal Control

For real vector functions $\mathbf{\Phi} = (u, v, \xi)$, $\hat{\mathbf{\Phi}} = (\hat{u}, \hat{v}, \hat{\xi}) \in (L_2(\Omega))^3$ we consider the inner product [4]

$$(\mathbf{\Phi}, \hat{\mathbf{\Phi}}) = \int_{\Omega'} R_E^2 \sin\varphi\left(H(u\hat{u} + v\hat{v}) + g\xi\hat{\xi}\right) d\lambda d\varphi$$

and the norm

$$\|\mathbf{\Phi}\| = (\mathbf{\Phi}, \mathbf{\Phi})^{1/2} < \infty.$$

For integral formulation of the problem (4) – (5) we take the inner product of system (4) in $(L_2(\Omega))^3$ by the arbitrary vector function $\widehat{\boldsymbol{\Phi}} = (\hat{u}, \hat{v}, \hat{\xi}) \in (L_2(\Omega))^2 \times H^1(\Omega) \equiv W$ and perform integration by parts taking into account the boundary condition (5).

Definition. *A vector function* $\boldsymbol{\Phi} = (u, v, \xi) \in (L_2(\Omega))^2 \times H^1(\Omega) \equiv W$ *is called a weak solution of the problem* (4) – (5) *if the integral identity*

$$a(\boldsymbol{\Phi}, \mathbf{W}) = f(\mathbf{W}) + b(d, \mathbf{W}) \qquad (7)$$

holds for any vector function $\mathbf{W} = (w^u, w^v, w^\xi) \in W$. *Here*

$$a(\boldsymbol{\Phi}, \mathbf{W}) = \int_{\Omega'} R_E^2 \sin\varphi \Big(\frac{1}{\tau} \left(H(uw^u + vw^v) + g\xi w^\xi \right) + Hl(uw^v - vw^u)$$

$$+ HR_f(uw^u + vw^v) \Big) d\lambda d\varphi$$

$$+ \int_{\Omega'} R_E Hg \left(\left(u\frac{\partial w^\xi}{\partial \lambda} - w^u \frac{\partial \xi}{\partial \lambda} \right) + \sin\varphi \left(v\frac{\partial w^\xi}{\partial \varphi} - w^v \frac{\partial \xi}{\partial \varphi} \right) \right) d\lambda d\varphi$$

$$+ \beta g \int_{\Gamma_2} \sqrt{gH} \xi w^\xi \, ds,$$

$$f(\mathbf{W}) = \int_{\Omega'} R_E^2 \sin\varphi \left(H(f_1 w^u + f_2 w^v) + g f_3 w^\xi \right) d\lambda d\varphi$$

$$+ \int_{\Omega'} \frac{1}{\tau} R_E^2 \sin\varphi \left(H(u^k w^u + v^k w^v) + g\xi^k w^\xi \right) d\lambda d\varphi,$$

$$b(d, \mathbf{W}) = g \int_{\Gamma_2} \sqrt{gH} dw^\xi \, ds.$$

Notice that the boundary condition (5) is natural for the problem (4), hence, it imposes no restriction on spaces of trial and test functions.

In [4] it has been proved for $\beta > 0$ the problem (7) has a solution.

Considering the bilinear forms $a(\boldsymbol{\Phi}, \cdot)$ and $b(d, \cdot)$ as bounded linear functionals defined for any functions $\boldsymbol{\Phi} \in W$ and $d \in L_2(\Gamma_2)$, respectively, we can write the equality (7) as an operator equation [7]:

$$A\boldsymbol{\Phi} = \tilde{F} + Bd \qquad (8)$$

with operators $A : W \to (L_2(\Omega))^3$ and $B : L_2(\Gamma_2) \to (L_2(\Gamma_2))^3$ and $\tilde{F} \in (L_2(\Omega))^3$ which is induced by the inner product on an element $\mathbf{F} = (f_1 + \frac{1}{\tau}u^k, \ f_2 + \frac{1}{\tau}v^k, \ f_3 + \frac{1}{\tau}\xi^k)$.

Rewrite the equation (6) on boundary Γ_0 in the following form:

$$C\boldsymbol{\Phi} = \xi_{obs} \qquad (9)$$

with a trace operator $C : W \to H^{1/2}(\Gamma_0)$ and $\xi_{obs} \in L_2(\Gamma_0)$ and consider (9) as a condition for the operator equation (8). Since Γ_0 is Lipchitz boundary the space $H^{1/2}(\Gamma_0)$ is compactly embedded into $L_2(\Gamma_0)$ [7]. As ξ_{obs} does not necessarily belong to $H^{1/2}(\Gamma_0)$, so the problems (8), (9) or (7), (6) are ill-posed.

According to [4, 7], the problem (8) – (9) is uniquely and densely solvable provided that mes$(\Gamma_0 \cap \Gamma_1) > 0$.

In order to solve this problem the technique presented in [7] is applied.

Thus consider the following family of optimal control problem.

Problem 2. *Let ξ_{obs} be given on Γ_0. For fixed $\alpha \geq 0$ find the boundary function d_α on Γ_2 and the vector-function $\boldsymbol{\Phi}_\alpha = (u_\alpha, v_\alpha, \xi_\alpha)$ which satisfy the following operator equation*

$$A\boldsymbol{\Phi}_\alpha = \tilde{F} + Bd_\alpha$$

and minimize the following cost functional:

$$J_\alpha(d_\alpha, \xi_\alpha(d_\alpha)) = \frac{1}{2}g\left(\alpha \int_{\Gamma_2} \sqrt{gH}\left(\frac{\partial d_\alpha}{\partial s}\right)^2 ds + \int_{\Gamma_0} \sqrt{gH}(\xi_\alpha - \xi_{obs})^2 ds\right). \quad (10)$$

Each solution d_α of this problem satisfies a system of variational equations (Euler optimality equations) which has the following form:

$$A\boldsymbol{\Phi}_\alpha = \tilde{F} + Bd_\alpha, \quad A^*\widehat{\boldsymbol{\Phi}}_\alpha = J'_{\alpha,\,\boldsymbol{\Phi}}(d_\alpha, \boldsymbol{\Phi}_\alpha), \quad J'_{\alpha,\,d}(d_\alpha, \boldsymbol{\Phi}_\alpha) + B^*\widehat{\boldsymbol{\Phi}}_\alpha = 0, \quad (11)$$

where $J'_{\alpha,\,d}, J'_{\alpha,\,\boldsymbol{\Phi}}$ mean the variations of the functional J_α with respect to d and $\boldsymbol{\Phi}$, respectively, operators $A^*:W \to (L_2(\Omega))^3$, $B^*:(L_2(\Gamma_2))^3 \to L_2(\Gamma_2)$ are a linear adjoint to operators A, B, respectively, vector-functions $\boldsymbol{\Phi}_\alpha \in W$, $\widehat{\boldsymbol{\Phi}}_\alpha \in W$ are solutions of direct and adjoint problems.

In our case, Euler equations (11) for Problem 2 generate the following family of problems:

$$a(\boldsymbol{\Phi}_\alpha, \mathbf{W}) = f(\mathbf{W}) + b(d_\alpha, \mathbf{W}) \qquad \forall \, \mathbf{W} = (w^u, w^v, w^\xi) \in W, \quad (12)$$

$$a(\widehat{\mathbf{W}}, \widehat{\boldsymbol{\Phi}}_\alpha) = g\int_{\Gamma_0} \sqrt{gH}(\xi_\alpha - \xi_{obs})\widehat{w}^\xi d\Gamma \quad \forall \, \widehat{\mathbf{W}} = (\widehat{w}^u, \widehat{w}^v, \widehat{w}^\xi) \in W, \quad (13)$$

$$\alpha\frac{\partial}{\partial s}\left(\sqrt{gH}\frac{\partial d_\alpha}{\partial s}\right) = \sqrt{gH}\widehat{\xi}_\alpha \quad \text{on } \Gamma_2, \quad d_\alpha(\gamma_0) = d_\alpha(\gamma_1) = 0. \quad (14)$$

Here (14) is a boundary value problem for an ordinary differential equation along the boundary Γ_2, and γ_0, γ_1 are ends of $\Gamma_2 \subset \Gamma$.

Thus, using (12) – (13), (14) to determine a solution u^{k+1}, v^{k+1}, ξ^{k+1} and d^{k+1} at $(k+1)$ time instant, we can apply the following iterative process.

Iterative Algorithm 1

1. Take some $d_\alpha^{(0)}$ on Γ_2. From here on, when describing the algorithm, a superscript in parentheses denotes the number of an iteration step. Put $u_\alpha^{(0)} = u^k$, $v_\alpha^{(0)} = v^k$, $\xi_\alpha^{(0)} = \xi^k$.

Let ε be a given accuracy.

2. While

$$\left(\int_{\Gamma_0} \sqrt{gH}(\xi^{(l)} - \xi_{obs}^{(l)})^2 \, d\Gamma \right)^{1/2} \geq \varepsilon \left(\int_{\Gamma_0} \sqrt{gH}(\xi_{obs})^2 \, d\Gamma \right)^{1/2} \tag{15}$$

an iteration step is performed:

2.1. Using $d_\alpha^{(l)}$, we solve the direct problem (12) and determine $u_\alpha^{(l)}$, $v_\alpha^{(l)}$, $\xi_\alpha^{(l)}$ (index $(k+1)$ of time step is omitted as usual).

2.2. Using a solution $\xi_\alpha^{(l)}$ of the direct problem in the boundary condition for the adjoint one, we solve the adjoint problem (13) and determine $\widehat{u}_\alpha^{(l)}$, $\widehat{v}_\alpha^{(l)}$, $\widehat{\xi}_\alpha^{(l)}$.

2.3. Using a solution $\widehat{\xi}_\alpha^{(l)}$ of the adjoint problem, we solve the boundary value problem (14) and determine $\tilde{d}_\alpha^{(l)}$. Then use this solution for the iterative refinement $d_\alpha^{(l)}$:

$$d_\alpha^{(l+1)} = d_\alpha^{(l)} + \gamma_l(\tilde{d}_\alpha^{(l)}). \tag{16}$$

Here γ_l, α are parameters of the method.

2.4. Put $d_\alpha^{(l)} = d_\alpha^{(l+1)}$, $l = l + 1$ and go to point **2**.

Basing on results of [4, 7] we can prove the following theorem.

Theorem 1. *Let* $\mathrm{mes}(\Gamma_0 \cap \Gamma_1) > 0$ *and* $\mathrm{mes}(\Gamma_2) > 0$. *Then*
(1) *the problem* (12) − (13), (14) *is well-posed for any* $\alpha > 0$;
(2) *if* $\boldsymbol{\Phi}$, d *are a weak solution of the problem* (12), (9) *for some* $\alpha > 0$, *then*

$$\|\boldsymbol{\Phi} - \boldsymbol{\Phi}_\alpha\| + \|\xi - \xi_{obs}\|_{L_2(\Gamma_0)} \to 0, \qquad as \ \alpha \to +0;$$

(3) *for sufficiently small* $\gamma_l > 0$ *the Iterative algorithm 1 converges, in addition,*

$$\|\boldsymbol{\Phi}_\alpha^{(l)} - \boldsymbol{\Phi}\| + \|d_\alpha^{(l)} - d\|_{L_2(\Gamma_2)} + \|\xi_\alpha^{(l)} - \xi_{obs}\|_{L_2(\Gamma_0)} \to 0$$

as $\alpha \to +0$, $k \to \infty$.

Thus, on each time interval (t_k, t_{k+1}) for sufficiently large $l = L \gg 0$ and sufficiently small $0 < \alpha \ll 1$, $u^{k+1} \approx u_\alpha^{(L)}$, $v^{k+1} \approx v_\alpha^{(L)}$, $\xi^{k+1} \approx \xi_\alpha^{(L)}$ can be taken as the solution of the differential problem (4) − (5), (6).

The numerical solution of the direct and adjoint problems is based on the finite elements method. Consider a consistent triangulation $\mathcal{T} = \{\omega_i\}|_{i=1}^{N_{el}}$ of the domain Ω' [8]. The Bubnov-Galerkin method is used for discretization of our problem with respect to space. Linear functions on triangular finite elements are used as trial and test functions. In [5] a priori stable estimation for the discrete analogue is derived and the second order of approximation in internal nodes for an uniform grid is shown.

3   Numerical Tests with Boundary Data Recovery

We consider the water area of the sea of Okhotsk and a part of the Pacific Ocean near the Kuril Islands as a computational domain. The domain is bounded by a "square": $\Omega = [42°, 62°]$ N. $\times [135°, 162°]$ E., its liquid boundary Γ_2 passes along $\lambda = 161,1°$ E. and along $\varphi = 41,5°$ N. From here on, for convenience, along the λ– and φ–axes instead of radian measure we use degrees of eastern longitude and northern latitude, respectively.

Test calculations for the water area of the Sea of Okhotsk were performed on grids constructed on the basis of the ETOPO2 open bathymetric data base [8].

Since in general case for a nonstationary problem initial data are unknown and the procedure does not assume initial data recovery, in Ω we consider the following problem.

Firstly, we solve a steady-state problem using the function d which is given on all "liquid" boundary, independent of time, and has the following form:

$$d(\lambda, \varphi) = A exp \left(- \left(\frac{\lambda - \lambda_0}{2D} \right)^2 - \left(\frac{\varphi - \varphi_0}{2D} \right)^2 \right) \quad \text{on} \quad \Gamma_2. \qquad (17)$$

The steady-state solution is used as initial data. The values ξ from the steady-state solution taking on some part of the boundary Γ_0 are considered as "observation" data. Then we "forgot" values of d.

The aim of the numerical test is recovery of the function d on the whole liquid boundary using our "observation" data. To this end, d is recovered everywhere on the "liquid" boundary with Iterative algorithm 1 starting with $d \equiv 0$.

Using the steady-state procedure we obtain a rather smooth function as "observation" data. However, actual observation data, as a rule, are not so smooth. In this connection, in one test "white noise" is superimposed on the values of ξ on the boundary Γ_0. Then "noisy" values of ξ are taken as observation data. Moreover, in another test observation data with gaps is considered, i.e. data is given on the part of the "liquid" boundary only (Fig. 1).

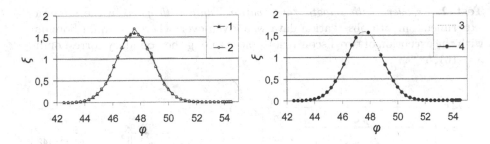

Fig. 1. The observation data on the east boundary part of the Pacific ocean: 1, 3 — smooth, 2 — "noisy"', 4 — with gaps

Thus, the problem (4)–(5) is solved with the cost functional (10) with two different type of observation data (6).

Test 1. *Recovery with "noisy" observation data.* The function of superimposition of "white noise" satisfies the following condition: the magnitude of "noise" which is added to each particular value of ξ, may not exceed

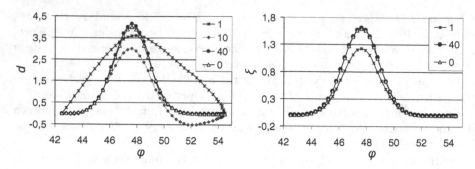

Fig. 2. Dependence of the functions d and ξ upon the number of iteration steps (1, 10, 40) on a liquid boundary of Ω with "noisy" observation data. 0 — exact d (on the left) and ξ (on the right).

given percent of the magnitude of ξ. The obtained perturbed function is used when recovering. Fig. 2 illustrates the recovery process of the functions d and ξ in the case that "noise" can introduce maximal deviation equal to 10 percent in "observation" data.

Fig. 2 shows that the recovered function d retains smoothness (17) despite the errors introduced in "observation" data and it is close to the exact solution. Moreover, the recovered free surface elevation ξ on the boundary practically coincides with the observation data without "noise". Thus, there is a smoothing of the errors introduced by "noise".

It is interesting to note that in [6] another regularizer in the functional of Problem 2 is considered. When using functional from [6], recovered free surface elevation ξ tends to noisy data.

Test 2. *Recovery with smooth "observation" data with gaps.* In the numerical experiment smooth observation data was given everywhere on liquid boundary with the exception of two discontiguous pieces along the boundary corresponding $\lambda = 161,5°$.

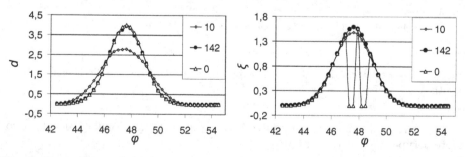

Fig. 3. Dependence of the functions d and ξ upon the number of iteration steps (10, 142) on a liquid boundary of Ω with observation data with gaps 0 (the graph of ξ on the right). Graph 0 on the left — exact d.

Results of the numerical experiments are shown in Fig. 3. Here we can see our method recovers free surface elevation ξ and d on the whole liquid boundary including points without observation data. Recovery passes in 142 iteration steps and there is good coincidence of the recovered values and the exact solution.

The numerical tests demonstrate some advantages of the boundary recovery method with the cost functional (10) in comparison with the functional proposed in [6]. The method recovers the boundary function d on the whole liquid boundary using observation data given on a part of the boundary only and moreover it is stable to perturbation and errors in data.

Test 3. *Dynamic recovery of d.* The aim is to recover d at each time instant for the problem of propagation of some initial perturbation of ξ. To get observation data the direct problem of waves propagation in time with some initial perturbation of ξ nearby the north part of the boundary Ω of the Pacific ocean (Pic. 4)

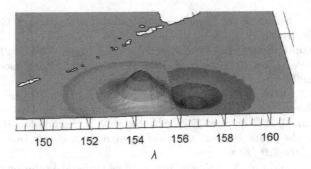

Fig. 4. The initial perturbation of ξ nearby the north part of the boundary Ω of the Pacific ocean

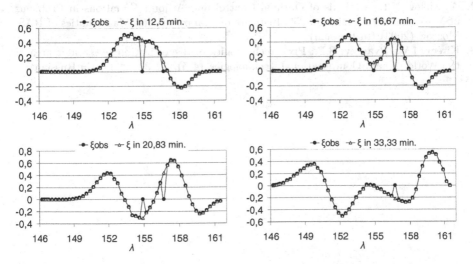

Fig. 5. The recovered ξ with observation data ξ_{obs} (with gaps) for some time instants on the north part of the boundary Ω of the Pacific ocean

is solved. Thus we obtain the observation data ξ_{obs} for each time instant. Then the boundary function d is "forgotten" ($d \equiv 0$ on the whole liquid boundary) and the inverse problem of recovery of d in time is solved at each instant t_k, $k = 1, ..., K$, of time. The results of recovery for some time instants are shown in fig. 5.

The numerical experiments are demonstrated good recovery of unknown boundary function from the boundary condition together with velocity vector and free surface elevation in whole computation domain both at a time instant and in dynamics case.

Acknowledgement. The work was supported by Russian Foundation of Fundamental Researches (grant 11-01-00224-a)

References

1. Marchuk, G.I., Kagan, B.A.: Dynamics of Ocean Tides. Leningrad, Gidrometizdat, 471 (1983) (in Russian)
2. Gill, A.E.: Atmosphere-Ocean Dynamics. Academic Press, 662 (1982)
3. Kowalik, Z., Polyakov, I.: Tides in the Sea of Okhotsk. J. Phys. Oceanogr. 28(7), 1389–1409 (1998)
4. Agoshkov, V.I.: Inverse problems of the mathematical theory of tides: boundary-function problem. Russ. J. Numer. Anal. Math. Modelling. 20(1), 1–18 (2005)
5. Kamenshchikov, L.P., Karepova, E.D., Shaidurov, V.V.: Simulation of surface waves in basins by the finite element method. Russian J. Numer. Anal. Math. Modelling. 21(4), 305–320 (2006)
6. Karepova, E., Shaidurov, V., Dementyeva, E.: The numerical solution of data assimilation problem for shallow water equations. Int. J. of Num. Analysis and Modeling, Series B. 2(2-3), 167–182 (2011)
7. Agoshkov, V.I.: Methods of Optimal Control and Adjoint Equations in Problems of Mathematical Physics, p. 256. Institute of Computational Mathematics of RAS, Moscow (2003) (in Russian)
8. Kireev, I.V., Pyataev, S.F.: Pixel technologies of discretization of a water area of the World Ocean. Computational Technologies 14(5), 30–39 (2009) (in Russian)

On a Mathematical Model of Adaptive Immune Response to Viral Infection

Mikhail Kolev[1,*], Ana Markovska[2], and Adam Korpusik[3]

[1] Faculty of Mathematics and Computer Science, University of Warmia and Mazury, Słoneczna 54, 10-710 Olsztyn, Poland
kolev@matman.uwm.edu.pl
[2] Faculty of Mathematics and Natural Sciences, South-West University "N. Rilski", Ivan Mihajlov 66, 2700 Blagoevgrad, Bulgaria
[3] Faculty of Technical Sciences, University of Warmia and Mazury, Oczapowskiego 11, 10-719 Olsztyn, Poland

Abstract. In this paper we study a mathematical model formulated within the framework of the kinetic theory for active particles. The model is a bilinear system of integro-differential equations (IDE) of Boltzmann type and it describes the interactions between virus population and the adaptive immune system. The population of cytotoxic T lymphocytes is additionally divided into precursor and effector cells. Conditions for existence and uniqueness of the solution are studied. Numerical simulations of the model are presented and discussed.

Keywords: numerical simulations, integro-differential equations, nonlinear dynamics, kinetic model.

1 Introduction

The application of mathematical and numerical approaches in life sciences is a promising area of research, which develops quickly during the last decades. It requires a close teamwork between specialists in mathematics, numerical analysis, computer programming, biology, medicine etc. In this way the traditional experimental biological approaches can be fruitfully complemented by mathematical modeling methods. This can help for better understanding of the way of functioning of the biological systems and for reduction of the amount of lengthy experiments needed for the design of therapeutical methods [5].

The purpose of the present paper is to present an example of application of mathematical and computational methods in the field of virology. The contents of our work are organized as follows. In Section 2 we present basic biological information on the viral infections and the defence mechanisms of the host organisms. In Section 3 we describe a mathematical model of adaptive immune response to virus. The model is a complicated system of partial integro-differential equations. A theorem for existence and uniqueness of its solution is proved.

---

* Corresponding author.

I. Dimov, I. Faragó, and L. Vulkov (Eds.): NAA 2012, LNCS 8236, pp. 355–362, 2013.

In Section 4 we describe the performed discretization of the model and present results of our simulations. Section 5 concludes the paper and suggests some future research directions.

2   Viruses and Immune System

An important feature of the viruses is that they do not possess their own metabolism [1, 9]. For their replication, they have to enter susceptible cells of the host organism and use their metabolic mechanism. After their reproduction inside the infected cells, the viral particles can leave and destroy the host cells, and start infecting other susceptible cells. Too great viral replication often leads to significant decreasing of the life span of the infected cells. Thus, the viral infections may lead to various diseases, some of which such as influenza, hepatitis C, AIDS can be very dangerous or even lethal for the infected individuals [9].

In order to protect the organisms against viruses and other foreign antigens, various mechanisms have evolved during the evolution. They form the immune system, which can be observed in some forms even in very simple species. The immune system of higher organisms is one of the most complicated their systems. It includes various molecules, cells, specialized tissues and organs.

The immune system may be divided into two main parts: (i) innate (natural) immune system and (ii) acquired (adaptive) immune system. The innate immunity is non-specific, it is not directed against any specific virus strain. It includes physical barriers (saliva, skin, stomach acid), changes (inflammation) and immune cells without immunological memory (e.g. macrophages, dendritic cells). The innate defense mechanisms provide the initial protection against infections.

The acquired immune system is a remarkably adaptive defense system that has evolved in vertebrates. It needs more time to develop than the innate immunity and mediates the later defenses against pathogens. The acquired immunity is virus-specific: it can specifically recognize the viral physical structure and perform responses depending on the identified virus strain. This important ability of the acquired immunity to adapt to the specific type of infection and to respond accordingly explains the second name of this very significant component of the immune system. The adaptive responses are performed by cells called lymphocytes. Besides recognition of virus particles, the adaptive immunity is also able to establish immunological memory, i.e. to produce higher amounts of virus-specific lymphocytes (even after achieving virus clearance) and thus to carry out better responses in future encounters with the same virus strain. The adaptive immunity can be subdivided into two main branches: humoral and cell-mediated (or cellular) immunity [1].

The humoral immune response is performed by B lymphocytes, which produce antibodies (AB) that are able to bind to free virus particles. Thus viruses are either neutralized or tagged for destruction by other cells of the immune system (e.g. macrophages). AB can be very effective in suppressing the virus population, but they are helpless against infected cells. This means that despite the ongoing humoral response, virus particles can still reproduce [1].

The cellular immunity is directed against the infected cells. The cell-mediated response is performed by T lymphocytes: T helper cells (Th) and cytotoxic T lymphocytes (CTL). The cytotoxic T lymphocytes are able to specifically recognize and destroy the infected cells. The Th cells, on the other hand, take part in the regulation of the immune responses, both cellular and humoral [9].

3   Mathematical Model

The mathematical model presented here is developed within the so-called "kinetic theory for active particles" (KTAP) [7]. The basic concept of KTAP is the functional activity of individuals (particles) taking part in the interactions between the composing subsystems of an overall complex system. This approach has been successfully applied to model phenomena in various areas [4–6].

The model describes the interactions between the populations of susceptible uninfected cells, infected cells, free virus particles, AB, precursor CTL (CTL_P) and effector CTL (CTL_E). Each population is denoted by the corresponding subscript $i = 1, \ldots, 6$. The functional states of the interacting individuals are characterized by a variable $u \in [0, 1]$ describing the specific biological activity of each individual.

We neglect the presence of internal degree of freedom of the population of uninfected cells labeled by $i = 1$. This population is assumed to be independent of its activation states. The activity of the population of infected cells labeled by $i = 2$ denotes the rate of reproduction of viruses inside the infected cell and the rate of destruction of the infected cells due to the viral activity. Infected cells with higher activation states are supposed to produce more virus particles, at the cost of faster destruction of the cells.

The activity of the population of free virus particles labeled by $i = 3$ describes its ability to infect susceptible uninfected cells. The activity of the population of antibodies labeled by $i = 4$ describes their ability to destroy free virus particles.

Populations of precursor CTL labeled by $i = 5$ and of effector CTL labeled by $i = 6$ participate in the cellular immune response. Upon contact with virus, precursor cells proliferate. This results in establishment of CTL memory and creation of effector cells. Hence, we assume that the activity of precursor CTL describes the rate of CTL_P proliferation. The activity of the population of CTL_E describes their ability to destroy infected cells.

Further, we introduce the following notation. Let

$$f_i = f_i(t, u) \quad : [0, \infty] \times [0, 1] \to R_+, \quad i = 1, \ldots, 6,$$

denotes the distribution function of the i-th population with activity state $u \in [0, 1]$ at time $t \geq 0$. The concentration of the i-th population at time $t \geq 0$ is then given by:

$$
\begin{aligned}
n_1(t) &= f_1(t, u), \quad \forall u \in [0, 1], \ t \geq 0, \\
n_i(t) &= \int_0^1 f_i(t, u) du \quad : [0, \infty) \to R_+, \ i = 2, \ldots, 6.
\end{aligned}
\tag{1}
$$

The difference of the equation for $n_1(t)$ in (1) is due to the assumed independency of the population $i = 1$ of its activation states u. Further, we denote:

$$n_i^*(t) = \int_0^1 u f_i(t, u) du \quad : [0, \infty) \to R_+, \quad i = 2, ..., 6. \tag{2}$$

The model of the competition between the virus particles and the adaptive immune system is given by the following system of partial IDE:

$$\frac{dn_1}{dt}(t) = S_1(t) - d_{11}n_1(t) - d_{13}n_1(t)n_3^*(t), \tag{3}$$

$$\frac{\partial f_2}{\partial t}(t, u) = p_{13}^{(2)}(1 - u)n_1(t)n_3^*(t) - d_{22}u f_2(t, u) - d_{26}f_2(t, u)n_6^*(t) \\ + c_{22}\left(2\int_0^u (u - v)f_2(t, v)dv - (1 - u)^2 f_2(t, u)\right), \tag{4}$$

$$\frac{\partial f_3}{\partial t}(t, u) = p_{22}^{(3)}n_2^*(t) - d_{33}f_3(t, u) - d_{34}f_3(t, u)n_4^*(t), \tag{5}$$

$$\frac{\partial f_4}{\partial t}(t, u) = p_{34}^{(4)}(1 - u)n_3(t)n_4(t) - d_{44}f_4(t, u), \tag{6}$$

$$\frac{\partial f_5}{\partial t}(t, u) = p_{25}^{(5)}(1 - u)n_2(t)n_5^*(t) - d_{55}f_5(t, u), \tag{7}$$

$$\frac{\partial f_6}{\partial t}(t, u) = p_{25}^{(6)}(1 - u)n_2(t)n_5^*(t) - d_{66}f_6(t, u), \tag{8}$$

supplemented by the following initial conditions

$$n_1(0) = n_1^{(0)}, \quad f_i(0, u) = f_i^{(0)}(u), \quad i = 2, ..., 6. \tag{9}$$

The initial conditions and all parameters of the model are assumed to be non-negative. Additionally, we suppose that $p_{13}^{(2)} = 2d_{13}$. The meanings of the parameters and corresponding terms participating in the model equations (3)-(8) are described in Table 1.

Table 1. Model parameters and variables

Par.	Description
u	activation state
t	time
$S_1(t)$	generation of new uninfected cells
d_{11}	natural death of uninfected cells
d_{13}	infectivity rate of uninfected cells
$p_{13}^{(2)}$	rate of appearance of new infected cells
d_{22}	natural death of infected cells
d_{26}	destruction of infected cells by effector CTLs
c_{22}	steady progress of infected cells towards increasing their activation states
$p_{22}^{(3)}$	production of new viruses
d_{33}	natural death of viruses
d_{34}	destruction of free viruses by antibodies
$p_{34}^{(4)}$	generation of new antibodies
d_{44}	natural death of antibodies
$p_{25}^{(5)}$	production of new precursor CTLs
d_{55}	natural death of precursor CTLs
$p_{25}^{(6)}$	production of new effector CTLs
d_{66}	natural death of effector CTLs

Now, let us consider some properties of the solution to the initial value problem (3)-(9). We introduce the following notation:

$$\mathbf{X} = \{\mathbf{f} = (n_1, f_2, ..., f_6) : \quad |n_1| < \infty, \quad \text{and} \quad f_i \in L_1(0,1), \text{ for } i = 2, ..., 6\},$$

$$\mathbf{X}^+ = \{\mathbf{f} = (n_1, f_2, ..., f_6) \in \mathbf{X} : \quad n_1 \geq 0, \quad \text{and} \quad f_i \geq 0, \, i = 2, ..., 6, \text{a.e.}\}.$$

We have the following theorem.

Theorem 1. *Let $S_1 \in C^0([0, \infty); R_+)$. For every $T > 0$ there exists a unique solution*

$$\mathbf{f} \in C^0([0, T]; \mathbf{X}) \cap C^1((0, T); \mathbf{X})$$

to system (3)-(8) with the initial datum $\mathbf{f}^{(0)} = (n_1^{(0)}, f_2^{(0)}, ..., f_6^{(0)})$, $\mathbf{f}^{(0)} \in \mathbf{X}^+$. The solution satisfies $\mathbf{f}(t) \in \mathbf{X}^+, \forall t \in [0, T]$.

Proof. It is easy to check that the operators defined by the right–hand sides of Eqs. (3)-(8) are Lipschitz–continuous in $\mathbf{X}$. From this, local existence and uniqueness follows. Standard proof can be performed for example by transforming system (3)-(8) into corresponding integral system and applying the Banach fixed point theorem, similarly to a proof of Picard's theorem, see, e.g. [3].

The nonnegativity of the solution can be easily proved by using the successive approximation method (see, e.g. [2]).

It remains to find *a priori* estimates for the solution. From the continuity of $S_1(t)$ and Eq. (3) it follows that:

$$\frac{d}{dt} n_1(t) \leq \sup_{[0,T]} S_1(t), \quad n_1(t) \leq t \sup_{[0,T]} S_1(t). \tag{10}$$

Therefore, the concentration $n_1(t)$ of uninfected cells is bounded on each finite time interval $[0, T]$.

Taking into account the equality

$$\int_0^1 c_{22} \left(2 \int_0^u (u - v) f_2(t, v) dv - (1 - u)^2 f_2(t, u) \right) du = 0,$$

the integration of Eq. (4) from 0 to 1 with respect to u yields:

$$\frac{d}{dt} n_2(t) = 0.5 p_{13}^{(2)} n_1(t) n_3^*(t) - d_{22} n_2^*(t) - d_{26} n_2(t) n_6^*(t) \tag{11}$$

Due to the relation $p_{13}^{(2)} = 2 d_{13}$, after summing up Eqs. (3) and (11) we have:

$$\frac{d}{dt} (n_1(t) + n_2(t)) = S_1(t) - d_{11} n_1(t) - d_{22} n_2^*(t) - d_{26} n_2(t) n_6^*(t) \leq \sup_{[0,T]} S_1(t) \tag{12}$$

Due to the continuity of $S_1(t)$, it follows from (12) that the sum of the concentrations of infected and uninfected cells $n_1(t) + n_2(t)$ is bounded on each finite time interval $[0, T]$. Since $n_1(t)$ is bounded, it follows that $n_2(t)$ is also bounded on $[0, T]$.

From the boundedness of $n_1(t)$ and $n_2(t)$, after integration of Eqs. (5)-(8) from 0 to 1 with respect to u the boundedness of $n_3(t)$, $n_4(t)$, $n_5(t)$ and $n_6(t)$ on $[0, T]$ follows. Therefore $f_i(t, u) \in L_1(0, 1)$, $i = 2, ..., 6$, with respect to u on $[0, T]$.

From the obtained *a priori* estimates it follows that the solution to the initial value problem (3)-(9) exists and it is unique on each finite time interval $[0, T]$, which finishes the proof.

4   Numerical Experiments and Discussion

The Cauchy problem (3)-(9) consisting of 6 nonlinear partial integro-differential equations is solved numerically. First, we discretize the equations (4)-(8) in the activation state variable $u \in [0, 1]$ by introducing a uniform mesh

$$u_i = i\Delta u, \quad i = 0, 1, \ldots, N, \tag{13}$$

where Δu and N are chosen such that $N\Delta u = 1$ and N is a positive integer. This yields a system of $5N + 6$ ordinary differential equations allowing to find approximate solutions to the model (3)-(8).

The system of ordinary differential equations corresponding to the discretized model (3)-(8) is solved by using the code **ode15s** from the Matlab ODE suite (see, e.g. [8]) with $RelTol = 10^{-3}$ and $AbsTol = 10^{-4}$. The participating integrals are approximated by the use of the composite Simpson's rule. The obtained numerical solutions of the discretized system are used to compute the approximations to the functions $n_2(t)$,..., $n_6(t)$, by the use of Eq. (1).

The aim of our numerical experiments is to study the role of the viral replication inside the infected cells (denoted by $p_{22}^{(3)}$) for the dynamics and outcome of the viral infection.

Let us denote by c the rate of total CTL production (both CTL_P and CTL_E) and by q - the fraction of the newly produced CTL that become effector cells (q is the CTL_E differentiation rate). Then

$$p_{25}^{(5)} = c(1 - q), \quad p_{25}^{(6)} = cq.$$

The role of parameter $p_{22}^{(3)}$ is studied for small and high rates of CTL differentiation into effector cells (denoted by q). The value of q is a tradeoff between strong CTL memory and less dangerous acute phase of infection [7].

The initial conditions and parameters of the model have been set to simulate a full adaptive (humoral and cellular) immune response to viral infection. We have assumed the initial presence of susceptible uninfected cells, virus particles, antibodies and precursor CTL, as well as the initial absence of infected cells and effector CTL. For every $j = 0, ..., N$, the initial values for populations and the parameters have been set as follows:

$$n_1(0) = 1, \quad f_{2,j}(0) = 0.0, \quad f_{3,j}(0) = 0.1,$$

$$f_{4,j}(0) = 0.1, \quad f_{5,j}(0) = 0.1, \quad f_{6,j}(0) = 0.0.$$

$$S_1(t) = 1, \quad d_{11} = 1, \quad d_{13} = 2.5, \quad d_{22} = 1.5, \quad d_{26} = 300,$$

$$c_{22} = 0.1, \quad d_{33} = 1, \quad p_{34}^{(4)} = 10, \quad d_{34} = 10, \quad d_{44} = 1,$$

$$d_{55} = 0.1, \quad d_{66} = 1, \quad c = 11.5, \quad \text{(i) } q = 0.1, \quad \text{(ii) } q = 0.6.$$

Further, for fixed value of q, we have changed the rate of virus production inside the infected cells (parameter $p_{22}^{(3)}$) in order to obtain the equilibrium and minimal concentrations of the uninfected cells $n_1(t)$. The equilibrium concentration determines the outcome of the infection, while the minimal concentration describes the severity of the acute phase of infection.

The results of our simulations have shown that the rate of virus production inside the infected cells influences both course and outcome of the viral infection (see Fig. 1). Lower values of parameter $p_{22}^{(3)}$ result in less onerous acute phase of infection and higher equilibrium concentration of uninfected cells.

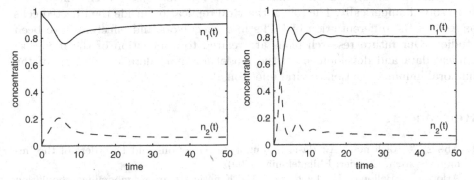

Fig. 1. Dynamics of the populations of uninfected and infected cells with CTL differentiation rate $q = 0.1$ and virus production rates: (i) $p_{22}^{(3)} = 5$, (ii) $p_{22}^{(3)} = 25$

An increase in the rate of virus production results in a decrease in the minimal value of $n_1(t)$. Moreover, initially (for low enough values of $p_{22}^{(3)}$) it results in a decrease of the equilibrium value of $n_1(t)$ as well. There exists a threshold p_{eq} after which the increase in parameter $p_{22}^{(3)}$ affects only the minimal concentration of uninfected cells - the equilibrium concentration remains the same. These results are illustrated in Fig. 2.

In the case of lower CTL differentiation rate ($q = 0.1$) the minimal value of $n_1(t)$ is lower than for $q = 0.6$ (see Fig. 2). On the other hand, the equilibrium value decreases slower in this case. The threshold p_{eq} is also higher for $q = 0.1$. Our results confirm the observations that lower rates of CTL differentiation into effector cells results in more onerous acute phase of infection, but stronger CTL memory (see e.g. [7]).

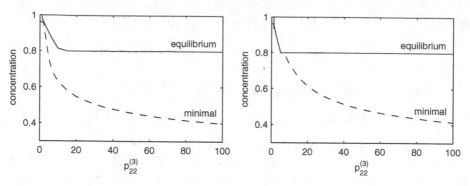

Fig. 2. Influence of virus production rate on the minimal and equilibrium values of $n_1(t)$ for CTL differentiation rates: (i) $q = 0.1$, (ii) $q = 0.6$

5   Conclusions

In the paper we have used a mathematical model to study some features of the competition between virus and adaptive immune system. The model has been solved numerically. The role of the viral replication inside the infected cells as well as the differentiation of CTLs into precursors and effectors have been studied. Our future research plans are related to application of the model to clinical data and development of the model for more detailed analysis of the humoral immune reaction to viral infections.

References

1. Abbas, A., Lichtman, A.: Basic Immunology: Functions and Disorders of the Immune System. Elsevier, Philadelphia (2009)
2. Arlotti, L., Bellomo, N., Lachowicz, M.: Kinetic equations modelling population dynamics. Transport Theory Statist. Phys. 29, 125–139 (2000)
3. Belleni-Morante, A.: Applied Semigroups and Evolution Equations. Oxford Univ. Press, Oxford (1979)
4. Bellomo, N., Carbonaro, B.: Toward a mathematical theory of living systems focusing on developmental biology and evolution: a review and perspectives. Physics of Life Reviews 8, 1–18 (2011)
5. Bellomo, N., Bianca, C.: Towards a Mathematical Theory of Complex Biological Systems. World Scientific, Singapore (2011)
6. Bianca, C.: Mathematical modelling for keloid formation triggered by virus: Malignant effects and immune system competition. Math. Models Methods Appl. Sci. 21, 389–419 (2011)
7. Kolev, M., Korpusik, A., Markovska, A.: Adaptive immunity and CTL differentiation - a kinetic modeling approach. Mathematics in Engineering, Science and Aerospace 3, 285–293 (2012)
8. Shampine, L., Reichelt, M.: The Matlab ODE suite. SIAM J. Sci. Comput. 18, 1–22 (1997)
9. Wodarz, D.: Killer Cell Dynamics. Springer, New York (2007)

Positivity Preserving Numerical Method
for Non-linear Black-Scholes Models

Miglena N. Koleva

FNSE, University of Rousse, 8 Studentska Str., Rousse 7017, Bulgaria
mkoleva@uni-ruse.bg

Abstract. A motivation for studying the nonlinear Black- Scholes equation with a nonlinear volatility arises from option pricing models taking into account e.g. nontrivial transaction costs, investors preferences, feedback and illiquid markets effects and risk from a volatile (unprotected) portfolio. In this work we develop positivity preserving algorithm for solving a large class of non-linear models in mathematical finance on the original (infinite) domain. Numerical examples are discussed.

1 Introduction and Model Formulation

The solution of the (linear) Black-Scholes equation (Black and Scholes, 1973) has been derived under several restrictive assumptions like e.g. frictionless, liquid and complete markets, etc. We also recall that the linear Black-Scholes equation provides a perfectly replicated hedging portfolio. In the last decades some of these assumptions have been relaxed in order to model, for instance, the presence of transaction costs (see e.g. Leland [19], Avellaneda and Parás [5]), feedback and illiquid market effects due to large traders choosing given stock-trading strategies (Frey and Patie [10]), imperfect replication and investors preferences (Barles and Soner [6]), risk from unprotected portfolio (Jandačka-Ševčovič, [15]).

This models defer from the classical Black-Scholes equation by a non-constant volatility term σ, which depends on time t, spot price S of the underlying and the second derivative (Greek Γ) of the option price $V(S,t)$. Hence, the model equation is the following nonlinear partial differential equation

$$V_t + \frac{1}{2}\sigma^2(t, S, V_{SS})S^2 V_{SS} + (r - q)SV_S - rV = 0, \quad 0 \leq S < \infty, \quad 0 \leq t \leq T, \quad (1)$$

with constant short rate r, dividend yield q, maturity T and volatility $\sigma^2(t, S, V_{SS})$ depending on the particular model.

We will study (1) for European Call option, i.e. the value $V(S, t)$ is the solution to (1), $q = 0$ on $0 \leq S < \infty$, $0 \leq t \leq T$ with the following terminal and boundary conditions ($E > 0$ is the exercise price):

$$V(S, T) = \max\{0, S - E\}, \quad 0 \leq S < \infty,$$
$$V(0, t) = 0, \quad 0 \leq t \leq T, \quad (2)$$
$$V(S, t) = S - Ee^{-r(T-t)}, \quad S \to \infty.$$

I. Dimov, I. Faragó, and L. Vulkov (Eds.): NAA 2012, LNCS 8236, pp. 363–370, 2013.

The model (1) can be written as the backward parabolic fully nonlinear PDE $V_t + S^2 F(S, V_S, V_{SS}) = 0$. At some conditions on F, the most used of which are $F(S, p, r) \in C^2$, $V(S, T) = V_T(S)$, $F_r(S, V_T'(S), V_T''(S)) > 0$, in [1,2,6,15] was obtained results for existence and uniqueness of solutions (classical or viscosity). It was checked that the model described above satisfies these conditions.

Let $G(S, V_S, V_{SS}) = S^2 F(S, V_S, V_{SS})$, $G \in C^k((0, T) \times R^3)$. We briefly discuss the maximum principle (MP) for (1). Let Π_T be the rectangle $\Pi_T = \{(S, t) : 0 < t < T, -\infty < a < b < +\infty\}$ and $G_r(S, p, r) \geq 0$ everywhere and $G_r(S, 0, 0) = 0$. Then, if V is a classical solution of $V_t + G(S, V_S, V_{SS}) = 0$ in Π_T we have: $\max_{\overline{\Pi}_T} V = \max_{\Gamma_T} u$ and $\min_{\overline{\Pi}_T} V = \min_{\Gamma_T} V$, where $\Gamma_T = I \cup II \cup III$, $I = \{0 < t < T, x = a\}$, $II = \{a < x < b, t = 0\}$, $III = \{0 < t < T, x = b\}$ is the parabolic part of the boundary. For the proof, see [2].

There exists many discretizations, algorithms and some numerical methods for different versions of the non-linear Black-Scholes equation [4,7,8,12,14]. In [20], authors develop positivity-preserving (i.e. the non-negativity of the numerical solution to be guaranteed) first-order fully implicit scheme for models arising from pricing European options under transaction costs. In our previous work [17] we developed a *fast*, second order both in space and time a kernel-based method for solving a large class of non-linear models in mathematical finance, computed on large enough truncated region. But the non-negativity of the numerical solution is not guaranteed. In this work, having in mind MP discussed above, we will present efficient, *positivity preserving* algorithm for solving the same non-linear Black-Scholes models on the original (infinite) interval. We develop implicit-explicit methods on quasi-uniform mesh (QUM), implementing the idea of van Leer flux limiter [11,13,21].

An often used approach to overcome the *degeneration* at $S = 0$ and to obtain a *forward* parabolic problem, is the variable transformation [1,8,12]

$$x(S) = \log\left(\frac{S}{E}\right), \quad \tau(t) = \frac{1}{2}\sigma_0^2(T - t), \quad u(x, \tau) = e^{-x}\frac{V}{E}.$$

Now, denoting $K = 2r/\sigma_0^2$ (σ_0 is the volatility of the underlying asset), the equation (1) transforms into

$$u_\tau - \tilde{\sigma}^2(\tau, x, u_x, u_{xx})(u_x + u_{xx}) - Ku_x = 0, \quad x \in \mathbb{R}, \quad 0 \leq \tau \leq \frac{\sigma_0^2 T}{2}, \quad (3)$$

$$\tilde{\sigma}_L^2 = 1 + f(u_x, u_{xx}), \qquad f(u_x, u_{xx}) = Le \cdot \text{sign}(u_x + u_{xx}),$$

$$\tilde{\sigma}_{BS}^2 = 1 + f(x, \tau, u_x, u_{xx}), \quad f(x, \tau, u_x, u_{xx}) = \Psi[a^2 E e^{K\tau + x}(u_x + u_{xx})],$$

$$\tilde{\sigma}_{JS}^2 = 1 + f(x, u_x, u_{xx}), \qquad f(x, u_x, u_{xx}) = \mu[E e^x(u_x + u_{xx})]^{1/3},$$

$$\tilde{\sigma}_{AP}^2 = f(u_x, u_{xx}), \qquad f(u_x, u_{xx}) = \begin{cases} \sigma_{max}^2, & u_x + u_{xx} \leq 0, \\ \sigma_{min}^2, & u_x + u_{xx} > 0, \end{cases}$$

$$\tilde{\sigma}_{FP}^2 = f(x, u_x, u_{xx}), \qquad f(x, u_x, u_{xx}) = [1 - \rho \cdot \lambda(E e^x)(u_x + u_{xx})]^{-2},$$

where $\tilde{\sigma}^2 = \tilde{\sigma}_{L,BS,JS,AP,FP}^2$ corresponds to Leland, Barles-Soner, Jandačka-Ševčovič, Avellaneda-Parás and Frey-Patie models, respectively. Here $0 < Le <$

1 is the Leland number, a is a parameter measure transaction cost and risk aversion, ρ is a parameter measuring the market liquidity and $\lambda(S)$ describes the liquidity profile in the dependence of the asset price. In $\tilde{\sigma}_{BS}^2$, $\Psi(s)$ solves an ODE [6] with implicit exact solution derived in [7]

$$\sqrt{s} = -\sinh^{-1}\sqrt{\Psi}/\sqrt{\Psi + 1} + \sqrt{\Psi}, \quad \text{for } s > 0, \quad \Psi(s) > 0,$$
$$\sqrt{-s} = \sin^{-1}\sqrt{-\Psi}/\sqrt{\Psi + 1} - \sqrt{-\Psi}, \quad \text{for } s < 0, \quad -1 < \Psi(s) < 0.$$

It is shown that $-1 < \Psi(s) < \infty$, $s \in \mathbb{R}$. Next, v^p ($p = 1/3$) in $\tilde{\sigma}_{JS}^2$ stands for the signed power function, i.e. $v^p = v|v|^{p-1}$, $\mu = 3(C^2 M/(2\pi))^{1/3}$, where $M \geq 0$ is the transaction cost measure, $C \geq 0$ is the risk premium measure and $SV_{SS} < \pi/(32C^2 R)$. In Avellaneda-Parás model the volatility is not known exactly, but is assumed that lie between extreme values σ_{max}^2 and σ_{min}^2. It is clear that for all models $\tilde{\sigma}^2 > 0$.

The problem (3) is computed by the following initial and boundary conditions, corresponding to (2)

$$u(x,0) = u^0(x) = \max(0, 1 - e^{-x}), \quad \lim_{x \to -\infty} u(x,\tau) = 0, \quad \lim_{x \to \infty} u(x,\tau) = 1. \quad (4)$$

Note that if $u \geq 0$ it follows that $V \geq 0$. Therefore we will concentrate on numerical solution of the problem (3)-(4).

The remaining part of this paper is organized as follows. In Section 2, we provide some basic tools for the numerical methods, which are presented in the next section. Finally, in Sections 4,5, we give numerical examples and concluding remarks.

2   Preliminaries

We start with some basic definitions, statements and notations. We also present meshes and derivative approximations.

Mesh in space. In the interval $[0, 1]$ we consider the uniform mesh $\omega_h = \{\xi_i = ih, \; h = 1/N, \; i = 0, \ldots, N\}$, where N is a positive number. Let $x(\xi)$, $\xi \in [0, 1]$, $x \in [a, b]$ is a strong monotone sufficiently smooth function. Following [3], we define the mesh $\omega_N = \{x_i = x(\frac{i}{N}), \; 0 \leq i \leq N\}$ in $[a, b]$.

Next, extending the idea of [3], we shall implement to our model problem (3)-(4) the mesh ω_N, defined on the infinite interval $(-\infty, +\infty)$, see Figure 1:

$$\omega_N = \begin{cases} x(\xi) = x^-(\xi), & x \leq 0, \\ x(\xi) = x^+(\xi), & x \geq 0 \end{cases}, \quad m_1 + m_2 = N, \quad x^-(1) = x^+(0) = 0,$$

$$x^-(\xi) = c_1 \ln(\xi), \quad h_{m_1-1}^- = x_{m_1}^- - x_{m_1-1}^- \simeq \frac{c_1}{m_1}, \quad x_1^- = c_1 \ln(m_1) \quad (5)$$

$$x^+(\xi) = -c_2 \ln(1 - \xi), \quad h_0^+ = x_1^+ - x_0^+ \simeq \frac{c_2}{m_2}, \quad x_{m_2-1}^+ = c_2 \ln(m_2), \quad (6)$$

where $c_1 > 0$ and $c_2 > 0$ are controlling (stretching) parameters. It is easy to check that the mesh ω_N consists of two quasi-uniform meshes: (5) and (6) [16].

The choice of c_1 and c_2 are coming from the fact that the half of intervals are in domain with length $\sim c_1 + c_2$. The first interval of (5): $[x_0^-, x_1^-]$ is infinite, but the point $x_{1/2}^-$ is finite, since the non-integer nodes are given by $x_{i+\alpha}^- = x^-(\frac{i+\alpha}{m_1})$, $|\alpha| < 1$. The same is for $x^+(\xi)$: the last interval of (6): $[x_{m_2-1}^+, x_{m_2}^+]$, is infinite, but the point $x_{m_2-1/2}^+$ is finite, since the non-integer nodes are given by $x_{i+\alpha}^+ = x^+(\frac{i+\alpha}{m_2})$, $|\alpha| < 1$. Therefore, the QUM transforms the infinite domain into finite number of intervals and *places the original boundary condition directly on the infinity*. The mesh ω_N consists of $N + 1$ grid nodes, $x_0 = -\infty$ and $x_{N+1} = \infty$.

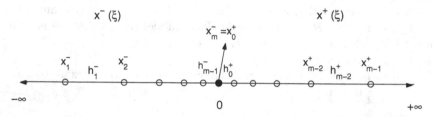

Fig. 1. QUM , $c_1 = c_2 = 1$, $m = m_1 = m_2 = 6$

In the discrete space ω_N we denote continuous time approximation to a function $u(x_i, t)$ by $u_i := u_i(t)$. We shall use the following derivative approximations

$$\left(\frac{\partial u}{\partial x}\right)_i \approx u_{\mathring{x},i} := \frac{u_{i+1/2} - u_{i-1/2}}{\hbar_i}, \quad \left(\frac{\partial u}{\partial x}\right)_{i+1/2} \approx u_{x,i} := \frac{u_{i+1} - u_i}{2\hbar_i^+}, \quad (7)$$

$$\left(\frac{\partial u}{\partial x}\right)_{i-1/2} \approx u_{\overline{x},i} := \frac{u_i - u_{i-1}}{2\hbar_i^-}, \quad \left(\frac{\partial^2 u}{\partial x^2}\right)_i \approx u_{\overline{x}x,i} := \frac{1}{\hbar_i} \left[u_{x,i+1/2} - u_{\overline{x},i-1/2}\right] (8)$$

with local truncation errors of order $O(N^{-2})$ and $\hbar_i = x_{i+1/2} - x_{i-1/2}$, $\hbar_i^+ = x_{i+3/4} - x_{i+1/4}$, $\hbar_i^- = x_{i-1/4} - x_{i-3/4}$.

Flux limiter. In this paper we will use the *van Leer limiter* [11,13,21]

$$\Phi(\theta) = \frac{|\theta| + \theta}{1 + |\theta|}, \quad (9)$$

where $\Phi(\theta)$ is Lipschitz continuous, continuously differentiable for all $\theta \neq 0$, and

$$\Phi(\theta) = 0, \quad \text{if } \theta \leq 0 \quad \text{and} \quad \Phi(\theta) \leq 2\min(1, \theta). \quad (10)$$

Note that at the extreme points of u, the slopes $(u_{i+1} - u_i)/2\hbar^+$ and $(u_i - u_{i-1})/2\hbar^-$ have opposite signs and $\Phi(\theta) = 0$.

Following [11,18] due to the symmetry property of the flux limiter

$$\Phi(\theta) = \theta\Phi(\theta^{-1}), \quad (11)$$

the numerical flux $F_{i+1/2} = F(u_{i+1/2})$ is constructed in a nonlinear way

$$F_{i+1/2} = u_i + \frac{1}{2}\Phi(\theta_{i+1/2})(u_i - u_{i-1}) \quad \text{with} \quad \theta_{i+1/2} = \frac{u_{i+1} - u_i}{u_i - u_{i-1}}. \quad (12)$$

For value u_{N+2} at outer grid node $x_{m_2}^+$, a second order fictitious extrapolation will be used: $u_{N+2} = 3u_{N+1} - 3u_N + u_{N-1}$ [11].

Time discretization. Grid points over $[0, \sigma_0^2 T/2]$ are defined by $\tau_{k+1} = \tau_k + \triangle\tau_k$, $k = 0, 1, 2, \ldots$, $\tau_0 = 0$. Approximation of $u(x_i, \tau_k)$ is denoted by u_i^k. It is well known [9] that for the non-negativity of the numerical solution $\mathbf{u}^{k+1} = [u_0^{k+1}, \ldots, u_N^{k+1}]^T$ of the approximation of a parabolic problem at time t_{k+1}, written in equivalent matrix form, $\mathcal{M}\mathbf{u}^{k+1} = \mathcal{K}^k\mathbf{u}^k$, the sufficient and necessary condition is that $\mathcal{M}$ is a M-matrix (it guaranties that its inverse is non-negative) and all elements of the vector $\mathcal{K}^k\mathbf{u}^k$ should be non-negative.

3   Numerical Method

In this section we develop an efficient explicit-implicit numerical methods, which preserves the positive property of the model problem (3)-(4) and provides $O(\tau + N^{-2})$ approximation to the solution of the continuous model problems. To this aim we use van Leer flux-limited technique [11,13,21], combined with quasi-uniform mesh in space and adaptive mesh in time.

Reflecting the indices that appear in u_i (see (12)) about $i+1/2$ [11] and from (7), (11), denoting for simplicity $\theta_{i+1/2}^{k}{}^{-1} := (\theta_{i+1/2}^k)^{-1}$, we find

$$u_{\bar{x},i}^k = \frac{1}{\hbar_i} A_i^k (u_{i+1} - u_i), \quad A_i^k = 1 - \frac{1}{2}\Phi(\theta_{i+3/2}^k) + \frac{1}{2}\Phi(\theta_{i+1/2}^{k}{}^{-1}), \quad (13)$$

where in view of (9), (10) we have $0 \le A_i^k \le 2$, $i = 1, \ldots, N$ and $k = 0, 1, \ldots$.

Let $\widetilde{\sigma}_i^2 := \widetilde{\sigma}^2(t^k, x_i, u_{\bar{x},i}^k, u_{\bar{x}\bar{x},i}^k)$, $k = 0, 1, \ldots$. Thus the full discretization of (3)-(4) is

$$u_0^{k+1} = 0, \quad u_{N+1}^{k+1} = 1, \quad \frac{u_i^{k+1} - u_i^k}{\triangle\tau_k} - \widetilde{\sigma}_i^2 u_{\bar{x}\bar{x},i}^{k+1} = (\widetilde{\sigma}_i^2 + K)u_{\bar{x},i}^k, \quad i = 1, \ldots, N,$$

or more in details, from (13),(8), for $i = 1, \ldots, N$ we have

$$\left(\frac{1}{\triangle\tau_k} + \frac{\widetilde{\sigma}_i^2}{2\hbar_i} \left(\frac{1}{\hbar_i^+} + \frac{1}{\hbar_i^-} \right) \right) u_i^{k+1} - \left(\frac{\widetilde{\sigma}_i^2}{2\hbar_i\hbar_i^+} \right) u_{i+1}^{k+1} - \left(\frac{\widetilde{\sigma}_i^2}{2\hbar_i\hbar_i^-} \right) u_{i-1}^{k+1}$$

$$= \left(\frac{1}{\triangle\tau_k} - \frac{\widetilde{\sigma}_i^2 + K}{\hbar_i} A_i^k \right) u_i^n + \frac{\widetilde{\sigma}_i^2 + K}{\hbar_i} A_i^k u_{i+1}^k. \quad (14)$$

The approximation (14) contains $u_0 = 0$ and $u_{N+1} = 1$, but not $x_0 = -\infty$ and $x_{N+1} = \infty$.

Note that the tridiagonal coefficient matrix $\mathcal{M}$ is strictly diagonally dominant with positive main diagonal entries and non-positive off-diagonal entries, which is a sufficient (but not necessary) condition for $\mathcal{M}$ to be an M-matrix. Next, from discussions in Section 2 follows that if $\mathbf{u}^n \ge 0$ and

$$\frac{1}{\triangle t_k} - \frac{\rho_i}{\hbar_i} \left[1 + \frac{1}{2}\Phi(\theta_{i+1/2}^{k}{}^{-1}) \right] \ge 0, \quad \rho_i := \widetilde{\sigma}_i^2 + K,$$

then $\mathbf{u}^{n+1} \geq 0$, $n = 0, 1, \ldots$. This leads to additional restriction of the van Leer limiter [22] and consequently for the time step size

$$\Phi(\theta_{i+1/2}^{k}{}^{-1}) = \min\left\{ \frac{|\theta_{i+1/2}^{k}{}^{-1}| + \theta_{i+1/2}^{k}{}^{-1}}{1 + |\theta_{i+1/2}^{k}{}^{-1}|}, \frac{2\hbar_i}{\rho_i \Delta t_k} - 2 \right\}, \quad \Delta t_k \leq \min_{1 \leq i \leq N-1}\left\{ \frac{\hbar_i}{\rho_i} \right\}. \quad (15)$$

With this approach, the time step restriction is relaxed (two times) at the expense of the restriction of the flux limiter.

Also, one can see that the estimate (15) for the time step could be considered as a restriction of the Courant number [21].

Note that the QUM stretches from 0 to $\pm\infty$, placing more points in the neighborhood of the grid point $(0,0)$ where the initial condition is non-differentiable (see Figure 1). Thus, owing to the use of QUM, we overcome the problem with non-smooth initial data for our transformed problem (3)-(4) at $u^0(0)$, see [23].

4   Numerical Experiments

We will test the accuracy of the presented difference schemes (14) for three typical examples: Leland, Avellaneda-Parás and Barles-Soner models. We deal with exact solution $u_{ex}(x,t)$. The error $E_i = u_{ex}(x_i, T) - u_i^T$, $i = 1, \ldots, N$ in maximal is given by $\|E^N\|_\infty = \max_{1 \leq i \leq N} |E_i|$ and the convergence rate is calculated using double mesh principle $CR_\infty = \log_2(\|E^N\|_\infty / \|E^{2N}\|_\infty)$.

In the computations we add a small positive number ($\sim 10^{-30}$) to both numerator and denominator of the gradient ratio in (12) in order to avoid division by zero in uniform flow regions.

The option and mesh parameters are: $r = 0.1$, $\sigma_0 = 0.2$, $E = 100$, $T = 2/\sigma_0^2$, $m_1 = m_2 = N/2$, experimentally found optimal value of controlling parameters $c_1 = c_2 = 2.5$ (i.e. $x_{1/2} = -x_{N-1/2}$), $\Delta\tau = h^2$.

Example 1 (Benchmark test with smooth initial condition). Let $u_{ex}(x,t) = e^{-Kt-x^2}$. Thus, instead of (4) we impose the corresponding to $u_{ex}(x,t)$ initial and boundary conditions and add appropriate residual term in the right-hand side of (3). For Leland model we take $Le = 0.5$, $\sigma^2 = \tilde{\sigma}_L^2$. In the case $\sigma^2 = \tilde{\sigma}_{AP}^2$, there is a small difference between the parameters σ_{max}^2, σ_{min}^2 ($\sigma_{max}^2 = 0.25$, $\sigma_{min}^2 = 0.15$ [12]) and we observe the same behavior of the solution as for Leland model. We test also Barles-Soner model for $a = 0.01$. Errors and convergence results in maximal discrete norm are summarized in Table 1.

Example 2 (Test with original initial and boundary conditions (4)). As exact solution of (3)-(4) we set the solution of linear Black-Scholes equation, adding appropriate function in the right-hand side of (3). The model parameters are the same as in Example 1. The results are given in Table 2. They are very satisfactory - the convergence rate in space is close to 2.

Table 1. Errors and convergence rates in maximal discrete norms, Examples 1

	Leland model		Avellaneda-Parás model		Barles-Soner model	
N	$\|E^N\|_\infty$	CR_∞	$\|E^N\|_\infty$	CR_∞	$\|E^N\|_\infty$	CR_∞
20	7.46962e-3		9.96842e-3		8.61912e-3	
40	2.53043e-3	1.5617	4.33341e-3	1.2019	2.98258e-3	1.5310
80	8.26893e-4	1.6136	1.66123e-3	1.3833	8.71307e-4	1.7753
160	2.18895e-4	1.9175	4.71415e-4	1.8172	2.38051e-4	1.8719
320	5.51499e-5	1.9888	1.24344e-4	1.9227	6.08203e-5	1.9686
640	1.36100e-5	2.0187	3.17048e-5	1.9716	1.53374e-5	1.9875

Table 2. QUM, errors and convergence rates, Example 2

				Avellaneda-Parás model		Barles-Soner model	
N	$\max\limits_{1\le i\le N} \hbar_i$	$\min\limits_{1\le i\le N} \hbar_i$	$x_{1/2}$	$\|E^N\|_\infty$	CR_∞	$\|E^N\|_\infty$	CR_∞
20	2.74653	2.56466e-1	-4.74280	5.92107e-2		7.22740e-2	
40	2.74653	1.26589e-1	-6.47567	1.46356e-2	2.0164	1.46073e-2	2.3069
80	2.74653	6.28939e-2	-8.20854	1.61228e-3	2.4861	4.36117e-3	1.7439
160	2.74653	3.13481e-2	-9.94140	7.54217e-4	1.7923	1.07680e-3	2.0179
320	2.74653	1.56495e-2	-11.67427	2.07501e-4	1.8619	2.66742e-4	2.0132
640	2.74653	7.81861e-3	-13.40714	5.68565e-5	1.8677	7.04872e-5	1.9200
1280	2.74653	3.90778e-3	-15.14001	1.48777e-5	1.9342	1.81406e-5	1.9581

5 Conclusions

In this paper we have presented efficient, positivity preserving algorithm for solving a large class of nonlinear models in mathematical finance. The solution is computed on infinite interval, taking into account the solution behavior at the infinity. Moreover, we overcome the problem with non-smooth initial function. The schemes proposed can be generalized to higher dimensional problem.

Acknowledgements. This work is supported by Project DID 02/37-2009.
  The author thanks to the referees for pointing out the drawbacks of the work, and contribute to enhance the quality of the paper.

References

1. Abe, R., Ishimura, N.: Existence of solutions for the nonlinear partial differential equation arising in the optimal investment problem. Proc. Japan Acad. Ser. A 84, 11–14 (2008)
2. Agliardi, R., Popivanov, P., Slavova, A.: Nonhypoellipticity and comparisson principle for partial differential equations of Black-Scholes type, Nonlin. Analisys: Real Word Appl. 12, 1429–1436 (2011)
3. Alshina, E., Kalitkin, N., Panchenko, S.: Numerical solution of boundary value problem in unlimited area. Math. Modelling 14(11), 10–22 (2002) (in Russian)

4. Ankudinova, J., Ehrhardt, M.: On tne numerical solution of nonlinear Black-Scholes equations. Int. J. of Comp. and Math. with Appl. 56, 779–812 (2008)
5. Avellaneda, M., Parás, A.: Managing the volatility risk of portfolios of derivative securities: the Lagrangian uncertain volatility model. Appl. Math. Fin. 3(1), 21–52 (1996)
6. Barles, G., Soner, H.M.: Option pricing with transaction costs and a nonlinear Black-Scholes equation. Finance and Stochastics 2(4), 369–397 (1998)
7. Company, R., Navarro, E., Pintos, J., Ponsoda, E.: Numerical solution of linear and nonlinear Black-Scholes option pricing equations. Int. J. of Comp. and Math. with Appl. 56, 813–821 (2008)
8. Düring, B., Fournié, M., Jüngel, A.: Convergence of a high-order compact finite difference scheme for a nonlinear Black-Scholes equation. ESAIM: M2AN 38(2), 359–369 (2004)
9. Faragó, I., Komáromi, N.: Nonnegativity of the numerical solution of parabolic problems, Colloquia matematica societatis János Bolyai. Numer. Meth. 59, 173–179 (1990)
10. Frey, R., Patie, P.: Risk management for derivatives in illiquid markets: a simulation-study. In: Adv. in Fin. and Stoch., pp. 137–159. Springer, Berlin (2002)
11. Gerisch, A., Griffiths, D.F., Weiner, R., Chaplain, M.A.J.: A positive splitting method for mixed hyperbolic-parabolic systems, Num. Meth. for PDEs 17(2), 152–168 (2001)
12. Heider, P.: Numerical Methods for Non-Linear Black-Scholes Equations. Appl. Appl. Math. Fin. 17(1), 59–81 (2010)
13. Hundsdorfer, W., Verwer, J.: Numerical Solution of Time-Dependent Advection-Diffusion-Reaction Equations. Springer, Heidelberg (2003)
14. Ishimura, N., Koleva, M.N., Vulkov, L.G.: Numerical solution of a nonlinear evolution equation for the risk preference. In: Dimov, I., Dimova, S., Kolkovska, N. (eds.) NMA 2010. LNCS, vol. 6046, pp. 445–452. Springer, Heidelberg (2011)
15. Jandačka, M., Ševčovič, D.: On the risk-adjusted pricing-methodology-based valuation of vanilla options and explanation of the volatility smile. J. of Appl. Math. 3, 235–258 (2005)
16. Knabner, P., Angerman, L.: Numerical Methods for Elliptic and Parabolic Partial Differential Equations. Springer (2003)
17. Koleva, M.N., Vulkov, L.G.: A kernel-based algorithm for numerical solution of nonlinear pDEs in finance. In: Lirkov, I., Margenov, S., Waśniewski, J. (eds.) LSSC 2011. LNCS, vol. 7116, pp. 566–573. Springer, Heidelberg (2012)
18. Kusmin, D., Turek, S.: High-resolution FEM-TVD schemes based on a fully multidimensional flux limiter. J. of Comp. Phys. 198(1), 131–158 (2004)
19. Leland, H.E.: Option pricing and replication with transactions costs. J. of Finance 40(5), 1283–1301 (1985)
20. Lesmana, D., Wang, S.: An upwind finite difference method for a nonlinear Black–Scholes equation governing European option valuation under transaction costs. Applied Math. and Comp. (in press)
21. LeVeque, R.J.: Numerical Methods for Conservation Laws, Birkhäuser (1992)
22. MacKinnon, R.J., Carey, G.F.: Positivity-preserving, flux-limited finite-difference and finite-element methods for reactive transport. Numer. Meth. Fluids 41, 151–183 (2003)
23. Oosterlee, C.W., Leentvaar, C.W., Huang, X.: Accurate American option pricing by grid stretching and high order finite dfferences, Tech. rep. Delft Institute of Appl. Math., Delft University of Technology, Delft, the Netherlands (2005)

A Multicomponent Alternating Direction Method for Numerical Solution of Boussinesq Paradigm Equation

Natalia Kolkovska and Krassimir Angelow

Institute of Mathematics and Informatics, Bulgarian Academy of Sciences,
Acad. Bonchev Str. bl.8, 1113 Sofia, Bulgaria
natali@math.bas.bg

Abstract. We construct and analyze a multicomponent alternating direction method (a vector additive scheme) for the numerical solution of the multidimensional Boussinesq Paradigm Equation (BPE). In contrast to the standard splitting methods at every time level a system of many finite difference schemes is solved. Thus, a vector of the discrete solutions to these schemes is found. It is proved that these discrete solutions converge to the continuous solution in the uniform mesh norm with $O(|h|^2 + \tau)$ order. The method provides full approximation to BPE and is efficient in implementation. The numerical rate of convergence and the altitudes of the crests of the traveling waves are evaluated.

Keywords: Boussinesq Equation, multicomponent ADI method, vector additive scheme, Sobolev type problem.

1 Introduction

Consider the Cauchy problem for the Boussinesq Paradigm Equation (BPE)

$$\frac{\partial^2 u}{\partial t^2} - \beta_1 \Delta \frac{\partial^2 u}{\partial t^2} = \Delta u - \beta_2 \Delta^2 u + \alpha \Delta f(u), \quad (x,y) \in \mathbb{R}^2, \ 0 < t \le T, \ T < \infty \quad (1)$$

on the unbounded region $\mathbb{R}^2$ with asymptotic boundary conditions

$$u(x,y,t) \to 0, \quad \Delta u(x,y,t) \to 0, \quad |(x,y)| \to \infty, \quad (2)$$

and initial conditions

$$u(x,y,0) = u_0(x,y), \quad \frac{\partial u}{\partial t}(x,y,0) = u_1(x,y). \quad (3)$$

Here $f(u) = u^p$, $p \in \mathbb{N}$, $p \ge 2$, Δ is the Laplace operator and the constants α, β_1 and β_2 are positive. It is shown in [6] how equation (1) could be derived from the original Boussinesq system. Note that equation (1) is unsolved relative to the time derivative $\frac{\partial^2}{\partial t^2}$ (the Laplace operator acts on the second time derivative). Thus, problem (1)–(3) is of Sobolev type according to the terminology of [16].

I. Dimov, I. Faragó, and L. Vulkov (Eds.): NAA 2012, LNCS 8236, pp. 371–378, 2013.
© Springer-Verlag Berlin Heidelberg 2013

A lot of papers are devoted to computational simulations of one dimensional BPE. In contrast the two dimensional problems are essentially less studied. The efficient algorithms for evaluation of the discrete approximation to the solution u of BPE presented in [5,7,10] are based on the representation of an implicit finite difference scheme as pair of an elliptic and a hyperbolic discrete equations. In [11,19] the regularization method is applied and the operator of the same finite difference scheme is factorized in order to reduce the evaluation of the numerical solution to a sequence of three simple schemes.

Numerous papers are dealing with the construction and investigation of splitting methods for numerical solution of second order evolutionary problems, see e.g. [8,9,13,14] and the references there. A multicomponent alternating direction method (ADI) for solving evolutionary problems is proposed and analyzed by Abrashin in [1]. In the method at each time step a system of finite difference equations is solved and a vector of discrete solutions to these schemes is found. This method is called a 'vector additive scheme' in [3,15,18]. Varying applications of the method can be found in [2,3,4,15,17]. The method has the following advantages. First, each finite difference scheme from the system approximates the initial continuous problem. Second, the method can be applied to equations with mixed derivatives and to problems posed on complicated domains. Third, the discrete solutions to the linear multicomponent ADI scheme satisfy a discrete identity which is an approximation to the conservation law valid for the solution of the linear initial problem. As a result the multicomponent method for linear problems is unconditionally stable. Thus, the method can be treated as a generalization of the classical ADI methods to cases of space dimensions $n > 2$. Fourth, the numerical implementation of the method is efficient. The main disadvantage of the vector additive schemes is that their implementation demands more computational resources (memory and time) compared to the standard schemes since at each time level two discrete equations have to be solved.

The aim of the paper is to construct and analyze a multicomponent ADI method for evaluation of the numerical solution to BPE (1)–(3). The algorithm for evaluation of the numerical solutions is proposed in Section 2. In Section 3 the numerical method is analyzed theoretically. First, it is proved for the particular case of linear BPE ((1)–(3) with $f(u) \equiv 0$) that the discrete solutions satisfy an identity, which is a proper discretization to the exact conservation law. Then the convergence of the method for the nonlinear BPE is established in the energy semi-norm. Important error estimates in the uniform and Sobolev mesh norms are derived and summarized. In Section 4 the evolution of 2D solitary waves with different velocities is computed with the multicomponent ADI scheme. The numerical rate of convergence and the altitudes of the crests of the traveling waves are also evaluated.

2   A Multicomponent ADI Finite Difference Scheme

We discretize BPE (1)–(3) on a sufficiently large space domain $\Omega = [-L_1, L_1] \times [-L_2, L_2]$. We assume that the solution and its derivatives are negligibly small outside Ω. For integers N_1, N_2 set the space steps $h_1 = L_1/N_1, h_2 = L_2/N_2$ and $h = (h_1, h_2)$. Let $\Omega_h = \{(x_i, y_j) : x_i = ih_1, i = -N_1, \ldots, N_1, y_j = jh_2, j = -N_2, \ldots, N_2\}$. Next, for integer K we denote the time step by $\tau = T/K$.

We consider mesh functions $v_{(i,j)}^{(k)}$ defined on $\Omega_h \times \{t^k\}$ on the time levels $t^k = k\tau$, $k = 0, 1, 2, \ldots, K$. Whenever possible the subscripts (i, j) of the mesh functions are omitted.

The discrete scalar product $\langle v, w \rangle = \sum_{i,j} h_1 h_2 v_{(i,j)}^{(k)} w_{(i,j)}^{(k)}$ and the corresponding $L_{2,h}$ discrete norm $\| \cdot \| = \langle v, v \rangle^{\frac{1}{2}}$ are associated with the space of mesh functions which vanish on the boundary of Ω_h.

The operators A_1 and A_2 are defined as second finite differences of the mesh functions in the x the direction and in the y direction, i.e. $A_1 v_{(l,m)}^{(k)} = -(v_{(l-1,m)}^{(k)} - 2v_{(l,m)}^{(k)} + v_{(l+1,m)}^{(k)})h_1^{-2}$. Then $A_1^2 v$ will stand for the fourth finite difference in the first space direction times $(h_1)^4$ and $A_1 A_2$ will be an approximation of the mixed fourth derivative $\frac{\partial^4}{\partial x^2 \partial y^2}$. The finite differences $v_t^{(k)} = (v^{(k+1)} - v^{(k)}) \tau^{-1}$ and $v_{\bar{t}t}^{(k)} = (v^{(k+1)} - 2v^{(k)} + v^{(k-1)}) \tau^{-2}$ are used for the approximation of the first and second time derivatives respectively.

We start with the construction of a multicomponent finite difference scheme for (1)–(3). At each time level k we consider two discrete approximations $v_{(i,j)}^{(1)(k)}$ and $v_{(i,j)}^{(2)(k)}$ to $u(ih_1, jh_2, k\tau)$. We deal with the following system of implicit finite difference schemes

$$
\begin{aligned}
&v_{\bar{t}t}^{(1)(k)} + \beta_1 A_1 v_{\bar{t}t}^{(1)(k)} + \beta_1 A_2 v_{\bar{t}t}^{(1)(k-1)} + A_1 v^{(1)(k+1)} + A_2 v^{(2)(k)} \\
&+ \beta_2 A_1^2 v^{(1)(k+1)} + \beta_2 A_1 A_2 v^{(1)(k)} + \beta_2 A_2^2 v^{(2)(k)} + \beta_2 A_1 A_2 v^{(2)(k)} \qquad (4)\\
&= -\alpha A_1 f(v^{(1)(k)}) - \alpha A_2 f(v^{(2)(k)}),
\end{aligned}
$$

$$
\begin{aligned}
&v_{\bar{t}t}^{(2)(k)} + \beta_1 A_1 v_{\bar{t}t}^{(1)(k)} + \beta_1 A_2 v_{\bar{t}t}^{(2)(k)} + A_1 v^{(1)(k+1)} + A_2 v^{(2)(k+1)} \\
&+ \beta_2 A_1^2 v^{(1)(k+1)} + \beta_2 A_1 A_2 v^{(1)(k+1)} + \beta_2 A_2^2 v^{(2)(k+1)} + \beta_2 A_1 A_2 v^{(2)(k)} \\
&= -\alpha A_1 f(v^{(1)(k)}) - \alpha A_2 f(v^{(2)(k)}). \qquad (5)
\end{aligned}
$$

Note that the nonlinear function $f(u)$ is evaluated on the time level k and the approximation $A_2 v_{\bar{t}t}^{(1)(k-1)}$ on the previous time level is used in (4). Thus, the scheme (4)–(5) is a four-level scheme and values of the numerical solution on the first three time levels $t = -\tau$, $t = 0$ and $t = \tau$ are required in order to start the method. We evaluate initial values for $v^{(1)}$ and $v^{(2)}$ on time levels $t = 0$ and $t = \tau$ using formulas

$$
v_{(i,j)}^{(m)(0)} = u_0(x_i, y_j), \quad m = 1, 2, \tag{6}
$$

$$
v_{(i,j)}^{(m)(1)} = u_0(x_i, y_j) + \tau u_1(x_i, y_j) \tag{7}
$$

$$
- 0.5\tau^2 (I + \beta_1 A)^{-1} (A u_0 + \beta_2 A^2 u_0 + \alpha A f(u_0))(x_i, y_j), \quad m = 1, 2,
$$

where $\mathcal{A} = A_1 + A_2$. The third initial value $v^{(m)(-1)}$ for $m = 1, 2$ at time level $t = -\tau$ is found from the equation

$$v^{(m)(0)}_{\bar{t}t(i,j)} = -(I + \beta_1 \mathcal{A})^{-1} \left(\mathcal{A}u_0 + \beta_2 \mathcal{A}^2 u_0 + \alpha \mathcal{A} f(u_0) \right)(x_i, y_j), \; m = 1, 2. \quad (8)$$

For approximation of the second boundary condition in (2) the mesh is extended outside the domain Ω_h by one line at each spatial boundary and the symmetric second-order finite difference is used for the approximation of the second derivatives in (2).

Suppose in the following that the exact solution to (1)–(3) is smooth enough, e.g. $u \in C^{6,6} \left(\mathbb{R}^2 \times [0, T] \right)$. Straightforward calculations via Taylor series expansion at point (x_i, y_j, t_k) show that the local approximation error of the discrete equations (4)–(5) is $O(h^2 + \tau)$. Also (6)–(7) approximate the initial conditions locally with $O(|h|^2 + \tau^2)$ error.

The numerical algorithm for evaluation of $v^{(1)}$ and $v^{(2)}$ is as follows. Suppose the values of $v^{(1)}$ and $v^{(2)}$ on the three consecutive time levels $(k-2)$, $(k-1)$ and (k) are known. Then (4) is an implicit scheme along the x direction and is explicit along the y direction. Thus, for each $j = -N_2, \cdots, N_2$ the vector $\{v^{(1)(k+1)}_{(i,j)}, i = -N_1, \cdots, N_1\}$ can be found from equation (4) as a solution of linear five-diagonal system. Analogously, (5) is an explicit scheme in the x direction and is implicit in the y direction with respect to $v^{(2)(k+1)}$. Thus, for each $i = -N_1, \cdots, N_1$ the vectors $\{v^{(2)(k+1)}_{(i,j)}, j = -N_2, \cdots, N_2\}$ can be evaluated from equation (5) as a solution of linear five-diagonal system. As a result the implementation of the schemes (4)–(5) can be done by efficient numerical algorithms.

Remark 1. *In order to achieve efficient algorithm in equation (4) we use $A_2 v^{(1)(k-1)}_{\bar{t}t}$ for approximation of $\frac{\partial^4 u}{\partial t^2 \partial y^2}$ instead of the straightforward $A_2 v^{(1)(k)}_{\bar{t}t}$. But we pay for this efficiency by having low order of approximation – the error of discretization of equation (4) is $O(\tau + |h|^2)$ only. In addition the scheme becomes a four-level one. Note that the straightforward approximation on level k could easily lead to a $O(\tau^2 + |h|^2)$ scheme, but the efficiency would be lost in this case.*

In the case of "good" (or "proper") Boussinesq equation the combined time-space derivative is removed from (1), i.e. $\beta_1 = 0$ is set in (1), and a three-level multicomponent ADI scheme with $O(|h|^2 + \tau^2)$ approximation error can be proposed and analyzed following the ideas of this paper.

3   Theoretical Analysis

First we consider the linear problem, i.e. (1)–(3) with $f \equiv 0$. We define operators $\Lambda_1(u) = -\frac{\partial^2 u}{\partial x^2}$ and $\Lambda_2(u) = -\frac{\partial^2 u}{\partial y^2}$ in the space of functions which vanish at infinity together with their second derivatives.

Let $|| \cdot ||$ stand for the standard norm in $L_2(\mathbb{R}^2)$. Denote by E the energy functional

$$E(u)(t) = \left\| \Lambda_1^{\frac{1}{2}} \frac{\partial u}{\partial t}(\cdot, t) \right\|^2 + \left\| \Lambda_2^{\frac{1}{2}} \frac{\partial u}{\partial t}(\cdot, t) \right\|^2 + \beta_2 \left\| (\Lambda_1 + \Lambda_2) \frac{\partial u}{\partial t}(\cdot, t) \right\|^2$$

$$+ \beta_1 \left\| \Lambda_1^{\frac{1}{2}} \frac{\partial^2 u}{\partial t^2}(\cdot, t) \right\|^2 + \beta_1 \left\| \Lambda_2^{\frac{1}{2}} \frac{\partial^2 u}{\partial t^2}(\cdot, t) \right\|^2 + \left\| \frac{\partial^2 u}{\partial t^2}(\cdot, t) \right\|^2. \tag{9}$$

It is straightforward to prove that the solution to problem (1)–(3) with $f = 0$ satisfies the identity $E(u)(t) = E(u)(0)$ for every $t \geq 0$ or, equivalently the energy functional $E(u)$ is preserved in time.

We shall obtain a similar discrete identity for the solution to (4)–(8) with $f(u) = 0$. First we define $\mathbf{v}^{(k)}$ as the couple of solutions $(v^{(1)(k)}, v^{(2)(k)})$ and the semi-norm (or the energy norm) $N(\mathbf{v}^{(k)})$ by

$$N(\mathbf{v}^{(k)}) = \|A_1^{\frac{1}{2}} v_t^{(1)(k)}\|^2 + \|A_2^{\frac{1}{2}} v_t^{(2)(k)}\|^2 + \beta_2 \|A_1 v_t^{(1)(k)} + A_2 v_t^{(2)(k)}\|^2$$

$$+ \beta_1 \|A_1^{\frac{1}{2}} v_{\bar{t}t}^{(1)(k)}\|^2 + \beta_1 \|A_2^{\frac{1}{2}} v_{\bar{t}t}^{(2)(k)}\|^2 + \|v_{\bar{t}t}^{(2)(k)}\|^2. \tag{10}$$

Following the proof of Theorem from page 318 of [1], it can be established

Theorem 1 (Discrete summation identity). *For every $Q = 1, 2, 3, \ldots, K$ the solution $\mathbf{v}^{(Q)}$ to problem (4)–(8) with $f(u) = 0$ satisfies the equality*

$$N(\mathbf{v}^{(Q)}) + \tau \sum_{k=1}^{Q} \tau \left(\|A_1^{\frac{1}{2}} v_{\bar{t}t}^{(1)(k)}\|^2 + \beta_2 \|A_1 v_{\bar{t}t}^{(1)(k)}\|^2 + \beta_1 \|A_1^{\frac{1}{2}} v_{\bar{t}t\bar{t}}^{(1)(k)}\|^2 \right)$$

$$+ \tau \sum_{k=1}^{Q} \tau \left(\|A_2^{\frac{1}{2}} v_{\bar{t}t}^{(2)(k)}\|^2 + \beta_2 \|A_2 v_{\bar{t}t}^{(2)(k)}\|^2 + \beta_1 \|A_2^{\frac{1}{2}} v_{\bar{t}t\bar{t}}^{(2)(k)}\|^2 \right)$$

$$+ \tau \sum_{k=1}^{Q} \tau \|A_1 v_t^{(1)(k)} + \beta_2 A_1^2 v_t^{(1)(k)} + \beta_2 A_1 A_2 v_t^{(2)(k-1)} + \beta_1 A_1 v_{\bar{t}t\bar{t}}^{(1)(k-1)}\|^2$$

$$+ \tau \sum_{k=1}^{Q} \tau \|A_2 v_t^{(2)(k)} + \beta_2 A_2^2 v_t^{(2)(k)} + \beta_2 A_1 A_2 v_t^{(1)(k)} + \beta_1 A_2 v_{\bar{t}t\bar{t}}^{(2)(k-1)}\|^2$$

$$= N(\mathbf{v}^{(0)}). \tag{11}$$

Consequently the energy norm of the numerical solution at each fixed time level deviates from the energy norm of the initial data by a small term of first order in time step, i.e. $N(\mathbf{v}^{(Q)}) - N(\mathbf{v}^{(0)}) = O(\tau)$ for $Q = 1, 2, \ldots, K$.

We state now our main theorem

Theorem 2 (Convergence of the Multicomponent ADI Scheme). *Assume that the solution u to BPE obeys $u \in C^{6,6} \left(\mathbb{R}^2 \times [0, T] \right)$ and the solutions $v^{(1)(k)}$, $v^{(2)(k)}$ to the multicomponent ADI scheme (4)–(8) are bounded in the* *maximum norm for every* $k = 1, 2, 3, \cdots, K$. *Then* $v^{(1)}$ *and* $v^{(2)}$ *converge to the exact solution u as* $|h|, \tau \to 0$ *and the energy norm estimate*

$$N(\mathbf{z}^{(k)}) \leq C \left(|h|^2 + \tau \right)^2, \quad k = 1, 2, \cdots, K$$

holds with a constant C independent on h and τ, where $z^{(1)(k)} = v^{(1)(k)} - u(\cdot, k\tau)$ *and* $z^{(2)(k)} = v^{(2)(k)} - u(\cdot, k\tau)$ *are the errors of the method.*

Sufficient conditions for global existence of bounded solution to (1)–(3) in $C^{6,6} \left(\mathbb{R}^2 \times [0, T] \right)$ are given in [21]. Stability and instability of solitary wave solutions to (1) are treated in many papers, see e.g. [20] and the references therein.

The problem for the boundedness of the discrete solutions to (4)–(8) imposed in Theorem 2 is still open. The boundedness (locally in time) of the discrete solution to a conservative finite difference scheme for BPE can be found in [12].

Corollary 1. *Under the assumptions of the Theorem 2 the multicomponent ADI scheme admits the following error estimates for every* $k = 1, 2, \cdots, K$, $m = 1, 2$

$$\|z^{(1)(k)}\| + \|z^{(2)(k)}\| + \|A_1^{\frac{1}{2}} z^{(1)(k)}\| + \|A_2^{\frac{1}{2}} z^{(2)(k)}\| \leq C \left(|h|^2 + \tau \right),$$

$$\|A_1 z^{(1)(k)} + A_2 z^{(2)(k)}\| \leq C \left(|h|^2 + \tau \right),$$

$$\|z^{(m)(k)}\|_{L_\infty} \leq C \left(|h|^2 + \tau \right), \quad \|z_t^{(m)(k)}\| + \|z_{\bar{t}t}^{(m)(k)}\| \leq C \left(|h|^2 + \tau \right).$$

4   Numerical Results

In this section some numerical tests concerning the convergence of the multicomponent ADI method and the evolution of the numerical solution are presented in the 2D case. The computational domain is $[-30, 30] \times [-30, 30]$. The numerical solutions are evaluated for parameters $\alpha = 3$, $\beta_1 = 3$, $\beta_2 = 1$, $p = 2$ and initial conditions u_0, u_1 given in [5]. These initial conditions correspond to a solitary wave which moves along the y-axis with velocity c.

Table 1. Dependence of the convergence rate on time step and space steps

τ	$h_1 = h_2$	Rate $v^{(1)}$	Rate $v^{(2)}$	τ	$h_1 = h_2$	Rate $v^{(1)}$	Rate $v^{(2)}$
0.08	0.075	-	-	0.02	0.3	-	-
0.04	0.075	0.9384	0.9450	0.02	0.15	2.5502	2.6853
0.02	0.075	-	-	0.02	0.075	-	-

Table 1 contains the numerical rate of convergence at time $T = 8$. The accuracy of the proposed schemes in the uniform norm is calculated by Runge method using three nested meshes. We observe that the experimental rate of convergence with respect to time step approximates the theoretical rate of convergence $O(\tau)$. Regarding the convergence with respect to spatial steps, the numerical rate of convergence is better than the rate of convergence $O(|h|^2)$ proved in Corollary 1.

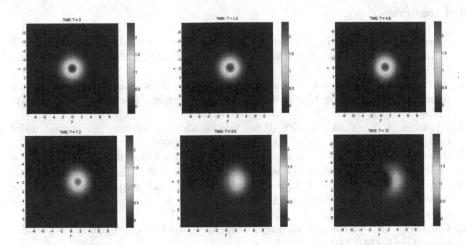

Fig. 1. (Color on-line) Evolution with velocity $c = 0.2$ of the numerical solution in time, $t = 0; 2.4; 4.8; 7.2; 9.6; 12$

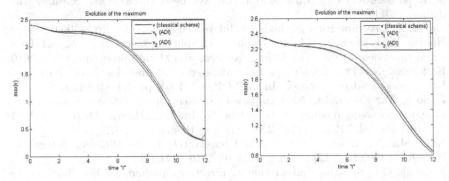

Fig. 2. (Color on-line) Evolution in time of the the altitudes of the crests of the solutions for velocity $c = 0$ (left) and velocity $c = 0.2$ (right)

On Figure 1 evolution of the numerical solution with velocity $c = 0.2$ is shown. For $t < 4.8$ the shape of the numerical solution is similar to the initial solution. For larger times the numerical solution changes its initial form and transforms into a diverging propagating wave. Evolutions of the altitudes of the crests of the solutions $v^{(1)}$ and $v^{(2)}$ in time are shown on Figure 2. For comparison the same quantity obtained by the conservative scheme from [12] is also plotted. It can be observed that the behavior of the altitudes of the crests of the numerical solutions obtained by the multicomponent ADI scheme is similar to the altitudes of the crests of the numerical solution given by the conservative scheme [7,12]. Thus, the proposed numerical method corresponds very well to the results evaluated by the well studied (theoretically and numerically) finite difference schemes from [7,12].

Acknowledgments. This work has been partially supported by the Bulgarian Science Found under grant DDVU 02/71.

References

1. Abrashin, V.N.: A variant of alternating directions method for solving of multidimensional problems of mathematical physics. Part I. Differ. Eq. 26, 314–323 (1990)
2. Abrashin, V.N., Dzuba, I.: A variant of alternating directions method for solving of multidimensional problems of mathematical physics. Part II. Differ. Eq. 26, 1179–1191 (1990)
3. Abrashin, V.N., Vabishchevich, P.N.: Vector additive schemes for second order evolutionary problems. Differ. Eq. 34, 1666–1674 (1998)
4. Abrashin, V.N., Zhadaeva, N.G.: Economical additive finite difference schemes for multidimensional nonlinear non stationary problems. Differ. Eq. 38, 960–971 (2002)
5. Chertock, A., Christov, C.I., Kurganov, A.: Central-Upwind Schemes for the Boussinesq Paradigm Equation. In: Computational Science and High Performance Computing IV, NNFM, vol. 113, pp. 267–281 (2011)
6. Christov, C.I.: An Energy-consistent Dispersive Shallow-water Model. Wave Motion 34, 161–174 (2001)
7. Christov, C.I., Kolkovska, N., Vasileva, D.: On the Numerical Simulation of Unsteady Solutions for the 2D Boussinesq Paragigm Equation. In: Dimov, I., Dimova, S., Kolkovska, N. (eds.) NMA 2010. LNCS, vol. 6046, pp. 386–394. Springer, Heidelberg (2011)
8. Geiser, J.: Decomposition methods for differential equations. CRC (2009)
9. Holden, H., Karlsen, K., Lie, K., Risebro, N.: Splitting methods for partial differential equations with rough solutions. Analysis and Matlab programs. EMS (2010)
10. Kolkovska, N.: Two families of finite difference schemes for multidimensional Boussinesq paradigm equation. In: AIP CP, vol. 1301, pp. 395–403 (2010)
11. Dimova, M., Kolkovska, N.: Comparison of some finite difference schemes for Boussinesq paradigm equation. In: Adam, G., Buša, J., Hnatič, M. (eds.) MMCP 2011. LNCS, vol. 7125, pp. 215–220. Springer, Heidelberg (2012)
12. Kolkovska, N., Dimova, M.: A New Conservative Finite Different Scheme for Boussinesq Paradigm Equation. CEJM 10, 1159–1171 (2011)
13. Marchuk, G.I.: Splitting and alternating directions method. In: Handbook of Numerical Analysis, vol. 1. Elsevier (1990)
14. Samarskii, A.A.: The theory of difference schemes. In: Pure and Applied Mathematics, vol. 240. Marcel Dekker
15. Samarskii, A.A., Vabishchevich, P.N.: Additive schemes for problems of mathematical physics. Nauka, Moscow (1999)
16. Sveshnikov, A., Al'shin, A., Korpusov, M., Pletner, Y.: Linear and nonlinear equations of Sobolev type. Fizmatlit, Moscow (2007) (in Russian)
17. Vabishchevich, P.N.: Regularized Additive Operator-Difference Schemes. Comp. Math. Math. Phys. 50, 449–457 (2010)
18. Vabishchevich, P.N.: On a new class of additive (splitting) operator-difference schemes. Math. Comp. 81, 267–276 (2012)
19. Vasileva, D., Dimova, M.: Comparison of two numerical approaches to Boussinesq paradigm equation. LNCS (submitted for publication in this volume)
20. Wang, Y., Mu, C., Deng, J.: Strong instability of solitary-wave solutions for a nonlinear Boussinesq equation. Nonlinear Analysis 69, 1599–1614 (2008)
21. Xu, R., Liu, Y.: Global existence and nonexistence of solution for Cauchy problem of multidimensional double dispersion equations. J. Math. Anal. Appl. 359, 739–751 (2009)

Finite Volume Approximations for Incompressible Miscible Displacement Problems in Porous Media with Modified Method of Characteristics

Sarvesh Kumar

Department of Mathematics,
Indian Institute of Space Science and Technology,
Thiruvananthapuram- 695 547, Kerala, India
http://www.iist.ac.in

Abstract. The incompressible miscible displacement problem in porous media is modeled by a coupled system of two nonlinear partial differential equations, the pressure-velocity equation and the concentration equation. The pressure-velocity is elliptic type and the concentration equations is convection dominated diffusion type. It is known that miscible displacement problems follow the natural law of conservation and finite volume methods are conservative. Hence, in this paper, we present a mixed finite volume element method (FVEM) for the approximation of the pressure-velocity equation. Since concentration equation is convection dominated diffusion type and most of the numerical methods suffer from the grid orientation effect and modified method of characteristics(MMOC) minimizes the grid orientation effect. Therefore, for the approximation of the concentration equation we apply a standard FVEM combined MMOC. *A priori* error estimates are derived for velocity, pressure and concentration. Numerical results are presented to substantiate the validity of the theoretical results.

Keywords: modified method of characteristics, mixed methods, finite volume element methods, miscible displacement problems, error estimates, numerical experiments.

1 Introduction

A mathematical model describing miscible displacement of one incompressible fluid by another in a horizontal porous medium reservoir $\Omega \subset \mathbb{R}^2$ with boundary $\partial\Omega$ of unit thickness over a time period of $J = (0, T]$ is given by

$$\mathbf{u} = -\frac{\kappa(x)}{\mu(c)}\nabla p \quad \forall (x, t) \in \Omega \times J \tag{1}$$

$$\nabla \cdot \mathbf{u} = q \qquad \forall (x, t) \in \Omega \times J, \tag{2}$$

$$\phi(x)\frac{\partial c}{\partial t} + \mathbf{u} \cdot \nabla c - \nabla \cdot (D(\mathbf{u})\nabla c) = g(x, t, c) = (\tilde{c} - c)q \quad \forall (x, t) \in \Omega \times J, \tag{3}$$

I. Dimov, I. Faragó, and L. Vulkov (Eds.): NAA 2012, LNCS 8236, pp. 379–386, 2013.

with conditions $\mathbf{u} \cdot \mathbf{n} = 0$, $D(\mathbf{u})\nabla c \cdot \mathbf{n} = 0$ $\forall (x,t) \in \partial\Omega \times J$, $c(x,0) = c_0(x)$ $\forall x \in \Omega$.

Here, Ω is bounded open set in $\mathbb{R}^2$ with Lipschitz boundary, $\mathbf{u}(x,t)$ and $p(x,t)$ are, respectively, the Darcy velocity and the pressure of the fluid mixture, $c(x,t)$ is the concentration of the fluid, $\tilde{c}$ is the concentration of the injected fluid, $\kappa(x)$ is the 2×2 permeability tensor of, the medium, $q(x,t)$ is the external source/sink term that accounts for the effect of injection and production wells, $\phi(x)$ is the porosity of the medium, $\mu(c)$ is the concentration dependent viscosity and $D(\mathbf{u}) = D(x,\mathbf{u})$ is the diffusion-dispersion tensor. For more details about this model, we refer to [5,12] and references therein. We assume that the functions ϕ, μ, κ and q are bounded, i.e., there exist positive constants ϕ_*, ϕ^*, μ_*, μ^*, κ_*, κ^*, q^*, D^* such that

$$0 < \phi_* \leq \phi(x) \leq \phi^*, \quad 0 < \mu_* \leq \mu(x,c) \leq \mu^*, \quad 0 < \kappa_* \leq \kappa(x) \leq \kappa^*, \tag{4}$$

$$|q(x)| \leq q^*, \quad D(x,\mathbf{u}) \leq D^*. \tag{5}$$

The mathematical theory for the system (1)-(3) under suitable assumptions on the data have been discussed [10,11] and for numerical approximation, we refer to [3,5,8] . Recently, Kumar [2] has discussed a mixed and discontinuous FVEM for incompressible miscible displacement problems.

We would like to mention that the model which describe the miscible displacement of one incompressible fluid by another in porous media is similar to reactive flows problem in porous media. The theory and numerical methods for reactive flow problems article well developed in the literature. Authors in [4] have discussed stability of the reactive flows and numerical approximation of reactive flows problems given in [7].

In this paper, we present a mixed FVEM for (1)-(2) and a standard FVEM combined with MMOC for (3). This paper is organized as follows: In Section 2, FVE approximation procedure is discussed. A priori error estimates for velocity, pressure and concentration are presented in Section 3. Finally in Section 4, the numerical procedure is discussed and some numerical experiments are presented.

The basic idea behind the MMOC is to set the hyperbolic part, i.e., $\phi\dfrac{\partial c}{\partial t} + \mathbf{u} \cdot \nabla c$, as a directional derivative.

Set $\psi(x,t) = (|\mathbf{u}(x,t)|^2 + \phi(x)^2)^{\frac{1}{2}} = (u_1(x,t)^2 + u_2(x,t)^2 + \phi(x)^2)^{\frac{1}{2}}$. The directional derivative of the concentration $c(x,t)$ in the direction of $\mathbf{s}$ is given by $\dfrac{\partial c}{\partial s} = \dfrac{\partial c}{\partial t}\dfrac{\phi(x)}{\psi(x,t)} + \dfrac{\mathbf{u}.\nabla c}{\psi(x,t)}$, where $\mathbf{s}(x,t) = \dfrac{(u_1(x,t),u_2(x,t),\phi(x))}{\psi(x,t)}$. Hence, (3) can be rewritten as

$$\psi(x,t)\frac{\partial c}{\partial s} - \nabla \cdot (D(\mathbf{u})\nabla c) = (\tilde{c} - c)q \quad \forall (x,t) \in \Omega \times J. \tag{6}$$

Since (6) is in the form of heat equation, the behavior of the numerical solution of (6) should be better than (3) if the derivative term $\dfrac{\partial c}{\partial s}$ is approximated properly.

Let $0 = t_0 < t_1 < \cdots t_N = T$ be a given partition of the time interval $[0,T]$ with the time step size Δt. For very small values of Δt, the characteristic direction starting from (x, t_{n+1}) crosses $t = t_n$ at (see Fig. 1)

$$\check{x} = x - \frac{\mathbf{u}^{n+1}}{\phi(x)}\Delta t, \tag{7}$$

This suggests us to approximate the characteristic directional derivative at $t = t_{n+1}$ as

$$\frac{\partial c}{\partial s}\big|_{t=t_{n+1}} \approx \frac{c^{n+1} - c(\check{x}, t^n)}{\Delta s} = \frac{c^{n+1} - c(\check{x}, t_n)}{((x - \check{x})^2 + (t_{n+1} - t_n)^2)^{1/2}}, \tag{8}$$

Using (7), we obtain $\psi(x,t)\frac{\partial c}{\partial s}\big|_{t=t_{n+1}} \approx \phi(x)\frac{c^{n+1} - \check{c}^n}{\Delta t}$, here $\check{c}^n = c(\check{x}, t_n)$.

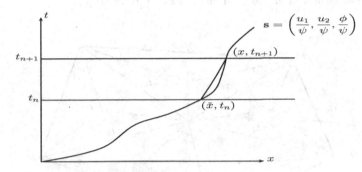

Fig. 1. An illustration of the definition $\check{x}$

2   Finite Volume Element Approximation

Weak Formulations
Define $U = \{\mathbf{v} \in H(\mathrm{div}; \Omega) : \mathbf{v} \cdot \mathbf{n} = 0 \text{ on } \partial\Omega\}$. Note that (1)-(2) with Neumann boundary condition has a solution for pressure, which is only unique up to an additive constant. The non-uniqueness of (1)-(2) may be avoided by considering the quotient space: $W = L^2(\Omega)/\mathbb{R}$. Multiply (1) and (2) by $\mathbf{v} \in U$ and $w \in W$, respectively, and integrate over Ω. A use of Green's formula and $\mathbf{v} \cdot \mathbf{n} = 0$ on $\partial\Omega$, yields the following weak formulation: Find $(\mathbf{u}, p) : \bar{J} \longrightarrow U \times W$ satisfying

$$(\kappa^{-1}\mu(c)\mathbf{u}, \mathbf{v}) - (\nabla \cdot \mathbf{v}, p) = 0 \quad \forall \mathbf{v} \in U, \tag{9}$$
$$(\nabla \cdot \mathbf{u}, w) = (q, w) \quad \forall w \in W. \tag{10}$$

Similarly, multiply (3) by $z \in H^1(\Omega)$, integrate over Ω to obtain a weak formulation for the concentration equation (3) as follows:
Find a map $c : \bar{J} \longrightarrow H^1(\Omega)$ such that for $t \in (0, T]$ and for $z \in H^1(\Omega)$

$$(\phi\frac{\partial c}{\partial t}, z) + (\mathbf{u} \cdot \nabla c, z) + a(\mathbf{u}; c, z) = (g(c), z) \quad c(x, 0) = c_0(x) \ \forall x \in \Omega, \tag{11}$$

where, $a(\mathbf{u}; \phi, \psi) = \int_\Omega D(\mathbf{u})\nabla\phi \cdot \nabla\psi dx \quad \forall \phi, \psi \in H^1(\Omega)$.

Trial Spaces for Velocity and Pressure

$$U_h = \{\mathbf{v_h} \in U : \mathbf{v_h}|_T = (a + bx, c + by) \ \forall T \in \mathcal{T}_h\},$$

$$W_h = \{w_h \in W : w_h|_T \text{ is a constant } \forall T \in \mathcal{T}_h\}.$$

Here, $\mathcal{T}_h$ be a regular partition of the domain $\bar{\Omega}$ into closed triangles T. The test space V_h for velocity is defined by

$$V_h = \{\mathbf{v_h} \in (L^2(\Omega))^2 : \mathbf{v_h}|_{T_M^*} \text{ is a constant } \forall T_M^* \in \mathcal{T}_h^* \text{ and } \mathbf{v_h} \cdot \mathbf{n} = 0 \text{ on } \partial\Omega\},$$

where T_M^* denote the dual element corresponding to the mid-side node M, see Fig 2. For construction of the dual elements, we refer to [6].

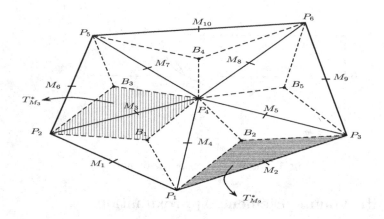

Fig. 2. Primal grid $\mathcal{T}_h$ and dual grid $\mathcal{T}_h^*$

The mixed FVE approximation corresponding to (1)-(2) can be written as: find $(\mathbf{u_h}, p_h) : \bar{J} \longrightarrow U_h \times W_h$ such that for $t \in (0, T]$, $\mathbf{v_h} \in V_h$ and $w_h \in W_h$ the following holds:

$$(\kappa^{-1}\mu(c_h)\mathbf{u_h}, \mathbf{v_h}) - \sum_{i=1}^{N_m} \mathbf{v_h}(M_i) \cdot \int_{\partial T_{M_i}^*} w_h \, \mathbf{n}_{T_{M_i}^*} \, ds = 0 \qquad (12)$$

$$(\nabla \cdot \mathbf{u_h}, w_h) = (q, w_h) \qquad \forall w_h \in W_h, \qquad (13)$$

where c_h is an approximation to c obtained from (18). Now, we introduce a dual mesh $\mathcal{V}_h^*$ based on $\mathcal{T}_h$ which will be used for the approximation of the concentration equation. For construction, we refer to [1].

For applying the standard FVEM to approximate the concentration, we define the trial space M_h on $\mathcal{T}_h$ and the test space L_h on $\mathcal{V}_h^*$.

Trial and Test Spaces for Concentration

$$M_h = \{z_h \in C^0(\bar{\Omega}) \ : \ z_h|_T \in P_1(T) \ \ \forall T \in \mathcal{T}_h\},$$

$$L_h = \{w_h \in L^2(\Omega) \ : \ w_h|_{V_P^*} \text{ is a constant} \ \ \forall V_P^* \in \mathcal{V}_h^*\},$$

where dual mesh $\mathcal{V}_h^*$ based on $\mathcal{T}_h$ and V_P^* is the control volume associated with node P, see [1]. Multiply (6) by $z_h \in L_h$, integrate over the control volumes, the FVE approximation c_h for $z_h \in L_h$ can be written as:

$$(\psi \frac{\partial c_h}{\partial s}, z_h) + a_h(\mathbf{u_h}; c_h, z_h) + (c_h q, z_h) = (\tilde{c}q, z_h) \quad \forall z_h \in L_h, \tag{14}$$

where,

$$a_h(\mathbf{v}; \chi, \phi_h) = -\sum_{j=1}^{N_h} \int_{\partial V_{P_j}^*} \left(D(\mathbf{v})\nabla\chi \cdot \mathbf{n}_{P_j} \right) \phi_h \, ds \ \ \forall v \in U, \ \chi \in H^1(\Omega), \ \phi_h \in L_h.$$

To approximate the concentration at any time say t_{n+1}, we use the approximation to the velocity at the previous time step. The fully discrete scheme corresponding to (12), (13) and (14) is defined as: For $n = 0, 1 \cdots N$, find $(c_h^n, p_h^n, \mathbf{u_h^n}) \in M_h \times W_h \times U_h$ such that $\forall \chi_h \in L_h$, $\mathbf{v_h} \in U_h$ and $w_h \in W_h$

$$c_h^0 = R_h c(0), \tag{15}$$

$$(\kappa^{-1}\mu(c_h^n)\mathbf{u_h^n}, \gamma_h \mathbf{v_h}) - \sum_{i=1}^{N_m} \mathbf{v_h}(M_i) \cdot \int_{\partial T_{M_i}^*} p_h^n \, \mathbf{n}_{T_{M_i}^*} \, ds = 0, \tag{16}$$

$$(\nabla \cdot \mathbf{u_h^n}, w_h) = (q^n, w_h), \tag{17}$$

$$\left(\phi \frac{c_h^{n+1} - \hat{c}_h^n}{\Delta t}, \chi_h\right) + a_h(\mathbf{u_h^n}; c_h^{n+1}, \chi_h) + (q^{n+1}c_h^{n+1}, \chi_h)$$
$$= (q^{n+1}\tilde{c}^{n+1}, \chi_h), \tag{18}$$

where $\hat{c}_h^n = c_h(x - \frac{\mathbf{u_h^n}}{\phi}\Delta t, t_n)$ and R_h is the projection of c defined as

$$A(\mathbf{u}; c - R_h c, \chi_h) = 0 \qquad \forall \chi_h \in L_h. \tag{19}$$

Here, $A(\mathbf{u}; \phi, \chi_h) = a_h(\mathbf{u}; \phi, \chi_h) + (q\phi, \chi_h) + (\lambda\phi, \chi_h) \quad \forall \chi_h \in L_h.$
The function λ will be chosen such that the coercivity of $A(\mathbf{u}; \cdot, \cdot)$ is assured.

3 Error Estimates

A. Error Estimates for Concentration
In order to derive error estimates, we make the following assumption on $\tilde{c}, c, p$ and $\mathbf{u}$:

$$\tilde{c}, q \in L^\infty(0, T; H^1), c_t \in L^\infty(0, T; H^2), c_{tt} \in L^\infty(0, T; H^1),$$
$$\nabla.\mathbf{u} \in L^\infty(0, T; H^1), (\nabla.\mathbf{u})_t \in L^2(0, T; L^2), p \in L^\infty(0, T; H^1) \tag{20}$$

Remark 1. Under the above assumption, we sate the following main theorem of this article and for a proof, we refer to [13].

Theorem 1. *Let c^n and c_h^n be the solutions of (11) and (18) at $t = t_n$ respectively, and let $c_h(0) = c_{0,h} = R_h c(0)$. Further assume that $\Delta t = O(h)$. Then, for sufficiently small h, there exists a positive constant $C(T)$ independent of h but dependent on the bounds of κ^{-1} and μ such that*

$$\max_{0 \leq n \leq N} \|c^n - c_h^n\|_{(L^2(\Omega))^2}^2 \leq C\left[h^4 + (\Delta t)^2\right]. \tag{21}$$

B. Error Estimates for Velocity and Pressure

Using Raviart-Thomas and L^2 projection, for a given c the following estimates for **u** and p can be shown. For proof, we refer [6].

Theorem 2. *Assume that the triangulation $\mathcal{T}_h$ is quasi-uniform. Let $(\mathbf{u}, p)$ and $(\mathbf{u_h}, p_h)$, respectively, be the solutions of (9)-(10) and (12)-(13). Then, there exists a positive constant C, independent of h, but dependent on the bounds of κ^{-1} and μ such that*

$$\|\mathbf{u} - \mathbf{u_h}\| + \|p - p_h\| \leq C \left[\|c - c_h\| + h(\|\mathbf{u}\|_{L^\infty(0,T;(H^1(\Omega))^2)} + \|p\|_1\right], \tag{22}$$

$$\|\nabla \cdot (\mathbf{u} - \mathbf{u_h})\| \leq Ch\|\nabla \cdot \mathbf{u}\|_1, \tag{23}$$

Here, $\|\cdot\| = \|\cdot\|_{L^\infty(0,T;L^2)}$ and $\|\cdot\|_1 = \|\cdot\|_{L^\infty(0,T;H^1)}$

Remark 2. Using Theorem 2, the similar estimates can be obtained for velocity and pressure as given in Theorem 1.

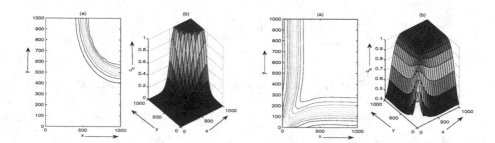

Fig. 3. Contour (a) and surface (b) plot at   **Fig. 4.** Contour (a) and surface (b) plot at
t=3 years   t=10 years

4   Numerical Experiments

For our numerical experiments, we take with $q = q^+ - q^-$ and $g(x,t,c) = \bar{c}q^+ - cq^-$, where $\bar{c}$ is the injection concentration and q^+ and q^- are the production and injection rates, respectively.

For the test problems, we have taken data from [9]. $\Omega = (0,1000) \times (0,1000)$ ft^2, $[0,T] = [0,3600]$ days, $\mu(0) = 1.0$ cp. The injection well is located at the upper right corner $(1000, 1000)$ with the injection rate $q^+ = 30\text{ft}^2/\text{day}$ and injection concentration $\bar{c} = 1.0$. The production well is located at the lower left corner with the production rate $q^- = 30\text{ft}^2/\text{day}$ and the initial concentration is $c(x, 0) = 0$. For time discretization, we take $\Delta t_p = 360$ days and $\Delta t_c = 120$ days, i.e., we divide each pressure time interval into three subintervals.

Test1: Take $\kappa = 80$, $\phi = .1$, $M = 1$, $d_m = 1$ and the dispersion coefficients are zero. In the numerical simulation for spatial discretization we divide in 20 number of divisions both along x and y axis. For time discretization, we take $\Delta t_p = 360$ days and $\Delta t_c = 120$ days, i.e., we divide each pressure time interval into three subintervals. The surface and contour plots for the concentration at $t = 3$ and $t = 10$ years are presented in Fig. 3 and Fig. 4, respectively.

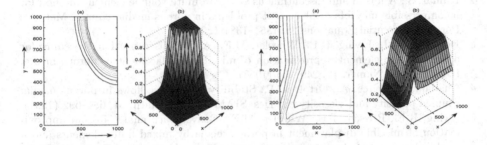

Fig. 5. Contour (a) and surface (b) plot t=3 years

Fig. 6. Contour (a) and surface (b) plot t=10 at years

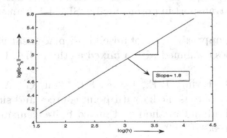

Fig. 7. Order of convergence in L^2- norm

Test 2: We take $\kappa = 80$ on the sub domain $\Omega_L := (0, 1000) \times (0, 500)$ and $\kappa = 20$ on the sub domain $\Omega_U := (0, 1000) \times (500, 1000)$. The contour and surface plot at $t = 3$ and $t = 10$ years are given in Fig. 5 and Fig. 6, respectively. We compute the order of convergence in L^2 norm. To discretize the time interval $[0, T]$, we take uniform time step $\Delta t = 360$ days for pressure and concentration equation. The computed order of convergence is given in Fig. 7. Note that the computed order of convergence matches with the theoretical order of convergence derived in Theorem 1.

Acknowledgments. The author would like to thank referees, Prof. Neela Nataraj and Prof. Amiya Kumar Pani (Department of Mathematics, IIT Bombay, India) for their valuable suggestions and the comments which helped to improve the manuscript.

I would like to thank IIST, Trivandrum for providing me financial support for attending the 5th International Conference on "Numerical Analysis and Application", June 15-20, 2012, Bulgaria.

References

1. Kumar, S., Nataraj, N., Pani, A.K.: Finite volume element method for second order hyperbolic equations. Int. J. Numerical Analysis and Modeling 5, 132–151 (2008)
2. Kumar, S.: A mixed and discontinuous Galerkin finite volume element method for incompressible miscible displacement problems in porous media. Numer. Methods Partial Differential Equations 28, 1354–1381 (2012)
3. Douglas Jr., J., Ewing, R.E., Wheeler, M.F.: A time-discretization procedure for a mixed finite element approximation of miscible displacement in porous media. RAIRO Anal. Numér. 17, 249–265 (1983)
4. Chadam, J., Peirce, A., Ortoleva, P.: Stability of reactive flows in porous media: coupled porosity and viscosity changes. SIAM J. Appl. Math. 51, 684–692 (1991)
5. Ewing, R.E., Russell, T.F., Wheeler, M.F.: Convergence analysis of an approximation of miscible displacement in porous media by mixed finite elements and a modified method of characteristics. Comput. Meth. Appl. Mech. Engrg. 47, 73–92 (1984)
6. Chou, S.H., Kwak, D.Y., Vassilevski, P.: Mixed covolume methods for elliptic problems on triangular grids. SIAM J. Numer. Anal. 35, 1850–1861 (1998)
7. Sun, S., Wheeler, M.F.: Symmetric and nonsymmetric discontinuous Galerkin methods for reactive transport in porous media. SIAM J. Numer. Anal. 43, 195–219 (2005)
8. Duran, R.G.: On the approximation of miscible displacement in porous media by a method characteristics combined with a mixed method. SIAM J. Numer. Anal. 14, 989–1001 (1988)
9. Wang, H., Liang, D., Ewing, R.E., Lyons, S.L., Qin, G.: An approximation to miscible fluid flows in porous media with point sources and sinks by an Eulerian-Lagrangian localized adjoint method and mixed finite element methods. SIAM J. Sci. Comput. 22, 561–581 (2000)
10. Chen, Z., Ewing, R.E.: Mathematical analysis for reservoir models. SIAM J. Numer. Anal. 30, 431–453 (1999)
11. Feng, X.: On existence and uniqueness results for a coupled system modeling miscible displacement in porous media. J. Math. Anal. Appl. 194, 883–910 (1995)
12. Russell, T.F.: An incompletely iterated characteristic finite element method for a miscible displacement problem. Ph.DThesis, University of Chicago, Illinois (1980)
13. Kumar, S., Yadav, S.: Modified Method of Characteristics Combined with Finite Volume Element Methods for Incompressible Miscible Displacement Problems in Porous Media (To be Communicated)

Buckling of Non-isotropic Plates with Cut-Outs

Alexandr V. Lebedev and Andrei L. Smirnov

St. Petersburg State University, Mathematics and Mechanics Faculty
Universitetsky Prospekt, 28, 198504 St. Petersburg, Russia
http://www.math.spbu.ru

Abstract. To consider the buckling of non-homogeneous elastic thin structures weakened by holes, we analyze the effect of the area of rectangular or circular holes on a critical buckling load under compression of rectangular or circular plates made of isotropic, orthotropic or transversally isotropic materials.

Keywords: Non-isotropic plate buckling, non-homogeneous plate buckling.

1 Introduction

This research is concerned with the buckling analysis of non-homogeneous (weakened by holes or cut-outs) isotropic or non-isotropic (orthotropic or transversely isotropic) thin-walled elastic plates. The purpose of the study is to examine the effect of the area and ratios of rectangular or circular holes on the critical loading of rectangular or circular plates. The effect of the boundary conditions and the plate side ratios are also analyzed. We limit ourselves to the analysis of plates under external compressive planar loadings. The plates are considered to be thin enough to apply the 2D Kirchhoff–Love theory [1]. Mathematically the buckling problems for plates with cutouts are reduced to the solution of boundary value problems for multiply connected domains, which are solved through analytical and/or numerical methods including the Bubnov–Galerkin method [2] and FEM.

2 Buckling of Isotropic Plates

2.1 Rectangular Plates

We start with the analysis of an isotropic thin plate under axial compressive load. The load is directed along the lateral faces of the plate of length a, with side ends of length b where $a \geqslant b$ and the side ratio is $k = a/b$. We consider only simply supported boundary conditions. The free edges of the central hole are parallel to the plate sides and the square hole has length d.

For homogeneous plates the buckling load may be derived analytically [1]

$$N_{cr} = \min_{m,n} \left[((nk)^2 + m^2) \frac{\pi^2 D}{a^2 m^2} \right]; \qquad k = \frac{a}{b}.$$

I. Dimov, I. Faragó, and L. Vulkov (Eds.): NAA 2012, LNCS 8236, pp. 387–394, 2013.
© Springer-Verlag Berlin Heidelberg 2013

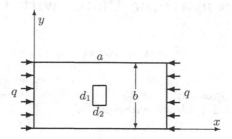

Fig. 1. Buckling of rectangular plate under compression

Here n is the number of waves in the axial direction, m is the number of waves in the transversal direction, and D is the cylindrical stiffness. For plates with holes, the results obtained by means of the Bubnov–Galerkin method are reported in [2]. Here we compare them with the results of the numerical analysis of the problem by means of the FEM package ANSYS. The most important and interesting point is the effect of the hole area on the critical buckling loading and the buckling modes.

In Fig. 2 one can see the effect of the plate sides ratio on the critical loading for a plate with relative thickness $h = 0.01$ and Poisson's ratio $\nu = 0.3$, where N_0 and N_{cr} are the critical buckling loads for a homogeneous plate (see [1]) and for a plate with a square hole respectively, $d = 0.1$.

It appeared that the critical buckling loadings may either increase or decrease. Presumably the effect when "mechanical buckling strengths of the perforated plates, contrary to expectation, increase rather than decrease as the hole sizes grow larger" was firstly reported in [3]. In our research it was found that, for

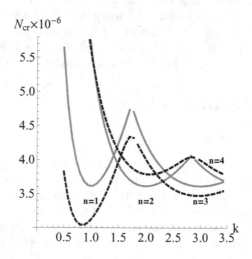

Fig. 2. Buckling of isotropic plates. Effect of the plate side ratio on the critical loading for plates with a hole with $d = 0.1$ (dashed lines) and a homogeneous plate.

example, for axially compressed rectangular plate for buckling nodes with odd wave numbers the critical loading decreases when the hole area decreases and for even wave numbers the buckling load increases when the hole area decreases [4]. The explanation of this phenomenon is in the initial compressive stresses developing in the narrow strips between the hole and the plate edges. One should remember that a hole not only affects the plate stiffness but also influences initial stress-strain state. These initial stresses are higher for the stronger supports of the lateral edges of the plate and they increase with Poisson's ratio (see [4] and [5]). That leads to the increase of the critical load.

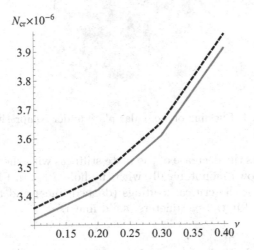

Fig. 3. Effect of Poisson's ratio on critical buckling loading for rectangular plate with a hole

The ratio of the hole sides plays an important role, which is revealed when we analyze plates with holes of equal areas. For buckling under axial loading for all cases, the extension of the hole in the axial direction leads to a decreasie of the critical loading. For a hole elongated in the transversal direction, the width of the strip is smaller and the intensity of the initial stresses is higher and the critical loading increases. The change of the hole side ratio may also cause a switch of the buckling modes. Clearly, this is valid only within some limits for the hole sides. When the width of the side strip becomes too small, the local buckling occurs and one of the strips buckles as a beam under compressive load.

2.2   Circular Plates

For the circular plate (radius R) with a central circular hole (radius r) under radial compressive load q the dependence of the critical load on the hole area is more predictable.

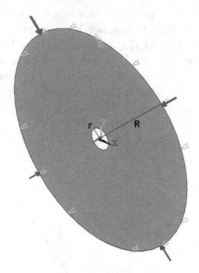

Fig. 4. Buckling of a circular plate under compresion

The main effect is the decrease of the plate stiffness with the hole area and the critical load goes down monotonically with the hole area. In Fig. 5, we compare numerical results for the critical loadings (dashed lines) and those obtained in [6] by the method of initial parameters (solid lines).

3   Buckling of Orthotropic Plates

The buckling behavior of non-isotropic plates has some specific features. As an example, we consider the buckling of a plate made of orthotropic material with Young's moduli E_x, and E_y, Poisson's ratios ν_{xy} and ν_{yx}, and shear modulus G. Since we wish to study the effect of non-isotropy on the buckling load we assume that

$$E_x = E_0(1 + |\epsilon|)^{\text{sgn } \epsilon}, E_y = E_0(1 + |\epsilon|)^{-\text{sgn } \epsilon},$$
$$\nu_{xy} = \nu_0(1 + |\epsilon|)^{\text{sgn } \epsilon},$$
$$E_x\nu_{yx} = E_y\nu_{xy} = E_0\nu_0, G = E_0/(2(1 + \nu_0)).$$

(1)

So, for small ϵ, this material is almost isotropic. For positive ϵ, the material is stiffer in the x-direction, for negative ϵ, it is stiffer in the y-direction. Note that for small $\epsilon > 0$, $E_x \approx E_0(1 + \epsilon)$ and $E_y \approx E_0(1 - \epsilon)$.

3.1   Rectangular Plates

The effect of orthotropy on buckling load of the rectangular plate with a central square hole with different hole area $S^* = d^2$ under axial compression is shown in Fig. 6, where N_0 and N_{cr} are the critical buckling loads for plates without and

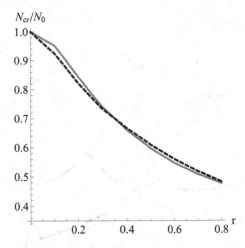

Fig. 5. Buckling of isotropic plates. Effect of the hole area on critical loadings for a circular plate with a circular hole of radius r.

with a hole, respectively. Again for a homogeneous plate we use the analytical formula [1]

$$N_{cr} = \pi^2 \left[D_{11} \left(\frac{n}{a}\right)^2 + 2H \left(\frac{m}{b}\right)^2 \left(\frac{n}{a}\right)^2 + D_{22} \left(\frac{m}{b}\right)^4 \right], \qquad (2)$$

where

$$D_{11} = \frac{E_x h^3}{12(1 - \nu_{xy}\nu_{yx})}, \quad D_{22} = \frac{E_y h^3}{12(1 - \nu_{xy}\nu_{yx})}, \quad H = \nu_{yx} D_{11} + \frac{G h^3}{6}.$$

Even for a relatively small hole, the effect of non-isotropy is very significant: if the plate becomes stiffer in the axial direction and softer in the transversal direction the critical buckling loading decreases very rapidly with $\epsilon > 0$ and for a plate stiffer in the transverse direction the critical buckling loading increases. Again, this underlines the crucial effect for buckling of the initial stresses "carried by the narrow side strips of material along the plate boundaries" [3]. It is interesting that for orthotropic plates even for a very small hole area the buckling mode does not change with the plate side ratio. It looks as the hole fixes the buckling mode wavenumbers that leads to rather monotone dependence of the critical load on the plate material stiffness ratio.

3.2   Circular Lates

Similarly, the critical loading of an orthotropic circular plate with a central circular hole under radial compression depends on the material stiffness ratio. Here the important effect comes from the initial stresses in the circumferential direction. For materials (1) stiffer in the circumferential direction ($\epsilon < 0$) the critical load increases (see Fig. 7).

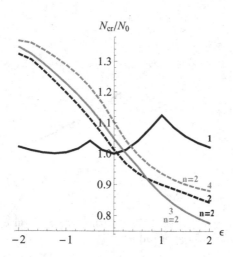

Fig. 6. Buckling of orthotropic plates. Rectangular plate with $k = 2$: 1 – a homogenous plate, and 2,3,4 – a plate with a central square hole of area $S^* = 0.01; 0.04; 0.09$, respectively.

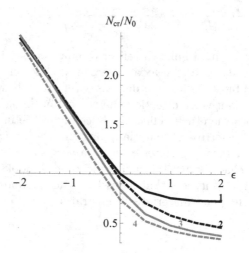

Fig. 7. Buckling of orthotropic plates. Circular plate with $R = 1$: 1 –a homogenous plate, and 2,3,4 – a plate with a central circular hole with $r = 0.1; 0.2; 0.3$, respectively.

4   Buckling of Transversally Isotropic Plates

Finally, we consider the buckling behavior of transversally isotropic plates with the following elastic moduli:

$$E_x = E_y = E_0(1 + |\epsilon|)^{\text{sgn}\ \epsilon}, E_z = E_0(1 + |\epsilon|)^{-\text{sgn}\ \epsilon},$$
$$\nu_{xy} = \nu_{yx} = \nu_0(1 + |\epsilon|)^{\text{sgn}\ \epsilon}, \tag{3}$$
$$E_x \nu_{yx} = E_y \nu_{xy} = E_z \nu_{xz} = E_0 \nu_0, G = E_0/(2(1 + \nu_0)).$$

For positive ϵ the material is stiffer in the x, y-directions (in the plane), and for negative ϵ, it is stiffer in the z-direction (along the thickness).

4.1   Rectangular Plates

For rectangular transversally isotropic plates, the effect of the material properties is shown in Fig. 8. For stiffer planar materials ($\epsilon > 0$), the buckling load is higher and the buckling mode essentially depends on the stiffness parameter. For small ϵ, the critical load increases with the hole area. Here a buckling mode switching exists. One can see the significant difference with the case of the orthotropic material. For large enough ϵ in absolute value, the critical buckling value becomes smaller than for the isotropic plate.

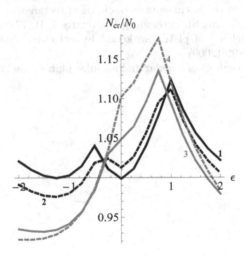

Fig. 8. Buckling of transversally isotropic plates. Rectangular plate with $k = 2$: 1 – homogenous plate, and $2,3,4$ – plate with central square hole of area $S^* = 0.01; 0.04; 0.09$, respectively.

4.2   Circular Plates

For transversally isotropic circular plates, the change of planar ratio and in thickness, Young's modulus leads to an increase of the critical buckling load which decreases monotonically with the hole area.

5   Conclusions

The presence of a hole or cut-outs may lead to either increasing or decreasing critical buckling load for compressed plates depending on the boundary conditions, geometric parameters of the plate and the hole and the material properties. For rectangular plates, the main effect comes from stresses in the lateral strips. For buckling of non-isotropic circular plates, the material stifness ratio plays the key role.

Acknowledgements This research was partly supported by the RFBR grant #10-01-00244-a.

References

1. Alfutov, N.A.: Fundamentals of buckling analysis of elastic systems. Machinostroyenie, Russia (1991) (in Russian)
2. Preobrazhensky, I.N.: Buckling and vibrations of plates and shells with the holes. Machinostroyenie, Russia (1981) (in Russian)
3. Ko, W.L.: Mechanical- and thermal-buckling behavior of rectangular plates with different central cutouts. Nasa report, Dryden Flight Research Center (1998)
4. Lebedev, A.V.: Effects of the cut-outs on buckling of rectangular elastic plates under axial compression. Vestnik St. Petersburg University 1(4), 77–83 (2009)
5. Lebedev, A.V.: Buckling of plates weakened by cut-outs. Vestnik St. Petersburg University 1(2), 94–99 (2009)
6. Chizhevsky, C.G.: Analysis of circular and annular plates. Machinostroyenie, Russia (1977) (in Russian)

On a Research of Hybrid Methods

Mehdiyeva Galina, Imanova Mehriban, and Ibrahimov Vagif

Baku State University,
Z.Khalilov str. 23, AZ1148, Baku, Azerbaijan
imn_bsu@mail.ru
http://www.bsu.az

Abstract. Constructed hybrid methods of the high accuracy the experts examined that's for solving integral and integro-differential equations. Using hybrid methods for solving integral equations belongs to Makroglou. Here, developing these idea, explored a more general hybrid method which is applied to solving Volterra integral equations and also constructed a concrete method with the degree $p = 8$. However, order of accuracy for the known corresponding methods is of level $p \leq 4$.

Keywords: Volterra integral equation, a hybrid method, stability and degree of hybrid method, multistep methods.

1 Introduction

As is known, the solving problems of natural science are reduced to solving integral equation among of which nonlinear integral equations with variable boundaries are most popular. The first application of such equation to applied problems was performed by Abel (see [1, p.12]). A wide application of integral equations with variable boundaries belongs to Volterra that applied such equations since 1887 (see. [2, p.67]).

Consider nonlinear Volterra integral equation:

$$y(x) = g(x) + \int_{x_0}^{x} K(x, s, y(s))ds, \quad x \in [fx_0, X]. \tag{1}$$

Suppose that equation (1) has a unique continuous solution determined on the segment $x \in [x_0, X]$.

Volterra has investigated equation (1) in the case when the kernel of the integral the function $K(x, z, y)$ is linear with respect to y (i.e. $K(x, z, y) = \varphi(x, z)y$). For solving these equation Volterra suggested the quadrature method that is used up to day. However, taking into account some shortcomings of the methods of quadrature methods the scientists modified these methods and at some cases suggested another methods as the methods of spline functions, collocation and etc. (see e.g. [3–5]). Unlike these methods, here to the numerical solution of integral equation (1) we apply hybrid methods. As remark above the application of hybrid methods to the solving integral equations belongs to Makroglou

I. Dimov, I. Faragó, and L. Vulkov (Eds.): NAA 2012, LNCS 8236, pp. 395–402, 2013.

(see [6]). This method was modified in the paper [7]. Here we attempt to generalize some known methods in the form of a hybrid method that may be written in the following form:

$$\sum_{i=0}^{k} \alpha_i y_{n+i} = \sum_{i=0}^{k} \alpha_i g_{n+i} + h \sum_{i=0}^{k} \sum_{j=0}^{k} \beta_i^{(j)} K(x_{n+j}, x_{n+i}, y_{n+i}) +$$

$$+ h \sum_{i=0}^{k} \sum_{j=0}^{k} \gamma_i^{(j)} K(x_{n+j}, x_{n+l_i}, y_{n+l_i}), \tag{2}$$

where $l_i = i + \nu_i$ ($|\nu_i| < 1$, $i = 0, 1, 2, ..., k$).

For the construction the methods of type (2), suppose that equation (1) has a continuous solution determined on the segment $x \in [x_0, X]$, and the kernel of the integral the function $K(x, z, y)$ is continuous on totality of arguments and is defined on the domain $G = \{x_0 \le z \le x + \varepsilon \le X + \varepsilon, |y| \le b\}$. Usually for $\varepsilon = 0$, some coefficients of method (2) are chosen in the form $\beta_i^{(j)} = 0$, $\gamma_i^{(j)} = 0$ for $j < i$. There are many methods for solving equation (1). One of them is the use of quadrature formulae after application of which to calculation of the integral participating in equation (1) we have:

$$y(x_n) = g_n + h \sum_{j=0}^{n} a_j K(x_n, x_j, y_j),$$

where the coefficients of the quadrature formula a_j ($j = 0, 1, ..., n$) are some real numbers, and $g_m = g(x_m)$ ($m = 0, 1, 2, ...$).

Note that the solution of equation (1) is compared with the solution of initial value problem for ordinary differential equations. In order to illustrate what has been said, assume that the kernel of the integral has the following form:

$$K(x, z, y) = \sum_{l=0}^{m} a_l(x) b_l(z, y).$$

Then the solving of equation (1) may be reduced to the solving of the following system of equations:

$$y(x) = g(x) + \sum_{l=0}^{m} a_l(x) \vartheta_l(x), \tag{3}$$

$$\vartheta_l' = b_l(x, y), \quad \vartheta_l(0) = 0 \ (l = 0, 1, 2, ..., m). \tag{4}$$

Thus,the solving of equation (1) is reduced to the solving the system ordinary differential equations of first order. Taking into account that the numerical solution of problem (4) has been studied well, because by using known methods we can solve the system consisted of equations (3) and (4). Application hybrid multistep methods to the solution of problem (4) began after the known papers of Gear and Butcher (see [8], [9]), but investigation of hybrid one-step methods began after the paper of Hammer and Hollingsworth (see [10]), that enters into the class of implicit Runge-Kutta methods (see e.g. [11, p.217]).

Here, for determined the numerical solution of problems (4) for $l = 0$, we apply the following hybrid method:

$$\sum_{i=0}^{k} \alpha_i \vartheta_{n+i} = h \sum_{i=0}^{k} \beta_i b_{n+i} + h \sum_{i=0}^{k} \gamma_i b_{n+i+\nu_i} \quad (|\nu_i| < 1, \ i = 0, 1, \ldots, k), \quad (5)$$

where $b_m = b(x_m, y_m) \ (m = 0, 1, 2, \ldots)$.

2 Construction of Hybrid Methods

Taking into account the advantage of hybrid methods, consider the investigation of method (5). Note that method (5) has the generalizes many of the known methods. Rewrite method (5) in the following form:

$$\sum_{i=0}^{k} \alpha_i z_{n+i} = h \sum_{i=0}^{k} \beta_i z'_{n+i} + h \sum_{i=0}^{k} \gamma_i z'_{n+i+\nu_i} \quad (|\nu_i| < 1, \ i = 0, 1, 2, \ldots, k). \quad (6)$$

It is easy to see that relation (6) is a difference equation with the constant coefficients of order k. Therefore, some authors call the k-step methods of type (6) for $\gamma_i = 0, \ (i = 0, 1, 2, \ldots, k)$ as the finite-difference methods (see e.g. [12], [13, p.483]).

Before considering the definition of the values of the quantities $\alpha_i, \beta_i, \gamma_i, \nu_i$ $(i = 0, 1, 2, \ldots, k)$, establish some restrictions imposed on the coefficients of the method (6). These conditions have the following form:

A: The values of the quantities $\alpha_i, \beta_i \ (i = 0, 1, 2, \ldots, k)$ are real numbers, moreover, $\alpha_k \neq 0$.
B: Characteristic polynomials

$$\rho(\lambda) \equiv \sum_{i=0}^{k} \alpha_i \lambda^i, \quad \sigma(\lambda) \equiv \sum_{i=0}^{k} \beta_i \lambda^i, \quad \gamma(\lambda) \equiv \sum_{i=0}^{k} \gamma_i \lambda^{i+\nu_i};$$

have no common multipliers different from the constant.
C: $\sigma(1) + \gamma(1) \neq 0$ and $p \geq 1$ are holds.

Note that the accuracy of the finite-difference methods of the type (6) is determined by means of the notion of degree, since the quantity k is an order of method (6). Because, the notion of degree for method (6), is determined as follows.

Definition 1. *Let the function $z(x)$ be determined on the segment $x \in [x_0, X]$ and be sufficiently smooth. Then, an integer quantity $p > 0$ calls as the degree of the method (6) if the following holds:*

$$\sum_{i=0}^{k} (\alpha_i z(x + ih) - h(\beta_i z'(x + ih) + \gamma_i z'(x + (i + \nu_i)h))) = O(h^{p+1}), \ h \to 0, \quad (7)$$

here $x = x_0 + nh$ fixed point.

One of the basic problems in researching method (6), is definition of the maximum value for the degree p. However, in the method (6), value of k is the given. Therefore, the specialists try to define the relation between the order and degree of method. For defining the relation between the quantities p and k, we use the method of undetermined coefficients, that uses the Taylor expansion of the functions $z(x + ih)$, $z'(x + ih)$ and $z'(x + (i + \nu_i)h)$ in the relation (7), that may be written in the following form:

$$z(x + ih) = z(x) + ihz'(x) + \frac{(ih)^2}{2!} z''(x) + \ldots + \frac{(ih)^p}{p!} z^{(p)}(x) + O(h^{p+1}), \quad (8)$$

$$z'(x + m_ih) = z'(x) + m_ihz''(x) + \frac{(m_ih)^2}{2!} z'''(x) + \ldots +$$

$$+ \frac{(m_ih)^{p-1}}{(p-1)!} z^{(p)}(x) + O(h^p). \tag{9}$$

Here the quantity m_i takes the values $m_i = i + \nu_i$ and $m_i = i$ $(i = 0, 1, 2, ..., k)$.

Assume that the method (6) has the degree p and allowing the expansions (8) and (9) in asymptotic relation (7) we have:

$$\sum_{i=0}^{k} \alpha_i = 0; \quad \sum_{i=0}^{k} i\alpha_i = \sum_{i=0}^{k} (\beta_i + \gamma_i),$$

$$\sum_{i=0}^{k} \frac{i^l}{l!} \alpha_i = \sum_{i=0}^{k} \left(\frac{i^{l-1}}{(l-1)!} \beta_i + \frac{(i+\nu_i)^{l-1}}{(l-1)!} \gamma_i \right), \quad l = 2, 3, \ldots, p. \tag{10}$$

It is easy to prove that for the existence of nontrivial solutions there should be $p + 1 < 4k + 4$. Hence it follows that $p_{max} = 4k + 2$. Usually the methods with the degree for are unstable. The stability of method (6) is similarly defined from the methods obtained from (6) for $\nu_i = 0$ $(i = 0, 1, ..., k)$.

Definition 2. *Method (6) is stable if the roots of its characteristic polynomial $\rho(\lambda)$ lie interior to a unique circle whose boundaries have no multiple roots.*

Now proof that conditions (7) and (10) are equivalent if the function $z(x)$ is sufficiently smooth. Consider the following lemma.

Lemma 1. *Let the function $z(x)$ be sufficiently smooth. Then, satisfaction of the quantities $\alpha_i, \beta_i, \gamma_i, \nu_i$ $(i = 0, 1, 2, \ldots, k)$ of the system algebraic equations (10) is necessary and sufficient be fulfilled asymptotic equalities (7).*

Proof. At first proof the sufficiency of conditions (10) for method (6) to have the degree p. To this end we investigate the following:

$$\sum_{i=0}^{k} (\alpha_i z(x + ih) - h\beta_i z'(x + ih) - h\gamma_i z'(x + (i + \nu_i)h)) =$$

$$= \sum_{i=0}^{k} \alpha_i z(x) + h \sum_{i=o}^{k} (i\alpha_i - \beta_i - \gamma_i)z'(x) + \sum_{i=0}^{k} \alpha_i z(x) +$$

$$+ h \sum_{i=o}^{k} (i\alpha_i - \beta_i - \gamma_i)z'(x) + h^2 \sum_{i=o}^{k} \left(\frac{i^2}{2}\alpha_i - i\beta_i - (i+\nu_i)\gamma_i \right) z''(x) + \ldots +$$

$$+ h^p \sum_{i=0}^{k} \left(\frac{i^p}{p!}\alpha_i - \frac{i^{p-1}}{(p-1)!}\beta_i - \frac{(i+\nu_i)^{p-1}}{(p-1)!}\gamma_i \right) z^{(p)}(x) + O(h^{p+1}). \qquad (11)$$

If we take into account that the quantities $\alpha_i, \beta_i, \gamma_i, \nu_i$ $(i = 0, 1, 2, ..., k)$ are the solution of system (10), then asymptotic equality (7) is follows from (11). Now we prove the necessity, i.e. prove that if the quantities $\alpha_i, \beta_i, \gamma_i, \nu_i$ $(i = 0, 1, 2, ..., k)$ satisfies condition (7), then they are the solution of system (10). Indeed, taking into account (7) in (11), we have:

$$\sum_{i=0}^{k} \alpha_i z(x) + h \sum_{i=o}^{k} (i\alpha_i - \beta_i - \gamma_i)z'(x) + \ldots +$$

$$+ h^p \sum_{i=0}^{k} \left(\frac{i^p}{p!}\alpha_i - \frac{i^{p-1}}{(p-1)!}\beta_i - \frac{(i+\nu_i)^{p-1}}{(p-1)!}\gamma_i \right) z^{(p)}(x) = 0. \qquad (12)$$

It is known that $1, x, x^2, ..., x^p$ or $z(x), z'(x), ..., z^{(p)}(x)$ for $z(x) \neq 0$ and $z^{(j)}(x) \neq 0$ $(i = 0, 1, ..., p)$ are linearly independent systems. Then from (12) it follows that the quantities $\alpha_i, \beta_i, \gamma_i, \nu_i$ $(i = 0, 1, 2, ..., k)$ satisfies system (10). Thus we proved the lemma. $\qquad\qquad\qquad\qquad\qquad\qquad\qquad\qquad\qquad\qquad\qquad\qquad$ □

Now consider the construction of specific methods with certain accuracy and assume that $k = 2$. Then from system (10) for $l_i = i + \nu_i (i = 0, 1, 2)$, we have:

$$\beta_2 + \beta_1 + \beta_0 + \gamma_2 + \gamma_1 + \gamma_0 = 2\alpha_2 + \alpha_1,$$
$$2\beta_2 + \beta_1 + l_2\gamma_2 + l_1\gamma_1 + l_0\gamma_0 = (2^2\alpha_2 + \alpha_1)/2,$$
$$2^2\beta_2 + \beta_1 + l_2^2\gamma_2 + l_1^2\gamma_1 + l_0^2\gamma_0 = (2^3\alpha_2 + \alpha_1)/3,$$
$$2^3\beta_2 + \beta_1 + l_2^3\gamma_2 + l_1^3\gamma_1 + l_0^3\gamma_0 = (2^4\alpha_2 + \alpha_1)/4,$$
$$2^4\beta_2 + \beta_1 + l_2^4\gamma_2 + l_1^4\gamma_1 + l_0^4\gamma_0 = (2^5\alpha_2 + \alpha_1)/5, \qquad (13)$$
$$2^5\beta_2 + \beta_1 + l_2^5\gamma_2 + l_1^5\gamma_1 + l_0^5\gamma_0 = (2^6\alpha_2 + \alpha_1)/6,$$
$$2^6\beta_2 + \beta_1 + l_2^6\gamma_2 + l_1^6\gamma_1 + l_0^6\gamma_0 = (2^7\alpha_2 + \alpha_1)/7,$$
$$2^7\beta_2 + \beta_1 + l_2^7\gamma_2 + l_1^7\gamma_1 + l_0^7\gamma_0 = (2^8\alpha_2 + \alpha_1)/8,$$
$$2^8\beta_2 + \beta_1 + l_2^8\gamma_2 + l_1^8\gamma_1 + l_0^8\gamma_0 = (2^9\alpha_2 + \alpha_1)/9.$$

If put $\alpha_2 = 1$, $\alpha_1 = 0$, $\alpha_0 = -1$, in this system, then by solving the obtained system of nonlinear algebraic equations, we have:

$$\beta_2 = 64/180, \quad \beta_1 = 98/180, \quad \beta_0 = 18/180,$$

$$\gamma_2 = 18/180, \quad \gamma_1 = 98/180, \quad \gamma_0 = 64/180,$$

$$l_2 = 1 + \sqrt{21}/14, \quad l_1 = 1, \quad l_0 = 1 - \sqrt{21}/14,$$

Hence we get the following method:

$$y_{n+2} = y_n + h(64y'_{n+2} + 98y'_{n+1} + 18y'_n)/180 +$$

$$+ h(18y'_{n+l_2} + 98y'_{n+1} + 64y'_{n+l_0})/180. \tag{14}$$

Obviously, for using method (14) the values of the quantities y_{n+l_0}, y_{n+l_2} should be known and these values should be calculated with the order $O(h^8)$.

Now consider the application of method (14) to the numerical solution of equation (1). Following results of the paper [14] while applying method (14) to solving equation (1), in one variant it will have the following form:

$$y_{n+2} = y_n + g_{n+2} - g_n + h(64K(x_{n+2}, x_{n+2}, y_{n+2}) + 49K(x_{n+2}, x_{n+1}, y_{n+1}) +$$

$$+ 49K(x_{n+1}, x_{n+1}, y_{n+1}) + 9K(x_{n+1}, x_n, y_n) + 9K(x_n, x_n, y_n))/180 +$$

$$+ h(9K(x_{n+2}, x_{n+l_2}, y_{n+l_2}) + 9K(x_{n+l_2}, x_{n+l_2}, y_{n+l_2}) +$$

$$+ 49K(x_{n+2}, x_{n+1}, y_{n+1}) + 49K(x_{n+1}, x_{n+1}, y_{n+1}) +$$

$$+ 32K(x_{n+2}, x_{n+l_0}, y_{n+l_0}) + 32K(x_{n+1}, x_{n+l_0}, y_{n+l_0}))/180. \tag{15}$$

3   Comparison of Some Methods Indented for Solving Integral Equations with Variable Boundaries

Consider to solving of equation(1)by the quadrature method. Then we have:

$$y(x_{n+k}) = g_{n+k} + h \sum_{j=0}^{n+k} a_j K(x_{n+k}, x_{n+j}, y_{n+j}) + R_n \quad (n = 0, 1, 2, \ldots). \tag{16}$$

Here the coefficients of the quadrature formulae a_j $(j = 0, 1, ..., n + k)$ are some real numbers, R_n is the remainder term of the quadrature formula, k is an entire constant. It is easy to see that while passing from one current point x_{n+k} to the next x_{n+k+1} it is necessary to calculate again the sum participating in the right side of equality (16) since the quantity $K(x_{n+k}, x_{n+j}, y_{n+j})$ is replaced by the quantity $K(x_{n+k+1}, x_{n+j}, y_{n+j})$. For removing the mentioned shortcoming of the quadrature method it is suggested to use one of the above-given schemes.

In order to illustrate the obtained results, here we investigated the solution of equation (1) by quadrature methods, which suggested in [14], and by the methods using the change of the kernel of the integral with degenerate function. And as a numerical method we used the trapezoid method since in the paper [3] the below-given integral equations were solved by the trapezoid methods.

1. $y(x) = \exp(-x) + \int\limits_0^x \exp(-(x - s))y^2(s)ds, \quad x \in [0; 0, 1]$, step $h = 0, 02$, the

   exact solution which represented in the form $y(x) \equiv 1$;

2. $y(x) = 1 + \frac{x^2}{2} + \int\limits_0^x \exp(x - s)(1 + \frac{s^2}{2})ds$, $x \in [0, 1]$, step $h = 0, 1$, but the

exact solution was represented in the form $y(x) = 2 \exp(x) - x - 1$;

3. $y(x) = x + \int\limits_0^x \sin(x - s)y(s)ds$, $x \in [0, 1]$, step $h = 0, 1$, the exact solution

was represented in the form $y(x) = x + x^3/6$.

First, the solution of integral equations use the trapezoid method, obtained from the method (2) for $\gamma_i^{(j)} = 0$ $(i, j = 0, 1, 2, ..., k)$. In the second method after applying the kernel of the integral method of degenerate kernel to their computing. The trapezoid method is used in all cases, the second method was more accurate than the methods used in [3]. To compare these results apply to the following table

number example	x	Accuracy of the method from [3]	Accuracy of the trapezoid method	Accuracy of the trapezoid method using degenerate kernels
I	0.02	1.E-2	6.6E-7	6.7E-7
	0.04	1.5E-2	3.9E-4	1.3E-6
	0.1	1.1E-2	4E-3	3.5E-6
II	0.1	-7.9E-2	8.3E-5	1.6E-4
	0.2	1.E-4	1.1E-2	3.3E-4
	1.0	7.5E-2	7.3E-1	1.8E-3
III	0.1	-1.0E-3	1.6E-4	8.2E-5
	0.2	3.E-3	8.3E-4	1.6E-4
	1.0	2.5E-1	1.4E-1	8.9E-4

Note that the method (15) has a higher accuracy, but its use has some difficulty, therefore, consider the following method:

$$y_{n+1} = y_n + h\left(5y'_{n+\frac{1}{2}+\alpha} + 8y'_{n+\frac{1}{2}} + 5y'_{n+\frac{1}{2}-\alpha}\right)/18, \quad (\alpha = \sqrt{15}/10). \quad (17)$$

This method is stable and has the degree $p = 6$. Consider to construct an algorithm for using method (17). Assume that y_1 is known, and consider the calculation of y_{n+2} $(n = 0, 1, 2, ...)$. For applying method (17) to calculation of the values of the quantities $y_{n\pm\alpha}$, should be known and they are determined by means of the following formula:

$$y_{n+\alpha} = y_n + 2hy'_n + \alpha^2 h((h^2 - 12\alpha + 6)y'_{n+3/2} -$$

$$- (3\alpha^2 - 48\alpha + 27)y'_{n+1} + (3\alpha^2 - 60\alpha + 54)y'_{n+1/2} - (\alpha^2 - 24\alpha + 33)y'_n)/18, \quad (18)$$

where $\alpha = 3/2 \pm \sqrt{15}/10$.

4   Conclusion

Here prove that stable hybrid methods are more accurate than the corresponding multi-step methods. There are stable methods of type (2) with a degree $p = 3k +$

1, but stable methods received from the formula (2) have a degree $p_{max} = k + 2$. Therefore, hybrid methods are more promising than ordinary multistep methods. Constructed, here stable method (15) is implicit, but the method (17) is explicit. But, the algorithm suggested for using method (17) is implicit. We have shown here that one can construct an algorithm for the use of hybrid methods with high accuracy. Note that the investigation of the solutions above mentioned examples show that, in possibility to solving the integral equation with variable boundaries, to apply the method of degenerate kernels.

Acknowledgement. This research has been supported by the Science Development Foundation of Azerbaijan (EIF-2011-1(3)-82/27/1).

References

1. Polishuk, Y.M.: Vito Volterra. Nauka, Leningrad (1977)
2. Volterra, V.: Theory of functional and of integral and integro-differensial equations. Nauka, Moscow (1982)
3. Verlan, A.F., Sizikov, V.S.: Integral equations: methods, algorithms, programs. Naukova Dumka, Kiev (1986)
4. Brunner, H.: Implicit Runge-Kutta Methods of Optimal oreder for Volterra integro-differential equation. Methematics of Computation 42(165), 95–109 (1984)
5. Lubich, C.: Runge-Kutta theory for Volterra and Abel integral equations of the second kind. Mathematics of Computation 41(163), 87–102 (1983)
6. Makroglou, A.: Hybrid methods in the numerical solution of Volterra integro-differential equations. Journal of Numerical Analysis 2, 21–35 (1982)
7. Mehdiyeva, G., Ibrahimov, V., Imanova, M.: On one application of hybrid methods for solving Volterra integral equations, Dubai, pp. 809–813. World Academy of Science, Engineering and Technology (2012)
8. Gear, C.S.: Hybrid methods for initial value problems in ordinary differential equations. SIAM, J. Numer. Anal. 2, 69–86 (1965)
9. Butcher, J.C.: A modified multistep method for the numerical integration of ordinary differential equations. J. Assoc. Comput. Math. 12, 124–135 (1965)
10. Hammer, P.C., Hollingsworth, J.W.: Trapezoildal methods of approximating solution of differential equations. MTAC 9, 92–96 (1955)
11. Hairier, E., Norsett, S.P., Wanner, G.: Solving ordinary differential equations. Mir, Moscow (1990) (Russian)
12. Imanova, M.N.: One the multistep method of numerical solution for Volterra integral equation. Transactions Issue Mathematics and Mechanics Series of Physical-technical and Mathematical Science VI(1), 95–104 (2006)
13. Bakhvalov, N.S.: Numerical methods. Nauka, Moscow (1973)
14. Mehdiyeva, G., Ibrahimov, V., Imanova, M.: On an application of the Cowell type method. News of Baku University. Series of Physico-mathematical sciences, (2), 92-99 (2010)
15. Imanova, M., Mehdiyeva, G., Ibrahimov, V.: Application of the Forward Jumping Method to the Solving of Volterra Integral Equation. In: Conference Proceedings NumAn 2010 Conference in Numerical Analysis, Chania, Greece, pp. 106–111 (2010)

Finite Difference Scheme for a Parabolic Transmission Problem in Disjoint Domains

Zorica Milovanović

University of Belgrade, Faculty of Mathematics
Studentski trg 16
zorica.milovanovic@gmail.com

Abstract. In this paper we investigate a parabolic transmission problem in disjoint domains. A priori estimate for its weak solution in appropriate Sobolev-like space is proved. A finite difference scheme approximating this problem is proposed and analyzed.

1 Introduction

In applications, especially in engineering, often are encountered composite or layered structure, where the properties of individual layers can vary considerably from the properties of the surrounding material. Layers can be structural, thermal, electromagnetic or optical role, etc. Mathematical models of energy and mass transfer in domains with layers lead to so called transmission problems. In this paper we consider a class of non-standard parabolic transmission problem in disjoint domain (see [4]). As model example it is taken an area consisting of two non-adjacent rectangles.In each subarea was given a initial-boundary problem of parabolic type, where the interaction between their solutions described by nonlocal integral conjugation conditions.

2 Formulation of the Problem

As a model example,we consider the following initial-boundary-value problem (IBVP): Find functions $u_1(x, y, t)$ and $u_2(x, y, t)$ that satisfy the system of parabolic equations for (1)-(11):

$$\frac{\partial u_1}{\partial t} - \frac{\partial}{\partial x}\left(p_1(x,y)\frac{\partial u_1}{\partial x}\right) - \frac{\partial}{\partial y}\left(q_1(x,y)\frac{\partial u_1}{\partial y}\right) + r_1(x,y)u_1 = f_1(x,y,t), \quad (1)$$

$$(x,y) \in \Omega_1 = (a_1,b_1) \times (c_1,d_1), \quad t > 0,$$

$$\frac{\partial u_2}{\partial t} - \frac{\partial}{\partial x}\left(p_2(x,y)\frac{\partial u_2}{\partial x}\right) - \frac{\partial}{\partial y}\left(q_2(x,y)\frac{\partial u_2}{\partial y}\right) + r_2(x,y)u_2 = f_2(x,y,t), \quad (2)$$

$$(x,y) \in \Omega_2 = (a_2,b_2) \times (c_2,d_2), \quad t > 0,$$

I. Dimov, I. Faragó, and L. Vulkov (Eds.): NAA 2012, LNCS 8236, pp. 403–410, 2013.

where $-\infty < a_1 < b_1 < a_2 < b_2 < +\infty$ and $c_2 < c_1 < d_1 < d_2$, the initial conditions

$$u_1(x,y,0) = u_{10}(x,y), \quad (x,y) \in \Omega_1, \quad u_2(x,y,0) = u_{20}(x,y), \quad (x,y) \in \Omega_2, \quad (3)$$

the simplest external Dirichlet boundary conditions

$$
\begin{aligned}
u_1(a_1,y,t) = 0, \quad y \in (c_1,d_1), \quad u_2(x,c_2,t) = 0, \quad x \in (a_2,b_2), \\
u_2(b_2,y,t) = 0, \quad y \in (c_2,d_2), \quad u_2(x,d_2,t) = 0, \quad x \in (a_2,b_2),
\end{aligned}
\quad (4)
$$

and the internal conjugation conditions of non-local Robin-Dirichlet type

$$p_1(b_1,y)\frac{\partial u_1}{\partial x}(b_1,y,t) + \alpha_1(y)u_1(b_1,y,t) = \int_{c_2}^{d_2} \beta_1(y,y')u_2(a_2,y',t)\,dy', \quad (5)$$

$$-p_2(a_2,y)\frac{\partial u_2}{\partial x}(a_2,y,t) + \alpha_2(y)u_2(a_2,y,t) =$$

$$
\begin{cases}
\int_{c_1}^{d_1} \beta_2(y,y')u_1(b_1,y',t)\,dy' + \int_{a_1}^{b_1} \check\beta_2(y,x')u_1(x',c_1,t)\,dx', & y \in (c_2,c_1), \quad (6) \\
\int_{c_1}^{d_1} \beta_2(y,y')u_1(b_1,y',t)\,dy', & y \in (c_1,d_1), \\
\int_{c_1}^{d_1} \beta_2(y,y')u_1(b_1,y',t)\,dy' + \int_{a_1}^{b_1} \hat\beta_2(y,x')u_1(x',d_1,t)\,dx', & y \in (d_1,d_2),
\end{cases}
$$

$$-q_1(x,c_1)\frac{\partial u_1}{\partial y}(x,c_1,t) + \check\alpha_1(x)u_1(x,c_1,t) = \int_{c_2}^{c_1} \check\beta_1(x,y')u_2(a_2,y',t)\,dy', \quad (7)$$

$$q_1(x,d_1)\frac{\partial u_1}{\partial y}(x,d_1,t) + \hat\alpha_1(x)u_1(x,d_1,t) = \int_{d_1}^{d_2} \hat\beta_1(x,y')u_2(a_2,y',t)\,dy'. \quad (8)$$

Throughout the paper we assume that the input data satisfy the usual regularity and ellipticity conditions

$$p_i(x,y),\ q_i(x,y),\ r_i(x,y) \in L_\infty(\Omega_i), \quad i = 1,2, \quad (9)$$

$$0 < p_{i0} \le p_i(x,y), \quad 0 < q_{i0} \le q_i(x,y), \quad 0 \le r_i(x,y), \quad \text{a.e in } \Omega_i, \quad i = 1,2 \quad (10)$$

and

$$
\begin{aligned}
&\alpha_i \in L_\infty(c_i,d_i), \quad i = 1,2, \qquad \check\alpha_1,\ \hat\alpha_1 \in L_\infty(a_1,b_1), \\
&\beta_i \in L_\infty((c_i,d_i) \times (c_{3-i},d_{3-i})), \quad i = 1,2, \\
&\check\beta_1 \in L_\infty((a_1,b_1) \times (c_2,c_1)), \quad \hat\beta_1 \in L_\infty((a_1,b_1) \times (d_1,d_2)), \\
&\check\beta_2 \in L_\infty((c_2,c_1) \times (a_1,b_1)), \quad \hat\beta_2 \in L_\infty((d_1,d_2) \times (a_1,b_1)).
\end{aligned}
\quad (11)
$$

In real physical problems (see [1]) we also often have

$$\alpha_i > 0, \quad \beta_i > 0, \quad \hat\alpha_1 > 0, \quad \check\alpha_1 > 0, \quad \hat\beta_i > 0, \quad \check\beta_i > 0, \quad i = 1,2.$$

3   Existence and Uniqueness of Weak Solutions

We introduce the product space

$$L = L_2(\Omega_1) \times L_2(\Omega_2) = \{v = (v_1, v_2) | v_i \in L_2(\Omega_i)\},$$

endowed with the inner product and associated norm

$$(u, v)_L = (u_1, v_1)_{L_2(\Omega_1)} + (u_2, v_2)_{L_2(\Omega_2)}, \quad \|v\|_L = (v, v)_L^{1/2},$$

where

$$(u_i, v_i)_{L_2(\Omega_i)} = \iint_{\Omega_i} u_i v_i \, dx dy, \quad i = 1, 2.$$

We also define the spaces

$$H^k = \{v = (v_1, v_2) | \ v_i \in H^k(\Omega_i)\}, \quad k = 1, 2, \ldots$$

endowed with the inner product and associated norm

$$(u, v)_{H^k} = (u_1, v_1)_{H^k(\Omega_1)} + (u_2, v_2)_{H^k(\Omega_2)}, \quad \|v\|_{H^k} = (v, v)_{H^k}^{1/2},$$

where $H^k(\Omega_i)$ are the standard Sobolev spaces. In particular, we set

$$H_0^1 = \left\{ v \in H^1 \, | \, v_1(a_1, y) = 0, \ v_2(b_2, y) = v_2(x, c_2) = v_2(x, d_2) = 0 \right\}.$$

Finally, with $u = (u_1, u_2)$ and $v = (v_1, v_2)$ we define the following bilinear form:

$$
\begin{aligned}
A(u, v) =& \iint_{\Omega_1} \left(p_1 \frac{\partial u_1}{\partial x} \frac{\partial v_1}{\partial x} + q_1 \frac{\partial u_1}{\partial y} \frac{\partial v_1}{\partial y} + r_1 u_1 v_1 \right) dx dy \\
&+ \iint_{\Omega_2} \left(p_2 \frac{\partial u_2}{\partial x} \frac{\partial v_2}{\partial x} + q_2 \frac{\partial u_2}{\partial y} \frac{\partial v_2}{\partial y} + r_2 u_2 v_2 \right) dx dy \\
&+ \int_{c_1}^{d_1} \alpha_1(y) \, u_1(b_1, y) v_1(b_1, y) \, dy + \int_{c_2}^{d_2} \alpha_2(y) \, u_2(a_2, y) v_2(a_2, y) \, dy \\
&+ \int_{a_1}^{b_1} \check{\alpha}_1(x) \, u_1(x, c_1) v_1(x, c_1) \, dx + \int_{a_1}^{b_1} \hat{\alpha}_1(x) \, u_1(x, d_1) v_1(x, d_1) \, dx \\
&- \iint_{c_1 c_2}^{d_1 d_2} \beta_1(y, y') u_2(a_2, y') v_1(b_1, y) \, dy' dy - \iint_{c_2 c_1}^{d_2 d_1} \beta_2(y, y') u_1(b_1, y') v_2(a_2, y) \, dy' dy \\
&- \iint_{a_1 c_2}^{b_1 c_1} \check{\beta}_1(x, y') u_2(a_2, y') v_1(x, c_1) \, dy' dx - \iint_{a_1 d_1}^{b_1 d_2} \hat{\beta}_1(x, y') u_2(a_2, y') v_1(x, d_1) \, dy' dx \\
&- \iint_{c_2 a_1}^{c_1 b_1} \check{\beta}_2(y, x') u_1(x', c_1) v_2(a_2, y) \, dx' dy - \iint_{d_1 a_1}^{d_2 b_1} \hat{\beta}_2(y, x') u_1(x', d_1) v_2(a_2, y) \, dx' dy
\end{aligned}
$$

$$(12)$$

Lemma 1. *Under the conditions (9) and (11) the bilinear form A, defined by (12), is bounded on $H^1 \times H^1$. If besides it the conditions (10) are fulfilled, this form satisfies the Gårding's inequality on H_0^1, i.e. there exist positive constants m and κ such that*

$$A(u,u) + \kappa \|u\|_L^2 \geq m\|u\|_{H^1}^2, \quad \forall u \in H_0^1.$$

Proof (comp. [5]). Boundedness of A follows from (9) and (11) and the trace theorem

$$\|u_i\|_{L_2(\partial\Omega_i)} \leq C\|u_i\|_{H^1(\Omega_i)}.$$

From (10) and Poincaré type inequalities we immediately obtain

$$\sum_{i=1}^{2} \iint_{\Omega_i} \left[p_i \left(\frac{\partial u_i}{\partial x}\right)^2 + q_i \left(\frac{\partial u_i}{\partial y}\right)^2 + r_i u_i^2 \right] dxdy \geq c_0 \|u\|_{H^1}^2,$$

where c_0 is a computable constant depending on p_{i0}, q_{i0}, $b_i - a_i$ and $d_i - c_i$. Further, using Cauchy-Schwartz and ϵ-inequalities we obtain,

$$\left| \int_{c_1}^{d_1} \alpha_1(y)\, u_1^2(b_1,y)\, dy - \int_{c_1}^{d_1}\int_{c_2}^{d_2} \beta_1(y,y')u_2(a_2,y')u_1(b_1,y)\, dy'dy \right.$$
$$+ \int_{c_2}^{d_2} \alpha_2(y)\, u_2^2(a_2,y)\, dy - \int_{c_2}^{d_2}\int_{c_1}^{d_1} \beta_2(y,y')u_1(b_1,y')u_2(a_2,y)\, dy'dy$$
$$+ \int_{a_1}^{b_1} \check{\alpha}_1(x)\, u_1^2(x,c_1)\, dx - \int_{a_1}^{b_1}\int_{c_2}^{c_1} \check{\beta}_1(x,y')u_2(a_2,y')u_1(x,c_1)\, dy'dx$$
$$+ \int_{a_1}^{b_1} \hat{\alpha}_1(x)\, u_1^2(x,d_1)\, dx - \int_{a_1}^{b_1}\int_{d_1}^{d_2} \hat{\beta}_1(x,y')u_2(a_2,y')u_1(x,d_1)\, dy'dx$$
$$- \int_{c_2}^{c_1}\int_{a_1}^{b_1} \check{\beta}_2(y,x')u_1(x',c_1)u_2(a_2,y)\, dx'dy$$
$$\left. - \int_{d_1}^{d_2}\int_{a_1}^{b_1} \hat{\beta}_2(y,x')u_1(x',d_1)u_2(a_2,y)\, dx'dy \right| \leq \epsilon\|u\|_{H^1}^2 + \frac{C}{\epsilon}\|u\|_L^2.$$

Taking $0 < \epsilon < c_0$ we obtain Gårding's inequality:

$$A(u,u) \geq (c_0 - \epsilon)\|u\|_{H^1}^2 - \frac{C}{\epsilon}\|u\|_L^2,$$

where $m = c_0 - \epsilon$ and $\kappa = \frac{C}{\epsilon}$.   □

Let Ω be a domain in $\mathbb{R}^n$ and $u(t)$ a function mapping Ω into Hilbert space H. In the usual manner [8] we set $L_2(\Omega, H) = H^0(\Omega, H)$. Further, we define $H^{1,1/2} = L_2((0,T), H^1) \cap H^{1/2}((0,T), L)$. Let $H^{-1} = (H_0^1)^*$ be the dual space for H_0^1. The spaces H_0^1, L and H^{-1} form a Gelfand triple $H_0^1 \subset L \subset H^{-1}$ (see [8]), with continuous and dense embeddings. We also introduce the space

$$W(0,T) = \left\{ u \,\middle|\, u \in L_2((0,T), H_0^1), \frac{\partial u}{\partial t} \in L_2((0,T), H^{-1}) \right\}$$

with inner product

$$(u, v)_{W(0,T)} = \int_0^T \left[(u(\cdot, t), v(\cdot, t))_{H^1} + \left(\frac{\partial u}{\partial t}(\cdot, t), \frac{\partial v}{\partial t}(\cdot, t) \right)_{H^{-1}} \right] dt$$

Multiplying equation (1) by $v_1(x, y)$ and equation (2) by $v_2(x, y)$, and integrating by parts using condition (4)-(8), we get the weak form of (1)-(8):

$$\left(\frac{\partial u}{\partial t}(\cdot, t), v \right)_L + A(u(\cdot, t), v) = (f(\cdot, t), v)_L, \quad \forall v \in H_0^1. \tag{13}$$

Applying Theorem 26.1 from [8] to (13) we obtain the following assertion.

Theorem 1. *Let the assumptions (9), (10) and (11) hold and suppose that $u_0 = (u_{10}, u_{20}) \in L$, $f = (f_1, f_2) \in L_2((0, T), H^{-1})$. Then for $0 < T < +\infty$ the initial-boundary-value problem (1)-(8) has a unique weak solution $u \in W(0, T)$, and this depends continuously on f and u_0.*

4 A Priori Estimate

Because the norm $\| \cdot \|_{H^{-1}}$ is not computable we restrict our investigations to problem (1)-(8) with right hand sides of the form

$$f_i(x, y, t) = f_{i0}(x, y, t) + \frac{\partial(\varrho_i(x) f_{i1}(x, y, t))}{\partial x} + \frac{\partial(\theta_i(y) f_{i2}(x, y, t))}{\partial y}$$

$$+ \int_0^T \frac{f_{i3}(x, y, t, t') - f_{i3}(x, y, t', t)}{|t - t'|} dt', \quad i = 1, 2, \tag{14}$$

where $f_{i0}, f_{i1}, f_{i2} \in L_2((0, T), L_2(\Omega_i)) = L_2(Q_i)$, $Q_i = \Omega_i \times (0, T)$, $f_{i3} \in L_2((0, T)^2, L_2(\Omega_i)) = L_2(R_i)$, $R_i = \Omega_i \times (0, T)^2$, $\varrho_i \in C([a_i, b_i])$, $\theta_1 \in C([c_1, d_1])$, $\theta_2(y) = 1$ and

$$c_1(b_1 - x) \leq \varrho_1 \leq C_1(b_1 - x), \quad x \in (a_1, b_1), \quad C_1 \geq c_1 > 0,$$

$$c_2(x - a_2) \leq \varrho_2 \leq C_2(x - a_2), \quad x \in (a_2, b_2), \quad C_2 \geq c_2 > 0, \tag{15}$$

$$s_1(d_1 - y)(y - c_1) \leq \theta_1 \leq S_1(d_1 - y)(y - c_1), \quad y \in (c_1, d_1), \quad S_1 \geq s_1 > 0.$$

Theorem 2. *Let the assumptions (9)-(11) and (15) hold and let $u_{i0} \in L_2(\Omega_i))$, $f_{i0}, f_{i1}, f_{i2} \in L_2(Q_i)$, $f_{i3} \in L_2(R_i)$, $i = 1, 2$. Then the initial-boundary-value problem (1)-(8),(14) has a unique weak solution $u = (u_1, u_2) \in H^{1,1/2}$ and the a priori estimate*

$$\|u\|_{H^{1,1/2}}^2 \leq C \sum_{i=1}^2 \left(\|u_{i0}\|_{L_2(\Omega_i)}^2 + \|f_{i0}\|_{L_2(Q_i)}^2 + \|f_{i1}\|_{L_2(Q_i)}^2 \right.$$

$$\left. + \|f_{i2}\|_{L_2(Q_i)}^2 + \|f_{i3}\|_{L_2(R_i)}^2 \right) \tag{16}$$

holds.

The proof is analogous to the proof of Theorem 1 in [5].

5   Finite Difference Approximation

Let $\overline{\omega}_{i,h_i}$ be a uniform mesh in $[a_i, b_i]$, with step size $h_i = (b_i - a_i)/n_i$, $i = 1, 2$. We denote $\omega_{i,h_i} := \overline{\omega}_{i,h_i} \cap (a_i, b_i)$, $\omega_{i,h_i}^- := \omega_{i,h_i} \cup \{a_i\}$, $\omega_{i,h_i}^+ := \omega_{i,h_i} \cup \{b_i\}$. Analogously we define a uniform mesh $\overline{\omega}_{i,k_i}$ in $[c_i, d_i]$, with the step size $k_i = (d_i - c_i)/m_i$, $i = 1, 2$ and its submeshes $\omega_{i,k_i} := \overline{\omega}_{i,k_i} \cap (c_i, d_i)$, $\omega_{i,k_i}^- := \omega_{i,k_i} \cup \{c_i\}$, $\omega_{i,k_i}^+ := \omega_{i,k_i} \cup \{d_i\}$. We assume that $h_1 \asymp h_2 \asymp k_1 \asymp k_2$. Finally, we introduce a uniform mesh $\overline{\omega}_\tau$ in $[0, T]$ with step size $\tau = T/n$ and set $\omega_\tau := \overline{\omega}_\tau \cap (0, T)$, $\omega_\tau^- := \omega_\tau \cup \{0\}$, $\omega_\tau^+ := \omega_\tau \cup \{T\}$. We will consider vector-functions of the form $v = (v_1, v_2)$ where v_i is a mesh function defined on $\overline{\omega}_{i,h_i} \times \overline{\omega}_{i,k_i} \times \overline{\omega}_\tau$, $i = 1, 2$. We define difference quotients in the usual way (see [2],[6],[7]):

$$v_{i,x} = \frac{v_i(x + h_i, y, t) - v_i(x, y, t)}{h_i} = v_{i,\overline{x}}(x + h_i, y, t),$$

$$v_{i,y} = \frac{v_i(x, y + k_i, t) - v_i(x, y, t)}{k_i} = v_{i,\overline{y}}(x, y + k_i, t),$$

$$v_{i,t} = \frac{v_i(x, y, t + \tau) - v_i(x, y, t)}{\tau} = v_{i,\overline{t}}(x, y, t + \tau).$$

We approximate the initial-boundary-value problem (1)-(8) with the following explicit finite difference scheme:

$$v_{1,t} - (\overline{p}_1 v_{1,\overline{x}})_x - (\overline{q}_1 v_{1,\overline{y}})_y + \overline{r}_1 v_1 = \overline{f}_1, \quad x \in \omega_{1,h_1}, \ y \in \omega_{1,k_1}, \ t \in \omega_\tau^- \tag{17}$$

$$v_{1,t}(b_1, y, t) + 2/h_1 \big[(\overline{p}_1(b_1, y)v_{1,\overline{x}}(b_1, y, t)) + \alpha_1(y)v_1(b_1, y, t)$$

$$-k_2 \sum_{y' \in \omega_{2,k_2}} \beta_1(y, y')v_2(a_2, y', t)\big] - (\overline{q}_1 v_{1,\overline{y}})_y(b_1, y, t) \tag{18}$$

$$+\overline{r}_1(b_1, y)v_1(b_1, y, t) = \overline{f}_1(b_1, y, t), \quad y \in \omega_{1,k_1}, \ t \in \omega_\tau^-$$

$$v_{1,t}(x, c_1, t) - 2/k_1 \big[(\overline{q}_1(x, c_1 + k_1)v_{1,y}(x, c_1, t)) - \check{\alpha}_1(x)v_1(x, c_1, t)$$

$$-k_2 \sum_{y' \in \omega_{2,k_2}} \check{\beta}_1(x, y')v_2(a_2, y', t)\big] - (\overline{p}_1 v_{1,\overline{x}})_x(x, c_1, t) \tag{19}$$

$$+\overline{r}_1(x, c_1)v_1(x, c_1, t) = \overline{f}_1(x, c_1, t), \quad x \in \omega_{1,h_1}, \ t \in \omega_\tau^-$$

$$v_{1,t}(x, d_1, t) + 2/k_1 \big[(\overline{q}_1(x, d_1)v_{1,\overline{y}}(x, d_1, t)) + \hat{\alpha}_1(x)v_1(x, d_1, t)$$

$$-k_2 \sum_{y' \in \omega_{2,k_2}} \hat{\beta}_1(x, y')v_2(a_2, y', t)\big] - (\overline{p}_1 v_{1,\overline{x}})_x(x, d_1, t) \tag{20}$$

$$+\overline{r}_1(x, d_1)v_1(x, d_1, t) = \overline{f}_1(x, d_1, t), \quad x \in \omega_{1,h_1}, \ t \in \omega_\tau^-$$

$$v_{2,t} - (\overline{p}_2 v_{2,\overline{x}})_x - (\overline{q}_2 v_{2,\overline{y}})_y + \overline{r}_2 v_2 = \overline{f}_2, \quad x \in \omega_{2,h_2}, \ y \in \omega_{2,k_2}, \ t \in \omega_\tau^- \tag{21}$$

$$v_{2,t}(a_2, y, t) - 2/h_2 \big[(\overline{p}_2(a_2 + h_2, y)v_{2,x}(a_2, y, t)) + \alpha_2(y)v_2(a_2, y, t)$$

$$+k_1 \sum_{y' \in \omega_{1,k_1}} \beta_2(y, y')v_1(b_1, y', t) + h_1 \sum_{x' \in \omega_{1,h_1}} \check{\beta}_2(y, x')v_1(x', c_1, t)\big] \tag{22}$$

$$-(\overline{q}_2 v_{2,\overline{y}})_y(a_2, y, t) + \overline{r}_2(a_2, y)v_2(a_2, y, t) = \overline{f}_2(a_2, y, t),$$

$$y \in (c_2, c_1), \ t \in \omega_\tau^-$$

$$v_{2,t}(a_2, y, t) - 2/h_2 \big[(\overline{p}_2(a_2 + h_2, y) v_{2,x}(a_2, y, t)) - \alpha_2(y) v_2(a_2, y, t)$$

$$+ k_1 \sum_{y' \in \omega_{1,k_1}} \beta_2(y, y') v_1(b_1, y', t) \big] - (\overline{q}_2 v_{2,\overline{y}})_y(a_2, y, t) \tag{23}$$

$$+ \overline{r}_2(a_2, y) v_2(a_2, y, t) = \overline{f}_2(a_2, y, t), \quad y \in (c_1, d_1), \ t \in \omega_\tau^-$$

$$v_{2,t}(a_2, y, t) - 2/h_2 \big[(\overline{p}_2(a_2 + h_2, y) v_{2,x}(a_2, y, t)) - \alpha_2(y) v_2(a_2, y, t)$$

$$+ k_1 \sum_{y' \in \omega_{1,k_1}} \beta_2(y, y') v_1(b_1, y', t) + h_1 \sum_{x' \in \omega_{1,h_1}} \hat{\beta}_2(y, x') v_1(x', d_1, t) \big] \tag{24}$$

$$- (\overline{q}_2 v_{2,\overline{y}})_y(a_2, y, t) + \overline{r}_2(a_2, y) v_2(a_2, y, t) = \overline{f}_2(a_2, y, t),$$

$$y \in (d_1, d_2), \ t \in \omega_\tau^-$$

$$v_1(a_1, y, t) = 0, \quad y \in \omega_{1,k_1}, \ t \in \overline{\omega}_\tau,$$

$$v_2(x, c_2, t) = 0, \quad x \in \omega_{2,h_2}, \ t \in \overline{\omega}_\tau$$

$$v_2(b_2, y, t) = 0, \quad y \in \omega_{2,k_2}, \ t \in \overline{\omega}_\tau, \tag{25}$$

$$v_2(x, d_2, t) = 0, \quad x \in \omega_{2,h_2}, \ t \in \overline{\omega}_\tau,$$

$$v_i(x, y, 0) = u_{i0}(x, y), \quad x \in \omega_{i,h_i}^\pm, \ y \in \omega_{i,k_i}, \quad i = 1, 2, \tag{26}$$

where denoted

$$\overline{p}_i(x, y) = \frac{1}{2}[p_i(x, y) + p_i(x - h_i, y)], \quad x \in \omega_{i,h_i}^+, \ y \in \omega_{i,k_i}, \quad i = 1, 2,$$

$$\overline{q}_i(x, y) = \frac{1}{2}[q_i(x, y) + q_i(x, y - k_i)], \quad x \in \omega_{i,h_i}^\pm, \ y \in \omega_{i,k_i}^+, \quad i = 1, 2,$$

$$\overline{r}_i(x, y) = r_i(x, y) \quad \text{and} \quad \overline{f}_i(x, y) = f_i(x, y), \quad i = 1, 2.$$

Finite difference scheme (17)-(26) is computationally efficient. It follows from the general theory of difference schemes [7], that the finite difference scheme (17)-(26) is stable under condition

$$\tau \leq c \ \min\{h_1^2, h_2^2, k_1^2, k_2^2\} \tag{27}$$

where c is computable constants depending on $\max p_i$ and $\max q_i$, $i = 1, 2$.

6   Numerical Examples

The test example is problem (1)-(8), with $a_1 = 1$, $b_1 = 2$, $c_1 = 0.2$, $d_1 = 0.8$, $a_2 = 3$, $b_2 = 4.5$, $c_2 = 0$, $d_2 = 1$, $t \in [0, 1]$. The coefficients are:

$$p_1(x, y) = e^{x+y}, \quad q_1(x, y) = \sin(x + y), \quad r_1(x, y) = x + y,$$

$$p_2(x, y) = x^2 + y^2, \quad q_2(x, y) = x(1 + y), \quad r_2(x, y) = x - y,$$

$$\alpha_1(y) = \alpha_2(y) = \check{\alpha}_1(x) = \hat{\alpha}_1(x) = \beta_1(y, y') = \beta_2(y, y') = 1,$$

$$\check{\beta}_2(y, x') = \hat{\beta}_2(y, x') = 0,$$

$$\check{\beta}_1(x, y') = 0.16 - y', \quad \hat{\beta}_1(x, y') = y' - 0.83.$$

In the right hand sides of equations (1),(2) we determine functions f_1 and f_2 in such a manner that $u = (u_1, u_2)$,

$$u_1(x, y, t) = 2 \left[\cos(10\pi y - \pi) + 1\right](x - a_1)^2(y - c_1)(d_1 - y)e^{-t},$$
$$u_2(x, y, t) = 2 \left[\cos(4\pi y - \pi) + 1\right](x - b_2)^2(y - c_2)(d_2 - y)e^{-t}$$

is the exact solution of the problem (1)-(8). Mesh parameters are $h_1 = h_2 = h$, $k_1 = k_2 = k$. The results are given in discrete max norm and the convergence rate (CR) is calculated using double mesh principle (see [3]):

$$E^h = \|u_h - u\|, \quad CR = \log_2[E^h/E^{h/2}].$$

Table 1. Error and convergence rate in max discrete norm

Mesh	Ω_1 Error (CR)	Ω_2 Error (CR)
h=0.5, k=0.3, τ=0.0002	0.0332	0.1932
h=0.25, k=0.15, τ=0.00005	0.0076 (2.1271)	0.0443 (2.1247)
h=0.125, k=0.075, τ=0.00000125	0.0016 (2.2479)	0.0103 (2.1047)

Acknowledgement. The research of author was supported by Ministry of Education and Science of Republic of Serbia under project 174015.

References

1. Amosov, A.A.: Global solvability of a nonlinear nonstationary problem with a nonlocal boundary condition of radiation heat transfer type. Differential Equations 41(1), 96–109 (2005)
2. Jovanović, B.S.: Finite difference method for boundary value problems with weak solutions. Posebna izdanja Mat. Instituta 16, Belgrade (1993)
3. Jovanović, B.S., Koleva, M.N., Vulkov, L.G.: Convergence of a FEM and two-grid algorithms for elliptic problems on disjoint domains. J. Comput. Appl. Math. 236, 364–374 (2011)
4. Jovanović, B.S., Vulkov, L.G.: Finite difference approximation of strong solutions of a parabolic interface problem on disconected domains. Publ. Inst. Math. 84(98), 37–48 (2008)
5. Jovanović, B.S., Vulkov, L.G.: Numerical solution of a two-dimensional parabolic transmission problem. Int. J. Numer. Anal. Model. 7(1), 156–172 (2010)
6. Süli, E.: Finite element methods for partial differential equations. University of Oxford (2007)
7. Samarskii, A.A.: The theory of difference schemes. Marcel Dekker (2001)
8. Wloka, J.: Partial differential equations. Cambridge University Press (1987)

An Efficient Hybrid Numerical Scheme for Singularly Perturbed Problems of Mixed Parabolic-Elliptic Type

Kaushik Mukherjee[1] and Srinivasan Natesan[2]

[1] Department of Mathematics, Indian Institute of Space Science and Technology, Thiruvananthapuram - 695547, India
kaushik@iist.ac.in, mathkaushik@gmail.com
[2] Department of Mathematics, Indian Institute of Technology Guwahati, Guwahati - 781039, India
natesan@iitg.ernet.in

Abstract. This article is dealt with the study of a hybrid numerical scheme for a class of singularly perturbed mixed parabolic-elliptic problems possessing both boundary and interior layers. The domain under consideration is partitioned into two subdomains. In the first subdomain, the given problem takes the form of parabolic reaction-diffusion type, whereas in the second subdomain elliptic convection-diffusion-reaction types of problems are posed. To solve these problems, the time derivative is discretized by the backward-Euler method, while for the spatial discretization the classical central difference scheme is used on the first subdomain and a hybrid finite difference scheme is proposed on the second subdomain. The proposed method is designed on a layer resolving piecewise-uniform Shishkin mesh and computationally it is shown that the method converges ε-uniformly with almost second-order spatial accuracy in the discrete supremum norm.

1 Introduction

Let us denote the domains for describing the model problem by

$$\Omega = (0,1), \ G^- = (0,\xi) \times (0,T], \ G^+ = (\xi,1) \times (0,T], \ G = \Omega \times (0,T].$$

We here consider the following class of singularly perturbed mixed parabolic-elliptic initial-boundary-value problems (IBVPs) posed on the domain $G^- \cup G^+$:

$$\begin{cases} L_{1,\varepsilon}u(x,t) \equiv \left(\dfrac{\partial u}{\partial t} - \varepsilon \dfrac{\partial^2 u}{\partial x^2} + b(x,t)u \right)(x,t) = f(x,t), \quad (x,t) \in G^-, \\[2mm] L_{2,\varepsilon}u(x,t) \equiv \left(-\varepsilon \dfrac{\partial^2 u}{\partial x^2} - a(x,t)\dfrac{\partial u}{\partial x} + b(x,t)u \right)(x,t) = f(x,t), \quad (x,t) \in G^+, \\[2mm] u(x,0) = s_0(x), \quad x \in \overline{\Omega} = [0,1], \\[2mm] u(0,t) = s_1(t), \quad u(1,t) = s_2(t), \quad t \in (0,T], \end{cases}$$

$$(1)$$

I. Dimov, I. Faragó, and L. Vulkov (Eds.): NAA 2012, LNCS 8236, pp. 411–419, 2013.
© Springer-Verlag Berlin Heidelberg 2013

where $0 < \varepsilon \ll 1$ is a small parameter, the convection coefficient a is sufficiently smooth on G^+, the source term f is sufficiently smooth on $G^- \cup G^+$ and the coefficient b is sufficiently smooth on $\overline{\Omega}$ such that

$$\begin{cases} b(x,t) \geq 0 \quad \text{on } \overline{\Omega}, \\ \alpha^* > a(x,t) > \alpha > 0, \, x > \xi, \end{cases} \tag{2}$$

and the solution $u(x,t)$ satisfies the following interface conditions

$$[u] = 0, \quad \left[\frac{\partial u}{\partial x}\right] = 0, \quad \text{at } x = \xi. \tag{3}$$

Here, we define the jump of u, denoted by $[u]$, across the point of discontinuity $x = \xi$ by $[u](\xi,t) = u(\xi^+,t) - u(\xi^-,t)$, where $u(\xi^\pm,t) = \lim_{x \to \xi \pm 0} u(x,t)$. Under sufficient smoothness and necessary compatibility conditions imposed on the data s_0, s_1 and s_2, the IBVP (1)-(3) admits a unique solution $u \in \mathbb{C}^{1+\lambda}(G) \cap \mathbb{C}^{2+\lambda}(G^- \cup G^+)$, which in general possesses a boundary layer at $x = 0$ and interior layers of different widths in the neighborhood of the point $x = \xi$ as discussed in [1]. These types of problems arise in the context of electromagnetic metal forming (see, e.g., [6]).

Singularly perturbed problems of mixed type have been studied mainly by Braianov for stationary case in [2] and for non-stationary case in [1]. In this article, we consider a class of singularly perturbed mixed parabolic-elliptic IBVPs whose solutions exhibit both boundary and interior layers. Here, our objective is to devise an efficient numerical scheme which is ε-uniformly convergent in the discrete supremum norm and also having almost second-order accuracy with respect to the spatial variable. In the first subdomain G^- we propose the classical central difference scheme for the spatial discretization of the problem of parabolic reaction-diffusion type, whereas in the second subdomain G^+ we propose a hybrid finite difference scheme for the spatial discretization of the problem of elliptic convection-diffusion-reaction type. The time derivative in the given problem is discretized by the classical backward-Euler method. To accomplish this purpose, we discretize the domian using a spacial rectangular mesh that consists a piecewise-uniform Shishkin mesh condensed closely to the boundary and interior layers in the spatial direction and a uniform mesh in the time direction. Note that in the recent paper [4], we have thoroughly studied a similar hybrid scheme for a class of singularly perturbed parabolic convection-diffusion IBVPs with discontinuous convection coefficient exhibiting strong interior layers.

The rest of the article is organized as follows: In Section 2, we describe the piecewise-uniform Shishkin mesh and also provide the detailed construction of the proposed numerical scheme. Finally, this section is ended up with a brief discussion on the stability criteria of the proposed scheme. In Section 3 we present the numerical results for a test example having known exact solution. Finally, we summarize the main conclusions in Section 4.

2 Numerical Approximation

In this section, we first give a description of the suitable mesh used for the discretization of the domain and then explicitly describe the difference scheme for the discretization of the problem (1)-(3). Finally, we briefly discuss the ε-uniform stability of the proposed scheme.

2.1 Discretization of the Domain

Consider the domain $\overline{G} = \overline{\Omega} \times [0,T] = [0,1] \times [0,T]$ and let $N \geq 16$ be a positive even integer. Here, we construct a rectangular mesh $\overline{G}_\varepsilon^{N,M} = \overline{\Omega}_x^{N,\varepsilon} \times \mathbb{S}_t^M$, which is a combination of the piecewise-uniform Shishkin mesh condensed near to the boundary and interior layers for the spatial variable and a uniform mesh for the temporal variable. Firstly, to define the piecewise-uniform Shishkin mesh we divide the spatial domain $\overline{\Omega}$ into five subintervals as

$$\overline{\Omega} = [0,\sigma_1] \cup [\sigma_1, \xi - \sigma_1] \cup [\xi - \sigma_1, \xi] \cup [\xi, \xi + \sigma_2] \cup [\xi + \sigma_2, 1],$$

for some σ_1, σ_2 that satisfy $0 < \sigma_1 \leq \xi/4$, $0 < \sigma_2 \leq (1-\xi)/2$. On the subintervals $[0,\sigma_1], [\xi - \sigma_1, \xi]$ a uniform mesh with $N/8$ mesh-intervals is placed and a uniform mesh with $N/4$ mesh-intervals is placed on the subintervals $[\sigma_1, \xi - \sigma_1], [\xi, \xi + \sigma_2], [\xi + \sigma_2, 1]$ such that

$$\Omega_x^{N,\varepsilon} = \{x_i : 1 \leq i \leq N/2 - 1\} \bigcup \{x_i : N/2 + 1 \leq i \leq N - 1\}$$

denotes the set of interior points of the mesh. Clearly $x_{N/2} = \xi$ and $\overline{\Omega}_x^{N,\varepsilon} = \{x_i\}_0^N$. Note that this mesh is uniform when $\sigma_1 = \xi/4$, $\sigma_2 = (1-\xi)/2$. It is fitted to the problem (1)-(3) by choosing σ_1 and σ_2 to be the following functions of N and ε

$$\sigma_1 = \min\left\{\frac{\xi}{4}, \tau_1 \sqrt{\varepsilon} \ln N\right\}, \quad \sigma_2 = \min\left\{\frac{1-\xi}{2}, \frac{2\tau_2\varepsilon}{\alpha} \ln N\right\},$$

where τ_1, τ_2 are constants. On the time domain $[0,T]$, we introduce the equidistant meshes in the temporal variable such that

$$\mathbb{S}_t^M = \{t_n = n\,\Delta t, \, n = 0, \ldots, M, \, \Delta t = T/M\},$$

where M denotes the number of mesh elements in the t-direction. Let us denote the step sizes in space by

$$h_i = x_i - x_{i-1}, \, i = 1, \ldots, N, \qquad \widehat{h}_i = h_i + h_{i+1}, \, i = 1, \ldots, N - 1.$$

Further, denote the mesh width h_i in the spatial direction as follows:

$$h_i = \begin{cases} h_{(l)} = 8\sigma_1/N, & \text{for } i = 1, \ldots, N/8, 3N/8 + 1, \ldots, N/2, \\ h_{(r)} = 4\sigma_2/N, & \text{for } i = N/2 + 1, \ldots, 3N/4. \end{cases}$$

2.2   The Backward-Euler Hybrid Finite Difference Scheme

On the domain G^-, the problem (1)-(3) is discretized in the spatial variable by the classical central difference scheme, whereas on the domain G^+ a hybrid scheme is proposed for the spatial discretization. The hybrid scheme consists of the midpoint upwind scheme in the outer region $[\xi + \sigma_2, 1]$ and the classical central difference scheme in the interior layer region $(\xi, \xi + \sigma_2)$, while at the point of discontinuity, second-order one-sided difference approximations are used to keep the continuity of the spatial derivative. On the other hand, we employ the backward-Euler method for discretizing the time derivative. Before describing the scheme, for a given mesh function $v(x_i, t_n) = v_i^n$, define the forward, backward and central difference operators D_x^+, D_x^- and D_x^0 in space and the backward difference operator D_t^- in time by

$$
\begin{cases}
D_x^+ v_i^n = \dfrac{v_{i+1}^n - v_i^n}{h_{i+1}}, \quad D_x^- v_i^n = \dfrac{v_i^n - v_{i-1}^n}{h_i}, \quad D_x^0 v_i^n = \dfrac{v_{i+1}^n - v_{i-1}^n}{\widehat{h}_i} \\[2mm]
\text{and} \quad D_t^- v_i^n = \dfrac{v_i^n - v_i^{n-1}}{\Delta t},
\end{cases} \tag{4}
$$

respectively, and we define the second-order finite difference operator δ_x^2 in space by

$$
\delta_x^2 v_i^n = \frac{2(D_x^+ v_i^n - D_x^- v_i^n)}{\widehat{h}_i}.
$$

Also, define $v_{i+1/2}^n = (v_{i+1}^n + v_i^n)/2$. Then the proposed numerical scheme takes the following form on the mesh $\overline{G}_\varepsilon^{N,M}$:

$$
\begin{cases}
U_i^0 = s_0(x_i), \quad \text{for } i = 0, \ldots, N, \\[1mm]
\begin{cases}
L_{1,cen}^{N,M} U_i^{n+1} = f_i^{n+1}, & \text{for } i = 1, \ldots, N/2 - 1, \\[1mm]
L_{2,cen}^{N,M} U_i^{n+1} = f_i^{n+1}, & \text{for } i = N/2 + 1, \ldots, 3N/4 - 1, \\[1mm]
L_{2,mu}^{N,M} U_i^{n+1} = f_{i+1/2}^{n+1}, & \text{for } i = 3N/4, \ldots, N - 1, \\[1mm]
D_x^F U_i^{n+1} - D_x^B U_i^{n+1} = 0, & \text{for } i = N/2, \\[1mm]
U_0^{n+1} = s_1(t_{n+1}), \quad U_N^{n+1} = s_2(t_{n+1}),
\end{cases} \\[1mm]
\text{for } n = 0, \ldots, M - 1,
\end{cases} \tag{5}
$$

where

$$
\begin{cases}
L_{1,cen}^{N,M} U_i^{n+1} = D_t^- U_i^{n+1} - \varepsilon \delta_x^2 U_i^{n+1} + b_i^{n+1} U_i^{n+1}, \\[2mm]
L_{2,cen}^{N,M} U_i^{n+1} = -\varepsilon \delta_x^2 U_i^{n+1} - a_i^{n+1} D_x^0 U_i^{n+1} + b_i^{n+1} U_i^{n+1}, \\[2mm]
L_{2,mu}^{N,M} U_i^{n+1} = -\varepsilon \delta_x^2 U_i^{n+1} - a_{i+1/2}^{n+1} D_x^+ U_i^{n+1} + b_{i+1/2}^{n+1} U_{i+1/2}^{n+1},
\end{cases} \tag{6}
$$

and

$$\begin{cases} D_x^F U_{N/2}^n = (-U_{N/2+2}^n + 4U_{N/2+1}^n - 3U_{N/2}^n)/2h_{(r)}, \\ D_x^B U_{N/2}^n = (U_{N/2-2}^n - 4U_{N/2-1}^n + 3U_{N/2}^n)/2h_{(l)}. \end{cases} \tag{7}$$

2.3 Stability Analysis

After doing a short calculation the difference scheme (5)-(7) can be transformed to a tridiagonal system of equations, which will be solved by a suitable solver to obtain the numerical solution at the $(n+1)$th level. Subsequently we consider the following form of the scheme (5)-(7) on the mesh $\overline{G}_\varepsilon^{N,M}$:

$$\begin{cases} U_i^0 = s_0(x_i), \quad \text{for } i = 0, \ldots, N, \\ \begin{cases} L_{hyb}^{N,M} U_i^{n+1} = \widetilde{f_{hyb,i}}^{n+1}, \quad \text{for } i = 1, \ldots, N-1, \\ U_0^{n+1} = s_1(t_{n+1}), \quad U_N^{n+1} = s_2(t_{n+1}), \end{cases} \\ \text{for } n = 0, \ldots, M-1. \end{cases} \tag{8}$$

Here, the difference operator $L_{hyb}^{N,M}$ and the right hand side term $\widetilde{f_{hyb}}^{n+1}$ are respectively defined as

$$\begin{cases} L_{hyb}^{N,M} U_i^{n+1} = [r_i^- U_{i-1}^{n+1} + r_i^0 U_i^{n+1} + r_i^+ U_{i+1}^{n+1}] + [p_i^- U_{i-1}^n + p_i^0 U_i^n + p_i^+ U_{i+1}^n], \quad \text{and} \\ \widetilde{f_{hyb,i}}^{n+1} = [m_i^- f_{i-1}^{n+1} + m_i^0 f_i^{n+1} + m_i^+ f_{i+1}^{n+1}], \end{cases} \tag{9}$$

for $i = 1, \ldots, N-1$, where the coefficients $r_i^-, r_i^0, r_i^+; p_i^-, p_i^0, p_i^+; m_i^-, m_i^0, m_i^+$ can be derived from the difference scheme (5)-(7).

Then using [Lemma 3.12, Part II] of the book of Roos et al. [5], we can show that the operator $L_{hyb}^{N,M}$ satisfies the following discrete minimum principle.

Let $G_\varepsilon^{N,M} = \overline{G}_\varepsilon^{N,M} \cap G$ and $\Gamma_\varepsilon^{N,M} = \overline{G}_\varepsilon^{N,M} \backslash G_\varepsilon^{N,M}$.

Lemma 1. *Assume that $N \geq N_0$, where*

$$\frac{N_0}{\ln N_0} \geq 4\tau_2 \frac{\alpha^*}{\alpha}, \tag{10}$$

$$\frac{\alpha N_0}{2} \geq \|b\|_\infty \quad \text{and} \quad (\|b\|_\infty + \Delta t^{-1}) \leq \frac{2\Bbbk N_0^2}{\ln^2 N_0}, \tag{11}$$

where $\Bbbk = (1/8\tau_1)^2$. Then if the mesh function Z satisfies $Z \geq 0$ on $\Gamma_\varepsilon^{N,M}$, then $L_{hyb}^{N,M} Z \geq 0$ in $G_\varepsilon^{N,M}$ implies that $Z \geq 0$ at each point of $\overline{G}_\varepsilon^{N,M}$.

Finally, applying the discrete minimum principle with a suitable barrier function one can obtain the following stability result. This implies the uniqueness of the numerical solution.

Lemma 2. *Let U be the solution of (8) and the assumptions (10) and (11) of Lemma 1 hold. Then we have the following stability bound*

$$\|U\|_{\infty,\overline{G}_\varepsilon^{N,M}} \leq \|U\|_{\infty,\Gamma_\varepsilon^{N,M}} + \frac{(1+T)}{\mu}\|\widetilde{f_{hyb}}\|_{\infty,\overline{G}_\varepsilon^{N,M}},$$

where $\mu = \min\{1, \alpha/(1-\xi)\}$.

It is to be noted that the detailed error analysis of the proposed numerical scheme will be carried out in our working paper [3].

3   Numerical Results

In this section, we present the numerical results obtained by the newly proposed scheme (5). To do this, we conduct the numerical experiments for the following test example on the piecewise-uniform rectangular mesh $\overline{G}_\varepsilon^{N,M}$. In all the cases, the numerical experiments are performed by choosing the constant $\tau_1 = \tau_2 = 2.4$, $\alpha = 0.9$ and $\Delta t = 0.8/N$, otherwise it is mentioned.

Example 1. Consider the following mixed parabolic-elliptic IBVP:

$$\begin{cases} \dfrac{\partial u}{\partial t} - \varepsilon\dfrac{\partial^2 u}{\partial x^2} = \exp(t/2)\sin(\pi x)\big(0.5 + \varepsilon\pi^2\big), & (x,t) \in (0,0.5) \times (0,1], \\[2mm] -\varepsilon\dfrac{\partial^2 u}{\partial x^2} - \dfrac{\partial u}{\partial x} = \pi\exp(t/2)\big(\sin(\pi x) + \varepsilon\pi\cos(\pi x)\big), & (x,t) \in (0.5,1) \times (0,1], \\[2mm] [u(x,t)] = 0, \quad \left[\dfrac{\partial u(x,t)}{\partial x}\right] = 0, \quad \text{at } x = 0.5, \\[2mm] u(x,0) = \varphi(x), \quad 0 \leq x \leq 1, \quad u(0,t) = u(1,t) = \exp(t/2), \quad 0 < t \leq 1, \end{cases} \tag{12}$$

where $\varphi(x)$ is the solution of the corresponding stationary problem such that $\varphi(0) = \varphi(1) = 1$. Then the exact solution is $u(x,t) = \exp(t/2)\varphi(x)$. As the exact solution of the IBVP (12) is known, for each ε, we calculate the maximum point-wise error by

$$e_\varepsilon^{N,\Delta t} = \max_{(x_i,t_n)\in\overline{G}_\varepsilon^{N,M}} \left| u(x_i,t_n) - U^{N,\Delta t}(x_i,t_n) \right|,$$

where $u(x_i,t_n)$ and $U^{N,\Delta t}(x_i,t_n)$ respectively denote the exact and the numerical solution obtained on the mesh $\overline{G}_\varepsilon^{N,M}$ with N mesh intervals in the spatial direction and M mesh intervals in the t-direction such that $\Delta t = T/M$ is the uniform time step. In addition, we determine the corresponding order of convergence by

$$p_\varepsilon^{N,\Delta t} = \log_2\left(\frac{e_\varepsilon^{N,\Delta t}}{e_\varepsilon^{2N,\Delta t/2}}\right).$$

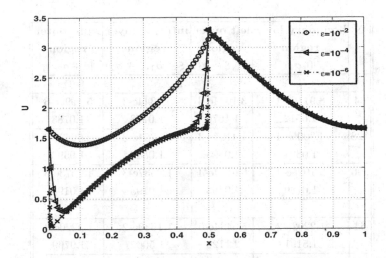

Fig. 1. Numerical solution at time $t = 1$, $N = 128$ for Example 1

Table 1. Maximum point-wise errors and the corresponding order of convergence for Example 1

ε	Number of mesh intervals N				
	32	64	128	256	512
1	1.2118e-2	5.4202e-3	2.5673e-3	1.2487e-3	6.1576e-4
	1.1607	1.0781	1.0399	1.0200	
10^{-2}	5.3114e-2	2.7075e-2	1.1657e-2	3.7649e-3	1.0985e-3
	0.9720	1.2158	1.6305	1.7771	
10^{-4}	2.7781e-2	1.3763e-2	5.2540e-3	1.7853e-3	5.8644e-4
	1.0133	1.3893	1.5573	1.6061	
10^{-5}	2.7934e-2	1.3900e-2	5.3205e-3	2.0099e-3	7.4457e-4
	1.0070	1.3854	1.4044	1.4327	
10^{-6}	2.7974e-2	1.3935e-2	5.3388e-3	2.0202e-3	7.4851e-4
	1.0053	1.3842	1.4020	1.4324	
$e^{N,\Delta t}$	**5.3114e-2**	**2.7075e-2**	**1.1657e-2**	**3.7649e-3**	**1.0985e-3**
$p^{N,\Delta t}$	**0.9720**	**1.2158**	**1.6305**	**1.7771**	

Table 2. Maximum point-wise errors and the corresponding order of convergence calculated for Example 1 by taking $M = N^2$

N	boundary layer region $[0, \sigma_1)$	left outer region $[\sigma_1, \xi - \sigma_1]$	interior layer region $(\xi - \sigma_1, \xi + \sigma_2)$	right outer region $[\xi + \sigma_2, 1]$
	$\varepsilon = 1$			
64	8.9116e-4	3.1684e-3	5.4344e-3	5.4990e-3
	1.0512	1.1676	1.0773	1.0783
128	4.3005e-4	1.4105e-3	2.5754e-3	2.6043e-3
	1.0324	1.0895	1.0386	1.0392
256	2.1025e-4	6.6284e-4	1.2537e-3	1.2673e-3
	1.0180	1.0463	1.0193	1.0198
	$\varepsilon = 10^{-4}$			
64	1.2202e-2	5.5484e-5	1.1630e-2	1.3270e-3
	1.5191	2.1120	1.5062	1.9769
128	4.2572e-3	1.2835e-5	4.0941e-3	3.3712e-4
	1.6975	2.0694	1.7439	1.9511
256	1.3125e-3	3.0580e-6	1.2224e-3	8.7187e-5
	1.9887	2.0011	2.0608	1.9037
	$\varepsilon = 10^{-6}$			
64	1.2200e-2	4.6652e-4	1.1802e-2	1.3245e-3
	1.5191	1.8868	1.4983	1.9999
128	4.2568e-3	1.2615e-4	4.1775e-3	3.3113e-4
	1.5283	2.0903	1.5285	1.9995
256	1.4758e-3	2.9625e-5	1.4481e-3	8.2811e-5
	1.6390	2.4329	2.0773	1.9989

Now, for each N and Δt, we define $e^{N, \Delta t} = \max_{\varepsilon} e_{\varepsilon}^{N, \Delta t}$ as the ε-uniform maximum point-wise error and the corresponding local ε-uniform order of convergence is defined by

$$p^{N, \Delta t} = \log_2 \left(\frac{e^{N, \Delta t}}{e^{2N, \Delta t/2}} \right).$$

The computed maximum point-wise errors $e_{\varepsilon}^{N, \Delta t}$ and the corresponding order of convergence $p_{\varepsilon}^{N, \Delta t}$ for Example 1 are precisely presented in Table 1 for $\varepsilon = 1, 10^{-2}, 10^{-4}, 10^{-5}, 10^{-6}$. Clearly, from the results given in Table 1 we observe that the computed ε-uniform errors $e^{N, \Delta t}$ decrease monotonically as N increases. This ensures that the proposed scheme (5) is ε-uniformly convergent. Despite this, for clarity of the presentation the presence of interior layers near the point of discontinuity $x = 0.5$ and that of the boundary layer at $x = 0$ in the numerical solution of the IBVP (12) have been displayed in Figure 1, as the parameter ε decreases.

Moreover, in order to justify the spatial order of convergence, we carried out the numerical experiments by taking $M = N^2$ and also displayed the computed maximum point-wise errors as well as the corresponding order of convergence for $\varepsilon = 1, 10^{-4}, 10^{-6}$ in Table 2. This in fact shows that the proposed scheme (5) is second-order spatial accurate outside the layer regions and almost second-order spatial accurate inside the layer regions, irrespective of the perturbation parameter ε (however small it may be). It is to be noted that as it can be proved theoretically that the method converges uniformly with almost second-order spatial accuracy provided $\varepsilon \leq C_0 N^{-1}$ for some fixed constant C_0, so for large value of ε the method merely shows first-order convergence in space.

4   Conclusion

In this article, an efficient hybrid numerical scheme has been proposed to solve a class of singularly perturbed mixed parabolic-elliptic problems using a piecewise-uniform Shishkin mesh resolving both boundary and interior layers. Through a precise stability analysis it has been shown that the proposed scheme is ε-uniformly stable, which leads to the uniqueness of the numerical solution. Further, we have experimented computationally through a test example that the hybrid scheme is ε-uniformly convergent and is of almost second-order accurate with respect to the spatial variable. Therefore, looking towards the accuracy and the ε-uniform convergence of the proposed scheme, it can be naturally concluded that a rigorous theoretical analysis of the proposed scheme can be pursued further as a challenging and interesting work to support the above computational analysis.

References

1. Braianov, I.A.: Numerical solution of a mixed singularly perturbed parabolic-elliptic problem. J. Math. Anal. and Appl. 320, 361–380 (2006)
2. Braianov, I.A.: Uniformly convergent difference scheme for singularly perturbed problem of mixed type. Electron. Trans. Numer. Anal. 23, 288–303 (2006)
3. Mukherjee, K., Natesan, S.: Uniform convergence analysis of hybrid numerical scheme for singularly perturbed problems of mixed type, working paper (2013)
4. Mukherjee, K., Natesan, S.: ε-Uniform error estimate of hybrid numerical scheme for singularly perturbed parabolic problems with interior layers. Numer. Alogrithms 58(1), 103–141 (2011)
5. Roos, H.G., Stynes, M., Tobiska, L.: Robust Numerical Methods for Singularly Perturbed Differential Equations, 2nd edn. Springer, Berlin (2008)
6. Stiemer, M.: A Galarkin method for mixed parabolic-elliptic partial differential equations. Numer. Math. 116, 435–462 (2010)

Numerical Experiments with a Shishkin Numerical Method for a Singularly Perturbed Quasilinear Parabolic Problem with an Interior Layer

E. O' Riordan and J. Quinn*

School of Mathematical Sciences, Dublin City University,
Glasnevin, Dublin 9, Ireland

Abstract. In *Russ. Acad. Dokl. Math., 48, 1994, 346–352*, Shishkin presented a numerical algorithm for a quasilinear time dependent singularly perturbed differential equation, with an internal layer in the solution. In this paper, we implement this method and present numerical results to illustrate the convergence properties of this numerical method.

Keywords: Singularly Perturbed, Shishkin mesh, Interior Turning Point.

1   Introduction

In [1], Shishkin presented a computational algorithm for a class of time-dependent quasilinear singularly perturbed differential equations, whose solutions contain an internal shock layer. The algorithm involves approximating the location of the shock layer, for all values of time, and splitting the global problem into two time dependent boundary turning point problems at the approximate shock location. The method outlined in the algorithm to approximate the shock location is intricate. A discrete approximation of the shock layer location is constructed to generate a smooth curve in time. No numerical experiments were presented in [1] to illustrate the details of the algorithm in practice. In [1], for sufficiently small values of the perturbation parameter, error bounds are presented in two cases. In the first case, a pointwise error bound of order 0.2 was established in a neighbourhood away from the approximate shock location. In the second case, a theoretical error bound of order 0.2 was established in a neighbourhood of the approximate shock location. The bound in the second case is theoretical since it requires knowledge of the exact shock location, which remains unknown.

In this paper, we implement the algorithm described in [1], as it applies to a one-dimensional time dependent singularly perturbed parabolic problem. The resulting numerical output indicates that the numerical algorithm in [1] is indeed practical and, in the main, the numerical results also support the theoretical error bounds established in [1].

---

* This research was supported by the Irish Research Council for Science, Engineering and Technology.

I. Dimov, I. Faragó, and L. Vulkov (Eds.): NAA 2012, LNCS 8236, pp. 420–427, 2013.

2   Outline of the Shishkin Algorithm

We describe the algorithm as it is applied to a sub-class of the problem class considered in [1]. We make minor modifications below for practical purposes to ease implementation. For a notable modification, a note is given to inform the reader. Consider the following problem class on the domain $G = \Omega \times (0, T]$, $\Omega = (0, 1)$, $\Gamma = (0, T)$; Find y_ε such that

$$\left(\varepsilon \frac{\partial^2}{\partial x^2} - 2y_\varepsilon \frac{\partial}{\partial x} - b - \frac{\partial}{\partial t}\right) y_\varepsilon(x, t) = f(x, t), \quad (x, t) \in G,$$

$$y_\varepsilon(0, t) = A > 0, \quad y_\varepsilon(1, t) = B < 0, \quad t \in [0, T], \tag{P_ε}$$

$$y_\varepsilon(x, 0) = u_\varepsilon(x), \quad x \in \Omega, \quad b(x, t) \geqslant 0, \quad (x, t) \in \bar{G},$$

where $u_\varepsilon(0) = A$, $u_\varepsilon(d) = 0$, $u_\varepsilon(1) = B$ and u_ε exhibits an ε-dependent interior layer at d for some $d \in (0, 1)$. On either side of d, the initial condition $u_\varepsilon(x)$ can be decomposed into a regular component v and a layer component w. On the interval $[0, d]$, $u_\varepsilon = v_L + w_L$ and on $[d, 1]$, $u_\varepsilon = v_R + w_R$, where v_L, v_R and a sufficient number of their derivatives are bounded independently of ε. Also it is assumed

$$|w_{L(R)}^{(k)}(x)| \leqslant C\varepsilon^{-k} e^{-C\frac{|d-x|}{\varepsilon}}, \quad 1 \leqslant k \leqslant 5, \quad x \in [0, d] \ ([d, 1]). \tag{2.1}$$

Furthermore, it is assumed that v_L and v_R are well defined over $[0, 1]$. We will use an explicit initial condition in our numerical experiment in Section 3 and hence know v_L and v_R explicitly.

The algorithm specifies the use of a simple difference scheme in the case when $\varepsilon > (N^{-2/5} + M^{-2/5})$ and a non-trivial scheme otherwise. If $\varepsilon > (N^{-2/5} + M^{-2/5})$ then we discretise the parabolic problem (P_ε) on a uniform grid as follows: Find Y_ε such that

$$(\varepsilon D_x^- D_x^+ Y_\varepsilon - D_x^- Y_\varepsilon^2 - D_t^- Y_\varepsilon - b Y_\varepsilon)(x_i, t_j) = f(x_i, t_j), \quad (x_i, t_j) \in G^{N,M},$$
$$\tag{$P_\varepsilon^{N,M}$}$$

$$Y_\varepsilon(0, t_j) = A, \quad Y_\varepsilon(1, t_j) = B, \quad t_j \in \bar{\Gamma}^M, \quad Y_\varepsilon(x_i, 0) = u_\varepsilon(x_i), \quad x_i \in \Omega^N,$$

where $\bar{\Omega}^N = \bar{\Omega} \cap \{x_i | x_i = \frac{i}{N}, \ i = 0, \ldots, N\}$, $\Omega^N = \Omega \cap \bar{\Omega}^N$, $\bar{\Gamma}^M = \bar{\Gamma} \cap \{t_j | t_j = \frac{j}{M}, \ j = 0, \ldots, M\}$, $\Gamma^M = \Gamma \cap \bar{\Gamma}^M$, $\bar{G}^{N,M} = \bar{\Omega}^N \times \bar{\Gamma}^M$, $G^{N,M} = G \cap \bar{G}^{N,M}$. In [1], Shishkin presents the pointwise error bound $|(y_\varepsilon - Y_\varepsilon)(x_i, t_j)| \leqslant C(N^{-1/5} + M^{-1/5})$ for $\varepsilon > (N^{-2/5} + M^{-2/5})$. Note that throughout the paper, D_x^-, D_x^+ and D_t^- denote the standard backward and forward difference operators in space and time i.e. for any mesh function $Z(x_i, t_j)$, we have $D_x^- Z(x_i, t_j) := (Z(x_i, t_j) - Z(x_{i-1}, t_j))/(x_i - x_{i-1})$, $D_x^+ Z(x_i, t_j) := D_x^- Z(x_{i+1}, t_j)$ and $D_t^- Z(x_i, t_j) = (Z(x_i, t_j) - Z(x_i, t_{j-1}))/(t_j - t_{j-1})$.

If $\varepsilon \leqslant (N^{-2/5} + M^{-2/5})$ then we follow the algorithm outlined below. There exists a function $s(t)$ for which $y_\varepsilon(s(t), t) = 0$, $(s(t), t) \in \bar{G}$. A transition layer appears in the vicinity of $s(t)$ for all $t \in [0, T]$. Consider the following reduced left and right boundary initial value problems:

$$(-2y_L\frac{\partial}{\partial x} - \frac{\partial}{\partial t} - b)y_L(x,t) = f(x,t), \quad (x,t) \in (0,1] \times (0,T], \qquad (2.2a)$$

$$y_L(0,t) = A, \quad t \in [0,T], \quad y_L(x,0) = v_L(x), \quad x \in (0,1] \cap \Omega, \qquad (2.2b)$$

$$(-2y_R\frac{\partial}{\partial x} - \frac{\partial}{\partial t} - b)y_R(x,t) = f(x,t), \quad (x,t) \in [0,1) \times (0,T], \qquad (2.2c)$$

$$y_R(x,0) = v_R(x), \quad x \in [0,1) \cap \Omega, \quad y_R(1,t) = B, \quad t \in [0,T]. \qquad (2.2d)$$

Define the reduced solution of (P_ε) for all $t \in [0,T]$ to be

$$y_0(x,t) := (y_L + y_R)(x,t). \qquad (2.3)$$

The leading term of the asymptotic expansion of the transition layer location $s(t)$ is the solution, s_0, of the nonlinear initial value problem

$$s_0'(t) = y_0(s_0(t),t), \quad t \in (0,T], \quad s_0(0) = d. \qquad (2.4)$$

In [1], Shishkin states that

$$|s(t) - s_0(t)| \leqslant C\varepsilon \ln(1/\varepsilon). \qquad (2.5)$$

We detail the construction of $s^*(t)$ (in (2.12)), an approximation to s_0, below. However, to keep the continuous and discrete subproblems separate in our presentation, we skip to the application of $s^*(t)$: we apply the transform $\xi(x,t) = \frac{1}{2}(1+\mu(x,t)(x-s^*(t)))$, $\mu(x,t) = 1/s^*(t)$ for $x \leqslant s^*(t)$ and $\mu(x,t) = 1/(1-s^*(t))$ for $x > s^*(t)$ to the domains G and Ω. The problem (P_ε) is transformed to the following problem:

$$\left(\frac{1}{\kappa_1^2}\left[\varepsilon\frac{\partial^2}{\partial\xi^2} + \kappa_1(\kappa_2 s^{*\prime}(t) - 2\tilde{y}_\varepsilon^\pm)\frac{\partial}{\partial\xi}\right] - \tilde{b} - \frac{\partial}{\partial t}\right)\tilde{y}_\varepsilon^\pm(\xi,t) = \tilde{f}(\xi,t), \quad (\xi,t) \in \tilde{G}^\pm,$$

$$\tilde{y}_\varepsilon^-(\xi,0) = \tilde{u}_\varepsilon(\xi), \; \xi \in (0,\tfrac{1}{2}), \qquad \tilde{y}_\varepsilon^+(\xi,0) = \tilde{u}_\varepsilon(\xi), \; \xi \in (0,\tfrac{1}{2}),$$

$$\tilde{y}_\varepsilon^-(0,t) = A, \; \tilde{y}_\varepsilon^-(\tfrac{1}{2},t) = 0, \quad \tilde{y}_\varepsilon^+(\tfrac{1}{2},t) = 0, \; \tilde{y}_\varepsilon^+(1,t) = B, \; t \in [0,T], \qquad (\tilde{P}_\varepsilon)$$

$$\tilde{G}^+ = ((0,\tfrac{1}{2}) \times [0,T]) \cap \tilde{G}, \quad \tilde{G}^- = ((\tfrac{1}{2},1) \times [0,T]) \cap \tilde{G},$$

$$\kappa_1 = \kappa(s^*(t)), \; \kappa_2 = \kappa(\xi), \; \kappa(\lambda) = \begin{cases} 2\lambda, & \xi < \tfrac{1}{2}, \\ 2(1-\lambda), & \xi > \tfrac{1}{2}, \end{cases}$$

where $\tilde{G}$ is the transformed domain $\xi(G)$ and the notation $\tilde{g}(\xi,t) := g(x^{-1}(\xi,t),t)$.

Note 1. The transformation proposed in [1] is $\xi(x,t) = \frac{1}{2}(1+x-s^*(t))$. That is, with $\mu(x,t) = 1$, which maps $s^*(t) \mapsto \frac{1}{2}$ for all t. However, this transformation does not generate a rectangular computational domain, which would be ideal to solve the corresponding discrete problem. Hence we modify the original prescribed transformation.

We now give details of the discretisations of (2.2)-(2.4) and of the transformed problem $(\tilde{P}_\varepsilon)$. Denote N and M as the space and time discretisation parameters respectively. Discretising (2.2), we solve the following reduced left and right discrete boundary initial value problems:

$$(-2Y_L D_x^- - D_t^- - b)Y_L(x_i, t_j) = f(x_i, t_j), \tag{2.6a}$$

$$(x_i, t_j) \in ((0, 1] \cap \bar{\Omega}^N) \times \bar{\Gamma}^M, \tag{2.6b}$$

$$Y_L(0, t_j) = A, \quad t_j \in \bar{\Gamma}^M, \quad Y_L(x_i, 0) = v_L(x_i), \quad x_i \in (0, d] \cap \Omega^N, \tag{2.6c}$$

$$(-2Y_R D_x^+ - D_t^- - b)Y_R(x_i, t_j) = f(x_i, t_j), \tag{2.6d}$$

$$(x_i, t_j) \in ([0, 1) \cap \bar{\Omega}^N) \times \bar{\Gamma}^M, \tag{2.6e}$$

$$Y_R(x_i, 0) = v_R(x_i), \quad x_i \in [d, 1) \cap \Omega^N, \quad Y_R(1, t_j) = B, \quad t_j \in \bar{\Gamma}^M. \tag{2.6f}$$

Define the discrete reduced solution for all $t_j \in \bar{\Gamma}^M$ to be

$$Y_0(\eta, t_j) := (Y_L + Y_R)(x_i, t_j) \quad \text{for all} \quad \eta \in [x_i, x_{i+1}), \quad x_i \in \bar{\Omega}^N. \tag{2.7}$$

Next, we discretise (2.4) and solve for S_0^M, a discrete approximation to s_0, where

$$D^- S_0^M(t_j) = Y_0(S_0^M(t_j), t_j), \quad t_j \in \bar{\Gamma}^M \setminus t_0, \quad S_0^M(0) = d. \tag{2.8}$$

Note 2. Later, for practical purposes, we need to define S_0^M beyond T. We extend Y_0 to $\bar{Y}_0$ s.t $\bar{Y}_0(\eta, t_j) = Y_0(\eta, t_j), t_j \leqslant T, t_j \in \bar{\Gamma}^M$ and $\bar{Y}_0(\eta, t) = Y_0(\eta, T), t > T$. This is suitable assuming Y_L and Y_R have reached steady-state by time T. We can then continue to solve iteratively for S_0^M beyond T as far as required on the extended mesh

$$\bar{\Gamma}^+ := \bar{\Gamma}^{\frac{M}{2}} \cup \{t_{i+M} = T/Mi \mid i = 1, 2, 3, \dots\}. \tag{2.9}$$

The algorithm now performs a "smoothing" routine on S_0^M to retrieve a continuous function $s^*(t)$. Construct the piecewise constant function $\bar{s}$ as follows

$$\bar{s}(\eta) = \begin{cases} d, & \eta \leqslant 0 \\ S_0^M(t_j), & \eta \in (t_{j-1}, t_j] \subset \bar{\Gamma}^+. \end{cases} \tag{2.10}$$

The algorithm requires the construction of a "piecewise-quadratic, nonnegative, compactly supported function", ω, with support on the set $[-L, L]$, $L = \frac{1}{\sqrt{M}}$, with $\omega \in C^1(-L, L)$ and $\int_{-L}^{L} \omega(\chi) \, d\chi = 1$ where M is the time discretisation parameter. An example of such a function is

$$\omega(\chi) = \begin{cases} \frac{2}{L^3}(\chi + L)^2, & \chi \in [-L, -\frac{L}{2}), \\ -\frac{2}{L^3}\chi^2 + \frac{1}{L}, & \chi \in [-\frac{L}{2}, \frac{L}{2}), \\ \frac{2}{L^3}(\chi - L)^2, & \chi \in [\frac{L}{2}, L]. \end{cases} \tag{2.11}$$

The function s^* is then constructed as follows

$$s^*(t) = \int_{t-L}^{t+L} \bar{s}(\eta)\omega(\eta - t) \, d\eta, \quad t \in [0, T]. \tag{2.12}$$

Note that $s^*(0) \neq d$. In [1], Shishkin states that

$$|s(t) - s^*(t)| \leqslant C(N^{-1/3} \ln(N) + M^{-1/3} \ln(N)).$$

The integral in (2.12) can be solved exactly by integrating over each subinterval (t_{j-1}, t_j) (2.10).

Remark 1. The reason why we extend the definition of S_0^M is so that $s^*(T)$ and $s^{*\prime}(T)$ are well defined.

We discretise $(\tilde{P}_\varepsilon)$ as follows:

$$\left(\frac{1}{\kappa_1^2} \left[\varepsilon D_\xi^{S_1} D_\xi^{-S_1} \tilde{Y}_\varepsilon^\pm + \kappa_1 (\kappa_2 s^{*\prime}(t_j)) D_\xi^{S_2} \tilde{Y}_\varepsilon^\pm - D_\xi^{S_1} (\tilde{Y}_\varepsilon^\pm)^2) \right] \right.$$
$$\left. - \tilde{b} - D_t^- \tilde{Y}_\varepsilon^\pm \right) (\xi_i, t_j) = \tilde{f}(\xi_i, t_j), \quad (\xi_i, t_j) \in \tilde{G}^{\frac{N}{2}, M\pm},$$

$$D_\xi^{S_1} = \begin{cases} D_\xi^-, & \xi_i \leqslant \frac{1}{2}, \\ D_\xi^+, & \xi_i > \frac{1}{2}, \end{cases} \quad D_\xi^{-S_1} = \begin{cases} D_\xi^+, & \xi_i \leqslant \frac{1}{2}, \\ D_\xi^-, & \xi_i > \frac{1}{2}, \end{cases} \quad D_\xi^{S_2} = D_\xi^{\mathrm{sgn}(\kappa_1 \kappa_2 s^{*\prime}(t_j))},$$

$$\tilde{G}^{\frac{N}{2}, M+} = \tilde{G}^+ \cap (\Omega_L^{\frac{N}{2}} \times \bar{\Gamma}^M), \quad \tilde{G}^{\frac{N}{2}, M-} = \tilde{G}^- \cap (\Omega_R^{\frac{N}{2}} \times \bar{\Gamma}^M), \qquad (\tilde{P}_\varepsilon^{N,M})$$

$$\theta > \tfrac{1}{5} \| y_L - y_R \|, \quad \kappa_1 = \kappa(s^*(t_j)), \quad \kappa_2 = \kappa(\xi_i),$$

$$\tilde{Y}_\varepsilon^-(0, t_j) = A, \quad \tilde{Y}_\varepsilon^\pm(\tfrac{1}{2}, t_j) = 0, \quad \tilde{Y}_\varepsilon^+(1, t_j) = B, \quad t_j \in \bar{\Gamma}^M,$$

$$\tilde{Y}_\varepsilon^\pm(\xi_i, 0) = u_\varepsilon^*(\xi_i), \quad \xi_i \in \Omega_{L \backslash R}^{\frac{N}{2}},$$

where the meshes $\Omega_L^{\frac{N}{2}}$ and $\Omega_R^{\frac{N}{2}}$ are defined by

$$\bar{\Omega}_L^{N/2} = \left\{ x_i \left| \begin{array}{ll} x_i = \frac{4(1/2 - \sigma)}{N} i, & 0 \leqslant i \leqslant \frac{N}{4}, \\ x_i = \frac{1}{2} - \sigma + \frac{4\sigma}{N}(i - \frac{N}{4}), & \frac{N}{4} < i \leqslant \frac{N}{2}, \end{array} \right. \right\}, \qquad (2.13)$$

$$\bar{\Omega}_R^{N/2} = \left\{ x_i \left| \begin{array}{ll} x_i = \frac{1}{2} + \frac{4\sigma}{N} i, & 0 \leqslant i \leqslant \frac{N}{4}, \\ x_i = \frac{1}{2} + \sigma + \frac{4(1/2 - \sigma)}{N}(i - \frac{N}{4}), & \frac{N}{4} < i \leqslant \frac{N}{2}, \end{array} \right. \right\}, \qquad (2.14)$$

$$\sigma = \min\{\tfrac{1}{4}, \theta \varepsilon \log N\}, \quad \Omega_{L \backslash R}^{N/2} = \bar{\Omega}_{L \backslash R}^{N/2} \backslash \{x_0, x_N\}, \quad \theta > 0.2|A - B|. \quad (2.15)$$

In [1], u_ε^* is not necessarily the initial condition u_ε of (P_ε) under the transform ξ but is prescribed as any function "somehow constructed" such that $|(u_\varepsilon^* - \tilde{u}_\varepsilon)(\xi_i)| \leqslant C(N^{-1} + M^{-1})$.

Remark 2. If $\tilde{x}_i \in \Omega$ is such that $\tilde{x}_i = \xi^{-1}(\xi_i)$ and Z is the inverse transform of $\tilde{Y}_\varepsilon^\pm$, that is $Z(\tilde{x}_i, t_j) = \tilde{Y}_\varepsilon^\pm(\xi^{-1}(\xi_i), t_j)$, then the error bounds presented in [1] for $\varepsilon \leqslant (N^{-2/5} + M^{-2/5})$ can be described as follows. For all $\tilde{x}_i \in \Omega$ such that $|\tilde{x}_i - s^*(t)| \geqslant C > 0$, the following pointwise error bound holds:

$$|(y_\varepsilon - Z)(\tilde{x}_i, t_j)| \leqslant C(N^{-1/5}(\ln N)^{1/2} + M^{-1/5}(\ln M)^{1/2}).$$

For all $\tilde{x}_i \in \Omega$ such that $|\tilde{x}_i - s^*(t)| \leqslant C$, the following error bound holds:

$$|y_\varepsilon(\tilde{x}_i + (s(t) - s^*(t)), t_j) - Z(\tilde{x}_i, t_j)| \leqslant C(N^{-1/5}(\ln N)^{1/2} + M^{-1/5}(\ln M)^{1/2}).$$

3   Numerical Example

As a sample initial condition for (P_ε), we take the following choice:

$$u_\varepsilon(x) = u_\varepsilon(x; d) = \begin{cases} (A - x)\tanh((d - x)/\varepsilon)/\tanh(\frac{d}{\varepsilon}), & x \leqslant d, \\ (B + 1 - x)\tanh((x - d)/\varepsilon)/\tanh(\frac{1-d}{\varepsilon}), & x > d. \end{cases} \quad (3.1)$$

Thus we take $v_L(x) = A - x$ and $v_R(x) = B + 1 - x$.

For the discrete problem $(P_\varepsilon^{N,M})$, we choose as an initial condition, $u_\varepsilon^*(\xi_i) = \tilde{u}_\varepsilon(\xi_i; s^*(0))$. That is $u_\varepsilon(x; s^*(0))$ as defined in (3.1) under the transform ξ. Note that since $\xi : s^*(0) \mapsto \frac{1}{2}$, this choice ensures $u_\varepsilon^*(\frac{1}{2}) = 0$ whereas $\tilde{u}_\varepsilon(\frac{1}{2}; d) \neq 0$. From the definition of s^* in (2.12), $s^*(0) \to d$ as $M \to \infty$ since $L \to 0$. Hence ω tends to the δ-function as $M \to \infty$. In this experiment, we do not establish the rate, in terms of N and M, at which $u_\varepsilon(x; s^*(0))$ converges to $u_\varepsilon(x; d)$.

In this example we solve $(P_\varepsilon^{N,M})$ and $(\tilde{P}_\varepsilon^{N,M})$ with $b(x,t) = 2x$, $f(x,t) = \sin(4x)$, $A = -B = 3$, $d = 0.5$ and $T = 1$. We choose $\theta = \frac{1}{4}|A - B|$ in $(\tilde{P}_\varepsilon^{N,M})$. A plot of S^M and s^* is included in Figure 1(a) along with a graph of a numerical solution of $(\tilde{P}_\varepsilon^{N,M})$ with the transform ξ reversed in Figure 1(c). We also include a plot of s^* for $T = 6$ computed using $N = M = 1024$ for a range of values of d, with all other problem data the same, in Figure 2(b). In Figure 2(b), we observed that $s^*(t)$ tends to a constant value as $t \to \infty$ and this constant is independent of the location d of the layer in the initial condition.

We solve for $U_\varepsilon^N(z_i, t_j)$ for $(z_i, t_j) \in \bar{\Lambda}^N$ where if $\varepsilon > N^{-2/5} + M^{-2/5}$ then

$$\bar{\Lambda}^N := \bar{G}^{N,M}, \qquad U_\varepsilon^N(z_i, t_j) := Y_\varepsilon(z_i, t_j),$$

and if $\varepsilon \leqslant N^{-2/5} + M^{-2/5}$ then

$$\bar{\Lambda}^N := \bar{\tilde{G}}^{\frac{N}{2}, N-} \cup \bar{\tilde{G}}^{\frac{N}{2}, N+}, \qquad U_\varepsilon^N(z_i, t_j) := \begin{cases} \tilde{Y}_\varepsilon^-(z_i, t_j), & z_i \leqslant \frac{1}{2}, \\ \tilde{Y}_\varepsilon^+(z_i, t_j), & z_i > \frac{1}{2}. \end{cases}$$

Table 1. Computed rates R_ε^N for sample values of N and ε

	R_ε^N					
ε	N=64	N=128	N=256	N=512	N=1024	N=2048
2^{-0}	0.94	0.85	0.94	0.96	0.97	0.99
2^{-1}	0.51	0.71	0.84	0.91	0.96	0.98
2^{-4}	0.94	0.84	0.94	0.96	0.98	.
2^{-5}	0.29	0.28	0.84	0.93	0.97	0.98
2^{-6}	0.39	0.47	0.56	0.78	0.78	0.82
2^{-7}	0.39	0.47	0.56	0.78	0.78	0.82
2^{-8}	0.39	0.47	0.56	0.78	0.78	0.82
$\vdots$	$\vdots$	$\vdots$	$\vdots$	$\vdots$	$\vdots$	
2^{-20}	0.43	0.47	0.56	0.78	0.78	0.82
R^N	0.59	0.35	0.83	0.78	0.78	0.82

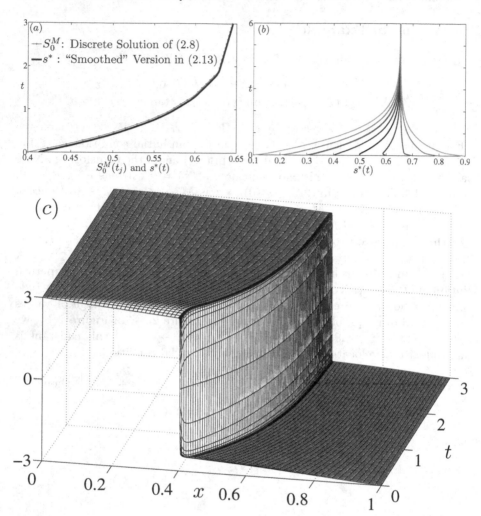

Fig. 1. Graphs: (a) Plot of t versus S^M and t versus $s^*(t)$ for $N = M = 32$ over $t \in [0,T]$.; (b) Plot of t versus $s^*(t)$ for $N = M = 1024$ over $t \in [0,6]$ for $d = 0.1, 0.2, \cdots, 0.9$.; (c) Numerical Solution of $(\tilde{P}_\varepsilon^N)$ under reverse of transform ξ for $\varepsilon = 2^{-10}$ and $N = M = 128$

Here Y_ε is the solution of $(P_\varepsilon^{N,N})$ defined over the grid $\bar{G}^{N,N}$ and $\tilde{Y}_\varepsilon^\pm$ are the numerical solutions of $(\tilde{P}_\varepsilon^{N,N})$ over the grids $\overline{\tilde{G}}^{\frac{N}{2},N\pm}$, all using N and M mesh intervals in space and time respectively with $N=M$.

Note the following linearisations we use to solve the nonlinear discrete problems. We replace $(Y_L D_x^- Y_L)(x_i,t_j)$ with $Y_L(x_{i-1},t_j)D_x^- Y_L(x_i,t_j)$ in (2.6a), we linearise $(Y_R D_x^+ Y_R)(x_i,t_j)$ in (2.6d) analogously. We replace $D_\xi^-(\tilde{Y}_\varepsilon^+)^2(\xi_i,t_j)$ in $(\tilde{P}_\varepsilon^{N,N})$ with $(\tilde{Y}_\varepsilon^+(\xi_i,t_{j-1}) + \tilde{Y}_\varepsilon^+(\xi_{i-1},t_{j-1}))D_\xi^- \tilde{Y}_\varepsilon^+(\xi_i,t_j)$.

We linearise $D_\xi^+(\tilde{Y}_\varepsilon^-)^2(\xi_i, t_j)$ in $(\tilde{P}_\varepsilon^{N,N})$ and $(D_x^- Y_\varepsilon^2)$ in $(P_\varepsilon^{N,M})$ analogously. We replace $Y_0(S_0^M(t_j), t_j)$ in (2.8) with $Y_0(S_0^M(t_{j-1}), t_j)$.

We compute differences D_ε^N and rates R_ε^N for various values of N and ε presented in Table 1 where

$$D_\varepsilon^N := \max_{\substack{(z_i, t_j) \in \Lambda^N \\ t_j \geqslant 0.05}} |(U_\varepsilon^N - \bar{U}_\varepsilon^{2N})(z_i, t_j)|, \quad D^N := \max_\varepsilon D_\varepsilon^N, \tag{3.2a}$$

$$R_\varepsilon^N := \log_2 \frac{D_\varepsilon^N}{D_\varepsilon^{2N}} \quad \text{and} \quad R^N := \log_2 \frac{D^N}{D^{2N}}, \tag{3.2b}$$

where $\bar{U}_\varepsilon^{2N}$ is the interpolation of U_ε^{2N} onto the grid Λ^N. Note that in our experiments, we did not observe convergence of $(U_\varepsilon^N - \bar{U}_\varepsilon^{2N})(z_i, t_j)$ in a small region after $t_0 = 0$. We compute the differences for all time $t_j \geqslant 0.05$ as in (3.2a). Note that in Table 1, entries in the rows for $\varepsilon = 2^{-2}, 2^{-3}$ and 2^{-4} have been excluded, because over this parameter range of (ε, N), two separate numerical algorithms are implemented either side of the constraint $\varepsilon = N^{-0.4}$. The bound in (2.5) is insufficient for such high test values of ε. The computed rates of uniform convergence R^N indicate a positive rate of convergence for this particular example over the range of test values for ε and N.

Reference

1. Shishkin, G.I.: Difference approximation of the Dirichlet problem for a singularly perturbed quasilinear parabolic equation in the presence of a transition layer. Russian Acad. Sci. Dokl. Math. 48(2), 346–352 (1994)

High Performance Computing of Data for a New Sensitivity Analysis Algorithm, Applied in an Air Pollution Model

Tzvetan Ostromsky[1], Ivan Dimov[1], Rayna Georgieva[1],
Pencho Marinov[1], and Zahari Zlatev[2]

[1] Institute of Information and Communication Technologies,
Bulgarian Academy of Sciences, Acad. G. Bonchev str., bl. 25-A, 1113 Sofia, Bulgaria
{ceco,rayna}@parallel.bas.bg, {ivdimov,pencho}@bas.bg
http://parallel.bas.bg/dpa/EN/index.htm
[2] National Centre for Environment and Energy, University of Århus,
Frederiksborgvej 399 P.O. Box 358, DK-4000 Roskilde, Denmark
zz@dmu.dk
http://www.dmu.dk/AtmosphericEnvironment

Abstract. Variance-based sensitivity analysis approach has been proposed for studying of input parameters' error contribution into the output results of a large-scale air pollution model, the Danish Eulerian Model, in its unified software implementation, called UNI-DEM. A three-stage sensitivity analysis algorithm, based on analysis of variances technique (ANOVA) for calculating Sobol global sensitivity indices and computationally efficient Monte Carlo integration techniques, has recently been developed and successfully used for sensitivity analysis study of UNI-DEM with respect to several chemical reaction rate coefficients.

As a first stage it is necessary to carry out a set of computationally expensive numerical experiments and to extract the necessary multidimensional sets of sensitivity analysis data. A specially adapted for that purpose version of the model, called SA-DEM, was created, implemented and run on an IBM Blue Gene/P supercomputer, the most powerful parallel machine in Bulgaria. Its capabilities have been extended to be able to perturb the 4 different input data sets with anthropogenic emissions by regularly modified perturbation coefficients. This is a complicated and challenging computational problem even for such a powerful supercomputer like IBM BlueGene/P. Its efficient numerical solution required optimization of the parallelization strategy and improvements in the memory management. Performance results of some numerical experiments on the IBM BlueGene/P machine will be presented and analyzed.

1 Introduction

The Danish Eulerian Model (DEM) [11] is a large-scale air pollution model, used to calculate the concentrations of various dangerous pollutants and other relevant chemical species over a large geographical region (4800 × 4800 km),

I. Dimov, I. Faragó, and L. Vulkov (Eds.): NAA 2012, LNCS 8236, pp. 428–436, 2013.

including whole Europe, the Mediterranean and neighbouring to them parts of Asia and Africa. The model takes into account the main physical and chemical processes between these species, the actual meteorological conditions, emissions, etc.. Among the input data sets of the Danish Eulerian Model, the anthropogenic emissions are some of the most important and expensive to be evaluated precisely. On the other hand, strong and costly measures are recently being taken in most developed countries to cut off these emissions, which results in a stable trend to reduction (with uneven rate for the different countries and species). That is why it is important to know with more certainty the effect of various emission changes on the expected output concentrations and particularly on those of the most dangerous pollutants. The sensitivity analysis study subject to this paper would help us to get more definite answers to that sort of questions.

The general concept of sensitivity analysis is described briefly in Section 2. The Danish Eulerian Model and some features of its software implementation are described in Section 3. In Section 4 its sensitivity study with respect to the input emissions and the specially developed for the purpose SA-DEM version are discussed. Some numerical experiments, performance and scalability results obtained on the IBM BlueGene/P, are presented in Section 5. Finally, some conclusions are drawn.

2   Sensitivity Analysis Concept - Sobol Approach

Sensitivity analysis (SA) is the study of how much the uncertainty in the input data of a model (due to any reason: inaccurate measurements or calculation, approximation, data compression, etc.) is reflected in the accuracy of the output results [8]. Two kinds of sensitivity analysis are present in the existing literature, local and global. Local SA studies how much some small variations of inputs around a given value can change the value of the output. Global SA takes into account all the variation range of the input parameters, and apportions the output uncertainty to the uncertainty in the input data. Subject to our study in this paper is the global sensitivity analysis.

Several sensitivity analysis techniques have been developed and used throughout the years [8]. In general, these methods rely heavily on special assumptions connected to the behaviour of the model (such as linearity, monotonicity and additivity of the relationship between input and output parameters of the model). Among the quantitative methods, variance-based methods are most often used. The main idea of these methods is to evaluate how the variance of an input or a group of inputs contributes to the variance of model output.

Assume that a model is represented by the model function $u = f(\mathbf{x})$, where the input parameters $\mathbf{x} = (x_1, x_2, \ldots, x_d) \in U^d \equiv [0, 1]^d$ are independent (non-correlated) random variables with a known *joint probability distribution function*. In this way the output u becomes also a random variable (as a function of the random vector $\mathbf{x}$) and let $\mathbf{E}$ be its mathematical expectation. Let $\mathbf{D}[\mathbf{E}(u|x_i)]$ be the variance of the conditional expectation of u with respect to x_i and $\mathbf{D}_u$ - the total variance according to u. This indicator is called *first-order sensitivity index* by Sobol [9] or sometimes *correlation ratio*.

Total Sensitivity Index (TSI) [9] of x_i, $i \in \{1, \ldots, d\}$ is the sum of the complete set of mutual sensitivity indices of any order:

$$S_{x_i}^{tot} = S_i + \sum_{l_1 \neq i} S_{il_1} + \sum_{l_1, l_2 \neq i, l_1 < l_2} S_{il_1 l_2} + \ldots + S_{il_1 \ldots l_{d-1}} \qquad (1)$$

where $S_{il_1 \ldots l_{j-1}}$ is j^{th} order sensitivity index for x_i $(1 \leq j \leq d)$, S_i is called the *main effect* of x_i. In most practical problems the high dimensional terms can be neglected, reducing significantly the number of summands in (1).

The Sobol approach, used in the current study, is based on a unique decomposition of the model function into orthogonal terms (summands) of increasing dimension and zero means. Its main advantage is computing in an uniform way not only the first order indices, but also those of higher order (in a similar way as the main effects are computed). The total sensitivity index can then be calculated with one Monte Carlo integral per factor [10]. A comparison with other techniques for global SA, used to study the variability of DEM output has been presented in [1].

3   The Danish Eulerian Model

In this section we describe shortly the Danish Eulerian Model (DEM) [11] and its current production version UNI-DEM. It is mathematically represented by the following system of partial differential equations, in which the unknown concentrations of a large number of chemical species (pollutants and other chemically active components) take part. The main physical and chemical processes (advection, diffusion, chemical reactions, emissions and deposition) are represented in that system.

$$\frac{\partial c_s}{\partial t} = -\frac{\partial(u c_s)}{\partial x} - \frac{\partial(v c_s)}{\partial y} - \frac{\partial(w c_s)}{\partial z} +$$

$$+ \frac{\partial}{\partial x}\left(K_x \frac{\partial c_s}{\partial x}\right) + \frac{\partial}{\partial y}\left(K_y \frac{\partial c_s}{\partial y}\right) + \frac{\partial}{\partial z}\left(K_z \frac{\partial c_s}{\partial z}\right) + \qquad (2)$$

$$+ E_s + Q_s(c_1, c_2, \ldots c_q) - (k_{1s} + k_{2s}) c_s, \quad s = 1, 2, \ldots q ,$$

where q is the number of equations (equals to the number of chemical species and groups of species, represented separately in the chemical scheme of the model);

- c_s – the concentrations of the chemical species;
- u, v, w – the wind components along the coordinate axes;
- K_x, K_y, K_z – diffusion coefficients;
- E_s – the emission functions in the domain;
- k_{1s}, k_{2s} – dry / wet deposition coefficients;
- $Q_s(c_1, c_2, \ldots c_q)$ – non-linear functions describing the chemical reactions between species under consideration.

3.1 Splitting into Submodels

This huge and rather complicated computational problem is not suitable for direct numerical treatment. For this purpose it is necessary to split it into submodels in accordance with the main physical and chemical processes in the atmosphere. The most straightforward sequential splitting [4] is used by now in the production version of the model, although other splitting methods have also been considered and implemented in some experimental versions [3]. Applying the sequential splitting scheme, the above system (2) can be decomposed into three submodels, defined by the following equations:

$$\frac{\partial c_s}{\partial t} = -\frac{\partial(uc_s)}{\partial x} - \frac{\partial(vc_s)}{\partial y} + \frac{\partial}{\partial x}\left(K_x \frac{\partial c_s}{\partial x}\right) + \frac{\partial}{\partial y}\left(K_y \frac{\partial c_s}{\partial y}\right) \quad \text{\textbf{hor. advection \& diffusion}}$$

$$\frac{\partial c_s}{\partial t} = E_s + Q_s(c_1^{(2)}, c_2^{(2)}, \ldots c_q^{(2)}) - (k_{1s} + k_{2s})c_s \quad \text{\textbf{chemistry, emissions \& deposition}}$$

$$\frac{\partial c_s}{\partial t} = -\frac{\partial(wc_s)}{\partial z} + \frac{\partial}{\partial z}\left(K_z \frac{\partial c_s}{\partial z}\right) \quad \text{\textbf{vertical transport}}$$

Spatial and time discretization of the above submodels on the EMEP[1] grid or its refinements (see Table 1) makes each of them a huge computational task, challenging for the most powerful supercomputers available nowadays. That is why the parallelization has always been a highly important issue in the computer implementation of DEM since its very early stages. A coarse-grain parallelism, based on partitioning of the spatial domain, is usually the primary parallelization strategy in the computer implementation of DEM. It was shown to be very efficient and well-balanced on widest class of nowadays parallel machines. Other parallelizations are also possible and suitable to certain classes of supercomputer platforms [6,7].

3.2 UNI-DEM Package and Some Essential Implementation Features

The development and improvements of DEM throughout the years has lead to a variety of different versions with respect to the grid-size/resolution, vertical layering (2D or 3D model respectively) and the number of species in the chemical scheme. The most prospective of them have been united in the package UNI-DEM. By setting a small number of parameters one can select the desired version in UNI-DEM and its running options. Some of these user-defined parameters and their optional values are shown in Table 1.

A coarse-grain parallelization strategy based on partitioning of the spatial domain in strips or blocks is currently used in UNI-DEM. The recent more advanced bidirectional block partitioning has been used in the experiments, presented in

---

[1] European Monitoring and Evaluation Programme.

Table 1. User-determined parameters for selecting an appropriate UNI-DEM version

Parameter	Description	Optional values		
NX = NY	Grid size (Grid step)	96×96 (50 km)	288×288 (16.7 km)	480×480 (10 km)
NZ	# layers (2D/3D)	1 or 10		
NEQUAT	# chem. species	35, 56 or 168		
NSIZE	chunk size	integer divisor of (NX $\times$ NY)		

this paper. It is based on partitioning of the horizontal grid into rectangular blocks and requires communication for boundary values exchange on each time step. Improving the data locality for more efficient cache utilization is achieved by using *chunks* to group properly the small tasks in the chemistry-deposition stage. The parameter NSIZE (see Table 1) determins the size of the chunks in dependense with the local cache size of the target machine. It should be chosen so that the amount of data in a chunk fits entirely in the cache.

4   Sensitivity Analysis of DEM with Respect to the Input Emissions by Using SA-DEM

Part of the large input data set of DEM are the anthropogenic emissions over the discretized spatial domain. They are defined as vectors with four different components $E = (E^N, E^C, E^S, E^A)$, given separately in the input data flow. These components correspond to four different groups of pollutants, i.e.: (i) nitrogen oxides ($NO + NO_2$); (ii) anthropogenic hydrocarbons; (iii) sulphur dioxide (SO_2); (iv) ammonia (NH_3).

Among the huge amount of output data, the mean monthly concentrations are one of the most commonly used. For the purpose of this particular research we consider the following pollutants (their mean monthly concentrations, more precisely said): (i) ozone (O_3); (ii) ammonia (NH_3) and (iii) ammonium sulphate and ammonium nitrate ($NH_4SO_4 + NH_4NO_3$). These are taken in 3 grid points of the computational domain, selected nearest to three European cities with different climate and level of pollution: Milan, Manchester and Edinburgh.

The full concept of this study can be represented by the scheme in Fig. 1. More detail on it can be found in [1,5].

The first stage of computations consists of generation of input data necessary for the particular sensitivity analysis study. In our case this means to perform a number of experiments with UNI-DEM by doing certain perturbations in the data for these emissions.

SA-DEM is a modification of UNI-DEM, specially adjusted to be used in the first stage (see Fig. 1) of our sensitivity analysis studies. It was created initially for studying sensitivity of DEM with respect to some chemical rate coefficients in the chemistry submodel [1]. There are additional input parameters in the main program, allowing the user to make some perturbation of the parameters subject to sensitivity analysis. Ratios of the following type

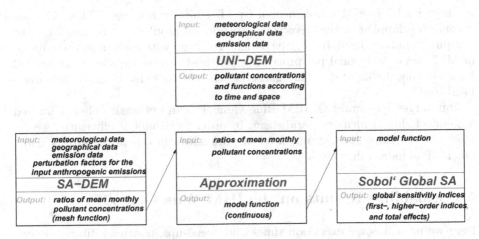

Fig. 1. Stages in the sensitivity studies of UNI-DEM

$$r_s(\alpha) = \frac{c_s^\alpha(a_i, b_i)}{c_s^{bas}(a_i, b_i)},$$

are calculated to form the necessary data (model mesh functions) for the particular SA study, where c_s^α is the "perturbed" mean monthly concentration of some pollutant at some point of the domain, while c_s^α is the corresponding basic value (without any perturbation). By now we used it to perform two kinds of perturbation (with respect to some chemical rate constants and with respect to the input anthropogenic emissions). A detailed discussion on sensitivity study of DEM with respect to the input emissions can be found in [2].

SA-DEM includes a new driver routine that automatically generates a set of tasks in order to produce the necessary results for the particular sensitivity analysis study. It allows to perform in parallel a large number of experiments (for different values of α). This is a typical SIMD[2] task, if considering the coarsest possible level of the structure of our algorithm. By using it we introduce a new, higher level of parallelism in SA-DEM on the top of the grid-partitioning level, the basis for distributed-memory MPI parallelization in UNI-DEM. Almost no communications are necessary on this (coarsest) level of parallelism, except (occasionally) synchronization 'by data', as certain temporary data files are used by all the processes. This is implemented (in MPI) mainly by global MPI barriers. A new communication subroutine is made to split the global communicator (MPI_COMMON_WORLD) and to define separate communicators for each of the top-level parallel tasks. The communicators are very useful on the lower level of parallelism (where intensive communications are performed on each time-step).

The second level of parallelism in SA-DEM is based on domain decomposition of the horizontal grid. This is the traditional distributed-memory parallelization

---

[2] Single Instruction Multiple Data, according to Flynn's taxonomy (1966).

strategy for UNI-DEM (implemented by MPI library routines). There is some domain overlapping of the advection-diffusion subproblems, a reason for some computational overhead. Its relative weight grows up with increasing the number of MPI tasks. Additional pre-processing and post-processing stages are needed for scattering the input data and gathering the results (which causes some overhead too).

Our target hardware, an IBM Blue Gene/P, can optionally offer a limited amount of shared memory parallelism. In order to exploit it efficiently, we introduced the third (finer-grain) level of parallelism in our algorithm by using OpenMP standard directives.

5   Scalability Results on the IBM Blue Gene/P

Here we present some execution times and speed-ups, obtained during experiments with the new version of SA-DEM, designed for SA studies with respect to the input emissions, on the Bulgarian IBM Blue Gene/P supercomputer. The IBM Blue Gene/P is a large high-performance system with 8192 CPU in total and peak performance more than 23 TFLOPS. It consists of 2048 compute cards (nodes), each of them being a quard core PowerPC 450 (4 CPU, 850 MHz, 2 GB RAM). A compute card is in fact a 4-CPU shared-memory computational unit with multithreading support via OpenMP. It can be used in 3 different modes: VN, DUAL and SMP. With respect to the MPI parallelism there are 4 MPI processes per node in VN mode, 2 - in DUAL mode, and one in VN mode. In the last two cases the machine offers some limited amount of shared memory parallelism. It can be exploited on the lowest level in our new 3-level parallel implementation of SA-DEM.

In accordance with the load managing policy of this machine and in order not to waste resource, our experiments with SA-DEM start from 20 CPU. Thus, for the sake of compatibility the speed-up in Table 2 is calculated as follows: $Sp(n) = 20 \frac{T(n)}{T(20)}$, where n is the number of processors (given in the first column). The time and the **speed-up (Sp)** of the main computational stages and in total are given in separate columns. The last column contains also the total efficiency E (in percent), where $E = 100 \, Sp(n)/n \, \%$.

The total time includes also the MPI communication time as well as the time for some I/O procedures, which are not parallelizable. Moreover, the larger the number of MPI tasks, the more I/O device conflicts arise, which results in a significant drop-down in the total efficiency. I/O device access appear to be the performance bottleneck in this case, partially avoided by using the lowest level OpenMP parallelization (see Table 2). On the other hand, the computational stages scale pretty well, even the speed-up of the chemistry stage tends to be slightly superlinear (due to the cache memory effects).

Table 2. Time (T) in seconds and **speed-up (Sp)** of SA-DEM for input emissions perturbation (block-partitioned version) on the IBM Blue Gene/P

Time and **speed-up** of SA-DEM (MPI+OpenMP) on the IBM Blue Gene/P $(96 \times 96 \times 1)$ grid, 35 species, CHUNKSIZE=48									
#	MPI p-s ×		Advection		Chemistry		TOTAL		
CPU	OMP thr.	MODE	T [s]	**(Sp)**	T [s]	**(Sp)**	T [s]	**(Sp)**	E [%]
20	20 × 1	SMP	6650	**(20)**	27924	**(20)**	31206	**(20)**	100%
40	40 × 1	SMP	3595	**(37)**	13621	**(41)**	16424	**(38)**	95%
80	40 × 2	DUAL	1797	**(74)**	6729	**(83)**	8790	**(71)**	89%
160	80 × 2	DUAL	911	**(146)**	3447	**(162)**	4334	**(144)**	90%
240	120 × 2	DUAL	643	**(207)**	2347	**(238)**	3448	**(181)**	75%
320	160 × 2	DUAL	506	**(263)**	1751	**(319)**	2916	**(214)**	67%
480	240 × 2	DUAL	358	**(371)**	1183	**(472)**	2517	**(248)**	52%
640	160 × 4	SMP	218	**(609)**	875	**(638)**	1914	**(326)**	51%
960	480 × 2	DUAL	209	**(635)**	579	**(964)**	1846	**(338)**	35%
960	240 × 4	SMP	153	**(867)**	585	**(955)**	1576	**(396)**	41%
1280	320 × 4	SMP	118	**(1125)**	440	**(1268)**	1418	**(440)**	34%
1920	480 × 4	SMP	94	**(1421)**	296	**(1885)**	1308	**(477)**	25%
2560	640 × 4	SMP	79	**(1674)**	290	**(1924)**	1160	**(538)**	21%
3840	960 × 4	SMP	51	**(2595)**	147	**(3807)**	957	**(652)**	17%

6   Concluding Remarks

The new 3-level parallel implementation of SA-DEM is a high performance tool for producing sensitivity analysis data, capable to exploit efficiently the computational power of the large and powerful supercomputer like IBM Blue Gene/P up to its full capacity.

Chemistry, the most computationally expensive stage of the model, scales almost perfectly in the whole range of experiments.

Advection stage scales pretty well in most of the experiments, with an expected modest slow-down in the efficiency. It is due to a significant boundary overlapping of the domain partitioning when approaching the inherent partitioning limitations.

With increasing the number of processors the time for I/O operations becomes strongly dominant. The problem comes from the insufficient number of I/O devices compared to the CPU number and other resources of the machine. This is the reason for the total efficiensy dropdown in the experiments with extremely high parallelism.

Acknowledgments. This research is partly supported by the Bulgarian NSF grants DCVP02/1/2010 (SuperCA++) and DTK 02/44/2009.

References

1. Dimov, I., Georgieva, R., Ivanovska, S., Ostromsky, T., Zlatev, Z.: Studying the Sensitivity of the Pollutants Concentrations Caused by Variations of Chemical Rates. Journal of Computational and Applied Mathematics 235(2), 391–402 (2010)
2. Dimov, I.T., Georgieva, R., Ostromsky, T., Zlatev, Z.: Sensitivity Studies of Pollutant Concentrations Calculated by UNI-DEM with Respect to the Input Emissions. Central European Journal of Mathematics, "Numerical Methods for Large Scale Scientific Computing" 11, 1531–1545 (2013)
3. Dimov, I., Ostromsky, T., Zlatev, Z.: Challenges in using splitting techniques for large-scale environmental modeling. In: Faragó, I., Georgiev, K., Havasi, Á. (eds.) Advances in Air Pollution Modeling for Environmental Security. NATO Science Series, vol. 54, pp. 115–132. Springer (2005)
4. Marchuk, G.I.: Mathematical modeling for the problem of the environment. Studies in Mathematics and Applications, vol. 16. North-Holland, Amsterdam (1985)
5. Ostromsky, T., Dimov, I., Georgieva, R., Zlatev, Z.: Air pollution modelling, sensitivity analysis and parallel implementation. Int. Journal of Environment and Pollution 46(1-2), 83–96 (2011)
6. Ostromsky, T., Zlatev, Z.: Parallel Implementation of a Large-scale 3-D Air Pollution Model. In: Margenov, S., Waśniewski, J., Yalamov, P. (eds.) LSSC 2001. LNCS, vol. 2179, pp. 309–316. Springer, Heidelberg (2001)
7. Ostromsky, T., Zlatev, Z.: Flexible Two-level Parallel Implementations of a Large Air Pollution Model, in: Numerical Methods and Applications. In: Dimov, I.T., Lirkov, I., Margenov, S., Zlatev, Z. (eds.) NMA 2002. LNCS, vol. 2542, pp. 545–554. Springer, Heidelberg (2003)
8. Saltelli, A., Tarantola, S., Campolongo, F., Ratto, M.: Sensitivity Analysis in Practice: A Guide to Assessing Scientific Models. Halsted Press (2004)
9. Sobol, I.M.: Sensitivity estimates for nonlinear mathematical models. Mathematical Modeling and Computational Experiment 1, 407–414 (1993)
10. Sobol, I.M.: Global Sensitivity Indices for Nonlinear Mathematical Models and Their Monte Carlo Estimates. Mathematics and Computers in Simulation 55(1-3), 271–280 (2001)
11. WEB-site of DEM: http://www.dmu.dk/AtmosphericEnvironment/DEM

Numerical Methods for Evolutionary Equations with Delay and Software Package PDDE

Vladimir Pimenov[1] and Andrey Lozhnikov[1,2]

[1] Ural Federal University, Ekaterinburg, Russia
[2] Institute of Mathematics and Mechanics, Ekaterinburg, Russia

Abstract. The paper gives a survey of the author's results on the grid-based numerical algorithms for solving the evolutionary equations (parabolic and hyperbolic) with the effect of heredity on a time variable. From uniform positions we construct analogs of schemes with weights for the one-dimensional heat conduction equation with delay of general form, analog of a method of variable directions for the equation of parabolic type with time delay and two spatial variables, analog of the scheme with weights for the equation of hyperbolic type with delay. For the one-dimensional heat conduction equation and the wave equation we obtained conditions on the weight coefficients that ensure stability on the prehistory of the initial function. Numerical algorithms are implemented in the form of software package Partial Delay Differential Equations (PDDE) toolbox.

Keywords: grid-based numerical methods, delay, partial differential equations, stability, convergence order, software package.

Introduction

The effects of delay of different kinds can arise in many mathematical models described by the evolutionary equations of parabolic and hyperbolic types [1]. One of well-known examples in ecology is the Hutchinson's equation with diffusion

$$\frac{\partial u}{\partial t} = a^2 \frac{\partial^2 u}{\partial x^2} + \alpha u(x,t)(1 - u(x, t - \tau)), \qquad \tau > 0$$

which can be also considered as the Kolmogorov-Petrovskii-Piskunov equation with delay. Such equations are difficult for analytical investigations and so the numerical methods for solving them are of great interest. Variants of the method of lines reduce the problem under consideration to the problem of numerical solving the systems of functional differential equations. The algorithms for these systems are elaborated well enough [2,3], but the main obstacle is the high stiffness of the systems.

In the paper we consider the methods of discretization both by time and by spatial independent variables. The methods are based on the idea of separating the discrete prehystory to the past and the present parts. By present part

I. Dimov, I. Faragó, and L. Vulkov (Eds.): NAA 2012, LNCS 8236, pp. 437–444, 2013.

(current state of required function) we construct complete analogs of the algorithms known for equations without delay [4]. For accounting the past part we use the interpolation with specified properties. These constructions lead to the systems of difference equations with the effect of heredity. The main difficulties connected with nonlinear dependance of difference equations from prehistory of discrete model are overcome by proposed earlier approach [3,5] to constructing a general scheme of numerical solution of functional differential equations, which allows to investigate local error, stability and convergence of these schemes. Using a common technique we construct and study the analog of difference scheme with weights for one-dimensional heat conduction equation with delay of general type, the analog of alternating direction method for equation of parabolic type with time delay and two spatial variables, the analog of difference scheme with weights for hyperbolic equation with delay.

It should be noted that other approaches to constructing numerical methods for evolutionary functional differential equations were studied in some papers (see, for example, [6,7]).

1 Heat Conduction Equation with Aftereffect

Let us consider parabolic-type equation with general time delay

$$\frac{\partial u}{\partial t} = a^2 \frac{\partial^2 u}{\partial x^2} + f(x, t, u(x,t), u_t(x, \cdot)); \tag{1}$$

here $u(x,t)$ is the required function, $x \in [0, X]$, $t \in [0, T]$, $u_t(x, \cdot) = \{u(x, t + s), -\tau \le s < 0\}$ is the prehistory for the required function by the time t, τ is the value of delay. We assume that the functional $f(x, t, u, v(\cdot))$ is given on $[0, X] \times [0, T] \times R \times Q$. We denote by $Q = Q[-\tau, 0)$ the set of functions $u(s)$ that are piecewise continuous on $[-\tau, 0]$ with a finite number of points of discontinuity of first kind and right continuous at the points of discontinuity, $\|u(\cdot)\|_Q = \sup_{s \in [-\tau, 0)} |u(s)|$.

Let the initial conditions $u(x, t) = \varphi(x, t)$, $x \in [0, X]$, $t \in [-\tau, 0]$, and the boundary conditions $u(0, t) = g_0(t)$, $u(X, t) = g_1(t)$, $t \in [0, T]$, be given.

We assume that the functional f and the functions g_1, g_2, φ are such that this problem has a unique solution $u(x, t)$ [1].

Let us make discretization of problem. Let $h = X/N$, introducing the points $x_i = ih$, $i = 0, \ldots, N$, and let $\Delta = T/M$, introducing the points $t_j = j\Delta$, $j = 0, \ldots, M$. We assume that $\tau/\Delta = K$ is a positive integer. Denote by u_j^i approximations of functions $u(x_i, t_j)$ at the nodes. For every fixed $i = 0, \ldots, N$, introduce the discrete prehistory by time t_j, $j = 0, \ldots, M$: $\{u_k^i\}_j = \{u_k^i, j - m \le k \le j\}$. The interpolation-extrapolation operator is, by definition, the operator defined on set of all admissible prehistories and acting by the rule $I : \{u_k^i\}_j \rightarrow v^i(\cdot) \in Q[-\tau, \Delta]$. Ways of creation of operators of interpolation-extrapolation are considered in [3].

For $0 \le s \le 1$ consider the family of methods

$$\frac{u^i_{j+1} - u^i_j}{\Delta} = sa^2 \frac{u^{i-1}_{j+1} - 2u^i_{j+1} + u^{i+1}_{j+1}}{h^2} + (1-s)a^2 \frac{u^{i-1}_j - 2u^i_j + u^{i+1}_j}{h^2} + F^i_j(v^i(\cdot)),$$
(2)

$$i = 1, \ldots, N-1, \quad j = 0, \ldots, M-1$$

with the initial conditions $u^i_0 = \varphi(x_i, 0)$, $i = 0, \ldots, N$, $v^i(t) = \varphi(x_i, t)$, $t < 0$, $i = 0, \ldots, N$, and the boundary conditions $u^0_j = g_0(t_j)$, $u^N_j = g_1(t_j)$, $j = 0, \ldots, M$. Here the functional $F^i_j(v^i(\cdot))$ is defined on $Q[-\tau, \Delta]$ and related to the functional $f(x_i, t_j, u^i_j, v^i(\cdot))$. The functions $v^i(\cdot)$ are images under the interpolation-extrapolation operator.

Denote $\varepsilon^i_j = u(x_i, t_j) - u^i_j$, $i = 0, \ldots, N$, $j = 0, \ldots, M$.

We will say that the method converges if $\varepsilon^i_j \to 0$ as $h \to 0$ and $\Delta \to 0$ for all $i = 0, \ldots, N$ and $j = 0, \ldots, M$. We will say that the method converges with order $h^p + \Delta^q$, if there exists a constant C such that $|\varepsilon^i_j| \le C(h^p + \Delta^q)$ for all $i = 0, \ldots, N$ and $j = 0, \ldots, M$.

The residual is, by definition, the grid function

$$\Psi^i_j = \frac{u(x_i, t_{j+1}) - u(x_i, t_j)}{\Delta} - sa^2 \frac{u(x_{i-1}, t_{j+1}) - 2u(x_i, t_{j+1}) + u(x_{i+1}, t_{j+1})}{h^2} -$$

$$(1-s)a^2 \frac{u(x_{i-1}, t_j) - 2u(x_i, t_j) + u(x_{i+1}, t_j)}{h^2} - F^i_j(u_{t_j}(x_i, (\cdot))).$$

Theorem 1. *Let the relation*

$$s \ge \frac{1}{2} - \frac{1}{4\sigma}, \qquad \sigma = \frac{a^2 \Delta}{h^2}$$

holds, the residual has order $\Delta^{p_1} + h^{p_2}$, the function F^i_j satisfies Lipschitz condition, the operator of interpolation-extrapolation I satisfies Lipschitz condition and has error order Δ^{p_0} [3]. Then the method converges with order $\Delta^{\min\{p_1, p_0\}} + h^{p_2}$.

The proof of this statement is based on a combination of methods of the general theory of difference schemes [4] and methods of the general scheme of numerical methods for functional-differential equations [3,5]. This proof is published in work [8], concrete algorithms and results of numerical experiments are given in the same work.

2 Alternating Direction Method

Let us consider two-dimensional parabolic equation with general time delay

$$\frac{\partial u}{\partial t} = a^2 \left(\frac{\partial^2 u}{\partial x^2} + \frac{\partial^2 u}{\partial y^2} \right) + f(x, y, t, u(x, y, t), u_t(x, y, \cdot)),$$
(3)

$$x \in [0, X], \, y \in [0, Y], \, t \in [0, T], \, u_t(x, y, \cdot) = \{u(x, y, t+s), -\tau \le s < 0\}.$$

Let the initial conditions $u(x, y, t) = \varphi(x, y, t)$, $x \in [0, X]$, $y \in [0, Y]$, $t \in [-\tau, 0]$, and the boundary conditions $u(0, y, t) = g_0(y, t)$, $u(X, y, t) = g_1(y, t)$, $y \in [0, Y]$, $t \in [0, T]$, $u(x, 0, t) = g_2(x, t)$, $u(x, Y, t) = g_3(x, t)$, $x \in [0, X]$, $t \in [0, T]$, be given. We assume that the functional $f(x, y, t, u, v(\cdot))$ is defined on $[0, X] \times [0, Y] \times [0, T] \times R \times Q$.

Let us make discretization of the problem and construct the difference scheme: $h_x = X/N_1$, $h_y = Y/N_2$, $x_i = ih_x$, $i = 0, \ldots, N_1$, $y_j = jh_y$, $j = 0, \ldots, N_2$, $\Delta = T/M$, $t_k = k\Delta$, $k = 0, \ldots, M$, $\tau/\Delta = m$ is integer, $u(x_i, y_j, t_k) \approx u_k^{i,j}$, $\{u_l^{i,j}\}_k = \{u_l^{i,j}, k - m \leq l \leq k\}$, $I : \{u_l^{i,j}\}_k \to v^{i,j}(\cdot) \in Q[-\tau, \Delta]$,

$$\frac{u_{k+\frac{1}{2}}^{i,j} - u_k^{i,j}}{\Delta/2} = \frac{a^2}{h_x^2}\left(u_{k+\frac{1}{2}}^{i+1,j} - 2u_{k+\frac{1}{2}}^{i,j} + u_{k+\frac{1}{2}}^{i-1,j}\right) + \frac{a^2}{h_y^2}\left(u_k^{i,j+1} - 2u_k^{i,j} + u_k^{i,j-1}\right) +$$

$$F_{k+\frac{1}{2}}^{i,j}(v^{i,j}(\cdot)),$$

$$\frac{u_{k+1}^{i,j} - u_{k+\frac{1}{2}}^{i,j}}{\Delta/2} = \frac{a^2}{h_x^2}\left(u_{k+\frac{1}{2}}^{i+1,j} - 2u_{k+\frac{1}{2}}^{i,j} + u_{k+\frac{1}{2}}^{i-1,j}\right) + \frac{a^2}{h_y^2}\left(u_{k+1}^{i,j+1} - 2u_{k+1}^{i,j} + u_{k+1}^{i,j-1}\right) +$$

$$F_{k+\frac{1}{2}}^{i,j}(v^{i,j}(\cdot)),$$

$F_k^{i,j}(v^{i,j}(\cdot)) = f(x_i, y_j, t_k, u_k^{i,j}, v_{t_k}^{i,j}(\cdot))$. Initial and boundary conditions are set respectively.

The proof of convergence of the method and results of numerical experiments are given in [9].

3   Wave Equation with Aftereffect

Let us consider wave equation with the effect of heredity

$$\frac{\partial^2 u}{\partial t^2} = a^2 \frac{\partial^2 u}{\partial x^2} + f(x, t, u(x, t), u_t(x, \cdot)), \tag{4}$$

$0 \leq t \leq T$, $0 \leq x \leq X$, with the initial conditions $u(0, t) = g_1(t)$, $u(X, t) = g_2(t)$, $0 \leq t \leq T$, and the boundary conditions $u(x, t) = \varphi(x, t)$, $0 \leq x \leq X$, $-\tau \leq t \leq 0$.

Let us make discretization of problem as for the equation of hyperbolic type and for $0 \leq s \leq 1$ let us consider the set of methods:

$$\frac{u_{j+1}^i - 2u_j^i + u_{j-1}^i}{\Delta^2} = sa^2 \frac{u_{j+1}^{i-1} - 2u_{j+1}^i + u_{j+1}^{i+1}}{h^2} + sa^2 \frac{u_{j-1}^{i-1} - 2u_{j-1}^i + u_{j-1}^{i+1}}{h^2} +$$

$$(1 - 2s)a^2 \frac{u_j^{i-1} - 2u_j^i + u_j^{i+1}}{h^2} + F_j^i(v^i(\cdot)),$$

with the boundary conditions $u_j^0 = g_1(t_j)$, $u_j^N = g_2(t_j)$, $j = 0, \ldots, M$, and the initial conditions $u_j^i = \varphi(x_i, t_j)$, $i = 0, \ldots, N$, $j \leq 0$. $F_j^i(v^i(\cdot))$ has the same sense as for the equation of parabolic type.

For $s = 0$ we obtain an explicit scheme, for $0 < s \leq 1$ and any fixed j we obtain a linear tridiagonal system with respect to u_{j+1}^i with diagonal dominance that can be effectively solved by sweep method.

The detailed analysis of errors of this set of methods is carried out in the article [10].

4   The Software Package PDDE and Some Examples

All described above numerical algorithms are implemented in MATLAB in the form of software package PDDE (Partial Delay Differential Equation) toolbox. The package contains the programs for numerical solution of parabolic equations with one and two spatial variables and equations of hyperbolic type. These equations can contain both lumped variable delays and some types of distributed (integral) delays. For example, the parabolic equation with one spatial variable can include two type integral delays:

$$\text{a)} \quad \int_{-\tau(t)}^{0} g_1(x, s)\, u(x, t + s)\, ds, \qquad \text{b)} \quad \int_{-\tau(t)}^{0} g_2(u(x, t + s))\, ds.$$

Further we consider the examples of numerical simulation of some equations using PDDE toolbox.

Example 1. The equation with variable delay

$$\frac{\partial u(x, t)}{\partial t} = 0.04 \frac{\partial^2 u(x, t)}{\partial x^2} - \frac{0.04 + x}{t^2} \sqrt{u(x, t - t/2)} \tag{5}$$

with initial and boundary conditions

$$u(x, r) = e^{x/r}, \quad 0.75 \leq r \leq 1.5, \ 1 \leq x \leq 5; u(1, t) = e^{1/t}, \ u(5, t) = e^{5/t}, \quad 1.5 \leq t \leq 6$$

has the exact solution $u(x, t) = e^{x/t}$. The approximate solution of this equation was obtained by grid method (2) with weight $s = 0.5 - 1/(12\sigma)$. Table 1 contains the comparison of the norms of difference between the matrices of the exact and approximate solutions of the equation (5) for different steps h and Δ.

Table 1. The norms of differences for different steps

N	10	10	10	10	20	50	100
M	10	20	50	100	10	10	10
$\|U\|$	1.3311	0.3508	0.0570	0.0143	2.7700	7.0006	14.029

Here N and M are the numbers of partition points corresponding to steps h and Δ. The norms of the difference were calculated by the formula

$$\|U\| = \max_{0 \leq j \leq M} \sum_{i=0}^{N} |u(x_i, t_j) - u_j^i|.$$

Table 1 shows that the ratio of steps strongly affect the accuracy of calculations.

Example 2. The following example demonstrates the effect of delay to solution. The equation

$$\frac{\partial u(x,t)}{\partial t} = \frac{\partial^2 u(x,t)}{\partial x^2} - 2\,u(x,t-\tau), \qquad \tau > 0 \tag{6}$$

with initial and boundary conditions

$$u(x,r) = 1, \quad 1-\tau \le r \le 1, \quad 0 \le x \le 4; \quad u(0,t) = u(4,t) = 1, \quad 1 \le t \le 10$$

has the solutions presented in Figs. 1–4 with different τ. The appearance of delay induces oscillations of solution. The greater the delay value, the greater the amplitude of the oscillations and the lower their frequency. With small delay values the solution is stabilizing over the time, the oscillations are damping. With large delay values the solution becomes unstable, the amplitude of oscillations are increasing over the time.

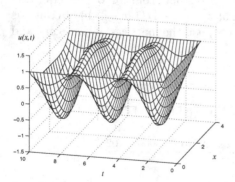

Fig. 1. Solution of (6) with $\tau = 0$

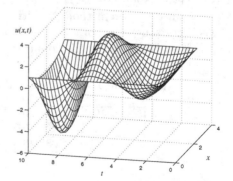

Fig. 2. Solution of (6) with $\tau = 0.5$

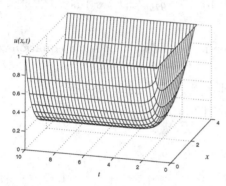

Fig. 3. Solution of (6) with $\tau = 1$

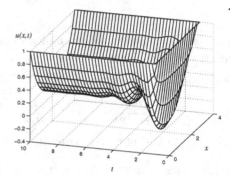

Fig. 4. Solution of (6) with $\tau = 2$

Example 3. Let us consider the solutions of the Hutchinson's equation with diffusion

$$\frac{\partial u}{\partial t} = \frac{\partial^2 u}{\partial x^2} + 2\,u(x,t)(1 - u(x, t - \tau)), \qquad \tau > 0 \tag{7}$$

with initial and boundary conditions

$$u(x,r) = 0.1, \quad 1 - \tau \leq r \leq 1, \quad 0 \leq x \leq 4; \quad u(0,t) = u(4,t) = 0.1, \quad 1 \leq t \leq 10.$$

Figures 5–8 show similar to the previous example effects due to the occurrence of delay. But unlike Example 2 here one can notice an interval of delay, within which the corresponding solutions come to steady oscillations.

The properties of solutions of equation (7) were studied in many papers (see, for example, [11] and references therein).

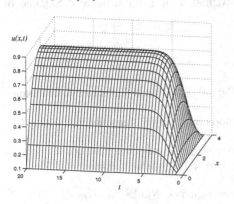

Fig. 5. Solution of (7) with $\tau = 0$

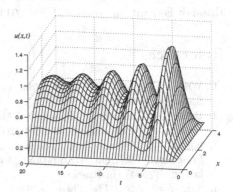

Fig. 6. Solution of (7) with $\tau = 0.9$

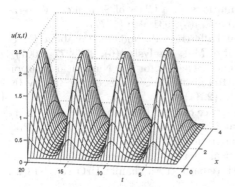

Fig. 7. Solution of (7) with $\tau = 1.4$

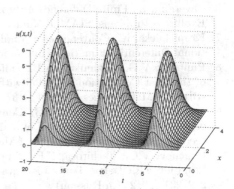

Fig. 8. Solution of (7) with $\tau = 2$

Conclusion

The paper presents some principles of construction of analogs of the classical grid schemes for the equations with delay of general type. The methods are based on the idea of separating the past and the present parts in the structure of the object state as well as on the use of interpolation and extrapolation with specified properties, for the low-order schemes the interpolation is piecewise-linear with extrapolation by continuation. The simplicity and efficiency of the method has allowed to create the appropriate software.

Further development of this technique assumes its application to more general multidimensional problems of parabolic and hyperbolic types with various boundary conditions, to the construction of the various grid schemes solving the advection equation with delay, mixed functional differential equations, and other evolutionary problems with heredity.

Acknowledgement. This work was supported by Russian Foundation for Basic Research (project 13-01-00089) and by Ministry of Education and Science of Russian Federation (project 1.994.2011).

References

1. Wu, J.: Theory and Applications of Partial Functional Differential Equations. Springer, New York (1996)
2. Bellen, A., Zennaro, M.: Numerical Methods for Delay Differential Equations. Oxford Science Publications, Oxford (2003)
3. Kim, A.V., Pimenov, V.G.: i-Smooth Analysis and Numerical Methods for Solving Functional Differential Equations. Regulyarn, Khaotichesk, Dinamika, Moscow-Izhevsk (2004) (in Russian)
4. Samarskii, A.A.: The Theory of Difference Schemes. Nauka, Moscow (1977)
5. Pimenov, V.G.: General Linear Methods for Numerical Solving Functional Differential Equations. Differential Equations 37(1), 105–114 (2001) (in Russian)
6. Tavernini, L.: Finite Difference Approximations for a Class of Semilinear Volterra Evolution Problems. SIAM J. Numer. Anal. 14(5), 931–949 (1977)
7. Kropielnicka, K.: Convergence of Implicit Difference Methods for Parabolic Functional Differential Equations. Int. Journal of Mat. Analysis 1(6), 257–277 (2007)
8. Pimenov, V.G., Lozhnikov, A.B.: Difference Schemes for the Numerical Solution of the Heat Conduction Equation with Aftereffect. Proceedings of the Steklov Institute of Mathematics 275(S. 1), 137–148 (2011)
9. Lekomtsev, A.V., Pimenov, V.G.: Convergence of the Alternating Direction Methods for the Numerical Solution of a Heat Conduction Equation with Delay. Proceedings of the Steklov Institute of Mathematics 272(1), 101–118 (2011)
10. Pimenov, V.G., Tashirova, E.E.: Numerical Methods for Solution of Hyperbolic Equation with Delay. Trudy Instituta Matematiki i Mekhaniki UrO RAN 18(2), 222–231 (2012) (in Russian)
11. Faria, T., Huang, W.: Stability of periodic solutions arising from Hopf bifurcation for a reaction-diffusion equation with time delay. Differential Equations and Dynamical Systems 31, 125–141 (2002)

Interior Layers in Coupled System of Two Singularly Perturbed Reaction-Diffusion Equations with Discontinuous Source Term

S. Chandra Sekhara Rao and Sheetal Chawla

Department of Mathematics, Indian Institute of Technology Delhi,
Hauz Khas, New Delhi-110 016, India
scsr@maths.iitd.ac.in, chawlaasheetal@gmail.com

Abstract. We study a coupled system of two singularly perturbed linear reaction-diffusion equations with discontinuous source term. A central difference scheme on layer-adapted piecewise-uniform mesh is used to solve the system numerically. The scheme is proved to be almost first order uniformly convergent, even for the case in which the diffusion parameter associated with each equation of the system has a different order of magnitude. Numerical results are presented to support the theoretical results.

Keywords: Singular perturbation, Coupled system, Discontinuous source term, Uniformly convergent, Shishkin mesh, Interior layers.

1 Introduction

Consider a coupled system of singularly perturbed reaction-diffusion equations with discontinuous source term on the unit interval $\Omega = (0,1)$, and assume a single discontinuity in the source term at a point $d \in \Omega$. Let $\Omega_1 = (0,d)$ and $\Omega_2 = (d,1)$, and let the jump at d in any function is given as $[\omega](d) = \omega(d+) - \omega(d-)$.

The corresponding boundary value problem is:

Find $u_1, u_2 \in C^0(\overline{\Omega}) \cap C^1(\Omega) \cap C^2(\Omega_1 \cup \Omega_2)$, such that

$$(\boldsymbol{L}\boldsymbol{u})_1(x) := -\epsilon_1 u_1''(x) + a_{11}(x)u_1(x) + a_{12}(x)u_2(x) = f_1(x), \qquad x \in \Omega_1 \cup \Omega_2. \quad (1)$$

$$(\boldsymbol{L}\boldsymbol{u})_2(x) := -\epsilon_2 u_2''(x) + a_{21}(x)u_1(x) + a_{22}(x)u_2(x) = f_2(x), \qquad x \in \Omega_1 \cup \Omega_2. \quad (2)$$

$$u_1(0) = p, \quad u_1(1) = q, \quad u_2(0) = r, \quad u_2(1) = s, \qquad (3)$$

where ϵ_1, ϵ_2 are small parameters such that $0 < \epsilon_1 \le \epsilon_2 \le 1$.

We assume that the coupling matrix satisfies the following positivity conditions.

$$a_{12} \le 0 \text{ and } a_{21} \le 0, \quad \text{for all } x \in \overline{\Omega}, \qquad (4)$$

I. Dimov, I. Faragó, and L. Vulkov (Eds.): NAA 2012, LNCS 8236, pp. 445–453, 2013.

and for some constant α we have

$$0 < \alpha := \min_{\overline{\Omega}}\{\alpha_1, \alpha_2\}, \quad \alpha_1 := \min_{x \in \overline{\Omega}}\{a_{11}(x) + a_{12}(x)\}, \quad \alpha_2 := \min_{x \in \overline{\Omega}}\{a_{21}(x) + a_{22}(x)\}$$

$$(5)$$

The source terms f_1, f_2 are sufficiently smooth on $\overline{\Omega} \setminus \{d\}$, and their derivatives have jump discontinuity at the same point. Due to that discontinuity, interior layers arise in the solution of the problem.

In the literature for the scalar case, interior layers were considered in addition to the boundary layers [1–4]. A numerical method for singularly perturbed coupled system of two second order ordinary differential equations with equal parameters having discontinuous source term was considered in [6].

This paper is organized as follows. Section 1, presents the properties of the exact solution of the problem. In Section 2, the discretization of the problem is given using central difference scheme and a piecewise uniform Shishkin mesh, which is fitted to both boundary and interior layers. In Section 3, error analysis of the present method is given. Results of numerical experiments are presented in Section 4.

Notations: We shall use C to denote a generic constant and $\boldsymbol{C} = (C, C)^T$ a generic positive constant vector such that both are independent of perturbation parameters ϵ_1, ϵ_2 and also of the discretization parameter N but may not be same at each occurrence. Define $\boldsymbol{v} \leq \boldsymbol{w}$ if $v_1(x) \leq w_1(x)$ and $v_2(x) \leq w_2(x)$ $x \in [0,1]$. We consider the maximum norm and denote it by $\|.\|_S$, where S is a closed and bounded subset in $[0,1]$. For a real valued function $v \in C(S)$ and for a vector valued function $\boldsymbol{v} = (v_1, v_2)^T \in C(S)^2$, we define $\|v\|_S := \max_{x \in S}|v(x)|$ and $\|\boldsymbol{v}\|_S := \max\{\|v_1\|_S, \|v_2\|_S\}$.

Theorem 1. *The problem (1)-(3) has a solution* $\boldsymbol{u} = (u_1, u_2)^T$ *with* $u_1, u_2 \in C^1(\Omega) \cap C^2(\Omega_1 \cup \Omega_2)$.

Proof. The result can be proved by following similar arguments considered in [6].

Theorem 2. *Suppose* $u_1, u_2 \in C^0(\overline{\Omega}) \cap C^2(\Omega_1 \cup \Omega_2)$. *Further suppose that* $\boldsymbol{u} = (u_1, u_2)^T$ *satisfies* $\boldsymbol{u}(0) \geq \boldsymbol{0}$, $\boldsymbol{u}(1) \geq \boldsymbol{0}$, $\boldsymbol{Lu(x)} \geq \boldsymbol{0}$ *in* $\Omega_1 \cup \Omega_2$ *and* $[\boldsymbol{u}'](d) \leq \boldsymbol{0}$. *Then* $\boldsymbol{u}(x) \geq \boldsymbol{0}$, *for all* $x \in \overline{\Omega}$.

Proof. Let $u_1(p) := \min_{x \in \overline{\Omega}}\{u_1(x)\}$ and $u_2(q) := \min_{x \in \overline{\Omega}}\{u_2(x)\}$. Assume without loss of generality $u_1(p) \leq u_2(q)$. If $u_1(p) \geq 0$, then there is nothing to prove. Suppose that $u_1(p) < 0$, then proof is completed by showing that this leads to contradiction. Note that $p \neq \{0,1\}$, So either $p \in \Omega_1 \cup \Omega_2$ or p=d. In the first case, $(\boldsymbol{Lu})_1(p) = -\epsilon_1 u_1''(p) + a_{11}(p)u_1(p) + a_{12}(p)u_2(p) < 0$.
In the other case proceed as in [2], with the assumption that there exists a neighborhood $N_h = (d - h, d)$ such that $u_1(x) < 0$ and $u_1(x) < u_2(x)$ for all $x \in N_h$. Now choose a point $x_1 \neq d$, $x_1 \in N_h$ such that $u_1(x_1) > u_1(d)$. It follows from the mean value theorem that, for some $x_2 \in N_h$, $u_1'(x_2) < 0$, and

for some $x_3 \in N_h$, $u_1''(x_3) > 0$; also note that $u_1(x_3) < 0$, since $x_3 \in N_h$. Thus $(\boldsymbol{Lu})_1(x_3) = -\epsilon_1 u_1''(x_3) + a_{11}(x_3)u_1(x_3) + a_{12}(x_3)u_2(x_3) < 0$, which leads to a contradiction.

Lemma 1. *Let $A(x)$ satisfy (4)-(5). If $\boldsymbol{u} = (u_1, u_2)^T$ be the solution of (1), then,*

$$\|\boldsymbol{u}\|_{\overline{\Omega}} \leq max\{\|\boldsymbol{u}(0)\|, \|\boldsymbol{u}(1)\|, \tfrac{1}{\alpha}\|\boldsymbol{f}\|_{\Omega_1 \cup \Omega_2}\}.$$

Proof. Define the functions
$$\Psi^{\pm}(x) := (\psi_1^{\pm}(x), \psi_2^{\pm}(x))^T \text{ with } \psi_1^{\pm}(x) := M \pm u_1(x) \text{ and } \psi_2^{\pm}(x) := M \pm u_2(x),$$
where $M = max\{\|\boldsymbol{u}(0)\|, \|\boldsymbol{u}(1)\|, \tfrac{1}{\alpha}\|\boldsymbol{f}\|_{\Omega_1 \cup \Omega_2}\}$ and $\Psi^{\pm}(x) = M\boldsymbol{e} \pm \boldsymbol{u}(x)$.

Note that $\boldsymbol{e} := (1,1)^T$ is the unit column vector.

Now, $\Psi^{\pm}(0) \geq \boldsymbol{0}$, $\Psi^{\pm}(1) \geq \boldsymbol{0}$, $\boldsymbol{L}\Psi^{\pm}(x) \geq \boldsymbol{0}$ for all $x \in \Omega_1 \cup \Omega_2$ and $[(\Psi^{\pm})'](d) = \pm[\boldsymbol{u}'](d) = \boldsymbol{0}$ as $\boldsymbol{u} \in C^1(\Omega)^2$, it follows from the maximum principle that $\Psi^{\pm}(x) \geq \boldsymbol{0}$ for all $x \in \overline{\Omega}$, which leads to the required bound on $\boldsymbol{u}(x)$.

To derive the sharper bounds on the solution, the solution be decomposed into a sum, composed of a discontinuous regular component $\boldsymbol{v}$, and a discontinuous singular component $\boldsymbol{w}$. That is, $\boldsymbol{u} = \boldsymbol{v} + \boldsymbol{w}$. These components are defined as the solutions of the following problems:

$$\boldsymbol{Lv}(x) = \boldsymbol{f}(x), \quad x \in \Omega_1 \cup \Omega_2, \quad \boldsymbol{v}(x) = A^{-1}(x)\boldsymbol{f}(x), \quad x \in \{0, d-, d+, 1\}. \quad (6)$$

By the assumptions (4)-(5) the matrix A is invertible, and

$$\boldsymbol{Lw}(x) = \boldsymbol{0}, \qquad\qquad x \in \Omega_1 \cup \Omega_2; \qquad\qquad (7)$$

$$\boldsymbol{w}(x) = \boldsymbol{u}(x) - \boldsymbol{v}(x), \quad x \in \{0,1\}, \quad [\boldsymbol{w}](d) = -[\boldsymbol{v}](d), \quad [\boldsymbol{w}'](d) = -[\boldsymbol{v}'](d). \quad (8)$$

The following layer functions are used in defining the bounds on derivatives:

$$B_{\epsilon_{l_1}}(x) := exp(-x\sqrt{\alpha/\epsilon_1}) + exp(-(d-x)\sqrt{\alpha/\epsilon_1}), \qquad\qquad (9)$$
$$B_{\epsilon_{r_1}}(x) := exp((d-x)\sqrt{\alpha/\epsilon_1}) + exp(-(1-x)\sqrt{\alpha/\epsilon_1}), \qquad\qquad (10)$$
$$B_{\epsilon_{l_2}}(x) := exp(-x\sqrt{\alpha/\epsilon_2}) + exp(-(d-x)\sqrt{\alpha/\epsilon_2}), \qquad\qquad (11)$$
$$B_{\epsilon_{r_2}}(x) := exp((d-x)\sqrt{\alpha/\epsilon_2}) + exp(-(1-x)\sqrt{\alpha/\epsilon_2}). \qquad\qquad (12)$$

Following the technique used in [5], the bounds on the discontinuous regular and discontinuous singular components; and on their derivarives are obtained.

Lemma 2. *The discontinuous regular component $\boldsymbol{v}$ and its derivatives satisfy the bounds given by $\|\boldsymbol{v}^{(k)}\|_{\Omega_1 \cup \Omega_2} \leq C$ for $k=0,1,2.$ with*

$$\|v_1'''\|_{\Omega_1 \cup \Omega_2} \leq C\epsilon_1^{-1/2} \quad and \quad \|v_2'''\|_{\Omega_1 \cup \Omega_2} \leq C\epsilon_2^{-1/2}.$$

Lemma 3. *The discontinuous singular component $\boldsymbol{w}$ and its derivatives satisfy the bounds given by*

$$|w_1(x)| \leq \begin{cases} CB_{\epsilon_{l_2}}(x), & x \in \Omega_1 \\ CB_{\epsilon_{r_2}}(x), & x \in \Omega_2, \end{cases} \qquad\qquad |w_2(x)| \leq \begin{cases} CB_{\epsilon_{l_2}}(x), & x \in \Omega_1 \\ CB_{\epsilon_{r_2}}(x), & x \in \Omega_2, \end{cases}$$

$$|w_1'(x)| \leq \begin{cases} C(\epsilon_1^{-1/2}B_{\epsilon l_1}(x) + \epsilon_2^{-1/2}B_{\epsilon l_2}(x)), \\ C(\epsilon_1^{-1/2}B_{\epsilon r_1}(x) + \epsilon_2^{-1/2}B_{\epsilon r_2}(x)), \end{cases} \quad |w_2'(x)| \leq \begin{cases} C(\epsilon_2^{-1/2}B_{\epsilon l_2}(x)), \ x \in \Omega_1 \\ C(\epsilon_2^{-1/2}B_{\epsilon r_2}(x)), \ x \in \Omega_2, \end{cases}$$

$$|w_1''(x)| \leq \begin{cases} C(\epsilon_1^{-1}B_{\epsilon l_1}(x) + \epsilon_2^{-1}B_{\epsilon l_2}(x)), \\ C(\epsilon_1^{-1}B_{\epsilon r_1}(x) + \epsilon_2^{-1}B_{\epsilon r_2}(x)), \end{cases} \quad |w_2''(x)| \leq \begin{cases} C(\epsilon_2^{-1}B_{\epsilon l_2}(x)), \ x \in \Omega_1 \\ C(\epsilon_2^{-1}B_{\epsilon r_2}(x)), \ x \in \Omega_2, \end{cases}$$

$$|w_1'''(x)| \leq \begin{cases} C(\epsilon_1^{-3/2}B_{\epsilon l_1}(x) + \epsilon_2^{-3/2}B_{\epsilon l_2}(x)), \ x \in \Omega_1 \\ C(\epsilon_1^{-3/2}B_{\epsilon r_1}(x) + \epsilon_2^{-3/2}B_{\epsilon r_2}(x)), \ x \in \Omega_2, \end{cases}$$

$$|w_2'''(x)| \leq \begin{cases} C\epsilon_2^{-1}(\epsilon_1^{-1/2}B_{\epsilon l_1}(x) + \epsilon_2^{-1/2}B_{\epsilon l_2}(x)), \quad x \in \Omega_1 \\ C\epsilon_2^{-1}(\epsilon_1^{-1/2}B_{\epsilon r_1}(x) + \epsilon_2^{-1/2}B_{\epsilon r_2}(x)), \ x \in \Omega_2. \end{cases}$$

To get more sharper bounds, a more precise decomposition of the function w is required. This is achieved by following the technique used in [5].

Lemma 4. *Suppose that* $\epsilon_1 < \frac{\epsilon_2}{4}$ *holds. Then the components* w_1 *and* w_2 *can be decomposed as follows:*

$$w_1(x) = w_{1,\epsilon_1}(x) + w_{1,\epsilon_2}(x), \qquad w_2(x) = w_{2,\epsilon_1}(x) + w_{2,\epsilon_2}(x),$$

where

$$|w_{1,\epsilon_1}''(x)| \leq \begin{cases} C\epsilon_1^{-1}B_{\epsilon l_1}(x), \ x \in \Omega_1 \\ C\epsilon_1^{-1}B_{\epsilon r_1}(x), \ x \in \Omega_2, \end{cases} \quad |w_{1,\epsilon_2}'''(x)| \leq \begin{cases} C\epsilon_2^{-3/2}B_{\epsilon l_2}(x), \ x \in \Omega_1 \\ C\epsilon_2^{-3/2}B_{\epsilon r_2}(x), \ x \in \Omega_2, \end{cases}$$

$$|w_{2,\epsilon_1}''(x)| \leq \begin{cases} C\epsilon_2^{-1}B_{\epsilon l_1}(x), \ x \in \Omega_1 \\ C\epsilon_2^{-1}B_{\epsilon r_1}(x), \ x \in \Omega_2, \end{cases} \quad |w_{2,\epsilon_2}'''(x)| \leq \begin{cases} C\epsilon_2^{-3/2}B_{\epsilon l_2}(x), \ x \in \Omega_1 \\ C\epsilon_2^{-3/2}B_{\epsilon r_2}(x), \ x \in \Omega_2. \end{cases}$$

2   Discretization of the Problem

We use a piecewise uniform Shishkin mesh which uses these transition parameters:

$$\sigma_{\epsilon l_2} := \min\left\{\frac{d}{4}, \sqrt{\frac{\epsilon_2}{\alpha}}\ln N\right\}, \qquad \sigma_{\epsilon r_2} := \min\left\{\frac{(1-d)}{4}, \sqrt{\frac{\epsilon_2}{\alpha}}\ln N\right\},$$

$$\sigma_{\epsilon l_1} := \min\left\{\frac{d}{8}, \frac{\sigma_{\epsilon l_2}}{2}, \sqrt{\frac{\epsilon_1}{\alpha}}\ln N\right\}, \quad \sigma_{\epsilon r_1} := \min\left\{\frac{(1-d)}{4}, \frac{\sigma_{\epsilon r_2}}{2}, \sqrt{\frac{\epsilon_1}{\alpha}}\ln N\right\}.$$

The interior points of the mesh are denoted by

$$\Omega^N = \{x_i : 1 \leq i \leq \tfrac{N}{2} - 1\} \cup \{x_i : \tfrac{N}{2} + 1 \leq i \leq N - 1\} = \Omega_1^N \cup \Omega_2^N.$$

Let $h_i = x_i - x_{i-1}$ be the ith mesh step and $\hbar_i = \frac{h_i + h_{i+1}}{2}$, clearly $x_{\frac{N}{2}} = d$ and $\overline{\Omega}^N = \{x_i : i = 0, 1, 2, \ldots..N\}$. Let N=$2^l$, $l \geq 5$ be any positive integer.

We divide $\overline{\Omega}_1^N$ into five sub-intervals $[0, \sigma_{\epsilon l_1}], [\sigma_{\epsilon l_1}, \sigma_{\epsilon l_2}], [\sigma_{\epsilon l_2}, d - \sigma_{\epsilon l_2}], [d - \sigma_{\epsilon l_2}, d - \sigma_{\epsilon l_1}]$ and $[d - \sigma_{\epsilon l_1}, d]$. The sub-interval $[0, \sigma_{\epsilon l_1}] [\sigma_{\epsilon l_1}, \sigma_{\epsilon l_2}], [d - \sigma_{\epsilon l_2}, d - \sigma_{\epsilon l_1}]$ and $[d - \sigma_{\epsilon l_1}, d]$ is divided into $\frac{N}{16}$ equidistant elements and the sub-interval $[\sigma_{\epsilon l_2}, d - \sigma_{\epsilon l_2}]$ is divided into $\frac{N}{4}$ equidistant elements. Similarly, in $\overline{\Omega}_2^N$ the sub-intervals $[d, d + \sigma_{\epsilon r_1}], [d + \sigma_{\epsilon r_1}, d + \sigma_{\epsilon r_2}], [1 - \sigma_{\epsilon r_2}, 1 - \sigma_{\epsilon r_1}]$ and $[1 - \sigma_{\epsilon r_1}, 1]$ is divided into $\frac{N}{16}$ equidistant elements and the sub-interval $[d + \sigma_{\epsilon r_2}, 1 - \sigma_{\epsilon r_2}]$ is divided into $\frac{N}{4}$ equidistant elements.

On the piecewise-uniform mesh Ω^N a standard centered finite difference operator is used. Then the fitted mesh method for the system is:

Define the discrete finite difference operator L^N as follows:

$$L^N U = f \qquad \text{for all } x_i \in \Omega^N \tag{13}$$

with boundary conditions $U(x_0) = p$, $\quad U(x_N) = q$, $\tag{14}$

where $L^N = -E\delta^2 + A(x)$, and at $x_{N/2} = d$ the scheme is given by

$$L^N U(d) \equiv -E\delta^2 U(d) + A(d) U(d) = \overline{f}(d) \tag{15}$$

where $\delta^2 Z(x_i) = (D^+ Z(x_i) - D^- Z(x_i))\frac{1}{\overline{h}_i}$, $\quad D^+ Z(x_i) = \frac{Z(x_{i+1}) - Z(x_i)}{h_{i+1}}$

$$D^- Z(x_i) = \frac{Z(x_i) - Z(x_{i-1})}{h_i} \quad \text{and} \quad \overline{f}(d) = \frac{f(d - h_{\frac{N}{2}}) + f(d + h_{\frac{N}{2}+1})}{2}.$$

Lemma 5. *Suppose that a mesh function* $\mathbf{W}(x_i)$ *satisfies:* $\mathbf{W}(0) \geq 0$, $\mathbf{W}(1) \geq 0$, $(\mathbf{L}^N \mathbf{W})(x_i) \geq 0$, *for all* $x_i \in \overline{\Omega}^N$, *then* $\mathbf{W}(x_i) \geq \mathbf{0}$ *for all* $x_i \in \overline{\Omega}^N$.

Proof. Let x_{p_1}, x_{p_2} be any two points at which $W_1(x_j)$ and $W_2(x_j)$ attain its minimum value on $\overline{\Omega}^N$. That is, $W_i(x_{p_i}) = \min\limits_{x_j \in \overline{\Omega}^N} \{W_i(x_j)\} \quad$ for i=1,2.

Without loss of generality, assume $W_1(x_{p_1}) \leq W_2(x_{p_2})$. If $W_1(x_{p_1}) \geq 0$, then there is nothing to prove. If we choose $W_1(x_p) < 0$ then this leads to contradiction.

3   Error Analysis

By the Taylor's series expansion on discontinuous regular and discontinuous singular components, we have

$$|\epsilon_k(\frac{d^2}{dx^2} - \delta^2)v_k(x_i)| \leq \begin{cases} C\epsilon_k(x_{i+1} - x_{i-1})|v_k|_3 & (16) \\ C\epsilon_k h^2 |v_k|_4, \quad x_{i+1} - x_i = x_i - x_{i-1} = h, & (17) \end{cases}$$

and

$$|\epsilon_k(\frac{d^2}{dx^2} - \delta^2)w_k(x_i)| \leq \begin{cases} C\epsilon_k(x_{i+1} - x_{i-1})|w_k|_3 & (18) \\ C\epsilon_k h^2 |w_k|_4, \quad x_{i+1} - x_i = x_i - x_{i-1} = h & (19) \\ C\epsilon_k \max\limits_{x \in [x_{i-1}, x_{i+1}]} |w_k''(x_i)|, & (20) \end{cases}$$

where $k = 1, 2$, $i \neq \frac{N}{2}$, $|z_k|_j := \max|\frac{d^j z}{dx^j}|$, $\forall j \in N$.

Note that if $\sigma_{\epsilon_{l_2}} = \frac{d}{4}$, $\sigma_{\epsilon_{r_2}} = \frac{(1-d)}{4}$, $\sigma_{\epsilon_{l_1}} = \frac{d}{8}$ and $\sigma_{\epsilon_{r_1}} = \frac{(1-d)}{8}$, then the mesh is uniform. In that case, N^{-1} is very small with respect to ϵ_1 and ϵ_2 and therefore a classical analysis could be used to prove the uniform convergence of the scheme. If $2\sigma_{\epsilon_{l_1}} = \sigma_{\epsilon_{l_2}} = \sqrt{\frac{\epsilon_2}{\alpha}} \ln N$ and $2\sigma_{\epsilon_{r_1}} = \sigma_{\epsilon_{r_2}} = \sqrt{\frac{\epsilon_2}{\alpha}} \ln N$ then $\epsilon_2 = O(\epsilon_1)$, which is discussed in [6]. The most interesting case to discuss is $\sigma_{\epsilon_{l_1}} = \sigma_{\epsilon_{r_1}} = \sqrt{\frac{\epsilon_1}{\alpha}} \ln N$ and $\sigma_{\epsilon_{l_2}} = \sigma_{\epsilon_{r_2}} = \sqrt{\frac{\epsilon_2}{\alpha}} \ln N$.

Using (16) and bounds on discontinuous regular components, we have

$$|(\boldsymbol{L}^N - \boldsymbol{L})\boldsymbol{v}(x_i)| \le \tfrac{1}{3}(h_i + h_{i+1}) \begin{pmatrix} \epsilon_1\|v_1'''\|_{\Omega_1\cup\Omega_2} \\ \epsilon_2\|v_2'''\|_{\Omega_1\cup\Omega_2} \end{pmatrix} \le C \begin{pmatrix} N^{-1} \\ N^{-1} \end{pmatrix}.$$

Now we evaluate the error estimates for the discontinuous singular components for different sub-intervals.

(i) For $x_i \in [\sigma_{\epsilon_{l_2}}, d-\sigma_{\epsilon_{l_2}}] \cup [d+\sigma_{\epsilon_{r_2}}, 1-\sigma_{\epsilon_{r_2}}]$. Consider first that $x_i \in [\sigma_{\epsilon_{l_2}}, \frac{d}{2}]$. Using (20), bounds on discontinuous singular components and (9), we have

$$| ((\boldsymbol{L}^N - \boldsymbol{L})\boldsymbol{w})_1(x_i) | \le C \parallel B_{\epsilon_{l_2}} \parallel_{[x_{i-1}, x_{i+1}]} = B_{\epsilon_{l_2}}(x_{i-1}).$$

Thus,

$$\parallel B_{\epsilon_{l_2}} \parallel_{[x_{i-1}, x_{i+1}]} \le 2e^{(-\sigma_{\epsilon_{l_2}}+16\sigma_{\epsilon_{l_2}}/N)\frac{\sqrt{\alpha}}{\sqrt{\epsilon_2}}} \le CN^{-1}.$$

We can prove a similar result when $x_i \in [\frac{d}{2}, d-\sigma_{\epsilon_{l_2}}]$. Similar arguments prove a similar result for the sub-interval $[d + \sigma_{\epsilon_{r_2}}, 1 - \sigma_{\epsilon_{r_2}}]$.
  Hence, for $x_i \in [\sigma_{\epsilon_{l_2}}, d-\sigma_{\epsilon_{l_2}}] \cup [d+\sigma_{\epsilon_{r_2}}, 1-\sigma_{\epsilon_{r_2}}]$ we have

$$\begin{pmatrix} | ((\boldsymbol{L}^N - \boldsymbol{L})\boldsymbol{w})_1(x_i) | \\ | ((\boldsymbol{L}^N - \boldsymbol{L})\boldsymbol{w})_2(x_i) | \end{pmatrix} \le \begin{pmatrix} CN^{-1} \\ CN^{-1} \end{pmatrix}.$$

(ii) For $x_i \in (0, \sigma_{\epsilon_{l_1}}) \cup (d - \sigma_{\epsilon_{l_1}}, d) \cup (d, d+\sigma_{\epsilon_{r_1}}) \cup (1 - \sigma_{\epsilon_{r_1}}, 1)$.

Using (18) and the bounds on the discontinuous singular components together with the inequality $h_i + h_{i+1} \le 32\sqrt{\frac{\epsilon_1}{\alpha}}N^{-1}\ln N$ yields

$$\begin{pmatrix} | ((\boldsymbol{L}^N - \boldsymbol{L})\boldsymbol{w})_1(x_i) | \\ | ((\boldsymbol{L}^N - \boldsymbol{L})\boldsymbol{w})_2(x_i) | \end{pmatrix} \le \frac{h_i + h_{i+1}}{3} \begin{pmatrix} \epsilon_1\|w_1'''\| \\ \epsilon_2\|w_2'''\| \end{pmatrix} \le C \begin{pmatrix} N^{-1}\ln N \\ N^{-1}\ln N \end{pmatrix}.$$

(iii) For $x_i \in (\sigma_{\epsilon_{l_1}}, \sigma_{\epsilon_{l_2}}) \cup (d-\sigma_{\epsilon_{l_2}}, d-\sigma_{\epsilon_{l_1}}) \cup (d+\sigma_{\epsilon_{r_1}}, d+\sigma_{\epsilon_{r_2}}) \cup (d-\sigma_{\epsilon_{r_2}}, d-\sigma_{\epsilon_{r_1}})$.
First assume that $\frac{\epsilon_2}{4} \le \epsilon_1 \le \epsilon_2$.

Using (18) and bounds on the discontinuous singular components, we have

$$\begin{pmatrix} | ((\boldsymbol{L}^N - \boldsymbol{L})\boldsymbol{w})_1(x_i) | \\ | ((\boldsymbol{L}^N - \boldsymbol{L})\boldsymbol{w})_2(x_i) | \end{pmatrix} \le \frac{h_i + h_{i+1}}{3} \begin{pmatrix} \epsilon_1\|w_1'''\| \\ \epsilon_2\|w_2'''\| \end{pmatrix} \le C \begin{pmatrix} N^{-1}\ln N \\ N^{-1}\ln N \end{pmatrix}.$$

Next, suppose that $\epsilon_1 < \frac{\epsilon_2}{4}$, then we have

$$\begin{pmatrix} | ((\boldsymbol{L}^N - \boldsymbol{L})\boldsymbol{w})_1(x_i) | \\ | ((\boldsymbol{L}^N - \boldsymbol{L})\boldsymbol{w})_2(x_i) | \end{pmatrix} \le \begin{pmatrix} \epsilon_1(D^2 - \frac{d^2}{dx^2})w_{1,\epsilon_1}(x_i) \\ \epsilon_2(D^2 - \frac{d^2}{dx^2})w_{2,\epsilon_1}(x_i) \end{pmatrix} + \begin{pmatrix} \epsilon_1(D^2 - \frac{d^2}{dx^2})w_{1,\epsilon_2}(x_i) \\ \epsilon_2(D^2 - \frac{d^2}{dx^2})w_{2,\epsilon_2}(x_i) \end{pmatrix}. \quad (21)$$

Consider the first part of (21). Using the analysis in (i), we obtain

$$\begin{pmatrix} \epsilon_1(D^2 - \frac{d^2}{dx^2})w_{1,\epsilon_1}(x_i) \\ \epsilon_2(D^2 - \frac{d^2}{dx^2})w_{2,\epsilon_1}(x_i) \end{pmatrix} \le 2 \begin{pmatrix} \epsilon_1\|w_{1,\epsilon_1}''\|_{[x_{i-1}, x_{i+1}]} \\ \epsilon_2\|w_{2,\epsilon_1}''\|_{[x_{i-1}, x_{i+1}]} \end{pmatrix} \le C \begin{pmatrix} N^{-1} \\ N^{-1} \end{pmatrix}.$$

For the second part of (21), and the analysis in (ii) we have

$$\begin{pmatrix} \epsilon_1(D^2 - \frac{d^2}{dx^2})w_{1,\epsilon_2}(x_i) \\ \epsilon_2(D^2 - \frac{d^2}{dx^2})w_{2,\epsilon_2}(x_i) \end{pmatrix} \le \frac{h_i + h_{i+1}}{3} \begin{pmatrix} \epsilon_1\|w_{1,\epsilon_2}'''\| \\ \epsilon_2\|w_{2,\epsilon_2}'''\| \end{pmatrix} \le C \begin{pmatrix} N^{-1}\ln N \\ N^{-1}\ln N \end{pmatrix}.$$

Combining the results of (i)-(iii) for discontinuous singular components, we have

$$\begin{pmatrix} |\ ((\boldsymbol{L}^N - \boldsymbol{L})\boldsymbol{w})_1(x_i)\ | \\ |\ ((\boldsymbol{L}^N - \boldsymbol{L})\boldsymbol{w})_2(x_i)\ | \end{pmatrix} \leq C \begin{pmatrix} N^{-1}\ln N \\ N^{-1}\ln N \end{pmatrix}.$$

The point $x_{N/2} = d$ and $h_{\frac{N}{2}} = h_{\frac{N}{2}+1} = h = \sqrt{\frac{\epsilon_1}{\alpha}}\ln N$. Consider

$$((\boldsymbol{L}^N(\boldsymbol{U} - \boldsymbol{u}))_1(d)) = \overline{f}_1(d) + \frac{\epsilon_1}{h^2}\int_{t=d}^{d+h}\int_{s=d}^{t} u_1''(s)\ ds\ dt - \frac{\epsilon_1}{h^2}\int_{t=d-h}^{d}\int_{s=d}^{t} u_1''(s)\ ds\ dt$$

$$- a_{11}(d)u_1(d) - a_{12}(d)u_2(d)$$

$$= \frac{1}{h^2}\int_{t=d}^{d+h}\int_{s=d}^{t}\int_{p=s}^{d+h} + \frac{1}{h^2}\int_{t=d-h}^{d}\int_{s=d}^{t}\int_{p=d-h}^{s}(f_1 - a_{11}u_1 - a_{12}u_2)'(p)\ dp\ ds\ dt$$

$$- a_{11}(d)u_1(d) - a_{12}(d)u_2(d) + \frac{1}{2}(a_{11}(d-h)u_1(d-h) + a_{12}(d-h)u_2(d-h))$$

$$+ \frac{1}{2}(a_{11}(d+h)u_1(d+h) + a_{12}(d+h)u_2(d+h)),$$

$$= \frac{1}{h^2}\int_{t=d}^{d+h}\int_{s=d}^{t}\int_{p=s}^{d+h} + \frac{1}{h^2}\int_{t=d-h}^{d}\int_{s=d}^{t}\int_{p=d-h}^{s}(f_1 - a_{11}u_1 - a_{12}u_2)'(p)\ dp\ ds\ dt$$

$$+ \frac{1}{2}\int_{t=d}^{d-h}(a_{11}(t)u_1(t) + a_{12}(t)u_2(t))'\ dt + \frac{1}{2}\int_{t=d}^{d+h}(a_{11}(t)u_1(t) + a_{12}(t)u_2(t))'\ dt,$$

which implies
$|(\boldsymbol{L}^N(\boldsymbol{U} - \boldsymbol{u}))_1(d))| \leq C(N^{-1}\ln N)$ and $|(\boldsymbol{L}^N(\boldsymbol{U} - \boldsymbol{u}))_2(d))| \leq C(N^{-1}\ln N)$.

Theorem 3. *Let $\boldsymbol{u}$ and $\boldsymbol{U}$ be the exact and the numerical solutions of the problem (1)-(3). Then, for N sufficiently large*

$$\max_{x_i \in \Omega^N} |\boldsymbol{U}(x_i) - \boldsymbol{u}(x_i)| \leq CN^{-1}\ln N.$$

Proof. Consider the mesh function $\Psi(x_i) = C(N^{-1}lnN) \pm (\boldsymbol{U} - \boldsymbol{u})(x_i)$. This function satisfies $\Psi(x_0) \geq 0$, $\Psi(x_N) \geq 0$, and $\boldsymbol{L}^N\Psi(x_i) \geq 0$, for all $x_i \in \overline{\Omega}^N$. Using discrete maximum principle we get $\Psi(x_i) \geq 0$ for all $x_i \in \overline{\Omega}^N$, and then the required result follows.

4 Numerical Results

Example 4.1 Consider the test Problem

$$- \epsilon_1 u_1''(x) + 2(x+1)^2 u_1(x) - (1+x^3)u_2(x) = f_1(x), \qquad x \in \Omega_1 \bigcup \Omega_2$$
$$- \epsilon_2 u_2''(x) - 2\cos(\tfrac{\Pi}{4}x)u_1(x) + 2.2e^{1-x}u_2(x) = f_2(x), \qquad x \in \Omega_1 \bigcup \Omega_2$$
$$u(0) = 0, \qquad u(1) = 0,$$

where

$$f_1(x) = \begin{cases} 2e^x & \text{for } 0 \leq x \leq 0.5 \\ 1 & \text{for } 0.5 < x \leq 1 \end{cases} \text{ and } f_2(x) = \begin{cases} 10x+1 & \text{for } 0 \leq x \leq 0.5 \\ 2 & \text{for } 0.5 < x \leq 1. \end{cases}$$

For the construction of piecewise-uniform Shishkin mesh $\overline{\Omega}_N$, we take $\alpha = 0.95$. The Exact solution of the Example 4.1 is not known. Therefore we estimate the error for $\boldsymbol{U}$ by comparing it to the numerical solution $\tilde{\boldsymbol{U}}$ obtained on the mesh $\tilde{x}_j$ that contains the mesh points of the original mesh and their midpoints, that is, $\tilde{x}_{2j} = x_j$, $j=0,\ldots\ldots,N$, $\tilde{x}_{2j+1} = (x_j + x_{j+1})/2$, $j=0,\ldots\ldots,N-1$.

For different values of N and ϵ_1, ϵ_2, we compute

$$D_{\epsilon_1,\epsilon_2}^N := \|(\mathbf{U} - \tilde{\mathbf{U}})(x_j)\|_{\overline{\Omega}^N}.$$

If $\epsilon_1 = 10^{-j}$ for some non-negative integer j , set

$$D_{\epsilon_1}^N := \max\{D_{\epsilon_1,1}^N, D_{\epsilon_1,10^{-1}}^N, D_{\epsilon_1,10^{-2}}^N, \ldots\ldots, D_{\epsilon_1,10^{-j}}^N\}.$$

Then the parameter-uniform error is computed as

$$D^N := \max\{D_1^N, D_{10^{-1}}^N, \ldots\ldots, D_{10^{-16}}^N\},$$

and the order of convergence is calculated using the formula $p^N := log_2(\frac{D^N}{D^{2N}})$.

Table 1. Maximum point-wise errors $D_{\epsilon_1}^N$, D^N and ϵ_1, ϵ_2−uniform rate of convergence p^N for Example 4.1

$\epsilon_1 = 10^{-j}$	N=64	N=128	N=256	N=512	N=1024	N=2048
$j = 0$	1.33E-04	3.32E-05	8.31E-06	2.08E-06	5.14E-07	1.29E-07
1	6.77E-04	1.69E-04	4.24E-05	1.06E-05	2.65E-06	6.47E-07
2	1.99E-03	5.01E-04	1.26E-04	3.14E-05	7.85E-06	1.96E-06
3	7.91E-03	2.08E-03	5.35E-04	1.34E-04	3.36E-05	8.39E-06
4	1.92E-02	7.98E-03	2.78E-03	8.50E-04	2.15E-04	5.38E-05
5	2.55E-02	8.07E-03	2.81E-03	8.99E-04	7.82E-05	2.24E-05
6	2.91E-02	1.09E-02	3.62E-03	1.11E-03	3.24E-04	9.18E-05
7	3.23E-02	1.50E-02	5.75E-03	1.95E-03	6.20E-04	1.89E-04
.	.	.	.	.	.	.
.	.	.	.	.	.	.
.	.	.	.	.	.	.
16	3.23E-02	1.50E-02	5.75E-03	1.95E-03	6.20E-04	1.89E-04
D^N	3.23E-02	1.50E-02	5.75E-03	1.95E-03	6.20E-04	1.89E-04
p^N	1.10	1.38	1.56	1.65	1.71	

Theorem 3 proves almost first order of convergence, but the results of the table 1 show almost second order of convergence. These results give us a scope to improve the theoretical results, which requires improvement on the bounds on the derivatives of the solution.

Acknowledgements. The authors gratefully acknowledge the valuable comments and suggestions from the anonymous referees.

References

1. de Falco, C., O'Riordan, E.: Interior Layers in a reaction-diffusion equation with a discontinuous diffusion coefficient. Int. J. Numer. Anal. Model. 7, 444–461 (2010)
2. Farrell, P.A., Miller, J.J.H., O'Riordan, E., Shishkin, G.I.: Singularly perturbed differential equations with discontinuous source terms. In: Miller, J.J.H., Shishkin, G.I., Vulkov, L. (eds.) Analytical and Numerical Methods for Convection-Dominated and Singularly Perturbed Problems, pp. 23–32. Nova Science, New York (2000)

3. Farrell, P.A., Hegarty, A.F., Miller, J.J.H., O'Riordan, E., Shishkin, G.I.: Singularly perturbed convection-diffusion problems with boundary and weak interior layers. J. Comput. Appl. Math. 166, 133–151 (2004)
4. Farrell, P.A., Hegarty, A.F., Miller, J.J.H., O'Riordan, E., Shishkin, G.I.: Global maximum norm parameter-uniform numerical method for a singularly perturbed convection-diffusion problem with discontinuous convection coeffcient. Math. Comput. Modelling 40, 1375–1392 (2004)
5. Madden, N., Stynes, M.: A Uniformly convergent numerical method for a coupled system of two singularly perturbed linear reaction-diffusion problems. IMA J. Numer. Anal. 23, 627–644 (2003)
6. Tamilselvan, A., Ramanujam, N., Shanthi, V.: A numerical method for singularly perturbed weakly coupled system of two second order ordinary differential equations with discontinuous source term. J. Comput. Appl. Math. 202, 203–216 (2007)

Finite Differences Method for the First Order 2-D Partial Equation

Mahir Rasulov[1], E. Ilhan Sahin[2], and M. Gokhan Soguksu[3]

[1] Beykent University, Department of Mathematics and Computing,
Sisli-Ayazaga Campus, 34396 Istanbul, Turkey
[2] Yildiz Technical University, Department of Physics, Istanbul, Turkey
[3] Zorlu Holding, Istanbul, Turkey
mresulov@beykent.edu.tr, shnethem@gmail.com, soguksu@zorlu.com

Abstract. In this study a new method for finding exact solution of the
Cauchy problem subject to a discontinuous initial profile for the two dimensional scalar conservation laws is suggested. For this aim, first, some
properties of the weak solution of the linearized equation are investigated.
Taking these properties into consideration an auxiliary problem having
some advantages over the main problem is introduced. The proposed
auxiliary problems also permit us to develop effective different numerical algorithms for finding the solutions. Some computer experiments are
carried out.

1 Introduction

It is known that theoretical investigation of many problems in sciences and
engineering, particularly in fluid dynamics requires to study the nonlinear hyperbolic equations of conservation laws. It is typical in such problems that their
solutions admit the points of discontinuity whose locations are unknown before-
hand. Therefore, even in one dimensional case they pose a special challenge for
theoretical and numerical analysis.

The mathematical theory of the Cauchy problem for one dimensional nonlinear hyperbolic equation is studied in [4], [7], [8], [10]. Profound results for the
mathematical theory of the initial boundary value problem for scalar conservation laws in several space dimensions can be found in [1], [2], [3], [6], [7] and
[12]. Most hyperbolic equations (or systems) of conservation laws in physics are
nonlinear and solving them analytically is often difficult, sometimes impossible.

There are many sensitive numerical methods for the solving of the Cauchy
problem of nonlinear hyperbolic equations [5], [8], and etc. Very relevant results
to the basis of the main themes are found [6], [9].

In this study the original method for finding the exact and numerical solution
of the Cauchy problem for the two dimensional scalar equation

$$u_t(x,y,t) + f_x(u(x,y,t)) + g_y(u(x,y,t)) = 0 \qquad (1)$$

is suggested. The conditions which the functions $f(u)$ and $g(u)$ must satisfy will
be expressed later.

I. Dimov, I. Faragó, and L. Vulkov (Eds.): NAA 2012, LNCS 8236, pp. 454–462, 2013.
© Springer-Verlag Berlin Heidelberg 2013

The Cauchy problem for equation (1), when $f(u) = g(u) = F(u)$ has been studied in [11] analytically as well as numerically. In order to study in full detail at first the case $f(u) = Au$ and $g(u) = Bu$ is considered. Here A and B are given real constants.

2 The Linear Equation and Some Properties of the Exact Solution

Let $R_+^3 = R^2 \times [0, T)$ be Euclidean space of the points (x, y, t), where $(x, y) \in R^2$, $t \in [0, T)$ and T is a given constant. In R_+^3 we consider the following Cauchy problem

$$u_t(x, y, t) + Au_x(x, y, t) + Bu_y(x, y, t) = 0, \tag{2}$$

$$u(x, y, 0) = u_0(x, y). \tag{3}$$

Here, $u_0(x, y)$ is any given function having both positive and negative slopes, or piecewise continuous. The problem (2), (3) later on will be called as the main problem. We support that $A \neq B$ and let D_{xy} be the domain defined as follows $D_{xy} = \{(\xi, \eta), \ a \leq \xi \leq x, \ b \leq \eta \leq y\} \subseteq R^2, \ t \in R_+^T = [0, T)$. By ∂D_{xy} we denote the boundary of the domain D_{xy}.

It is easily shown that the function

$$u(x, y, t) = u_0(x - At, y - Bt) \tag{4}$$

is the exact solution of the main problem, and such solution is called a soft solution of the problem (2), (3).

Definition 1. *The function $u(x, y, t)$ satisfying the initial condition (3) is called a weak solution of the problem (2), (3) if the following integral relation*

$$\int_{R^2} \int_{R_+^T} \{u(x, y, t) [\varphi_t(x, y, t) + A\varphi_x(x, y, t) + B\varphi_y(x, y, t)]\} \, dx dy dt$$

$$+ \int_{R^2} \varphi(x, y, 0) u_0(x, y) dx dy = 0 \tag{5}$$

holds for every test functions $\varphi(x, y, t)$ defined and differentiable in R_+^3 and vanishes for $\sqrt{x^2 + y^2} + t$ sufficiently large.

Theorem 1. *If the $u(x, y, t)$ is a continuous weak solution of the main problem, then the function $u(x, y, t) = u_0(x - At, y - Bt)$ is a soft solution of the main problem.*

Proof. According to the Definition 1 we have

$$\int_{R^2} \int_{R_+^T} \{u(x, y, t) [v_t(x, y, t) + Av_x(x, y, t) + Bv_y(x, y, t)]\} \, dx dy dt$$

$$+ \int_{R^2} v(x, y, 0) u_0(x, y) dx dy = 0 \tag{6}$$

for every smooth function $v(x, y, t)$ which vanishes for $\sqrt{x^2 + y^2} + t$ sufficiently large.

Applying the following transformation of variables

$$\xi = x - At, \quad \eta = y - Bt, \quad \tau = t, \tag{7}$$

the equality (6) can be expressed as

$$\int_{R^2} \left\{ \int_0^\infty u(\xi, \eta, \tau) v_\tau(\xi, \eta, \tau) d\eta + v(\xi, \eta, 0) u(\xi, \eta, 0) \right\} d\xi = 0$$

for every smooth $v(x, y, t)$ with appropriate support. If we define $F(\xi, \eta) = \int_0^\infty u(\xi, \eta, \tau) v_\tau(\xi, \eta, \tau) d\tau$ then the previous relation can be rewritten as $\int_{R^2} [F(\xi, \eta) + u(\xi, \eta, 0) v(\xi, \eta, 0)] d\xi = 0$. This implies that $F(\xi, \eta) + u(\xi, \eta, 0) v(\xi, \eta, 0) = 0$ for, if there were and domain $[\xi_1, \xi_2,] \times [\eta_1, \eta_2]$ where this is different from zero (say positive), we could define a new $\bar{v}(\xi, \eta)$ coinciding with v in $[\xi_1, \xi_2,] \times [\eta_1, \eta_2]$ and zero elsewhere (smoothing it out as necessary). Then $\int_{R^2} [\bar{F}(\xi, \eta) + u(\xi, \eta, 0) \bar{v}(\xi, \eta, 0)] d\xi > 0$, which is a contradiction. We conclude that

$$\int_{R_+^T} u(\xi, \eta, \tau) v_\tau(\xi, \eta, \tau) d\tau + v(\xi, \eta, 0) u(\xi, \eta, 0) = 0.$$

Since, $\int_{R_+^T} v_\tau(\xi, \eta, \tau) d\tau = -v(\xi, \eta, 0)$, then $\int_{R_+^T} v_\tau(\xi, \eta, \tau) u(\xi, \eta, 0) d\tau = -v(\xi, \eta, 0)$ $u_0(\xi, \eta, 0)$, and we get $\int_{R_+^T} [u(\xi, \eta, \tau) - u(\xi, \eta)] v_\tau(\xi, \eta, \tau) d\tau = 0$. From the continuity of u we conclude that $u(\xi, \eta, \tau) = u(\xi, \eta) = u_0(x - At, y - Bt)$.

Theorem 2. *If $u_0(x - At, y - Bt)$ is integrable, then the function $u(x, y, t)$ defined by the formula (4) is a weak solution of the main problem.*

Proof. Let v be a smooth function which vanishes for $\sqrt{x^2 + y^2} + t$ sufficiently large. Consider the expression

$$\int_{R^2} \int_{R_+^T} u_0(x - At, y - Bt) \left\{ v_t(x, y, t) + A v_x(x, y, t) + B v_y(x, y, t) \right\} dx dy dt$$

$$+ \int_{R^2} v(x, y, 0) u_0(x, y) dx dy. \tag{8}$$

Taking account into consideration the (7), we have

$$\int_{R^2} \int_{R_+^T} u_0(\xi, \eta) v_\tau(\xi, \eta, \tau) d\xi d\eta d\tau + \int_{R^2} v(x, y, 0) u_0(x, y) dx dy.$$

During the integration with respect to τ, since $v = 0$ for sufficiently large τ, the sum of these integrals vanishes. Hence, the function $u(x, y, y) = u_0(x - At, y - Bt)$ is a weak solution of the main problem.

Integrating the equation (2) on the domain D_{xy} with respect to x, y and using the Green's formula we get

$$\frac{\partial}{\partial t} \int_a^x \int_b^y u(\xi, \eta, t) d\xi d\eta + A \int_b^y u(x, \eta, t) d\eta$$

$$-B \int_a^x u(\xi, y, t) d\xi = \varphi(a, y, t) - \psi(x, b, t). \tag{9}$$

Here, $\varphi(a, y, t) = A \int_b^y u(a, \eta, t) d\eta$, $\quad \psi(x, b, t) = B \int_a^x u(\xi, b, t) d\xi$.

The problem finding the solution of the equation (9) subject to (3) will be called as first auxiliary problem. We introduce the following operator $\Im(\cdot) = (\cdot)_{xy}$ and the function defined as

$$v(x, y, t) = \int_a^x \int_b^y u(\xi, \eta, t) d\xi d\eta + \Phi_1(a, y, t) - \Psi_1(x, b, t), \tag{10}$$

where $\Psi_1(x, b, t)$ and $\Phi_1(a, y, t)$ are the integrals of $\psi(x, b, t)$ and $\varphi(a, y, t)$ with respect to t, respectively. It is easily shown, that the function $\varphi(a, y, t) - \psi(x, b, t) \in$ ker$\Im$. Indeed,

$$\Im \left[\varphi(a, y, t) - \psi(x, b, t) \right] = \Im \left[A \int_b^y u(a, \eta, t) d\eta + B \int_a^x u(\xi, b, t) d\xi \right] = 0.$$

For the sake of simplicity, we denote $H(a, b, x, y, t) = \varphi(a, y, t) - \psi(x, b, t)$ and it is obvious that $\Im H(a, b, x, y, t) = 0$. Taking into consideration (10), the equation (9) takes the form

$$v_t(x, y, t) + A v_x(x, y, t) + B v_y(x, y, t) = 0. \tag{11}$$

The initial condition for the equation (11) is

$$v(x, y, 0) = v_0(x, y), \tag{12}$$

here the function $v_0(x, y)$ is any differentiable solution of the equation

$$\Im (v_0(x, y)) = u_0(x, y). \tag{13}$$

From (10) we have

$$\Im (v(x, y, t)) = u(x, y, t). \tag{14}$$

Indeed, if we differentiate the relation (10) with respect to x, and then with respect to y we prove the validity of (14). The auxiliary problem (11), (12) has the following advantages:

- The differentiability property of the function $v(x, y, t)$ with respect to x and y is higher than $u(x, y, t)$
- The function $u(x, y, t)$ may be discontinuous.
- By obtaining the solution $u(x, y, t)$ of the problem (11), (12), we does not use the derivatives u_x, u_y, u_t, which does not already exist, usually.

It is obvious that the solution of the auxiliary problem is not unique.

Theorem 3. *If the function $v(x, y, t)$ is soft solution of the auxiliary problem (11), (12), then the function $u(x, y, t)$ defined by (14) is a weak solution of the main problem*

Proof. Let the function $\varphi(x, y, t)$ is a test function and we consider the following expression

$$0 = \int_{R_+^3} \Im\varphi(x, y, t) \{v_t + Av_x + Bv_y\} \, dxdydt.$$

After some simple manipulation we get

$$\int_{R_+^3} \varphi(x, y, t) \{\Im v_t + A\Im v_x + B\Im v_y\} \, dxdydt = 0.$$

Taking into consideration of (14) and applying integration by part to the last equality with respect to t, x, y respectively we get the relation (6), which shown that u is weak solution of the main problem.

3   Two Dimensional Scalar Conservation Law

In this section the two dimensional scalar equation which describes a certain conservation law as

$$u_t + f_x(u) + g_y(u) = 0 \tag{15}$$

is considered.

Relatively $f(u)$ and $g(u)$ we assume that

(i) $f(u)$ and $g(u)$ are continuous differentiable functions,
(ii) $f'(u) \geq 0$, $g'(u) \geq 0$ for $u \geq 0$,
(iii) $f''(u)$ and $g''(u)$ have an alternative signs, i.e. the $f(u)$ and $g(u)$ functions have the concave and convex parts.

The solution of the problem (15), (3) obtained using the characteristics method is

$$u(x, y, t) = u_0(\xi, \eta), \tag{16}$$

where $\xi = x - f'(u)t$ and $\eta = y - g'(u)t$ are the special coordinates moving with speed of $f'(u)$ and $g'(u)$, respectively. From (16) we get

$$u_x = \frac{(u_0)_\xi}{1 + [f''(u)(u_0)_\xi + g''(u)(u_0)_\eta]t}, \quad u_y = \frac{(u_0)_\eta}{1 + [f''(u)(u_0)_\xi + g''(u)(u_0)_\eta]t},$$

$$u_t = \frac{f'(u)(u_0)_\xi + g'(u)(u_0)_\eta}{1 + [f''(u)(u_0)_\xi + g''(u)(u_0)_\eta]t}.$$

As it is seen from these formulas, if $(u_0)_\xi < 0$, $(u_0)_\eta < 0$ and $f''(u) > 0$, $g''(u) > 0$ or $((u_0)_\xi > 0$, $(u_0)_\eta > 0$ and f''(u) < 0, g''(u) < 0) at the value $t > T_0 \equiv -[f''(u)(u_0)_\xi + g''(u)(u_0)_\eta]^{-1}\big|_{\min}$ the derivatives u_x, u_y and u_t are approaching to infinity. Therefore the classical solution of the problem (15), (3) does not exist.

Definition 2. *The function $u(x, y, t)$ satisfying the initial condition (3) is called a weak solution of the problem (15), (3) if the following integral relation*

$$\iint_{R^2}\int_{R_+^T} \{u\varphi_t + f(u)\varphi_x + g(u)\varphi_y\}\, dxdydt + \iint_{R^2} u_0(x, y)\varphi(x, y, 0)dxdy = 0$$

(17)

holds for every test functions $\varphi(x, y, t)$ defined and differentiable in R_+^3 and vanishes for $\sqrt{x^2 + y^2} + t$ sufficiently large.

In order to find the weak solution of the problem (15),(3) in sense (17) we will introduce the auxiliary problem as above. Integrating the equation (15) on the region D_{xy} we get

$$0 = \iint_{D_{xy}} \{u_t + f_\xi(u) + g_\eta(u)\}\, d\xi d\eta = \frac{\partial}{\partial t} \int_a^x \int_b^y u(\xi, \eta, t)d\xi d\eta$$

$$+ \int_b^y [f(u(x, \eta, t)) - f(u(a, \eta, t))]\, d\eta + \int_a^x [g(u(\xi, y, t)) - g(u(\xi, b, t))]\, d\xi.$$

The last equality can be rewritten as

$$\frac{\partial}{\partial t} \int_a^x \int_b^y u(\xi, \eta, t)d\xi d\eta + \int_b^y f(u(x, \eta, t))\, d\eta + \int_a^x g(u(\xi, y, t))\, d\xi = \Phi(a, y, t)$$

$+\Psi(x, b, t)$. Here $\Phi(a, y, t) = \int_b^y f(u(a, \eta, t))\, d\eta$, $\Psi(x, b, t) = \int_a^x g(u(\xi, b, t))\, d\xi$. It is clearly seen that $\Phi(a, y, t) + \Psi(x, b, t) \in \ker \Im$. We denote by $v(x, y, t)$ following expression

$$v(x, y, t) = \int_a^x \int_b^y u(\xi, \eta, t)d\xi d\eta + H_1(a, b, x, y, t),$$

(18)

where $H_1(a, b, x, y, t) \in \ker \Im$. From (18) we have $u(x, y, t) = \Im(v(x, y, t))$. Taking into consideration to (18) we get

$$v_t(x, y, t) + \int_b^y f(u(x, \eta, t))\, d\eta + \int_a^x g(u(\xi, y, t))\, d\xi = 0.$$

(19)

The initial condition for the (19) is

$$v(x, y, 0) = v_0(x, y).$$

(20)

Here the function $v_0(x, y)$ is any differentiable solution of the equation (13).

Theorem 4. *If the function $v(x, y, t)$ is solution of the auxiliary problem (19), (20), then the function $u(x, y, t)$ expressed by $u(x, y, t) = \Im v(x, y, t)$ is a weak solution of the main problem (15) and (3).*

4   Finite Differences Schemes in a Class of Discontinuous Functions

In this section, we intended to develop the numerical method for finding the solution of the problem (2), (3), and investigate some properties of it. For this aim, we will use the auxiliary problem (9), (3). In order to demonstrate the effectiveness of the proposed method in this study we will consider only linear case.

4.1   The Finite Differences Scheme for Cauchy Problem

In order to construct the numerical algorithm, the domain of definition of the problem (2), (3) is covered by the following grid

$$\omega_{h_1,h_2,\tau} = \left\{(x_i, y_j, t_k),\quad x_i = ih_1,\quad y_j = jh_2,\quad t_k = k\tau,\quad i,j,k = 0,1,2,...\right\}$$

where $h_1 > 0$, $h_2 > 0$ and $\tau > 0$ are steps of the grid with respect to x, y and t, respectively.

In order to approximate of the equation (9) by the finite differences, the integrals leaving in (9) are approximated as follows

$$\int_a^x u(\xi, y, t)d\xi = h_1 \sum_{\nu=1}^{i} U_{\nu,j,k}, \quad \int_b^y u(x, \eta, t)d\eta = h_2 \sum_{\mu=1}^{j} U_{i,\mu,k} \tag{21}$$

and

$$\int_a^x \int_b^y u(\xi, \eta, t)d\xi d\eta = h_1 h_2 \sum_{\nu=1}^{i} \sum_{\mu=1}^{j} U_{\nu,\mu,k}. \tag{22}$$

Here $U_{i,\mu,k}$ are approximate values of the function $u(x, y, t)$ at points (x_i, y_μ, t_k). Taking into consideration (21), (22) the equation (9) at any point (i, j, k) of the grid $\omega_{h_1,h_2,\tau}$ is approximated as follows

$$U_{i,j,k+1} = (1 - \frac{\tau}{h_1}A + \frac{\tau}{h_2}B)U_{i,j,k} - \frac{\tau}{h_1}A\sum_{\mu=1}^{j-1} U_{i,\mu,k} + \frac{\tau}{h_2}B\sum_{\nu=1}^{i-1} U_{\nu,j,k}$$

$$- \sum_{\nu=1}^{i-1}\sum_{\mu=1}^{j-1} (U_{\nu,\mu,k+1} - U_{\nu,\mu,k}) + \frac{\tau}{h_1}A\sum_{\mu=1}^{j-1} U_{0,\mu,k} - \frac{\tau}{h_2}B\sum_{\nu=1}^{i-1} U_{\nu,n,k}, \tag{23}$$

$$(i = 0,1,2,....N;\quad j = 0,1,2,...,M,\quad k = 0,1,2,...,).$$

The initial condition for (23) is $U_{i,j,0} = u_0(x_i, y_j)$, $(i = 0,1,2,...;\ j = 0,1,2,...)$.

Now, we approximate the problem (11), (12) by the finite differences. For this aim, we introduce the following notations $U(x_i, y_j, t_k) = U_{i,j}$, $U(x_i, y_j, t_{k+1}) = \hat{U}_{i,j}$, $\Delta_{\bar{x}}U_{i,j} = (U_{i,j} - U_{i-1,j})$, $\Delta_{\bar{y}}U_{i,j} = (U_{i,j} - U_{i,j-1})$, $\Delta_{\bar{x}\bar{y}}^2 U_{i,j} = (\Delta_{\bar{x}}U_{i,j} - $

$\Delta_{\bar{x}} U_{i,j-1}$), $\Delta_{\bar{y}\bar{x}}^2 U_{i,j} = (\Delta_{\bar{y}} U_{i,j} - \Delta_{\bar{y}} U_{i-1,j})$. In this notations the problem (11), (12) is approximated by the finite differences scheme as follows

$$\hat{V}_{i,j} = V_{i,j} - A\frac{\tau}{h_1}\Delta_{\bar{x}} V_{i,j} - B\frac{\tau}{h_2}\Delta_{\bar{y}} V_{i,j}, \tag{24}$$

$$V_{i,j,0} = V_{i,j}^{(0)}. \tag{25}$$

Here, the function $V_{i,j}^{(0)}$ is any continuous solution of the equation $V_{\bar{x}\bar{y}}^{(0)} = u_0(x_i, y_j)$, and the grid functions $U_{i,j}$ and $V_{i,j}$ represent approximate values of the functions $u(x, y, t)$ and $v(x, y, t)$ at point (i, j, k) respectively.

It is easy to prove that, if grid function $\hat{V}_{i,j}$ is the solution the problem (24), (25), then the grid function $\hat{U}_{i,j}$ defined by $\hat{U}_{i,j} = \frac{1}{h_1 h_2}\Delta_{\bar{x}\bar{y}}^2 \hat{V}_{i,j}$ is the solution of the equation

$$\hat{U}_{i,j} = U_{i,j} - A\frac{\tau}{h_1}\Delta_{\bar{x}}^j U_{i,j} - B\frac{\tau}{h_2}\Delta_{\bar{y}}^i U_{i,j}, \tag{26}$$

Similarly algorithms can be written for the problem (19), (20). Using the algorithms (24), (25) some computer experiments were carried out.

5   Conclusion

In this study an original method for finding the exact and numerical solutions of the Cauchy problem for the first order 2-D nonlinear partial equations in a class of discontinuous functions is proposed. The properties of the exact solution of the linearized equation are also studied.

The special auxiliary problem whose solution is more smoother than the solution of the main problem is introduced, which makes possible to develop efficient and sensitive algorithms that describe all physical properties of the investigated problem accurately.

References

1. Kurganov, A., Tadmor, E.: Solution of Two-Dimensional Riemann Problems for Gas Dynamics without Riemann Problem. Numerical Methods for Partial Differential Equations 18, 548–608 (2002)
2. Levy, D., Puppo, G., Russo, G.: A third order central WENO scheme for 2D conservation laws. Applied Numerical Mathematics 33, 415–421 (2000)
3. Yoon, D., Hwang, W.: Two- Dimensional Riemann Problem for Burger's Equation. Bull. Korean Math. Soc. 45(1), 191–205 (2008)
4. Godlewski, E., Raviart, R.E.: Hyperbolic Systems of Conservation Laws. Mathematiques & Applications (1991)
5. Godunov, S.K.: A Difference Scheme for Numerical Computation of Discontinuous Solutions of Equations of Fluid Dynamics. Mat. Sb. 47(89), 271–306 (1959)
6. Kroner, D.: Numerical Schemes for Conservation Laws. J.Wiley & Sons Ltd. and B.G. Tenber (1997)

7. Kruzkov, S.N.: First Order Quasilinear Equation in Several Independent Variables. Math. Sbornik 81, 217–243 (1970)
8. Lax, P.D.: Weak Solutions of Nonlinear Hyperbolic Equations and Their Numerical Computations
9. LeVeque, R.J.: Finite Volume Methods for Hyperbolic Problems, 558 p. Cambridge University Press (2002)
10. Oleinik, O.A.: Discontinuous Solutions of Nonlinear Differential Equations. Usp. Math. Nauk 12 (1957)
11. Rasulov, M.A., et al.: Identification of the Saturation Jump in the Process of Oil Displacement by Water in a 2D Domain. Dokl RAN, USSR 319(4), 943–947 (1991)
12. Toro, E.F., Titarev, V.A.: ADER Schemes for Scalar Non-linear Hyperbolic Conservation Laws with Source Terms in Three-space Dimensions. J. Comput. Phys. 202(1), 196–215 (2005)

Numerical Analysis of a Hydrodynamics Problem with a Curved Interface

Alexey V. Rukavishnikov

Institute of Applied Mathematics, Far-Eastern Branch,
Russian Academy of Sciences, Dzerzhinsky St. 54, Khabarovsk, 680000 Russia
Far Eastern State Transport University, Serysheva St. 47, Khabarovsk, 680021 Russia
alexeyruk@mail.ru

Abstract. In the paper, we present a two-dimensional problem that is obtained by sampling in time and linearising a problem regarding two-phase flow of a viscous fluid without mixing. The fluid satisfies the incompressible Navier-Stokes equations, and it is assumed that there is a time-varying curved interface Γ between liquid phases of different densities and viscosities. The primary result of this paper is an estimate of the convergence rate of an approximate solution to the exact solution of a problem regarding special norms. The results of numerical experiments agree with theoretical estimates of the convergence rate of the approximate solution to the exact solution in the special norms of grid spaces.

Keywords: Curved interface, domain decomposition, mortar method.

1 Introduction

In the paper, we present a two-dimensional problem that is obtained by sampling in time and linearising a problem regarding two-phase flow of a viscous fluid without mixing. The fluid satisfies the incompressible Navier-Stokes equations, and it is assumed that there is a time-varying curved interface Γ between liquid phases of different densities and viscosities (see [1]).

Regarding the discontinuity of the coefficients, we divide the original domain Ω into subdomains Ω_i, such that on each of the subdomains, the coefficients are constant; accordingly, we determine the variational formulation of the problem separately for each Ω_i and coordinate the solutions on Γ using the conditions of weak continuity. Because Γ between the subdomains Ω_i is a curve, we can produce a partition Ω_{ih} for each Ω_i composed of triangles such that Γ can be interpolated in a piecewise linear manner. An approximate variational formulation of the problem is defined independently on each Ω_{ih} in conjunction with the identification of mortar functions on Γ (see [2]). Earlier, this method was considered in [3] and [4] for elliptical problems. For a saddle point problem, different methods have been theoretically investigated (see [5] and [6]), but the solution has been numerically realised in [7] only for the case of a straight interface.

The primary result of this paper is an estimate of the convergence rate of an approximate solution to the exact solution of a problem in special norms. We also present numerical simulation results for the model problem.

I. Dimov, I. Faragó, and L. Vulkov (Eds.): NAA 2012, LNCS 8236, pp. 463–470, 2013.
© Springer-Verlag Berlin Heidelberg 2013

2   Problem Statement

Let $\Omega \subset \mathbf{R}^2 = \{\bar{\mathbf{x}} : \bar{\mathbf{x}} = (x_1, x_2)\}$ be a convex polygonal domain with boundary $\partial\Omega\,(\bar{\Omega} = \Omega \cup \partial\Omega)$. Consider simply connected subdomains Ω_1 and Ω_2 such that $\bar{\Omega}_1 \cup \bar{\Omega}_2 = \bar{\Omega}, \bar{\Omega}_1 \cap \bar{\Omega}_2 = \Gamma$. Let Γ be a sufficiently smooth (not a closed, self-avoiding) curve, the ends of which belong to $\partial\Omega$.

The problem is to find a velocity $\mathbf{u} = (u^1, u^2)$ and a pressure P such that

$$\frac{\varrho}{\tau}\mathbf{u} - \operatorname{div} \sigma(\mathbf{u}, P) + \varrho(g \times \mathbf{u}) = \mathbf{F}, \qquad \operatorname{div} \mathbf{u} = 0 \quad \text{in } \Omega,$$
$$[\mathbf{u}] = 0, \quad [\sigma(\mathbf{u}, P) \cdot \mathbf{n}] = 0 \quad \text{on } \Gamma, \qquad \mathbf{u} = 0 \quad \text{on } \partial\Omega, \tag{1}$$

$\mathbf{F} = (F^1, F^2), g$ are functions defined in Ω, $\sigma(\mathbf{u}, P) = \{2\,\mu\,\varepsilon_{ij}(\mathbf{u}) - \delta_{ij}P\}_{i,j=1,2}$ is a stress tensor, and $\varepsilon(\mathbf{u}) = \{\varepsilon_{ij}(\mathbf{u})\}_{i,j=1,2} = \{\frac{1}{2}(\frac{\partial u^j}{\partial x_i} + \frac{\partial u^i}{\partial x_j})\}_{i,j=1,2}$ is a strain tensor. The positive coefficients of viscosity μ and density ϱ of (1) are piecewise constants: $\mu = \begin{cases} \mu_1, \bar{\mathbf{x}} \in \Omega_1, \\ \mu_2, \bar{\mathbf{x}} \in \Omega_2, \end{cases} \varrho = \begin{cases} \varrho_1, \bar{\mathbf{x}} \in \Omega_1, \\ \varrho_2, \bar{\mathbf{x}} \in \Omega_2, \end{cases} g \times \mathbf{u} = \begin{pmatrix} -gu^2 \\ gu^1 \end{pmatrix}, \delta_{ij} = \begin{cases} 1, i = j, \\ 0, i \neq j, \end{cases}$

$[z] := z_1|_{\Gamma \cap \bar{\Omega}_1} - z_2|_{\Gamma \cap \bar{\Omega}_2}$, where $z_k|_{\Gamma \cap \bar{\Omega}_k}$ is a trace function of z_k on Γ, $\tau > 0$, $\mathbf{n}$ is a unit normal vector to Γ. $H_*^m(\Omega) = \{z \in L_2(\Omega) : z_k = z|_{\Omega_k} \in H^m(\Omega_k), m \in \mathbf{N}\}, \|z\|_{m,\Omega}^* = (\sum_{k=1}^2 \|z_k\|_{m,\Omega_k}^2)^{\frac{1}{2}}$; $\mathbf{H}_*^m(\Omega) = \{\mathbf{v} = (v^1, v^2) : v^l \in H_*^m(\Omega)\}$, $\|\mathbf{v}\|_{\mathbf{m},\Omega}^* = (\sum_{l=1}^2 (\|v^l\|_{m,\Omega}^*)^2)^{\frac{1}{2}}$; $\mathbf{H}_{00}^{\frac{1}{2}}(\Gamma) = \{\boldsymbol{\rho} : \rho^l \in H^{\frac{1}{2}}(\Gamma), \bar{\rho}^l \in H^{\frac{1}{2}}(\partial\Omega_k)\}$,

$\bar{\rho}^l = \begin{cases} \rho^l, \text{ on } \Gamma, \\ 0, \text{ on } \partial\Omega_k \backslash \Gamma, \end{cases} \|\boldsymbol{\rho}\|_{\mathbf{H}_{00}^{\frac{1}{2}}(\Gamma)} = (\sum_{l=1}^2 \inf_{\substack{z \in H^1(\Omega_k) \\ z^l|_{\Gamma} = \rho^l, z^l|_{\partial\Omega_k \backslash \Gamma} = 0}} \|z^l\|_{1,\Omega_k}^2)^{\frac{1}{2}}$;

$\mathbf{V}(\Omega) = \{\mathbf{v} \in \mathbf{H}_*^1(\Omega) : [\mathbf{v}] \in \mathbf{H}_{00}^{\frac{1}{2}}(\Gamma), \mathbf{v} = 0 \text{ on } \partial\Omega\}, \|\mathbf{v}\|_{\mathbf{V}(\Omega)} = (\sum_{k=1}^2 \|\mathbf{v}_k\|_{1,\Omega_k}^2 +$

$+ \|[\mathbf{v}]\|_{\mathbf{H}_{00}^{\frac{1}{2}}(\Gamma)}^2)^{\frac{1}{2}}$; $\mathbf{M}(\Gamma)$ is a dual space to $\mathbf{H}_{00}^{\frac{1}{2}}(\Gamma)$, $\|\boldsymbol{\nu}\|_{\mathbf{M}(\Gamma)} = \sup_{\boldsymbol{\mu} \in \mathbf{H}_{00}^{\frac{1}{2}}(\Gamma)} \frac{\int_\Gamma \boldsymbol{\nu} \cdot \boldsymbol{\mu}\, d\Gamma}{\|\boldsymbol{\mu}\|_{\mathbf{H}_{00}^{\frac{1}{2}}(\Gamma)}}$;

$\mathbf{Y}(\Omega) = \{\mathbf{v} \in \mathbf{V}(\Omega) : \int_\Gamma \chi[\mathbf{v}] \cdot \boldsymbol{\nu}\, d\Gamma = 0\ \forall \boldsymbol{\nu} \in \mathbf{M}(\Gamma)\}$ with the norm of $\mathbf{V}(\Omega)$ ($\chi \in \in C^1(\Gamma)$ is a weight function defined by $\chi = \frac{1}{\det(\gamma' \circ \gamma^{-1})}$; γ is a parametrization of Γ such that $\gamma \in C^2(\hat{I}, \mathbf{R}^2)$ and $|\gamma(y_1) - \gamma(y_2)| \succeq |y_1 - y_2|$ $\forall y_1, y_2 \in \hat{I} = [a, b]$); $X(\Omega) = \{Q : Q_k = Q|_{\Omega_k} \in L_2(\Omega_k)\}, \|Q\|_{X(\Omega)} = (\sum_{k=1}^2 \|Q_k\|_{0,\Omega_k}^2)^{\frac{1}{2}}$.

Here and below, we use the notation $\mathbf{A} \succeq \mathbf{B}$ ($\mathbf{A} \preceq \mathbf{B}$) to denote that there exists a constant $C > 0$ such that $\mathbf{A} \geq C\mathbf{B}$ ($\mathbf{A} \leq C\mathbf{B}$), and $\mathbf{A} \cong \mathbf{B}$, denotes that both $\mathbf{A} \succeq \mathbf{B}$ and $\mathbf{A} \preceq \mathbf{B}$. Let $F_k^j \in L_2(\Omega_k)$ and $g_k \in L_2(\Omega_k)$ be Lipschitz continuous functions.

The variational problem given in (1) is as follows: *find* $(\mathbf{u}, P, \boldsymbol{\lambda}) \in \mathbf{V}(\Omega) \times X(\Omega) \times \mathbf{M}(\Gamma)$ *such that for all* $(\boldsymbol{\varphi}, Q, \boldsymbol{\nu}) \in \mathbf{V}(\Omega) \times X(\Omega) \times \mathbf{M}(\Gamma)$:

$$a(\mathbf{u}, \boldsymbol{\varphi}) + b(\boldsymbol{\varphi}, P) + d(\boldsymbol{\varphi}, \boldsymbol{\lambda}) = l(\boldsymbol{\varphi}), \ b(\mathbf{u}, Q) = 0, \quad d(\mathbf{u}, \boldsymbol{\nu}) = 0, \tag{2}$$

$l(\boldsymbol{\varphi}) = \sum_{m=1}^2 \int_{\Omega_m} \mathbf{F}_m \cdot \boldsymbol{\varphi}_m\, d\bar{\mathbf{x}}$; $b(\boldsymbol{\varphi}, P) = -\sum_{m=1}^2 \int_{\Omega_m} \operatorname{div} \boldsymbol{\varphi}_m P_m\, d\bar{\mathbf{x}}$; $a(\mathbf{u}, \boldsymbol{\varphi}) =$

$= \sum_{m=1}^2 \int_{\Omega_m} (\frac{\varrho_m}{\tau}\mathbf{u}_m \cdot \boldsymbol{\varphi}_m + 2\,\mu_m\,\varepsilon(\mathbf{u}_m) : \varepsilon(\boldsymbol{\varphi}_m) - \varrho_m(\mathbf{u}_m \cdot (g_m \times \boldsymbol{\varphi}_m)))\, d\bar{\mathbf{x}}$;

$d(\boldsymbol{\varphi}, \boldsymbol{\lambda}) = \sum\limits_{m=1}^{2} \int_\Gamma (-1)^{m+1} \chi \, \boldsymbol{\lambda} \cdot \boldsymbol{\varphi}_m|_{\Gamma \cap \bar{\Omega}_m} \, d\Gamma$. To match the solution on Γ in (2) we use following conditions: $\int_\Gamma \chi \, (u_1 - u_2)^i \, \nu^i \, d\Gamma = 0 \quad \forall \boldsymbol{\nu} \in \mathbf{M}(\Gamma)$,

$$\int_\Gamma (\sigma_{ij}(\mathbf{u}_1, P_1)(n_1)^j) \varphi^i d\Gamma = \int_\Gamma -(\sigma_{ij}(\mathbf{u}_2, P_2)(n_2)^j) \varphi^i d\Gamma \ \forall \boldsymbol{\varphi} \in \mathbf{H}_{00}^{\frac{1}{2}}(\Gamma), \quad (3)$$

where $(n_2)^j = -(n_1)^j$ and $(n_m)^j$ – j-th component of the outward normal $\mathbf{n}_m$ on Γ with respect to Ω_m, $m = 1, 2$. For the closure by (3) obtained in the Ω_1 and Ω_2 equations (2), we introduce the supporting vector $\boldsymbol{\lambda} = (\lambda^1, \lambda^2)$ which is defined by the relation $\int_\Gamma \lambda^i \, \varphi^i \, d\Gamma = \int_\Gamma -\frac{1}{\chi} \, (\sigma_{ij}(\mathbf{u}_1, P_1)(n_1)^j) \, \varphi^i \, d\Gamma$.

3   The Scheme of the Finite Element Method (FEM)

First, we perform a triangulation Υ_h of the domain Ω. For each Ω_j, we have a quasi-uniform partition (see [9]) of Ω_{jh} into triangles with sides of order h_j. We denote the triangles by K and refer to them as finite elements. Their set is denoted by $\Upsilon_h^{(j)}$. The partition $\Omega_{jh} = \bigcup_{K \in \Upsilon_h^{(j)}} K$ approximates the subdomain Ω_j such that $\partial \Omega_{jh}$ is a piecewise linear interpolation $\partial \Omega_j$. The ends of Γ are interpolation nodes. Thus for Γ, we construct two piecewise linear interpolations Γ_{1h} and Γ_{2h}, which are parts of $\partial \Omega_{1h}$ and $\partial \Omega_{2h}$, respectively. We parameterise Γ_{1h} using continuous piecewise linear functions $\gamma_h^j : \hat{I} \to \mathbf{R}^2$, $\hat{I} = [a, b]$, $\gamma_h^1(a) = \gamma_h^2(a), \gamma_h^1(b) = = \gamma_h^2(b)$. As an approximation for the velocity components, we choose the vertices and midpoints of the sides K, and for the pressure, we use the vertices of K. We denote the set of nodes that belong to Γ_{1h}, by $\mathcal{N}$; these nodes are vertices of K and define a set $\hat{\mathcal{N}} = \{y \in \hat{I} : \gamma_h^1(y) = r; r \in \mathcal{N} \text{ on } \Gamma_h \equiv \Gamma_{1h}\}$. We note that a subset of the nodes of the approximations Ω_{1h} and Ω_{2h} does not coincide on Γ. The partition of Ω is denoted by $\Omega_h = \Omega_{1h} \cup \Omega_{2h}$.

We introduce the FE spaces on Ω_{jh}: for the velocity components, $V_h^{(j)}(\Omega_{jh})$ is a subspace of continuous, quadratic on K, FE functions in $H^1(\Omega_{jh})$, which vanish on $\partial \Omega \cap \partial \Omega_j$; for the pressure, $X_h^{(j)}(\Omega_{jh})$ is a subspace of continuous, linear on K, FE functions in $L_2(\Omega_{jh})$. The method presented here is a first-order Taylor-Hood method (see [10]). We denote the space of Lagrange multipliers (mortar functions) on $\hat{I}$ by $\hat{M}_h$, which is constructed on a grid with nodes of $\hat{\mathcal{N}}$ such that $\forall \hat{\varpi}_h \in \hat{M}_h$, the following conditions hold: $\hat{\varpi}_h$ is a continuous function of L_2, and $\hat{\varpi}_h$ is quadratic on each inner segment and linear on the end segments of $\hat{I}$. Hence on Γ_h, define a space M_h such that $M_h = \hat{M}_h \circ \gamma^{-1}$. In this case, we can express the solution $\lambda_{jh}^m \in M_h, m, j = 1, 2$, as a linear combination of coefficients and basic functions of M_h. As in [6], $\lambda_h^m = \lambda_{1h}^m = -\lambda_{2h}^m$, where $\boldsymbol{\lambda}_h = (\lambda_h^1, \lambda_h^2)$ is a grid analogue to the vector $\boldsymbol{\lambda}$. Define the FE spaces on Ω_h and Γ_h:
$\mathbf{V}_h(\Omega_h) = \{\mathbf{v}_h : v_{lh}^j \in V_h^{(l)}(\Omega_{lh})\}, \|\mathbf{v}_h\|_{\mathbf{V}_h} = (\|\mathbf{v}_h\|_{1,\Omega_h}^2 + \|[\mathbf{v}_h]_h\|_{\frac{1}{2},h,\Gamma}^2)^{\frac{1}{2}}$;
$\|\mathbf{v}_h\|_{1,\Omega_h} = (\sum\limits_{j,l=1}^{2} \sum\limits_{K \in \Upsilon_h^{(l)}} \|v_{lh}^j|_K\|_{1,K}^2)^{\frac{1}{2}}, \|[\mathbf{v}_h]_h\|_{\frac{1}{2},h,\Gamma} = h^{-\frac{1}{2}} \|[\mathbf{v}_h]_h\|_{0,\Gamma}$; here $[\mathbf{v}_h]_h$ is a discrete jump of a function $\mathbf{v}_h$ on Γ, which we define later; $X_h(\Omega_h) = \{Q_h :$

$$Q_{lh} \in X_h^{(l)}(\Omega_{lh})\}, \ \|Q_h\|_{X_h} = (\sum_{l=1}^2 \|Q_{lh}\|_{0,\Omega_{lh}}^2)^{\frac{1}{2}} = (\sum_{l=1}^2 \sum_{K \in \Upsilon_h^{(l)}} \|Q_{lh}|_K\|_{0,K}^2)^{\frac{1}{2}};$$

$\mathbf{M}_h(\Gamma_h) = \{\boldsymbol{\nu}_h : \nu_h^j \in M_h(\Gamma_h)\}, \ \|\boldsymbol{\nu}_h\|_{-\frac{1}{2},h,\Gamma} = h^{\frac{1}{2}} \|\boldsymbol{\nu}_h\|_{0,\Gamma}; \ \mathbf{Y}_h(\Omega_h) = \{\boldsymbol{\omega}_h \in \mathbf{V}_h : \int_\Gamma \chi[\boldsymbol{\omega}_h]_h \cdot \boldsymbol{\nu}_h d\Gamma = 0 \forall \boldsymbol{\nu}_h \in \mathbf{M}_h\}$ with the norm of $\mathbf{V}_h(\Omega_h)$.

Remark 1. In all propositions, we use a common parameter to denote the grid step, h.

Definition 1. *The approximate problem (1) expressed using the FEM is as follows:find* $(\mathbf{u}_h, P_h, \boldsymbol{\lambda}_h) \in \mathbf{V}_h \times X_h \times \mathbf{M}_h$ *such that* $\forall (\boldsymbol{\varphi}_h, Q_h, \boldsymbol{\nu}_h) \in \mathbf{V}_h \times X_h \times \mathbf{M}_h$:

$$\begin{aligned} a_h(\mathbf{u}_h, \boldsymbol{\varphi}_h) + b_h(\boldsymbol{\varphi}_h, P_h) + d_h(\boldsymbol{\varphi}_h, \boldsymbol{\lambda}_h) &= l_h(\boldsymbol{\varphi}_h), \\ b_h(\mathbf{u}_h, Q_h) &= 0, d_h(\mathbf{u}_h, \boldsymbol{\nu}_h) = 0, \end{aligned} \quad (4)$$

$$l_h(\boldsymbol{\varphi}_h) = \sum_{i=1}^2 \int_{\Omega_{ih}} \bar{\mathbf{F}}_i \cdot \boldsymbol{\varphi}_{ih} \, d\bar{\mathbf{x}}, \ d_h(\mathbf{u}_h, \boldsymbol{\nu}_h) = \int_\Gamma \chi[\mathbf{u}_h]_h \cdot \boldsymbol{\nu}_h \, d\Gamma; \ a_h(\mathbf{u}_h, \boldsymbol{\varphi}_h) =$$

$$= \sum_{i=1}^2 \sum_{K \in \Upsilon_h^{(i)}} \int_K \left(\frac{\varrho_i}{\tau} \mathbf{u}_{ih} \cdot \boldsymbol{\varphi}_{ih} + 2\mu_i \, \varepsilon(\mathbf{u}_{ih}) : \varepsilon(\boldsymbol{\varphi}_{ih}) - \varrho_i (\mathbf{u}_{ih} \cdot (\bar{g}_i \times \boldsymbol{\varphi}_{ih})) \right) dK;$$

$$b_h(\mathbf{u}_h, Q_h) = -\sum_{i=1}^2 \sum_{K \in \Upsilon_h^{(i)}} \int_K div_K \mathbf{u}_{ih} \, Q_{ih} dK,$$

where $\bar{g}_i$ and $\bar{F}_i^j$ are Lipschitz continuous functions on $\Omega_i \cup \Omega_{ih}$ such that $\bar{g}_i = g_i$, $\bar{F}_i^j = F_i^j$ on $\Omega_i \cap \Omega_{ih}$, $\|\bar{g}_i\|_{W_\infty^1(\Omega_i \cup \Omega_{ih})} \le C_{g_i}, \|\bar{F}_i^j\|_{W_\infty^1(\Omega_i \cup \Omega_{ih})} \le C_{F_i^j}$.

4   Convergence Analysis

We do not analyze problem (4) directly. To obtain a priori bounds for the discretization error, we proceed in two steps. In the first step, we introduce and analyze a variational problem based on blending elements (BE). In the second step, we interprete (4) as a perturbed blending approach.

We perform a triangulation $\tilde{\Upsilon}_h$ of the domain Ω. Each Ω_j is divided into triangles $\tilde{K}$ such that if two vertices of $\tilde{K}$ belong to Γ, then the side (curve) connecting them lies on Γ and is parameterised by γ. The above triangulation of Ω_j is denoted by $\tilde{\Upsilon}_h^{(j)}$. Let $\tilde{\Omega}_h = \bigcup_{\tilde{K} \in \tilde{\Upsilon}_h} \tilde{K}$, $\tilde{\Omega}_{jh} = \bigcup_{\tilde{K} \in \tilde{\Upsilon}_h^{(j)}} \tilde{K}$ be partitions of Ω, Ω_j, respectively. The FE $\tilde{K}$ is called BE, and the method is BEM. We assume that $\tilde{\Upsilon}_h$ and $\tilde{\Upsilon}_h^{(j)}$ differ from Υ_h and $\Upsilon_h^{(j)}$ only in that the sides of $K \in \Upsilon_h^{(j)}$ (if they have common points) approximate Γ, whereas the sides of $\tilde{K} \in \tilde{\Upsilon}_h^{(j)}$ coincide with part of Γ. The relationships between $\tilde{K}, K$ and the basic element $\hat{K}$ is defined by one-to-one correspondences $\tilde{F}_{\tilde{K}}, F_K$ ($\tilde{F}_{\tilde{K}} : \hat{K} \to \tilde{K}$, $F_K : \hat{K} \to K$).

Furthermore, we define spaces on $\tilde{\Omega}_{jh}$: $\check{V}_h^{(j)}(\tilde{\Omega}_{jh})$ is a subspace of continuous functions $\check{v} \in H^1(\tilde{\Omega}_{jh})$, which vanish on $\partial\Omega \cap \partial\tilde{\Omega}_{jh}$ such that for all $\tilde{K} \in \tilde{\Upsilon}_h^{(j)}$: $\check{v}|_{\tilde{K}} \circ \tilde{F}_{\tilde{K}}$ is a polynomial of second degree on $\hat{K}$, and $\tilde{X}_h^{(j)}(\tilde{\Omega}_{jh})$ is a subspace of continuous functions $\tilde{q} \in L_2(\tilde{\Omega}_{jh})$ such that for all $\tilde{K} \in \tilde{\Upsilon}_h^{(j)}$: $\tilde{q}|_{\tilde{K}} \circ \tilde{F}_{\tilde{K}}$ is a polynomial of first degree on $\hat{K}$. If $\check{v}_{jh}^l \in \check{V}_h^{(j)}(\tilde{\Omega}_{jh})$, then $\check{v}_h^l \in \check{V}_h(\tilde{\Omega}_h)$ and $\check{\mathbf{v}}_h \in$ $\in \check{\mathbf{V}}_h(\tilde{\Omega}_h) = \check{V}_h(\tilde{\Omega}_h) \times \check{V}_h(\tilde{\Omega}_h)$; if $\tilde{q}_{jh} \in \tilde{X}_h^{(j)}(\tilde{\Omega}_{jh})$, then $\tilde{q}_h \in \tilde{X}_h(\tilde{\Omega}_h), j, l = 1, 2$.

Here, we note that $\tilde{X}_h(\tilde{\Omega}_h)$ is a space for the pressure and $\check{\mathbf{V}}_h(\tilde{\Omega}_h)$ is not a final space for the velocity field. The final space for it is $\tilde{\mathbf{V}}_h(\tilde{\Omega}_h)$.

Now, we define the correspondence between functions q_h and $\tilde{q}_h$ of spaces X_h and $\tilde{X}_h$ ($\mathbf{v}_h$ and $\tilde{\mathbf{v}}_h$ of spaces $\mathbf{V}_h$ and $\tilde{\mathbf{V}}_h$). Consider $\tilde{K} \in \tilde{\Upsilon}_h^{(m)}$, which have a side parameterised by γ. Each $\tilde{K}$ corresponds to $K \in \Upsilon_h^{(m)}$. We have the equality $\tilde{q}_h \circ \tilde{F}_{\tilde{K}} = q_h \circ F_K$. Then, we define a mapping $T : X_h \to \tilde{X}_h$ such that for all $K \in$ $\in \Upsilon_h^{(m)} : \tilde{q}_h = Tq_h = q_h \circ G_K$, where $G_K = F_K \circ \tilde{F}_{\tilde{K}}^{-1}$ is transformation $\tilde{K}$ to K. Let $\nabla_{\tilde{K}}$ be the gradient operator on $\tilde{K} \in \tilde{\Upsilon}_h^{(m)}$, then, we have the relation for the gradient operator ∇_K on the appropriate $K \in \Upsilon_h^{(m)} : \nabla_K = ((G'_K)^{-1})^T \nabla_{\tilde{K}}$, where G'_K is the Jacobi matrix of the transformation G_K. Moreover, $div_{\tilde{K}} \equiv \nabla_{\tilde{K}}\cdot$ and $div_K \equiv \nabla_K\cdot$. Let $\sum_{m=1}^{2} \sum_{\tilde{K} \in \tilde{\Upsilon}_h^{(m)}} \int_{\tilde{K}} \tilde{q}_{mh} div_{\tilde{K}} \tilde{\mathbf{v}}_{mh} d\tilde{K} = 0 \ \forall \tilde{q}_h \in \tilde{X}_h$. Then, we construct a mapping $\mathbf{S} : \mathbf{V}_h \to \tilde{\mathbf{V}}_h$ such that $\forall K \in \Upsilon_h^{(m)} : \tilde{\mathbf{v}}_h = \mathbf{S}\mathbf{v}_h$ and

$$\sum_{i=1}^{2} \sum_{K \in \Upsilon_h^{(i)}} \int_K q_{ih} \, div_K \, \mathbf{v}_{ih} \, dK = 0 \quad \forall q_h \in X_h. \tag{5}$$

Proposition 1 ([11]). *Let* $G_K : \tilde{K} \to K$ *–* C^2*-diffeomorphism and* $\mathbf{w} \in \mathbf{H}^1(K)$ *be a differentiable vector field on* K. *Then the following identity holds:* $\det G'_K \, div_K \mathbf{w} = div_{\tilde{K}} (\det G'_K (G'_K)^{-1} \check{\mathbf{w}}), \ \check{\mathbf{w}} = \mathbf{w} \circ G_K.$

If we use Proposition 1 and go from K to $\tilde{K}$ in (5), then $\forall \tilde{q}_h \in \tilde{X}_h$:

$$\sum_{i=1}^{2} \sum_{\tilde{K} \in \tilde{\Upsilon}_h^{(i)}} \int_{\tilde{K}} \tilde{q}_{ih} div_{\tilde{K}} (\det G'_K (G'_K)^{-1} (\mathbf{v}_{ih} \circ G_K)) d\tilde{K} = 0. \tag{6}$$

The identity (6) is satisfied if only if on each $K \in \Upsilon_h^{(m)} : \tilde{q}_h = Tq_h$ and $\tilde{\mathbf{v}}_h = \mathbf{S}\mathbf{v}_h = (G'_K)^{-1}(\mathbf{v}_h \circ G_K) \det G'_K$, where $\mathbf{v}_h \circ G_K$ is a second-degree polynomial on $\tilde{K}$. We define a discrete jump of the function $\mathbf{v}_h \in \mathbf{V}_h$ on Γ: $[\mathbf{v}_h]_h = [\mathbf{S}\mathbf{v}_h]$. Then, we have the identities $b(\tilde{\mathbf{v}}_h, \tilde{q}_h) = b_h(\mathbf{v}_h, q_h), \quad d(\tilde{\mathbf{v}}_h, \boldsymbol{\nu}_h) = d_h(\mathbf{v}_h, \boldsymbol{\nu}_h)$. $\tilde{\mathbf{V}}_h(\tilde{\Omega}_h)$ is a space for the velocity field $\tilde{\mathbf{v}}_h$, $\|\tilde{\mathbf{v}}_h\|_{\tilde{\mathbf{V}}_h} = (\|\tilde{\mathbf{v}}_h\|_{1,\tilde{\Omega}_h}^2 + \|[\tilde{\mathbf{v}}_h]\|_{\frac{1}{2},h,\Gamma}^2)^{\frac{1}{2}}$, where $\|\tilde{\mathbf{v}}_h\|_{1,\tilde{\Omega}_h} = (\sum_{j,l=1}^{2} \|\tilde{v}_{jh}^l\|_{1,\tilde{\Omega}_{jh}}^2)^{\frac{1}{2}} = (\sum_{j,l=1}^{2} \sum_{\tilde{K} \in \tilde{\Upsilon}_h^{(j)}} \|\tilde{v}_{jh}^l\|_{1,\tilde{K}}^2)^{\frac{1}{2}}$; $\tilde{X}_h(\tilde{\Omega}_h)$ is a space for a pressure $\tilde{q}_h$, $\|\tilde{q}_h\|_{\tilde{X}_h} = (\sum_{j=1}^{2} \|\tilde{q}_{jh}\|_{0,\tilde{\Omega}_{jh}}^2)^{\frac{1}{2}}$; $\tilde{\mathbf{Y}}_h(\tilde{\Omega}_h) =$ $= \{\tilde{\mathbf{v}}_h \in \tilde{\mathbf{V}}_h : \int_\Gamma \chi [\tilde{\mathbf{v}}_h] \cdot \boldsymbol{\nu}_h \, d\Gamma = 0 \ \forall \boldsymbol{\nu}_h \in \mathbf{M}_h\}$ with the norm of $\tilde{\mathbf{V}}_h(\tilde{\Omega}_h)$.

Definition 2. *The approximation of problem (1) solved by the BEM is as follows:find* $(\tilde{\mathbf{u}}_h, \tilde{P}_h, \tilde{\boldsymbol{\lambda}}_h) \in \tilde{\mathbf{V}}_h \times \tilde{X}_h \times \mathbf{M}_h$ *such that* $\forall (\tilde{\boldsymbol{\varphi}}_h, \tilde{Q}_h, \boldsymbol{\nu}_h) \in \tilde{\mathbf{V}}_h \times \tilde{X}_h \times \mathbf{M}_h$:

$$a(\tilde{\mathbf{u}}_h, \tilde{\boldsymbol{\varphi}}_h) + b(\tilde{\boldsymbol{\varphi}}_h, \tilde{P}_h) + d(\tilde{\boldsymbol{\varphi}}_h, \tilde{\boldsymbol{\lambda}}_h) = l(\tilde{\boldsymbol{\varphi}}_h), b(\tilde{\mathbf{u}}_h, \tilde{Q}_h) = 0, d(\tilde{\mathbf{u}}_h, \boldsymbol{\nu}_h) = 0. \tag{7}$$

Theorem 1 ([12]). *Let* $\mathbf{u} \in \mathbf{Y}(\Omega) \cap \mathbf{H}_*^2(\Omega)$, $P \in X(\Omega) \cap H_*^1(\Omega)$, $\boldsymbol{\lambda} \in \mathbf{H}^{\frac{1}{2}}(\Gamma)$. *Let* $(\tilde{\mathbf{u}}_h, \tilde{P}_h, \tilde{\boldsymbol{\lambda}}_h)$ *be an approximate solution that satisfies (7),* $\tilde{\mathbf{u}}_h \in \tilde{\mathbf{Y}}_h$, $\tilde{P}_h \in$

$\in \tilde{X}_h, \tilde{\lambda}_h \in \mathbf{M}_h$. *Then, the following convergence estimate holds:*
$$\|\mathbf{u} - \tilde{\mathbf{u}}_h\|_{\tilde{\mathbf{V}}_h} + \|P - \tilde{P}_h\|_{\tilde{X}_h} + \|\lambda - \tilde{\lambda}_h\|_{-\frac{1}{2}, h, \Gamma} \preceq h\, (\|\mathbf{u}\|_{2,\Omega}^* + \|P\|_{1,\Omega}^*).$$

In practice, the calculation of integrals over curved elements near the Γ may cause some technical difficulties. It is much easier to realize the numerical method if we replace integrals over the curved $\tilde{K}$ by the integrals over the corresponding straight K. Let $\mathbf{v}_h, \mathbf{w}_h \in \mathbf{V}_h$ and $\tilde{\mathbf{w}}_h = \mathbf{S}\,\mathbf{w}_h, \tilde{\mathbf{v}}_h = \mathbf{S}\,\mathbf{v}_h \in \tilde{\mathbf{V}}_h$ and $\bar{a}(\tilde{\mathbf{w}}_h, \tilde{\mathbf{v}}_h) :=$ $= a_h(\mathbf{w}_h, \mathbf{v}_h)$ on $\tilde{\mathbf{V}}_h \times \tilde{\mathbf{V}}_h$. We denote by $\mathbf{u}'_h$ and P'_h the vector $\mathbf{S}\mathbf{u}_h$ and the function $T P_h$, respectively. Then, the approximate formulation (4) is defined as follows: *find* $(\mathbf{u}'_h, P'_h, \lambda_h) \in$ $\in \tilde{\mathbf{V}}_h \times \tilde{X}_h \times \mathbf{M}_h$, *such that* $\forall (\tilde{\boldsymbol{\varphi}}_h, \tilde{Q}_h, \boldsymbol{\nu}_h) \in \tilde{\mathbf{V}}_h \times \tilde{X}_h \times \mathbf{M}_h$ $(\tilde{\boldsymbol{\varphi}}_h = \mathbf{S}\boldsymbol{\varphi}_h)$:

$$\bar{a}(\mathbf{u}'_h, \tilde{\boldsymbol{\varphi}}_h) + b(\tilde{\boldsymbol{\varphi}}_h, P'_h) + d(\tilde{\boldsymbol{\varphi}}_h, \lambda_h) = l_h(\boldsymbol{\varphi}_h), b(\mathbf{u}'_h, \tilde{Q}_h) = 0, d(\mathbf{u}'_h, \boldsymbol{\nu}_h) = 0. \quad (8)$$

Theorem 2 ([12]). *Let* $\mathbf{u} \in \mathbf{Y}(\Omega) \cap \mathbf{H}_*^2(\Omega), P \in X(\Omega) \cap H_*^1(\Omega), \lambda \in \mathbf{H}^{\frac{1}{2}}(\Gamma)$. *Let* $(\mathbf{u}'_h, P'_h, \lambda_h)$ *be an approximate solution that satisfies* (8), $\mathbf{u}'_h \in \tilde{\mathbf{Y}}_h, P'_h \in$ $\in \tilde{X}_h, \lambda_h \in \mathbf{M}_h$. *Then, the following convergence estimate holds:*
$$\|\mathbf{u} - \mathbf{u}'_h\|_{\tilde{\mathbf{V}}_h} + \|P - P'_h\|_{\tilde{X}_h} + \|\lambda - \lambda_h\|_{-\frac{1}{2}, h, \Gamma} \preceq h\, (\|\mathbf{u}\|_{2,\Omega}^* + \|P\|_{1,\Omega}^*).$$

It is known (see, for example, [13]) that any function $z_i \in H^s(\Omega_i), s = 1, 2$, can be extended to $\mathbf{R}^2$ with preservation of the class H^s, i.e., exists an operator E_i^s that satisfies the following:

$$E_i^s : H^s(\Omega_i) \to H^s(\mathbf{R}^2),\ E_i^s z_i = z_i\ \text{in}\ \Omega_i,\ \|E_i^s z_i\|_{s, \mathbf{R}^2} \preceq \|z_i\|_{s, \Omega_i}. \quad (9)$$

Let $z \in H_*^s(\Omega)$. We say that E^s operates to z $(E^s z)$ if for every restriction $z_i = z|_{\Omega_i} \in H^s(\Omega_i)$, the operator E_i^s acts as defined in (9). If $q \in H_*^1(\Omega), q_h \in X_h$ and $\mathbf{w} \in \mathbf{H}_*^2(\Omega), \mathbf{w}_h \in \mathbf{V}_h$, then $\|E^1 q - q_h\|_{X_h}^2 := \sum_{k=1}^{2} \|E_1^k q_k - q_{kh}\|_{0, \Omega_{kh}}^2,$

$$\|\mathbf{E}^2 \mathbf{w} - \mathbf{w}_h\|_{\mathbf{V}_h}^2 := \sum_{k=1}^{2} \|\mathbf{E}_k^2 \mathbf{w}_k - \mathbf{w}_{kh}\|_{1, \Omega_{kh}}^2 + \|[\mathbf{E}^2 \mathbf{w} - \mathbf{w}_h]_h\|_{\frac{1}{2}, h, \Gamma}^2.$$

Theorem 3 ([12]). *Let* $\mathbf{u} \in \mathbf{Y}(\Omega) \cap \mathbf{H}_*^2(\Omega), P \in X(\Omega) \cap H_*^1(\Omega)$. *Let* $(\mathbf{u}_h, P_h)$ *be an approximate solution that satisfies* (4), $\mathbf{u}_h \in \mathbf{Y}_h, P_h \in X_h$. *Then, the following convergence estimate holds:*
$$\|\mathbf{E}^2 \mathbf{u} - \mathbf{u}_h\|_{\mathbf{V}_h} + \|E^1 P - P_h\|_{X_h} \preceq h\, (\|\mathbf{u}\|_{2,\Omega}^* + \|P\|_{1,\Omega}^*).$$

5 Numerical Results

Let $\bar{\Omega} = [-\pi, \pi] \times [-\pi, \pi]$, and let Γ be defined such that $x = a\sin(\frac{3}{2}(t - \pi))$, $y = t, t \in [-\pi, \pi], a > 0$. The subdomain Ω_1 lies to the left of the Γ, and Ω_2 lies to the right. In the numerical experiments, we set $\varrho = \begin{cases} 1, \bar{\mathbf{x}} \in \Omega_1, \\ 5, \bar{\mathbf{x}} \in \Omega_2, \end{cases}$ $\mu = \begin{cases} 1, \bar{\mathbf{x}} \in \Omega_1, \\ 3, \bar{\mathbf{x}} \in \Omega_2, \end{cases}$ $g = b \cdot \text{rot}\,\mathbf{u}; b = 0.995; c = 1/3; \tau = 1$. The parameter a, which is responsible for the location of the interface Γ, will be changed. As a solution of the problem

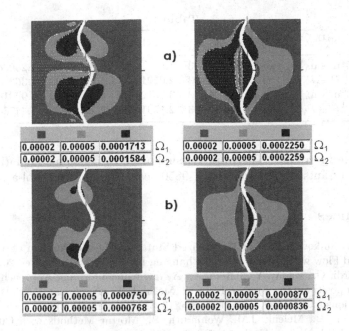

| 0.00002 | 0.00005 | 0.0001713 | Ω_1 |
| 0.00002 | 0.00005 | 0.0001584 | Ω_2 |

| 0.00002 | 0.00005 | 0.0002250 | Ω_1 |
| 0.00002 | 0.00005 | 0.0002259 | Ω_2 |

| 0.00002 | 0.00005 | 0.0000750 | Ω_1 |
| 0.00002 | 0.00005 | 0.0000768 | Ω_2 |

| 0.00002 | 0.00005 | 0.0000870 | Ω_1 |
| 0.00002 | 0.00005 | 0.0000836 | Ω_2 |

Fig. 1. The errors for the velocity components on the norm of $C(\bar{\Omega}_h)(a = 0.4)$: a)$N_{X_1}^{(1)} = 48$; b) $N_{X_1}^{(1)} = 96$

(1), we take
$$u^1(\bar{\mathbf{x}}) = \begin{cases} \frac{3}{2}\,a\,(x_1 - a\sin(\frac{3}{2}(x_2 - \pi)))\cos(\frac{3}{2}(x_2 - \pi)), \bar{\mathbf{x}} \in \Omega_1, \\ \frac{3}{2}\,c\,a\,(x_1 - a\sin(\frac{3}{2}(x_2 - \pi)))\cos(\frac{3}{2}(x_2 - \pi)), \bar{\mathbf{x}} \in \Omega_2, \end{cases}$$

$$u^2(\bar{\mathbf{x}}) = \begin{cases} x_1 - a\sin(\frac{3}{2}(x_2 - \pi)), \bar{\mathbf{x}} \in \Omega_1, \\ c\,(x_1 - a\sin(\frac{3}{2}(x_2 - \pi))), \bar{\mathbf{x}} \in \Omega_2, \end{cases} \qquad P(\bar{\mathbf{x}}) = \begin{cases} \sin x_1 \sin x_2, \ \bar{\mathbf{x}} \in \Omega_1, \\ c\sin x_1 \sin x_2, \ \bar{\mathbf{x}} \in \Omega_2. \end{cases}$$

We note that the solution of the model problem (1) is chosen such that the velocity field $\mathbf{u}$ is continuous on Γ, and that the pressure P and derivatives of the velocity components $\mathbf{u}$ are discontinuous on Γ.

Let $N_{X_1}^{(j)}$ and $N_{X_2}^{(j)}$ be the numbers of segments of the partition Ω_j on the axis Ox_1 and Ox_2, respectively. Step $h_{X_2}^{(j)}$ in direction of Ox_2 is equal to $2\pi/N_{X_2}^{(j)}$. For each $(x_2)_k^{(j)} = k \cdot h_{X_2}^{(j)}$, $k = 0, \ldots, N_{X_2}^{(j)}$, we define a uniform step of partition Ω_j in the direction of the axis Ox_1 and consider the case $N_{X_2}^{(j)} = 2\,N_{X_1}^{(j)}$, $N_{X_1}^{(2)} = 2\,N_{X_1}^{(1)}$. Using the incomplete Uzawa algorithm [14], we perform numerical calculations. In each iteration, we use GMRES method (see [15]). In Table 1, we show values of the errors for the different parameters a and numbers $N_{X_1}^{(1)}$.

The results of the numerical experiments demonstrate that the errors of the solutions for a velocity field in the norm of space $\mathbf{V}_h(\Omega_h)$ and a pressure in the norm of space $X_h(\Omega_h)$ decrease as the first power h for each a.

Table 1.

$N_{X_1}^{(1)}$	a	48	64	80	96
$\|\mathbf{E}^2\mathbf{u} - \mathbf{u}_h\|_{\mathbf{V}_h}$	0.2	2.4980e-03	1.8791e-03	1.5032e-03	1.2503e-03
$\|E^1 P - P_h\|_{X_h}$	0.2	2.7268e-03	2.0521e-03	1.6451e-03	1.3639e-03
$\|\mathbf{E}^2\mathbf{u} - \mathbf{u}_h\|_{\mathbf{V}_h}$	0.4	3.9778e-03	3.0012e-03	2.3941e-03	1.9901e-03
$\|E^1 P - P_h\|_{X_h}$	0.4	2.8997e-03	2.1791e-03	1.7502e-03	1.4476e-03

Acknowledgements. This work was supported by Russian Foundation of Basic Research (grants 10-01-00060, 11-01-98502-east, 12-01-31018-mol-a).

References

1. Rukavishnikov, A.V.: The Generalized Statement of the Problem of a Two-phase Liqued Flow with a Continuously Changing Interface. Math. Model. 20, 3–8 (2008)
2. Bernardi, C., Maday, Y., Patera, A.: A New Nonconforming Approach to Domain Decomposition: the Mortar Element Method. In: Nonlinear Partial Differential Equations and their Applications, pp. 13–51. Pitman (1994)
3. Flemisch, B., Melenk, J.M., Wohlmuth, B.: Mortar Methods with Curved Interfaces. Appl. Numer. Math. 54, 339–361 (2005)
4. Huang, J., Zou, J.: A Mortar Element Method for Elliptic Problems with Discontinuous Coefficients. IMA Jour. of Numer. Anal. 22, 549–576 (2002)
5. Ben Belgacem, F.: The Mixed Mortar Finite Element Method for the Incompressible Stokes Problem: Convergence Analysis. SIAM J. Numer. Anal. 37, 1085–1100 (2000)
6. Rukavishnikov, A.V., Rukavishnikov, V.A.: On the Nonconformal Finite Element Method for the Stokes Problem with a Discontinuous Coefficient. Sib. Zh. Ind. Mat. 10, 104–117 (2007) (in Russian)
7. Rukavishnikov, A.V.: On Construction of a Numerical Method for the Stokes Problem with a Discontinuous Coefficient of Viscosity. Vychisl. Tekhnol. 14, 110–123 (2009)
8. Gordon, W.J., Hall, C.A.: Transfinite Element Methods: Blending Functions Interpolation over Arbitrary Curved Element Domains. Numer. Math. 21, 109–129 (1973)
9. Ciarlet, P.: The Finite Element Method for Elliptic Problems. North-Holland, Holland (1978)
10. Brezzi, F., Fortin, M.: Mixed and Hybrid Finite Element Methods. Springer, New York (1991)
11. Richter, T.: Numerical Methods for Fluid-structure Interaction Problems. Course of Lectures (2010), http://www.numerik.uni-hd.de/richter/SS10/fsi/fsi.pdf
12. Rukavishnikov, A.V.: The Nonconformal Finite Element Method for a One Problem of Hydrodynamics with a Curved Interface. Comp. Math. and Math. Physics 52, 2072–2094 (2012) (in Russian)
13. Grisvard, P.: Elliptic Problems in Nonsmooth Domains. Pitman, Boston (1985)
14. Bramble, J.H., Pasciak, J.E., Vassilev, A.T.: Analysis of the Inexact Uzawa Algorithm for Saddle Point Problems. SIAM J. Numer. Anal. 34, 1072–1092 (1997)
15. Saad, Y.: Iterative Methods for Sparse Linear Systems. PWS, New Jersey (1996)

Parallel Program Systems for Problems of the Dynamics of Elastic-Plastic and Granular Media

Oxana V. Sadovskaya

Institute of Computational Modeling SB RAS
Akademgorodok 50/44, 660036 Krasnoyarsk, Russia
o_sadov@icm.krasn.ru

Abstract. Parallel program systems for numerical solution of 2D and 3D problems of the dynamics of deformable media with constitutive relationships of rather general form on the basis of universal mathematical model describing small strains of elastic, elastic-plastic and granular materials are worked out. Computational algorithm is based on the splitting methods with respect to physical processes and spatial variables. Some computations of dynamic problems with and without taking into account the moment properties of a material were performed on clusters.

Keywords: Dynamics, elasticity, plasticity, granular medium, Cosserat continuum, shock-capturing method, parallel computational algorithm.

1 Introduction

Under modeling the processes of propagation of the stress waves in geomaterials (granular and porous media, soils and rocks) it is necessary to take into account two main factors connected with their structural inhomogeneity. The first of them is a sharp decrease of the compliance in compression at the time of the collapse of pores with corresponding increase of the wave velocity. The second factor is the conversion of a part of the wave energy into the energy of rotational motion of particles in the material microstructure. Rotational motion can be considered within the framework of a mathematical model of the Cosserat continuum [1], in which along with the velocity of translational motion the angular velocity of a particle is introduced, and along with the stress tensor the couple stress tensor, determining the rotation, is introduced. To make possible the description of deformation of materials with different resistance to tension and compression, the rheological method was supplemented by a new element, a rigid contact, which serves for imitation of perfectly granular material with rigid particles. By using rigid contact in combination with conventional rheological elements (a spring simulating elastic properties of a material, a viscous damper, and a plastic hinge) one can construct constitutive equations of granular materials and soils with elastic-plastic particles and of porous materials like metal foams, see [4].

I. Dimov, I. Faragó, and L. Vulkov (Eds.): NAA 2012, LNCS 8236, pp. 471–478, 2013.

In order to obtain correct numerical solutions for structurally inhomogeneous materials by means of finite-difference methods, computations must be performed on a grid whose meshes are smaller than the characteristic size of the particles of a material. To solve 2D and 3D dynamic problems, parallel algorithms can be efficiently used because they make it possible to distribute the computational load between a lot of number of the cluster nodes, increasing thereby the accuracy of numerical solutions.

2   Mathematical Models

The model for description of the process of deformation of elastic bodies is given by the system of equations

$$A\frac{\partial U}{\partial t} = \sum_{i=1}^{n} B^i \frac{\partial U}{\partial x_i} + QU + G, \tag{1}$$

where U is unknown vector–function, A is a symmetric positive definite matrix of coefficients under time derivatives, B^i are symmetric matrices of coefficients under derivatives with respect to the spatial variables, Q is an antisymmetric matrix, G is a given vector, n is the spatial dimension of a problem (2 or 3). The dimension of the system (1) and concrete form of matrices–coefficients are determined by the used mathematical model.

When taking into account the plastic deformation of a material, the system of equations (1) is replaced by the variational inequality

$$(\tilde{U} - U)\left(A\frac{\partial U}{\partial t} - \sum_{i=1}^{n} B^i \frac{\partial U}{\partial x_i} - QU - G\right) \geq 0, \qquad \tilde{U}, \, U \in F, \tag{2}$$

where F is a given convex set, by means of which some constraints are imposed on possible states of a medium, $\tilde{U}$ is an arbitrary admissible element of F.

In the problems of mechanics of granular media with plastic properties a more general variational inequality

$$(\tilde{V} - V)\left(A\frac{\partial U}{\partial t} - \sum_{i=1}^{n} B^i \frac{\partial V}{\partial x_i} - QV - G\right) \geq 0, \qquad \tilde{V}, \, V \in F, \tag{3}$$

takes place, where the vector–functions V and U are related by the equations

$$V = \lambda U + (1 - \lambda) U^\pi, \qquad U = \frac{1}{\lambda} V - \frac{1-\lambda}{\lambda} V^\pi. \tag{4}$$

Here $\lambda \in (0, 1]$ is the parameter of regularization of the model characterizing the ratio of elastic moduli in tension and compression, U^π is the projection of the vector of solution onto the given convex cone K, by means of which the different resistance of a material to tension and compression is described.

The set F of admissible variations, included in (2) and (3), is defined by the Mises yield condition

$$F = \left\{ U \,\middle|\, \tau(\sigma) \leq \tau_s \right\},$$

where σ is the stress tensor, $\tau(\sigma)$ is the intensity of tangential stresses, τ_s is the yield point of particles. As a convex cone K of stresses, allowed by the strength criterion, the Mises–Schleicher circular cone

$$K = \left\{ U \mid \tau(\sigma) \le æ p(\sigma) \right\}$$

is used, where $p(\sigma)$ is the hydrostatic pressure, æ is the parameter of internal friction. In plane problems of the dynamics of elastic-plastic media the vector–function U consists of 6 unknown functions – 2 components of the velocity vector and 4 components of the symmetric stress tensor:

$$U = \left(v_1, v_2, \sigma_{11}, \sigma_{22}, \sigma_{33}, \sigma_{12} \right).$$

In spatial problems there are 9 unknown functions:

$$U = \left(v_1, v_2, v_3, \sigma_{11}, \sigma_{22}, \sigma_{33}, \sigma_{23}, \sigma_{13}, \sigma_{12} \right).$$

When taking into account the rotational degrees of freedom of particles in the microstructure of a material in the Cosserat continuum model [1], this vector–function along with the components of the velocity vector v and of the stress tensor σ contains also the components of the angular velocity ω and of the nonsymmetric couple stress tensor m. In 2D problems the vector–function U consists of 12 unknown functions:

$$U = \left(v_1, v_2, \sigma_{11}, \sigma_{22}, \sigma_{33}, \sigma_{12}, \sigma_{21}, \omega_3, m_{23}, m_{32}, m_{31}, m_{13} \right),$$

and in 3D problems it consists of 24 unknown functions:

$$U = \Big(v_1, v_2, v_3, \sigma_{11}, \sigma_{22}, \sigma_{33}, \sigma_{23}, \sigma_{32}, \sigma_{31}, \sigma_{13}, \sigma_{12}, \sigma_{21},$$
$$\omega_1, \omega_2, \omega_3, m_{11}, m_{22}, m_{33}, m_{23}, m_{32}, m_{31}, m_{13}, m_{12}, m_{21} \Big).$$

When taking into account the granularity of elastic-plastic materials, the unknown vector–function V involves the nonzero components of the velocity vector v and the actual stress tensor σ, and in the vector–function U instead of σ the conditional stress tensor s, defined by the Hooke law, is included. In the most general model of a granular medium with different resistance to tension and compression, which takes into account rotation of the particles, the vector–function U is as follows:

$$U = \Big(v_1, v_2, v_3, s_{11}, s_{22}, s_{33}, s_{23}, s_{32}, s_{31}, s_{13}, s_{12}, s_{21},$$
$$\omega_1, \omega_2, \omega_3, n_{11}, n_{22}, n_{33}, n_{23}, n_{32}, n_{31}, n_{13}, n_{12}, n_{21} \Big),$$

and the vector–function V is obtained from U by replacing the components of tensors s and n on σ and m, respectively.

3 Computational Algorithm

In the framework of considered mathematical models the parallel computational algorithm was worked out for numerical solution of problems on the propagation of stress and strain waves in media with complex rheological properties [4]. The system of equations (1) is solved by means of the splitting method with respect to spatial variables. Variational inequalities (2) and (3) are solved by splitting with physical processes, which leads to the system (1) at each time step.

Technology of parallelization of computational algorithm is based on the method of two-cyclic splitting with respect to spatial variables [3]. In the three-dimensional case, on the time interval $(t_0, t_0 + \Delta t)$ the splitting method involves 7 stages: the solution of a one-dimensional problem in the x_1 direction on the interval $(t_0, t_0 + \Delta t/2)$, similar stages in the x_2 and x_3 directions, the solution of a system of linear ordinary differential equations with the matrix Q, the recalculation of a problem in the x_3 direction on the interval $(t_0 + \Delta t/2, t_0 + \Delta t)$, and the recalculation in the x_2 and x_1 directions. The two-cyclic splitting method is of second-order accuracy provided that at its stages second-order schemes are used. Besides, it ensures the stability of a numerical solution in spatial case provided that stability conditions for one-dimensional systems are fulfilled.

To maintain second-order accuracy, at the 4th stage, when a system of ordinary differential equations is solved, in linear problems (where the vector–functions V and U are equal to one another) the Crank–Nicholson implicit nondissipative difference scheme is applied. The more general scheme is used for the nonlinear case, and the solution is constructed by the method of successive approximations [4]. For the solution of remaining one-dimensional systems of equations an explicit monotone finite-difference ENO-scheme of the "predictor–corrector" type with piecewise-linear distributions of velocities and stresses over meshes, based on the principles of grid-characteristic methods [2], is applied.

In the most general model, described by the inequality (3), at the "corrector" step the relationships

$$U^{j+1/2} = U_{j+1/2} + \frac{\Delta t}{2} A^{-1} \left(B^i \frac{V_{j+1} - V_j}{\Delta x_i} + G^i \right)$$

are used (in the case of constant matrix–coefficients). Here the index $j + 1/2$ is related to the center of a mesh of a spatial difference grid, a superscript corresponds to an actual time level and a subscript corresponds to a previous level. The vector $V^{j+1/2}$ is calculated from $U^{j+1/2}$ by the formula (4). If the matrices are variable then the corresponding terms of the conservative approximation are taken as a difference derivative with respect to x. At the "predictor" step a system of differential equations arises (equations on characteristics for the model of an elastic medium):

$$Y_l A \frac{\partial U}{\partial t} = c_l Y_l A \frac{\partial V}{\partial x_i} + Y_l G^i \, ,$$

hence, after approximation

$$\left(I_l^{j+1/2} \right)^{\pm} = I_{l\,j+1/2} \pm \alpha_{l\,j+1/2} \frac{\Delta x_i}{2} + \left(c_l \beta_l + Y_l G^i \right)_{j+1/2} \frac{\Delta t}{4} \, ,$$

where Y_l and c_l are left eigenvectors and eigenvalues of the matrix $B^i A^{-1}$, α_l and β_l are derivatives of coefficients of the decomposition of U and V in the basis Y_l: $I_l = (Y_l A)_{j+1/2} U$ and $J_l = (Y_l A)_{j+1/2} V$, received by means of the iterative procedure of limit reconstruction, indices "−" and "+" mark the values of these coefficients on the left and on the right boundaries of a mesh. The procedure of limit reconstruction consists in the construction of monotone piecewise-linear splines which approximate I_l and J_l with minimal discontinuities on boundaries of neighbouring meshes of a grid.

To take into account plastic properties of materials, the variational inequalities (2) and (3) after approximation of the time derivative by a finite difference on the interval $(t_0, t_0 + \Delta t)$ at each mesh of a spatial grid are reduced to

$$(\widetilde{V} - V) A (U - \bar{U}) \geq 0 , \qquad \widetilde{V}, V \in F ,$$

where $\bar{U}$ is the solution of elastic problem at given fixed instant t_0 ($V = U$ for the inequality (2)). Then, taking into account (4), this inequality takes the form: $(\widetilde{V} - V) A (V - (1 - \lambda) V^\pi - \lambda \bar{U}) \geq 0$. Hence, by definition of a projection, $V = ((1 - \lambda) V^\pi + \lambda \bar{U})^\Pi$. The mapping, defined by the right-hand side, is contractive for the positive regularization parameter λ. Thus, the solution correction procedure is in determining a fixed point of the contractive mapping and can be implemented by method of successive approximations. For the case of the Prandtl–Reuss elastic-plastic material this procedure does not require iterations and coincides with the well-known Wilkins procedure for stresses correction.

Numerical algorithm allows to simulate the propagation of elastic-plastic waves produced by external mechanical effects in a medium body, aggregated of arbitrary number of heterogeneous blocks with curvilinear boundaries. Examples of regular decomposition of a computational domain into blocks bounded by curvilinear surfaces are presented in Fig. 1: in 2D case a medium body consists of 12 blocks, in 3D case it includes 24 blocks. The block structure of a body is regular in the sense that its separated blocks consisting of homogeneous materials

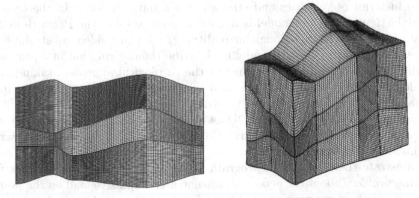

Fig. 1. Finite-difference grids in block media with curvilinear interfaces: 2D case (*left*), 3D case (*right*)

can be numbered with three indices along the axes of a Cartesian coordinate system. This numbering can be introduced only if interfaces of blocks are consistent with each other. If there exist inconsistent interfaces, then it is necessary to extend these interfaces and to perform a fictitious regular decomposition of a medium body with a large number of blocks of the same material.

The parallelizing of computations is carried out using the MPI library, the programming language is Fortran. The data exchange between processors occurs at step "predictor" of the finite-difference scheme by means of the function MPI_Sendrecv. At first each processor exchanges with neighboring processors the boundary values of their data, and then calculates the required quantities in accordance with the difference scheme. Mathematical models are embedded in programs by means of software modules that implement the constitutive relationships, the initial data and boundary conditions of problems, and also conditions of pasting together of solutions on inconsistent grids of neighboring blocks. The universality of programs is achieved by a special packing of the variables, used at each node of the cluster, into large one-dimensional arrays. Computational domain is distributed between the cluster nodes by means of 1D, 2D or 3D decomposition so as to load the nodes uniformly and to minimize the number of passing data. Detailed description of the parallel algorithm one can found in [4].

4 Parallel Program Systems

Using this computational algorithm, the program systems for numerical solution of two-dimensional and three-dimensional elastic-plastic problems of the dynamics of granular media (2Dyn_Granular, 3Dyn_Granular) and for solution of two-dimensional and three-dimensional dynamic problems of the Cosserat elasticity theory (2Dyn_Cosserat, 3Dyn_Cosserat) were worked out.

Parallel program systems 2Dyn_Granular and 3Dyn_Granular are intended for numerical realization of the universal mathematical model describing small strains of elastic, plastic and granular materials. In the case of elastic material this model is reduced to the system of equations (1), hyperbolic by Friedrichs, written in terms of velocities and stresses in a symmetric form. In the case of elastic-plastic material the model is a special formulation of the Prandtl–Reuss theory in the form of variational inequality (2) with one-sided constraints on stresses. Generalization of the model to describe the deformation of a granular material (3), (4) is obtained by means of the rheological approach, taking into account different resistance of a material to tension and compression. Computational domain may have a block structure, composed of an arbitrary number of layers and blocks in each layer (2D case), or an arbitrary number of layers, strips in a layer and blocks in a strip (3D case) from different materials with self-consistent curvilinear interfaces.

To illustrate the efficiency of program systems, numerical computations for the Lamb problem about the action of concentrated impulsive load on the boundary of elastic medium were carried out. The level curves of normal stress for 2D problem at different time moments are shown in Fig. 2, and the level surfaces of normal stress for 3D problem are shown in Fig. 3.

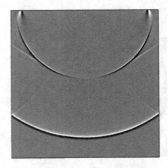

Fig. 2. 2D Lamb's problem about the action of concentrated load on the boundary of a half-plane for a momentless medium: level curves of normal stress σ_{11}, 250th time step (*left*) and 500th time step (*right*)

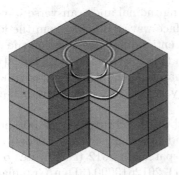
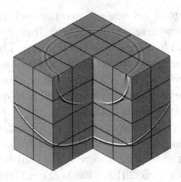

Fig. 3. 3D Lamb's problem about the action of concentrated load on the surface of a half-space for a momentless medium: level surfaces of normal stress σ_{11}, 100th time step (*left*) and 200th time step (*right*)

Program systems 2Dyn_Cosserat and 3Dyn_Cosserat allow to solve plane and spatial dynamic problems of the moment elasticity, taking into account rotations of the particles of microstructure of a material within the framework of the theory of small strains. The model is formulated as the system of equations (1), hyperbolic by Friedrichs, written in terms of the vectors of velocities of translational and rotational motion, as well as the tensors of stresses and couple stresses. The initial data of boundary-value problem are formulated in terms of displacements, rotation angles, velocities of translational and rotational motion, stresses and couple stresses. On interblock boundaries the conditions of continuity of the velocity and angular velocity vectors, the stress and couple stress vectors are placed. On external boundaries the boundary conditions in velocities of translational and rotational motion, stresses and couple stresses, as well as mixed boundary conditions and symmetry conditions, ensuring mathematical correctness of a problem, can be specified. Numerical results for the Lamb problem in the case of a moment medium are shown in Fig. 4 (2D problem).

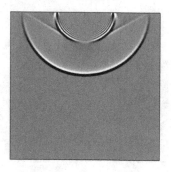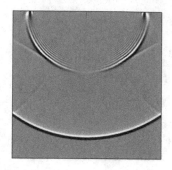

Fig. 4. 2D Lamb's problem about the action of concentrated load on the boundary of a half-plane for a moment medium: level curves of normal stress σ_{11}, 250th time step (*left*) and 500th time step (*right*)

In Figs. 2 – 4 one can see the incident longitudinal and transverse waves, conical transverse waves and the Raleigh surface waves, moving from the loading point inside computational domain. The essential difference between the results of computations for the Cosserat continuum (Figs. 4) and for the classical elasticity theory (Figs. 2, 3) is that in a moment material behind the front of transverse wave the additional system of high-frequency waves, caused by rotational motion of the particles, is observed.

Numerical computations were performed on the cluster MVS–100k of Joint Supercomputer Center of the Russian Academy of Sciences (Moscow). Parallel program systems were registered by Rospatent in 2012 (certificates of state registration no. 2012613989 (2Dyn_Granular), 2012613990 (3Dyn_Granular) and 2012614823 (2Dyn_Cosserat), 2012614824 (3Dyn_Cosserat)).

Acknowledgments. This work was supported by the Complex Fundamental Research Program no. 18 "Algorithms and Software for Computational Systems of Superhigh Productivity" of the Presidium of RAS and the Russian Foundation for Basic Research (grant no. 11–01–00053).

References

1. Cosserat, E., Cosserat, F.: Théorie des Corps Déformables. In: Chwolson, O.D. (ed.) Traité Physique, Librairie Scientifique, pp. 953–1173. A. Hermann et Fils, Paris (1909)
2. Kulikovskii, A.G., Pogorelov, N.V., Semenov, A.Y.: Mathematical Aspects of Numerical Solution of Hyperbolic Systems. Monographs and Surveys in Pure and Applied Mathematics, vol. 118. Chapman & Hall/CRC, Boca Raton, London (2001)
3. Marchuk, G.I.: Splitting Methods. Nauka, Moscow (1988) (in Russian)
4. Sadovskaya, O., Sadovskii, V.: Mathematical Modeling in Mechanics of Granular Materials. Advanced Structured Materials, vol. 21. Springer, Heidelberg (2012)

On Thermodynamically Consistent Formulations of Dynamic Models of Deformable Media and Their Numerical Implementation

Vladimir M. Sadovskii

Institute of Computational Modeling SB RAS
Akademgorodok 50/44, 660036 Krasnoyarsk, Russia
sadov@icm.krasn.ru

Abstract. Mathematical models of the dynamics of elastic-plastic and granular media are formulated as variational inequalities for hyperbolic operators with one-sided constraints describing the transition of a material in plastic state. On this basis a priori integral estimates are constructed in characteristic cones of operators, from which follows the uniqueness and continuous dependence on initial data of solutions of the Cauchy problem and of the boundary-value problems with dissipative boundary conditions. With the help of an integral generalization of variational inequalities the relationships of strong discontinuity in dynamic models of elastic-plastic and granular media are obtained, whose analysis allows us to calculate velocities of shock waves and to construct discontinuous solutions. Original algorithms of solution correction are developed which can be considered as a realization of the splitting method with respect to physical processes.

Keywords: Dynamics, granular medium, elasticity, plastic shock wave, discontinuous solution, variational inequality, computational algorithm.

1 Introduction

Thermodynamically consistent systems of conservation laws were firstly obtained by Godunov and his followers for the models of reversible thermodynamics – elasticity theory, gas dynamics and electrodynamics [3–5]. Such form of equations assumes the setting so-called generating potentials $\Phi(U)$ and $\Psi_j(U)$ ($j = 1, ..., n$, where n is the spatial dimension of a model) depending on the vector U, whose components are projections of the velocity vector, components of the stress tensor and other thermodynamic state parameters. By means of generating potentials the system is written in divergent form as follows:

$$\frac{\partial}{\partial t}\frac{\partial \Phi}{\partial U} = \sum_{j=1}^{n}\frac{\partial}{\partial x_j}\frac{\partial \Psi_j}{\partial U} , \tag{1}$$

I. Dimov, I. Faragó, and L. Vulkov (Eds.): NAA 2012, LNCS 8236, pp. 479–486, 2013.
© Springer-Verlag Berlin Heidelberg 2013

or in a more general form, including terms that are independent of derivatives. The additional conservation law

$$\frac{\partial}{\partial t}\left(U\frac{\partial\Phi}{\partial U} - \Phi\right) = \sum_{j=1}^{n}\frac{\partial}{\partial x_j}\left(U\frac{\partial\Psi_j}{\partial U} - \Psi_j\right) \tag{2}$$

is valid for the divergent system (1). The equation (2) may be a conservation law of energy or of entropy.

Thermodynamically consistent systems of conservation laws of the form (1), (2) turn out to be very useful in justification of the mathematical correctness of models. Based on such formulation a priori estimates of solutions in characteristic cones of the operator can be obtained, from which it follows the uniqueness and continuous dependence on initial data for the Cauchy problem and for boundary value problems with dissipative boundary conditions. It is intended for the integral generalization of the model, which allows to construct discontinuous solutions. For numerical analysis of the system (1), (2) the effective shock-capturing methods, such as Godunov's method [6], adapted to the computation of solutions with discontinuities, caused by concentrated and impulsive perturbations, may be applied.

The present paper addresses to generalization and application of this approach for the analysis of thermodynamically irreversible models of mechanics of deformable media taking into account plastic deformation of materials.

2   Special Formulations

In geometrically linear approximation the system of equations of the dynamic elasticity can be written in common notations in the next form:

$$\rho\frac{\partial v}{\partial t} = \nabla\cdot\sigma + \rho g, \qquad \frac{\partial}{\partial t}\frac{\partial\Phi_0}{\partial\sigma} = \frac{1}{2}\left(\nabla v + \nabla v^*\right). \tag{3}$$

In this case the vector U consists of velocities and stresses only. The dimension of U depends on the spatial dimension n. Generating potentials are given by $\Phi = \rho v^2/2 + \Phi_0(\sigma)$, $\Psi_j = (\sigma\cdot v)_j$, where the projection of a vector onto the axis x_j is marked by the index j. The system (3) itself is a thermodynamically consistent system of conservation laws. It can also be represented in the form

$$A\frac{\partial U}{\partial t} = \sum_{j=0}^{n}B^j\frac{\partial U}{\partial x_j} + QU + G, \tag{4}$$

where $A = \partial^2\Phi/\partial U^2$, $B^j = \partial^2\Psi_j/\partial U^2$ are square matrices composed of mechanical parameters of a material, G is a vector containing the mass forces, Q is a matrix, which is nonzero in the curvilinear coordinate system.

If the strain potential Φ_0 is a strongly convex twice differentiable function, then the system (4) belongs to the well-studied class of systems, which are hyperbolic by Friedrichs [7]. In particular, in the linear elasticity theory

$\Phi_0 = \sigma : a : \sigma/2$ is a quadratic form, whose coefficients are given by the tensor a of elastic compliance of a material – a positive definite fourth-rank tensor, possessing a special symmetry.

In a more general nonlinear model of a granular material having different resistance to tension and compression [10], the condition of differentiability of a strain potential is violated. For example, for an ideal (not cohesive) granular material with elastic particles $\Phi_0 = \sigma : a : \sigma/2 + \delta_K(\sigma)$, where K is the cone of admissible stresses (the Coulomb–Mohr cone or the Mises–Schleicher cone) and δ_K is the indicator function, equals to zero on K and equals to infinity in the exterior of K. Such form of a potential indicates that tensile stresses are not possible in an ideal granular medium. This model is reduced to the system

$$\rho \frac{\partial v}{\partial t} = \nabla \cdot \sigma + \rho g\,, \qquad a : \frac{\partial s}{\partial t} = \frac{1}{2}\left(\nabla v + \nabla v^*\right)\,, \qquad \sigma = s^\pi\,,$$

where s^π is a projection of the conditional stress tensor s onto the cone K in the norm $|s| = \sqrt{s : a : s}$. The last system can be represented in the form

$$A\frac{\partial U}{\partial t} = \sum_{j=0}^{n} B^j \frac{\partial U^\pi}{\partial x_j} + Q\,U^\pi + G\,, \tag{5}$$

but this system is not reduced to the Godunov system (1), written in terms of generating potentials.

When taking into account the irreversible deformation, in the models of viscoelastic media the dissipative potential $H(\sigma)$ is introduced. The derivative of this potential is equal to the viscous strain rate tensor. The constitutive equation in (2) is replaced by the next equation:

$$a : \frac{\partial \sigma}{\partial t} = \frac{1}{2}\left(\nabla v + \nabla v^*\right) - \frac{\partial H}{\partial \sigma}\,. \tag{6}$$

It is usually assumed that the dissipative potential is a convex function. In this case for any symmetric tensor $\tilde{\sigma}$ the following inequality is valid

$$H(\tilde{\sigma}) - H(\sigma) \geq (\tilde{\sigma} - \sigma) : \frac{\partial H}{\partial \sigma}\,,$$

which is equivalent to the definition of convexity. If the indicator function of the set F, whose boundary describes the yield surface, is taken as a dissipative potential in (6), then we obtain the variational formulation of constitutive relationships of the Prandtl–Reuss theory of elastic-plastic flow [13]:

$$(\tilde{\sigma} - \sigma) : \left(a : \frac{\partial \sigma}{\partial t} - \nabla v\right) \geq 0\,, \qquad \sigma,\ \tilde{\sigma} \in F\,. \tag{7}$$

It can be shown that the system of equations of the dynamics of a viscoelastic medium with the constitutive equation (6) is reduced to the semilinear system (4), in which the matrix Q depends on the solution. In the case of an elastic-plastic medium (7) variational inequality of general form

$$(\tilde{U} - U)\left(A\frac{\partial U}{\partial t} - \sum_{j=0}^{n} B^j \frac{\partial U}{\partial x_j} - QU - G\right) \geq 0, \qquad U, \tilde{U} \in F, \qquad (8)$$

takes place. It should be noted that under our assumptions relative to matrices–coefficients the differential operator of variational inequality is hyperbolic by Friedrichs and that it can be written with the help of generating potentials.

Let us consider a more general variational inequality, describing the process of dynamic deformation of a granular medium with elastic-plastic particles:

$$(\tilde{V} - U^\pi)\left(A\frac{\partial U}{\partial t} - \sum_{j=0}^{n} B^j \frac{\partial U^\pi}{\partial x_j} - QU^\pi - G\right) \geq 0, \qquad U^\pi, \tilde{V} \in F, \qquad (9)$$

from which the above models can be obtained as special cases when K and F coincide with the whole stress space.

3   A Priori Estimates and Weak Solutions

For the difference of two sufficiently smooth solutions of the variational inequality (8) a priori estimates in characteristic cones, generalizing estimates for solutions of the hyperbolic system of equations, can be obtained. Let U and $\bar{U}$ be two such solutions. Assuming that $\tilde{U} = \bar{U}$ in the inequality (8) and $\tilde{U} = U$ in the analogous inequality written for $\bar{U}$, and summing results we obtain

$$(\bar{U} - U)\left(A\frac{\partial(\bar{U} - U)}{\partial t} - \sum_{j=0}^{n} B^j \frac{\partial(\bar{U} - U)}{\partial x_j} - Q(\bar{U} - U) - \bar{G} + G\right) \leq 0.$$

This inequality, after reduction to the divergent form, is integrated over the set of a truncated cone type in the space of variables $(x_1, ..., x_n, t)$ with the bases $t = t_0$ and $t = t_1$, which is built by solving the equation or inequality of Hamilton–Jacobi [3, 13]. Transforming by Green's formula we obtain the estimate:

$$\|\bar{U} - U\|(t_1) \leq \|\bar{U} - U\|(t_0)\exp\alpha(t_1 - t_0) +$$

$$+ \beta\int_{t_0}^{t_1} \|\bar{G} - G\|(t)\exp\alpha(t_1 - t)\,dt. \qquad (10)$$

Here $\|U\|(t)$ is the energy norm, equals to the integral of $\sqrt{UAU}$ over the cross-section of cone by the hyperplane $t = $ const, α and β are the coefficients, independent on both solutions. Uniqueness of a solution of the Cauchy problem

$$U\big|_{t=t_0} = U^0$$

and its continuous dependence on initial data and right-hand side in a small time interval follows from the estimate (10).

A similar estimate can be obtained in a truncated cone, adjacent to a part of boundary, where the dissipative boundary conditions are given. Dissipativity is understood in a usual sense: the fulfillment of these conditions for two vector–functions $\bar{U}$ and U must ensure the fulfillment of the inequality

$$(\bar{U} - U) \sum_{j=1}^{n} \nu_j \, B^j (\bar{U} - U) \leq 0 \, ,$$

where ν_j are the projections of an external normal vector.

It is impossible to deduce a similar estimate for the difference of solutions of the variational inequality (9) because of the nonlinear projector onto the cone K, included in the inequality. In this case the energy estimate can be obtained, by means of which the boundness of solutions in cones and the stability of trivial solution with respect to small perturbations of initial data and right-hand side can be proved. In the monograph [10] the estimate is obtained for difference of two solutions of the regularized inequality (9), where the projection U^π is replaced by the vector $V = \varepsilon U + (1 - \varepsilon) U^\pi$ with a small parameter $\varepsilon > 0$. However, the limit under $\varepsilon \to 0$ in this estimate does not give the desired result. It is essential that estimate (10) is valid for the variational inequality (8) in the presence of discontinuity surfaces of solutions [13].

The problem of constructing weak (discontinuous) solutions in the theory of plasticity is not completely solved till nowadays. Under the assumption of continuity of the stress deviator on the plastic shock wave Mandell in [8] has constructed a system of equations of a strong discontinuity in the elastic-plastic flow theory with the Mises yield condition. Bykovtsev and Kretova in [2] have obtained similar relationships with the Tresca–Saint-Venant yield condition under the assumption of immutability of principal axes of the stress tensor onto the sides of wave front. The author of this paper proposed a method of integral generalization of variational inequalities, which allowed to investigate the plastic shock waves for a wide range of models without any additional assumptions. This method was applied to the analysis of discontinuous solutions in perfectly elastic-plastic media [12, 15], in materials with linear and nonlinear hardening [14, 17]. In more general form it has been applied to the analysis of shock-wave motion of granular media [9, 16]. The method occurs to be extremely fruitful in the study of models, which are formulated as variational inequalities with linear differential operators. In the nonlinear case, the system of relationships of a strong discontinuity, obtained with the help of this method, as a rule, is not closed, and the construction of the realizability conditions must be accompanied by the use of additional considerations about the evolutionarity and the stability of a shock.

We describe the method of integral generalization on an example of variational inequality with operator, expressed in terms of generating potentials:

$$(\tilde{U} - U)\left(\frac{\partial}{\partial t} \frac{\partial \Phi}{\partial U} - \sum_{j=1}^{n} \frac{\partial}{\partial x_j} \frac{\partial \Psi_j}{\partial U} \right) \geq 0 \, , \qquad U, \, \tilde{U} \in F \, . \tag{11}$$

At first this inequality must be written in the divergent form:

$$\tilde{U}\left(\frac{\partial}{\partial t}\frac{\partial\Phi}{\partial U} - \sum_{j=1}^{n}\frac{\partial}{\partial x_j}\frac{\partial\Psi_j}{\partial U}\right) \geq \frac{\partial}{\partial t}\left(U\frac{\partial\Phi}{\partial U} - \Phi\right) - \sum_{j=1}^{n}\frac{\partial}{\partial x_j}\left(U\frac{\partial\Psi_j}{\partial U} - \Psi_j\right).$$

Then both its parts are multiplied by a compactly supported test function $\chi \geq 0$ and integrated over the space–time domain using Green's formula. As a result, we obtain the inequality, which does not contain derivatives of the solution:

$$\int\int\left(-\frac{\partial\Phi}{\partial U}\frac{\partial(\chi\tilde{U})}{\partial t} + \sum_{j=1}^{n}\frac{\partial\Psi_j}{\partial U}\frac{\partial(\chi\tilde{U})}{\partial x_j}\right)dx\,dt \geq$$

$$\geq \int\int\left(-\left(U\frac{\partial\Phi}{\partial U} - \Phi\right)\frac{\partial\chi}{\partial t} + \sum_{j=1}^{n}\left(U\frac{\partial\Psi_j}{\partial U} - \Psi_j\right)\frac{\partial\chi}{\partial x_j}\right)dx\,dt\,.$$

Taking into account arbitrariness in the choice of the function χ, the next variational inequality at the points of surface of strong discontinuity can be obtained:

$$\tilde{U}\left[c\frac{\partial\Phi}{\partial U} + \sum_{j=1}^{n}\nu_j\frac{\partial\Psi_j}{\partial U}\right] \geq c\left[U\frac{\partial\Phi}{\partial U} - \Phi\right] + \sum_{j=1}^{n}\nu_j\left[U\frac{\partial\Psi_j}{\partial U} - \Psi_j\right], \qquad (12)$$

where $c \geq 0$ is the velocity of the wave front, square brackets are used to denote a jump of function under transition through the discontinuity. This inequality allows to analyze admissible discontinuities of a solution in specific models. In a similar way one can obtain more general relationships of a strong discontinuity for the solution of variational inequality (9), which does not reduced to (11).

The inequality (12) possesses comprehensive information about the discontinuities in the case of quadratic generating potentials, when a differential operator is linear. As it turned out, jumps of velocities and stresses on the fronts of plastic shock waves, velocities of waves and even their number are essentially dependent on the form of a plasticity condition. In the presence of flat portions of the yield surface the number of shock waves increases. In the geometrically linear models the velocities of shock waves are calculated in terms of given phenomenological parameters of a material and do not depend on the stress state in the vicinity of a front. Conversely, in the nonlinear models the theorem of definiteness of a shock wave holds: the state behind the front is uniquely determined by the state before the wave front and by given velocity of the wave. In perfectly media the class of discontinuous solutions depends essentially on the type of state function, connecting the hydrostatic pressure with the density and temperature, on the nature of dependence of the yield point on temperature, as well as the presence of plane faces on the yield surface [15]. In the models of a nonlinearly hardening the number of shock waves is determined by the number of convex down parts on the diagram of shear [9]. In the models of granular media the number of possible discontinuities increases in comparison with usual elastic-plastic media. Shock-wave transitions of a material from the loosened state to the compressed one, so-called signotons, are added to these discontinuities. Shock adiabats of plane signotons of low intensity are analyzed in [9, 10].

4 Numerical Approaches

Besides notable successes in the investigation of a problem of discontinuous solutions, the formulations of models in the form of variational inequalities turned
out to be useful in the constructing computational algorithms. The approximation of differential operator and constraint on an example of the inequality (11)
leads to the following discrete problem:

$$(\tilde{U} - \hat{U}^{k+1})\left(\frac{\partial \Phi^{k+1}}{\partial U} - \frac{\partial \Phi^k}{\partial U} - \Delta t \sum_{j=1}^{n} \Lambda_j \frac{\partial \Psi_j^k}{\partial U}\right) \geq 0, \qquad \hat{U}^{k+1}, \tilde{U} \in F,$$

where $\hat{U}^{k+1}$ is a special combination of U^{k+1} and U^k, Δt is the time step of
a grid, and Λ_j is the differential operator approximating the partial derivative
with respect to the spatial variable x_j. In the case of $\hat{U}^{k+1} = U^{k+1}$ there is the
most simple problem. Its solution can be found in two steps: at first the vector

$$\frac{\partial \bar{\Phi}^{k+1}}{\partial U} = \frac{\partial \Phi^k}{\partial U} - \Delta t \sum_{j=1}^{n} \Lambda_j \frac{\partial \Psi_j^k}{\partial U},$$

implementing the explicit finite-difference scheme on the time step for the system of equations (1), is calculated and then the solution correction is made in
accordance with the variational inequality

$$(\tilde{U} - \hat{U}^{k+1})\left(\frac{\partial \Phi^{k+1}}{\partial U} - \frac{\partial \bar{\Phi}^{k+1}}{\partial U}\right) \geq 0, \qquad U^{n+1}, \tilde{U} \in F,$$

which is equivalent by the convexity of $\Phi(U)$ to the problem of conditional
minimization of the function $\Phi(U^{k+1}) - U^{k+1}\partial\bar{\Phi}^{k+1}/\partial U$ under the constraint
$U^{k+1} \in F$. These steps can be considered as a result of application to the variational inequality (11) of the splitting method with respect to physical processes:
at first an elastic problem is solved, then its solution is corrected by means of
the plastic constitutive relationships.

If the generating potential $\Phi(U)$ is a quadratic function, then the solution
correction is reduced to determining projection of the vector $\bar{U}^{k+1}$ onto the
convex set F with respect to corresponding norm. This method of correction
was used by Wilkins [18] under numerical solution of elastic-plastic problems,
it is widespread now. The algorithm, corresponding to $\hat{U}^{k+1} = (U^{k+1} + U^k)/2$,
possesses a more accuracy. It is realized rather simple in [1].

More general variants of the solution correction algorithm in the problems
of the dynamics of granular media taking into account the plastic deformation
of the particles are developed in the monograph [10]. The method of splitting
with respect to spatial variables, which is applied for the solution of multidimensional problems in the first stage of the computational algorithm, and the
finite-difference schemes for the solution of one-dimensional systems of equations
under the realization of the splitting method are described there, too.

Acknowledgments. This work was supported by the Complex Fundamental Research Program no. 18 "Algorithms and Software for Computational Systems of Superhigh Productivity" of the Presidium of RAS and the Russian Foundation for Basic Research (grant no. 11–01–00053).

References

1. Annin, B.D., Sadovskii, V.M.: The numerical realization of a variational inequality in the dynamics of elastoplastic bodies. Comput. Math. Math. Phys. 36 (9), 1313–1324 (1996)
2. Bykovtsev, G.I., Kretova, L.D.: On shock wave propagation in elastoplastic media. J. Appl. Math. Mech. 36 (1), 106–116 (1972)
3. Godunov, S.K.: Equations of Mathematical Physics. Nauka, Moscow (1979) (in Russian)
4. Godunov, S.K., Peshkov, I.M.: Symmetric hyperbolic equations in the nonlinear elasticity theory. Comput. Math. Math. Phys. 48 (6), 975–995 (2008)
5. Godunov, S.K., Romenskii, E.I.: Elements of Continuum Mechanics and Conservation Laws. Kluwer Academic / Plenum Publishers, New York, Boston (2003)
6. Godunov, S.K., Zabrodin, A.V., Ivanov, M.Y., Kraiko, A.N., Prokopov, G.P.: Numerical Solution of Multidimensional Problems of Gas Dynamics. Nauka, Moscow (1976) (in Russian)
7. Friedrichs, K.O.: Symmetric hyperbolic linear differential equations. Commun. Pure Appl. Math. 7(2), 345–392 (1954)
8. Mandel, J.: Ondes plastiques dans un milieu indéfini à trois dimensions. J. de Mécanique 1(1), 119–141 (1962)
9. Sadovskaya, O.V., Sadovskii, V.M.: Elastoplastic waves in granular materials. J. Appl. Mech. Tech. Phys. 44(5), 741–747 (2003)
10. Sadovskaya, O., Sadovskii, V.: Mathematical Modeling in Mechanics of Granular Materials. Advanced Structured Materials, vol. 21. Springer, Heidelberg (2012)
11. Sadovskii, V.M.: Hyperbolic variational inequalities. J. Appl. Math. Mech. 55(6), 927–935 (1991)
12. Sadovskii, V.M.: To the problem of constructing weak solutions in dynamic elastoplasticity. Int. Ser. Numer. Math. 106, 283–291 (1992)
13. Sadovskii, V.M.: Discontinuous Solutions in Dynamic Elastic-Plastic Problems. Fizmatlit, Moscow (1997) (in Russian)
14. Sadovskii, V.M.: Elastoplastic waves of strong discontinuity in linearly hardening media. Mech. Solids 32(6), 104–111 (1997)
15. Sadovskii, V.M.: On the theory of shock waves in compressible plastic media. Mech. Solids 36(5), 67–74 (2001)
16. Sadovskii, V.M.: On the theory of the propagation of elastoplastic waves through loose materials. Doklady Phys. 47(10), 747–749 (2002)
17. Sadovskii, V.M.: To the analysis of the structure of finite-amplitude transverse shock waves in a plastic medium. Mech. Solids 38(6), 31–39 (2003)
18. Wilkins, M.L.: Calculation of elastic-plastic flow. In: Methods in Computational Physics. Fundamental Methods in Hydrodynamics, vol. 3, pp. 211–263. Academic Press, New York (1964)

An Effective Method of Electromagnetic Field Calculation

Alexey Shamaev and Dmitri Knyazkov

IPMech RAS, Prosp. Vernadskogo 101, Block 1, Moscow, 119526 Russia
knyaz@ipmnet.ru

Abstract. The problem of calculation of electromagnetic field from a large number (10^8 and more) of elementary radiating objects for the needs of holographic lithography is considered. The specially designed big pixel method is proposed. This method was implemented as a part of a parallel software package and was used on the MVS-100K JSCC RAS and the MIIT T4700 clusters (100 TFLOPS and 4.7 TFLOPS correspondingly). The big pixel method allows calculating of Gabor holograms for images of real-size chip topologies on modern clusters. A parallel efficiency of the algorithm was investigated. It is shown an example of Gabor hologram synthesis for an image of topology consisting of $1.6*10^9$ elements.

Keywords: HPC, parallel algorithms, cluster computations, holography simulation, highly oscillatory integrals.

1 Introduction

In this paper we are considering a calculation of electromagnetic fields for one specific optical scheme (fig. 1) used for construction of holographic images. It is assumed that all desired images consist of simple geometrical shapes composed from rectangles and squares with sides being parallel to coordinate axes and having multiple sizes (see gray areas at fig. 2).

Our problem arises from a holographic lithography [1–3] where there is a need to solve creation, optimization and reconstruction problems for such images. For images of small-size and medium-size structures or so-called "topologies" (squares, strips, test miras) such problems could be solved with the use of cluster supercomputers having relatively low performance [4], but when dealing with images of full-size chip layer topologies it is indispensable to develop fast methods of electromagnetic field calculation suitable for the considered optical scheme.

The main computational problem in the holographic lithography is to calculate such a hologram mask which produces the desired image. While it is rather simple to calculate a mask in ordinary projection lithography [5] (in most cases this mask is just a homothetic copy of an image of an initial topology), here at holographic lithography, on the contrary, holographic mask calculation is very time consuming, obligatory technological step. At the same time calculating holographic mask by the means of standard methods demands computational

I. Dimov, I. Faragó, and L. Vulkov (Eds.): NAA 2012, LNCS 8236, pp. 487–494, 2013.

power exceeding existing HPC-computers performance in order of magnitudes. That is why developing of a method of fast calculation of electromagnetic fields needed for Gabor hologram synthesis is a key issue in the holographic lithography and this particular problem will be taken in mind through the course of the current paper.

2   Problem Statement

A scheme of a holographic lithography system [1, 2] is shown at fig. 1. A radiation W illuminates a holographic plate, which is denoted as Ω_H and is situated in $O_H xy$ plane. The plate's transmissivity is described by the a real-value function $T(x,y)$: $\Omega_H \to [0,1]$. Domains $\{(x,y) \mid (x,y) \in \Omega_H, T(x,y) = 1\}$ are totally transparent for the radiation, while domains $\{(x,y) \mid (x,y) \in \Omega_H, T(x,y) = 0\}$ don't transmit the radiation W at all. The resulting aerial image (intensity distribution) is formed at $O_I \xi \eta$ plane.

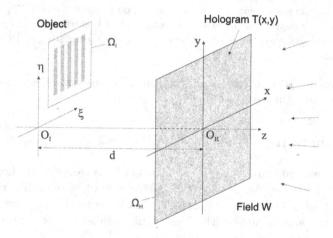

Fig. 1. Optical scheme

Let $q_0(\xi, \eta) : \Omega_I \to \{0,1\}$ be some function defining the required image, which corresponds to the desired paths configuration (topology) on a future chip layer. Let's consider an object field - a complex function $g(\xi, \eta) : \Omega_I \to \mathbb{C}$ defined at Ω_I domain in $O_I \xi \eta$ plane. When simulating an illumination by a plane wave R we could assume $g(\xi, \eta) = q_0(\xi, \eta) R(\xi, \eta)$. A Gabor hologram for the object $g(\xi, \eta)$ is a function $T_g(x,y) : \Omega_H \to [0,1]$ defined by the following formula:

$$T_g(x,y) = \left| \iint\limits_{\Omega_I} K(x,y,\xi,\eta) g(\xi,\eta) d\xi d\eta + \overline{W}(x,y) \right|^2 , \tag{1}$$

where $K(x, y, \xi, \eta) = \frac{e^{i\tilde{k}r}}{r}$, $r = \sqrt{(x - \xi)^2 + (y - \eta)^2 + d^2}$, $\tilde{k} = \frac{2\pi}{\lambda}$ is wave number, λ is a wave length and $\overline{W}(x, y)$ is a field complex-conjugated to the reconstructing field $W(x, y)$ (fig. 1).

Radiation intensity ("reconstructed image") $I(\xi, \eta)$ is registered at observation domain Ω_I laying at object plane $O_I \xi \eta$ at distance d from the hologram. If the field at $O_H xy$ plane before propagating through the hologram with the transparency function $T(x, y)$ is described by $W(x, y)$, then radiation intensity at object plane could be calculated in scalar approximation with the use of Kirchhoff integral in the following way:

$$I_T(\xi, \eta) = \left| \iint\limits_{\Omega_H} K(x, y, \xi, \eta) W(x, y) T(x, y) dx dy \right|^2 . \tag{2}$$

A reconstructed image $q(\xi, \eta) = I_{T_g}(\xi, \eta)$ created by Gabor hologram $T_g(x, y)$ will be rather close to the given image $q_0(\xi, \eta)$ for the case when $g(\xi, \eta) = q_0(\xi, \eta) R(\xi, \eta)$.

It should be noted, that it is necessary to use vector radiation model when dealing with high aperture angles. In this case formula (1) is not changing, and image reconstruction calculation could be reduced to calculation of several integrals having the same type as (2) using, for example, Stratton-Chu formulas [6]. Therefore, the main issue in the hologram creation problem (1) (as well as in the problem of its illumination modelling (2)) is calculation of the following integral:

$$I(x, y) = \iint\limits_{\Omega} K(x, y, \xi, \eta) \varphi(\xi, \eta) d\xi d\eta \tag{3}$$

with a rapidly oscillating kernel

$$K(x, y, \xi, \eta) = \frac{e^{i \frac{2\pi}{\lambda} \sqrt{(x - \xi)^2 + (y - \eta)^2 + d^2}}}{\sqrt{(x - \xi)^2 + (y - \eta)^2 + d^2}} \tag{4}$$

at points $(-\frac{b}{2} + \gamma(i + \frac{1}{2}), -\frac{b}{2} + \gamma(j + \frac{1}{2}))$, where b is a hologram size, γ is a hologram cell size, $i, j = 1, ..., M$, while $M = [\frac{b}{\gamma}]$.

As the grid size M could be up to 10^6 as well as an image grid size N ($N \approx M$), it is critically important to develop a technique (or so-called fast summation method [7]) allowing to produce the calculation (3), (4) in $O(N^2 log N)$ operations instead of N^4 given by direct summation. Later in this paper it is described a method specially designed for the calculation of Gabor holograms in the holographic lithography specific optical scheme (fig. 1). It substantially uses some peculiarities of the problem. The first is that the domain where $\varphi(\xi, \eta)$ is non-zero consists of rectangles with sides parallel to coordinate axis, because metal paths on a resulting chip layer do have such configuration, and therefore the initial image has the same structure. That is why we could decompose the integration domain into rectangles or squares without loosing accuracy. The second one is that the transparency function $T_g(x, y)$ of the Gabor hologram oscillates very slow (it is done intentionally by using special optical scheme shown at fig. 1). Therefore we need to calculate electromagnetic field $I(x, y)$ on rather sparse grid, that allows to considerably reduce number of operations but still allows to implement FFT-based approach. As a result the proposed big pixel method allows to obtain $C_{BP}N^2logN$ operations asymptotic with such a constant C_{BP} that calculating Gabor holograms for real chip layer topologies becomes possible on the state-of-the-art cluster computational systems.

3 Big Pixel Method

Let's decompose an image domain Ω_I (square $a \times a$ having center at (ξ_0, η_0)) into big pixels, i.e. squares $\Box_{kl}$ having size $\sigma \times \sigma$ with centers at $(\xi_0 - \frac{a}{2} + \sigma(k + \frac{1}{2}), \eta_0 - \frac{a}{2} + \sigma(l + \frac{1}{2}))$, $k, l = 0, ..., N - 1$, where $N = [\frac{a}{\sigma}]$. Thus, on every $\Box_{kl}$ the function of topology image q has a constant value q_{kl}: 1, if the square $\Box_{kl}$ belongs to topology and 0 otherwise (fig. 2). We could also add some non-zero phase ψ (for example, neighbor elements are assumed to have phases 0 and π, when implementing a phase-shift technology [3]) to certain squares. This additional phase will also be constant on every $\Box_{kl}$ and will be accounted for by multiplying q by $e^{i\psi}$. The illumination R of the object practically doesn't vary on a big pixel area, thus it could be assumed that object field function $g = qe^{i\psi}R$ is constant on every $\Box_{kl}$ and is equal to g_{kl} on it. The initial integral (3) could be transformed in the following way:

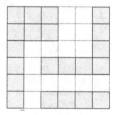

Fig. 2. Integration domain is decomposed into big pixels. Gray areas are radiating (or transparent for radiation), while white ones are not.

$$I(x, y) = \iint_O g(\xi, \eta)K(x, y, \xi, \eta)d\xi d\eta = \sum_{k,l=0}^{N-1} \varphi(k, l) \iint_{\Box_{kl}} K(x, y, \xi, \eta)d\xi d\eta. \quad (5)$$

Let's denote $S_{kl}(x, y) = \iint_{\Box_{kl}} K(x, y, \xi, \eta)d\xi d\eta$. It is a result of diffraction of electromagnetic field on the square hole (big pixel) $\Box_{kl}$. This integral could be found analytically in the far-field approximation [8]. Let a hologram calculation mesh step to be equal to the step σ on the topology. Then number of points in hologram mesh is equal to $H \times H$, where $H = [\frac{b}{\sigma}]$. Let's notice that the integral $S_{kl}(x,y)$ is invariant when shifting along an object and a hologram simultaneously: $\forall \delta_k, \delta_l \in \mathbb{Z}$ we have $S_{k+\delta_k\, l+\delta_l}(x + \delta_k\sigma, y + \delta_l\sigma) = S_{kl}(x,y)$. Using this property and denoting $S(i,j) = S_{00}(i\sigma, j\sigma)$, the sum (5) could be rearranged in the following way:

$$I(i,j) = \sum_{k,l=0}^{N-1} \varphi(k,l)S(i-k, j-l), \quad i,j = 0, ..., H-1, \tag{6}$$

where $I(i,j) = I(-\frac{b}{2} + \sigma(i + \frac{1}{2}), -\frac{b}{2} + \sigma(j + \frac{1}{2}))$.

Direct computation of the convolution (6) requires $N^2 H^2 = [\frac{a}{\sigma}]^2 [\frac{b}{\sigma}]^2$ operations. Therefore, as $a \sim b$, the number of operations will increase like the forth power when increasing the image size a or decreasing the characteristic size σ, it will be impossible to perform the calculation for big-size image in appropriate time even when supercomputer is used. However the convolution (6) could be calculated by 3 two-dimensional Fourier transforms [9]:

$$\mathbf{I} = F^{-1}[F(\varphi) \otimes F(\mathbf{S})], \tag{7}$$

where F and F^{-1} are direct and inverse two-dimensional discrete Fourier transforms, $\otimes$ is element-wise multiplication, and $\mathbf{I}$, φ, $\mathbf{S}$ are corresponding matrices. Using fast Fourier transform gives the following number of operations: $\Upsilon_{BP} = 3 \cdot 2C_1((k+1)N - 1)^2 log_2((k+1)N - 1)$. Throwing away unimportant terms and assuming $C_1 = 5$ [9] we will have

$$\Upsilon_{BP} = 30(k+1)^2 N^2 log_2((k+1)N). \tag{8}$$

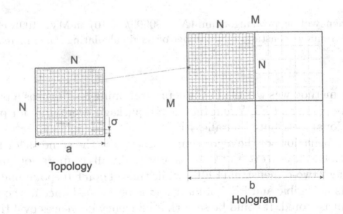

Fig. 3. Computational meshes for hologram synthesis when the big pixel method with decomposition of a hologram into areas is used

3.1   Cluster Realization of the Big Pixel Method

When realizing the big pixel method on a cluster, a hologram is divided into uniform square areas. Each area has the same size as topology (fig. 3) and is calculated separately. Total number of such areas is k^2. A number of operations is $\Upsilon_{BPA} = (2k^2 + 1)2C_1(2N - 1)^2 log_2(2N - 1)$. For $C_1 = 5$ we will have

$$\Upsilon_{BP} = (80k^2 + 40)N^2 log_2(2N).\tag{9}$$

This is a little more than for original big pixel method (8) but for this modification of the big pixel algorithm the requirements for memory is decreasing by two orders (precisely by $\frac{(k+1)^2}{4}$ times).

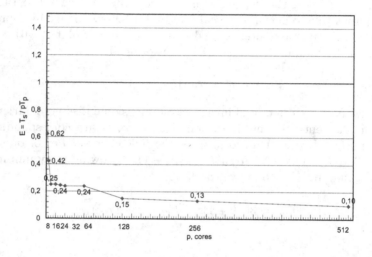

Fig. 4. Efficiency of big pixel algorithm ($N = 8000$, $k = 10$) on MVS-100K cluster. The problem size remains constant as a number of used calculating cores increases from 1 to 512.

Big pixel method was written in C++ programming language as a part of Bin-Net software package. FFTW and Intel MKL packages were used for performing two-dimensional distributed parallel FFT.

Efficiency behaviour of the algorithm is practically the same as for FFT algorithm (fig. 4), because on every of k^2 steps practically all computation time is spent by computing 2 two-dimensional FFTs, while time of point-to-point matrices multiplication is negligible and FFT of matrix φ is performed once in the beginning and its result is stored. It could be seen that efficiency decreases by 4 times while a number of processes increases from 1 to 8 on one node (one MVS-100K computational node has two quad-core processors), then it is at about constant level of 25% when a number of cores increases up to 64 (8 nodes) and then continues to

Fig. 5. Result of reconstruction of an image fragment for the topology consisting of test miras

decrease reaching 10% on 512 cores (64 nodes). This poor efficiency is caused by excessive interprocessor communication needed to perform a transposition of the distributively stored matrix during a calculation of 2-d FFT.

Numerical Computation Example. The test topology consisted of 40000 × 40000 computational pixels filled with test miras having size 57 × 37. Gabor hologram was 10 times bigger than the object, i.e. $k = 10$. This computation was performed on 640 calculating cores of MVS100-K JSCC RAS supercomputer and took 100 minutes. Execution time for every FFT of matrix having 79999 × 79999 double precision complex numbers was equal to 29 seconds. For controlling the appropriateness of hologram synthesis it was calculated a small fragment of a reconstructed image (2) containing 9 miras. Result of this reconstruction is shown at fig. 5.

4   Conclusion

In the paper it is described the big pixel method allowing to calculate Gabor holograms for modern chips with arbitrary topology on the state-of-the-art computing systems. The modified variant (allowing decreasing memory amount requirements by two orders) of the big pixel method was implemented as a part of BinNet parallel software package.

In a case when IC topology has periodic (regular) structure, it is possible to increase big pixel algorithm performance by considering these regular structures as elementary pixels.

In the case when $\gamma > \sigma$, there is another way to increase method performance, that is based on using multiple grids. Each FFT-grid should have step equal to the hologram cell size γ, while a big pixel size on an image remains the same σ as for ordinary big pixel method. This approach allows to use RAM more efficiently and to decrease the needed number of operations. The key issue allowing such improvement is that in this case, which is really important for applications [1–3], we need to calculate electromagnetic field on a hologram on very sparse grid. Its step is bigger than a period of field oscillation and is bigger than the minimal characteristic topology size σ on an image. Ability to use such big step γ appears when the transparency function $T_g(x, y)$ of a hologram is sufficiently smooth. In some ways such an algorithm should be an optimal one for the problem (2), (3), (4) when considering an optical scheme like the one shown on fig. 1.

It could be seen on fig. 5 that the reconstructed image q is rather far from the given topology image q_0 (which is equal only to 0 or 1). An optimization problem $\rho(q_0, q) \xrightarrow[T,g]{} \min$ could be considered to improve resulting image quality [10]. A variation of diversity functional for $\rho = L_2$ could be obtained analytically:

$$\delta J_{L_2}(T) = 4Re(V(x, y), \overline{\delta T(x, y)})_{(x,y)}, \tag{10}$$

where

$$V(x, y) = \iint K(\xi - x, \eta - y) \left[|K * T|^2 \cdot K * T - K * T \cdot q_0^2 \right] (\xi, \eta) d\xi d.\eta$$

Therefore, a gradient method of optimization could be proposed, where for computing variation (10) of the corresponding functional it is enough to calculate several convolutions having the same type as (3), so the big pixel method could be applied for solving such optimization problems.

References

1. Borisov, M.V., Knyazkov, D.Y., Shamaev, A.S., et al.: Methods of the Development and Correction of the Quality of Holographic Images of Geometry Objects with Subwave-size Elements. Doklady Physics 55(9), 436–440 (2010)
2. Borisov, M.V., Knyazkov, D.Y., Shamaev, A.S., et al.: Method of Producing Holographic Images. RF Patent No 2396584 C1 (2009)
3. Borisov, M.V., Knyazkov, D., Yu., S.A.S., et al.: Phase-shift at Subwavelength Holographic Lithography (SWHL). In: Proceedings of SPIE, vol. 8352 (2012)
4. Knyazkov, D.: Simulation of holography using multiprocessor systems. In: Adam, G., Buša, J., Hnatič, M. (eds.) MMCP 2011. LNCS, vol. 7125, pp. 270–275. Springer, Heidelberg (2012)
5. Singh, V., Hu, B., Bollepalli, S., Wagner, S., Borodovsky, Y.: Making a Trillion Pixels Dance. In: Proc. of SPIE, vol. 6924, pp. 0S1–0S12 (2008)
6. Stratton, J.A., Chu, L.J.: Diffraction Theory of Electromagnetic Waves. Phys. Rev. 56, 99–107 (1939)
7. Greengard, L.: Fast Algorithms for Classical Physics. Science 265, 909–914 (1994)
8. Bass, F.G., Fux, I.M.: Wave Scatter on a Statistically Uneven Surface. Nauka, Moscow (1972) (in Russian)
9. Nussbaumer, H.J.: Fast Fourier Transform and Convolution Algorithms, 2nd edn. Springer, Heidelberg (1982)
10. Knyazkov, D.Y.: Optimization of Electromagnetic Fields in Holographic Lithography Using the Method of Local Variations. Journal of Computer and Systems Sciences International 50(6), 953–963 (2011)

Finite Differences Method for One Dimensional Nonlinear Parabolic Equation with Double Nonlinearity

Bahaddin Sinsoysal[1] and Turgay Coruhlu[2]

[1] Beykent University, Department of Mathematics and Computing
[2] Beykent University, School of Vocational Studies,
Sisli-Ayazaga Campus, 34396 Istanbul, Turkey
{bsinsoysal,turgayc}@beykent.edu.tr

Abstract. The finite differences scheme for finding a numerical solution of the parabolic equation with double nonlinearity is suggested. For this purpose a special auxiliary problem having some advantages over the main problem is introduced. Some properties of the numerical solution are studied and using the advantages of the proposed auxiliary problem, the convergence of the numerical solution to the exact solution in the sense of mean is proven.

Keywords: Parabolic equation with double nonlinearity, auxiliary problem, finite differences scheme in a class of discontinuous functions.

1 Introduction

It is known that many problems of hydrodynamics, such as the motion of a non-newtonian fluid in a porous medium, the motion of water in natural watercourse and etc are modeled by nonlinear partial differential equations of the parabolic type with double nonlinearity under the corresponding initial and boundary conditions. Investigation of these kinds of problems are important practically and theoretically both, [2], [4], [6].

Let $D_T = \{(x,t) \mid 0 \leq x \leq \ell,\ 0 < t < T\} \subset R^2$. In D_T we will investigate the initial-boundary problem

$$\frac{\partial u}{\partial t} = \frac{\partial}{\partial x} K \left(\frac{\partial \sigma(u)}{\partial x} \right) - \frac{\partial Q(u)}{\partial x}, \tag{1}$$

$$u(x,0) = u_0(x), \tag{2}$$

$$u(0,t) = u_1(t), \quad u(\ell,t) = u_2(t). \tag{3}$$

Here, $u_0(x)$, $u_1(t)$ and $u_2(t)$ are given functions, $u(x,t)$ is an unknown function, $K(s)$, $\sigma(u)$ and $Q(u)$ are known functions with respect to s, u, respectively which satisfy the following conditions:

- $K(s)$, $\sigma(u)$ and $Q(u)$ are non negative and bounded functions for the bounded s and u respectively;

I. Dimov, I. Faragó, and L. Vulkov (Eds.): NAA 2012, LNCS 8236, pp. 495–501, 2013.
© Springer-Verlag Berlin Heidelberg 2013

- $K'(s) > 0$ and $\sigma'(u) > 0$ and $Q'(u) > 0$ for $u > 0$ and are finite, $K(|s|) - K(0) \geq m_0 > 0, |s| \geq m_1$;
- $K(0) = K'(0) = \sigma(0) = \sigma'(0) = Q(0) = Q'(0) = 0$;
- $K''(s)$ changes its sign.

In [6] the first type initial boundary value problem for the equation (1) was investigated and the existence of a weak solution in an appropriate space was proven, when $Q(u) = 0$, $\sigma(u) = u$, and $K(s) = s^{p-2}$, $p > 2$. It should be noted that the construction of the exact solution of the equation (1) is impossible since the equation (1) is nonlinear. But, finding the exact solution of the equation is possible in a special case of $K(s)$, $\sigma(u)$ and $Q(u)$, (see [4]). When $K(s) = s^n$, $(n > 2)$ and $\sigma(u) = u$, (that is, sign of $K''(s)$ does not change), the equation (1) has the solution in the traveling wave, [10], [12]. It is proven that the solution is continuous, but its first and second derivatives have points of discontinuities. On the other hand, the function $\frac{\partial u^n}{\partial x}$ is continuous. The existence of global solution of the Cauchy problem for the parabolic equation with double nonlinearity are studied in paper [6], [14], [5], [11], [1].

Note 1. In the case when $K'_s(s) < 0$, $Q(u) = 0$ the equation (1) models the heat and mass exchange in stratified turbulent shear flow, [3].

In the case when $K(s) \equiv 0$ and $Q(u)$ is the Buckley-Leverett's function, the equation (1) models simultaneous microscopical motion of two phase incompressible fluids in a porous medium, regardless of the capillary pressure and have been investigated in detail in [7], [13], [15].

Definition 1. *A function $u(x,t)$ satisfying the conditions (2) and (3) is called to be a weak solution of the problem (1)-(3), if the following integral relation*

$$\int_{D_T^\ell} \left\{ u \frac{\partial f}{\partial t} + \left[K\left(\frac{\partial \sigma(u)}{\partial x} \right) - Q(u) \right] \frac{\partial f}{\partial x} \right\} dx dt +$$

$$\int_0^T \left[K\left(\frac{\partial \sigma(u(0,t))}{\partial x} \right) - Q(u(0,t)) \right] f(0,t) dx + \int_0^\infty u_0(x) f(x,0) dx = 0$$

holds for every test function $f(x,t) \in \overset{o}{W}{}^2_{1,1}(D_T^\ell)$ and $f(x,T) = 0$, where $\overset{o}{W}{}^2_{1,1}(D_T^\ell)$ is a Sobolev space on $\overset{o}{D}{}_T^\ell$.

In order to find the weak solution of the problem (1)-(3), according to [8], [9] we introduce the following auxiliary problem

$$\frac{\partial v(x,t)}{\partial t} = K\left(\frac{\partial \sigma(u(x,t))}{\partial x} \right) - Q(u), \tag{4}$$

$$v(x,0) = v_0(x), \tag{5}$$

$$u(0,t) = u_1(t), \quad u(\ell,t) = u_2(t). \tag{6}$$

Here $v_0(x)$ is any differentiable solution of the equation $\frac{dv_0(x)}{dx} = u_0(x)$.

Theorem 1. *If $v(x,t)$ is the solution of the auxiliary problem (4)-(6), then the function defined by*

$$u(x,t) = \frac{\partial v(x,t)}{\partial x} \qquad (7)$$

is a weak solution of the main problem (1)-(3).

2   Numerical Solution and Its Properties

In order to develop a numerical algorithm for the problem (1), (2) and (3), at first, we cover the region D_T^ℓ by the grid

$$\Omega_{h,\tau}^\ell = \{(x_i, t_k) \mid x_i = ih,\ t_k = k\tau,\ i = 0,1,2,...n;\ k = 0,1,2,...;\ h > 0, \tau > 0\}.$$

Here, h and τ are the steps of the grid $\Omega_{h,\tau}^\ell$ respect to x and t variables, respectively, $\bar{\Omega}_{h,\tau}^\ell = \Omega_{h,\tau}^\ell + \gamma_{h,\tau}^\ell$, where $\gamma_{h,\tau}^\ell$ is a set of boundary nodes of the grid $\Omega_{h,\tau}^\ell$, and $\Omega_{h,\tau}^\ell = \Omega_h^\ell \times \Omega_\tau^\ell$. We denote by $U_{i,k}$ the approximate value of grid function $u(x_i, t_k)$ which is defined in $\Omega_{h,\tau}^\ell$.

The problem (4)-(6) is approximated at any points (x_i, t_k) of the grid $\Omega_{h,\tau}^\ell$ as follows

$$\frac{V_{i,k+1} - V_{i,k}}{\tau} = K\left(\frac{\sigma\left(U_{i,k+1}\right) - \sigma\left(U_{i-1,k+1}\right)}{h}\right) - Q(U_{i,k+1}), \qquad (8)$$

$$V_{i,0} = v_0(x_i), \qquad (9)$$

$$V_x\mid_{i=0} = u_1(t_k), \quad V_{\bar{x}}\mid_{i=n} = u_2(t_k). \qquad (10)$$

Theorem 2. *If $V_{i,k}$ is the numerical solution of the auxiliary problem (8)-(10), then the grid function defined by*

$$U_{i,k+1} = \frac{V_{i,k+1} - V_{i-1,k+1}}{h}$$

is a numerical solution of the problem

$$\frac{U_{i,k+1} - U_{i,k}}{\tau} = \frac{1}{h}\left[K\left(\frac{\sigma\left(U_{i+1,k+1}\right) - \sigma\left(U_{i,k+1}\right)}{h}\right)\right.$$

$$\left. - K\left(\frac{\sigma\left(U_{i,k+1}\right) - \sigma\left(U_{i-1,k+1}\right)}{h}\right)\right] - \frac{Q(U_{i,k+1}) - Q(U_{i-1,k+1})}{h}, \qquad (11)$$

$$U_{i,0} = u_0(x_i), \qquad (12)$$

$$U_{0,k} = u_1(t_k), \quad U_{n,k} = u_2(t_k). \qquad (13)$$

Now we will prove some properties of the numerical solution $U_{i,k}$.

1. *The solution of the problem (11)-(13) takes the greatest (or smallest) value at the boundary nodes, that is*

$$0 \le m \le U_{i,k} \le M,$$

where $m = \min\{u_0, u_1, u_2\}$ and $M = \max\{u_0, u_1, u_2\}$.

Proof. For the sake of simplicity we introduce the notations $U_{i,k} = U$, $U_{i,k+1} = \hat{U}$ and $\hat{U}_\pm = U_{i\pm1,k+1}$ and hence, we can rewrite the equation (11) as

$$\hat{U} - U = \frac{\tau}{h} K' \left(\frac{\partial \sigma}{\partial x}\right) \left[\frac{\sigma(\hat{U}_+) - \sigma(\hat{U})}{h} - \frac{\sigma(\hat{U}) - \sigma(\hat{U}_-)}{h}\right] - \frac{Q(\hat{U}) - Q(\hat{U}_-)}{h}.$$

(14)

Assume that $\hat{U} \ne \text{const}$ and $\hat{U}$ takes the greatest value at some points of the grid $\Omega^\ell_{h,\tau}$ rather than at nodes of $\gamma^\ell_{h,\tau}$. Then, there is such a point $(x_1, t_1) \in \Omega^\ell_{h,\tau}$ that $\hat{U}$ takes the maximal value and at least even at some neighborhood points $U(x_1, t_1)$ is less than $\hat{U}(x_1, t_1)$.

If $\hat{U}(x_1, t_1) > U(x_1, t_1)$, since the function $\sigma(u)$ is monotone, the left part of the relation (14) is positive, but the right part is negative. Hence, we arrive to inconsistency. In just the same way we arrive to inconsistency, if $\hat{U}_\pm(x_1, t_1) > \hat{U}(x_1, t_1)$. Similarly, we can prove that $\hat{U}$ does not take a minimal value at the inner nodes of the grid $\Omega^\ell_{h,\tau}$ as well.

2. *If*

$$\max\{|\alpha_1|, |\alpha_2|\} \le \text{const}$$

and

$$\max\left\{|\alpha_1|, |\alpha_2|, \left|\frac{d\alpha_1}{dt}\right|, \left|\frac{d\alpha_2}{dt}\right|\right\} \le \text{const},$$

then the estimation for the solution of the problem (11)-(13)

$$\left(\sigma_x(\hat{U}), [K(\sigma_x(\hat{U})) - K(0) - Q(u)]\right) \le \text{const}$$

(15)

is hold.

2.1   The Convergence of the Numerical Solution

In order to prove the convergence of the numerical solution to exact solution, let $\epsilon_{i,k}$, $\eta_{i,k}$ and $\delta_{i,k}$ be errors of approximation of the functions $\frac{\partial \sigma(u)}{\partial x}$, $\frac{\partial v}{\partial x}$ and $\frac{\partial v}{\partial t}$, respectively. Then the equation (4) can be rewritten in the form

$$v_t = K\left(\sigma_x\left(\hat{U}_{\bar{x}}\right)\right) - Q(v_{\bar{x}}) + \nu_{i,k}^{(1)} + \nu_{i,k}^{(2)}.$$

(16)

Since the function $K(s)$ is continuous

$$\nu_{i,k}^{(1)} = K\left(\sigma_x\left(\frac{\partial v}{\partial x}\right)\right) - K(\sigma_x(v_{\bar{x}})) = K(\sigma_x(u(x_i, t_k))) - K(\sigma_x(u(x_i^*, t_k)))$$

$$= K \left(\frac{\partial \sigma(u)}{\partial x} \Big|_{x=x^{**}} \right) - K \left(\frac{\partial \sigma(u)}{\partial x} \Big|_{x=x^{***}} \right) \longrightarrow 0.$$

Here, $| x_i - x_i^{**} | \le h$ and $| x_i - x_i^{***} | \le h$, and $\nu_{i,k}^{(2)} = \delta_{i,k} + \epsilon_{i,k} K'(s) - \eta_{i,k} Q'(s)$, $\nu_{i,k}^{(3)} = \nu_{i,k}^{(1)} + \nu_{i,k}^{(2)}$.

Subtracting the equation (16) from (4) and denoting R by $R = v - V$ we get

$$R_t = K'(s) \left[\sigma'(s^*) \hat{R}_x \right]_{\bar{x}} - Q'(s^*) R_{\bar{x}} + \nu_{i,k}^{(3)}, \tag{17}$$

$$R_{i,0} = h \sum_{j=1}^{i-1} \Delta_{0,j} = \tilde{\Delta}(0), \quad \left(\frac{\partial v}{\partial x} - V_x \right) \Big|_{i=0} = 0, \quad \left(\frac{\partial v}{\partial x} - V_{\bar{x}} \right) \Big|_{i=n} = 0.$$

Theorem 3. *For any τ and h the following inequalities are hold*

$$| v - V | \le \max_i | \tilde{\Delta}(0) | + 2T \max_{i,k} | \nu_{i,k}^{(3)} |, \tag{18}$$

$$\tau h \sum_{i,k} \left[\sigma(\hat{u}) - \sigma(\hat{U}) \right] (u - U) + \tau h \sum_{i,k} \left[K(\hat{u}) - K(\hat{U}) \right]_x (v - V)$$

$$\le \text{const} \left(\max_{i,k} | v - V | + \max_i | \eta_{i,k} | \right). \tag{19}$$

Proof. At first, the inequality (18) will be proven. Let us introduce the function ρ such that $v - V = \rho + A(t_k)$, here $A(t_k)$ will be chosen later. It is easily seen that the function ρ satisfies the equation

$$\rho_t = K'(s) \left[\sigma'(s^*) \hat{\rho}_{\bar{x}} \right]_x - Q'(s^*) \rho_{\bar{x}} + \nu_{i,k}^{(3)} - A_t' \tag{20}$$

with the following conditions

$$A(0) + \rho_{i,0} = (v - Y)_{i,0} = \tilde{\Delta}^{(0)}, \tag{21}$$

$$\left[\rho_x - \left(\frac{\partial v}{\partial x} - V_x \right) \right] \Big|_{i=0} = 0, \quad \left[\rho_{\bar{x}} - \left(\frac{\partial v}{\partial x} - V_{\bar{x}} \right) \right] \Big|_{i=n} = 0.$$

At the boundary points the equation (20) takes the following forms, for $i = 1$

$$\hat{\rho}_1 - \rho = \frac{\tau}{h} K'(s_1) \left[\sigma'(s_2) \left(\frac{\partial v}{\partial x} \Big|_{i=1} - \frac{\hat{V}_2 - \hat{V}_1}{h} \right) \right] + \nu_{1,k}^{(3)} - A_t'$$

and for $i = n$

$$\hat{\rho}_n - \hat{\rho}_{n-1} = \frac{\tau}{h} K'(s_n) \left[\sigma'(s_n) \left(\frac{\partial v}{\partial x} \Big|_{i=n} - \frac{\hat{V_{n+1}} - \hat{V}_n}{h} \right) \right] + \nu_{n,k}^{(3)} - A_t'.$$

Now we can apply the principle of maximum, if $A(t_k)$ is chosen as $A(t_k) +$ $\max_{i,k} \mid \nu_{i,k}^{(2)} \mid < 0$, for example, $A(t_k) = -2t_k \max_{i,k} \mid \nu_{i,k}^{(2)} \mid$. Therefore, for any k, $\mid \rho_{i,k} \mid \leq \max_i \mid \Delta_{i,k}^{(0)} \mid$ and for $v - V$ is valid

$$\mid v - V \mid \leq \mid \rho_{i,k} \mid + \mid A(t_k) \mid \leq \max_i \mid \Delta^{(0)} \mid + 2T \max_{i,k} \mid \nu_{i,k}^{(3)} \mid.$$

Now we will prove the inequality (19). For this aim, we consider

$$J = \tau h \sum_{i,k} \left\{ [\sigma(\hat{u}) - \sigma(\hat{U})](\hat{u} - \hat{U}) + (Q(\hat{u}) - Q(\hat{U})) \left(\hat{v} - \hat{V} \right) \right\}$$

we can rewrite the last relation as,

$$J = \tau h \sum_{i,k} \left\{ [\sigma(\hat{u}) - \sigma(\hat{U})] \left(\frac{\partial v}{\partial x} - \hat{V}_{\bar{x}} \right) + \left(Q(\hat{u}) - Q(\hat{U}) \right) \left(\hat{v} - \hat{V} \right) \right\} =$$

$$-\tau h \sum_{i,k} \left\{ [\sigma(\hat{u}) - \sigma(\hat{U})]_x - \left(Q(\hat{u}) - Q(\hat{U}) \right) \right\} \left(\hat{v} - \hat{V} \right) +$$

$$+\tau h \sum_{i,k} [\sigma(\hat{u}) - \sigma(\hat{U})]\eta_{i,k} = J_1 + J_2 + J_3,$$

where, $J_1 = -\tau h \sum_{i,k} [\sigma_x(\hat{u}) - Q(\hat{u})] \left(\hat{v} - \hat{V} \right)$, $J_2 = -\tau h \sum_{i,k} [\sigma_x(\hat{U}) - Q(\hat{U})]$ $\left(\hat{v} - \hat{V} \right)$, $J_3 = -\tau h \sum_{i,k} [\sigma(\hat{U}) - Q(\hat{U})]\eta_{i,k}$.

Now we shall separate the grid $\Omega_{\tau h}$ into two sub grids $\Omega_{\tau h}^{(1)}$ and $\Omega_{\tau h}^{(2)}$ where $\Omega_{\tau h}^{(1)} = \{(x_i, t_k) \mid\mid \sigma_x(\hat{u}) - Q(\hat{u}) \mid \leq 1\}$, $\Omega_{\tau h}^{(2)} = \{(x_i, t_k) \mid\mid \sigma_x(\hat{u}) - Q(\hat{u}) \mid > 1\}$. Then $\mid J_1 \mid_{\Omega_{\tau h}^{(1)}} \leq \tau h \sum_{i,k} \mid v - V \mid$. Let us estimate J_1 on the grid $\Omega_{\tau h}^{(2)}$. Under $\mid s \mid > 1$ and supposition $\mid K(s) - K(0) \mid \geq m_0 > 0$ we get

$$\left| J_1 \right|_{\Omega_{\tau h}^{(2)}} \leq \frac{1}{2} \tau h \sum_{i,k} \left| \sigma_x(u) \left[K(\sigma_x(u)) - K(0) - Q(u) \right] (v - V) \right| \leq \frac{c_1}{m_0} \max_{i,k} \left| v - V \right|.$$

Similarly, the $J_2 \Big|_{\Omega_{\tau h}^{(1)}}$ and $J_2 \Big|_{\Omega_{\tau h}^{(2)}}$ are estimated. Finally, for J we have

$$\left| J \right| \leq c_1 \max_{i,k} \left| \eta_{i,k} \right| + 2\tau h \sum_{i,k} \left| v - V \right| + c_2 \max_{i,k} \left| v - V \right|,$$

where c_j, $(j = 1, 2)$ are constants. The validity of the inequality (19) should follow from the last relation. On the basis of the suggested algorithms some numerical experiments were carried out.

3   Conclusion

The special auxiliary problem having the some advantages over the main problem is proposed, whose solution describes all physical properties of the investigated problem accurately.

The higher resolution numerical scheme for the solution of the nonlinear parabolic equation with double degeneration in a class of discontinuous functions is suggested. Using the auxiliary problem, the convergence of the numerical solution to the exact solution in sense of mean is proven.

References

1. Antontsev, S., Shmarev, S.: Parabolic Equations with Double Variable Nonlinearities. Mathematics and Computers in Simulation 81, 2018–2032 (2011)
2. Baklanovskaya, V.F.: A Study of the Method of Nets for Parabolic Equations with Degeneracy. Z. Vychisl. Mat. Mat. Fiz. 17(6), 1458–1473 (1977)
3. Barenblatt, G.I., Bertsch, M., Dal Passo, R., Prostokishin, V.M., Ughi, M.: A Mathematical Model of Turbulent Heat and Mass Exchange in Stratified Turbulent shear Flow. Journal Fluid Mechanics 253, 341–358 (1984)
4. Fokas, A.S., Yortsos, Y.C.: On the Exactly Solvable Equation $S_t = [(\beta S + \gamma)^{-2} S_x]_x + (\beta S + \gamma)^{-2} S_x$ Occurring in Two-Phase Flow in Porous Media. SIAM, J. Appl. Math. 42(2), 318–332 (1982)
5. Jager, W., Kacur, J.: Solution of Doubly Nonlinear and Degenerate Parabolic Problems by Relaxation Schemes. Modlisation Mathmatique et Analyse Numrique 29(5), 605–627 (1995)
6. Lions, J.L.: Some Methods of Solutions of Nonlinear Boundary Problem, Moskow (1972)
7. Rasulov, M.A.: On a Method of Solving the Cauchy Problem for a First Order Nonlinear Equation of Hyperbolic Type with a Smooth Initial Condition. Soviet Math. Dokl. 316(4), 777–781 (1991)
8. Rasulov, M.A., Ragimova, T.A.: A Numerical Method of the Solution of one Nonlinear Equation of a Hyperbolic Type of the First -Order. Dif. Equations 28(7), 2056–2063 (1992)
9. Rasulov, M.A.: A Numerical Method of Solving a Parabolic Equation with Degeneration. Dif. Equations 18(8), 1418–1427 (1992)
10. Samarskii, A.A.: Duty with Peaking in Problems for Quasi Linear Equations of Parabolic Type. Nauka, Moskow (1987)
11. Shang, H.: On the Cauchy Problem for the Singular Parabolic Equations with Gradient Term. Journal of Math. Anal. and Appl. 378, 578–591 (2011)
12. Sinsoysal, B.: The Analytical and a Higher-Accuracy Numerical Solution of a Free Boundary Problem in a Class of Discontinuous Functions. Mathematical Problems in Engineering (2012), doi:10.1155/2012/791026
13. Smoller, J.A.: Shock Wave and Reaction Diffusion Equations. Springer, New York (1983)
14. Wang, S., Deng, J.: Doubly Nonlinear Parabolic Equation with Nonlinear Boundary Conditions. Journal of Math. Anal. and Appl. 225, 109–121 (2001)
15. Whitham, G.B.: Linear and Nonlinear Waves. Wiley Int., New York (1974)

Numerical Solving of Functional Differential Equations in Partial Derivatives on a Remote Server through a Web Interface

Svyatoslav I. Solodushkin[1,2]

[1] Ural Federal University, Ekaterinburg, Russia
[2] Institute of Mathematics and Mechanics, Ekaterinburg, Russia

Abstract. The paper describes an architecture of information and computing server that allow one to carry out numerical modeling of evolutionary systems in partial derivatives with time delay. Algorithms implemented as m-files for MATLAB were compiled into dynamic linking libraries. Front end was elaborated with ASP.NET. Brief user guide and examples of a numerical modeling of certain systems are presented.

Introduction

Evolutionary systems are most generally described in scientific and engineering terms with respect to three-dimensional space and time. At times such systems are complicated with aftereffect, i.e. system dynamics depends not only on the present state but also on the prior state. Partial functional differential equations (PFDE) also known as delay partial differential equations provide a mathematical description of these systems. These equations include at least two independent variables (one of them usually is interpreted as time), an unknown function of the independent variables and partial derivatives of the unknown function with respect to the independent variables; description of the system dynamics involves value of the unknown function at some previous time.

PFDE are widely used for describing and mathematical modeling various processes and systems with delay [1]. Models that involve PFDE arise in biomedicine, geology, control theory etc. Their independent variables are time t and one or more dimensional variable x , which represents either position in space (for physical models [2]) or size of cells, their maturation level (for biomedical models [3]), or something other. The solutions (unknown function) of delay partial differential equations may represent temperature, or concentrations/densities of various particles, for example cells, bacteria, chemicals, animals and so on.

At present theoretical aspects of PFDE are studied with almost the same completeness as the corresponding parts differential equations in partial derivatives.

As usual even for the simplest types of PFDE are not known methods for finding solutions in an explicit form. So elaboration of numerical methods and their programm realization for PFDE is a very important problem.

Numerical methods are elaborated quite well for differential equations in partial derivatives; moreover these methods are presented in the form of standard

I. Dimov, I. Faragó, and L. Vulkov (Eds.): NAA 2012, LNCS 8236, pp. 502–508, 2013.

packets and toolboxes for use with MATLAB, Mathematica and other standard packets. Mathematica allow one to solve systems of DDEs in ordinary derivatives of any order, and offers such features as automatic selection of efficient and reliable DDE integration methods, direct input of DDEs in standard mathematical notation, fully automated lag function computation, arbitrary numerical-precision DDE solutions [19]. MATLAB provides functions to solve DDEs with general delays and DDEs of neutral type [20].

At the same time numerical methods for PFDE are elaborated much more poor and aren't presented in widely known software packets. Several approaches are used to solve PFDE. Method of lines [4–6] reduces PFDE to the system of FDE in ordinal derivatives witch could be solved by special method [7–9]; unfortunately after discretization with respect to two variables stiff systems appears. Implicit difference methods [10–15] allow to avoid stiffness by appropriate step choice. Various monotone iterative methods are used too.

The aim of this work is to present information and computing server that allow one to solve PFDE of parabolic and hyperbolic types; this paper continues [16]. This server is a result of efforts of chair of Computational Mathematics of the Ural Federal University.

1   The Class of Solvable Equations and Corresponding Algorithms

The class of equations which our server solves consist of parabolic functional differential equation (also known as heat conduction equation with aftereffect)

$$\frac{\partial u}{\partial t} = a^2 \frac{\partial^2 u}{\partial x^2} + f(x, t, u(x, t), u_t(x, \cdot)) \tag{1}$$

and hyperbolic functional differential equation

$$\frac{\partial^2 u}{\partial t^2} = a^2 \frac{\partial^2 u}{\partial x^2} + f(x, t, u(x, t), u_t(x, \cdot)), \tag{2}$$

here $x \in [x_0; X]$ — spacial and $t \in [t_0; \theta]$ — time independent variables; $u(x, t)$ — unknown function; $u_t(x, \cdot) = \{u(x, t+\xi), -\tau \leq \xi < 0\}$ — prehistory-function of the unknown function to the moment t.

In conjunction with equation (1) or (2) initial and boundary conditions are set

$$u(x, t) = \varphi(x, t), \quad x \in [x_0, X], t \in [t_0 - \tau; t_0], \tag{3}$$

$$u(x_0, t) = g_0(t), \ u(X, t) = g_1(t), \quad t \in [t_0; \theta], \tag{4}$$

consentaneous conditions are satisfied

$$\varphi(x_0, t_0) = g_0(t_0), \quad \varphi(X, t_0) = g_1(t_0). \tag{5}$$

Tasks (1), (3)–(5) and (2), (3)–(5) — are the typical boundary tasks. Let assume that functions φ, g_0, g_1 and functional f are such that appropriate task has unique solution $u(t, x)$ in classical sense [1].

Systems with delays (1), (2) are infinite dimensional systems because of the presence of the functional component $u_t(x, \cdot)$, which characterizes delays. One of the problem of simulations these systems consist in describing of functional $f(x, t, u(x, t), u_t(x, \cdot))$ by finite number of parameters, because for computer simulations usually only finite algorithms with finite number of input parameters can be used. Analyzing the structure of PFDE one can realize that in certain cases right-hand sides of these equations are combinations of finite dimensional functions and integrals. Our server implements numerical methods, which allow one to model systems with distributed and lumped delays; and functional $f(x, t, u(x, t), u_t(x, \cdot))$ may be of the following forms

$$\psi_1(x, t, u(x, t), u_t(x, t - \tau(t))),$$

$$\int_{\tau(t)}^{0} \psi_2(x, t, u(x, t), u_t(x, t + s)) \, ds.$$

Thus the right-hands sides of equations (1), (2) can be defined just by finite numbers of functions.

To obtain numerical solutions algorithms described in [12–14] are used.

2   Examples and User's Guide

Let us consider the following example [13]

$$\frac{\partial u}{\partial t} = a^2 \frac{\partial^2 u}{\partial x^2} + \frac{\sqrt{0.25t^2 + x^2}(2a^2t - t^2 - x^2)}{(t^2 + x^2)^2} \sin(u(x, t - 0.5t))$$

here $a = 0.2$, initial and boundary conditions are set

$$u(x, s) = \arcsin \frac{x}{\sqrt{s^2 + x^2}}, \quad 0.5 \leq s \leq 1, \ 0.5 \leq x \leq 4;$$

$$u(0.5, t) = \arcsin \frac{0.5}{\sqrt{t^2 + 0.5^2}}, \quad u(4, t) = \arcsin \frac{4}{\sqrt{t^2 + 4^2}}, \quad 1 \leq t \leq 5.$$

This task has an exact solution

$$u(t, x) = \arcsin \frac{x}{\sqrt{t^2 + x^2}}.$$

On the page "Parabolic equations" user should fill the form according the following instructions

1. in the textarea "Inhomogeneous function" `sin(ut)*` `(sqrt(t*t/4+x*x))*(2*0.2*0.2*t-t*t-x*x)/(t*t+x*x)^2`
2. in the text area "Initial condition function" `asin(x/sqrt(x^2+t^2))`,
3. in the text area "Border condition Left function" `asin(0.5/sqrt(0.5^2+t^2))`,
4. in the text area "Border condition Right function" `asin(4/sqrt(4^2+t^2))`,

5. in the text area "Time delay function" $t/2$,
6. in the text area "Parameter a" 0.2.

After that it's necessary to define grid characteristics

7. in the fields "Start X", "Finish X" and "Amount of nodes in the grid X"
   0.5, 4, 32 respectively,
8. in the fields "Start T", "Finish T" and "Amount of nodes in the grid T"
   1, 4, 18 respectively.

Receiving the number of partition points N and M for $[x_0; X]$ and $[t_0; \theta]$ respectively, programm defines steps of partitions $h = (X - x_0)/N$ and $\Delta = (\theta - t_0)/M$ and uniform grid $\{t_j, x_i\}_{j=0 i=0}^{M\ N}$, where $t_j = t_0 + j\Delta$, $j = 0, ..., M$, and $x_i = x_0 + ih$, $i = 0, ..., N$.

As a result user obtains an interactive plot and table $(M + 1) \times (N + 1)$ with numerical solution. A cell in the i-th row and j-th column contains approximate value of $u(t_i, x_j)$.

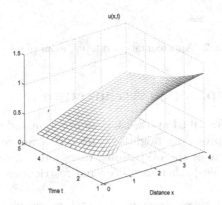

Fig. 1. Approximate solutions, example 1

Let us consider the second example

$$\frac{\partial^2 u}{\partial t^2} = \frac{1}{2}\frac{\partial^2 u}{\partial x^2} - 2u(x, t - 2\pi),$$

initial and boundary conditions are set

$$u(x, s) = e^{-x^2} \sin t, \ -2\pi \leq s \leq 0, \ 0 \leq x \leq 1;$$

$$u(0, t) = \sin t, \ u(1, t) = e^{-1} \sin t, \ 0 \leq t \leq 3\pi.$$

This task has an exact solution

$$u(t, x) = e^{-x^2} \sin t.$$

On the page "Hyperbolic equation" user must fill out a form in almost the same way as described above for the parabolic equation and click the button "Calculate".

The plot is interactive, i.e it allows rotation by means of the mouse in order to better review. The processing of these user manipulations performed on the server side; AJAX is used for the organization of client-server interaction without refreshing the web page.

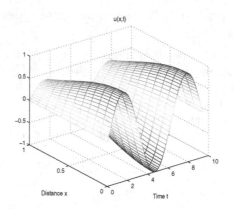

Fig. 2. Approximate solutions, example 2

3   Developed Application Architecture

Numerical algorithms described in [12, 13] were programmed as m-files for MAT-LAB. It's difficult to provide straightforward access for remote users to the MATLAB server because of several reasons: users must have high qualification (at least write own m-files), there are licensing restrictions and some security problems take place.

For the elimination of these problems, our system was elaborated — its fron-tend is IIS and backend is MATLAB Run Time Server. User needs modern web browser only, any additional plugins aren't required; system we describe here has simple and clear interface and doesn't imply user's knowledge of MATLAB programming. Remote user interact with systems by means of his web browser over HTTP and his right are constrained and controlled by web server.

Web interface was created in ASP.NET4/C#. To deal with MATLAB from C#-programs we use MATLAB Builder NE. MATLAB Builder NE lets you cre-ate .NET components from MATLAB programs that include MATLAB math and graphics developed with MATLAB. Developers can integrate these com-ponents into large Web applications and deploy them royalty-free to comput-ers that do not have MATLAB installed. Using MATLAB Compiler, MATLAB Builder NE encrypts certain MATLAB programs and then generates .NET wrap-pers around them so that they can be accessed just like native .NET compo-nents [17, 18].

The Builder NE support type-safe automatic conversion to and from managed .NET and MATLAB data types; to do this the Builder NE provides a set of data conversion classes derived from the abstract class `MWArray`.

Using the WebFigures feature in Builder NE one can display MATLAB figures on a Web site for graphical manipulation by end users. This enables them to use their graphical applications from anywhere on the Web without the need to download MATLAB or other tools that can costs considerably.

To use the component assembly `pfde` generated using the Builder NE from the C#-program it's should reference the namespaces for the MATLAB data conversion and webfigures assemblies as well as the namespace for the builder assembly generated for our component `pfde`, as shown:

```
using MathWorks.MATLAB.NET.Utility;
using MathWorks.MATLAB.NET.WebFigures;
using MathWorks.MATLAB.NET.Arrays;
using pfde;
```

The listing below shows string and number data handling in C#-program
```
MWArray inpICFun = ICFun.Text;
MWArray inpParamA = Convert.ToDouble(ParamA.Text);
```

To create interactive plot we use such code
```
WebFigureControl.WebFigure =
new WebFigure(mywebfig.pdep(inpICFun, inpBCFun,
inpInhomFun, inpTimeDelayFun, inpParamA,
inpX0, inpXn, inpXNumNod, inpT0, inpTm, inpTNumNod));
```

Acknowledgement. This work was supported by Russian Foundation for Basic Research (project 13-01-00089) and by Ministry of Education and Science of Russian Federation (project 1.994.2011).

Conclusion

Elaborated system allows one to easily add new algorithms, which are implemented as M-files for MATLAB. For that it's sufficient to design an suitable Web page with Web form, where user can input a description of his equation. The system has a modular structure that will allow in further work on the parallel MATLAB, while the Web interface does not change.

References

1. Wu, J.: Theory and applications of partial functional differential equations. Springer, New York (1996)
2. Wang, P.K.C.: Asymptotic stability of a time-delayed diffusion system. J. Appl. Mech. 30, 500–504 (1963)
3. Jorcyk, C.L., Kolev, M., Tawara, K., Zubik-Kowal, B.: Experimental versus numerical data for breast cancer progression. Nonlinear Analysis: Real World Applications 13, 78–84 (2012)

4. Tavernini, L.: Finite difference approximation for a class of semilinear volterra evolution problems. SIAM J. Numer. Anal. 14(5), 931–949 (1977)
5. Zubik-Kowal, B.: Stability in the Numerical Solution of Linear Parabolic Equations with a Delay Term. Bit Numerical Mathematics 41(1), 191–206 (2001)
6. Kamont, Z., Netka, M.: Numerical method of lines for evolution functional differential equations. J. Numer. Math. 19(1), 63–89 (2011)
7. Shampine, L.F., Tompson, S.: Solvin DDEs in MATLAB. Appl. Numer. Math. 37, 441–458 (2001)
8. Bellen, A., Zennaro, M.: Numerical methods for delay differential equations, numerical mathematics and scientific computation. Oxford University Press, Clarendon Press, New York (2003)
9. Kim, A.V., Pimenov, V.G.: i-Smooth analysis and numerical methods of solving finctional-differential equations. Regular and Chaotic Dynamics, Moscow (2004)
10. Kamont, Z., Leszczynski, H.: Stability of difference equations generated by parabolic differential-functional problems. Rend. Mat. Appl. 16(2), 265–287 (1996)
11. Kropielnicka, K.: Convergence of Implicit Difference Methods for Parabolic Functional Differential Equations. Int. Journal of Math. Analysis 1(6), 257–277 (2007)
12. Lekomtsev, A.V., Pimenov, V.G.: Convergence of the alternating direction method for the numerical solution of a heat conduction equation with delay. Proceedings of Institute of Mathematics and Mechanics 16(1), 102–118 (2010)
13. Pimenov, V.G., Lozhnikov, A.B.: Difference schemes for the numerical solution of the heat conduction equation with aftereffect. Proceedings of the Steklov Institute of Mathematics 275(1), 137–148 (2011)
14. Pimenov, V.G., Tashirova, E.E.: Numerical methods for the equation of hyperbolic type with delay. Trudi Instituta Matematiki i Mehaniki 18(2), 222–231 (2012)
15. Czernous, W., Kamont, Z.: Comparison of explicit and implicit difference methods for quasilinear functional differential equations. Appl. Math. 38(3), 315–340 (2011)
16. Solodushkin, S.I.: Access to the Solver of Functional Differential Equations Through the Web Interface. In: Recent Advances in Applied & Biomedical Informatics and Computational Engineering in Systems Applications, pp. 135–137 (2011)
17. MATLAB Builder NE User's Guide,
   http://www.mathworks.com/help/pdf_doc/dotnetbuilder/dotnetbuilder.pdf
18. MATLAB Application Deployment Web Example Guide,
   http://www.mathworks.com/help/pdf_doc/compiler/example_guide.pdf
19. Delay Differential Equations in Mathematica,
   http://www.wolfram.com/products/mathematica/
   newin7/content/DelayDifferentialEquations/
20. Delay Differential Equations in MATLAB,
   http://www.mathworks.com/help/matlab/delay-differential-equations.html

Adaptive Artificial Boundary Conditions for 2D Schrödinger Equation

Vyacheslav A. Trofimov and Anton D. Denisov

Lomonosov Moscow State University, Faculty of Computational Mathematics and
Cybernetics
vatro@cs.msu.ru, antondnsv@gmail.com

Abstract. We develop adaptive artificial boundary conditions for 2D
linear and nonlinear Schrödinger equation. These conditions are adaptive ones to the problem solution near the artificial boundary. This approach allows us to increase many times the efficiency of application of
the artificial boundary conditions. In this connection we discuss an influence of both round-off error and small value of intensity on accuracy
of computation of wave number of the beam in the vicinity of boundary.

Keywords: adaptive artificial boundary conditions, 2D Schrödinger
equation, wave number near the boundary.

1 Introduction

Nowadays, various problems which are described by Schrödinger equation (or
equations) are of great interest. Therefore, developing of corresponding artificial
boundary conditions is relevant also. As it is well-known, various types of such
conditions have previously been proposed [1–8]. Nevertheless, they have not both
sufficient simplicity and easy realization in some cases. Especially, this problem
is actual one for the case of multidimensional problems because the proposed artificial (non-reflecting) boundary conditions have only recently begun to explore
for 2D case and, in our opinion, they do not possess sufficient efficiency at computer modeling of nonlinear propagation of laser radiation. Below we develop
adaptive artificial boundary conditions for 2D linear and nonlinear Schrödinger
equation and discuss their approximation at angular points of considered rectangular domain. It should be stressed that early we have developed the artificial
boundary conditions which are adaptive to boundaries for 1D problem.

2 State of Problem

As it is well-known the laser beam propagation in a medium with cubic nonlinear response is described by the following nonlinear Schrödinger dimensionless
equation with respect to slowly varying envelope of laser beam $A(z, x, y)$:

$$\frac{\partial A}{\partial z} + iD_x\frac{\partial^2 A}{\partial x^2} + iD_y\frac{\partial^2 A}{\partial y^2} + i\gamma|A|^2 A = 0, \ \ 0 < z < L_z, \ \ 0 < x < L_x, \ \ 0 < y < L_y$$

$$(1)$$

I. Dimov, I. Faragó, and L. Vulkov (Eds.): NAA 2012, LNCS 8236, pp. 509–516, 2013.

with the initial condition

$$A|_{z=0} = e^{-\left(\frac{x-\theta_x L_x}{a_x}\right)^2 - \left(\frac{y-\theta_y L_y}{a_y}\right)^2}, \quad 0 \le x \le L_x, \quad 0 \le y \le L_y, \quad \theta_x, \theta_y \in [0,1] \ ,$$
(2)

where z – coordinate along which the beam propagates, x, y – transverse coordinates, D_x, D_y characterize the diffraction of beam along the corresponding coordinates, γ is a coefficient of nonlinearity. Parameters a_x, a_y denote the beam radius along x and y coordinates. Parameters θ_x, θ_y define the center of beam's position.

The domain in transverse coordinates is shown in Fig.1(a). Here, parameters $\{\Omega_{xL}, \Omega_{xR}, \Omega_{yL}, \Omega_{yR}\}$ characterize components of wave-vector of the light beam near the boundary in the directions those are perpendicular to the corresponding boundary. They are used in the boundary conditions. The dotted arrows show possible directions of the beam propagation through boundary and angular points.

Let us write boundary conditions for the linear problem. Using well-known solution of the linear Schrödinger equation (parameter γ in (1) is equal to zero) for initial Gaussian spatial distribution of complex amplitude, it is easy to write the components of wave-vector:

$$\{\Omega_{xL}, \Omega_{xR}\} = \frac{8 D_x z L_x}{a_x \left(1 + (4 D_x z)^2\right)} \{-\theta_x, 1 - \theta_x\},$$

$$\{\Omega_{yL}, \Omega_{yR}\} = \frac{8 D_y z L_y}{a_y \left(1 + (4 D_y z)^2\right)} \{-\theta_y, 1 - \theta_y\} \ .$$
(3)

In general case the corresponding component of wave vector can be computed using the beam phase which is calculated using the function $\psi = \arctan\left|\frac{A_L}{A_R}\right|$. However, the function ψ is discontinuous one. Therefore, for points of its discontinuity it is necessary to use special rule for computation of the beam phase. At domain points, in which the intensity is close to zero, one needs to smooth the function, using the interpolation of function. For this aim we use Lagrange polynomial of second order.

Secondly, because of round-off errors one can calculate the phase of optical beam with the intensity which is greater than $10^{-\delta}$, value of the parameter δ is defined by data format. In opposite case the phase of the complex amplitude is distorted as it will be demonstrated below. Therefore, to calculate the phase correctly it is necessary to make a computation at certain distance from the boundary where the intensity is larger than this crucial value.

We demonstrate the solution of this problem at considering the axially symmetric Gaussian beam propagation $D_x = D_y$ (see Fig.1(b)). In this case, for example the wave-vector Ω at the left or right boundaries on x-axis can be represented as $\Omega = \{\Omega \cos\alpha, \Omega \sin\alpha\}$, where $\Omega = \Omega_{xR}, \Omega_{xL}$. So, let us write the adaptive artificial boundary conditions:

$$\left(\frac{\partial A}{\partial z} \mp 2D_x \Omega \cos\alpha \frac{\partial A}{\partial x} \mp 2D_y \Omega \sin\alpha \frac{\partial A}{\partial y} + iD_x(\Omega \cos\alpha)^2 A+\right.$$

$$\left. +iD_y(\Omega \sin\alpha)^2 A + i\gamma|A|^2 A\right)_{x=0,L_x} = 0, \Omega = \{\Omega_{xL}, \Omega_{xR}\},$$

$$\left(\frac{\partial A}{\partial z} \mp 2D_y \Omega \sin\beta \frac{\partial A}{\partial y} \mp 2D_x \Omega \cos\beta \frac{\partial A}{\partial x} + iD_y(\Omega \sin\beta)^2 A+\right.$$

$$\left. +iD_x(\Omega \cos\beta)^2 A + i\gamma|A|^2 A\right)_{y=0,L_y} = 0, \Omega = \{\Omega_{yL}, \Omega_{yR}\}. \tag{4}$$

Values of $\cos\alpha$, $\cos\beta$ are calculated using the coordinates of the point at which the boundary condition is written (see Fig.1(b)):

$$\cos\alpha = (L_x - x_0)\left/\sqrt{(L_x - x_0)^2 + (y_1 - y_0)^2}\right.,$$

$$\cos\beta = (x_2 - x_0)\left/\sqrt{(x_2 - x_0)^2 + (L_y - y_0)^2}\right.. \tag{5}$$

Using the split-step method for solution of the linear problem, the angles α, β are chosen to be equal to zero. In this case one can write the following boundary conditions:

$$\left(\frac{\partial A}{\partial z} \mp 2D_x \Omega \frac{\partial A}{\partial x} + iD_x\Omega^2 A\right)_{x=0,L_x} = 0, \quad \Omega = \{\Omega_{xL}, \Omega_{xR}\},$$

$$\left(\frac{\partial A}{\partial z} \mp 2D_y \Omega \frac{\partial A}{\partial y} + iD_y\Omega^2 A\right)_{y=0,L_y} = 0, \quad \Omega = \{\Omega_{yL}, \Omega_{yR}\}. \tag{6}$$

The main difficulty of these conditions consists in definition of components $\{\Omega_{xL}, \Omega_{xR}, \Omega_{yL}, \Omega_{yR}\}$ of wave-vector in corresponding boundary points because the analytical solution is absent in the general case, especially for nonlinear propagation of laser radiation. In particular, the parameters $\{\Omega_{xL}, \Omega_{xR}, \Omega_{yL}, \Omega_{yR}\}$ in the corner points are defined by (3) for the analytical expression of the wave-vector components, and by (5) for adaptive boundary conditions. In computer simulation we use both types of boundary conditions for comparison of efficiency of the adaptive boundary conditions.

The problem, formulated above, possess some invariants. Usual Energy invariant (I_1), and Hamiltonian (I_3) are examined. These invariants are used for control of computer simulation results.

3   Finite-Difference Scheme for the Linear Schrödinger Equation

In the domain $W = (0, L_z) \times (0, L_x) \times (0, L_y)$ let us introduce the following mesh $\omega = \omega_z \times \omega_x \times \omega_y$ for the mesh function U and the mesh $\tilde{\omega} = \tilde{\omega}_k \times \omega_x \times \omega_y$ for

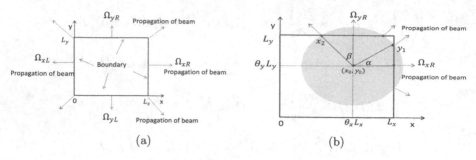

Fig. 1. Domain in transverse coordinate (a) and intensity beam profile in some section on longitudinal coordinate (b)

the mesh function $\tilde{U}$:

$$\omega_z = \{z_k = kh_z, k = 0, 1, ..., N_z, h_z = L_z/N_z\},$$
$$\tilde{\omega}_z = \{z_{k+0.5} = h_z(k + 0.5), k = 0, 1, ..., N_z - 1, h_z = L_z/N_z\},$$
$$\omega_x = \{x_m = mh_x, m = 0, 1, ..., N_x, h_x = L_x/N_x\},$$
$$\omega_y = \{x_n = nh_y, n = 0, 1, ..., N_y, h_y = L_y/N_y\}.$$

Let us introduce the following index-free notation:

$$U = U_{m,n} = U(z_k, x_m, y_n), \ \tilde{U} = U(z_{k+0.5}, x_m, y_n), \ \hat{U} = U(z_{k+1}, x_m, y_n),$$

$$\overset{0.5}{U} = 0.5(\hat{U} + U), \ \overset{0.5}{|U|^2} = 0.5(|\hat{U}|^2 + |U|^2).$$

The Laplace operator is approximated with the second order along the corresponding coordinate:

$$\Lambda_{\bar{x}x} U = \frac{U_{m+1} - 2U_m + U_{m-1}}{h_x^2}, \ \Lambda_{\bar{y}y} U = \frac{U_{n+1} - 2U_n + U_{n-1}}{h_y^2}.$$

Thus, we write the following finite-difference scheme in the inner nodes of the mesh for the equation (1) in linear case of laser beam propagation:

$$\frac{\hat{U} - U}{h_z} + iD_x\Lambda_{\bar{x}x} \overset{0.5}{U} + iD_y\Lambda_{\bar{y}y} \overset{0.5}{U} = 0, \ m = \overline{1, N_x - 1}, \ n = \overline{1, N_y - 1}. \quad (7)$$

Using the well-known split-step method, one can solve the following sequence of equations:

$$\frac{\tilde{U} - U}{h_z/2} + iD_x\Lambda_{\bar{x}x}\tilde{U} + iD_y\Lambda_{\bar{y}y}U = 0, \ m = 1, ..., N_x - 1, n = 1, ..., N_y - 1,$$

$$\frac{\hat{U} - \tilde{U}}{h_z/2} + iD_x\Lambda_{\bar{x}x}\tilde{U} + iD_y\Lambda_{\bar{y}y}\hat{U} = 0, \ n = 1, ..., N_y - 1, m = 1, ..., N_x - 1 \quad (8)$$

instead the equations (7). The last set of equations can be represented as three-diagonal algebraic equations that are solved by the sweep method.

At computer simulation of nonlinear propagation of laser beam ($\gamma \neq 0$) we use conservative two-stage iterative process which is written in following manner:

$$\frac{\overset{s+1}{\hat{U}}-U}{h_z} + iD_x\Lambda_{\bar{x}x}\overset{s+1}{\underset{0.5}{U}} + iD_y\Lambda_{\bar{y}y}\overset{s}{\underset{0.5}{U}} + i\gamma|\overset{s}{\underset{0.5}{U}}|^2\overset{s}{\underset{0.5}{U}} = 0, \quad \overset{s=0}{\hat{U}} = U,$$

$$\frac{\overset{s+2}{\hat{U}}-U}{h_z} + iD_x\Lambda_{\bar{x}x}\overset{s+1}{\underset{0.5}{U}} + iD_y\Lambda_{\bar{y}y}\overset{s+2}{\underset{0.5}{U}} + i\gamma|\overset{s+1}{\underset{0.5}{U}}|^2\overset{s+1}{\underset{0.5}{U}} = 0. \tag{9}$$

Let us write conditions for the right and top boundaries (for brevity). Expressions for the left and bottom boundaries look similar, excepting the sign before the finite-difference derivative in corresponding coordinate:

$$\frac{\overset{s+1}{\hat{U}}_{N_x,n} - U_{N_x,n}}{h_z} + 2D_x\kappa_x\frac{\overset{s+1}{\underset{0.5}{U}}_{N_x,n} - \overset{s+1}{\underset{0.5}{U}}_{N_x-1,n}}{h_x} + 2D_y\kappa_y\frac{\overset{s}{\underset{0.5}{U}}_{N_x,n} - \overset{s}{\underset{0.5}{U}}_{N_x,n-1}}{h_y} +$$

$$iD_x\kappa_x^2\overset{s+1}{\underset{0.5}{U}}_{N_x,n} + iD_y\kappa_y^2\overset{s}{\underset{0.5}{U}}_{N_x,n} + i\gamma|\overset{s}{\underset{0.5}{U}}_{N_x,n}|^2\overset{s}{\underset{0.5}{U}}_{N_x,n} = 0, n = \overline{1, N_y},$$

$$\frac{\overset{s+2}{\hat{U}}_{m,N_y} - U_{m,N_y}}{h_z} + 2D_y\nu_y\frac{\overset{s+2}{\underset{0.5}{U}}_{m,N_y} - \overset{s+2}{\underset{0.5}{U}}_{m,N_y-1}}{h_y} + 2D_x\nu_x\frac{\overset{s+1}{\underset{0.5}{U}}_{m,N_y} - \overset{s+1}{\underset{0.5}{U}}_{m-1,N_y}}{h_x} +$$

$$+iD_y\nu_y^2\overset{s+2}{\underset{0.5}{U}}_{m,N_y} + iD_x\nu_x^2\overset{s+1}{\underset{0.5}{U}}_{m,N_y} + i\gamma|\overset{s+1}{\underset{0.5}{U}}_{m,N_y}|^2\overset{s+1}{\underset{0.5}{U}}_{m,N_y} = 0, m = \overline{1, N_x}, \tag{10}$$

$$\kappa_x = \Omega_{xR}\cos\alpha, \quad \kappa_y = \Omega_{yR}\sin\alpha, \quad \nu_x = \Omega_{xR}\cos\beta, \quad \nu_y = \Omega_{yR}\sin\beta .$$

The criterion of stopping the iterations is:

$$\max_{x_m,y_n}\left|\overset{s+2}{\hat{U}} - \overset{s}{\hat{U}}\right| < \varepsilon_1 \max_{x_m,y_n}\left|\overset{s}{\hat{U}}\right| + \varepsilon_2, \quad \varepsilon_1 = 10^{-3}, \quad \varepsilon_2 = 10^{-2}\varepsilon_1 .$$

We apply this iterative process for solution of the linear problem also for its comparison with split-step method.

Efficiency of adaptive artificial boundary conditions is estimated using the following expressions:

$$\xi = \max_{z,x,y\in\omega}\left||U|^2 - |A|^2\right|, \quad c = \max_{z,x,y\in\omega}||U_R - A_R| + |U_I - A_I||,$$

$$l_2 = \sqrt{\sum_{m=0}^{N_x} \sum_{n=0}^{N_y} \left(|U_R - A_R|^2 + |U_I - A_I|^2 \right) h_x h_y} \, ,$$

$$I' = \max_{z \in \omega} \left\{ \sum_{m=0}^{N_x} \left(|U|^2 - |A|^2 \right)\big|_{N_y} h_x + \sum_{n=0}^{N_y} \left(|U|^2 - |A|^2 \right)\big|_{N_x} h_y \right\} \, ,$$

$$I'' = \max_{z \in \omega} \left\{ \sum_{m=0}^{N_x} \left(|U_R - A_R|^2 + |U_I - A_I|^2 \right)\big|_{N_y} h_x + \right.$$

$$\left. + \sum_{n=0}^{N_y} \left(|U_R - A_R|^2 + |U_I - A_I|^2 \right)\big|_{N_x} h_y \right\} \, . \tag{11}$$

4   Computer Simulation Results

First of all let us consider linear propagation of input Gaussian beam with parameters $a_x = a_y = 1$ in the domain $L_z = 5, L_x = L_y = 20$. In this case we use the split-step method. The corresponding computer simulation results are shown in the Table 1 and in Fig.2. With respect to the Table 1 one should be stressed that the lines with parameter Ω corresponds to computer simulation results at using the analytical expression for wave-vector components in artificial boundary conditions and presence of parameter $\tilde{\Omega}$ in the lines of Table 1 means that the wave-vector components is calculated numerically. In the last case one needs to make a calculation at some distance from the boundary to avoid the influence of round-off error on the phase distribution.

From the Table 1 it is clearly seen that the order of accuracy of the finite-difference scheme and the order of round-off are in agreement each with other. Comparing the Fig.2(a,b) we see high efficiency of proposed artificial boundary conditions with adaptive choice of local wave number near the boundary. Nevertheless, the distortion of beam phase can be present and it depends on way of computation. Comparison of Fig.2(c,d) clearly illustrates this statement. It is well-seen the distortion of the phase if we calculate the corresponding components of local wave vector numerically in each point of the right boundary.

High efficiency of developing method illustrates Fig.2(e,f) and Fig.3 also. In Fig.2(e,f) the intensity distribution is shown. In Fig.3 we see the distortion of the phase distribution in dependence of the z for $\tilde{\Omega}$. It is important to stress that the distortion of laser radiation phase, which appears in section $z = 1$, disappears in the section $z = 2.5$ and does not influence on the intensity profile and phase distribution at further propagation of laser beam.

Table 1. Invariants and deviation of numerical solution, obtained at using of the adaptive boundary conditions, from explicit solution of linear problem. The row with Ω corresponds to results of computer simulation for boundary conditions (3), $\tilde{\Omega}$ – for phase of laser beam, computed using the numerical solution.

	$10^3 I_1$	$10^3 I_3$	$10^3 \xi$	$10^3 l_2$	$10^3 c$	$10^3 I''$	$10^3 I'$
$\theta_x = 0.75, \theta_y = 0.5, h_z = h_x = h_y = 0.05$							
$\{I_1(0) = 1.5708, I_3(0) = -0.7850\}$							
Ω 0.943	1.97	1.15	8.32	2.64	1.24	0.013	
$\tilde{\Omega}$ 0.942	1.90	9.06	8.32	2.89	1.16	0.013	
$\theta_x = \theta_y = 0.75$							
Ω 1.88	3.54	1.15	9.26	2.63	2.40	0.026	
$\tilde{\Omega}$ 1.87	3.42	1.15	10.4	2.88	2.24	0.025	

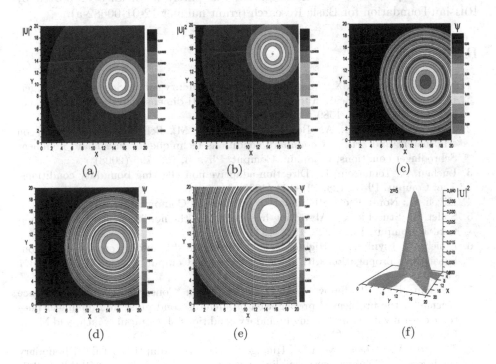

Fig. 2. Intensity distribution of laser beam in longitudinal section $z = 5$ for $\theta_x = 0.75, \theta_y = 0.5$ (a); for $\theta_x = \theta_y = 0.75$ (b). Phase distribution at $z = 2$(c), $z = 5$ (d) if the corresponding components of local wave vector is computed according with (3) (c) or numerically in each point of the right boundary (d). Phase distribution (e) and intensity profile (f) of the laser beam in section $z = 5$ at using the parameter $\tilde{\Omega}$ in the adaptive artificial boundary conditions for $\theta_x = \theta_y = 0.75$.

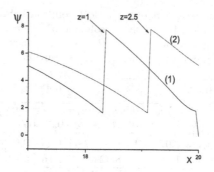

Fig. 3. Phase distribution of complex function $U = U(x, \theta_y L_y)$ for $\theta_x = 0.75, \theta_y = 0.5$ at section $z = 1$ (line 1), 2.5 (line 2)

Acknowledgements. The investigation was partly financially supported by Russian Foundation for Basic Research (grant number 12-01-00682-a).

References

1. Xu, Z., Han, H., Wu, X.: Adaptive absorbing boundary conditions for Schrödinger-type equations: application to nonlinear and multi-dimensional problems. J. Comput. Phys. 225, 1577–1589 (2007)
2. Antoine, X., Arnold, A., Besse, C., Ehrhardt, M., Schadle, A.: A review on transparent and artificial boundary conditions technique for linear and nonlinear Schrödinger equations. Commun. Comput. Phys. 4, 729–796 (2008)
3. Luchini, P., Tognaccin, R.: Direction-adaptive nonreflecting boundary conditions. J. of Comput. Phys. 128, 121–133 (1996)
4. Givoli, D.: Non-reflecting boundary conditions. J. Comput. Phys. 94(1) (1991)
5. Soffer, A., Stucchio, C.: Absorbing boundaries for the nonlinear Schrödinger equation. J. Comput. Phys. 225, 1218–1232 (2007)
6. Fibich, G., Tsynkov, S.: High order two-way artificial boundary conditions to non-linear wave propagation with back-scattering. J. Comput. Phys. 171(2), 632–677 (2001)
7. Tereshin, E.B., Trofimov, V.A., Fedotov, M.V.: Conservative finite-difference scheme for the problem of propagation of a femtosecond pulse in a nonlinear photonic crystal with nonreflecting boundary conditions. J. Comput. Math. and Math. Phys. 46(1), 161–171 (2006)
8. Trofimov, V.A., Denisov, A.D., Huang, Z., Han, H.: Adaptive artificial boundary conditions for 2D nonlinear Schrödinger equation. In: Proceedings of CMMSE 2011, Alicante, Spain, vol. 3, pp. 1150–1157 (2011)

Numerical Solution of Dynamic Problems in Block Media with Thin Interlayers on Supercomputers with GPUs

Maria Varygina

Institute of Computational Modeling
Siberian Branch of Russian Academy of Sciences,
Akademgorodok, 660036, Krasnoyarsk, Russia

Abstract. Parallel computational algorithms for the modeling of dynamic interaction of elastic blocks through thin viscoelastic interlayers in structurally inhomogeneous media such as rock are worked out. Numerical algorithms are based on monotonous grid- characteristic schemes with the balanced number of time steps in layers and interlayers. The numerical results to demonstrate qualitative characteristics of wave propagation in materials with microstructure are shown.

Introduction

Several nature materials such as rock have distinct structurally inhomogeneous block-hierarchical structure. Block structure appears on different scale levels from the size of crystal grains to the blocks of rock. Blocks are connected to each other with thin interlayers of rock with significantly weaker mechanical properties [1].

One of the most important technological problems of coal mining is the prognosis of sudden collapse of the coal mining roof. This process is preceded by a weakening of the mechanical contact between the blocks: the rock gains weakened microstructure. Such state of the media can be detected with inducing elastic waves of small amplitude and recording the response to these disturbances. This method can be used to develop special technical devices for the forehanded prediction and prevention of the emergency situations.

The purpose of this paper is to apply reliable computational algorithms for the numerical modeling of dynamic interactions of elastic blocks through thin viscoelastic interlayers in structurally inhomogeneous medium such as rock [2]. Parallel computation algorithms for the calculation of elastic wave propagation in rock media based on mathematical models that take into account complex rheological properties of layered materials are developed as a program complex for the multiprocessor computers with graphics processing units. The computations for the large number of layers allow to analyze specific waves related to the structural inhomogeneity, i.e. the so-called "pendulum" waves [3–5].

I. Dimov, I. Faragó, and L. Vulkov (Eds.): NAA 2012, LNCS 8236, pp. 517–523, 2013.

1   Mathematical Model

An approximate scheme of hierarchical structure of rock media is shown on Fig. 1. In the ideal case this is a nested layered structure with invariant ratio of the characteristic sizes of blocks and interlayers. Let us consider at first a fragment of the structure in one dimension, i.e. the interleaved system of n elastic layers of thickness h and elastic interlayers of thickness δ.

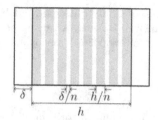

Fig. 1. Hierarchical structure of rock

Let ρ and ρ_0, c and c_0, $a = 1/(\rho c^2)$ and $a_0 = 1/(\rho_0 c_0^2)$ be the densities, velocities and elastic compliances of materials in the layer and the interlayer respectively. One-dimensional equations of elasticity theory inside the kth layer are given:

$$\rho \frac{\partial v^k}{\partial t} = \frac{\partial \sigma^k}{\partial x}, \quad a \frac{\partial \sigma^k}{\partial t} = \frac{\partial v^k}{\partial x}. \tag{1}$$

Here v^k is the longitudinal velocity in x-direction (x varies from 0 to h in each layer) and σ^k is the normal stress.

The behavior of the interlayer material is described by the system of ordinary differential equations:

$$\rho_0 \frac{d}{dt} \frac{v^{k+1} + v^k}{2} = \frac{\sigma^{k+1} - \sigma^k}{\delta}, \quad a_0 \frac{d}{dt} \frac{\sigma^{k+1} + \sigma^k}{2} = \frac{v^{k+1} - v^k}{\delta}. \tag{2}$$

This system contains boundary conditions for the mentioned above velocities and stresses, i.e. the left boundary condition is for the $(k+1)^{th}$ interlayer and the right is for the kth interlayer. Such system can be obtained by averaging the motion equations of elastic media considering thin interlayer ($\delta << h$). The inertial properties of interlayer are taken into account.

Systems of equations (1), (2) are complemented with the initial conditions and the boundary conditions

$$v^k = \sigma^k = 0, \quad (k = 1, ..., n), \quad \sigma^1(0, t) = -p(t), \quad v^n(h, t) = 0.$$

Here $p(t)$ is given external pressure. Statement correctness of the initial-boundary value problem can be proved by the methods based on integral estimates derived from the energy conservation law [6]:

$$\frac{1}{2}\frac{\partial}{\partial t}\sum_{k=1}^{n}\int_{0}^{h}\left(\rho\left|v^{k}(x,t)\right|^{2}+a\left|\sigma^{k}(x,t)\right|^{2}\right)dx+$$

$$+\frac{\delta}{2}\frac{d}{dt}\sum_{k=1}^{n}\left(\frac{\rho_{0}}{2}\left|v^{k+1}(0,t)+v^{k}(h,t)\right|^{2}+\frac{a_{0}}{2}\left|\sigma^{k+1}(0,t)+\sigma^{k}(h,t)\right|^{2}\right)=$$

$$=\sigma^{n}(h,t)v^{n}(h,t)-\sigma^{1}(0,t)v^{1}(0,t), \tag{3}$$

that is the direct consequence of systems (1), (2). The conservation law (3) also proves the thermodynamic consistency of the mathematical model.

Analysis of experimental data of wave propagation in layered media shows that the interlayers behave non-elastically even under small wave amplitudes. In Maxwell's model the strain of the interlayers is formed from elastic and viscous components. Viscoelastic interaction according to this model is described by system (2) after replacing the second equation by the more general equation:

$$a_{o}\frac{d}{dt}\frac{\sigma^{k+1}+\sigma^{k}}{2}+\frac{\sigma^{k+1}+\sigma^{k}}{2\eta}=\frac{v^{k+1}+v^{k}}{\delta}, \tag{4}$$

where η is the viscosity coefficient of interlayer material.

2   Numerical Algorithm

Numerical solution of the problem is based on the collapse of the gap Godunov scheme on a uniform grid with a time step $\tau=\Delta x/c$ admissible by the Courant-Friedrichs-Levy condition. In this case the scheme in the layer does not possess an artificial energy dissipation. Piece-wise linear ENO-reconstruction of the second-order accuracy is used with the lower values of time step [2].

Consistency conditions of the form (2) on the boundaries between layers and interlayers are calculated with the Godunov scheme as well. For this purpose in each artificially introduced cell simulating a single interlayer the collapse of the gap scheme is implemented. The independent time step in the interlayer in this scheme $\tau_{0}=\delta/c_{0}<<\tau$ is admissible by the Courant-Friedrichs-Levy condition. In the interlayer the number of time steps necessary to achieve the next time step $t+\tau$ of the main scheme is calculated. This scheme significantly reduces the effect of smoothing the numerical solution peaks with corresponding refinement of the obtained results.

Grid-characteristic interpretation of the method is shown schematically on Fig. 2. At the stage of solving system (2) with time step τ_{0} the equations of collapse of the gap are used on the boundaries of interlayers (stage "predictor" in the interlayer):

$$z_{0}v_{+}-\sigma_{+}=z_{0}v-\sigma, \quad zv_{+}+\sigma_{+}=zv^{k+1}+\sigma^{k+1},$$

$$z_{0}v_{-}+\sigma_{-}=z_{0}v+\sigma, \quad zv_{-}-\sigma_{+}=zv^{k}-\sigma^{k}.$$

Here $z=\rho c$ and $z_{0}=\rho_{0}c_{0}$ are the acoustic impedances of materials in layer and interlayer respectively. The values with upper indexes correspond to the

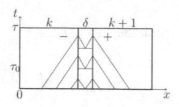

Fig. 2. Scheme of the grid-characteristic method in the boundary between the layers

boundary cells of interacting layers. The values with lower indexes "+" and "–" correspond to the right and the left boundaries of the interlayer respectively. The subsequent recalculation of the solution (stage "corrector") is done according to the following formulas:

$$\hat{v} = v + (\sigma_+ - \sigma_-) \frac{\tau_0}{\rho_0\,\delta}, \qquad \hat{\sigma} = \sigma + (v_+ - v_-) \frac{\tau_0}{a_0\,\delta}. \tag{5}$$

Here $\hat{v}$ and $\hat{\sigma}$ correspond to the next time step. Predictor values on the boundary between layers in the main scheme with time step τ are calculated by averaging values related to the cell boundaries on small mesh steps. Stage "corrector" of the main scheme in layers is performed in the usual way based on the integral analogues of differential equations (1).

In case of viscoelastic interlayers at stage "corrector" of the numerical algorithm Crank-Nicholson scheme is used instead of (5):

$$\rho_0 \frac{\hat{v} - v}{\tau_0} = \frac{\sigma_+ - \sigma_-}{\delta}, \quad a_0 \frac{\hat{\sigma} - \sigma}{\tau_0} + \frac{\hat{\sigma} + \sigma}{\eta} = \frac{v_+ - v_-}{\delta}, \tag{6}$$

which is also implemented with an explicit computational algorithm.

The scheme possess at least the first-order approximation relative to the introduced parameters of time step and mesh step. It can be determined that the difference analogue of the energy conservation law (3) is true for the scheme. The equality sign in (3) is replaced by the sign of non-strict inequality generally meaning that the scheme possess an artificial energy dissipation.

Numerical algorithm to solve planar problems is based on the space-variable two-cycling decomposition method. At first two stages one-dimensional problems in x_1 and x_2 directions in the time interval $(t; t + \tau/2)$ are solved. Third and fourth stages are the stages of recalculation of the problem in directions x_2 and x_1 in the time interval $(t + \tau/2; t + \tau)$. The two-cycling decomposition method ensures the stability of the numerical solution provided the Courant-Friedrichs-Levy stability condition for one-dimensional systems is fulfilled.

3   Numerical Results

The algorithms for calculation of thin elastic and viscoelastic interlayers are implemented for the multiprocessor computer systems with graphics accelerators with CUDA technology (Compute Unified Device Architecture). The calculations were performed on the 8-core computer with a graphics card NVIDIA Tesla C2050.

The computations of planar waves propagation induced by short and long Λ−impulses on the boundary of layered medium were performed. The layered medium consists of 512 layers of rock with microfractured elastic interlayers. Calculations were performed after the transition the system of equations to dimensionless variables with the following parameter values: $\rho_0/\rho = 0.76$, $a_0/a = 7.17$, $\delta/h = 0.027$. A uniform finite difference mesh in the layer consists of 16 cells and one cell is used in each interlayer.

On Fig. 3 and Fig. 4 the dependencies of dimensionless velocities of particles v of space coordinate x, divided by the layer thickness h in a problem of the load of Λ−impulse of pressure are shown. The impulse with a unit amplitude was induced on the left boundary of computational domain, the right boundary was fixed. Fig. 3 corresponds to the impulse duration equal to the time that elastic wave passes through one layer, Fig. 4 corresponds to the duration two and a half times longer. Fig. 3.a and Fig. 4.a show the dependencies of velocities at the time the incident wave passes approximately 370 layers (6000th time step of the main scheme). On Fig. 3.b and Fig. 4.b the incident wave passes approximately 200 layers in inverse direction (12000th time step).

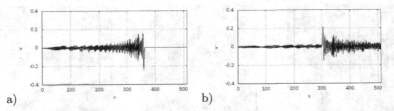

a)   b)

Fig. 3. Velocity behind front wave of incident (a) and reflected (b) waves induced by the short impulse in layered medium

These results demonstrate a qualitative difference between the wave pattern in layered media as compared with a homogeneous medium. This difference at the initial stage is revealed in the appearance of waves reflected from the interlayers, i.e. the characteristic oscillations behind the loading wave front as it passes through the interface. Eventually stationary wave pattern appears after multiple reflections behind the head wave front, i.e. the so-called pendulum wave predicted in papers [3–5]. The head wave amplitude increases with the impulse duration increasing and the amplitude of the oscillations behind the wavefront decreases and tends to zero. It is related to the fact that waves with wavelength considerably greater than the thickness of the layer are nearly not reflected from the interlayers. Thus it is possible to detect the weakened microstructure of layered or block medium with only sufficiently short wavelengths.

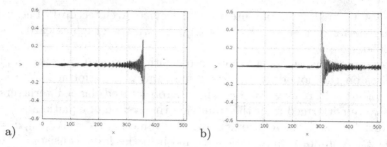

Fig. 4. Velocity behind front wave of incident (a) and reflected (b) waves induced by the long impulse in layered medium

Fourier analysis of the displacement of layers seismograms allows to identify the characteristic frequency of the pendulum wave due to the compliances of interlayers and their thickness.

The numerical results of Lamb problem on the action of instant concentrated load in tangent direction on a central part of a half-space are presented on Fig. 5. These results are obtained on a grid consisting of 512×1024 cells for different number of thin interlayers. Black lines are the positions of thin interlayers.

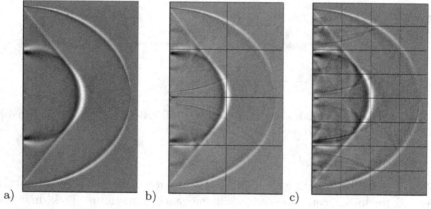

Fig. 5. Lamb problem: contour levels of σ_{12}. a) domain without interlayers, b) domain with 2×4 interlayers, c) domain with 4×8 interlayers.

All characteristic waves for the solution of Lamb's problem in the framework of the classical elasticity theory (incident longitudinal and transverse waves, two conical transverse waves, the Rayleigh surface waves rapidly damped with depth) are clearly distinguished on the level curves. The distinction is that for the domain with thin interlayers multiple re-reflections from the interlayers arised.

Acknowledgement. This work was supported by the Complex Fundamental Research Program no. 18.2 "Algorithms and Software for Computational Systems of Superhigh Productivity" of the Presidium of RAS and the Russian Foundation for Basic Research (grant no. 11-01-00053).

References

1. Sadovskii, M.A.: Natural lumpiness of rocks. DAN USSR 247(4), 829–831 (1979) (in Russian)
2. Varygina, M.P., Pohabova, M.A., Sadovskaya, O.V., Sadovskii, V.M.: Numerical algorithms for the analysis of elastic wave propagation in block media with thin interlayers. Computational Methods and Programming (12), 435–442 (2011) (in Russian)
3. Kurlenya, M.V., Oparin, V.N., Vostrikov, V.I.: On generation of elastic wave packages under impulse inducement of block media. Pendulum waves. DAN USSR 333(4), 3–13 (1993) (in Russian)
4. Aleksandrova, N.I., Chernikov, A.G., Sher, E.N.: Experimental testing of the one-dimensional model of wave propagation in block media. Physical and Technical Problems of Mining (3), 46–55 (2005) (in Russian)
5. Aleksandrova, N.I., Sher, E.N., Chernikov, A.G.: The impact of viscosity of interlayers on low-frequent pendulum waves propagation in block hierarchical media. Physical and Technical Problems of Mining (3), 3–13 (2008) (in Russian)
6. Sadovskaya, O., Sadovskii, V.: Mathematical Modeling in Mechanics of Granular Materials. Advanced Structured Materials, vol. 21, p. 390. Springer, Heidelberg (2012)

Asymptotic-numerical Investigation of Generation and Motion of Fronts in Phase Transition Models

Vladimir Volkov and Nikolay Nefedov

Department of Mathematics, Faculty of Physics,
Lomonosov Moscow State University, 119991 Moscow, Russia
nefedov@phys.msu.ru

Abstract. We propose an effective asymptotic-numerical approach to the problem of moving front type solutions in nonlinear reaction-diffusion-advection equations. The dimension of spatial variables for the location of a moving front is lower per unit then the original problem. This fact gives the possibility to save computing resources in numerical experiments and speed up the process of constructing approximate solutions with a suitable accuracy.

Keywords: singularly perturbed problems, moving fronts, reaction-diffusion-advection equations.

1   Statement of the Problem

The purpose of this paper is to develop effective asymptotic-numerical approaches to study solutions with internal transition layers – moving fronts in a mathematical model of reaction-diffusion-advection type. We demonstrate our method for the following problem.

Consider the equation

$$\varepsilon^2 \left(\Delta u + \nu(x,y) \frac{\partial u}{\partial x} \right) - \varepsilon^2 \frac{\partial u}{\partial t} = f(u, x, y, \varepsilon), \qquad (1)$$

$$x \in (0, a), y \in (-\infty, +\infty), t > 0$$

with the boundary, space periodicity and initial conditions

$$\left. \frac{\partial u}{\partial x} \right|_{x=0, x=a} = 0, \quad u(x, y, t, \varepsilon) = u(x, y + L, t, \varepsilon) \qquad (2)$$

$$u(x, y, t, \varepsilon)|_{t=0} = u^0(x, y). \qquad (3)$$

In equation (1) $\varepsilon > 0$ is a small parameter, which is a consequence of the parameters in the underlying physical problem. It should be noted that the appearance of the small parameter in front of the time derivative determines only the scale of the time, convenient for the further consideration.

The functions $\nu(x, y)$ and $f(u, x, y, \varepsilon)$ are assumed to be sufficiently smooth and L - periodic in the variable y.

I. Dimov, I. Faragó, and L. Vulkov (Eds.): NAA 2012, LNCS 8236, pp. 524–531, 2013.
© Springer-Verlag Berlin Heidelberg 2013

2   The Problem of Moving Front Formation

Suppose the following conditions are satisfied.

Condition (A_1). *The function $f(u, x, y, \varepsilon)$ is such that the equation*
$f(u, x, y, 0) = 0$ *has exactly three roots* $\varphi^{(\pm)}(x, y), \varphi^{(0)}(x, y)$. *Assume that*
$\varphi^{(-)}(x, y) < \varphi^{(0)}(x, y) < \varphi^{(+)}(x, y)$ *for all* $(x, y) \in \bar{D} = \{[0, a] \times (-\infty, +\infty)\}$
and $f_u\left(\varphi^{(\pm)}(x, y), x, y, 0\right) > 0$, $f_u\left(\varphi^{(0)}(x, y), x, y, 0\right) < 0$.

To simplify the further presentation, without a lost of generality we assume
that $\varphi^{(0)}(x, y) \equiv 0$, that can always be achieved by a suitable change of variables.

Condition (A_2). *There exists sufficiently smooth L- periodic curve* $x = h^0(y)$,
such that $u^0(x, y) < 0$ *for* $(x, y) \in D_0^{(-)} :=$
$\left\{(0 \le x < h^0(y)) \times (-\infty < y < +\infty)\right\}$ *and* $u^0(x, y) > 0$ *for* $(x, y) \in D_0^{(+)} :=$
$\left\{(h^0(y) < x \le a) \times (-\infty < y < +\infty)\right\}$.

Under the conditions (A_1) and (A_2) the following statement is valid.

Theorem 1. *Suppose that the conditions $(A_1))$ and $((A_2)$ are valid. Then there
exists a positive constant B such that for sufficiently small ε at time $t = t_B(\varepsilon) = B\varepsilon^2 |\ln \varepsilon|$ the solution $u(x, y, t, \varepsilon)$ of the problem (1) – (3) satisfies the following
estimates*

$$u(x, y, t_B, \varepsilon) = \varphi^{(-)}(x, y) + O(\varepsilon) \text{ for } (x, y) \in D_0^{(-)},$$
$$u(x, y, t_B, \varepsilon) = \varphi^{(+)}(x, y) + O(\varepsilon) \text{ for } (x, y) \in D_0^{(+)}.$$
excluding a fixed neighborhood of the curve $x = h^0(y)$.

The proof of this theorem is based on the works [1] and [2] with slight
modifications.

¿From Theorem 1 it follows that at the initial stage the solution of the problem
(1) – (3) quickly generates a sharp internal layer in the neighborhood of the curve
$x = h^0(y)$. For the following we assume that the front already exists, i.e. we put
$t_B = 0$.

In what follows we suppose that $u(x, y, t_B, \varepsilon) \equiv u^0(x, y, \varepsilon)$, and choose

$$u(x, y, t, \varepsilon)|_{t=0} = u^0(x, y, \varepsilon), \tag{4}$$

as initial conditions. Thus, we will consider the problem (1), (2), (4).

3   Description of the Moving Front

We define the function $I(x, y)$ by
$$I(x, y) := \int_{\varphi^{(-)}(x,y)}^{\varphi^{(+)}(x,y)} f(u, x, y, 0)\, du.$$

Condition (A_3). $I(x, y) \equiv 0$ *for* $(x, y) \in D$, *i.e. nonlinearity* $f(u, x, y, \varepsilon)$ *is
balanced.*

An asymptotic expansion of the front type solution of (1), (2), (4) can be
built by using the ideas of asymptotic theory of contrast structures (see e.q. [3]),
developed to describe moving fronts in [4]. Using this scheme to build the asymptotics let us consider two domains $D^{(-)}$ and $D^{(+)}$, separated by some sufficiently

smooth curve $x = h(y, t, \varepsilon)$, in the neighborhood of which the solution changes rapidly from $\varphi^{(-)}(x, y)$ to $\varphi^{(+)}(x, y)$. The curve $x = h(y, t, \varepsilon)$ is not known a priori and should be found in the process of the construction of the asymptotics. We define the location of the transition layer by the condition of the intersection of the solution (1) – (4) with the root of the degenerate equation $\varphi^{(0)}(x, y) \equiv 0$, therefore it holds $u(h(y, t, \varepsilon), y, t, \varepsilon) = 0$.

Let us consider the equation

$$\varepsilon^2 \left(\Delta u + \nu(x, y) \frac{\partial u}{\partial x} - \frac{\partial u}{\partial t} \right) = f(u, x, y, \varepsilon)$$

in two domains $D^{(-)} := \{0 < x < h(y, t, \varepsilon), y \in (-\infty, +\infty)\}$ and $D^{(+)} := \{h(y, t, \varepsilon) < x < a, y \in (-\infty, +\infty)\}$ with the conditions

$$\frac{\partial u}{\partial x}(0, y, t, \varepsilon) = 0, \quad u(h(y, t, \varepsilon), y, t, \varepsilon) = 0, \quad y \in (-\infty, +\infty)$$

$$u(x, y, 0, \varepsilon) = u^0(x, y, \varepsilon), \qquad (x, y) \in D^{(-)}$$

and

$$u(h(y, t, \varepsilon), y, t, \varepsilon) = 0, \quad \frac{\partial u}{\partial x}(a, y, t, \varepsilon) = 0, \quad y \in (-\infty, +\infty)$$

$$u(x, y, 0, \varepsilon) = u^0(x, y, \varepsilon), \qquad (x, y) \in D^{(+)}.$$

We seek the unknown curve $x = h(y, t, \varepsilon)$ in the form of a power series in ε

$$h(y, t, \varepsilon) = h_0(y, t) + \varepsilon h_1(y, t) + ...,$$

where terms will be defined in the process of the construction of the asymptotics.

Asymptotic expansions of the solutions of each of these problems contain regular and boundary series $U^{(\pm)}(x, y, t, \varepsilon) = u^{(\pm)}(x, y, t, \varepsilon) + P^{(\pm)}(\rho, y, t, \varepsilon) + Q^{(\pm)}(\xi, y, t, \varepsilon)$ and can be constructed using standard method of boundary functions. Boundary series $P^{(\pm)}(\rho, y, t, \varepsilon)$ and $Q^{(\pm)}(\xi, y, t, \varepsilon)$ describe the solution near the boundaries of D and the internal transition layer near the curve $x = h(y, t, \varepsilon)$. Regular series $u^{(\pm)}(x, y, t)$ in domains $D^{(-)}$ and $D^{(+)}$, and boundary series $P^{(\pm)}(\rho, y, t, \varepsilon)$ near the boundaries of D can be determined by the standard scheme [3]. Note that the boundary series $P^{(\pm)}(\rho, y, t, \varepsilon)$ are significant only in a small area near the boundaries of the area D, rapidly exponentially decrease and do not have a significant influence on the behavior of the internal transition layer. The boundary functions $Q^{(\pm)}(\xi, y, t, \varepsilon)$ describing the transition layer near the curve $x = h(y, t, \varepsilon)$ are defined also by the standard scheme [3].

The curve $x = h(y, t, \varepsilon)$ can be determined from the conditions of continuous matching for the functions $U^{(-)}(x, y, t, \varepsilon)$, $U^{(+)}(x, y, t, \varepsilon)$ (asymptotic expansions built in the domains $D^{(\pm)}$) and their first derivatives on the curve $x = h(y, t, \varepsilon)$ ($C^{(1)}$-matching conditions) :

$$\varepsilon \frac{\partial U^{(-)}}{\partial x} = \varepsilon \frac{\partial U^{(+)}}{\partial x} \text{ for } x = h(y, t, \varepsilon). \tag{5}$$

We briefly describe some detail of the procedure of finding the main terms of the asymptotic expansions, describing the moving front. For the construction of the functions $Q^{(\pm)}$ we introduce local coordinates $(r, \theta(y))$ and use them for the parabolic operator in the right part of (1), where $r = (x - h_0(y, t)) \cdot \sqrt{1 + h_{0y}^2}$ is the distance from the curve $x = h_0(y, t)$ along the normal to the curve with the sign "+" in the domain $D^{(+)}$ and with "–" in $D^{(-)}$, $\theta(y)$ – coordinate of the point on the curve $h_0(y, t)$ from which this normal is going, $h_{0y} = \frac{\partial h_0(y, t)}{\partial y}$.

As a result, using the stretched variable $\xi = \dfrac{r}{\varepsilon} \equiv \dfrac{(x - h_0(y, t)) \cdot \sqrt{1 + h_{0y}^2}}{\varepsilon}$ the operator $Lu = \varepsilon^2 \left(\Delta u + \nu(x, y) \dfrac{\partial u}{\partial x} - \dfrac{\partial u}{\partial t} \right)$ can be transformed into the form

$$Lu = \left(1 + \varepsilon^2 \alpha(\varepsilon\xi, y, \varepsilon)\right) \frac{\partial^2 u}{\partial \xi^2} + \varepsilon L_1 u + \varepsilon^2 L_2 u + \dots,$$

where the operators L_1, L_2 are known and the function α is also defined. In particular, important for the further part $L_1 u$ is

$$L_1 u = k(y) \frac{\partial u}{\partial \xi} + \left(1 + h_{0y}^2\right)^{1/2} \left(\nu(h_0(y, t), y) - \frac{\partial h_0}{\partial t} \right) \frac{\partial u}{\partial \xi},$$

where $k(y) = \dfrac{h_{0yy}}{\left(1 + h_{0y}^2\right)^{3/2}}$ is the curvature of the curve $x = h_0(y, t)$ calculated at the point $(h_0(y, t), y)$.

For the main terms of transition layer functions $Q_0^{(\pm)}(\xi, t)$ we obtain the problems

$$\frac{\partial^2 Q_0^{(\pm)}}{\partial \xi^2} = f\left(\varphi^{(\pm)}(h_0(y, t), y) + Q_0^{(\pm)}, h_0(y, t), y, 0 \right),$$

$$Q_0^{(\pm)}(\pm\infty, t) = 0, \quad Q_0^{(\pm)}(0, t) = -\varphi^{(\pm)}(h_0(y, t), y).$$

Taking into consideration the continuous function $\tilde{u} = \varphi^{(\pm)}(h_0(y, t), y) + Q_0^{(\pm)}$, and using (A_1), we can rewrite these problems as

$$\frac{\partial^2 \tilde{u}}{\partial \xi^2} = f(\tilde{u}, h_0(y, t), y, 0), \quad \xi \in R,$$

$$\tilde{u}(-\infty, t) = \varphi^{(-)}(h_0(y, t), y), \quad \tilde{u}(\infty, t) = \varphi^{(+)}(h_0(y, t), y), \quad \tilde{u}(0, t) = 0.$$

It is well known that for a nonlinearity $f(u, x, y, \varepsilon)$ satisfying the condition (A_3), this problem has a unique solution and the condition (5) is fulfilled in zeroth order

$$\left. \frac{\partial Q_0^{(+)}}{\partial \xi} \right|_{\xi=0} = \left. \frac{\partial Q_0^{(-)}}{\partial \xi} \right|_{\xi=0}$$

automatically for the curve $x = h_0(y, t)$. This curve – the main term in the asymptotic expansion for the front localization – will be determined in the next step by using the matching condition of first order.

For the transition layer functions $Q_1^{(\pm)}(\xi, t)$ we have

$$\frac{\partial^2 Q_1^{(\pm)}}{\partial \xi^2} - f_u(\tilde{u}, h_0(y,t), y, 0) \cdot Q_1^{(\pm)} = f_1^{(\pm)}(\xi, y, t), \quad \xi \in R,$$

$$Q_1^{(\pm)}(0, t) = 0, \quad Q_1^{(\pm)}(\pm\infty, t) = 0,$$

where $f_1^{(\pm)}(\xi, y, t)$ are known functions. For example,

$$f_1^{(-)}(\xi, y, t) = (1 + h_{0y}^2)^{1/2}$$

$$\left[-\frac{\partial h_0}{\partial t} \frac{\partial Q_0^{(-)}}{\partial \xi} + \frac{h_{0yy}(y,t)}{(1 + h_{0y}^2(y,t))^2} \frac{\partial Q_0^{(-)}}{\partial \xi} + \nu(h_0(y,t), y) \frac{\partial Q_0^{(-)}}{\partial \xi} \right] +$$

$$+ \left(\tilde{f}_u^{(-)} \cdot \varphi_x^{(-)}(h_0(y,t), y) + \tilde{f}_x^{(-)} \right) \cdot (h_1(y,t) + \xi) +$$

$$\tilde{f}_u^{(-)} \cdot u_1^{(-)}(h_0(y,t), y, t) + \tilde{f}_\varepsilon^{(-)}, \tag{6}$$

where the derivatives $\tilde{f}_u^{(-)}, \tilde{f}_x^{(-)}, \tilde{f}_\varepsilon^{(-)}$ are calculated at the point $(\tilde{u}^{(-)}, h_0(y,t), y, 0)$. The structure of the function $f_1^{(+)}(\xi, y, t)$ is fully identical. The solutions of these problems can be written explicitly (see [3]):

$$Q_1^{(\pm)}(\xi, y, t) = -\frac{\partial Q_0^{(\pm)}}{\partial \xi} \cdot \int_0^\xi \left(\frac{\partial Q_0^{(\pm)}}{\partial \tau} \right)^{-2} \cdot \int_\tau^{\pm\infty} \frac{\partial Q_0^{(\pm)}}{\partial \sigma} f_1^{(\pm)}(\sigma, y, t) \, d\sigma d\tau \quad . \tag{7}$$

¿From the first order of $C^{(1)}$-matching condition (5) we get

$$\frac{\partial Q_1^{(+)}}{\partial \xi}(0, y, t) = \frac{\partial Q_1^{(-)}}{\partial \xi}(0, y, t) \tag{8}$$

Using (6) and (7), the equality (8) can be transformed into the form

$$(1 + h_{0y}^2)^{1/2} \cdot \left[\frac{\partial h_0}{\partial t} - \frac{h_{0yy}}{(1 + h_{0y}^2)^2} - \nu(h_0(y,t), y) \right] \times$$

$$\int_{-\infty}^{+\infty} \left(\frac{\partial \tilde{u}}{\partial \xi} \right)^2 d\xi - \int_{-\infty}^{+\infty} \left(\tilde{f}_x^{(-)} \cdot \xi + \tilde{f}_\varepsilon^{(-)} \right) d\xi = 0.$$

Hence, finally we obtain the following problem to determine the location of the front in zeroth order approximation

$$\frac{\partial h_0}{\partial t} - \frac{\partial^2 h_0}{\partial y^2} \cdot [1 + h_{0y}^2]^{-2} - \nu(h_0(y,t), y)$$

$$= \frac{m(t,y)}{\sqrt{(1 + h_{0y}^2)}} \cdot \int_{-\infty}^{+\infty} \left(\tilde{f}_x^{(-)} \cdot \xi + \tilde{f}_\varepsilon^{(-)} \right) d\xi,$$

$$h_0(y, 0) = h^0(y)$$

with L- periodic condition in the variable y, where $m(y, t) = \left[\int\limits_{-\infty}^{+\infty} \left(\frac{\partial \tilde{u}}{\partial \xi} \right)^2 d\xi \right]^{-1}$.

Thus we obtain a nonlinear parabolic equation which determines the location of the moving front in zeroth order approximation. Note that the dimension of the spatial variables is lower per unit as we have at the original problem (1–3). This fact can be used in order to save significantly computing resources in the numerical experiment by replacing the original problem (1–3) by the simplified problem above, which can adequately describe the dynamics of the front. Thus, the use of such asymptotic-numerical approach gives the possibility to speed up the process of constructing approximate solutions with a suitable accuracy, and we have more efficient numerical calculations.

The higher construction of terms of the asymptotic expansion follows by the same way as in [4].

Theorem 2. *Under the conditions* (A_1), (A_2), (A_3) *there exist the solution* $u(x, y, t, \varepsilon)$ *of (1)–(3) and satisfies the estimate*

$$|u(x, y, t, \varepsilon) - U_n(x, y, t, \varepsilon)| \leq C\varepsilon^{n+1}.$$

Here, $U_n(x, y, t, \varepsilon) = U_n^{(\pm)}(x, y, t, \varepsilon)$ *and* $U_n^{(\pm)}(x, y, t, \varepsilon)$ *are the n-th partial sum of the asymptotic series*

$$U^{(\pm)}(x, y, t, \varepsilon) = u^{(\pm)}(x, y, t, \varepsilon) + P^{(\pm)}(\rho, y, t, \varepsilon) + Q^{(\pm)}(\xi, y, t, \varepsilon)$$

in the domains $D^{(-)}$ *and* $D^{(+)}, \xi = \dfrac{x - \sum\limits_{i=0}^{i=n+1} h_i(y, t)}{\varepsilon} \cdot \sqrt{1 + h_{0y}^2}$.

The proof of this theorem is based on the method of differential inequalities and the fact that all operators producing the asymptotic series are positively invertible (see [7]).

4 Numerical Experiment

The theoretical results on a moving front presented in the section before are compared with the results of numerical solution of the problem (1)-(3). For this purpose we use standard finite-difference scheme for the problem (1)-(3) and for the equation of front localization. Calculations were done in D, representing a rectangle with the sides $a = b = 1$, the advection coefficient has the form $\nu(x, y) = \nu_0 + \nu_L(y)$, $\nu_0 = const$ and $f(u, x, y, \varepsilon) = u^3 - u$.

In case $\nu_L(y) = 0$, a sharp transition layer moves along the axis X with the constant speed. This behavior is consistent with experimental [5], [6] and theoretical studies that show that the burning front, extending in a homogeneous environment in the absence of a hydrodynamic perturbation moves at a constant speed without changing its shape.

To illustrate the behavior of the front that occurs in the problem (1)-(3), we provide three numerical calculations. In the first experiment the initial condition is in the form of a linear front parallel to the axis y, and the advection parameters are $\nu_0 = const, \nu_L(y) = 0$. In this case, the front is moving with a constant speed ν_0 without changing its shape (Figure 1a.).

In the second case, the initial condition is selected as the front line, parallel to the axis y and the advection parameters are $\nu_0 = const, \nu_L = const > 0$ for $y \in [0.25; 0.75]$. In this case, the front begins to move in the direction of the advection, and after some time its form is stabilized, then the front continues to move (Figure 1b.) with the fixed form.

In the third case, the initial condition is selected such, that the front is specified by the equation $x = p\{\tanh(q(y - 0.25)) - \tanh(q(y - 0.75))\}$, where p and q are constants and the advection parameters are $\nu_0 = const, \nu_L(y) = 0$. As a result, under the effect of the curvature the front will move along the axis x, gradually losing the form and in the limit it is turning into a line parallel to the side of the square, as shown in Figure 1c.

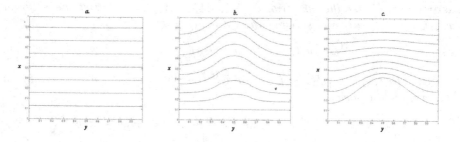

Fig. 1. Behavior of the moving front in the three numerical experiments

The analysis of the numerical calculations show a good agreement between the above qualitative descriptions of the front behavior and numerical calculations for (1)-(3). Figure 2. shows several sequent positions, found in the numerical experiments. From the picture one can see that the behavior of the solutions of (1)-(3) and of front motion problem coincides with a high accuracy.

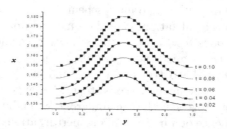

Fig. 2. Sequent positions of the front at times $t = 0.01, 0.06, 0.04, 0.08$ and 0.1. Points show the solution of (1)-(3). Solid lines show the solution of front motion problem.

Therefore, the asymptotic representation of the solution of (1)-(3) allows fully to describe the process of formation and dynamics of the transition layer (front), provides the estimates of its width and the time of its formation, as well as determines the shape of the front. Note that the asymptotic consideration is simple enough and extremely important for the effective estimates of the various system parameters.

Acknowledgement. This work is supported by RFBR, pr. N 13-01-00200.

References

1. Volkov, V.T., Grachev, N.E., Nefedov, N.N., Nikolaev, A.N.: On the formation of sharp transition layers in two-dimensional reaction-diffusion models. Journ. of Comp. Math. and Math. Phys. 47(8), 1301–1309 (2007)
2. Butuzov, V.F., Nefedov, N.N., Schneider, K.R.: On generation and propagation of sharp transition layers in parabolic problems. Vestnik MGU. Serija 3(1), 9–13 (2005) (in Russian)
3. Vasilieva, A.B., Butuzov, V.F., Nefedov, N.N.: Contrast structures in singularly perturbed problems. Fundamentalnaja i Prikladnala Matematika 4(3), 799–851 (1998) (in Russian)
4. Bozhevolnov, Y.V., Nefedov, N.N.: Front motion in parabolic reaction-diffusion problem. Computational Mathematics and Mathematical 50(2), 276–285 (2010)
5. Fife, P.C., Penrose, O.: Interfacial dynamics for thermodynamically consistent phase-field models with nonconserved order parameter. Electronic J. of Diff. Equations 1995 (16), 1–49 (1995)
6. Burger, J.G., Sahuquet, B.C.: Chemical Aspects of In Situ Combustion: Heat of Combustion and Kinetics. SPEJ 3599 (1972)
7. Nefedov, N.N.: The Method of Differential Inequalities for Some Classes of Nonlinear Singularly Perturbed Problems with Internal Layers. Differ. Uravn. 31(7), 1142–1149 (1995)

hp Finite Element Methods for Fourth Order Singularly Perturbed Boundary Value Problems

Christos Xenophontos[1], Markus Melenk[2], Niall Madden[3], Lisa Oberbroeckling[4], Pandelitsa Panaseti[1], and Antri Zouvani[1]

[1] University of Cyprus, PO BOX 20537, Nicosia 1678, Cyprus
xenophontos@ucy.ac.cy
[2] TU Wien, Wiedner Hauptstrasse 8-10, A-1040 Wien, Austria
[3] NUI Galway, Ireland
[4] Loyola University, Baltimore, MD, USA

Abstract. We consider fourth order singularly perturbed boundary value problems (BVPs) in one-dimension and the approximation of their solution by the hp version of the Finite Element Method (FEM). If the given problem's boundary conditions are suitable for writing the BVP as a second order system, then we construct an hp FEM on the so-called *Spectral Boundary Layer Mesh* that gives a robust approximation that converges exponentially in the energy norm, provided the data of the problem is analytic. We also consider the case when the BVP is not written as a second order system and the approximation belongs to a finite dimensional subspace of the Sobolev space H^2. For this case we construct suitable C^1−conforming hierarchical basis functions for the approximation and we again illustrate that the hp FEM on the *Spectral Boundary Layer Mesh* yields a robust approximation that converges exponentially. A numerical example that validates the theory is also presented.

1 Introduction: The Model Problems

Singularly Perturbed Problems (SPPs) arise in numerous applications from science and engineering, such as electrical networks, vibration problems, and in the theory of hydrodynamic stability [2]. In such problems, the highest derivative in the differential equation is multiplied by a very small positive parameter. This causes the solution to contain *boundary layers*, which are rapidly varying solution components that have support in a narrow neighbourhood of the boundary of the domain. Their numerical approximation must be carefully constructed in order for their effects to be accurately captured [8]. In the context of the FEM, this requires the use of *graded meshes* which include refinement near the boundary layer region that depends on the singular perturbation parameter. Examples of such meshes include the Bakhvalov [1] and Shishkin [10] meshes, which are used with finite differences and the h version of the FEM, as well as the *Spectral Boundary Element Mesh* [3] which is used with the p and hp versions of the FEM.

Most SPPs that have been studied in the literature concern second order differential operators; notable exceptions are the works [2,4,6,9]. In this paper we

I. Dimov, I. Faragó, and L. Vulkov (Eds.): NAA 2012, LNCS 8236, pp. 532–539, 2013.

consider fourth order elliptic boundary value problems (BVP) with two different types of boundary conditions: one that is suitable for writing the BVP as a second order system and one that is not. The model problem we study here is a simplified version of the well-known Orr-Sommerfeld equation from hydrodynamics (cf. [2]), and reads: Find $u(x)$ such that

$$\varepsilon^2 u^{(4)}(x) - \alpha(x)u''(x) + \beta(x)u(x) = f(x) \text{ in } I = (0,1), \tag{1}$$

where $\alpha \geq 0, \beta \geq 0$ and f are given (sufficiently smooth) functions and $\varepsilon \in (0,1]$ is a given parameter (that can approach 0). The notation u'' means the second derivative of u with respect to x and $u^{(n)}, n \in \mathbb{N}$, denotes the nth derivative of u with respect to x. Equation (1) is supplemented with one of the following two types of boundary conditions (which for simplicity are chosen as homogeneous):

$$u(0) = u'(0) = u(1) = u'(1) = 0, \tag{2}$$

or

$$u(0) = u''(0) = u(1) = u''(1) = 0. \tag{3}$$

As we will see in the next section, the BVP (1), (3) can be written as a second order system and its approximation will follow the work of [3]. For the BVP (1), (2) we will construct a C^1 approximation from a finite dimensional subspace of H^2, using a hierarchical basis from [7]. The results presented here summarize the analysis found in [3] and [5].

Throughout the paper we will utilize the notation $H^k(I)$ to mean the usual Sobolev space of functions defined on I, whose $0, 1, ..., k$ (generalized) derivatives belong to $L^2(I)$, with associated norm and seminorm $\|\cdot\|_{k,I}, |\cdot|_{k,I}$, respectively; the Lebesgue spaces $L^p(I), 1 \leq p \leq \infty$, are defined in the usual way and $\langle \cdot, \cdot \rangle_I$ denotes the usual $L^2(I)$ inner-product. We will also use the space

$$H_0^1(I) = \{u \in H^1(I) : u|_{\partial I} = 0\},$$

where ∂I denotes the boundary of I. Finally, the letters c, C (with or without subscripts) will denote generic positive constants independent of the solution and any discretization parameters.

2   Second Order Systems

We first consider the problem (1), (3), which can be written as the following second order system: Find $\mathbf{U}(x) = [u_1(x), u_2(x)]^T (= [u''(x), u(x)]^T)$ such that

$$-\mathbf{E}\mathbf{U}''(x) + \mathbf{A}(x)\mathbf{U}(\mathbf{x}) = \mathbf{F}(x) \text{ in } I, \ \mathbf{U}(a) = \mathbf{U}(b) = \mathbf{0}, \tag{4}$$

where

$$\mathbf{E} = \begin{bmatrix} \varepsilon^2 & 0 \\ 0 & 1 \end{bmatrix}, \mathbf{A}(x) = \begin{bmatrix} \alpha(x) & -\beta(x) \\ 1 & 0 \end{bmatrix}, \mathbf{F}(x) = \begin{bmatrix} -f(x) \\ 0 \end{bmatrix}. \tag{5}$$

For the remainder of the paper, we assume that the data of our problem are analytic and that there exist positive constants $C_\alpha, C_\beta, C_f, \gamma_\alpha, \gamma_\beta, \gamma_f$ independent of ε such that $\forall\, n = 0, 1, 2, \ldots$

$$\left\|\alpha^{(n)}\right\|_{L^\infty(I)} \le C_\alpha \gamma_\alpha^n n!\,, \quad \left\|\beta^{(n)}\right\|_{L^\infty(I)} \le C_\beta \gamma_\beta^n n!\,, \quad \left\|f^{(n)}\right\|_{L^\infty(I)} \le C_f \gamma_f^n n!\,. \quad (6)$$

Moreover, we assume that the matrix valued function $\mathbf{A}(x)$ in (5) is pointwise positive definite (but not necessarily symmetric), i.e. for some fixed $\delta > 0$

$$\boldsymbol{\xi}^T \mathbf{A} \boldsymbol{\xi} \ge \delta^2 \boldsymbol{\xi}^T \boldsymbol{\xi} \quad \forall\, \boldsymbol{\xi} \in \mathbb{R}^2\,, \quad \forall\, x \in \bar{I}. \quad (7)$$

The variational formulation of (4) reads: Find $\mathbf{U} := (u_1, u_2) \in \left[H_0^1(I)\right]^2$ such that

$$\mathcal{B}(\mathbf{U}, \mathbf{V}) = \mathcal{F}(\mathbf{V}) \quad \forall\, \mathbf{V} = (v_1, v_2) \in \left[H_0^1(I)\right]^2\,, \quad (8)$$

where

$$\mathcal{B}(\mathbf{U}, \mathbf{V}) = \varepsilon^2 \langle u_1', v_1'\rangle_I + \langle u_2', v_2'\rangle_I + \langle \alpha u_1 - \beta u_2, v_1\rangle_I + \langle u_1, v_2\rangle_I\,, \quad (9)$$
$$\mathcal{F}(\mathbf{V}) = \langle -f, v_2\rangle_I\,. \quad (10)$$

It follows that the bilinear form $\mathcal{B}(\cdot, \cdot)$ given by (9), is coercive with respect to the *energy norm*

$$\|\mathbf{U}\|_{E,I}^2 \equiv \|(u_1, u_2)\|_{E,I}^2 := \varepsilon^2 |u_1|_{1,I}^2 + |u_2|_{1,I}^2 + \delta^2 \left(\|u_1\|_{0,I}^2 + \|u_2\|_{0,I}^2\right), \quad (11)$$

i.e.,

$$\mathcal{B}(\mathbf{V}, \mathbf{V}) \ge \|\mathbf{V}\|_{E,I}^2 \quad \forall\, \mathbf{V} \in \left[H_0^1(I)\right]^2\,. \quad (12)$$

This fact, along with the continuity of $\mathcal{B}(\cdot, \cdot)$ and $\mathcal{F}(\cdot)$ imply the unique solvability of (8).

The discrete version of (8) reads: Find $\mathbf{U}^N := (u_1^N, u_2^N) \in [S_N]^2 \subset \left[H_0^1(I)\right]^2$ such that

$$\mathcal{B}(\mathbf{U}^N, \mathbf{V}) = \mathcal{F}(\mathbf{V}) \quad \forall\, \mathbf{V} = (v_1, v_2) \in [S_N]^2\,, \quad (13)$$

where S_N is a finite dimensional subspace of $H_0^1(I)$, to be defined shortly. The unique solvability of (13) follows from (7), and by the well-known Galerkin orthogonality, we have

$$\left\|\mathbf{U} - \mathbf{U}^N\right\|_{E,I}^2 \le C \inf_{\mathbf{V} \in [S_N]^2} \|\mathbf{U} - \mathbf{V}\|_{E,I}^2 \quad \forall\, \mathbf{V} \in [S_N]^2\,. \quad (14)$$

Before we define the space S_N we comment on the regularity of the solution to (8) and in particular we quote a relevant result from [3] that shows that the solution can be decomposed into an outer (smooth) part, a boundary layer part and a remainder that is exponentially (in ε) small.

Theorem 1. *[3] Assume (6) and (7) hold. Then there exist constants $C, \gamma, q, \nu > 0$ independent of $\varepsilon \in (0,1]$ such that the following assertions are true for the solution of (8):*

(I) $\left\| \mathbf{U}^{(n)} \right\|_{L^\infty(I)} \leq C \varepsilon^{-1/2} \gamma^n \max\{n, \varepsilon^{-1}\}^n \ \forall \ n = 0, 1, 2, \dots$

(II) $\mathbf{U}$ *can be written as* $\mathbf{U} = \mathbf{W} + \mathbf{U}_{BL} + \mathbf{R}$, *with*

$$\left\| \mathbf{W}^{(n)} \right\|_{L^\infty(I)} \leq C\gamma^n n^n \ \forall \ n = 0, 1, 2, \dots,$$

$$\left| \mathbf{U}_{BL}^{(n)}(x) \right| \leq C\gamma^n (\nu\varepsilon)^{-n} e^{-dist(x, \partial I)/\nu\varepsilon} \ \forall \ n = 0, 1, 2, \dots,$$

$$\|\mathbf{R}\|_{L^\infty(\partial I)} + \|\mathbf{R}\|_{E,I} \leq C e^{-q/\varepsilon}.$$

Moreover, the second component u_2^{BL} of $\mathbf{U}_{BL}$, satisfies the stronger estimate

$$\left| (u_2^{BL})^{(n)}(x) \right| \leq C\gamma^n \varepsilon^2 (\nu\varepsilon)^{-n} e^{-dist(x, \partial I)/\nu\varepsilon} \ \forall n = 0, 1, 2, \dots$$

Practically speaking, the above theorem states that for ε relatively large, the solution of (8) is analytic if the data α, β, f are analytic. If, on the other hand ε is small, then the solution may be decomposed into the three aforementioned parts and estimates on the derivatives of each part are given. This information is the key ingredient for the proof of (exponential) convergence of the proposed method.

We now define the space S_N: Let $\Delta = \{0 = x_0 < x_1 < \dots < x_M = 1\}$ be an arbitrary partition of $I = (0,1)$ and set $I_j = (x_{j-1}, x_j), h_j = x_j - x_{j-1}, j = 1, \dots, M$. We also define the master (or standard) element $I_{ST} = (-1, 1)$ and note that it can be mapped onto the jth element I_j by the linear mapping

$$x = Q_j(t) = \frac{1}{2}(1 - t)x_{j-1} + \frac{1}{2}(1 + t)x_j.$$

With $\Pi_p(I_{ST})$ denoting the space of polynomials of degree $\leq p$ on I_{ST}, we define the (finite dimensional) subspace

$$S^P(\Delta) = \left\{ \mathbf{V} \in \left[H_0^1(I)\right]^2 : \mathbf{V} \circ Q_j^{-1} \in \left(\Pi_{p_j}(I_{ST})\right)^2, j = 1, \dots, M \right\},$$

where $\circ$ denotes composition of functions. Then, we set

$$S_N \equiv S_0^p(\Delta) = S^P(\Delta) \cap \left[H_0^1(I)\right]^2. \tag{15}$$

We restrict our attention here to constant polynomial degree p for all elements, i.e. $p_j = p \ \forall \ j$, but clearly more general settings with variable polynomial degrees are possible. The following *Spectral Boundary Layer mesh* is essentially the minimal mesh that yields exponential convergence.

Definition 1. *(Spectral Boundary Layer mesh) For $\kappa > 0, p \in \mathbb{N}$ and $0 < \varepsilon \leq 1$, define the spaces $S(\kappa, p)$ of piecewise polynomials by*

$$S(\kappa, p) := \begin{cases} S_0^p(\Delta); \Delta = \{0, 1\} & \text{if } \kappa p \varepsilon \geq \frac{1}{2} \\ S_0^p(\Delta); \Delta = \{0, \kappa p \varepsilon, 1 - \kappa p \varepsilon, 1\} & \text{if } \kappa p \varepsilon < \frac{1}{2} \end{cases}$$

The parameter κ is user specific and depends on the problem under consideration as well as the length scales of the boundary layers – we refer to [8] for a more
detailed discussion of this issue and we note that in practice the value $\kappa = 1$
yields satisfactory results for most problems. We also note that the method we
are considering is not a true hp FEM since the location and not the number of
the elements changes; a more correct characterization would be a p version FEM
on a moving mesh, but in order to be consistent with the bibliography we utilize
the term hp FEM for our method. Obviously, additional refinement and/or using
a true hp version would yield better results but at the cost of using more degrees
of freedom – see [11] for a numerical comparison.

The main result is the following:

Theorem 2. *[3] Let* $\mathbf{U}$ *be the solution to (8) and let* $\mathbf{U}^N \in S(\kappa, p)$ *be the
solution of (13) with* $S(\kappa, p)$ *given by Definition 1. Then there exist constants*
$\kappa_0, C, \sigma > 0$ *depending only on the data* α, β, f, *such that for any* $0 < \kappa \leq \kappa_0$

$$\left\| \mathbf{U} - \mathbf{U}^N \right\|_{E,I} \leq C e^{-\sigma \kappa p}.$$

3   A C^1 Approximation

In this section we consider the BVP (1), (2) whose variational formulation reads:
Find $u \in H_0^2(I) := \left\{ u \in H^2(I) : u(0) = u'(0) = u(1) = u'(1) = 0 \right\}$, such that

$$\overline{\mathcal{B}}(u, v) = \overline{\mathcal{F}}(v) \ \forall \ v \in H_0^2(I), \tag{16}$$

where

$$\overline{\mathcal{B}}(u, v) = \int_0^1 \left\{ \varepsilon^2 u''(x) v''(x) + \alpha(x) u'(x) v'(x) + \beta(x) u(x) v(x) \right\} dx, \tag{17}$$

$$\overline{\mathcal{F}}(v) = \int_0^1 f(x) v(x) dx. \tag{18}$$

We continue to assume analyticity of the input data, i.e. (6), and coercivity of
the bilinear form $\overline{\mathcal{B}}(\cdot, \cdot)$ holds in the *energy norm*

$$|||u|||_{E,I}^2 := \overline{\mathcal{B}}(u, u). \tag{19}$$

Existence and uniqueness follow from the Lax-Milgram lemma as usual. The
discrete version of (16) reads: Find $u_N \in \overline{S}_N \subset H_0^2(I)$ such that

$$\overline{\mathcal{B}}(u_N, v) = \overline{\mathcal{F}}(v) \ \forall \ v \in \overline{S}_N, \tag{20}$$

and we have

$$|||u - u_N|||_{E,I} \leq C \inf_{v \in \overline{S}_N} |||u - v|||_{E,I} \ \forall \ v \in \overline{S}_N.$$

In order to define the space $\overline{S}_N$, we introduce, for $t \in I_{ST}$, the four *nodal* basis functions (cf. [7])

$$N_1(t) = \frac{1}{4}(1-t)^2(2+t), \; N_2(t) = \frac{1}{4}(1+t)^2(2-t),$$

$$N_3(t) = \frac{1}{4}(1-t)(2+t)^2, \; N_4(t) = \frac{1}{4}(1+t)(2-t)^2,$$

as well as the $p-3$ *internal* basis functions

$$N_i(t) = \frac{1}{\sqrt{2(i-5)}}\left(\frac{1}{2i-7}P_{i-5}(t) - \frac{2(i-5)}{(2i-7)(2i-3)}P_{i-3}(t) + \frac{1}{2i-3}P_{i-1}(t)\right),$$

for $i \geq 5$, where $P_i(t)$ is the Legendre polynomial of degree i. The (first two) nodal basis functions are equal to 1 at one endpoint of I_{ST} and 0 at the other. The internal basis functions are 0 at both endpoints of I_{ST}. There holds

$$\Pi_p(I_{ST}) = \text{span}\{N_1(t), ..., N_{p+1}(t)\},$$

and, with Δ, Q_j as in the previous section, we define the space

$$S^p(\Delta) := \{w \in H_0^2(I) : w \circ Q_j^{-1} \in \Pi_{p_j}(I_{ST}), j = 1, ..., m\}.$$

Hence, the subspace $\overline{S}_N \subset H_0^2(I)$ is chosen as

$$\overline{S}_N \equiv S_0^p(\Delta) = S^p(\Delta) \cap H_0^2(I). \tag{21}$$

The following proposition gives bounds on the nth derivative of the solution.

Proposition 1. *[5] Assume (6) and (7) hold. Then there exist constants C, γ, ν, $q > 0$ independent of $\varepsilon \in (0,1]$ such that the following assertions are true for the solution of (16):*

(I) $\left\|u^{(n)}\right\|_{L^\infty(I)} \leq C\gamma^n \max\{n^n, \varepsilon^{1-n}\} \; \forall \, n = 0, 1, 2, ...$

(II) u can be written as $u = w + u_{BL} + r$, with

$$\left\|w^{(n)}\right\|_{L^\infty(I)} \leq C\gamma^n n^n \; \forall \, n = 0, 1, 2, ...,$$

$$\left|u_{BL}^{(n)}(x)\right| \leq C\gamma^n \varepsilon^2 (\nu\varepsilon)^{-n} e^{-dist(x,\partial I)/\nu\varepsilon} \; \forall \, n = 0, 1, 2, ...,$$

$$\|r\|_{L^\infty(\partial I)} + \||r\||_{E,I} \leq Ce^{-q/\varepsilon}.$$

Using the above result we can prove the following.

Proposition 2. *[5] Let u be the solution to (16) and let $u_N \in S(\kappa, p)$ be the solution of (20) with $S(\kappa, p)$ given by Definition 1. Then there exist constants $\kappa_0, C, \sigma > 0$ depending only on the data α, β, f, such that for any $0 < \kappa \leq \kappa_0$*

$$\||u - u_N\||_{E,I} \leq Ce^{-\sigma\kappa p}.$$

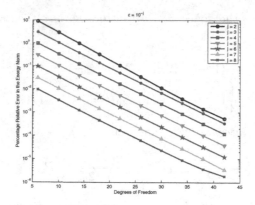

Fig. 1. Energy norm convergence for the *hp* version

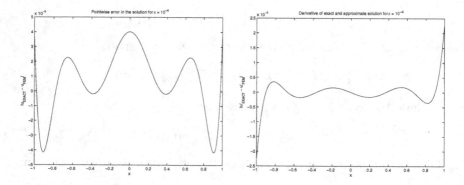

Fig. 2. Pointwise error in the approximation (left) and in the derivatives (right), for $\varepsilon = 10^{-4}$

4   Numerical Results

In this section we present the results of numerical computations for the problem studied in the previous section; ample numerical results pertaining to the problem studied in Section 2 may be found in [3] and [11]. We consider the BVP (1), (2) when the data is $\alpha = \beta = f = 1$; an exact solution is available, hence our reported results are reliable. Figure 1 shows the percentage relative error in the energy norm,

$$Error := 100 \times \frac{|||u - u_N|||_{E,I}}{|||u|||_{E,I}}, \qquad (22)$$

versus the number of degrees of freedom, N, in a semi-log scale. As the figure shows, the error curves are straight, something that verifies the exponential convergence of the proposed method. Moreover, as $\varepsilon \to 0$ the method not only does not deteriorate, but it actually performs better. This suggests that in the error estimate of Proposition 2, there is a positive power of ε present, something that is due to the fact that the problem has constant coefficients and the right hand side belongs to the subspace (see [8] for more details on this for second order scalar problems).

Figure 2 shows the pointwise error between the exact and *hp* approximation (computed with $p = 8$) as well as their derivatives, for $\varepsilon = 10^{-4}$. The high accuracy of the computed solution is readily visible from these figures.

References

1. Bakhvalov, N.S.: Towards optimization of methods for solving boundary value problems in the presence of boundary layers. Zh. Vychisl. Mat. Mat. Fiz. 9, 841–859 (1969) (in Russian)
2. Roos, H.-G., Stynes, M.: A uniformly convergent discretization method for a fourth order singular perturbation problem. Bonn. Math. Schr. 228, 30–40 (1991)
3. Melenk, M.J., Xenophontos, C., Oberbroeckling, L.: Robust exponential convergence of *hp*-FEM for singularly perturbed systems of reaction-diffusion equations with multiple scales. Ima. J. Num. Anal., 1–20 (2012), doi:10.1093/imanum/drs013
4. Pan, Z.-X.: The difference and asymptotic methods for a fourth order equation with a small parameter. In: BAIL IV Proceedings, pp. 392–397. Book Press, Dublin (1986)
5. Panaseti, P., Zouvani, A., Madden, N., Xenophontos, C.: A C^1–conforming *hp* Finite Element Method for fourth order singularly perturbed boundary value problems. Submitted to App. Numer. Math. (2013)
6. Roos, H.-G.: A uniformly convergent discretization method for a singularly perturbed boundary value problem of the fourth order. In: Review of Research, Faculty of Science, Mathematics Series, Univ. Novi Sad, vol. 19, pp. 51–64 (1989)
7. Schwab, C.: *p*- and *hp*-Finite Element Methods. Oxford Science Publications (1998)
8. Schwab, C., Suri, M.: The *p* and *hp* versions of the finite element method for problems with boundary layers. Math. Comp. 65, 1403–1429 (1996)
9. Shishkin, G.I.: A difference scheme for an ordinary differential equation of the fourth order with a small parameter at the highest derivative. Differential Equations 21, 1743–1742 (1985) (in Russian)
10. Shishkin, G.I.: Grid approximation of singularly perturbed boundary value problems with a regular boundary layer. Sov. J. Numer. Anal. Math. Model. 4, 397–417 (1989)
11. Xenophontos, C., Oberbroeckling, L.: A numerical study on the finite element solution of singularly perturbed systems of reaction-diffusion problems. Appl. Meth. Comp. 187, 1351–1367 (2007)

Quadrature Formula with Five Nodes for Functions with a Boundary Layer Component

Alexander Zadorin and Nikita Zadorin

Omsk Filial of Sobolev Mathematics Institute SB RAS,
Pevtsova 13, Omsk, 644099, Russia
zadorin@ofim.oscsbras.ru

Abstract. Quadrature formula for one variable functions with a boundary layer component is constructed and studied. It is assumed that the integrand can be represented as a sum of regular and boundary layer components. The boundary layer component has high gradients, therefore an application of Newton-Cotes quadrature formulas leads to large errors. An analogue of Newton-Cotes rule with five nodes is constructed. The error of the constructed formula does not depend on gradients of the boundary layer component. Results of numerical experiments are presented.

Keywords: function, numerical integration, boundary layer component, nonpolynomial interpolation, quadrature rule, uniform accuracy.

1 Introduction

The construction of Newton-Cotes formulas is based on the approximation of the integrand by a Lagrange polynomial. It is well known that the error of such composite quadrature formulas can be large for high-gradient functions. The construction of quadrature formulas for functions with large gradients was carried out in [1, 2] and in works of many other authors. Below, it is assumed that the integrand can be represented as a sum of a regular component having bounded derivatives up to some order and a given boundary layer component with large gradients in certain parts of the integration interval. In this article a quadrature formula with five nodes, exact for the boundary layer component, is constructed. Before we constructed similar formulas with two and three nodes [3]. We construct the quadrature formula for the integral

$$I(u) = \int_a^b u(x)\,dx, \tag{1.1}$$

where $u(x)$ can be represented as

$$u(x) = p(x) + \gamma\Phi(x),\ x \in [a, b]. \tag{1.2}$$

We assume that $u, \Phi \in C^6[a, b]$, the boundary layer component $\Phi(x)$ is a given bounded function with high gradient areas, the regular component $p(x)$ is

I. Dimov, I. Faragó, and L. Vulkov (Eds.): NAA 2012, LNCS 8236, pp. 540–546, 2013.
© Springer-Verlag Berlin Heidelberg 2013

bounded together with some derivatives, and γ is an unknown constant. The representation (1.2) holds for a solution of a singularly perturbed boundary value problem [4]. Without loss of generality we suppose that

$$\Phi^{(4)}(x) > 0, \ x \in [a, b]. \tag{1.3}$$

2   Construction of the Quadrature Formula

Let Ω be an uniform mesh of the interval $[a, b]$ with nodes $\{x_n\}, 0 \le n \le N$ and with a step h, $h = (b - a)/N$. We suppose that $N/4$ is integer, $u_n = u(x_n)$, $n = 0, 1, \ldots, N$.

We construct the quadrature formula for any interval $[x_{n-2}, x_{n+2}]$, where $n = 2, 6, \ldots, N - 2$. To calculate the integral

$$I_n(u) = \int_{x_{n-2}}^{x_{n+2}} u(x) \, dx \tag{2.1}$$

we consider Newton - Cotes formula with five nodes:

$$S_n(u) = \frac{2}{45} h \left[7u_{n-2} + 32u_{n-1} + 12u_n + 32u_{n+1} + 7u_{n+2} \right]. \tag{2.2}$$

Using (2.2), we write the composite quadrature formula for the integral (1.1)

$$S(u) = \frac{2}{45} h \sum_{n=2,4}^{N-2} \left(7u_{n-2} + 32u_{n-1} + 12u_n + 32u_{n+1} + 7u_{n+2} \right). \tag{2.3}$$

The error of the formula (2.3) is estimated in [2] as

$$|I(u) - S(u)| \le \frac{2}{945} (b - a) \max_{x \in [a,b]} |u^{(6)}(x)| h^6. \tag{2.4}$$

According to (2.4), if the derivative $u^{(6)}(x)$ is uniformly bounded, then the error of the formula (2.3) is the quantity of the order $O(h^6)$. If the derivative of the integrand has large values, then the accuracy of the formula (2.3) can degrade. Let $u(x) = \exp(-\varepsilon^{-1}x)$. We can easy verify that the error of the formula (2.3) is a quantity of the order $O(h)$ if $\varepsilon \le h$.

We modify the formula (2.2) so that the error does not depend on $\Phi(x)$. In [5–8] we constructed spline interpolation formulas exact for $\Phi(x)$. We found an application of such interpolation formulas in the construction of the two-grid method for a singularly perturbed problem [9]. Now we apply such interpolation formula to the construction of the quadrature formula.

In the interval $[x_{n-2}, x_{n+2}]$ we construct the interpolant

$$u_\Phi(x) = u_n + \beta_1(x - x_n) + \beta_2(x - x_n)^2 + \beta_3(x - x_n)^3 + G\Big[\Phi(x) - \Phi_n\Big] \tag{2.5}$$

for the function $u(x)$ with interpolation conditions at nodes $\{x_n, x_{n\pm1}, x_{n\pm2}\}$. We use the interpolant (2.5) in (2.1) and obtain the quadrature formula

$$S_{\Phi,n}(u) = \int_{x_{n-2}}^{x_{n+2}} u_\Phi(x)\,dx.$$

Integrating, we obtain

$$S_{\Phi,n}(u) = \frac{4h}{3}(2u_{n+1} - u_n + 2u_{n-1}) + G\left[\int_{x_{n-2}}^{x_{n+2}} \Phi(x)\,dx - \frac{4h}{3}(2\Phi_{n+1} - \Phi_n + 2\Phi_{n-1})\right],$$

(2.6)

where

$$G = \frac{u_{n+2} - 4u_{n+1} + 6u_n - 4u_{n-1} + u_{n-2}}{\Phi_{n+2} - 4\Phi_{n+1} + 6\Phi_n - 4\Phi_{n-1} + \Phi_{n-2}}.$$

(2.7)

The formula (2.6) is exact for the function $\Phi(x)$.

For a function $f \in C^4[x_{n-2}, x_{n+2}]$ the next relation is known

$$f_{n+2} - 4f_{n+1} + 6f_n - 4f_{n-1} + f_{n-2} = h^4 f^{(4)}(s_1), \quad \exists s_1 \in (x_{n-2}, x_{n+2}). \quad (2.8)$$

Using (1.3) and (2.8), we obtain that the expression (2.7) is correct.

The formula (2.6) contains Milne's rule for the integral (2.1):

$$I_n(f) \approx \frac{4h}{3}\left[2f_{n-1} - f_n + 2f_{n+1}\right].$$

It is known [10] that if $f \in C^4[a, b]$, then there is $s_2 \in (x_{n-2}, x_{n+2})$ such that

$$\int_{x_{n-2}}^{x_{n+2}} f(x)\,dx - \frac{4h}{3}\left[2f_{n-1} - f_n + 2f_{n+1}\right] = \frac{14}{45}h^5 f^{(4)}(s_2). \quad (2.9)$$

Lemma 1. *The following estimation is true*

$$|I_n(u) - S_{\Phi,n}(u)| \le \frac{44}{45}\max_x |p^{(4)}(x)|h^5, \quad x \in [x_{n-2}, x_{n+2}]. \quad (2.10)$$

Proof. Write (2.6) in a form:

$$S_{\Phi,n}(u) = \frac{4h}{3}(2u_{n+1} - u_n + 2u_{n-1}) + 4M_n h(u_{n+2} - 4u_{n+1} + 6u_n - 4u_{n-1} + u_{n-2}),$$

(2.11)

where

$$M_n = \frac{\int_{x_{n-2}}^{x_{n+2}} \Phi(x)\,dx - \frac{4}{3}h(2\Phi_{n+1} - \Phi_n + 2\Phi_{n-1})}{4h(\Phi_{n+2} - 4\Phi_{n+1} + 6\Phi_n - 4\Phi_{n-1} + \Phi_{n-2})}.$$

(2.12)

Now we prove that

$$0 < M_n < \frac{1}{6}. \quad (2.13)$$

Using relations (1.3), (2.9), we obtain that the numerator in the expression (2.12) is positive. The denominator in this expression is positive according to (1.3) and (2.8). Thus, $M_n > 0$. The condition $M_n < 1/6$ is equivalent to the inequality

$$\int\limits_{x_{n-2}}^{x_{n+2}} \Phi(x)\,dx < \frac{2h}{3}\left(\Phi_{n+2} + 4\Phi_n + \Phi_{n-2}\right).$$

This inequality is true because the next relation for Simpson rule is correct

$$\frac{2h}{3}\left(\Phi_{n+2} + 4\Phi_n + \Phi_{n-2}\right) = \int\limits_{x_{n-2}}^{x_{n+2}} \Phi(x)\,dx + \frac{1}{90}(2h)^5\Phi^{(4)}(s_3),\ \exists s_3 \in (x_{n-2}, x_{n+2}),$$

where $\Phi^{(4)}(s_3) > 0$ according to (1.3).

We proved (2.13). Now we take into account that the formula (2.11) is exact for $\Phi(x)$ and obtain

$$|I_n(u) - S_{\Phi,n}(u)| = |I_n(p) - S_{\Phi,n}(p)| \le \left|\frac{4}{3}h(2p_{n+1} - p_n + 2p_{n-1}) - \int\limits_{x_{n-2}}^{x_{n+2}} p(x)\,dx\right| +$$

$$+ 4M_n h\left|p_{n+2} - 4p_{n+1} + 6p_n - 4p_{n-1} + p_{n-2}\right|. \tag{2.14}$$

Using (2.8), (2.9), (2.13), we obtain from (2.14)

$$|I_n(u) - S_{\Phi,n}(u)| \le \left[\frac{14}{45} + \frac{4}{6}\right]\max_x |p^{(4)}(x)|h^5,\quad x \in [x_{n-2}, x_{n+2}].$$

Hence, the inequality (2.10) is true. The lemma is proved.   ◇

According to Lemma 1, the error of the constructed quadrature formula (2.6) does not depend on the boundary layer component.

Using (2.11), we write the composite quadrature formula for the integral (1.1)

$$S_\Phi(u) = 4h \sum_{n=2,4}^{N-2} \left[M_n u_{n-2} + \left(\frac{2}{3} - 4M_n\right)u_{n-1} + \left(6M_n - \frac{1}{3}\right)u_n + \right.$$

$$\left. + \left(\frac{2}{3} - 4M_n\right)u_{n+1} + M_n u_{n+2}\right]. \tag{2.15}$$

Using the estimate (2.10), we obtain

$$|I(u) - S_\Phi(u)| \le \frac{11}{45}(b-a)\max_{x\in[a,b]}|p^{(4)}(x)|h^4. \tag{2.16}$$

According to (2.16), the composite quadrature formula (2.15) has the error of the order $O(h^4)$ uniformly in a boundary layer component $\Phi(x)$.

We can obtain Newton-Cotes formula (2.3) from the constructed formula (2.15), setting $M_n = 7/90$.

If we know the location of a boundary layer, we can improve the accuracy of the composite quadrature formula. We offer to use the formula (2.11) in a boundary layer and the classical formula (2.2) out of the boundary layer.

To be definite, we assume that the boundary layer is located near the left endpoint of the interval $[a, b]$ and $|u^{(6)}(x)| \leq C_0$ for all $x \geq a + \sigma$. Let

$$m = \min_{n}\{n : x_{n-2} \geq a + \sigma, \quad (n - 2)/4 \text{ is integer }\}.$$

Consider the composite quadrature formula:

$$\tilde{S}_{\Phi}(u) = 4h \sum_{n=2,4}^{m-4} \left[M_n u_{n-2} + \left(\frac{2}{3} - 4M_n\right) u_{n-1} + \left(6M_n - \frac{1}{3}\right) u_n + \left(\frac{2}{3} - 4M_n\right) u_{n+1} + \right.$$

$$\left. + M_n u_{n+2}\right] + \frac{2}{45}h \sum_{n=m,4}^{N-2} \left[7u_{n-2} + 32u_{n-1} + 12u_n + 32u_{n+1} + 7u_{n+2}\right]. \quad (2.17)$$

Taking into account the estimates of an accuracy of formulas (2.2) and (2.11), we obtain:

$$|I(u) - \tilde{S}_{\Phi}(u)| \leq \frac{11}{45}(m - 2) \max_{x} |p^{(4)}(x)|h^5 + C_0(N - m + 2)\frac{2}{945}h^7. \quad (2.18)$$

If $m \ll N$, then the error of the formula (2.17) is a quantity of the order $O(h^5)$.

As example, we consider an exponential boundary layer near the point $x = 0$ of the interval $[0, 1]$. Let $\Phi(x) = \exp(-\varepsilon^{-1}x)$. We suppose that $\Phi^{(6)}(\sigma) = 1$ and obtain $\sigma = -6\varepsilon \ln(\varepsilon)$, where $0 < \varepsilon < 1$.

3   Numerical Results

Consider the integral

$$I(u) = \int_0^1 u(x)\, dx, \quad u(x) = \cos(\pi x/2) + \exp(-\varepsilon^{-1}x).$$

Here $\Phi(x) = \exp(-\varepsilon^{-1}x)$, $\varepsilon \in (0, 1]$.

Define computed order of an accuracy for a quadrature rule. Let

$$CR_{N,\varepsilon} = log_2(E_{N,\varepsilon}/E_{2N,\varepsilon}),$$

where $E_{N,\varepsilon}$ is the error of a tested composite quadrature formula with N mesh intervals.

In tables $e \pm m$ means $10^{\pm m}$.

Table 1 presents the error $E_{N,\varepsilon}$ and computed order $CR_{N,\varepsilon}$ of the composite rule (2.3) for various ε and N. As ε decreases, computed order of an accuracy decreases from six to one. It corresponds to the estimate (2.4) and to the remark, following after (2.4), about the accuracy of the rule (2.3) for small values of ε.

Table 1. The errors and computed orders of Newton-Cotes formula (2.3)

ε	N					
	$3*2^3$	$3*2^4$	$3*2^5$	$3*2^6$	$3*2^7$	$3*2^8$
10^{-1}	$1.01e-6$	$1.69e-8$	$2.69e-10$	$4.22e-12$	$6.54e-14$	$7.77e-16$
	5.9	6.0	6.0	6.0	6.4	
10^{-2}	$3.89e-3$	$4.04e-4$	$1.66e-5$	$3.69e-7$	$6.37e-9$	$1.02e-10$
	3.3	4.6	5.5	5.9	6.0	
10^{-3}	$1.20e-2$	$5.48e-3$	$2.24e-3$	$6.61e-4$	$9.33e-5$	$5.06e-6$
	1.1	1.3	1.8	2.8	4.2	
10^{-4}	$1.29e-2$	$6.38e-3$	$3.14e-3$	$1.52e-3$	$7.10e-4$	$3.05e-4$
	1.0	1.0	1.0	1.1	1.2	
10^{-5}	$1.29e-2$	$6.47e-3$	$3.23e-3$	$1.61e-3$	$8.00e-4$	$3.95e-4$
	1.0	1.0	1.0	1.0	1.0	

Table 2. The errors and computed orders of the constructed formula (2.15)

ε	N					
	$3*2^3$	$3*2^4$	$3*2^5$	$3*2^6$	$3*2^7$	$3*2^8$
10^{-1}	$4.39e-9$	$6.87e-11$	$1.07e-12$	$1.69e-14$	$6.66e-16$	$5.55e-16$
	6.0	6.0	6.0	4.7	$-$	
10^{-2}	$2.91e-7$	$6.03e-9$	$1.02e-10$	$1.63e-12$	$2.56e-14$	$1.33e-15$
	5.6	5.9	6.0	6.0	4.3	
10^{-3}	$8.40e-7$	$4.81e-8$	$2.46e-9$	$9.24e-11$	$2.17e-12$	$3.80e-14$
	5.6	4.1	4.7	5.4	5.8	
10^{-4}	$9.03e-7$	$5.59e-8$	$3.44e-9$	$2.08e-10$	$1.22e-11$	$6.53e-13$
	4.0	4.0	4.0	4.1	4.2	
10^{-5}	$9.10e-7$	$5.67e-8$	$3.54e-9$	$2.20e-10$	$1.37e-11$	$8.45e-13$
	4.0	4.0	4.0	4.0	4.0	

Table 3. The errors and computed orders of the combined formula (2.17)

ε	N					
	$3*2^3$	$3*2^4$	$3*2^5$	$3*2^6$	$3*2^7$	$3*2^8$
10^{-1}	$2.57e-9$	$3.89e-11$	$3.90e-13$	$3.55e-15$	$4.44e-16$	$5.55e-16$
	6.0	6.6	6.8	3.0	$-$	
10^{-2}	$7.51e-8$	$1.54e-9$	$4.19e-11$	$1.06e-12$	$1.85e-14$	$3.33e-16$
	5.6	5.2	5.3	5.8	5.8	
10^{-3}	$2.18e-7$	$6.27e-9$	$1.61e-10$	$2.43e-12$	$3.12e-14$	$2.66e-15$
	5.1	5.3	6.1	6.3	3.6	
10^{-4}	$2.34e-7$	$7.30e-9$	$2.25e-10$	$6.81e-12$	$1.99e-13$	$5.88e-15$
	5.0	5.0	5.0	5.1	5.1	
10^{-5}	$2.36e-7$	$7.41e-9$	$2.31e-10$	$7.21e-12$	$2.24e-13$	$7.54e-15$
	5.0	5.0	5.0	5.0	4.9	

Table 2 presents the error and the computed order of the composite formula (2.15) for various ε and h. Results support the estimate (2.16). As ε decreases, the computed order of an accuracy decreases from six to four. In a case $N = 3*2^8$ the error of the quadrature rule has the order of calculative errors, it has the influence on the value of $CR_{N,\varepsilon}$.

Table 3 presents the error and the computed order of the combined composite formula (2.17) for various ε and h. The results confirm the estimate (2.18). As ε decreases, the computed order of an accuracy decreases from six to five.

4   Conclusion

The quadrature formula with five nodes for the numerical integration of a function with a boundary layer component is constructed. It is proved that the error of proposed formula does not depend on a boundary layer component. Numerical experiments confirmed that constructed formula is more accurate in compare with Newton - Cotes formula.

Acknowledgements. Supported by Russian Foundation for Basic Research under Grants 10-01-00726, 11-01-00875.

References

1. Berezin, I.S., Zhidkov, N.P.: Computing Methods. Nauka, Moskow (1966) (in Russian)
2. Bakhvalov, N.S.: Numerical Methods. Nauka, Moskow (1975) (in Russian)
3. Zadorin, A.I., Zadorin, N.A.: Quadrature formulas for functions with a boundary-layer component. Comput. Math. Math. Phys. 51(11), 1837–1846 (2011)
4. Miller, J.J.H., O'Riordan, E., Shishkin, G.I.: Fitted Numerical Methods for Singular Perturbation Problems. World Scientific, Singapore (1996)
5. Zadorin, A.I.: Method of interpolation for a boundary layer problem. Sib. J. of Numer. Math. 10(3), 267–275 (2007) (in Russian)
6. Zadorin, A.I.: Interpolation Method for a Function with a Singular Component. In: Margenov, S., Vulkov, L.G., Waśniewski, J. (eds.) NAA 2008. LNCS, vol. 5434, pp. 612–619. Springer, Heidelberg (2009)
7. Zadorin, A.I., Zadorin, N.A.: Spline interpolation on a uniform grid for functions with a boundary-layer component. Comput. Math. Math. Phys. 50(2), 211–223 (2010)
8. Zadorin, A.I.: Spline interpolation of functions with a boundary layer component. Int. J. Numer. Anal. Model., series B 2(2-3), 562–579 (2011)
9. Vulkov, L.G., Zadorin, A.I.,: Two-grid algorithms for an ordinary second order equation with exponential boundary layer in the solution. Int. J. Numer. Anal. Model. 7(3), 580–592 (2010)
10. Dahlquist, G., Bjorck, A.: Numerical Methods in Scientific Computing, vol. 1. SIAM, Philadelphia (2008)

Numerical Study of Travelling Multi-soliton Complexes in the Ac-Driven NLS Equation

Elena Zemlyanaya[1] and Nora Alexeeva[2]

[1] Laboratory of Information Technologies, Joint Institute for Nuclear Research
141980 Dubna, Moscow Region, Russia
elena@jinr.ru
[2] Department of Mathematics, University of Cape Town,
Rondebosch 7701, South Africa
nora.alexeeva@uct.ac.za

Abstract. Travelling solitons of the undamped, externally driven nonlinear Schrödinger equation (NLS) are investigated by the numerical solution of the reduced ordinary differential equation and the corresponding linearized eigenvalue problem. This numerical approach is complemented by direct numerical simulations of the partial differential NLS equation. We show that in the small driving case, travelling solitons can form stably travelling bound states.

Keywords: Nonlinear Schrödinger equation, travelling solitons, Newtonian iteration, numerical continuation, stability, bifurcations.

1 Introduction

We consider the nonlinear Schrödinger equation (NLS) driven by a constant external force

$$i\psi_t + \psi_{XX} + 2|\psi|^2\psi - \psi = -h - i\gamma\psi, \quad \psi_X(\pm\infty) = 0, \tag{1}$$

where $h > 0$ and $\gamma > 0$ are, respectively, dimensionless parameters of external driving and linear damping; t and X are. respectively. dimensionless time and space. The externally driven nonlinear Schrödinger equation has undergone an extensive mathematical analysis because of a wide range of physical applications (physical phenomena associated with Eq.(1) are overviewed in [4]).

Two types of stationary soliton solutions of Eq.(1), ψ_- and ψ_+, are well studied, see [8,2,7]. The soliton ψ_- is known to be stable for small driving strengths but unstable otherwise; the other one, ψ_+, unstable independently of the choice of the forcing amplitude.

Strongly damped stationary multi-soliton complexes are investigated in [4,2,9]. Existence of stationary complexes in the undamped case (denoted "twist") is proved in the recent paper [4]. Two of twist solutions (denoted $T2$ and $T3$) have been reproduced numerically.

Travelling undamped waves of Eq.(1) are obtained in [4] by reducing the partial differential equation (1) to an ordinary differential equation where the

I. Dimov, I. Faragó, and L. Vulkov (Eds.): NAA 2012, LNCS 8236, pp. 547–554, 2013.

fixed constant velocity V plays a role of additional parameter of equation. Using two stationary solitons ψ_- and ψ_+ and two stationary complexes $T2$ and $T3$ as starting points, classes of localized travelling waves were obtained by the numerical continuation in the parameter space. Two families of stable single solitons have been identified: one family is stable for sufficiently low velocities while solitons from the second family stabilize when travelling faster than a certain critical speed. It was also shown that travelling ψ_- solitons can form stably travelling bound states.

This contribution extends numerical results of [4] by providing additional branches of travelling complexes and results of direct numerical simulation of Eq.(1). Our numerical approach is described in Sect.2. Numerical results are presented in Sect.3 and summarized in Sect.4.

Here we consider the undamped case $\gamma = 0$.

2   Formulation of Problem and Numerical Approach

We search localized travelling waves of the form $\psi(X, t) = \psi(X - Vt)$ by the numerical solution of the ordinary differential equation

$$- iV\psi_x + \psi_{xx} + 2|\psi|^2\psi - \psi = -h, \quad , \psi_x(\pm L) = 0, \quad L \to \infty. \quad (2)$$

where $x = X - Vt$ and V is a constant velocity of the travelling soliton.

We pathfollow stationary single solitons and complexes in nonzero velocity and classify stability and bifurcations of travelling waves as the parameters h and V are varied.

For graphical representation of results we calculate, at each step of continuation process, the momentum integral

$$P = \frac{i}{2} \int (\psi_x^*\psi - \psi_x\psi^*)dx \quad (3)$$

that is an integral of motion for equation (1), i.e. P is a physically meaningful characteristic of solutions.

Stability and bifurcations are analyzed on the basis of numerical solution of the corresponding linearized eigenvalue problem [4] as follows:

$$\mathcal{H}y = \lambda Jy, \quad \mathcal{H} = \begin{pmatrix} -\partial_x^2 + 1 - 2(3\mathcal{R}^2 + \mathcal{I}^2) & -V\partial_x - 4\mathcal{R}\mathcal{I} \\ V\partial_x - 4\mathcal{R}\mathcal{I} & -\partial_x^2 + 1 - 2(3\mathcal{I}^2 + \mathcal{R}^2) \end{pmatrix}. \quad (4)$$

Here y is a two-component vector-function $y(x) = \begin{pmatrix} u \\ v \end{pmatrix}$; $\mathcal{R} + i\mathcal{I} = \psi_s(x)$ is a solution of Eq.(2); J is a constant skew-symmetric matrix $J = \begin{pmatrix} 0 & -1 \\ 1 & 0 \end{pmatrix}$.

We routinely evaluate the spectrum of eigenvalues of Eq.(4) as we continue localized solutions in V. If there is at least one eigenvalue λ with Re$\lambda > 0$, the solution is unstable.

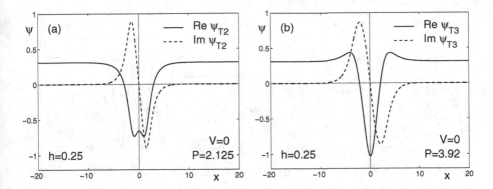

Fig. 1. Twist solutions T2 (a) and T3 (b) for $h = 0.25$, $V = 0$. The solid line shows the real and dashed imaginary part.

Localized solutions of Eq.(2) are investigated on the basis of the predictor-corrector algorithm with Newtonian iteration at each step of numerical continuation. We applied the 4th order accuracy Numerov's discretization. Details of numerical continuation technique are given in [9,10]. Typically, the calculations have been performed with $L = 500$ and the spatial stepsize $\Delta x = 0.005$.

For numerical solution of the eigenvalue problem (4) the Fourier discretization is applied. The resulting algebraic eigenvalue problem is solved by means of the standard EISPACK code.

For direct numerical simulation of Eq.(1) we employed the split-step method [1] with Fourier discretization in space. As an initial condition for the direct simulations we chosen the solution of Eq.(2) with the fixed value V.

3   Numerical Results

Scenarios of transformations of travelling solitons depend on the value of h. As representative cases, we consider three values of the driving parameter: $h = 0.05$, $h = 0.2$, and $h = 0.25$. We start with the case $h = 0.25$

3.1   The Case $h = 0.25$

The case $h = 0.25$ is the representative value of parameter region $h \geq 0.25$. As we have already mentioned, in addition to the simple solitons ψ_- and ψ_+ the equation (2) has two localized solutions in a form of multi-soliton bound states (denoted "twist" in [4]). Two twist solutions $T2$ and $T3$ are shown in the Fig.1 for $h = 0.25$.

Result of numerical continuation of $T2$ and $T3$ in nonzero velocity is shown on the Fig.2(a).

The continuation of the $T2$ twist (shown in Fig.1(a)) to the negative velocities proceeds according the following scenario: the twist transforms into a complex of two solitons ψ_-. At some negative V the curve turns back and connects to

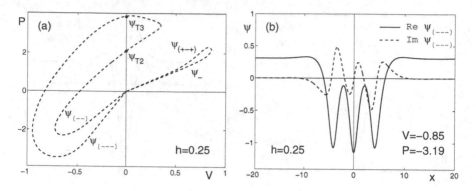

Fig. 2. (a) The $P(V)$ curve resulting from the continuation of the $T2$, $T3$, and ψ_- solutions for $h = 0.25$. All branches shown in this figure are unstable. More solution branches can be obtained by the reflection $V \to -V$, $P \to -P$. (b) A $\psi_{(---)}$ solution on the lower branch in (a). Here $h = 0.25$, $V = -0.85$, $P = -3.19$. The solid line shows the real and dashed imaginary part.

the origin on the (V, P) plane, with the distance between the two solitons bound in the complex increasing to infinity.

The continuation of $T2$ to positive V produces the following outcome. the curve $P(V)$ turns counterclockwise and crosses through the P-axis once again. The solution arising at the point $V = 0$ is nothing but the $T3$ twist shown in Fig.2(b). The subsequent continuation produces a hook-shaped curve which leads to the origin on the (V, P)-plane. The corresponding solution is a complex of three ψ_- solitons. Representative solution of the $\psi_{(---)}$ branch is shown in Fig.2(b). As $V, P \to 0$, the distance between the solitons grows to infinity.

The other like-loop branch shown in Fig.2(a) is associated with ψ_- soliton. As we continue the stationary ψ_- solution to nonzero velocity the $P(V)$ curve turns up and returns to the origin of the (V, P)-plane while solution gradually transforms into a three-soliton $\psi_{(+-+)}$. The separation between solitons in the complex is growing without limit as $V \to 0$.

All branches on the Fig.2 are unstable.

3.2   The Case $h = 0.2$

Result of continuation of complexes $T2$ and $T3$ in nonzero velocity for $h = 0.2$ is given in Fig.3a.

In the case $0.6 \leq h \leq 0.2$ the branch of travelling twist-solution $T2$ is connected with the branch of travelling single wave ψ_-. As we continue the $T2$ complex in negative velocity, the solution transforms to a complex of two well-separated ψ_- solitons. The corresponding curve makes a turn and eventually returns to the point $V = P = 0$. As $V \to 0$, the distance between the solitons in the complex tends to infinity.

The $P(V)$ curve associated with the $T3$ twist, transforms to three-soliton solution $\psi_{(+-+)}$ as we continue in positive velocity. Two-soliton complex $\psi_{(-+)}$ arises as the pitchfork bifurcation of the $\psi_{(+-+)}$ branch. When pathfollowing

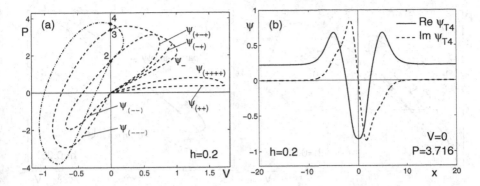

Fig. 3. (a) The full $P(V)$ bifurcation diagram for the ψ_-, $T2$, $T3$, and $T4$ solitons for $h = 0.2$. Solid points numbered 2,3,4 indicate the twist complexes $T2$, $T3$, $T4$. Also shown is the continuation of the $\psi_{(++)}$ branch. All branches shown in this figure are unstable. More solution branches can be obtained by the reflection $V \to -V$, $P \to -P$. (b) The twist solution $T4$ for $h = 0.2$, $V = 0$. The solid line shows the real and dashed imaginary part.

in negative velocity the $T3$-twist gradually transforms to a complex of three well-separated ψ_- solitons.

Besides the twist solutions $T2$ and $T3$, one more complex has been obtained numerically at $V = 0$. This solution (denoted $T4$) is shown in fig.3(b). The branch $P(V)$ emanating from the $T4$ twist is plotted in Fig.3(a) by dash-pointed line. When pathfollowing to $V < 0$ the $T4$ twist gradually transforms to the complex of two well-separated solitons ψ_+ and $T2$-soliton between them. As we continue the $T4$ twist in $V > 0$ it transforms to the three-soliton complex with $T2$ in the middle and two ψ_- solitons on the left and on the right hand. Distances between constituents of those complexes go infinity as V tends zero. Because $P_{\psi_-} = P_{\psi_+} = 0$ at $V = 0$, in both cases of positive and negative velocities the $P(V)$ curve ends exactly at the point $V = 0$, $P = P_{T2}$. It does not mean, however, that the $T4$-branch joins the $T2$-branch at this point.

The last branch $P(V)$ shown in fig.3(a) is associated with travelling complexes $\psi_{(++)}$ and $\psi_{(++++)}$. Both branches come to the point $V = P = 0$ where solutions have the form of two and four free-standing $\psi_{(+)}$ solitons, respectively.

As in the case $h = 0.25$, all branches in the Fig.3(a) are unstable.

3.3 The Case $h = 0.05$

The $P(V)$ diagram for $h = 0.05$ is shown in Fig.4(a). Continuing the $T2$ twist in the negative-V direction, it transforms into a complex of two ψ_- solitons. At some point along the curve ($V = -0.451$, $P = -0.76$), the complex *stabilizes*. Continuing further in the direction of negative V, the branch turns back; shortly after that the momentum reaches its minimum where the solution loses its stability. When continued beyond the turning point ($V = -0.513$, $P = -1.042$) and the point of minimum of momentum ($V = -0.503$, $P = -1.08$), the curve

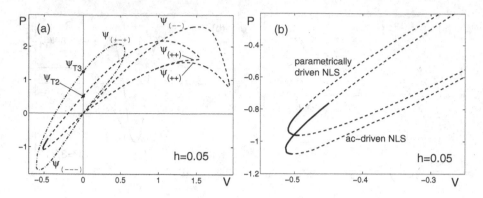

Fig. 4. (a) Continuation of the twist solitons $T2$ and $T3$ for $h = 0.05$. Also shown is the continuation of the $\psi_{(--)}$ branch. More solution branches can be obtained by the reflection $V \to -V$, $P \to -P$. (b) The stable portions of the $P(V)$ curves corresponding the travelling two-soliton complexes of Eq.(1) and Eq.(5). Stable branches are given by solid and unstable ones by dashed lines.

connects to the origin on the (V, P) plane (Fig.4(a)). The distance between the two ψ_- solitons in complex grows without bound as $V, P \to 0$.

It is interesting to note a similarity between the bifurcation diagram resulting from the continuation of the small-h twist in the externally driven NLS and the corresponding diagram for the case of parametrically driven NLS equation

$$i\psi_t + \psi_{xx} + 2|\psi|^2\psi - \psi = h\psi^* - i\gamma\psi. \tag{5}$$

Fig.4(b) reproduces stable portions of $P(V)$ branches corresponding the travelling complexes of Eq.(1) and Eq.(5). The latter is reproduced from [5] .

Note, in both case external and parametrical driving the $P(V)$ curve has a narrow interval where two different complexes (corresponding the top and low branches of the $P(V)$ graph) coexist and stably travel with the same velocity. For the Eq.(1) the bistability interval is $V \in [-0.513, -0.503]$, and for the Eq.(5) $V \in [-0.509, -0.4593]$.

Pathfollowing to $V > 0$ the $T2$ twist transforms into a $\psi_{(++)}$ complex. The $P(V)$ curve turns down and returns to the origin of the (V, P)-plane while separation between solitons in complex goes infinity.

Continuing the $T3$ twist to positive velocities, the solution transforms into a $\psi_{(+-+)}$ complex. If we, instead, continue to negative velocities, the $T3$ twist transforms into a triplet of ψ_- solitons. Both $V > 0$ and $V < 0$ parts of the curve turn and connect to the origin on the (V, P)-plane, see Fig.4(a). As V and P approach the origin on either side, the distance between the three solitons bound in the complex grows without limit.

Figure 4(a) also shows the $P(V)$ branch associated with complexes $\psi_{(++)}$ and $\psi_{(--)}$. Note, we have two coexisting complexes $\psi_{(++)}$ here, with close values of momentum but with different distance between solitons bounded in the complex.

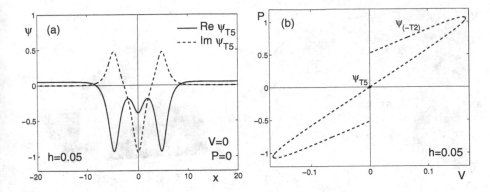

Fig. 5. (a) The $T5$ complex for $h = 0.2$, $V = 0$. The solid line shows the real and dashed imaginary part. (b) The $P(V)$ diagram associated with the $T5$ complex.

Beside of twist solutions $T2$, $T3$, $T4$ we have numerically obtained one more complex (denoted $T5$) where both real and imaginary parts are even, and $P = 0$. This complex is shown in Fig.5(a) for $h = 0.05$, $V = 0$. Corresponding branch of travelling complexes $P(V)$ is given in Fig.5(b). In both cases of $V < 0$ and $V > 0$ the symmetric solution $T5$ transforms to nonsymmetric complex of ψ_- and $T2$-solutions. As $V > 0$ the $P(V)$ curve turns up and tends the point $V = 0\ P = P_{T2}$ while the distance between ψ_- and $T2$ bounded in the complex is growing to infinity. The negative part of the $P(V)$ curve is symmetric with respect to origin of (V, P)-plane, i.e. $P(V) = -P(-V)$.

4 Summary

The stable domain of travelling complexes is shown on the diagram in Fig.6(a) (reproduced from [4]). In this region stably travelling complexes coexist with stable travelling single waves ψ_-. Fig.6(b) demonstrates stably travelling soliton for h=0.05 obtained in direct simulation. As an initial condition we have chosen the solution of eq.(2) with $V = -0.48$.

As in the case of parametrically driven NLS, ac-driven travelling complexes are stable only for small h. No stable travelling bound states has been found in our numerical study when $h \geq 0.06$.

Note that undamped stationary solutions with $P = 0$ can be continued to nonzero parameter of damping γ [5]. Our analysis confirms that the complex $T5$ can be continued to the $\gamma > 0$ direction indeed. Numerical investigation of weakly damped multi-soliton complexes is object of the further research.

EZ was partially supported by a grant under the JINR/RSA Research Collaboration Programme and by RFBR (grant No.12-01-00396). NA was supported by the National Research Foundation (grant No.67982).

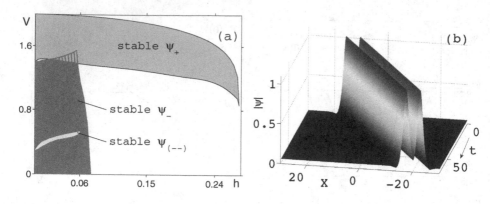

Fig. 6. Color online: (a) Stability chart of travelling solitons at the (V, h)-plane. Light (yellow online) inclusion reproduces the stability region of travelling complexes where stably travelling complexes coexist with stable travelling single waves ψ_-. The bistability region where travelling ψ_- and ψ_+ solitons coexist, is shaded with parallel lines. (b) Stably travelling complex. Result of direct numerical simulation for $h = 0.05$.

References

1. Alexeeva, N.V., Barashenkov, I.V., Pelinovsky, D.E.: Dynamics of the parametrically driven NLS solitons beyond the onset of the oscillatory instability. Nonlinearity 12, 103–140 (1999)
2. Barashenkov, I.V., Smirnov, Y.S.: Existence and stability chart for the ac-driven, damped nonlinear Schrodinger solitons. Phys. Rev. E 54, 5707–5725 (1996)
3. Barashenkov, I.V., Smirnov, Y.S., Alexeeva, N.V.: Bifurcation to multisoliton complexes in the ac-driven, damped nonlinear Schrodinger equation. Phys. Rev. 57, 2350–2364 (1998)
4. Barashenkov, I.V., Zemlyanaya, E.V.: Travelling solitons in the externally driven nonlinear Schroedinger equation. J. Phys. A: Math. Theor. 44, 465211 (2011)
5. Barashenkov, I.V., Zemlyanaya, E.V., Bär, M.: Travelling solitons in the parametrically driven nonlinear Schrödinger equation. Phys. Rev. E 64, 016603 (2001)
6. Barashenkov, I.V., Zemlyanaya, E.V.: Travelling solitons in the damped driven nonlinear Schrödinger equation. SIAM J. Appl. Math. 64(3), 800–818 (2004)
7. Barashenkov, I.V., Zemlyanaya, E.V.: Existence threshold for the ac-driven nonlinear Schrödinger solitons. Physica D 132, 363–372 (1999)
8. Barashenkov, I.V., Zhanlav, T., Bogdan, M.M.: Instabilities and Soliton Structures in the Driven Nonlinear Schrödinger Equation. In: Bar'yakhtar, V.G., et al. (eds.) Nonlinear World. IV International Workshop on Nonlinear and Turbulent Processes in Physics, Kiev, pp. 3–9. World Scientific, Singapore (1990)
9. Zemlyanaya, E.V., Barashenkov, I.V.: Numerical study of the multisoliton complexes in the damped-driven NLS. Math. Modelling 16(10), 3–14 (2004) (Russian)
10. Zemlyanaya, E.V., Barashenkov, I.V.: Numerical analysis of travelling solitons in the the parametrically driven NLS. Math. Modelling 17(1), 65–78 (2005) (Russian)

Asynchronous Differential Evolution
with Restart

Evgeniya Zhabitskaya[1,2] and Mikhail Zhabitsky[1,3]

[1] Joint Institute for Nuclear Research, Joliot-Curie 6, 141980, Dubna, Russia
[2] University Dubna, University Str. 19, 141980, Dubna, Russia
[3] Rock Flow Dynamics, Moscow, Russia

Abstract. Asynchronous Differential Evolution (ADE) [1] is a derivative-free method to solve global optimization problems. It provides effective parallel realization. In this work we derive ADE with restart (ADE-R). By increasing population size after each restart, new strategy enhances its chances to locate the global minimum. The ADE-R algorithm has convergence rate comparable or better than ADE with fixed population sizes. Performance of the ADE-R algorithm is demonstrated on a set of benchmark functions.

Keywords: global optimization, derivative-free optimization, differential evolution, evolution strategy.

1 Introduction

Many optimization problems can be reduced to the finding of the optimal set of parameters x^* from a real-value domain Ω which minimizes an objective function $f(x^*)$:

$$ f(x^*) \leqslant f(x), \quad \forall x \in \Omega, \quad x = \{x_j\}|_{j=0,\ldots,D-1} . \tag{1} $$

The Asynchronous Differential Evolution algorithm is efficient to solve possible nonlinear and non-differentiable global optimization problems [1]. It contains a few control parameters, but one of them — population size N_p — is crucial for DE: large population sizes ensure a high probability to locate the global minimum, at the same time small populations are characterized by faster convergence rate. Reasonable choice of N_p by a practitioner requires an intuition and considerable efforts.

In the next part of this paper we remind basic principles of the Asynchronous Differential Evolution algorithm. In the part three we introduce Asynchronous Differential Evolution with restart. As soon as new algorithm has diagnosed stagnation in progress to find a better candidate solution, restart is performed. Algorithm increases the population size with each restart, which enhances probability to locate the global minimum. Its performance is verified on a set of benchmark functions characteristic for real-parameter multidimensional optimization.

I. Dimov, I. Faragó, and L. Vulkov (Eds.): NAA 2012, LNCS 8236, pp. 555–561, 2013.

2   Asynchronous Differential Evolution

The ADE represents an asynchronous strategy derived from Classical Differential Evolution (CDE) algorithm [4, 5]. Differential evolution is a popular algorithm to solve black-box optimization problems. It is very competitive in the domain of ill-conditioned and separable multidimensional problems [3]. Unlike CDE, ADE incorporates mutation, crossover and selection operations into an asynchronous strategy. The ADE performance is competitive to the commonly used variants of Classical DE, while for parallel calculations ADE clearly outperforms the synchronous CDE strategies [1]. ADE algorithm contains a few control parameters, the constraints on their reasonable values were derived analytically [2].

The general scheme of the Asynchronous Differential Evolution algorithm is presented in Fig. 1. The algorithm operates over a population $P_x = \{x_i\}$ of size N_p, each population member is represented by a vector in the search domain $x \in \Omega$. The population is initialized by vectors randomly selected from Ω. Then ADE iteratively improves the population until some termination criterion is not met.

// Initialize a population $P_x = \{x_i\}|_{i=0,\ldots N_p-1}$, $x_i = \{x_{i,j}\}|_{j=0,\ldots D-1} \in \Omega$

P_x = initialize_population();

do {

    i = choose_target_vector(); // Choose target vector x_i

    // Mutation:

    $v_i = x_r + F(x_p - x_q)$; // mutant vector, $r \neq p \neq q$; p, q are random indices

    // Crossover (recombination):

    for $(j = 0;\ j < D;\ j = j + 1)$

$$u_{i,j} = \begin{cases} v_{i,j} & \mathrm{rand}(0,1) < C_r \text{ or } j = j_{\mathrm{rand}} \\ x_{i,j} & \text{otherwise} \end{cases} \quad \text{// trial vector}$$

    // Selection:

    if $(f(u_i) < f(x_i))$

        $x_i = u_i$;

} while (termination criterion not met);

Fig. 1. C-style scheme for the Asynchronous Differential Evolution algorithm

Population members are sorted according to their objective function value, for minimization problems the best one corresponds to the smallest value. The algorithm selects some vector, called *target* vector, from a population. In this article we will consider two variants: either a target vector is randomly chosen from the population ("rand") or the worst population vector is chosen as a target vector ("worst"). Another vector x_r is chosen as a *base* vector ("rand" or "best"). Two other randomly chosen vectors form a *difference* vector. The sum of the base vector with the difference vector multiplied by a *scale* factor F is a *mutant* vector:

$$v_i = x_r + F(x_p - x_q). \qquad (2)$$

Coordinate recombination (crossover) of the mutant vector with the target one results a *trial* vector. To ensure that the trial vector is different from the target one, at least one coordinate (randomly selected j_{rand}) is taken from the mutant vector. The objective function value of the trial vector is confronted to the function value of the target vector. If and only if a resulting trial vector provides a better value of the optimization function, it will replace its parent target vector in the population.

For ADE variants we introduce an ADE/w/x/y/z notation [1] similar to the CDE classification [5]. Here w corresponds to the choice of a *target* vector, this feature is specific of ADE. The choice of a *base* vector is denoted as x; y and z indicate the number of difference vectors added and crossover type respectively. In this article numerical results are obtained by ADE/rand/rand/1/bin and ADE/worst/best/1/bin strategies. The first one is characterized by high exploration ability, while the latter one has faster convergence to a local minimum.

The performance of the ADE is evaluated on the CEC2005 set of benchmark functions [6]. We selected four functions, which represent different classes of real-parameter optimization problems: f_1 — sphere function (unimodal, separable), f_6 — Rosenbrock function (a few minima, nonseparable), f_9 — Rastrigin function (multimodal, separable) and f_{11} — rotated Weierstrass function (multimodal, nonseparable). All functions are defined in a 10-dimensional parameter space.

A problem is considered to have been solved if its optimizer reached the global minimum within the predefined precision ε: $f(x) - f(x^*) < \varepsilon$. For the sphere function the precision was set to 10^{-6}, while for other functions $\varepsilon = 10^{-2}$. One trial represents 100 optimization executions. For each trial the probability of

Table 1. Convergence rate of ADE strategies for four benchmark functions

		ADE/rand/rand/1/bin			ADE/worst/best/1/bin		
	N_p	$\langle N_{\text{feval}} \rangle$	Median N_{feval}	P_{succ}	N_p	$\langle N_{\text{feval}} \rangle$	Median N_{feval}	P_{succ}
f_1	10	3641	3633	1.00	10	2986	2977	0.98
					15	4344	4333	1.00
f_6	20	2.38e+04	2.32e+04	0.92	50	1.92e+04	1.84e+04	0.85
	50	8.30e+04	8.20e+04	0.98	150	5.65e+04	5.59e+04	0.85
	70	1.28e+05	1.25e+05	1.00	600	2.12e+05	2.11e+05	0.91
f_9	15	3963	4012	0.93	20	4317	4289	0.92
	30	7976	7988	1.0	50	10138	10108	1.0
f_{11}	20	0.406e+05	0.381e+05	0.59	50	0.347e+05	0.333e+05	0.55
	50	2.939e+05	2.913e+05	0.83	100	0.870e+05	0.842e+05	0.73
	100	1.168e+06	1.122e+05	0.90	200	1.952e+05	1.887e+05	0.84

success P_{succ}, which is the number of executions which converged to the global minimum divided by the total number of executions, the average number of function evaluations $\langle N_{feval} \rangle$ and median N_{feval} for converged executions are recorded.

Performance of ADE for the selected functions is presented in Tab. 1. For both strategies the crossover rate C_r was set to 0.1 for separable functions and to 0.9 otherwise. The scale factor $F = 0.9$ was used. We performed trials for different population sizes $N_p = 10, 15, 20, 30, 50, 70, 100, 150, 200, 600$. Results with highest probability of success are reported.

Unlike CEC2005 testbench we have increased the limit on maximal number of function evaluations from 10^5 to 10^7. Even if we don't limit the maximal number of function evaluations, ADE will fail to locate the global minimum in many cases. The analysis of failed attempts reveals either degeneration of the population into a subspace or that the search ended up in a local minimum. Both above reasons can be overcome if one chooses sufficiently large population size N_p. From the other side, the search with a large population can consume exceedingly long computer time. For practical problems one should either adjust the population size to complexity of the function to be minimized or perform multiple searches with different population sizes, which requires efforts by a practitioner and additional CPU time.

3   Asynchronous Differential Evolution with Restart

To balance fast convergence rate with high probability of convergence we will introduce Asynchronous Differential Evolution with restart (ADE-R). The modified algorithm starts with some small population size N_p^{min}. We will use $N_p^{min}=10$ for numerical tests. As soon as ADE-R diagnoses stagnation in the search towards a better solution, the algorithm performs a restart by increasing the population size by a predefined factor k and starting an independent search.

To diagnose stagnation ADE-R tracks the maximal spread for population members in each coordinate Δx_j and in function values Δf:

$$\Delta x_j = \max_{i=0,...,N_p-1}\{x_{i,j}\} - \min_{i=0,...,N_p-1}\{x_{i,j}\}, \tag{3}$$

$$\Delta f = \max_{i=0,...,N_p-1}\{f_i\} - \min_{i=0,...,N_p-1}\{f_i\}. \tag{4}$$

As the factor F in the mutation operator (2) is of the order of one, a difference in the operator should be larger then the coordinate of the base vector multiplied by the machine epsilon. Otherwise adding this difference will have no effect due to rounding in floating point operations. Therefore we introduce the following criterion for restart

$$\exists j \quad \Delta x_j < \varepsilon_x \max_i\{|x_{i,j}|\}. \tag{5}$$

In the current realization of ADE float values are 8-bytes long (IEEE 754 standard) and corresponding machine epsilon is about 10^{-16}. Typically, ε_x should exceed 10^{-15}. At the same time one can try to use larger values of ε_x. In the latter case stagnation will be indicated ealier, which will potentially accelerate convergence, but too large ε_x will lead to excessive restarts and the algorithm will fail to locate the minimum. One should notice that the lower bound on feasible values of ε_x is due to an algorithm numerical representation and doesn't depend on the function to be minimized. The upper bound is problem-dependent and should be chosen with care.

Due to the comparison in the selection operator, we introduce another criterion for restart, based on the spread in function values within population:

$$\Delta f < \varepsilon_f \max_{i=0,\ldots,N_p-1} \{|f_i|\}. \tag{6}$$

Again the lower bound on ε_f is due to the algorithm numerical representation, while the upper bound is problem dependent. One should note that the upper bound on ε_f can be estimated *a priori* for the important class of generalized least-squares problems.

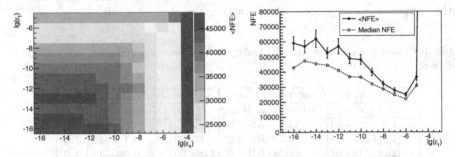

Fig. 2. Mean number of function evaluations to solve Rosenbrock problem f_6 ($D = 10$) as a function of ε_x and ε_f (left). Mean and median N_{feval} as a function of ε_f without criterion on Δ_x (5) (right).

We performed a scan in the $(\varepsilon_x, \varepsilon_f)$ plane to analyze the convergence rate as a function of applied criteria for the ADE/worst/best/1/bin strategy on Rosenbrock function f_6 (Fig. 2). Multiplier $k = 2$ was used to increase the population size after restart. For all points $P_{\text{succ}} = 1.0$ is ensured by ADE-R. Higher ε_x and ε_f values lead to earlier diagnostic of stagnation in evolution progress, which results in restart, and thus to faster convergence. But too large values cause excessive restarts. Upper limit on ε_f at 10^{-4} is defined by the objective function value in minimum $f^* = 360$ and required precision of 10^{-2}. From Fig. 2 one can conclude that even small problem-independent values of ε_x and ε_f, defined as the machine epsilon multiplied by a factor of $10^3 \ldots 10^4$, guarantee a reasonable convergence rate.

Performance of ADE with restart for four test functions is presented in Tab. 2. Factor $k = 2$ was used to increase population size after restart. For all functions $P_{\text{succ}} = 1$ has been achieved. Compared to ADE with fixed population sizes (Tab. 1), the improvement in the convergence rate for complex functions is observed. ADE-R automatically adapts the population size to match the complexity of the optimization problem. In this way strategy with restart eliminates time

Table 2. Convergence rate of ADE-R strategies for four benchmark functions for $\varepsilon_x = 10^{-12}$ and different ε_f

	ε_f	ADE-R/rand/rand/1/bin			ADE-R/worst/best/1/bin		
		$\langle N_{feval}\rangle$	Median N_{feval}	P_{succ}	$\langle N_{feval}\rangle$	Median N_{feval}	P_{succ}
f_1	10^{-12}	3641	3633	1.0	3070	2979	1.0
f_6	10^{-7}	4.91e+04	4.20e+04	1.0	2.88e+04	2.30e+04	1.0
	10^{-12}	5.11e+04	4.03e+04	1.0	4.09e+04	3.22e+04	1.0
f_9	10^{-7}	5339	2977	1.0	5865	6913	1.0
	10^{-12}	6121	2977	1.0	6942	8315	1.0
f_{11}	10^{-7}	3.22e+05	0.75e+05	1.0	1.54e+05	1.14e+05	1.0
	10^{-12}	2.02e+05	0.78e+05	1.0	1.78e+05	1.32e+05	1.0

consuming tuning of the optimal population size N_p for a selected optimization problem. All results in Tab. 2 were obtained with $N_p^{min} = 10$. In the case of complex objective functions, to be solved only with large population sizes, one can further improve convergence rate by starting the algorithm with higher N_p^{min}.

Table 3. Convergence rate of ADE-R strategies for different population size multipliers k after restart, $\varepsilon_x = 10^{-12}$, $\varepsilon_f = 10^{-12}$

	k	ADE-R/rand/rand/1/bin			ADE-R/worst/best/1/bin		
		$\langle N_{feval}\rangle$	Median N_{feval}	P_{succ}	$\langle N_{feval}\rangle$	Median N_{feval}	P_{succ}
f_6	1.2	4.88e+04	3.93e+04	1.0	8.56e+04	7.81e+04	1.0
	1.5	4.85e+04	3.93e+04	1.0	4.97e+04	4.14e+04	1.0
	2.0	5.11e+04	4.03e+04	1.0	4.09e+04	3.22e+04	1.0
	3.0	5.76e+04	4.06e+04	1.0	4.60e+04	3.42e+04	1.0
	5.0	7.42e+04	4.06e+04	1.0	5.60e+04	3.51e+04	1.0
f_{11}	1.2	2.15e+05	1.79e+05	1.0	2.37e+05	2.18e+05	1.0
	1.5	1.75e+05	1.14e+05	1.0	1.64e+05	1.27e+05	1.0
	2.0	2.02e+05	0.78e+05	1.0	1.78e+05	1.32e+05	1.0
	3.0	2.99e+05	1.16e+05	1.0	2.11e+05	1.14e+05	1.0
	5.0	1.00e+06	3.15e+05	0.96	2.66e+05	0.55e+05	1.0

The effect on the convergence rate by the population size multiplier k is illustrated in Tab. 3. To our experience reasonable values belong to the range $1.2 \ldots 3.0$, for many problems $k \in [1.5, 2.0]$ is optimal. Smaller values lead to finer steps in population growth thus to more frequent restarts.

With the help of ADE-R several practical optimization problems have been solved [7–9]. These problems have objective functions with complex terrain or multiple minima, therefore local-searching optimization algorithms failed.

4 Conclusions

We have constructed an algorithm of Asynchronous Differential Evolution with restart (ADE-R) to solve real-value global optimization problems.

Through introduction of simple criteria on population spread in parameter space and corresponding objective function values, the algorithm diagnoses stagnation in evolution towards a better solution and performs restart for an independent search with larger population size. ADE with restart automatically increases the population size according to the complexity of the problem to be solved. This new algorithm performs well on a standard test bench of real-value optimization problems, ensures high probability to locate the global extremum and has competitive convergence rate.

Acknowledgement. E.Z. was partially supported by the Program for collaboration between JINR (Dubna) and Bulgarian scientific centers.

References

1. Zhabitskaya, E., Zhabitsky, M.: Asynchronous Differential Evolution. In: Adam, G., Buša, J., Hnatič, M. (eds.) MMCP 2011. LNCS, vol. 7125, pp. 328–333. Springer, Heidelberg (2012)
2. Zhabitskaya, E.: Constraints on Control Parameters of Asynchronous Differential Evolution. In: Adam, G., Buša, J., Hnatič, M. (eds.) MMCP 2011. LNCS, vol. 7125, pp. 322–327. Springer, Heidelberg (2012)
3. Das, S., Suganthan, P.N.: Differential evolution: A Survey of the State-of-the-Art. IEEE Trans. Evol. Comput. 15, 4–31 (2011)
4. Price, K.V., Storn, R.M.: A Simple and Efficient Heuristic for Global Optimization over Continuous Spaces. J. of Global Optimization 11, 341–359 (1997)
5. Price, K.V., Storn, R.M., Lampinen, J.A.: Differential Evolution: A Practical Approach to Global Optimization. Springer, Heidelberg (2005)
6. Suganthan, P.N., et al.: Problem definitions and evaluation criteria for the CEC05 special session on real-parameter optimization. Technical report, Nanyang Technological University, Singapore (2005),
   http://www.ntu.edu.sg/home/epnsugan/index_files/CEC-05/CEC05.htm
7. Zhabitskaya, E.I.: Applying Asynchronous Differential Evolution method for estimation of the parameters of microscopic optical potential π-mesons elastic scattering on nuclei. In: Proceedings of the XVI Conference of Young Scientists and Specialists, Dubna, pp. 46–49 (2012),
   http://omus.jinr.ru/conference2012/conference_proceedings.pdf
8. Zhabitskaya, E.I., Zhabitsky, M.V., Zemlyanaya, E.V., Lukyanov, K.V.: Calculation of the parameters of microscopic optical potential for pion-nuclei elastic scattering by Asynchronous Differential Evolution algorithm. Computer Research and Modeling 4(3), 585–595 (2012) ISSN 2077-6853,
   http://crm.ics.org.ru/journal/article/1927/
9. Zemlyanaya, E.V., Lukyanov, V.K., Lukyanov, K.V., Zhabitskaya, E.I., Zhabitsky, M.V.: Pion-Nucleus Microscopic Optical Potential at Intermediate Energies and In-Medium Effect on the Elementary $\pi\pi N$ Scattering Amplitude. Nuclear Theory 31, 175–184 (2012) ISSN 1313-2822, http://arxiv.org/abs/1210.1069

Uniform Grid Approximation of Nonsmooth Solutions of a Singularly Perturbed Convection - Diffusion Equation with Characteristic Layers

Umar Kh. Zhemukhov

Moscow State University, Faculty of Computational
Mathematics and Cybernetics, 119991, Moscow, Russia
zhemukhov-u@yandex.ru

Abstract. A mixed boundary value problem for a singularly perturbed elliptic convection-diffusion equation with constant coefficients is considered in a square with the Dirichlet conditions imposed on the two sides, which are orthogonal to the flow direction, and with the Neumann conditions on the other two sides. Sufficient smoothness of the right-hand side and that of the boundary functions is assumed, which ensures the required smoothness of the solution in the considered domain, except the neighborhoods of the corner points. At the corner points themselves, zero order compatibility conditions alone are assumed to be satisfied.

For the numerical solution to the posed problem a nonuniform monotonous difference scheme is used on a rectangular piecewise uniform Shishkin grid. Non-uniformity of the scheme means that the form of the difference equations, which are used for the approximation, is not the same in different grid points but it depends on the value of the perturbing parameter.

Under assumptions made a uniform convergence with respect to ε of the numerical solution to the precise solution is proved in a discrete uniform metric at the rate $O(N^{-3/2} \ln^2 N)$, where N is the number of the grid points in each coordinate direction.

Keywords: singularly perturbed problems, condensing mesh, characteristic boundary layer, corner singularity, uniform convergence.

1 Introduction

A mixed boundary problem for a singularly perturbed convection-diffusion equation [1,2] is considered in the square $\Omega = (0,1)^2$ with the boundary $\partial\Omega$:

$$Lu \equiv -\varepsilon \Delta u + a \frac{\partial u}{\partial x} + qu = f(x,y), \quad (x,y) \in \Omega, \tag{1}$$

$$\frac{\partial u}{\partial \mathbf{n}} = \varphi(x,y), \quad (x,y) \in \partial\Omega_N, \tag{2}$$

$$u = g(x,y), \quad (x,y) \in \partial\Omega_D, \tag{3}$$

where $a = const > 0$, $q = const > 0$; $\mathbf{n}$ is the unit vector of an outer normal to $\partial\Omega_N$, and $\varepsilon \in (0,1]$ is a small parameter. The domain boundary consists

I. Dimov, I. Faragó, and L. Vulkov (Eds.): NAA 2012, LNCS 8236, pp. 562–570, 2013.
© Springer-Verlag Berlin Heidelberg 2013

of the parts $\partial\Omega_D = \Gamma_1 \cup \Gamma_3$ and $\partial\Omega_N = \Gamma_2 \cup \Gamma_4$, where Γ_k are the sides of the square Ω, ordered in the counterclockwise direction starting from $\Gamma_1 = \{(x,y) \in \partial\Omega \mid x = 0\}$, whereas $a_k = (x_k, y_k)$ are its vertices, ordered in the similar way being $a_1 = (0,0)$.

In the corner points a_k the boundary functions g and φ are assumed to be the subject to the following conditions

$$\frac{dg_1}{dy}(0) = -\varphi_1(0), \quad \frac{dg_1}{dy}(1) = \varphi_2(0), \quad \frac{dg_2}{dy}(0) = -\varphi_1(1), \quad \frac{dg_2}{dy}(1) = \varphi_2(1), \quad (4)$$

which are named [3] zero-order compatibility conditions, where $g_1(y) = g(0,y)$, $g_2(y) = g(1,y)$ and $\varphi_1(x) = \varphi(x,0)$, $\varphi_2(x) = \varphi(x,1)$.

The solution to problem (1)–(4) has a compound structure (see [2,4] e.g. and the literature cited there), which includes the regular boundary layer of width $O(\varepsilon)$ in a neighborhood of the right-hand boundary Γ_3, the two characteristic layers of width $O(\sqrt{\varepsilon})$ in neighborhoods of the upper and lower boundaries Γ_2 and Γ_4, the corner layers with corner singularities in neighborhoods of the vertices a_2, a_3 and the corner singularities in neighborhoods of the inflow vertices a_1, a_4. All this makes it difficult to solve problem (1)–(4) numerically. It is also known (see [3]) that the presence of the corner points adversely affects the smoothness of the solution.

Actually it is problem (1)–(3) though in more general setting (with variable coefficients and boundary conditions of the third kind imposed on $\partial\Omega_N$) that paper [1] is devoted to. In this paper a nonuniform monotonous scheme was made up for the equation and the scheme formally has the second order approximation with $\varepsilon \leq CN^{-1}$ (compare with [6,7] in one dimensional case) and, under the assumption that the compatibility conditions at the corner points are satisfied up to the 2nd order, the convergency at the rate $O(N^{-3/2} \ln N)$ on Shishkin mesh is proved.

The purpose of this work is to enhance the results [1] in the following directions: getting rid of the excessive compatibility conditions, which increased the smoothness of the solution up to $u(x,y) \in C^{4,\lambda}(\bar{\Omega})$, $\lambda \in (0,1)$; working out the uniform with respect to ε estimate of the convergency rate of order $O(N^{-3/2} \ln^2 N)$ for all $\varepsilon \in (0,1]$, instead of the former estimate for $\varepsilon \leq CN^{-1}$.

In the course of the paper we denote by $C, \tilde{C}$ and c some positive constants, which are different in any individual case and depend on input data, but not on N or ε.

2 Statement of the Finite Difference Problem

In paper [2] by means of the decomposition of the solution to problem (1)–(3) into smooth and boundary layer components some point-wise estimates to the solution and its derivatives are obtained, which we use here; their dependance on a small parameter ε and on the data compatibility conditions at the domain corners is shown. Before we introduce the difference scheme, let us write this solution as the following sum:

$$u(x,y) = S + E + w_1 + w_2 + w_3 + w_4 + \tilde{u}, \quad (x,y) \in \Omega, \quad (5)$$

where S is a smooth component, E is a regular boundary layer, w_1, w_4 and w_2, w_3 are two characteristic layers (an upper one and a lower one respectively) and two corner layers in neighborhoods of the corners a_2, a_3; $\tilde{u}$ is the remainder term, which is exponentially small.

Thus, on the set $\bar{\Omega}$ we introduce the grid $\bar{\Omega}^h = \bar{\omega}_x \times \bar{\omega}_y$, which is a tensor product of two piecewise unform Shishkin meshes. The mesh $\bar{\omega}_x$ contains $N/2$ points on each of the half-intervals $(0, 1 - \sigma_x]$ and $(1 - \sigma_x, 1]$ with the steps $\mathfrak{H}_1$ and $\mathfrak{h}_1$, respectively, and the mesh $\bar{\omega}_y$ contains $N/4$ points on the half-intervals $(0, \sigma_y], (1 - \sigma_y, 1]$ and $N/2$ points on the $(\sigma_y, 1 - \sigma_y]$ with steps $\mathfrak{h}_2$ and $\mathfrak{H}_2$ respectively, where

$$\mathfrak{H}_1 = \frac{2(1 - \sigma_x)}{N}, \ \mathfrak{h}_1 = \frac{2\sigma_x}{N}, \ \sigma_x = \min\left\{\frac{1}{2}; \frac{4\varepsilon \ln N}{a}\right\}; \ \mathfrak{H}_2 = \frac{2(1 - 2\sigma_y)}{N},$$

$$\mathfrak{h}_2 = \frac{4\sigma_y}{N}, \ \sigma_y = \min\left\{\frac{1}{4}; \frac{4\sqrt{\varepsilon} \ln N}{\beta}\right\}; \ \beta = \min\{a/12, q/2a, \sqrt{q}\}.$$

Therefore, to find the numerical solution of the problem (1)–(4) we use the inhomogeneous monotone difference scheme, in which for small values of ε convective term is approximated by the usual directional difference at the middle point $x_{i-1/2}$ out of the regular layer, while in the regular layer we use for this the central difference.

Before we write the difference scheme, we will introduce the following notation: $\Omega^h = \bar{\Omega}^h \cap \Omega$, $\Omega_1^h = \bar{\Omega}_1^h \cap \Omega^h$, $\Omega_2^h = \bar{\Omega}_2^h \cap \Omega^h$, $\bar{\Omega}_1^h = \{0 \le x_i \le 1 - \sigma_x, 0 \le y_j \le 1\}$, $\bar{\Omega}_2^h = \{1 - \sigma_x < x_i \le 1, 0 \le y_j \le 1\}$, $u_{\bar{x},i} = (u_i - u_{i-1})/h_i$, $u_{\hat{x},i} = (u_{i+1} - u_i)/\hbar_i$, $u_{\mathring{x},i} = (u_{i+1} - u_{i-1})/2\hbar_i$, $u_{x,i} = u_{\bar{x},i+1}$, $\hbar_i = (h_{i+1} + h_i)/2$, $f_{ij}^h = f(x_i, y_j)$, and u_{ij}^h is an approximative solution to problem (1)–(4).

On the mesh $\bar{\Omega}^h$ we assign a difference problem (see [1]) to problem (1)–(4)

$$L^h u_{ij}^h \equiv -\varepsilon \left(u_{\bar{x}\hat{x};ij}^h + u_{\bar{y}\hat{y};ij}^h\right) + L_1^h u_{ij}^h + q u_{ij}^h = F_{ij}^h, \quad \Omega^h \tag{6}$$

where

$$L_1^h u_{ij}^h \equiv \begin{cases} \left(a - \frac{q\mathfrak{H}_1}{2}\right) u_{\bar{x};ij}^h, \ \varepsilon < a\mathfrak{H}_1/2, \ \Omega_1^h, \\ a u_{\hat{x};ij}^h, \qquad\qquad \varepsilon < a\mathfrak{H}_1/2, \ \Omega_2^h, \\ a u_{\mathring{x};ij}^h, \qquad\qquad \varepsilon \ge a\mathfrak{H}_1/2, \ \Omega^h, \end{cases} \quad F_{ij}^h = \begin{cases} f_{i-1/2,j}^h, \ \varepsilon < a\mathfrak{H}_1/2, \ \Omega_1^h, \\ f_{ij}^h, \qquad \varepsilon < a\mathfrak{H}_1/2, \ \Omega_2^h, \\ f_{ij}^h, \qquad \varepsilon \ge a\mathfrak{H}_1/2, \ \Omega^h. \end{cases}$$

$$L^h u_{i0}^h \equiv \begin{cases} -u_{y;i0}^h + \dfrac{\mathfrak{h}_2}{2}\left[-u_{\bar{x}\hat{x};i0}^h + \dfrac{1}{\varepsilon}\left(a - \dfrac{q\mathfrak{H}_1}{2}\right)u_{\bar{x};i0}^h + \dfrac{q}{\varepsilon}u_{i0}^h\right] \\ -u_{y;i0}^h + \dfrac{\mathfrak{h}_2}{2}\left[-u_{\bar{x}\hat{x};i0}^h + \dfrac{a}{\varepsilon}u_{\hat{x};i0}^h + \dfrac{q}{\varepsilon}u_{i0}^h\right] \end{cases} =$$

$$= \begin{cases} \varphi_1(x_i) + \dfrac{\mathfrak{h}_2}{2\varepsilon}f_{i-1/2,0}^h, \ \varepsilon < \dfrac{a\mathfrak{H}_1}{2}, \ 0 < i \le N/2, \\ \varphi_1(x_i) + \dfrac{\mathfrak{h}_2}{2\varepsilon}f_{i0}^h, \qquad \varepsilon < \dfrac{a\mathfrak{H}_1}{2}, \ N/2 < i < N \text{ or } \varepsilon \ge \dfrac{a\mathfrak{H}_1}{2}, \ 0 < i < N, \end{cases} \tag{7}$$

$$L^h u_{iN}^h \equiv \begin{cases} u_{\bar{y};iN}^h + \dfrac{\mathfrak{h}_2}{2}\left[-u_{\bar{x}\hat{x};iN}^h + \dfrac{1}{\varepsilon}\left(a - \dfrac{q\mathfrak{H}_1}{2}\right)u_{\overset{\circ}{x};iN}^h + \dfrac{q}{\varepsilon}u_{iN}^h\right] \\[4mm] u_{\bar{y};iN}^h + \dfrac{\mathfrak{h}_2}{2}\left[-u_{\bar{x}\hat{x};iN}^h + \dfrac{a}{\varepsilon}u_{\overset{\circ}{x};iN}^h + \dfrac{q}{\varepsilon}u_{iN}^h\right] \end{cases} = $$

$$= \begin{cases} \varphi_2(x_i) + \dfrac{\mathfrak{h}_2}{2\varepsilon}f_{i-1/2,N}^h, \ \varepsilon < \dfrac{a\mathfrak{H}_1}{2}, \ 0 < i \le \dfrac{N}{2}, \\[4mm] \varphi_2(x_i) + \dfrac{\mathfrak{h}_2}{2\varepsilon}f_{iN}^h, \quad \varepsilon < \dfrac{a\mathfrak{H}_1}{2}, \ \dfrac{N}{2} < i < N \text{ or } \varepsilon \ge \dfrac{a\mathfrak{H}_1}{2}, \ 0 < i < N, \end{cases} \tag{8}$$

$$u_{0j}^h = g_1(y_j), \quad u_{Nj}^h = g_2(y_j), \qquad\qquad 0 \le j \le N. \tag{9}$$

In order to get the convergence rate estimate let us represent the solution to difference problem (6)–(9), by analogy with the differential problem, in the form

$$u^h = S^h + E^h + \sum_{k=1}^{4} w_k^h + \tilde{u}^h, \quad (x_i, y_j) \in \bar{\Omega}^h \tag{10}$$

where each component satisfies the following discrete problems

$$\begin{cases} L^h S_{ij}^h = F_{ij}^h, \quad \Omega^h, \\[3mm] L^h S_{i0}^h = -\dfrac{\partial S}{\partial y}(x_i, 0) + \dfrac{\mathfrak{h}_2}{2\varepsilon}F_{i0}^h, \quad 0 < i < N, \\[3mm] L^h S_{iN}^h = \dfrac{\partial S}{\partial y}(x_i, 1) + \dfrac{\mathfrak{h}_2}{2\varepsilon}F_{iN}^h, \quad 0 < i < N, \end{cases} \tag{11}$$

$$S_{0j}^h = S(0, y_j), \quad S_{Nj}^h = S(1, y_j), \qquad 0 \le j \le N. \tag{12}$$

$$\begin{cases} L^h E_{ij}^h = 0, \quad \Omega^h, \\[3mm] L^h E_{i0}^h = -\dfrac{\partial E}{\partial y}(x_i, 0), \quad 0 < i < N, \\[3mm] L^h E_{iN}^h = \dfrac{\partial E}{\partial y}(x_i, 1), \quad 0 < i < N, \end{cases} \tag{13}$$

$$E_{0j}^h = E(0, y_j), \quad E_{Nj}^h = E(1, y_j), \quad 0 \le j \le N. \tag{14}$$

$$\begin{cases} L^h w_{k;ij}^h = 0, \quad \Omega^h, \\[3mm] L^h w_{k;i0}^h = -\dfrac{\partial w_k}{\partial y}(x_i, 0), \quad 0 < i < N, \\[3mm] L^h w_{k;iN}^h = \dfrac{\partial w_k}{\partial y}(x_i, 1), \quad 0 < i < N, \end{cases} \tag{15}$$

$$w_{k;0j}^h = w_k(0, y_j), \quad w_{k;Nj}^h = w_k(1, y_j), \qquad k = 1, 2, 3, 4, \ \ 0 \le j \le N. \tag{16}$$

The estimate for the convergence of the numerical solution to the exact one will be obtained as a sum of the estimates for each term from (10). To this effect we represent the approximation error also as a sum

$$\psi_{ij}^h = \psi_{S;ij}^h + \psi_{E;ij}^h + \sum_{k=1}^4 \psi_{w_k;ij}^h + \psi_{\bar{u};ij}^h, \ (x_i, y_j) \in \bar{\Omega}^h, \text{ where } \psi_{S;ij}^h = L^h S(x_i, y_j) -$$

$$-L^h S_{ij}^h, \ \psi_{E;ij}^h = L^h \left(E(x_i, y_j) - E_{ij}^h \right), \ \psi_{w_k;ij}^h = L^h \left(w_k(x_i, y_j) - w_{k;ij}^h \right).$$

Further in the whole article we shall use the comparison principle [1] while working out the convergence rate estimates for each component from (10).

Theorem 1 (Comparison principle). *Let V_{ij}^h and W_{ij}^h be arbitrary mesh functions, defined on the mesh $\bar{\Omega}^h$, so that $\left| L^h V_{ij}^h \right| \leq L^h W_{ij}^h$ в $\Omega^h \cup \partial \Omega_N^h$, where $\Omega_N^h = \bar{\Omega}^h \cap \partial \Omega_N$ и $\left| V_{ij}^h \right| \leq W_{ij}^h$ on $\partial \Omega_D^h = \bar{\Omega}^h \cap \partial \Omega_D$. Then in $\bar{\Omega}^h$ the following estimate holds true $\left| V_{ij}^h \right| \leq W_{ij}^h$.*

The next theorem contains the main result of this paper.

Theorem 2. *Let $u(x_i, y_j)$ be a solution to the original problem (1)-(4), and u_{ij}^h be a solution to the discrete problem (6)-(9) on a piecewise uniform Shishkin mesh. Then for $\varepsilon \in (0, 1]$ the following rate convergence estimate holds true*

$$\left| u(x_i, y_j) - u_{ij}^h \right| \leq CN^{-3/2} \ln^2 N, \qquad (x_i, y_j) \in \bar{\Omega}^h. \qquad (17)$$

Proof. The proof follows from Theorem 3 and Remark 1, which are given in the next section. □

3 The Uniform Convergence of Numerical Solutions

The reasoning we use in this section when proving convergence for the smooth component S_{ij}^h, of the regular layer E_{ij}^h, and also for boundary layer components $w_{k;ij}^h$ $(k = \overline{1,4})$ outside the domains of the characteristic layer $\Omega_{w_1}^h = \{0 \leq x_i \leq 1, 0 \leq y_j < \sigma_y\}$ and of the corner layer $\Omega_{w_2}^h = \{1 - \sigma_x < x_i \leq 1, 0 \leq y_j < \sigma_y\}$ ($\Omega_{w_4}^h$ и $\Omega_{w_3}^h$ are dealt in the similar way), do not differ from those, which are given in [1].

Hence, proceeding as in [1] and taking the estimates for derivatives [2], we arrive at the following estimates

$$\left| S(x_i, y_j) - S_{ij}^h \right| \leq CN^{-2}, \qquad (x_i, y_j) \in \bar{\Omega}^h, \qquad (18)$$

$$\left| E(x_i, y_j) - E_{ij}^h \right| \leq C \begin{cases} N^{-2}, & (x_i, y_j) \in \bar{\Omega}_1^h, \\ N^{-2} \ln^2 N, & (x_i, y_j) \in \bar{\Omega}_2^h, \end{cases} \qquad (19)$$

$$\left| w_k(x_i, y_j) - w_{k;ij}^h \right| \leq CN^{-2}, \qquad (x_i, y_j) \in \bar{\Omega}^h \backslash \Omega_{w_k}^h. \qquad (20)$$

Some additional investigation is needed for the convergence rate estimates for $w_{1;ij}^h$ and $w_{2;ij}^h$ (estimates for $w_{4;ij}^h$ and $w_{3;ij}^h$ are obtained by analogy) in the domains of the characteristic layer and the corner layer respectively, including the corner singularities as well.

3.1   The Characteristic Layer

Let us consider the discrete function $w_{1;ij}^h$ of the lower characteristic layer. Using the derivative estimates [2], taking into consideration the approximation error of problem (6)–(9) and estimate (20) and also proceeding as in [5], we obtain

$$\left|\psi_{w_1;ij}^h\right| \le C \begin{cases} N^{-3/2}r_{1;ij}^{-1}\ln N, & x_i \in (0, 1-\sigma_x], \, y_j \in [0,\sigma_y), \\ N^{-2}\ln^2 N, & x_i \in (1-\sigma_x, 1), \, y_j \in [0,\sigma_y), \end{cases} \tag{21}$$

$$\left|w_1(x_i, y_{N/4}) - w_{1;iN/4}^h\right| = O(N^{-2}), \qquad 0 < i < N, \tag{22}$$

$$\left|w_1(0, y_j) - w_{1;0j}^h\right| = \left|w_1(1, y_j) - w_{1;Nj}^h\right| = 0, \quad 0 \le j \le N.$$

We choose the barrier function (see [5]) in view of the corner singularity in a neighborhood of the point $a_1 = (0,0)$. to estimate the convergence rate of $w_{1;ij}^h$. Thus, let us consider the function

$$\tilde{B}_{w_1}(x,y) = N^{-3/2}\ln N\left(CB_1(x,y) + \tilde{C}b_1(y)\ln N\right) + CN^{-2}, \quad (x,y) \in \Omega_{w_1}^h,$$

where $B_1(x,y) = \ln(r_1'/\mathfrak{H}_1) + \left(-\varphi'^2 - \varphi' + \pi/4 + \pi/2 + 1\right), b_1(y) = e^{-\frac{\beta y}{2\sqrt{\varepsilon}}}, y' = y,$

$x' = x + b\mathfrak{H}_1, \, r_1' = \sqrt{x'^2 + y'^2}, \, \varphi' = \arctan\frac{y}{x}, \, \tilde{C} = 64a/27(2\pi+7)^3\beta^2, \, b = const > 1$. The barrier function $\tilde{B}_{w_1}(x,y)$, unlike [5], contains an additional term $b_1(y)$ to enhance. The condition of choice for the constant b is given below.

Lemma 1. *If $w_{1;ij}^h$ is a solution to difference problem (15)–(16) at $k = 1$ and $\varepsilon < a\mathfrak{H}_1/2$, while $w_1(x_i, y_j)$ is a solution to the corresponding differential problem, then the following estimate is valid*

$$\left|w_1(x_i, y_j) - w_{1;ij}^h\right| \le C\left(N^{-3/2}\ln^2 N + N^{-2}\right), \quad (x_i, y_j) \in \Omega_{w_1}^h. \tag{23}$$

Proof. Validity of estimate (23) follows from the validity of inequality (see [5])

$$L^h\tilde{B}_{w_1;ij} \ge C(N^{-3/2}\ln N/r_{1;ij}) + CN^{-2}, \quad \Omega_{w_1}^h \backslash \partial\Omega_D^h, \tag{24}$$

and this implies the result by applying comparison principle for the approximation error in $\Omega_{w_1}^h$, in view of the estimates (21)–(22). Estimate (24) at $0 < j < N/4$ is obtained by the same method as in ([5], see (3.18), (3.19)), if we assume that $Lb_{1;j} \ge C(q,\beta)$ holds true and instead of (3.19) from [5] we require the fulfilment of condition
$r_{1;ij}' \ge \max_{ij}\left\{\left(6\mathfrak{H}_1\sqrt{2\pi+5}\right)/\sqrt{3}; (4q\mathfrak{H}_1^2)/a; 6\mathfrak{H}_1(2\pi+7)\right\}$. Since $r_{1;ij}' \ge (1+b)\mathfrak{H}_1$
in domain $\Omega_{w_1}^h\backslash\partial\Omega_D^h$, then the last inequality certainly holds true provided $6\mathfrak{H}_1(2\pi+7) \le (1+b)\mathfrak{H}_1$, by which b is determined. Similar reasoning also holds true at $y = 0$ if we take into consideration the derivative estimates from [2], choice of the constant $\tilde{C}$ and fulfilment of the following relations

$$LB_{1;i0} = \frac{1}{r_{1;i0}'}, \, Lb_{1;i0} \ge \frac{\beta}{2\sqrt{\varepsilon}}, \, 0 < i < N; \, \left|\frac{\partial^3 b_1(y)}{\partial y^3}\right| \le \frac{C}{\varepsilon\sqrt{\varepsilon}}, \, \left|\frac{\partial^4 b_1(y)}{\partial y^4}\right| \le \frac{C}{\varepsilon^2}.$$

And the following expression will be an analogue for inequality (3.19) from [5], which represents the Neumann boundary condition. And the expression $r'_{1;i0} \geq \frac{4}{\sqrt{3}} \max_{0<i<N} \left\{ \mathfrak{h}_2\sqrt{2(\pi+3)}; 2\sqrt{2\mathfrak{h}_1\mathfrak{h}_2} \right\}$, will be an analogue of the mentioned earlier inequality (3.19) from [5], which stands for the boundary the Neumann condition and holds true with our choice of b.

In case $N/2 < i < N$ the approximation error of the difference scheme contains the fourth order derivatives, but at the expense of multipliers $\mathfrak{h}_1^2, \mathfrak{h}_2^2$ all the reasoning holds true. We finish proving the lemma with the estimates $b_1(y) = e^{-\frac{\beta y}{2\sqrt{\varepsilon}}} \leq 1$, $C \leq B_1(x,y) \leq C(\ln N + 1)$ at $0 < x < 1$, $0 \leq y < \sigma_y$, the second of which is the implication of inequalities $N^{-1} < \mathfrak{H}_1 < 2N^{-1}$, $(1+b)\mathfrak{H}_1 \leq r'_{1;ij} \leq \sqrt{2}(1+b\mathfrak{H}_1)$. $\qquad\square$

3.2   The Corner Layer

Let us do convergence rate estimates for the function $w_{2;ij}^h$ of the corner layer and analyze the approximation error. Proceeding by analogy with characteristic layers, we obtain

$$\left| \psi_{w_2;ij}^h \right| \leq C \frac{N^{-2} \ln^2 N}{\varepsilon}, \qquad (x_i, y_j) \in \Omega_{w_2}^h \setminus \partial \Omega^h, \qquad (25)$$

$$\left| \psi_{w_2;i0}^h \right| \leq C \begin{cases} \dfrac{N^{-2} \ln^2 N}{r_{2;i0}} + \dfrac{N^{-3} \ln^3 N}{\varepsilon}, & r_{2;i0} < \varepsilon, \\[2mm] \dfrac{N^{-3} \ln^3 N}{\varepsilon}, & r_{2;i0} \geq \varepsilon, \end{cases} \quad N/2 < i < N, \qquad (26)$$

$$\left| w_2(x_{N/2}, y_j) - w_{2;N/2j}^h \right| = O(N^{-2}), \quad \left| w_2(1, y_j) - w_{2;Nj}^h \right| = 0, \quad 0 \leq j \leq \frac{N}{4},$$

$$\left| w_2(x_i, y_{N/4}) - w_{2;iN/4}^h \right| = O(N^{-2}), \qquad \frac{N}{2} \leq i \leq N, \qquad (27)$$

where $r_{2;ij} = \sqrt{(1 - x_i)^2 + y_j^2}$.

So the following lemma gives the convergence rate estimate for the $w_{2;ij}^h$.

Lemma 2. *If $w_{2;ij}^h$ is a solution to difference problem (15)–(16) at $k = 2$ and $\varepsilon < a\mathfrak{H}_1/2$, and $w_2(x_i, y_j)$ is a solution to the corresponding differential problem, then the following estimate is valid*

$$\left| w_2(x_i, y_j) - w_{2;ij}^h \right| \leq CN^{-2} \ln^3 N, \quad (x_i, y_j) \in \Omega_{w_2}^h. \qquad (28)$$

Proof. Let us introduce the notation $z_{ij}^h = w_2(x_i, y_j) - w_{2;ij}^h$ and assume $z_{ij}^h = z_{1;ij}^h + z_{2;ij}^h$, where $z_{1;ij}^h$, $z_{2;ij}^h$ satisfies the conditions (27) and the following inequalities (see (25), (26))

$$\left| L^h z_{1;ij}^h \right| \leq C \begin{cases} \dfrac{N^{-2} \ln^2 N}{\varepsilon}, & N/2 < i < N, \ 0 < j < N/4, \\[2mm] \dfrac{N^{-3} \ln^3 N}{\varepsilon}, & j = 0, \end{cases} \qquad (29)$$

$$|L^h z^h_{2;ij}| \leq C \frac{N^{-2} \ln^2 N}{r_{2;ij}}, \qquad N/2 < i < N, \ 0 \leq j < N/4. \tag{30}$$

Therefore the error estimate for z^h_{ij} will be obtained as the sum of the estimates for $z^h_{1;ij}, z^h_{2;ij}$.

We shall begin our investigation with $z^h_{1;ij}$. Let $B^h_{z_1;ij} = C(N^{-2} \ln^2 N) B^h_{E;ij}$, where

$$B^h_{E;ij} = \begin{cases} \prod\limits_{s=i+1}^{N} (1 + ah_s/2\varepsilon)^{-1}, & 0 \leq i < N, \ 0 \leq j \leq N, \\ 1, & i = N, \quad 0 \leq j \leq N. \end{cases}$$

Applying the difference operator from (6)–(8) to the barrier $B^h_{E;ij}$ and, using the approximation error bounds (27), (29), we shall get the inequalities

$$L^h B^h_{z_1;ij} \geq \frac{C(a)}{\varepsilon} B^h_{z_1;ij} \geq |L^h z^h_{1;ij}|, \qquad (x_i, y_j) \in \Omega^h_{w_2} \backslash \partial \Omega^h,$$

$$L^h B^h_{z_1;i0} \geq \frac{C(a)}{\sqrt{\varepsilon}} B^h_{z_1;i0} \geq |L^h z^h_{1;i0}|, \qquad N/2 < i < N,$$

$$B^h_{z_1;N/2j} \geq |z^h_{1;N/2j}|, \ B^h_{z_1;Nj} \geq |z^h_{1;Nj}|, \quad 0 \leq j \leq N/4,$$

$$B^h_{z_1;iN/4} \geq |z^h_{1;iN/4}|, \qquad N/2 \leq i \leq N.$$

Hence, by virtue of comparison principle we have

$$|z^h_{1;ij}| \leq B^h_{z_1;ij} \leq C N^{-2} \ln^2 N, \quad (x_i, y_j) \in \Omega^h_{w_2}. \tag{31}$$

In order to estimate the second term $z^h_{2;ij}$, which contains a corner singularity, by analogy with a characteristic layer we shall choose the following barrier function

$$B_{z_2}(x, y) = C(N^{-2} \ln^2 N) B_{w_2}(x, y), \qquad (x, y) \in \Omega^h_{w_2},$$

where $B_{w_2}(x, y) = 4 \ln(c\sigma_y/r'_2) + \left(-\varphi'^2 + 4\varphi' + \pi^2/4 - 2\pi + 1\right)$, $r'_2 = \sqrt{x'^2 + y'^2}$, $\varphi' = \arctan(y'/x')$, $x' = 1 - x$, $y' = y + \tilde{b}\mathfrak{h}_2$, $\tilde{b} = const > 1$.

Following the same reasoning as in the case of the characteristic layer and taking into account the following inequalities $r_{2;ij} \geq \max\{c\mathfrak{h}_1; c\mathfrak{h}_2; 2c\mathfrak{h}_1/a\}$, $B_{w_2} > 1$ in domain $\Omega^h_{w_2} \backslash \partial \Omega^h_D$, which hold true in the case of our choice of $\tilde{b}$, we can obtain the estimate

$$L^h B^h_{w_2;ij} \geq C(1/r_{2;ij}), \qquad \Omega^h_{w_2} \backslash \partial \Omega^h_D. \tag{32}$$

Then, in virtue of estimates (27), (30), (32) and of the choice of the barrier, also in virtue of application of comparison principle, the following estimate is valid

$$|z^h_{2;ij}| \leq C(N^{-2} \ln^2 N) B^h_{w_2;ij}, \qquad (x_i, y_j) \in \Omega^h_{w_2}. \tag{33}$$

Since in $\Omega^h_{w_2}$ the inequalities $c\mathfrak{h}_2 \leq r'_{2;ij} \leq c\sigma_y$ hold true, then the estimate $B_{w_2} \leq C \ln N$ is valid. Substituting this estimate into (33), we obtain

$$|z^h_{2;ij}| \leq C N^{-2} \ln^3 N, \qquad (x_i, y_j) \in \Omega^h_{w_2}. \tag{34}$$

Combining (31) and (34), in $\Omega^h_{w_2}$ we obtain the final error estimate. $\qquad \square$

3.3   The Final Results

Thus we have all the necessary information to obtain the rate estimate for the uniform in ε convergence of the solution to scheme (6)–(9) to the exact solution.

Theorem 3. *Let $u(x_i, y_j)$ be the solution to problem (1)–(4), and let u_{ij}^h be the solution to difference problem (6)–(9) at $\varepsilon < a\mathfrak{H}_1/2$. Then, for $N > N_0$, where N_0 is a positive integer, which does not depend on ε, the following estimates are valid*

$$\left| u(x_i, y_j) - u_{ij}^h \right| \leq C \begin{cases} N^{-3/2} \ln^2 N, & \Omega_{w_1}^h \cup \Omega_{w_4}^h, \\ N^{-2} \ln^3 N, & N/2 < i \leq N, \ N/4 \leq j \leq 3N/4, \\ N^{-2}, & 0 \leq i \leq N/2, \ N/4 \leq j \leq 3N/4. \end{cases}$$

Proof. The proof follows from (10), (18)–(20), (23) and (28). $\square$

Remark 1. The case $\varepsilon \geqslant a\mathfrak{H}_1/2$ is not investigated here in details. However we should note, that in this case the solution to problem (6)–(9) is also uniformly convergent, but this time the rate is $O(N^{-2} \ln^3 N)$. The general proof scheme for this fact remains the same as in the case of small values of ε. Only S_{ij}^h in Ω^h and $w_{1;ij}^h$, $w_{2;ij}^h$ require some additional investigation ($w_{4;ij}^h$ and $w_{3;ij}^h$ are examined by analogy) in corresponding boundary layer domains.

In conclusion we shall note that numerical calculations were performed, which corroborate the theoretical results.

Acknowledgments. The author expresses sincere acknowledgement to Professor V. B. Andreev for his valuable advice and helpful discussions.

References

1. Clavero, C., Gracia, J.L., Lisbona, F., Shishkin, G.I.: A robust method of improved order for convection-diffusion problems in a domain with characteristic boundaries. Zamm. Z. Angew. Math. Mech. 82, 631–647 (2002)
2. Naughton, A., Stynes, M.: Regularity and derivative bounds for a convection-diffusion problem with the Neumann boundary conditions on characteristic boundaries. Z. Anal. Anwend. 29, 163–181 (2010)
3. Volkov, E.A.: On differential properties of solutions of boundary value problems for the Laplace and Poisson equations on a rectangle. Trudy Mat. Inst. Steklov 77, 89–112 (1965) (in Russian)
4. Shih, S., Kellogg, R.B.: Asymptotic analysis of a singular perturbation problem. SIAM J. Math. Anal. 18, 1467–1511 (1987)
5. Andreev, V.B.: Pointwise approximation of corner singularities for singularly perturbed elliptic problems with characteristic layers. Internat. J. of Num. Analysis and Modeling 7, 416–428 (2010)
6. Stynes, M., Roos, H.-G.: The midpoint upwind scheme. Appl. Numer. Math. 23, 361–374 (1997)
7. Andreev, V.B., Kopteva, N.V.: On the convergence, uniform with respect to a small parameter, of monotone three-point difference schemes. Differ. Uravn. 34, 921–928 (1998) (in Russian); translation in Differential Equations, 34, 921 (1998)

Author Index